PIC Microcontroller:

An Introduction to Software and Hardware Interfacing

PIC Microcontroller:

An Introduction to Software and Hardware Interfacing

Han-Way Huang
Minnesota State University · Mankato

DELMAR
CENGAGE Learning™   Australia • Canada • Mexico • Singapore • Spain • United Kingdom • United States

DELMAR
CENGAGE Learning

PIC Microcontroller: An Introduction to Software and Hardware Interfacing

Han-Way Huang

Vice President, Technology and Trades SBU:

Alar Elken

Editorial Director:

Sandy Clark

Senior Acquisitions Editor:

Steve Helba

Senior Development Editor:

Michelle Ruelos Cannistraci

Senior Editorial Assistant:

Dawn Daugherty

Marketing Director:

Dave Garza

Channel Manager:

Fair Huntoon

Marketing Coordinator:

Casey Bruno

Production Director:

Mary Ellen Black

Production Manager:

Andrew Crouth

Production Editor:

Stacy Masucci

Technology Project Manager:

Kevin Smith

Technology Project Specialist:

Linda Verde

> For product information and technology assistance, contact us at
> **Cengage Learning Customer & Sales Support, 1-800-354-9706**
>
> For permission to use material from this text or product,
> submit all requests online at **cengage.com/permissions**
> Further permissions questions can be emailed to
> **permissionrequest@cengage.com**

ExamView® and ExamView Pro® are registered trademarks of FSCreations, Inc. Windows is a registered trademark of the Microsoft Corporation used herein under license. Macintosh and Power Macintosh are registered trademarks of Apple Computer, Inc. Used herein under license.

© 2007 Cengage Learning. All Rights Reserved. Cengage Learning WebTutor™ is a trademark of Cengage Learning.

ISBN-13: 978-1-4018-3967-3

ISBN-10: 1-4018-3967-3

Delmar Cengage Learning
5 Maxwell Drive
Clifton Park, NY 12065-2919
USA

Cengage Learning products are represented in Canada by Nelson Education, Ltd.

For your lifelong learning solutions, visit **delmar.cengage.com**

Visit our corporate website at **www.cengage.com**

Notice to the Reader

Publisher does not warrant or guarantee any of the products described herein or perform any independent analysis in connection with any of the product information contained herein. Publisher does not assume, and expressly disclaims, any obligation to obtain and include information other than that provided to it by the manufacturer. The reader is expressly warned to consider and adopt all safety precautions that might be indicated by the activities described herein and to avoid all potential hazards. By following the instructions contained herein, the reader willingly assumes all risks in connection with such instructions. The publisher makes no representations or warranties of any kind, including but not limited to, the warranties of fitness for particular purpose or merchantability, nor are any such representations implied with respect to the material set forth herein, and the publisher takes no responsibility with respect to such material. The publisher shall not be liable for any special, consequential, or exemplary damages resulting, in whole or part, from the readers' use of, or reliance upon, this material.

Printed in the United States of America
6 7 11 10

Contents

Chapter 2   PIC18 Assembly Language Programming   37

Chapter 3   PIC18 Development Tools   89

Chapter 4 Advanced Assembly Programming 125

Chapter 5 A Tutorial to the C Language and the Use of the C Compiler 183

Chapter 6 Interrupts, Resets, and Configuration   243

Chapter 9 Addressable Universal Synchronous Asynchronous Receiver Transceiver   407

Chapter 10 Serial Peripheral Interface   445

Chapter 11 Interintegrated Circuit Interface   503

Chapter 13 Controller Area Network   607

Chapter 15 System Configuration and Protection   741

Preface

Over the past few years, we have seen many changes in the 8-bit microcontroller market. Motorola stopped new development on its popular 8-bit 68HC11 microcontroller, which was one of the most taught microcontrollers in universities. Instead, Motorola developed the new 8-bit 68HC908 to take over the low-end applications of the 68HC11 and introduced the 16-bit 68HC12 and HCS12 to take over the high-end applications of the 68HC11. Several new players in the 8-bit microcontroller field have gained significant market share. Among them, Microchip has overtaken Motorola to become the new market leader in this competitive 8-bit microcontroller market.

In the past few years, more and more universities and colleges became interested and started to teach the Microchip PIC® microcontrollers. Microchip designs and manufactures several families of 8-bit microcontrollers. Among them, the PIC16 family and the PIC18 family are the two most important families. The PIC18 family was the latest development and has several advantages over the PIC16 microcontroller family:

1. The PIC18 family supports much larger program memory space. The PIC18 family supports 2 MB of program memory space, whereas the PIC16 family supports only 8 KB of program memory.

2. The PIC18 family supports external program memory. The PIC16 family does not.

3. The PIC18 family has much larger on-chip data memory to support the application.

4. The PIC18 family provides the access bank to minimize the data memory bank-switching overhead.

5. The PIC18 family provides more instructions, which sometimes makes assembly language programming a little easier.

6. The PIC18 family supports more peripheral functions than the PIC16 family does.

7. The PIC18 family devices can run at a higher clock rate and achieve better performance.

The PIC18 family provides a wide range of pin count from as few as 18 pins to as many as 80 pins at the time of this writing. In addition to the normal parallel I/O ports, the PIC18 family provides a wide spectrum of peripheral functions to satisfy the needs of different applications:

1. *Multiple serial interfaces.* The PIC18 family supports industrial standard serial interfaces, including the USART, SPI, I²C, and CAN bus. The SPI and I²C allow the PIC18 microcontroller to interface with numerous peripheral devices with serial interfaces, such as LED drivers, LCD drivers, A/D converters, D/A converters, real-time clock chips, SRAM, EEPROM, and phase-locked loop. The CAN bus has been widely used in automotive applications and factory automation and control.

2. *Sophisticated timer system.* A PIC18 microcontroller may provide input-capture, output-compare, pulse width modulation (PWM), real-time interrupt, and watchdog timer functions. Some PIC18 members have their PWM modules enhanced to support motor applications.

3. *A/D converter.* The PIC18 family provides a 10-bit resolution A/D converter.

4. *In-system programming capability.* Most PIC18 members provide on-chip flash program memory and allow the software to be upgraded in the system.

5. *In-circuit debug capability.* The PIC18 family implements a background debug mode and provides an in-circuit debug (ICD) interface, which allows the inexpensive in-circuit debugger to be implemented.

These features appear to be most desired by the end user. With these features, the PIC18 microcontroller is very suitable for those who want to learn about modern microcontroller interfacing and programming.

Intended Audience

This book is written for two groups of readers:

1. Students in electrical and computer engineering and technology who are taking an introductory course of microprocessor interfacing and applications. For this group of readers, this book provides a broad and systematic introduction to the microcontrollers. The material of this book would be adequate for two semesters.

2. Senior electrical engineering and computer engineering students and working engineers who want to learn about the PIC18 microcontroller and use it in a design project. For this group of readers, this book provides numerous more complicated examples to explore the functions and applications of the PIC18 microcontroller. For those readers who have learned at least one assembly language before, they might want to use C language to program their applications as much as possible to gain the productivity advantage provided by the C language.

Prerequisites

This book has been written with the assumption that the reader has taken a course on digital logic design and has been exposed to high-level language programming. The knowledge in digital logic will greatly facilitate the learning of the PIC18 microcontroller. The experience in high-level language programming will enable the reader to quickly implement a program construct using an appropriate assembly instruction sequence. The knowledge in assembly language programming is not required because one of the writing goals of this book is to teach the PIC18 assembly language programming.

Approach

Using a high-level language to program the application would be much more productive than using an assembly language because a high-level language such as C allows the user to work on the program logic at a much higher level. However, our experience shows that learning the assembly language programming can give the learner much insight into the functioning of hardware because the assembly instructions allow the user to have full control of the hardware. Based on these two considerations, this text uses both the assembly and the C languages to teach the programming of each peripheral function.

The main writing goal of this text is to facilitate the learning of the microcontroller interfacing and programming. Each subject is started with background issues and general concepts followed by the specific implementation in the PIC18 microcontroller. Numerous examples are used to illustrate the programming and interfacing of the PIC18 microcontroller.

Textbook Organization

Chapter 1 presents the basic concepts of computer hardware and software, microcontroller applications, the PIC18 architecture and addressing modes, a subset of PIC18 instructions, and a survey of the 8-bit microcontroller market.

Chapter 2 introduces the PIC18 instruction set, assembler directives, and the basic assembly language programming skills.

Chapter 3 provides a review of the PIC18 software and hardware development tools provided by Microchip and Shuan-Shizu Enterprise. Microchip provides a full spectrum of hardware and software development tools, whereas Shuan-Shizu developed several PIC18 demo boards for learning the PIC18 microcontrollers. Several tutorials have been provided to demonstrate the use of development tools and demo boards.

Chapter 4 is about more advanced topics in assembly language programming, including writing subroutines and making subroutine calls. A salient feature of this chapter is the implementation of software stack data structure. The PIC18 does not support the stack structure in hardware. This chapter puts forth great effort to illustrate how to use software techniques to implement the stack data structure. The stack structure has been heavily used in passing incoming parameters, returning computation result, and allocating local variables during subroutine calls.

Chapter 5 provides a tutorial on the C language and then examines the features of the Microchip C compiler. A tutorial on the use of the Microchip C compiler is also given in this chapter. A tutorial on using the Hi-Tech PICC-18 C compiler is given in Appendix C. The Hi-Tech PICC-18 C compiler appears to be easy to use and user friendly.

Chapter 6 discusses the concept and programming of interrupts and resets. The debounced key switches available on the demo boards from Shuan-Shizu Enterprise are especially suitable for experimenting with the interrupt handling and interrupt-driven I/O.

Chapter 7 introduces the basic concept of parallel I/O. This chapter also covers the interfacing and programming of simple I/O devices, including DIP switches, keypad scanning and debouncing, LEDs, LCDs, D/A converters, and the parallel slave port.

Chapter 8 explores the operation and applications of the timer system, including input capture, output compare, real-time interrupt, and pulse width modulation and the applications of these functions on the measurement of pulse width, period, duty cycle, and frequency; the creation of time delays, sirens, and songs; and motor control.

Chapter 9 deals with the serial communication interface USART. This interface can operate in both the asynchronous and the synchronous mode. However, this interface is often used to communicate with a PC using the EIA-232 interface.

Chapter 10 covers the synchronous serial interface SPI and the applications of this interface, including parallel port expansion with shift registers, multidigit display with seven-segment display driver chips, and time-of-day tracking.

Chapter 11 introduces the characteristics of the interintegrated circuit (I²C) and the applications of this interface, including EEPROM interfacing and ambient temperature measurement.

Chapter 12 presents the topic of analog-to-digital conversion and its applications in temperature, humidity, and barometric pressure measurement.

Chapter 13 presents the CAN 2.0 protocol and the PIC18 CAN module. Examples and programming of CAN bus have been provided.

Chapter 14 describes the PIC18 internal SRAM, data EEPROM, and flash memory. The issues related to external memory expansion are also explored: address space assignment, decoder design, and timing verification.

Chapter 15 describes the system configuration issue and miscellaneous issues of the PIC18 microcontroller, including the configuration registers, watchdog timer, low-power mode, program memory protection, data EEPROM protection, background debug mode, and in-circuit serial programming.

Pedagogical Features

Each chapter opens with a list of objectives. Every subject is presented in a step-by-step manner. Background issues and general concepts are presented before the specifics related to each PIC18 peripheral function are discussed. Numerous examples are then presented to demonstrate the use of each PIC18 peripheral function. Algorithmic steps and flowcharts are used to help the reader understand the program logic in most examples. Each chapter concludes with a summary along with exercise problems and lab assignments. Many of the exercise problems can also be used as lab assignments. A separate lab manual is not needed to support the learning of the PIC18 microcontroller.

Development Tools

Development tools are important in the learning of the PIC18 microcontroller. The MPLAB® IDE from Microchip is an integrated development environment, which consists of a context-sensitive text editor, cross assemblers for all Microchip microcontrollers, a linker, a librarian, a simulator, and drivers for several in-circuit emulators and the in-circuit debugger (ICD 2). The MPLAB IDE is updated several times each year and can be downloaded from the Microchip Web site free of charge (www.microchip.com).

The ICD 2 is an inexpensive tool for the user to download programs (hex files) onto the demo board for execution. This device also supports source-level debugging under the control of MPLAB IDE. The user connects the ICD 2 to a PC using the USB port or serial communication port and plugs into the target demo board via a connector.

Chapter 3 examines three demo boards for learning the PIC18 microcontroller made by Shuan-Shizu Enterprise (www.evb.com.tw). The SSE452 uses the PIC18F452 as its controller and allows the user to experiment with different types of PIC18 devices by providing optional 28-pin and 40-pin DIP ZIF sockets. The SSE8720 uses the PIC18F8720, whereas the SSE8860 uses the PIC18F8680 as their controllers. The SSE8680 includes the CAN transceiver chip MCP2551 and allows the user to experiment with CAN bus interfacing. All three demo boards include a 2 × 20 LCD kit, digital waveform outputs with frequency ranging from 1 Hz up to 16 MHz, DIP switches, debounced key switches, LEDs, an EEPROM with serial interface, a time-of-day chip, a temperature sensor, a rotary encoder, a connector for the speaker and the keypad, and connectors for accessing microcontroller signals. Microchip also produces several PIC18 demo boards for learning the PIC18 microcontroller.

This text uses the PIC18 C compiler from Microchip to compile all the C programs. This compiler does not have a student version. However, it has a demo version that can be run for a month to allow the user to experiment with the supported C language features. The demo version of the C compiler can be downloaded from the Microchip Web site. For those who are interested in the C compiler from Hi-Tech Inc., a tutorial is provided in Appendix C.

Complimentary CD

This text includes a complimentary CD that contains all appendices, including the source code of all example programs in the text and also the PDF files of datasheets of the PIC18 micro-controllers and peripheral chips discussed in the text.

Supplements

A CD dedicated to instructors who adopt this text is also available from the publisher. This CD contains solutions to all exercise problems and the lecture notes in PowerPoint format. Instructors are encouraged to modify the PowerPoint lecture notes to suit their teaching needs.

(ISBN #: 1401839681)

Acknowledgments

This book would not be possible without the help of a number of people. I would like to thank Carol Popovich of Microchip for her support during the development of this text. She has supported all my requests for the chips and development tools. I would also like to thank Professor Shujen Chen of DeVry University in Tinley Park, Illinois, for his useful suggestions to clarify several questions regarding the PIC18 microcontroller. I would like to thank Greg Clayton, of Thomson Delmar Learning, and Michelle Ruelos Cannistraci, senior developmental editor, for their enthusiastic support during the preparation of this book. I also appreciate the outstanding work for the production staff led by Stacy Masucci.

I would like to express my thanks for the many useful comments and suggestions provided by the following colleagues who reviewed this text during the course of its development:

Harold Broberg, Indiana University and Purdue University

Shujen Chen, DeVry University

Gerard Gambs, Pennsylvania Institute of Technology

Richard Helps, Brigham Young University

Greg Osborn, DeVry University

I am grateful to my wife, Su-Jane, and my sons, Craig and Derek, for their encouragement and support during the entire preparation of this book.

Han-Way Huang
Mankato, Minnesota

Trademark Information

MPLAB, PIC, PICmicro, PICSTART, PRO MATE, the Microchip logo and the Microchip name and logo are registered trademarks, and ICSP, In-Circuit Serial Programming, Microchip, Microchip in Control, PICC-18, and MPASM are trademarks of Microchip Technology Incorporated in the United States and other countries.

Avenue of Feedback

The author welcomes the report of errors and suggestions for improvement. Your input will be greatly appreciated. Error reports and suggestions can be sent directly to the author at han-way.huang@mnsu.edu or to the publisher.

About the Author

Dr. Han-Way Huang received his B.S. in electrical engineering from National Taiwan University and M.S. and Ph.D. degrees in computer engineering from Iowa State University. He has taught microprocessor and microcontroller applications extensively for 17 years. Before teaching at Minnesota State University, Mankato, Dr. Huang worked for four years in the computer industry. In addition to this book, Dr. Huang has also authored books on the Zilog Z80 (in Chinese), the Motorola 68HC11 (an introduction to the 68HC11 from Delmar Learning), the 8051 microcontroller (using the MCS-51 microcontroller from Oxford University Press), and the Motorola 68HC12 (an introduction to the 68HC12 from Delmar Learning).

1

Introduction to the PIC18 Microcontroller

1.1 Objectives

After completing this chapter, you should be able to:

- Define or explain the following terms: computer, processor, microprocessor, microcontroller, hardware, software, cross assembler, cross compiler, RAM, SRAM, ROM, EPROM, EEPROM, flash memory, byte, nibble, bus, KB, MB, GB, mnemonic, opcode, and operand

- Explain the differences among all of the PIC18 addressing modes

- Use appropriate PIC18 addressing mode to access operands

- Explain the banking operation of PIC18 data memory

- Perform simple operations using add, subtract, and data movement instructions

1.2 What Is a Computer?

A computer is made up of hardware and software. The hardware of a computer consists of four types of components:

- *Processor.* The processor is responsible for performing all of the computational operations and the coordination of the usage of resources of a computer. A computer system may consist of one or multiple processors. A processor may perform general-purpose computations or special-purpose computations, such as graphical rendering, printing, or network processing.

- *Input devices.* A computer is designed to execute programs that manipulate certain data. Input devices are needed to enter the program to be executed and data to be processed into the computer. There are a wide variety of input devices: keyboards, keypads, scanners, bar code readers, sensors, and so on.

- *Output devices.* No matter if the user uses the computer to do certain computation or to find information from the Internet or a database, the end results must be displayed or printed on paper so that the user can see them. There are many media and devices that can be used to present the information: CRT displays, flat-panel displays, seven-segment displays, printers, light-emitting diodes (LEDs), and so on.

- *Memory devices.* Programs to be executed and data to be processed must be stored in memory devices so that the processor can readily access them.

1.2.1 The Processor

A processor is also called the *central processing unit* (CPU). The processor consists of at least the following three components:

Registers. A register is a storage location inside the CPU. It is used to hold data and/or a memory address during the execution of an instruction. Because the register is very close to the CPU, it can provide fast access to operands for program execution. The number of registers varies greatly from processor to processor.

Arithmetic logic unit (ALU). The ALU performs all the numerical computations and logical evaluations for the processor. The ALU receives data from the memory, performs the operations, and, if necessary, writes the result back to the memory. Today's supercomputer can perform trillions of operations per second. The ALU and registers together are referred to as the *datapath* of the processor.

Control unit. The control unit contains the hardware instruction logic. The control unit decodes and monitors the execution of instructions. The control unit also acts as an arbiter as various portions of the computer system compete for the resources of the CPU. The activities of the CPU are synchronized by the system clock. The clock rates of modern microprocessors have exceeded 3.0 GHz at the time of this writing, where

1 GHz = 1 billion cycles per second

The period of a 1-GHz clock signal is 1 ns (10^{-9} second). The control unit also maintains a register called the *program counter* (PC) that keeps track of the address of the next instruction to be executed. During the execution of an instruction, the occurrence of an overflow, an addition carry, a subtraction borrow, and so forth are flagged by the system and stored in another register called a *status register*. The resultant flags are then used by the programmer for program-flow control and decision making.

1.2.2 The Microprocessor

The advancement of semiconductor technology allows the circuitry of a complete processor to be placed in one integrated circuit (also called a *chip*). A *microprocessor* is a processor packaged in a single integrated circuit. A *microcomputer* is a computer that uses a microprocessor as its CPU. A personal computer (PC) is a microcomputer. Early microcomputers were very slow. However, many personal computers manufactured in 2003 run at a clock rate higher than 3.0 GHz and are faster than some supercomputers of a few years ago.

Depending on the number of bits that a microprocessor can manipulate in one operation, a microprocessor is referred to as 4-bit, 8-bit, 16-bit, 32-bit, or 64-bit. This number is the *word length* (or *datapath length*) of the microprocessor. Currently, the most widely used microprocessors are 8-bit.

Although the clock rate of the microprocessor has been increased dramatically, the improvement in the *access time* (or simply called the speed) of the high-capacity memory chips (especially the most widely used DRAM chips to be discussed in Section 1.2.4) has been moderate at best. The microprocessor may complete one arithmetic operation in one clock cycle; however, it may take many clock cycles to access data from the memory chip. This disparity in speed makes the high clock rate of the microprocessor alone useless for achieving high throughput. The solution to this issue is adding a small high-speed memory to the CPU chip. This on-chip memory is called *cache memory*. The CPU can access data from the on-chip cache memory in one or two clock cycles because it is very close to the ALU. The cache memory is effective in improving the average memory access time because the CPU demonstrates *locality* in its access behavior. Within a short period of time, the CPU tends to access a small area in the memory repeatedly. Once the program segment or data has been brought into the cache, it will be referenced many times. This results in an average memory access time very close to that of the access time of the cache memory.

Microprocessors and input/output (I/O) devices have different characteristics and speed. A microprocessor is not designed to deal with I/O devices directly. Instead, peripheral chips (also called *interface chips*) are needed to make up the difference between the microprocessor and the I/O devices. For example, the Intel i8255 was designed to interface the 8-bit 8080 microprocessor from Intel, and the M6821 was designed to interface the 8-bit 6800 from Motorola with I/O devices.

Microprocessors have been widely used in many applications since they were invented. However, there are several limitations in the initial microprocessor designs that led to the development of microcontrollers:

- External memory chips are needed to hold programs and data because the early microprocessors did not have on-chip memory.
- Glue logic (such as address decoder and buffer chips) is required to interface with the memory chips.
- Peripheral chips are needed to interface with I/O devices.

Because of these limitations, a product designed with microprocessors cannot be made as compact as might be desirable. The development of microcontrollers has not only eliminated most of these problems but also enabled the design of many low-cost microprocessor-based products.

1.2.3 Microcontrollers

A microcontroller, or MCU, is a computer implemented on a single very large scale integrated (VLSI) circuit. In addition to those components contained in a microprocessor, an MCU also contains some of the following peripheral components:

- Memory
- Timers, including event counting, input capture, output compare, real-time interrupt, and watchdog timer
- Pulse-width modulation (PWM)
- Analog-to-digital converter (ADC)
- Digital-to-analog converter (DAC)
- Parallel I/O interface
- Asynchronous serial communication interface (UART)
- Synchronous serial communication interfaces (SPI, I2C, and CAN)
- Direct memory access (DMA) controller
- Memory component interface circuitry
- Software debug support hardware

The discussion of the functions and applications of these components is the subject of this text. Most of these functions are discussed in details in later chapters.

Since their introduction, MCUs have been used in almost every application that requires certain amount of intelligence. They are used as controllers for displays, printers, keyboards, modems, charge card phones, palm-top computers, and home appliances, such as refrigerators, washing machines, and microwave ovens. They are also used to control the operation of engines and machines in factories. One of the most important applications of MCUs is probably the automobile control. Today, a luxurious car may use more than 100 MCUs. Today, most homes have one or more MCU-controlled consumer electronics appliances. In these applications, people care about only the functionality of the end product rather than the MCUs being used to perform the control function. Products of this nature are often called *embedded systems*.

1.2.4 Memory

Programs and data are stored in memory in a computer system. A computer may contain semiconductor, magnetic, and/or optical memories. Only semiconductor memory is discussed in this text because magnetic and optical memories are seldom used in 8-bit MCU applications. Semiconductor memory can be further classified into two major types: *random-access memory* (RAM) and *read-only memory* (ROM).

Random-Access Memory

Random-access memory is *volatile* in the sense that it cannot retain data in the absence of power. RAM is also called *read/write memory* because it allows the processor to read from and write into it. Both read and write accesses to a RAM chip take roughly the same amount of time. As long as the power is on, the microprocessor can write data into a location in the RAM chip and read back the same contents later. Reading memory is nondestructive. When the microprocessor writes data to memory, the old data is written over and destroyed.

There are two types of RAM technologies: *static RAM* (SRAM) and *dynamic RAM* (DRAM). SRAM uses from four to six transistors to store one bit of information. As long as power is on, the information stored in the SRAM will not be degraded. Dynamic RAM uses one transistor and one capacitor to store one bit of information. The information is stored in the capacitor in the form of electric charge. The charge stored in the capacitor will leak away over time, so a periodic refresh operation is required to maintain the contents of DRAM.

RAM is mainly used to store *dynamic* programs and data. A computer user often wants to run different programs on the same computer, and these programs usually operate on different sets of data. The programs and data must therefore be loaded into RAM from hard disk or other secondary storage, and for this reason they are called *dynamic.*

READ-ONLY MEMORY

ROM is nonvolatile. If power is removed from ROM and then reapplied, the original data will still be there. As its name implies, ROM data can only be read. This is not exactly true. Most ROM technologies require special algorithm and voltage to write data into the chip. Without using the special algorithm and voltage, any attempt to write to the ROM memory will not be successful. There are many different kinds of ROM technologies in use today:

Masked-programmed read-only memory (MROM) is a type of ROM that is programmed when it is manufactured. The semiconductor manufacturer places binary data in the memory according to customer's specification. To be cost effective, many thousands of MROM memory chips, each containing a copy of the same data (or program), must be sold. Many people simply call MROM as ROM.

Programmable read-only memory (PROM) is a type of read-only memory that can be programmed in the field (often by the end user) using a device called a PROM programmer or PROM *burner.* Once a PROM has been programmed, its contents cannot be changed. PROMs are fuse-based; that is, end users program the fuses to configure the contents of memory.

Erasable programmable read-only memory (EPROM) is a type of read-only memory that can be erased by subjecting it to strong ultraviolet light. The circuit design of EPROM requires the user to erase the contents of a location before a new value can be written into it. A quartz window on top of the EPROM integrated circuit permits ultraviolet light to be shone directly on the silicon chip inside. Once the chip is programmed, the window can be covered with dark tape to prevent gradual erasure of the data. If no window is provided, the EPROM chip becomes *one-time programmable* (OTP) only. EPROM is often used in prototype computers where the software may be revised many times until it is perfected. EPROM does not allow erasure of the contents of an individual location. The only way to make change is to erase the entire EPROM chip and reprogram it. The programming of an EPROM chip is done electrically by using a device called an *EPROM programmer.* Today, most programmers are universal in the sense that they can program many different types of devices including EPROM, EEPROM, flash memory, and *programmable logic devices.*

Electrically erasable programmable read-only memory (EEPROM) is a type of nonvolatile memory that can be erased and reprogrammed by electrical signals. Like EPROM, the circuit design of EEPROM also requires the user to erase the contents of a memory location before writing a new value into it. EEPROM allows each individual location to be erased and reprogrammed. Unlike EPROM, EEPROM can be erased and programmed using the same programmer. However, EEPROM pays the price for being so flexible in its erasability. The cost of an EEPROM chip is much higher than that of an EPROM chip of comparable density.

Flash memory was invented to incorporate the advantages and avoid the disadvantages of both EPROM and EEPROM technologies. Flash memory can be erased and reprogrammed in the system without using a dedicated programmer. It achieves the density of EPROM, but it does not require a window for erasure. Like EEPROM, flash memory can be programmed and erased electrically. However, it does not allow the erasure of an individual memory location—the user can only erase a section or the entire chip. Today, more and more MCUs are incorporating on-chip flash memory for storing programs and data. The flash-based PIC18 MCUs allow you to erase one block of 64 bytes at a time.

1.3 The Computer Software

Programs are known as *software*. A program is a set of instructions that the computer can execute. The program is stored in the computer's memory in the form of binary numbers called *machine instructions*.

The length of a machine instruction of a computer may be fixed or variable. Fixing the instruction length makes instruction decoding simpler and hence can simplify the design of the processor. However, it has one potential drawback. The program length may be longer because of the inefficiency of instruction encoding. Most of the PIC18 instructions are 16 bits, whereas four of them are 32 bits. For example, the PIC18 machine instruction

0010 0100 0010 0000   (or 2420 in base 16)

adds the contents of the data register at the hexadecimal address 20 to the WREG register and leaves the sum in WREG. The machine instruction

0110 1010 0000 0101   (or 6A05 in base 16)

clears the contents of the data register located at the address 5 to 0.

When a machine instruction is fetched from the memory, it will be decoded in the control unit of the CPU. Appropriate control signals will then be generated to trigger the desired operation.

1.3.1 Assembly Language

It is not difficult to conclude that software development in machine language is extremely hard:

1. *Program entering*. The programmer must use the binary patterns of every machine instruction in order to enter a machine instruction. Before the user can memorize the binary pattern of each instruction, he or she must consult a lookup table constantly. In addition, the programmer must work on the program logic at a very low level, which will hinder the programming productivity.

2. *Program debugging*. Whenever a program does not perform as expected, the programmer will have a hard time to identify the instruction that caused the problem. A programmer will need to identify each machine instruction and then think about what operation is performed by that instruction. This is not an easy task.

3. *Program maintenance*. Most programs will need to be maintained in the long run. A programmer who did not write the program will have a hard time reading the program and figuring out the program logic.

Assembly language was invented to simplify the programming job. An *assembly program* consists of assembly instructions. An assembly instruction is the mnemonic representation of a machine instruction. For example, in the PIC18 MCU,

decf fp_cnt,F,A stands for "decrement the variable *lp_cnt* located at the access bank by 1"
addwf sum,F,A stands for "add the contents of the WREG register and the variable *sum* in the access bank and leaves the result in *sum*.

where the meaning of *access bank* is explained in Section 1.5.2.

With the invention of the assembly language, a programmer no longer needs to scan through the sequence of 0s and 1s in order to identify what instructions are in the program. This is a significant improvement over machine language programming.

The assembly program that the programmer enters is called *source program* or *source code*. The user needs to invoke an *assembler* program to translate the source program into machine language so that the computer can execute it. The output of an assembler is also called *object code*. There are two types of assemblers: *native assembler* and *cross assembler*. A native assembler runs on a computer and generates the machine code to be executed on the same computer or a different computer having the same instruction set. A cross assembler runs on a computer but generates machine code that will be executed by computers that have a different instruction set. The Microchip MPASM® is a cross assembler designed to run on a PC to translate assembly programs for the PIC MCUs.

1.3.2 High-Level Languages

There are a few drawbacks for assembly language programming:

- The programmer must be familiar with the hardware architecture on which the program is to be executed.

- A program (especially a long one) written in assembly language is difficult to understand for anyone other than the author.

- Programming productivity is not satisfactory for large programming projects because the programmer needs to work on the program logic at a very low level.

For these reasons, high-level languages such as C, C++, and Java were invented to avoid the problems of assembly language programming. High-level languages are close to plain English, and hence a program written in a high-level language becomes easier to understand. A statement in high-level language often needs to be implemented by tens or even hundreds of assembly instructions. The programmer can now work on the program logic at a much higher level, which makes the programming job much easier. A program written in a high-level language is also called a *source code*, and it requires a software program called a *compiler* to translate it into machine instructions. A compiler compiles a program into *object code*. Just as there are cross assemblers, there are *cross compilers* that run on one machine but translate programs into machine instructions to be executed on a computer with a different instruction set.

Some high-level languages are *interpreted*; that is, they use an *interpreter* to scan the user's source code and perform the operations specified. Interpreters do not generate object code. Programming languages that use this approach include Basic, Lisp, and Prolog. The Java language is partially compiled and partially interpreted. A program written in Java language is first compiled into byte code and then interpreted. The design purpose of this language is "compiled once, run everywhere."

High-level languages are not perfect, either. One of the major problems with high-level languages is that the machine code compiled from a program written in a high-level language is much longer and cannot run as fast as its equivalent in the assembly language. For this reason, many time-critical programs are still written in assembly language.

C language has been used extensively in MCU programming in the industry, and most MCU software tool developers provide cross C compilers. Both the C and the PIC18 assembly languages will be used throughout this text. The C programs in this text are compiled by the Microchip C cross compiler and tested on the PIC18 demo board.

1.4 Overview of the PIC18 MCU

Microchip has introduced six different lines of 8-bit MCUs over the years:

1. PIC12XXX: 8-pin, 12- or 14-bit instruction format
2. PIC14000: 28-pin, 14-bit instruction format (same as PIC16XX)
3. PIC16C5X: 12-bit instruction format
4. PIC16CXX: 14-bit instruction format
5. PIC17: 16-bit instruction format
6. PIC18: 16-bit instruction format

Each line of the PIC MCUs support different number of instructions with slightly different instruction formats and different design in their peripheral functions. This makes products designed with a different family of PIC MCUs incompatible. The members of the PIC18 family share the same instruction set and the same peripheral function design and provide from eight to more than 80 signal pins. This makes it possible to upgrade the PIC18-based product without changing the MCU family. One of the design goals of the PIC18 MCU is to eliminate the design flaws of other earlier MCU families and provide a better upgrade path to other families of MCUs. In terms of cost, the PIC18 MCUs are not more expensive than those in other families with similar capability.

The PIC18 MCUs provide the following peripheral functions:

1. Parallel I/O ports
2. Timer functions, including counters, input capture, output compare, real-time interrupt, and watchdog timer
3. Pulse width modulation (PWM)
4. SPI and I^2C serial interface
5. Universal Synchronous/Asynchronous Receiver Transmitter (USART)
6. A/D converter with 10-bit resolution
7. Analog comparator
8. Low-power operation mode
9. SRAM and EEPROM
10. EPROM or flash memory
11. Controller Area Network (CAN)

These peripheral functions are studied in detail in the following chapters. By October 2003, 40 devices in the PIC18 family have been in production, and more devices will be introduced in the coming few years. All these MCUs implement 77 instructions. Among them, 73 instructions are 16 bits, and the remaining four are 32 bits.

The features of all PIC18 devices are shown in Table 1.1. This table mentions many acronyms that may not make any sense at this point. However, all of them are explained in detail in later chapters.

Feature	PIC18C242	PIC18C252	PIC18C442	PIC18C452	PIC18C601	PIC18C801	PIC18C658
Operating frequency	DC-40 MHz	DC-40 MHz	DC-40 MHz	DC-40 MHz	DC-25 MHz	DC-25 MHz	DC-40 MHz
Program memory	16 KB	32 KB	16 KB	32 KB	0 KB	0 KB	32 KB
Data memory	512 Bytes	1.5 KB	512 Bytes	1.5 KB	1.5 KB	1.5 KB	1.5 KB Data
EEPROM	0	0	0	0	0	0	0
External program memory	No	No	No	No	256 KB	2 MB	No
Inerrupt sources	16	16	17	17	15	15	21
I/O ports	A..C	A..C	A..E	A..E	A..G	A..H, J	A..G
Timers	4	4	4	4	4	4	4
Capture/Compare/PWM modules	2	2	2	2	2	2	2
Serial Communication	MSSP, USART	MSSP, USART	MSSP, USART	MSSP, USART	MSSP, USART	MSSP, USART	MSSP, USART, CAN
Parallel Communication	No	No	PSP	PSP	No	No	PSP
10-bit A/D	5 channels	5 channels	8 channels	8 channels	8 channels	12 channels	12 channels
Low voltage detect	Yes	Yes	Yes	Yes	Yes	Yes	Yes
Brown out reset	Yes	Yes	Yes	Yes	No	No	Yes
Instruction set	77	77	77	77	77	77	77
8-bit external memory	No	No	No	NO	YES	YES	NO
8-bit external demuxed memory	No	No	No	NO	No	YES	NO
8-bit external memory	No	No	No	NO	YES	YES	NO
On-chip chip select signal	No	No	No	NO	$\overline{CSI}$	$\overline{CSI}$, $\overline{CS2}$	NO
Packages	28-pin PDIP 28-pin SOIC 28-pin JW	28-pin DIP 28-pin SOIC 28-pin JW	40-pin DIP 40-pin PLCC 40-pin TQFP 40-pin JW	40-pin DIP 44-pin PLCC 44-pin TQFP 40-pin JW	64-pin TQFP 68-pin PLCC	80-pin TQFP 84-pin PLCC	64-pin TQFP 68-pin CERQUAD 68-pin PLCC

Feature	PIC18C858	PIC18F242	PIC18F252	PIC18F442	PIC18F452	PIC18F258	PIC18F458
Operating frequency	DC-40 MHz	DC-40 MHz	DC-40 MHz	DC-40 MHz	DC-40 MHz	DC-40 MHz	DC-40 MHz
Program memory	32 KB	16 KB	32 KB	16 KB	32 KB	32 KB	32 KB
Data memory	1.5 KB	768 Bytes	1.5 KB	768 Bytes	1.5 KB	1.5 KB	1.5 KB
Data EEPROM	0	256	256	256	256	256	256
External program memory	No	No	No	No	No	No	No
Interrupt sources	21	17	17	18	18	17	21
I/O ports	A..H, J, K	A..C	A..C	A..E	A..E	A..C	A..E
Timers	4	4	4	4	4	4	4
Capture/Compare/PWM modules	2	2	2	2	2	1	1
Enhanced Capture/Compare/PWM modules	0	0	0	0	0	0	1
Serial Communication	MSSP, USART, CAN	MSSP, USART	MSSP, USART	MSSP, USART	MSSP, USART	MSSP, USART, CAN	MSSP, USART, CAN
Parallel Communication	PSP	No	No	PSP	PSP	No	PSP
10-bit A/D	16 channels	5 channels	5 channels	8 channels	8 channels	5 channels	8 channels
Low voltage detect	Yes	Yes	Yes	Yes	Yes	Yes	Yes
Brown out reset	Yes	Yes	Yes	Yes	Yes	Yes	Yes
Instruction set	77	77	77	77	77	77	77
Packages	80-pin TQFP 84-pin CERQUAD 80-pin PLCC	28-pin DIP 28-pin SOIC	28-pin DIP 28-pin PLCC	40-pin DIP 40-pin PLCC 44-pin TQFP	40-pin TQFP 44-pin PLCC 44-pin TQFP	28-pin SPDIP 28-pin SOIC	40-pin PDIP 44-pin PLCC 44-pin TQFP

Table 1.1 ■ Features of the PIC18 Microcontrollers (continued)

Feature	PIC18F248	PIC18F448	PIC18F6620	PIC18F6720	PIC18F8620	PIC18F8720	PIC18F1220
Operating frequency	DC-40 MHz	DC-25 MHz	DC-25 MHz	DC-25 MHz	DC-25 MHz	DC-250 MHz	DC-40 MHz
Program memory	16 KB	16 KB	64 KB	128 KB	64 KB	128 KB	4 KB
Data memory	768 Bytes	768 KB	3840 Bytes	3840 Bytes	3840 Bytes	3840 Bytes	256 Bytes
Data EEPROM	256 Bytes	256 Bytes	1024 Bytes	1024 Bytes	1024 Bytes	1024 Bytes	256 Bytes
External program memory	No	No	No	No	Yes	Yes	No
Interrupt sources	17	21	17	17	18	18	15
I/O ports	A..C	A..E	A..G	A..G	A..H, J	A..H, J	A..B
Timers	4	4	5	5	5	5	4
Capture/Compare/ PWM modules	1	1	5	5	5	5	1
Enhanced Capture/ Compare/PWM modules	0	1	0	0	0	0	0
Serial Communication	MSSP, USART, CAN	MSSP, USART, CAN	MSSP, USART	MSSP, USART	MSSP, USART	MSSP, USART	USART
Parallel Communication	No	PSP	PSP	PSP	PSP	PSP	No
10-bit A/D module	5 channels	8 channels	12 channels	12 channels	16 channels	16 channels	7 channels
Low voltage detect	Yes	Yes	Yes	Yes	Yes	Yes	Yes
Brown out reset	Yes	Yes	Yes	Yes	Yes	Yes	Yes
Instruction set	77	77	77	77	77	77	77
Packages	28-pin SPDIP 28-pin SOIC	40-pin PDIP 44-pin PLCC 44-pin TQFP	64-pin TQFP	64-pin TQFP	80-pin TQFP	80-pin TQFP	18-pin SDIP 18-pin SOIC 20-pin SSOP 28-pin QFN

Feature	PIC18F1320	PIC18F2220	PIC18F2320	PIC18F2439	PIC18F2539	PIC18F8520	PIC18F8525
Operating frequency	DC-40 MHz	DC-40 MHz	DC-40 MHz	DC-40 MHz	DC-40 MHz	DC-40 MHz	DC-40 MHz
Program memory	8 KB	4 KB	8 KB	12 KB	24 KB	32 KB	48 KB
Data memory	256 Bytes	512 Bytes	512 Bytes	640 Bytes	1408 Bytes	2048 Bytes	3840 Bytes
Data EEPROM	256 Bytes	256 Bytes	256 Bytes	256 Bytes	256 Bytes	1024 Bytes	1024 Bytes
External program memory	No	No	No	No	No	Yes	Yes
Interrupt sources	15	19	19	15	15	18	17
I/O ports	A..B	A..C	A..C	A..C	A..C	A..H, J	A..H, J
Timers	4	4	4	3	3	5	5
Capture/Compare/PWM modules	1	2	2	2	2	5	2
Enhanced Capture/ Compare/PWM modules	0	0	0	0	0	0	3
Serial Communication	EUSART	MSSP, USART	MSSP, USART	MSSP, USART	MSSP, USART	MSSP, USART (2)	MSSP, USART (2)
Parallel Communication	No	No	No	No	No	PSP	PSP
10-bit A/D module	7 channels	10 channels	10 channels	5 channels	5 channels	16 channels	12 channels
Low voltage detect	Yes	Yes	Yes	Yes	Yes	Yes	Yes
Brown out reset	Yes	Yes	Yes	Yes	Yes	Yes	Yes
Instruction set	77	77	77	77	77	77	77
Packages	18-pin SDIP 18-pin SOIC 20-pin SSOP 28-pin QFN	28-pin SDIP 28-pin SOIC	28-pin SDIP 28-pin SOIC	28-pin DIP 28-pin SOIC	28-pin DIP 28-pin SOIC	80-pin TQFP	80-pin TQFP

Table 1.1 ■ Features of the PIC18 Microcontrollers (continued)

Feature	PIC18F4439	PIC18F4539	PIC18F6520	PIC18F6525	PIC18F6585	PIC18F6621	PIC18F6680
Operating frequency	DC-40 MHz	DC-40 MHz	DC-40 MHz	DC-40 MHz	DC-40 MHz	DC-40 MHz	DC-40 MHz
Program memory	12 KB	24 KB	32 KB	48 KB	48 KB	64 KB	64 KB
Data memory	640 Bytes	1408 Bytes	2048 Bytes	3840 Bytes	3328 Bytes	3840 Bytes	3328 Bytes
Data EEPROM	256 Bytes	256 Bytes	1024 Bytes	1024 Bytes	1024 Bytes	1024 Bytes	1024 Bytes
External program memory	No	No	No	No	No	No	No
Interrupt sources	16	16	17	17	29	17	29
I/O ports	A..E	A..E	A..G	A..G	A..G	A..G	A..G
Timers	3	3	5	5	4	5	4
Capture/Compare/PWM modules	2 PWM	2 PWM	5	2	1	2	1
Enhanced Capture/Compare/PWM modules	0	0	0	3	1	3	1
Serial Communication	USART, MSSP	USART, MSSP	MSSP, USART (2)	MSSP, EUSART (2)	MSSP, EAUSART, ECAN	MSSP, EUSART (2)	MSSP, EUSART, ECAN
Parallel Communication	PSP	PSP	PSP	PSP	PSP	PSP	PSP
10-bit A/D module	8 channels	8 channels	12 channels	12 channels	12 channels	12 channels	12 channels
Low voltage detect	Yes	Yes	Yes	Yes	Yes	Yes	Yes
Brown out reset	Yes	Yes	Yes	Yes	Yes	Yes	Yes
Instruction set	77	77	77	77	77	77	77
Packages	40-pin DIP 44-pin TQFP 44-pin QFN	40-pin DIP 44-pin TQFP 44-pin QFN	64-pin TQFP	64-pin TQFP	64-pin TQFP 68-pin PLCC	64-pin TQFP	64-pin TQFP

Feature	PIC18F8585	PIC18F8621	PIC18F8680	PIC18F4220	PIC18F4320
Operating frequency	DC-40 MHz	DC-40 MHz	DC-40 MHz	DC-40 MHz	DC-40 MHz
Program memory	48 KB	64 KB	64 KB	4 KB	8 KB
Data memory	3328 Bytes	3840 Bytes	3328 Bytes	512 Bytes	512 Bytes
Data EEPROM	1024 Bytes	1024 Bytes	1024 Bytes	256 Bytes	256 Bytes
External program memory	Yes	Yes	Yes	No	No
Interrupt sources	16	16	17	17	20
I/O ports	A..H, J	A..H, J	A..H, J	A..E	A..E
Timers	4	5	4	4	4
Capture/Compare/PWM modules	1	2	1	1	1
Enhanced Capture/Compare/PWM modules	1	3	1	1	1
Serial Communication	EAUSART MSSP ECAN	AUSART (2) MSSP	MSSP, USART ECAN	MSSP, USART	MSSP, USART
Parallel Communication	PSP	PSP	PSP	PSP	PSP
10-bit A/D module	16 channels	16 channels	16 channels	13 channels	13 channels
Low voltage detect	Yes	Yes	Yes	Yes	Yes
Brown out reset	Yes	Yes	Yes	Yes	Yes
Instruction set	77	77	77	77	77
Packages	80-pin TQFP	80-pin TQFP	80-pin TQFP	40-pin DIP 44-pin TQFP 44-pin QFN	40-pin DIP 44-pin TQFP 44-pin QFN

Note.
Both PIC18F8585 and PIC18F8680 can only work with 25 MHz crystal when external memory is enabled.

Table 1.1 ■ Features of the PIC18 Microcontrollers (concluded)

The block diagram of the PIC18 member PIC18F8720 is shown in Figure 1.1. This Figure shows the peripheral functions implemented in the PIC18F8720. These are also discussed later in more details.

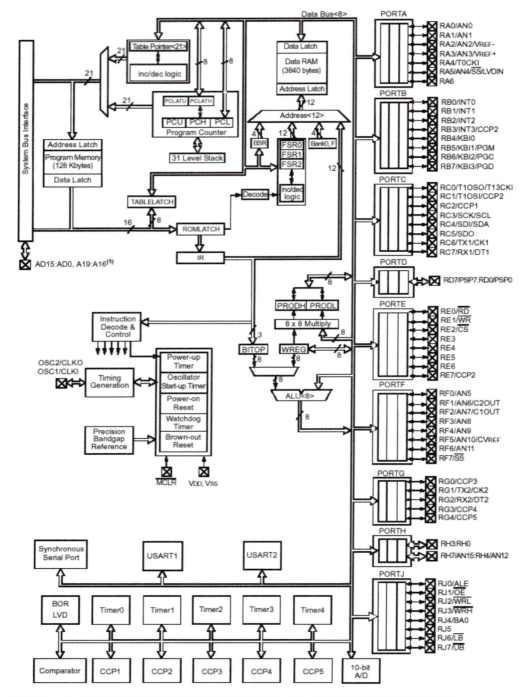

Figure 1.1 ■ Block diagram of the PIC18F8720 (reprint with permission of Microchip)

1.5 The PIC18 Memory Organization

Memory consists of a sequence of directly addressable "locations." A memory location is referred to as an *information unit*. A memory location in the PIC18 holds eight bits of information. Eight bits of information are called a *byte*. Sometimes one must deal with four bits of information at a time. Four bits of information are called a *nibble*. A memory location can be used to store data, instruction, the status of peripheral devices, and so on. An information unit has two components: its *address* and its *contents*, shown in Figure 1.2.

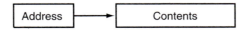

Figure 1.2 ■ The components of a memory location

Each location in memory has an address that must be supplied before its contents can be accessed. The CPU communicates with memory by first identifying the address of the location and then passing this address on the address bus. This is similar to the fact that a mail carrier needs an address in order to deliver a letter. The data is transferred between memory and the CPU along the data bus.

To differentiate the contents of a register or memory location from the address of a register or the memory location, the following notations are used throughout this text:

- *[register address]:* Refers to the contents of the register. For example, [WREG] refers to the contents of the WREG register; [0x20] refers to the contents of the general-purpose register at address 0x20. The prefix *0x* indicates that the number is represented in hexadecimal format. A number without a prefix is decimal.

- *address:* Refers to the register or memory location. For example, 0x10 refers to special function register at address 0x10.

1.5.1 Separation of Data Memory and Program Memory

As shown in Figure 1.3, the PIC18 MCU assigns data and program to different memory spaces and provides separate buses to them so that both are available for access at the same time.

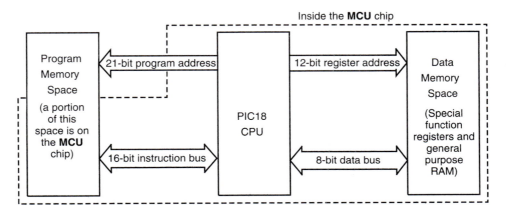

Figure 1.3 ■ The PIC18 memory spaces

The PIC18 MCU has a 21-bit program counter that is divided into three registers: PCU, PCH, and PCL. Among them, only the PCL register is directly accessible to the user. Both the PCH and the PCL are eight bits, whereas the PCU is five bits.

In the following discussion, memory addresses are referred to by using hex or decimal numbers. A hex number is indicated by adding a suffix *H* (or *h*) or prefix *0x* to the number. For example, 10H or 10h indicates hex 10 or decimal 16, and 0x20 represents the hex 20 or decimal 32. The MPASM also allows us to use *H'xx'* to specify a hex number *xx* and to use *D'yy'* to specify a decimal number *yy*. Hex refers to *hexadecimal* in the rest of this text.

Many digital systems have a large amount of memory. Therefore, special terms are often used to refer to the size of memory system. Among them, KB, MB, and GB are most often used:

- 1 KB refers to 2^{10} (1,024) bytes of memory.
- 1 MB refers to 2^{20} (1,048,576) bytes of memory.
- 1 GB refers to 2^{30} (1,073,741,824) bytes of memory.

1.5.2 PIC18 Data Memory

The PIC18 data memory is implemented as SRAM. Each location in the data memory is also referred to as a *register* or *file register*. The PIC18 MCU supports 4096 bytes of data memory. It requires 12 bits of address to select one of the data registers. The data memory map of the PIC18 MCU is shown in Figure 1.4.

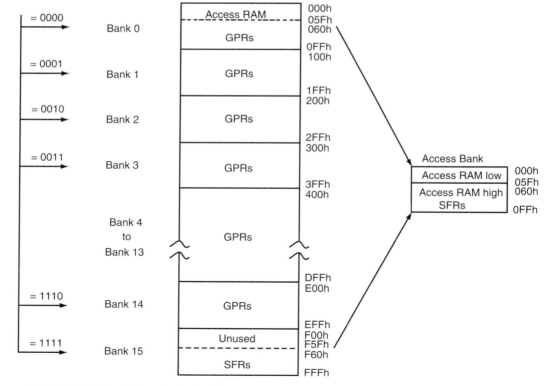

Note. 1. **BSR** is the 4-bit bank select register.

Figure 1.4 ■ Data memory map for PIC18 devices (redraw with permission of Microchip)

Because of the limited length of the PIC18 instruction (most instructions are 16 bits), only eight bits of the PIC18 instruction are used to specify the file register to be operated on. As a result, the PIC18 designers divided the 4096 file registers into 16 banks. Only one bank of 256 file registers is active at any time. An additional four bits are placed in a special register called bank select register (BSR) to select the bank to be active. The user needs to change the contents of the BSR register in order to change the active bank.

There are two types of registers: general-purpose registers (GPRs) and special-function registers (SFRs). GPRs are used to hold dynamic data when the PIC18 CPU is executing a program. SFRs are registers used by the CPU and peripheral modules for controlling the desired operation of the MCU. These registers are implemented as SRAM. A summary of these SFRs is listed in Appendix A.

The SFRs are assigned from the highest addresses and downward, whereas GPRs start from address 0 and upward. Depending on the device, some of the GPRs in the middle are not implemented. For example, the PIC18F452 has 1536 (six banks) bytes of data memory, and then banks 0 to 5 and bank 15 are implemented. The first 96 bytes (in bank 0, 0x000-0x05F) of the GPRs and the last 160 bytes (in bank 15, 0xF60-0xFFF) of the SFRs are grouped into a special bank called *access bank*. The functioning of the access bank is explained in Section 1.8. For the PIC18F242/252/442/452 MCUs, the access bank comprises of the upper 128 bytes in bank 15 and the lower 128 bytes in bank 0.

1.5.3 EEPROM Data Memory

At the time of this writing, all the PIC18 devices that have on-chip flash program memory also have either 256 bytes or 1024 bytes of data EEPROM. The data EEPROM is readable and writable during normal operation over the entire power supply range. The data EEPROM memory is not directly mapped in the register file space. Instead, it is indirectly addressed through the special function register. The operation of the data EEPROM is discussed in Chapter 14.

1.5.4 Program Memory Organization

Each PIC18 member has a 21-bit program counter and hence is capable of addressing the 2-MB program memory space. Accessing a nonexistent memory location will cause a read of all 0s.

Different members of the PIC18 family have different memory configurations. The PIC18CXX2 and PIC18CXX8 devices have on-chip EPROM program memory only and cannot access external memory. The PIC18C601 and PIC18C801 do not have on-chip memory. The PIC18C601 is capable of accessing 256 KB of external program memory, whereas the PIC18C801 can access 2 MB of external program memory. The PIC18FXX2, PIC18FXX8, and the PIC18F6620/6720 devices have on-chip flash program memory only. The PIC18F8585/8680/8621/8620/8720 can also access external program memory in addition to their on-chip flash program memory.

The PIC18 MCU has a 31-entry return address stack that is used to hold return addresses for subroutine call (to be discussed in Chapter 4) and interrupt processing (to be discussed in Chapter 6). This return address stack is not part of the program memory space. The program memory map is illustrated in Figure 1.5.

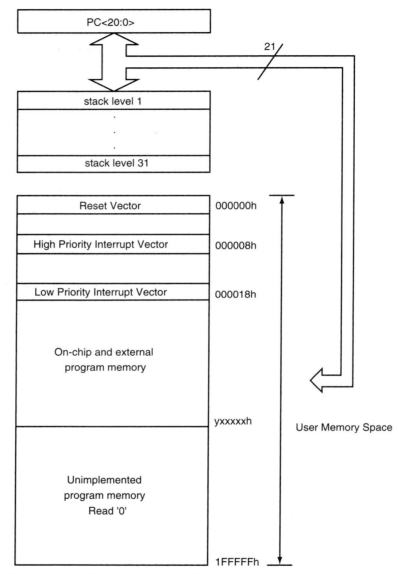

Note. **y** can be 0 or 1 whereas **x** can be 0-F

Figure 1.5 ■ PIC18 memory organization (redraw with permission of Microchip)

As shown in Figure 1.5, the address 000000h is assigned to the *reset vector*, which is the program-starting address after power-on or manual reset. The address 000008h is the starting address of the high-priority interrupt service routine. Sixteen bytes are allocated to the high-priority interrupt service routine by default. The address 000018h is the starting address for the low-priority interrupt service routine, and there is no default size for this service routine. The user program should follow the low-priority interrupt service routine. Reset is also discussed in Chapter 6.

1.6 The PIC18 CPU Registers

The PIC18 MCU has a group of registers, from 0xFD8 to 0xFFF (listed in Table 1.2), in the data memory space that are dedicated to the general control of the CPU operation. This group of registers can be referred to as *CPU registers.* Each of these CPU registers is discussed in an appropriate chapter of this book.

address	Name	Description
0xFFF	TOSU	Top of stack (upper)
0xFFE	TOSH	Top of stack (high)
0xFFD	TOSL	Top of stack (low)
0xFFC	STKPTR	Stack pointer
0xFFB	PCLATU	Upper program counter latch
0xFFA	PCLATH	High program counter latch
0xFF9	PCL	Program counter low byte
0xFF8	TBLPTRU	Table pointer upper byte
0xFF7	TBLPTRH	Table pointer high byte
0xFF6	TBLPTRL	Table pointer low byte
0xFF5	TABLAT	Table latch
0xFF4	PRODH	High product register
0xFF3	PRODL	Low product register
0xFF2	INTCON	Interrupt control register
0xFF1	INTCON2	Interrupt control register 2
0xFF0	INTCON3	Interrupt control register 3
0xFEF	INDF0 [1]	Indirect file register pointer 0
0xFEE	POSTINC0 [1]	Post increment pointer 0 (to GPRs)
0xFED	POSTDEC0 [1]	Post decrement pointer 0 (to GPRs)
0xFEC	PREINC0 [1]	Preincrement pointer 0 (to GPRs)
0xFEB	PLUSW0 [1]	Add WREG to FSR0
0xFEA	FSR0H	File select register 0 high byte
0xFE9	FSR0L	File select register 0 low byte
0xFE8	WREG	Working register
0xFE7	INDF1 [1]	Indirect file register pointer 1
0xFE6	POSTINC1 [1]	Post increment pointer 1 (to GPRs)
0xFE5	POSTDEC1 [1]	Post decrement pointer 1 (to GPRs)
0xFE4	PREINC1 [1]	Preincrement pointer 1 (to GPRs)
0xFE3	PLUSW1 [1]	Add WREG to FSR1
0xFE2	FSR1H	File select register 1 high byte
0xFE1	FSR1L	File select register 1 low byte
0xFE0	BSR	Bank select register
0xFDF	INDF2 [1]	Indirect file register pointer 2
0xFDE	POSTINC2 [1]	Post increment pointer 2 (to GPRs)
0xFDD	POSRDEC2 [1]	Post decrement pointer 2 (to GPRs)
0xFDC	PREINC2 [1]	Preincrement pointer 2 (to GPRs)
0xFDB	PLUSW2 [1]	Add WREG to FSR2
0xFDA	FSR2H	File select register 2 high byte
0xFD9	FSR2L	File select register 2 low byte
0xFD8	STATUS	Status register

Note. This is not a physical register

Table 1.2 ■ PIC18 CPU registers

The STATUS register, shown in Figure 1.6, contains the arithmetic status of the ALU. As with any other register, the STATUS register can be the destination of any instruction. If the STATUS register is the destination for an instruction that affects the Z, DC, C, OV, or N bits, then the write to these five bits is disabled. These bits are set or cleared according to the device logic. Therefore, the result of an instruction with the STATUS register as the destination may be different than intended. It is recommended, therefore, that only BCF, BSF, SWAPF, MOVFF, and MOVWF instructions be used to alter the STATUS register because these instructions do not affect the Z, C, DC, OV, or N bits of the STATUS register.

7	6	5	4	3	2	1	0
–	–	–	N	OV	Z	DC	C

N: Negative bit

    1 = arithmetic result is negative

    0 = arithmetic result is positive

OV: Overflow bit

    1 = Overflow occurred for signed arithmetic

    0 = No overflow occurred

Z: Zero flag

    1 = The result of an arithmetic or logic operation is zero.

    0 = The result of an arithmetic or logic operation is not zero.

DC: Digit carry/borrow bit

    For ADDWF, ADDLW, SUBLW, SUBWF instructions.

    1 = A carry-out from the 4th low-order bit of the result occurred.

    0 = No carry-out from the 4th low-order bit of the result occurred.

    For borrow, the polarity is reversed. For rotate (RRF, RLF)

    instructions, this bit is loaded with either the bit 4 or bit 3 of the

    source register.

C: Carry/borrow bit

    For ADDWF, ADDLW, SUBLW, SUBWF instructions.

    1 = A carry-out from the most significant bit of the result occurred.

    0 = No carry-out from the most significant bit of the result has

        occurred.

    For borrow, the polarity is reversed. For rotate (RRF, RLF)

    instructions, this bit is loaded with either the high or low order bit

    of the source register.

Figure 1.6 ■ The STATUS register (0xFD8) (redraw with permission of Microchip)

The WREG register (referred to as *working register*) is a special register that is involved in the execution of many instructions and can be the destination of many instructions.

1.7 The PIC18 Pipelining

The PIC18 designer divided the execution of most of the PIC18 instructions into two stages (*instruction fetch* and *instruction execution*) and then overlapped the execution of two consecutive instructions. Each stage takes one instruction clock cycle to complete. The result of the overlap of instruction execution is that most instructions take one instruction clock cycle to complete. This scheme is called *instruction pipelining.* An example of instruction pipelining is illustrated in Figure 1.7.

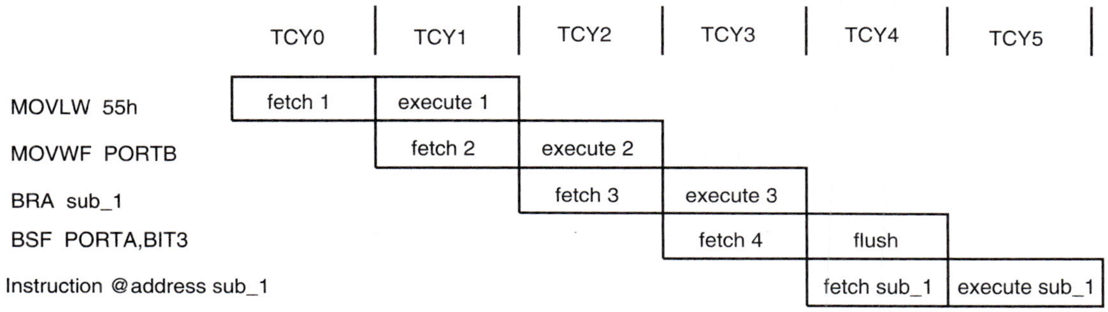

	TCY0	TCY1	TCY2	TCY3	TCY4	TCY5
MOVLW 55h	fetch 1	execute 1				
MOVWF PORTB		fetch 2	execute 2			
BRA sub_1			fetch 3	execute 3		
BSF PORTA,BIT3				fetch 4	flush	
Instruction @address sub_1					fetch sub_1	execute sub_1

Note: All instructions are single cycle, except for any program branches.

Figure 1.7 ■ An example of instruction pipeline flow

There are two problems caused by instruction pipelining: *data dependency hazard* and *control hazard.* In a program, it is common for one instruction to perform further operation on the result produced by the previous instruction. If the pipeline is designed in a way that the earlier instruction cannot write the result back to the register or memory location before it is used by the following instruction(s), then the data-dependency hazard has occurred. Most of the data-dependency hazards can be solved by *result forwarding.* However, if an instruction reads from a memory location (e.g., a load instruction) whereas the following instruction will use the returned value to perform certain operation, then result forwarding cannot resolve the hazard. This problem is usually solved by rearranging the instruction sequence to avoid this type of data dependency or inserting a no-op instruction. The dependency hazard problem will occur on pipelined processors with more than two stages. The PIC18 instruction pipeline has only two stages and does not have data-dependency hazard problems.

Control hazard is caused by branch instructions. Whenever a branch instruction reaches the execution stage and the branch is taken, then the following instructions in the pipeline need to be flushed because they are not allowed to take any effect by the program logic. In Figure 1.7, the instruction **BSF PORTA, BIT3** is flushed when it reaches the execution stage for this reason. There are several options to deal with control hazards in a pipelined processor. However, this issue is beyond the scope of this text.

The PIC18 MCU needs to access program memory during the instruction *fetch* stage and needs to access data memory during the instruction *execute* stage. When pipelining the execution of instructions, the PIC18 MCU needs to access the program memory and the data memory in the same clock cycle. This requirement is satisfied by separating the program memory from the data memory and providing separate buses to them.

The pipelined processor is explained clearly in Patterson and Hennessy's book *Computer Organization* published by Morgan Kaufman.

1.8 PIC18 Instruction Format

It was mentioned in Section 1.5.2 that data memory is divided into banks. Why would the banking scheme be used to control the access of data memory? The instruction format must be defined in order to understand this issue.

The instruction set is grouped into five basic categories:

1. *Byte-oriented operations.* The format of byte-oriented instructions is shown in Figure 1.8. The 6-bit field *opcode* specifies the operation to be performed by the ALU.

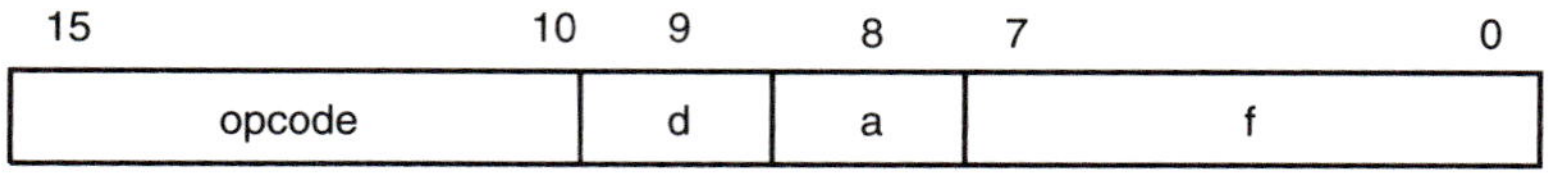

15		10	9	8	7		0
	opcode		d	a		f	

d = 0 for result destination to be WREG register.
d = 1 for result destination to be file register (f)
a = 0 to force Access Bank
a = 1 for BSR to select bank
f = 8-bit file register address

Figure 1.8 ■ Byte-oriented file register operations (redraw with permission of Microchip)

2. *Byte-to-byte operations (two-word).* The format of the instruction in this category is shown in Figure 1.9. There is only one instruction that uses this format: **movff f1, f2.** This instruction allows one to move data from one file register to another.

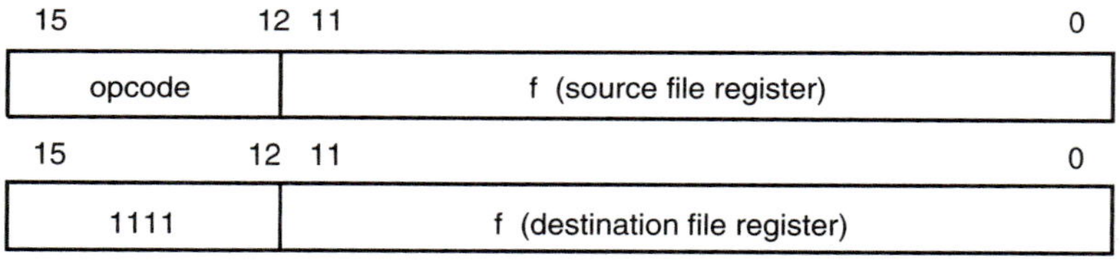

15		12	11		0
	opcode			f (source file register)	

15		12	11		0
	1111			f (destination file register)	

f = 12-bit file register address

Figure 1.9 ■ Byte-to-byte move operations (2 words) (redraw with permission of Microchip)

3. *Bit-oriented file register operations.* The format of instructions in this category is shown in Figure 1.10. This format uses an 8-bit field (f) to specify a file register as the operand.

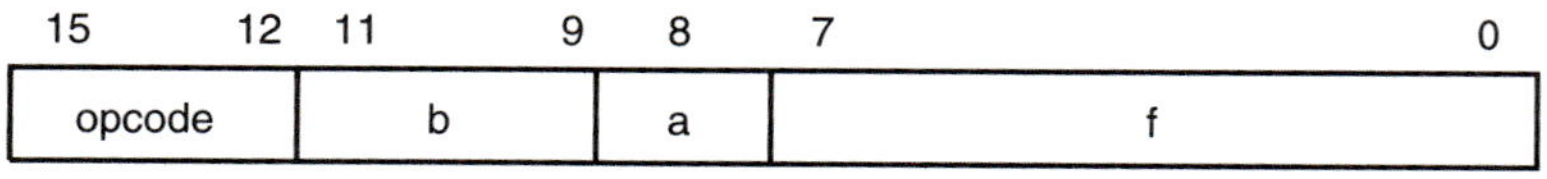

15		12	11		9	8	7		0
	opcode			b		a		f	

b = 3-bit position of bit in the file register (f).
a = 0 to force Access Bank
a = 1 for BSR to select bank
f = 8-bit file register address

Figure 1.10 ■ Bit-oriented file register operations (redraw with permission of Microchip)

4. *Literal operations.* Instructions in this category specify a *literal* (a *number*) as an operand. The format of instructions in this category is shown in Figure 1.11.

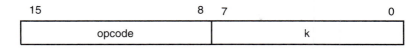

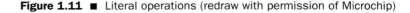

k = 8-bit immediate value

Figure 1.11 ■ Literal operations (redraw with permission of Microchip)

5. *Control operations.* The format of instructions in this category is shown in Figure 1.12. The notation n<7:0> stands for the bit 7 to bit 0 of the number n, whereas the notation n<19:8> stands for the bit 19 to bit 8 of the number n. The notation n<10:0> means that the number n is an 11-bit number. There are four different variations in their formats.

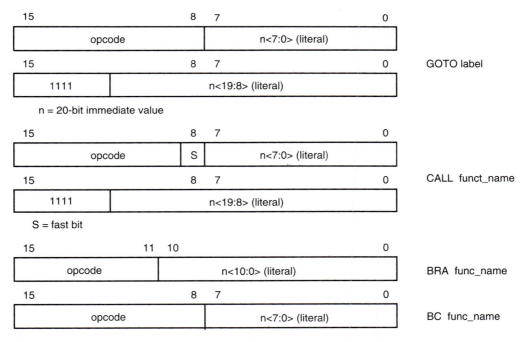

Figure 1.12 ■ Control operations (redraw with permission of Microchip)

As shown in Figures 1.8 to 1.12, all the PIC18 instructions use eight bits to specify the data register operand. A data register (excluding WREG) is also called a *file register*. Only 256 different registers can be specified by eight bits. However, all the PIC18 devices have more than 256 file registers, and hence additional information is needed to pinpoint the exact register to be operated on. This additional information is stored in the BSR register. The designer of the PIC18 MCU divided data memory into 16 (maximum) banks, with each bank having 256 data registers. The BSR register specifies the bank, and the f field in the instruction specifies the register number within the bank.

The banking scheme has been used in the PIC12, the PIC14000, the PIC16, and the PIC17 MCUs. This scheme allows a PIC MCU to incorporate more than 256 data registers on the CPU. However, it also adds a significant amount of overhead to the software because of the need to switch from one bank to another. In addition, it is easy to forget about bank switching, which will cause the software to fail.

In order to solve the problem caused by the banking scheme, the designers of the PIC18 MCU incorporated the *access bank*. The access bank consists of the lowest 96 GPRs and the highest 160 SFRs. As long as an instruction specifies a data register in the access bank, banking is ignored, and bank switching is unnecessary. All SFRs except a subset in the CAN module are in the access bank (CAN stands for *controller area network*). This makes bank switching unnecessary in many cases.

When bank switching is needed, the **movlb k** instruction can be used. This instruction places the value of k in the lower four bits of the BSR register. The result of the execution of this instruction is that it caused the data registers in bank k to become active.

In Figures 1.8 and 1.10, the a field in the PIC18 instruction allows the user to select the access bank. When writing program in assembly language, the assembler (MPASM) allows the user to use the letter A (a = 0) to specify the access bank. When the access bank is not chosen, one should use the word BANKED (a = 1) to allow the BSR register to do the bank selection.

The d field in Figure 1.8 allows the user to choose either the WREG or the file register as the destination of the instruction. The assembler allows the user to use the letter F (d = 1) to specify a file register and use the letter W (d = 0) to specify the WREG register as the destination. For example,

```
addwf    sum, F, A    ; sum is a GPR
```

adds the WREG register and sum in the access bank and places the result in sum.

```
addwf    sum,W, A
```

performs the same operation but leaves the result in the WREG register.

1.9 Addressing Modes

All MCUs use addressing modes to specify the operand to be operated on. The PIC18 MCU provides *register direct, immediate, inherent, indirect,* and *bit-direct* addressing modes for specifying instruction operands. As discussed in Chapter 2, assembler directives allow the user to use symbols to refer to memory locations. Using symbols to refer to memory locations makes the user program more readable. During the following discussion, symbols are used to refer to memory locations when appropriate.

1.9.1 Register Direct

The PIC18 device uses an 8-bit value to specify a data register as an operand. The register may be in the access bank or other banks. In the first case, the 8-bit value is used to select a register in the access bank, and the bank value in the BSR register is ignored. If the access bank is not selected, then the access is completed from the memory of the bank specified in the BSR register. The following instructions illustrate the register direct addressing mode:

```
movwf 0x1A, BANKED
```

copies the contents of the WREG register to the memory location 0x1A in the bank specified by the BSR register. The word BANKED (must be in uppercase) informs the assembler that the BSR register must be included in specifying the data register to be operated on.

movwf 0x45, A

copies the contents of the WREG register to the memory location 0x45 in the access bank.

movff reg1, reg2

copies the contents of the register reg1 to the register reg2. Both reg1 and reg2 are 12-bit values. The value of BSR is ignored.

1.9.2 Immediate Mode

In the immediate addressing mode, the actual operand is provided in the instruction. There is no need to access any memory location. The following instructions illustrate the immediate addressing mode:

addlw 0x20

adds the hex value 20 to the WREG register and places the sum in the WREG register.

movlw 0x15

loads the hex value 15 into the WREG register.

movlb 3

places the decimal value 3 in the lower four bits of the BSR register. The lower four bits become 0011. This instruction makes bank 3 the active bank. The value to be operated on directly is often called *literal*.

1.9.3 Inherent Mode

In the inherent mode, the operand is implied in the opcode field. The instruction opcode does not provide the address of the implied operand. The following instructions illustrate the inherent mode:

movlw 0x20

places the hex value 20 (decimal 32) in the WREG register. In this example, the value 0x20 is specified in the instruction machine code. The destination WREG is implied in the opcode field. No other address information for the WREG register is supplied.

andlw 0x13

performs an AND operation on the corresponding bits of the hex number 13 and the WREG register (i.e., bit i of WREG and with bit i of the value 0x13; i = 0 . . . 7). In this example, only the immediate value 0x13 is specified in the instruction machine code. The address of the WREG register 0xFE8 is not specified.

1.9.4 Indirect Mode

In this mode, a special function register is used as a pointer to the data memory location that is to be read and written. Since this register is in SRAM, the contents can be modified by the program. This can be useful for data tables in data memory and for *software stacks*. The software stack will be explained in Chapter 4.

There are three indirect addressing registers: FSR0, FSR1, and FSR2. To address the entire data memory space (4096 bytes), 12 bits are required. To store the 12-bit address information, two 8-bit registers are used. These indirect addressing registers are the following:

1. FSR0: composed of FSR0H and FSR0L
2. FSR1: composed of FSR1H and FSR1L
3. FSR2: composed of FSR2H and FSR2L

After placing the address of the data in one of the FSR registers, one needs to read from or write into one of the three registers that are not physically implemented in order to activate indirect addressing. These three registers are INDF0, INDF1, and INDF2.

If an instruction writes a value to INDF0, the value will be written to the data register with the address indicated by the register pair FSR0H:FSR0L. A read from INDF1 reads the data from the data register with the address indicated by the register pair FSR1H:FSR1L. INDFn can be used in a program anywhere an operand can be used. The process of indirect addressing is illustrated in Figure 1.13.

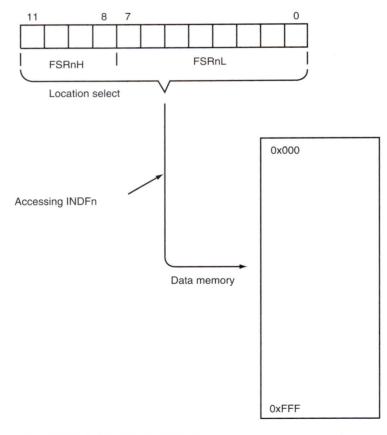

Figure 1.13 ■ Indirect addressing

Each FSR register has an INDF register associated with it plus four additional register addresses. Performing an operation on one of these five registers determines how the FSR will be modified during indirect addressing:

1. Do nothing to FSRn after an indirect access. This access is specified by using the register INDFn (n = 0 . . . 2).

2. Auto-decrement FSRn after an indirect access (postdecrement). This access is specified by using the register POSTDECn (n = 0 . . . 2).

3. Auto-increment FSRn after an indirect access (postincrement). This access is specified by using the register POSTINCn (n = 0 . . . 2).

4. Auto-increment FSRn before an indirect access (preincrement). This access is specified by using the register PREINCn (n = 0 . . . 2).

5. Use the value in the WREG register as an offset to FSRn. The signed value in WREG is added to the value in FSR to form an address before performing an indirect access. Neither the WREG nor the FSRn is modified after the access. This access is specified by using the register PLUSWn.

The following examples illustrate the usage of indirect addressing modes:

```
movwf INDF0
```

copies the contents of the WREG register to the data memory location specified by the FSR0 register. After the execution of this instruction, the contents of the FSR0 register are not changed.

```
movwf POSTDEC0
```

copies the contents of the WREG register to the data memory location specified by the FSR0 register. The contents of FSR0 are decremented by 1 after the operation.

```
movwf PREINC0
```

first increments the FSR0 register by 1 and then copies the contents of the WREG register to the data memory location specified by the FSR0 register.

```
clrf    PLUSW0
```

clears the memory location at the address equal to the sum of the value in the WREG register and that in the FSR0 register.

In the previous examples, one does not need to specify whether the register is in the access bank because the complete 12-bit data register address is taken from one of the FSR registers.

1.9.5 Bit-Direct Addressing Mode

The PIC18 MCU has five instructions to deal with an individual bit. These instructions use three bits to specify the bit to be operated on. For example,

```
BCF PORTB,3,A      ; integer 3 specifies the bit to be cleared
```

clears bit 3 of the data register PORTB, which will then pull the port B pin RB3 to low.

```
BSF PORTA,4,A      ; integer 4 specifies the bit to be set
```

sets bit 4 of the data register PORTA, which will then pull the port A pin RA4 to high.

1.10 A Sample of PIC18 Instructions

The PIC18 has 77 instructions. Four of these are 32-bit instructions, whereas the others are all 16 bits. A subset of the PIC18 instructions is examined in this section.

1.10.1 Data Movement Instructions

Memory data must be placed in appropriate registers before useful operations can be performed. Data movement instructions are provided for this purpose. A subset of the data movement instructions is listed in Table 1.3.

Mnemonic	Description	16-bit instruction word			Status affected
lfsr f, k	load FSR	1110 1110	00ff	k_{11}kkk	None
		1111 0000	$k_7$kkk	kkkk	
movf f, d, a	Move f	0101 00da	ffff	ffff	Z, N
movff fs, fd	Move fs (source) to f	1100 ffff	ffff	ffff	None
		1111 ffff	ffff	ffff	
movwf, f,a	Move WREG to f	0110 111a	ffff	ffff	None
swapf f, d, a	Swapp nibbles in f	0011 10da	ffff	ffff	None
movlb k	Move literal to BSR<3:0>	0000 0001	kkkk	kkkk	None
movlw k	Move literal to WREG	0000 1110	kkkk	kkkk	None

Note. Both **LFSR f, k** and **MVFF fs, fd** are 32-bit instructions

Table 1.3 ■ A sample of PIC18 data movement instructions

The instruction **lfsr f, k,** a 32-bit instruction, allows the user to place a 12-bit value in the FSR register specified by the value f. Two bits are provided for selecting the FSR registers (FSR0–FSR2). A 12-bit value (k) is contained in the instruction. The upper four bits (represented as k_{11}kkk in Table 1.3) of k are contained in the first word of the instruction, whereas the lower eight bits (represented as $k_7$kkk kkkk in Table 1.3) are contained in the second word.

The instruction **movf f, d, a** in Table 1.3 is provided for easy migration from the PIC16 family to the PIC18 family because the PIC16 family also has the same instruction. By setting the d field to 0 (represented by the letter W), this instruction will copy the contents of a file register to the WREG register. For example, the instruction

    movf 0x20,W,A

will copy the contents of the data register at 0x20 to the WREG register.

The **movff** instruction, a 32-bit instruction, can copy a file register in one bank to a file register in another bank without referring to the BSR register. Both the source and the destination registers are specified in 12 bits.

The **movlb k** instruction sets the bank k as the active bank. The **movlw k** instruction places the value k in the WREG register.

Example 1.1

▼

Write a PIC18 instruction (or instruction sequence) to transfer data from (a) WREG to data register at 0x30, (b) the data register at 0x30 to the data register at 0x40, (c) the data register at 0x40 to WREG, and (d) load the value 0x200 into FSR0.

Solution:

```
(a)  movwf 0x30,A        ; force access bank
(b)  movff 0x30, 0x40    ;
(c)  movf 0x40,W,A       ; force access bank and copy register 0x40 to WREG
(d)  lfsr FSR0, 0x200    ; load the value 0x200 into FSR0
```

1.10.2 ADD Instructions

ADD is the generic name of a group of instructions that perform the addition operation. The ADD instruction may have two or three operands. A three-operand ADD instruction includes the carry flag in the STATUS register as one of the operand. The PIC18 MCU has 3 ADD instructions:

```
addwf f, d, a     ; add WREG and f
addwfc f, d, a    ; add WREG, carry bit, and f
addlw k           ; add literal k to WREG
```

Example 1.2

Write an instruction to perform the following operations:

(a) Add the content of WREG and that of the data register at 0x40 (in access bank) and leave the sum in WREG.
(b) Increment the WREG register by 5.
(c) Add the WREG register, the register with the name of sum, and carry and leave the result in sum. The variable sum is in access bank.

Solution:

```
(a)  addwf 0x40,W,A
(b)  addlw 5
(c)  addwfc sum, F, A
```

Example 1.3

Write an instruction sequence to increment the contents of three registers 0x30–0x32 by 3.

Solution: The procedure for incrementing the value of a register by 3 is as follows:

Step 1
Place the value 3 in the WREG register.

Step 2
Execute the **addwf f, d, a** instruction.

The following instruction will increment the specified three registers by 3:

```
movlw 0x3
addwf 0x30, F, A    ; increment the register at 0x30 by 3
addwf 0x31, F, A    ; increment the register at 0x31 by 3
addwf 0x32, F, A    ; increment the register at 0x32 by 3
```

Example 1.4

▼

Write an instruction sequence to add the contents of three data registers located at 0x40–0x42 and store the sum at 0x50.

Solution: The required operation can be achieved by the following procedure:

Step 1
Load the contents of the register at 0x40 into the WREG register.

Step 2
Add the contents of the register at 0x41 into the WREG register.

Step 3
Add the contents of the register at 0x42 into the WREG register.

Step 4
Store the contents of the WREG register in the register at 0x50.

The following instructions will perform the desired operation:

```
movf 0x40, W, A      ; WREG ← [0x40]
addwf 0x41, W, A     ; add the contents of the register at 0x41 to WREG
addwf 0x42, W, A     ; add the contents of the register at 0x42 to WREG
movwf 0x50, A        ; 0x50 ← [WREG]
```

▲

Example 1.5

▼

Write an instruction sequence to add 10 to the data registers at 0x300–0x303 using the indirect postincrement addressing mode.

Solution: The procedure for solving this problem is as follows:

Step 1
Load the value 0x300 into the FSR0 register.

Step 2
Load the value 10 into the WREG register.

Step 3
Add the value in WREG to the data register pointed by FSR0 using the indirect postincrement mode.

Step 4
Repeat Step 3 three more times.

The instruction sequence that carries the operations from Step 1 to Step 3 is as follows:

```
movlw     0x0A
lfsr      FSR0,0x300      ; place 0x300 in FSR0
addwf     POSTINC0,F
addwf     POSTINC0,F
addwf     POSTINC0,F
addwf     POSTINC0,F
```

▲

1.10.3 SUB Instructions

SUB is the generic name of a group of instructions that perform the subtraction operation. The SUB instruction may have two or three operands. A three-operand SUB instruction includes the carry flag in the STATUS register as one of the operands. The PIC18 MCU provides four SUB instructions:

```
subfwb    f, d, a      ; subtract f from WREG with borrow
subwf     f, d, a      ; subtract WREG from f
subwfb    f, d, a      ; subtract WREG from f with borrow
sublw     k            ; subtract WREG from literal
```

Example 1.6
▼

Write an instruction sequence to subtract 9 from the data registers located at 0x50–0x53.

Solution: This operation can be implemented by placing 9 in the WREG register and then executing the **subwf f,F,A** instruction. The following instruction sequence will implement the required operation:

```
movlw 0x09          ; place 9 in the WREG register
subwf 0x50,F,A      ; decrement the contents of the register at 0x50 by 9
subwf 0x51,F,A      ; decrement the contents of the register at 0x51 by 9
subwf 0x52,F,A      ; decrement the contents of the register at 0x52 by 9
subwf 0x53,F,A      ; decrement the contents of the register at 0x53 by 9
```
▲

Example 1.7
▼

Write an instruction to perform each of the following operations:

Subtract the WREG register from the file register at 0x30 and leave the difference in the file register at 0x30.
Subtract the file register at 0x30 and borrow from the WREG register.
Subtract the WREG register from the file register at 0x50 and leave the difference in WREG.

Solution: The following instructions will perform the specified operations:

```
(a)      subwf 0x30, F, A
(b)      subfwb 0x30, W, A
(c)      subwf 0x50,W,A
```
▲

1.11 Overview of the 8-Bit MCU Market

There are many different 8-bit MCUs in the market today. Some of them follow the RISC design approach, whereas others follow the CISC design approach. This section briefly reviews the history of RISC and CISC and then provides a brief overview of the 8-bit MCU market.

1.11.1 CISC versus RISC

In the past, processor design followed either the Complex Instruction Set Computer (CISC) approach or the Reduced Instruction Set Computer (RISC) approach. The instructions of a processor designed with the CISC philosophy tend to be more complicated and perform more functions. The processor designers in the CISC camp believed that the instruction set designed with this philosophy in mind could make the machine code shorter and better match the syntax of high-level languages and hence could support high-level languages better. Experience proved that the resultant machine code was indeed shorter. However, many of the complex instructions are rarely used by the compiler, and the performance of the resultant machine was poor. In addition, the CISC computer took a longer time to design, which would likely cause the market window to be missed.

On the contrary, instructions of a RISC processor perform only simple operations. The addressing modes supported by a RISC computer tend to be simple as well. The followers of the RISC philosophy believed that a simple instruction set and simple addressing modes could simplify the design of the processor and that the resultant processor could run at a faster clock rate. More important, the RISC computer takes a shorter time to design and can be put on the market sooner. The earliest RISC processors have the following common features:

- Pipelined instruction execution.
- Fixed-length instructions (all instructions are either 32 or 16 bits long).
- Simple instruction set, simple instruction formats, and simple addressing modes.
- Large number of registers to support register-to-register operations (such operations are faster than register-to-memory or memory-to-memory operations).
- Delayed branch instruction. The earliest RISC processor allowed the instruction that followed the branch instruction to take effect regardless of whether the branch was taken. The purpose of this technique is to reduce the performance reduction caused by the branch instruction. This technique is no longer supported in the current generation RISC processors.
- Separation of the data memory and the program memory to allow simultaneous access of two memories. The computer architecture with separate data memory and program memory is often called *Harvard architecture.*

The compiled program for the RISC processor tends to be longer than its counterpart for the CISC processor. However, the resultant program for a RISC processor can run faster than its counterpart in a CISC processor. The current practice of processor design is to combine the strengths of both the RISC and the CISC processor.

Computer makers like to benchmark the performance of one another's machines. The adopted metric for comparing the performance of different machines is the running times of the same set of application programs on different machines. Because of the differences in design, machine A may run certain applications faster than machine B but run other applications slower than machine B. The selection of the set of application programs for comparison is arguable. Other metrics were also used to compare the performance of different machines. A common but inaccurate metric is MIPS (million instructions per second). It is obvious that MIPS means different things for processors with different instruction sets. The drawbacks of this metric have been detailed in Patterson and Hennessy's books. For computers made of the same microprocessor, the MIPS is a good metric for performance comparison if the same compiler is used.

1.11.2 Major 8-Bit MCUs

The 8-bit MCU market has experienced significant growth in the past decade. More than a billion 8-bit MCUs have sold annually in the past few years. Microchip started to fabricate 8-bit MCUs in late 1980s and by 1999 had sold its first billion MCUs. Thirty months later (in May 2002), Microchip had sold another billion MCUs. This statistic shows the strength of the 8-bit MCU market. In July 2003, Microchip became the number one vendor of the 8-bit MCU in terms of units of shipment. The success of the Microchip 8-bit MCUs can be attributed to the following factors:

- *Cost effectiveness.* MCUs made by Microchip follow the RISC design philosophy. They provide a simple instruction set and a fixed instruction length and use the pipelining technique to achieve high instruction execution throughput. Many Microchip devices have flash memory and provide in-system programming (ISP) capability, which allows the end user to upgrade product software without removing the MCU from the end product, an attractive feature. In addition, Microchip sells its MCUs at a competitive price.

- *Rich peripheral functions.* MCUs from Microchip have implemented all the peripheral functions that can be found in MCUs from other vendors. These peripheral functions include (1) parallel I/O ports; (2) timer functions, such as input capture, output compare, pulse-width modulation (PWM), counters, real-time interrupt (RTI), and watchdog timer; (3) 8-bit to 12-bit A/D converter; (4) multiple serial interface protocols (USART, SPI, and I²C); and (5) controller area network (CAN) controller.

- *Technical support.* Microchip provides sample designs and design consultation service to help its customers. Customers can ask for technical questions via phone calls or e-mail.

- *Development tools support.* The MPLAB· IDE from Microchip is undoubtedly the most important component in this area. MPLAB IDE is essentially an integrated development environment that incorporates a text editor, cross assemblers, simulators, programmer drivers, and a source-level debugger. It allows the user to perform source-level debugging in assembly or C language. Microchip makes MPLAB IDE even more attractive: it is free. Microchip also provides demo boards, programmers, in-circuit emulators, C compilers for the PIC17 MCU and the PIC18 MCU, and low-cost in-circuit debuggers to help customers develop their products. Many of these development tools are reviewed in Chapter 3.

- *University program.* Universities can request (so do other commercial users) free samples of MCUs and other development tools. Universities and students can purchase Microchip products at a discounted price. The web address for Microchip is www.microchip.com.

The PIC18 MCU can run with a 40-MHz crystal oscillator. With pipelining, most of the PIC18 instructions will take one instruction cycle (four crystal oscillator cycles) to execute. Therefore, the PIC18 MCU's performance is close to 10 MIPS.

Other major 8-bit MCUs include the following:

- The 68HC05, 68HC08, and 68HC11 from Motorola
- The 8051/8052 variants from more than 45 companies
- The AVR from Atmel

1.11.3 Motorola 8-Bit MCUs

Motorola is competing with Microchip for being the leader in terms of units of shipment of 8-bit MCUs. The 68HC05 and the 68HC08 MCUs are designed to serve low-end applications. The 68HC11 is the high-end 8-bit MCU from Motorola and was initially designed to serve the automotive market. The 68HC11 follows the CISC design philosophy. Its rich instruction set and addressing modes illustrate that. The rich instruction set and addressing modes make assembly programming easy. However, this is not an advantage for C programmers. Motorola has a good university support program. They provided free samples, low-cost demo boards, and free software (cross assembler) to help promote the 68HC11 and made it the most taught 8-bit MCU in U.S. universities. Most 68HC11 MCUs do not have on-chip flash memory. Serial interface protocols, such as UART and SPI, are standard. The 68HC11 provides a rich set of timer functions, including input capture, output compare, real-time interrupt (RTI), computer operate properly (COP) timer, and pulse accumulator. Most of the 68HC11 members have 8-bit A/D resolution only and do not implement the PWM function, which is an important feature for automotive and motor control applications. Motorola realizes these drawbacks and hence is pushing 68HC08 to take over the low-end applications of the 68HC11 and introduced the 16-bit HCS12 to take over 68HC11's high-end applications. Most of the 68HC08 members and all HCS12 members have on-chip flash memory and provide in-system programming capability. Motorola 8-bit MCUs have a combined 64-KB program and data memory.

1.11.4 Intel 8051/8052 Variants

The 8051 was designed by Intel and is the oldest 8-bit MCU. The 8051 has separate data memory and program memory spaces and follows the CISC design philosophy. Each memory space is 64 KB. The original 8051 has 4 KB of ROM, two 8-bit timers, four 8-bit I/O ports, and an asynchronous serial port. Later, Intel introduced 8052 as an enhancement to the 8051, adding another 8-bit timer to it. The initial 8051 MCU was very slow. It divides the crystal oscillator signal by 12 and uses it as the instruction clock signal to control the instruction execution. One instruction clock cycle takes 12 crystal oscillator cycles. An instruction may take from one to four instruction cycles to execute. The original 8051 MCU can run with a 12-MHz crystal oscillator. Therefore, the throughput of the original 8051 MCU is less than 1 MIPS. The 8051 provides one data pointer DPTR for accessing data memory. In many cases, this is inadequate.

Intel licensed the 8051 design to many semiconductor companies. To be competitive, most of the 8051 vendors add many enhancements to their implementations of the 8051. Many useful peripheral functions have been added by the 8051 variants. Some vendors even modified the 8051 design to shorten the instruction cycle time from 12 to 6 or 4 or even to a single oscillator clock cycle. Many vendors also add one or more data pointers to facilitate the access of data memory. On-chip flash program memory is also added by many vendors.

Philips produces the most 8051 variants, whereas Atmel produces the most flash-memory-based 8051 variants. Dallas/MAXIM Semiconductor shortens the 8051-instruction clock cycle time from 12 to 4 crystal oscillator cycles. Silicon Laboratory takes this approach to the limit and pushes the instruction clock cycle to one crystal oscillator cycle and provides devices running at 100 MHz. Silicon Laboratory 8051 variants are also pipelined. The 8051 variants from Silicon Laboratory provide (1) eight to over 100 I/O pins; (2) A/D converter with 8- to 16-bit resolutions; (3) serial communication interfaces, such as UART, SPI, I²C, and CAN controller; (4) on-chip flash program memory and in-system programming capability; (5) timer functions, such as input capture, output compare, counter, and PWM; (6) on-chip temperature sensor; and (7) free development software tools and inexpensive demo boards. The peak performance of Silicon Laboratory's 8051 variants could range from 20 to 100 MIPS.

Analog Devices Inc. focuses on the data acquisition market. Analog Devices call their 8051-based devices *MicroConverters* (ADuC). The ADuC provides A/D converters with 12- to 24-bit resolution and D/A converters with 12-bit resolution. Other peripheral functions, such as UART and SPI serial interface protocol, timers, power management, and COP, are also standard. The MSC1210 MCU from TI is another 8051 variant that follows an approach similar to that of Analog Devices. The MSC1210 has 32 KB of on-chip flash program memory, a 24-bit A/D converter, a 32-bit accumulator, three 8-bit timers, UART and SPI ports, and a watchdog timer. The MSC1210 uses four crystal oscillator cycles for one instruction cycle, as does the Dallas/MAXIM semiconductor.

Philips and Infineon are two major producers of 8051 variants. Many of their devices have flash program memory and in-system programming capability. Their 8051 variants also provide peripheral functions available in other 8051 vendors.

Because of the large number of 8051 variants and vendors, many hardware and software development tools are available from many vendors. The huge spectrum of the 8051 variants provides the user with the most choices in vendors, features, and development tools.

1.11.5 The Atmel AVR

In addition to producing 8051 variants, Atmel also produces an 8-bit RISC MCU, the AVR. The AVR has devices from 8 pin to over 64 pin. These devices are pipelined to achieve high throughput just like Microchip PIC18 MCU and Silicon Laboratory 8051 variants. The AVR provides up to 128 KB of on-chip flash program memory and optional 64-KB external memory and has in-system programming capability. Most peripheral functions available in the PIC18 are also available in the AVR. The AVR RISC MCU provides more instructions (133) than the PIC18 MCU does. The AVR has 32 general-purpose 8-bit registers and many other registers that are dedicated to the configuration and control of peripheral functions. Atmel also provides free software development tools and low-cost demo boards. Atmel's AVR MCU can run with a 16-MHz crystal oscillator and achieve the peak performance of 16 MIPS.

It is apparent that many 8-bit MCUs can be used in the same application because many devices have similar peripheral functions. The choice is often made on the basis of the following considerations:

- Device cost and system cost
- Availability of development tools and technical support
- Availability of engineering expertise on the specific MCU
- Device performance
- Personal preference

1.12 Summary

A computer system consists of hardware and software. The hardware consists of four major components: a processor (called the CPU), an input unit, an output unit, and memory. The processor can be further divided into three major parts: (1) registers, (2) the arithmetic logic unit, and (3) the control unit. All CPU activities are synchronized by the system clock. The clock frequency of the current generation of microprocessors (mainly Intel Pentium 4 and AMD Athlon-64) has exceeded 3 GHz. The CPU maintains a register called program counter (PC), which controls the memory address of the next instruction to be executed. During the execution of an instruction, the occurrence of an overflow, an addition carry, a subtraction borrow, and so forth are flagged by the system and stored in another register called the status register. The PIC18 MCU uses the STATUS register to record the occurrence of these conditions.

A microprocessor is a processor fabricated on a single integrated circuit. A microcomputer is a computer that uses a microprocessor as its CPU. Although microprocessors have been widely used since their invention, there are several limitations that led to the development of microcontrollers. First, a microprocessor requires external memory to store data and programs. Second, a microprocessor cannot interface directly to I/O devices; peripheral chips are needed. Third, glue logic is needed to interconnect external memory and peripheral interface chips to the microprocessor.

A microcontroller is a computer implemented on a VLSI chip. It contains everything contained in a microprocessor along with one or more of the following components:

- Memory
- Timer
- Pulse-width modulation (PWM) module
- Analog-to-digital converter
- Digital-to-analog converter
- Direct memory access (DMA) controller
- Parallel I/O interface
- Serial I/O interface
- Memory component interface circuitry
- Software debug circuitry

Software and data are stored in memory. Semiconductor memory chips can be classified into two major categories: random-access memory (RAM) and read-only memory (ROM).

There are many different types of ROM. MROM is a type of ROM that is programmed when it is fabricated. PROM is a type of fuse-based ROM that can be programmed in the field by the end user. EPROM is a type of ROM that is programmed electrically and erased by ultraviolet light. EEPROM is a type of ROM that can be erased and programmed electrically. The user can erase a location, a row, or the whole EEPROM chip in one operation. Flash memory can be erased and programmed electrically. However, flash memory can be erased only one row at a time or in bulk.

Programs are known as software. A program is a set of instructions that the computer hardware can execute. Programmers write a program in some kind of programming language. Only machine language was available during the early days of computers. A machine language program consists of a sequence of machine instructions. A machine instruction is a combination of 0s and 1s that inform the CPU to perform certain operation. Using machine language to write programs is difficult, and hence assembly language was developed to improve the productivity of programmers. Programs written in assembly language consist of a sequence of assembly instructions. An assembly instruction is the mnemonic representation of some machine instruction. Programs written in assembly language are still difficult to understand, and programming productivity is not high. High-level languages, such as C, C++, and Java, were invented to avoid the drawback of the assembly language. Programs written in assembly or high-level languages are called source code. Source code must be translated before it can be executed. The translator of a program written in assembly language is called an assembler, whereas the translator of a program written in a high-level language is called a compiler.

A memory location has two components: its contents and its address. When accessing a memory location, the CPU sends out the address on the address bus, and the memory components will place the requested value on the data bus.

The PIC18 MCU separates the program memory and data memory spaces and provides separate buses to access them. This Harvard architecture allows the PIC18 MCU to access the instruction and data in the same clock cycle. The group of data registers located between 0xFD8 and 0xFFF are used to control the general operation of the CPU and should be referred to as CPU registers. The program counter is 21 bits, which allows the PIC18 MCU to access up to 2 MB of program memory.

The PIC18 MCU provides 77 instructions. Seventy-three instructions are 16 bits. Four are 32 bits. These instructions are pipelined so that most of them will take one clock cycle to complete.

A data memory location is referred to as a data register. A small set of them provides the control and records the status of peripheral functions. This group is often referred to as special-function registers (SFRs). SFRs are located in the highest data memory space. The remaining data registers are used to store data or address information and are referred to as general-purpose registers (GPRs).

Each data register is associated with a 12-bit address. However, all (except one) PIC18 instructions provide 8-bit addresses for selecting the data register operand. An additional four bits of register address are stored in the bank select register (BSR). This addressing scheme divides data registers into 16 banks. The user needs to specifically change the value in the BSR register in order to switch to different bank.

Bank switching increases program complexity and may cause software errors. The designer of the PIC18 MCU minimized the problem of bank switching by creating the access bank. The access bank consists of the first 96 bytes of GPRs and the last 160 bytes of SFRs. Most of the PIC18 instructions have an "a" field for the user to force the access of access bank, which will ignore the BSR register during the addressing process of data registers. This scheme eliminates the need for bank switching in most cases.

The PIC18 provides five different addressing modes for accessing instruction operands:

- Register direct
- Immediate
- Inherent
- Indirect
- Bit direct

The indirect mode is used mainly to access data arrays and matrices. Three indirect pointers (FSR0–FSR2) are provided to support indirect access mode. Indirect pointers can be incremented or decremented before or after the access.

1.13 Exercises

E1.1 What is a processor? What components does it have?

E1.2 What makes a microprocessor different from the processor of a large computer?

E1.3 What makes an MCU different from a general-purpose microprocessor?

E1.4 What is the length (number of bits) of a PIC18 instruction?

E1.5 What is the length of the PIC18 program counter? How many different program memory locations can be addressed by the PIC18 MCU?

E1.6 How many bits are used to select a data register?

E1.7 What is the access bank? What benefits does the access bank provide?

E1.8 What is an assembler? What is a compiler?

E1.9 What is a cross assembler? What is a cross compiler?

E1.10 Explain the features of EPROM, EEPROM, and flash memory.

E1.11 What is instruction pipelining? What benefits does it provide?

E1.12 Write an instruction sequence to swap the contents of data registers at 0x300 and 0x200.

E1.13 Write an instruction sequence to load the value of 0x39 into data memory locations 0x100–0x103.

E1.14 Write an instruction sequence to subtract 10 from data memory locations 0x30–0x34.

E1.15 Write an instruction to store the contents of the WREG register in the data register located at 0x25 of bank 4.

E1.16 Write an instruction sequence to copy the contents of data memory at 0x100–0x103 to 0x200–0x203 using the postincrement addressing mode.

E1.17 Write an instruction sequence to copy the contents of data memory at 0x100–0x103 to 0x203–0x200, that is, in the reverse order by combining the use of postincrement and postdecrement modes.

E1.18 Write an instruction sequence to add 5 to data registers at $300–$303.

E1.19 Write an instruction sequence to add the contents of the data registers at $10 and $20 and store the sum at $30.

E1.20 Write an instruction sequence to subtract the contents of the data register at 0x20 from that of the data register at 0x30 and store the difference in the data register at 0x10.

2

PIC18 Assembly Language Programming

2.1 Objectives

After completing this chapter, you should be able to

- Explain the structure of an assembly language program

- Use assembler directives to allocate memory blocks and define constants

- Write assembly programs to perform simple arithmetic operations

- Write program loops to perform repetitive operations

- Use a flowchart to describe program flow

- Create time delays of any length using program loops

2.2 Introduction

Assembly language programming is a method of writing programs using instructions that are the symbolic equivalent of machine code. The syntax of each instruction is structured to allow direct translation to machine code.

This chapter begins the formal study of Microchip PIC18 assembly language programming. The format rules, specification of variables and data types, and the syntax rules for program statements are introduced in this chapter. The rules for the Microchip MPASM® assembler will be followed. The rules discussed in this chapter also apply to all other Microchip families of MCUs.

2.3 Assembly Language Program Structure

A program written in assembly language consists of a sequence of statements that tell the computer to perform the desired operations. From a global point of view, a PIC18 assembly program consists of three types of statements:

- *Assembler directives.* Assembler directives are assembler commands that are used to control the assembler: its input, output, and data allocation. An assembly program must be terminated with an END directive. Any statement after the END directive will be ignored by the assembler.
- *Assembly language instructions.* These instructions are PIC18 instructions. Some are defined with labels. The PIC18 MCU allows us to use up to 77 different instructions.
- *Comments.* There are two types of comments in an assembly program. The first type is used to explain the function of a single instruction or directive. The second type explains the function of a group of instructions or directives or the whole routine.

The source code of an assembly program can be created using any ASCII text file editor. Each line of the source file may consist of up to four fields:

- Label
- Mnemonic
- Operand(s)
- Comment

The order and position of these four fields are important. Labels must start in column 1. Mnemonics may start in column 2 or beyond. Operands follow the mnemonic. Comments may follow the operands, mnemonics, or labels and can start in any column. The maximum column width is 256 characters.

One should use space(s) or a colon to separate the label and the mnemonic and use space(s) to separate the mnemonic and the operand(s). Multiple operands must be separated by commas.

2.3.1 The Label Fields

A label must start in column 1. It may be followed by a colon (:), space, tab, or the end of line. Labels must begin with an alphabetic character or an underscore (_) and may contain alphanumeric characters, the underscore, and the question mark.

Labels may be up to 32 characters long. By default, they are case sensitive, but case sensitivity can be overridden by a command line option. If a colon is used when defining a label, it is treated as a label operator and not part of the label itself.

Example 2.1

▼

The following instructions contain valid labels:

(a) loop   addwf 0x20,F,A
(b) _again   addlw 0x03
(c) c?gtm   andlw 0x7F
(d) may2_june   bsf 0x07, 0x05,A

The following instructions contain invalid labels:

(e) isbig   btfsc 0x15,0x07,B   ;label starts at column 2
(f) 3or5   clrf 0x16,A   ;label starts with a digit
(g) three-four   cpfsgt 0x14,A   ;label contains illegal character "-"

▲

2.3.2 The Mnemonic Field

This field can be either an assembly instruction mnemonic or an assembler directive and must begin in column 2 or greater. If there is a label on the same line, instructions must be separated from that label by a colon or by one or more spaces or tabs.

Example 2.2

▼

Examples of mnemonic field:

(a) false   equ   0   ;**equ** is an assembler directive
(b)   goto   start   ;**goto** is the mnemonic
(c) loop: incf   0x20,W,A   ;**incf** is the mnemonic

▲

2.3.3 The Operand Field

If an operand field is present, it follows the mnemonic field. The operand field may contain operands for instructions or arguments for assembler directives. Operands must be separated from mnemonics by one or more spaces or tabs. Multiple operands are separated by commas. The following examples include operand fields:

(a)   cpfseq 0x20,A   ; "0x20" is the operand
(b) true   equ   1   ; "1" is the operand
(c)   movff 0x30,0x65 ; "0x30" and "0x65" are operands

2.3.4 The Comment Field

The comment field is optional and is added for documentation purpose. The comment field starts with a semicolon. All characters following the semicolon are ignored through the end of the line. The two types of comments are illustrated in the following examples.

(a)   decf   0x20,F,A   ;decrement the loop count
(b) ;the whole line is comment

Example 2.3

▼

Identify the four fields in the following source statement:

too_low   addlw   0x02   ; increment WREG by 2

Solution: The four fields in the given source statement are as follows:

(a) *too_low* is a label
(b) *addlw* is an instruction mnemonic
(c) *0x02* is an operand
(d) *;increment WREG by 2* is a comment

▲

2.4 Assembler Directives

Assembler directives look just like instructions in an assembly language program. Most assembler directives tell the assembler to do something other than creating the machine code for an instruction. Assembler directives provide the assembly language programmer with a means to instruct the assembler how to process subsequent assembly language instructions. Directives also provide a way to define program constants and reserve space for dynamic variables. Each assembler provides a different set of directives. In the following discussion, [ ] is used to indicate that a field is optional.

MPASM® provides five types of directives:

- *Control directives.* Control directives permit sections of conditionally assembled code.

- *Data directives.* Data directives are those that control the allocation of memory and provide a way to refer to data items symbolically, that is, by meaningful names.

- *Listing directives.* Listing directives are those directives that control the MPASM® listing file format. They allow the specification of titles, pagination, and other listing control.

- *Macro directives.* These directives control the execution and data allocation within macro body definitions.

- *Object directives.* These directives are used only when creating an object file.

2.4.1 Control Directives

The control directives that are used most often are listed in Table 2.1. Directives that are related are introduced in a group in the following:

if <expr>
else
endif

Directive	Description	Syntax
CODE	Begin executable code section	[<name>] code [<address>]
#DEFINE	Define a text substition section	#define <name> [<value>]
		#define <name> [<arg>, . . . <arg>] <value>
ELSE	Begin alternative assembly block to IF	else
END	End program block	end
ENDIF	End conditional assembly block	endif
ENDW	End a while loop	endw
IF	Begin conditionally assembled code block	if <expr>
IFDEF	Execute if symbol has been defined	ifdef <label>
IFNDEF	Execute if symbol has not been defined	ifndef <label>
#INCLUDE	Include additional source code	#include <<include_file>> | "<include_file>"
RADIX	Specify default radix	radix <default_radix>
#UNDEFINE	Delete a substition label	#undefine <label>
WHILE	Perform loop while condition is true	while <expr>

Table 2.1 ■ MPASM control directives

The **if** directive begins a conditionally assembled code block. If **<expr>** evaluates to true, the code immediately following **if** will assemble. Otherwise, subsequent code is skipped until an **else** directive or an **endif** directive is encountered. An expression that evaluates to 0 is considered logically false. An expression that evaluates to any other value is considered logically true.

The **else** directive begins alternative assembly block to **if.** The **endif** directive marks the end of a conditional assembly block. For example,

```
if version == 100 ; check current version
        movlw 0x0a
        movwf io_1,A
else
        movlw 0x1a
        movwf io_2,A
endif
```

will add the following two instructions to the program when the variable *version* is 100:

```
movlw 0x0a
movwf io_1,A
```

Otherwise, the following two instructions will be added instead:

```
movlw 0x1a
movwf io_2,A
end
```

This directive indicates the end of the program. An assembly language program looks like the following:

```
list p=xxxx      ; xxxx is the device name such as pic18F452
.                ; executable code
.                ; "
end              ; end of program
```

[<label>] code [<ROM address>]

This directive declares the beginning of a section of program code. If *<label>* is not specified, the section is named ".**code**". The starting address is initialized to the specified address or will be assigned at link time if no address is specified. For example,

```
reset   code      0x00
        goto      start
```

creates a new section called **reset** starting at the address 0x00. The first instruction of this section is **goto start.**

#define <name> [<string>]

This directive defines a text substitution string. Whenever **<name>** is encountered in the assembly code, **<string>** will be substituted. Using this directive with no <string> causes a definition of <name> to be noted internally and may be tested for using the **ifdef** directive. The following are examples for using the **#define** directive:

```
#define      length     20
#define      config     0x17,7,A
#define      sum3(x,y,z)           (x + y + z)

             .
             .
test         dw     sum3(1, length, 200)      ; place (1 + 20 + 200) at this location
             bsf       config       ; set bit 7 of the data register 0x17 to 1
```

#undefine <label>

This directive deletes a substitution string.

ifdef <label>

If *<label>* has been defined, usually by issuing a **#define** directive or by setting the value on the MPASM command line, the conditional path is taken. Assembly will continue until a matching **else** or **endif** directive is encountered. For example,

```
#define      test_val 2
             .
             .
             ifdef test_val
             <execute test code>        ; this path will be executed
             endif
```

ifndef <label>

If *<label>* has not been previously defined or has been undefined by issuing an **#undefine** directive, then the code following the directive will be assembled. Assembly will be enabled or disabled until the next matching **else** or **endif** directive is encountered. The following examples illustrate the use of this directive:

```
#define      lcd_port 1       ; set time_cnt on
             .
             .
             .
#undefine    lcd_port         ; set time_cnt off
             .
             .
             .
             infdef led_port
             .                 ; execute this
             .                 ; "
             endif
             end
```

#include "<include_file>"

This directive includes additional source file. The specified file is read in as source code. The effect is the same as if the entire text of the included file were inserted into the file at the location of the include statement. Up to six levels of nesting are permitted. *<include_file>* may be enclosed in quotes or angle brackets. If a fully qualified path is specified, only that path will be searched. Otherwise, the search order is current working directory, source file directory, MPASM executable directory. The following examples illustrate the use of this directive:

```
#include "p18F8720.inc"      ;search the current working directory
#include <p18F452.inc>
```

radix <default_radix>

This directive sets the default radix for data expressions. The default radix is hex. Valid radix values are: hex, dee, or oct.

while <expr>
endw

The lines between **while** and **endw** are assembled as long as **<expr>** evaluates to **true.** An expression that evaluates to zero is considered logically false. An expression that evaluates to any other value is considered logically true. A **while** loop can contain at most 100 lines and be repeated a maximum of 256 times. The following example illustrates the use of this directive:

```
test_mac    macro chk_cnt
            variable i
i = 0
            while i < chk_cnt
            movlw i
i + = 1
            endw
            endm
start
            test_mac 6
            end
```

The directives related to macro will be discussed later.

2.4.2 Data Directives

The MPASM data directives are listed in Table 2.2.

Directive	Description	Syntax
CBLOCK	Define a block of constant	cblock [<expr>]
CONSTANT	Declare symbol constant	constant <label> [=<expr>, . . ., <label>[=<expr>]
DA	Store strings in program memory	[<label>] da <expr>[,<expr>, . . ., <expr>]
DATA	Create numeric and text data	[<label>] data <expr>[,<expr>, . . ., <expr>]
		[<label>] data "<text string>"[, "<text_string>", . . .]
DB	Declare data of one byte	[<label>] db <expr>[,<expr>, . . ., <expr>]
		[<label>] db "<text string>"[, "<text_string>", . . .]
DT	Define table	[<label>] dt <expr>[,<expr>, . . ., <expr>]
		[<label>] dt "<text string>"[, "<text string>", . . .]
DW	Declare data of one word	[<label>] dw <expr>[,<expr>, . . ., <expr>]
		[<label>] dw "<text_string>"[,"<text_string>", . . .]
ENDC	End an automatic constant block	endc
EQU	Define an assembly constant	<label> equ <expr>
FILL	Fill memory	[<label> fill <expr>,<count>
RES	Reserve memory	[<label>] res <mem_units>
SET	Define an assembler variable	<label> set <expr>
VARIABLE	Declare symbol variable	variable <label>[=<expr>, . . ., <label>[=<expr>]]

Table 2.2 ■ MPASM* data directives

The user can choose his or her preferred radix to represent the number. MPASM supports the radices listed in Table 2.3.

```
cblock      [<expr>]
            <label>[:<increment>][,<label>[:<increment>]]
endc
```

Type	Syntax	Example
Decimal	D'<decimal digits>'	D'1000'
Hexadecimal	H'<hex_digits>' or	H'234D'
	0x<hex_digits>	0xC000
Octal	O'<oetal_digits>'	O'1357'
Binary	B'<binary_digits>'	B'01100001'
ASCII	'<character>'	'T'
	A'<character>'	A'T'

Table 2.3 ■ MPASMD radix specification

The *cblock* directive defines a list of named constants. Each <label> is assigned a value of one higher than the previous <label>. The purpose of this directive is to assign address offsets to many labels. The list of names ends when an **endc** directive is encountered.

<expr> indicates the starting value for the first name in the block. If no expression is found, the first name will receive a value one higher than the final name in the previous **cblock**. If the first **cblock** in the source file has no <expr>, assigned values start with zero.

If *<increment>* is specified, then the next *<label>* is assigned the value of <increment> higher than the previous *<label>*. The following examples illustrate the use of these two directives:

```
cblock 0x50
        test1, test2, test3, test4 ;test1..test4 get the value of 0x50..0x53
endc
cblock 0x30
        twoByteVal: 0, twoByteHi, twoByteLo
        queue: 40
        queuehead, queuetail
        double 1:2, double2:2
endc
```

The values assigned to symbols in the second cblock are the following:

- twoByteVal: 0x30
- twoByteHi: 0x30
- twoByteLo: 0x31
- queue: 0x32
- queuehead: 0x5A
- queuetail: 0x5B
- double1: 0x5C
- double2: 0x5E

constant <label> = <expr> [. . .,<label> = <expr>]

This directive creates symbols for use in MPASM expressions. Constants may not be reset after having once been initialized, and the expression must be fully resolvable at the time of the assignment. For example,

```
constant        duty_cyle = D'50'
```

will cause 50 to be used whenever the symbol *duty_cycle* is encountered in the program.

[<label>] data <expr>,[,<expr>, . . ., <expr>]
[<label>] data "<text_string>"[, "<text_string>"]

The **data** directive can also be written as **da**. This directive initializes one or more words of program memory with data. The data may be in the form of constants, relocatable or external labels, or expressions of any of the above. Each **expr** is stored in one word. The data may also consist of ASCII character strings, <text_string>, enclosed in single quotes for one character or double quotes for strings. Single character items are placed into the low byte (higher address) of the word, while strings are packed two bytes into a word. If an odd number of characters are given in a string, the final byte is zero. The following examples illustrate the use of this directive:

```
data    1,2,3               ; constants
data    "count from 1,2,3"  ; text string
data    'A'                 ; single character
data    main                ; relocatable label
```

[<label>] db <expr>[, <expr>, . . ., <expr>]

This directive reserves program memory words with packed 8-bit values. Multiple expressions continue to fill bytes consecutively until the end of expressions. Should there be an odd number of expressions, the last byte will be zero. When generating an object file, this directive can also be used to declare initialized data values. An example of the use of this directive is as follows:

```
        db      'b',0x22, 't', 0x1f, 's', 0x03, 't', '\n'
```

[<label>] de <expr>[, <expr>, . . ., <expr>]

This directive reserves memory words with 8-bit data. Each <expr> must evaluate to an 8-bit value. The upper eight bits of the program word are zeroes. Each character in a string is stored in a separate word. Although designed for initializing EEPROM data on the PIC16C8X, the directive can be used at any location for any processor. An example of the use of this directive is as follows:

```
        org     0x2000
        de      "this is my program", 0      ; 0 is used to terminate the string
```

[<label>] dt <expr> [, <expr>, . . ., <expr>]

This directive generates a series of **retlw** instructions, one instruction for each *<expr>*. Each *<expr>* must be an 8-bit value. Each character in a string is stored in its own **retlw** instruction. The following examples illustrate the use of this directive:

```
        dt      "A new era is coming", 0
        dt      1,2,3,4
```

[<label> dw <expr> [, <expr>, . . ., <expr>]

This directive reserves program memory words for data, initializing that space to specific values. Values are stored into successive memory locations, and the location is incremented by one. Expressions may be literal strings and are stored as described in the data directive. Examples on the use of this directive are as follows:

```
        dw      39, 24, "display data"
        dw      array_cnt-1
```

<label> equ <expr>

The **equ** directive defines a constant. Wherever the label appears in the program, the assembler will replace it with **<expr>**. Some examples of this directive are as follows:

```
true    equ     1
false   equ     0
four    equ     4
```

[<label>] fill <expr>, <count>

This directive specifies a memory fill value. The value to be filled is specified by **<expr>**, whereas the number of words that the value should be repeated is specified by **<count>.** The following example illustrates the use of this directive:

```
        fill    0x2020, 5        ;fill five words with the value of 0x2020 in program memory
```

[<label>] res <mem_units>

The MPASM® uses a memory location pointer to keep track of the address of the next memory location to be allocated. This directive will cause the memory location pointer to be advanced from its current location by the value specified in **<mem_units>.** In nonrelocatable code, *<label>* is assumed to be a program memory address. In relocatable code (using the MPLINK®), **res** can also be used to reserve data storage. For example, the following directive reserves 64 bytes:

```
buffer      res     64
```

<label>   set   <expr>

Using this directive, *<label>* is assigned the value of the valid MPASM expression specified by *<expr>*. The **set** directive is functionally equivalent to the **equ** directive except that a **set** value may be subsequently altered by other **set** directive. The following examples illustrate the use of this directive:

```
length        set       0x20
width         set       0x21
area_hi       set       0x22
area_lo       set       0x23
```

variable <label> [=<expr>][,<label>[=<expr>] . . .]

This directive creates symbols for use in MPASM expressions. Variables and constants may be used interchangeably in expressions. The **variable** directive creates a symbol that is functionally equivalent to those created by the **set** directive. The difference is that the **variable** directive does not require that symbols be initialized when they are declared.

2.4.3 Macro Directives

A *macro* is a name assigned to one or more assembly statements. There are situations in which the same sequence of instructions need to be included in several places. This sequence of instructions may operate on different parameters. By placing this sequence of instructions in a macro, the sequence of instructions need be typed only once. The macro capability not only makes us more productive but also makes the program more readable. The MPASM assembler macro directives are listed in Table 2.4.

<label> macro [<arg>, . . ., <arg>]
endm

Directive	Description	Syntax
ENDM	End of a macro definition	endm
EXITM	Exit from a macro	exitm
MACRO	Declare macro definition	<label> macro [<arg>, . . ., <arg>]
EXPAND	Expand macro listing	expand
LOCAL	Declare local macro variable	local <label> [, <label>]
NOEXPAND	Turn off macro expansion	noexpand

Table 2.4 ■ MPASM macro directives

A name is required for the macro definition. To invoke the macro, specify the name and the arguments of the macro, and the assembler will insert the instruction sequence between the **macro** and **endm** directives into our program. For example, a macro may be defined for the PIC18 as follows:

```
sum_of_3      macro     arg1,arg2, arg3
              movf      arg1,W,A
              addwf     arg2,W,A
              addwf     arg3,W,A
              endm
```

If the user wants to add three data registers at 0x20, 0x21, and 0x22 and leave the sum in WREG, he or she can use the following statement to invoke the previously mentioned macro:

```
sum_of_3 0x20,0x21,0x22
```

When processing this macro call, the assembler will insert the following instructions in the user program:

```
movf    0x20,W,A
addwf   0x21,W,A
addwf   0x22,W,A
```

exitm

This directive forces immediate return from macro expansion during assembly. The effect is the same as if an **endm** directive had been encountered. An example of the use of this directive is as follows:

```
test    macro arg1
        if arg1 == 3            ; check for valid file register
exitm
        else
error "bad file assignment"
        endif
        endm
```

expand
noexpand

The **expand** directive tells the assembler to expand all macros in the listing file, whereas the **noexpand** directive tells the assembler to do the opposite.

local <label>[,<label> . . .]

This directive declares that the specified data elements are to be considered in local context to the macro. *<label>* may be identical to another label declared outside the macro definition; there will be no conflict between the two. The following example illustrates the use of this directive:

```
<main code segment>
        .
        .
len     equ     5             ; global version
width   equ     8             ;
abc     macro   width
        local   len, label    ; local len and label
len     set     width         ; modify local len
label   res     len           ; reserve buffer
len     set     len-10
        endm                  ; end macro
```

2.4.4 Listing Directives

Listing directives are used to control the MPASM listing file format. The listing directives are listed in Table 2.5.

Directive	Description	Syntax
ERROR	Issue an error message	error "<text_string>"
ERRORLEVEL	Set error level	errorlevel 0 | 1 | 2 <+ | –><message number>
LIST	Listing options	list [<list_option>, . . ., <list_option>]
MESSG	Create user defined message	messg "<message_text>"
NOLIST	Turn off listing options	nolist
PAGE	Insert listing page eject	page
SPACE	Insert blank listing lines	space <expr>
SUBTITLE	Specify program subtitle	subtitle "<sub_text>"
TITLE	Specify program title	title "<title_text>"

Table 2.5 ■ MPASM assembler listing directives

error "<text_string>"

This directive causes *<text_string>* to be printed in a format identical to any MPASM error message. The error message will be output in the assembler list file. *<text_string>* may be from 1 to 80 characters. The following example illustrates the use of this directive:

```
bnd_check    macro   arg1
             if arg1 >= 0x20
                     error "argument out of range"
             endif
             endm
```

errorlevel {0 | 1 | 2 + <msgnum> | – <msgnum>} [, . . .]

This directive sets the types of messages that are printed in the listing file and error file. The meanings of parameters for this directive are listing in Table 2.6.

Setting	Effect
0	Messages, warnings, and errors printed
1	Warnings and errors printed
2	Errors printed
– <msgnum>	Inhibits printing of message <msgnum>
+ <msgnum>	Enables printing of message <msgnum>

Table 2.6 ■ Meaning of parameters for ERRORLEVEL directive

For example,

```
errorlevel 1, –202
```

enables warnings and errors to be printed and inhibits the printing of message number 202.

list [<list_option>, . . ., <list_option>]
nolist

The **list** directive has the effect of turning listing output on if it had been previously turned off. This directive can also supply the options listed in Table 2.7 to control the assembly process or format the listing file. The **nolist** directive simply turns off the listing file output.

messg "message_text"

Option	Default	Description
b = nnn	8	Set tab spaces
c = nnn	132	Set column width
f = <format>	INHX8M	Set the hex file output, <format> can be INHX32, INHX8M, or INHX8S
free	Fixed	Use free-format parser. Provided for backward compatibility
fixed	Fixed	Use fixed format paper
mm = {ON | OFF}	ON	Print memory map in list file
n = nnn	60	Set lines per page
p = <type>	None	Set processor type; for example, PIC18F8720
r = <radix>	hex	Set default radix; hex, dec, oct.
st = {ON | OFF}	ON	Print symbol table in list file
t = {ON | OFF}	OFF	Truncate lines of listing (otherwise wrap)
w = {0 | 1 | 2}	0	Set the message level. See ERRORLEVEL.
x = {ON | OFF}	ON	Turn macro expansion on or off

Table 2.7 ■ List directive options

This directive causes an informational message to be printed in the listing file. The message text can be up to 80 characters. The following example illustrates the use of this directive:

```
msg_macro macro
        messg "this is an messg directive"
        endm
```

page

This directive inserts a page eject into the listing file.

space <expr>

This directive inserts <expr> number of blank lines into the listing file. For example,

```
space 3
```

will insert three blank lines into the listing file.

title "<title_text>"

This directive establishes the text to be used in the top line of each page in the listing file. <title_text> is a printable ASCII string enclosed in double quotes. It must be 60 characters or less. For example,

```
title "prime number generator, rev 2.0"
```

causes the string "prime number generator, rev 2.0" to appear at the top of each page in the listing file.

subtitle "<sub_text>"

This directive establishes a second program header line for use as a subtitle in the listing output. <sub_text> is an ASCII string enclosed in double quotes, 60 characters or less in length.

2.4.5 Object File Directives

There are many MPASM directives that are used only in controlling the generation of object code. A subset of these directives is shown in Table 2.8.

Directive	Description	Syntax
BANKSEL	Generate RAM bank selecting code	banksel <label>
CODE	Begin executable code section	[<name> code [<address>]
__ CONFIG	Specify configuration bits	__config <expr> OR __ config <addr>, <expr>
EXTERN	Declare an external label	extern <label>[, <label>]
GLOBAL	Export a defined label	global <label>[,<label>]
IDATA	Begin initialized data section	[<name>] idata [<address>]
ORG	Set program origin	<label> org <expr>
PROCESSOR	Set processor type	processor <processor_type>
UDATA	Begin uninitialized data section	[<name>] udata [<address>]
UDATA_SHR	Begin shared uninitialized data section	[<name>] udata_shr [<address>]

Table 2.8 ■ MPASM object file directives

banksel <label>

This directive is an instruction to the linker to generate bank-selecting code to set the active bank to the bank containing the designated *<label>*. Only one *<label>* should be specified. In addition, *<label>* must have been previously defined. The instruction **movlb k** will be generated, and *k* corresponds to the bank in which *<label>* resides. The following example illustrates the use of this directive:

```
          udata
var1      res       1

          . . .
vark      res       1

          . . .
          code

          . . .
          banksel   var1
          movwf     var1
          banksel   vark
          movwf     vark
          pagesel   sub_x      ; to be discussed later
          call      sub_x

          . . .
sub_x     clrw

          . . .
          retlw     0
```

[<label>] code [<ROM address>]

This directive declares the beginning of a section of program code. If *<label>* is not specified, the section is named **".code"**. The starting address is initialized to the specified address or will be assigned at link time if no address is specified. The following example illustrates the use of this directive:

```
reset   code    0x00
        goto    start
```

__config <expr> or
__config <addr>, <expr>

This directive sets the processor's configuration bits to the value described by *<expr>*. Before this directive is used, the processor must be declared through the *processor* or *list* directive. The hex file output format must be set to INHX32 with the *list* directive when this directive is used with the PIC18 family. The following example illustrates the use of this directive:

```
list p = 18F8720, f = INHX32
__config 0xFFFF                ; default configuration bits
```

extern <label> [, <label>, . . .]

This directive declares symbols names that may be used in the current module but are defined as *global* in a different module. The *external* statement must be included before *<label>* is used. At least one label must be specified on the line. The following example illustrates the use of this directive:

```
        extern       function_x
                        . . .
        call         function_x
```

global <label> [, <label>, . . .]

This directive declares symbol names that are defined in the current module and should be available to other modules. This directive must be used after *<label>* is defined. At least one label must be included in this directive. The following example illustrates the use of this directive:

```
             udata
ax           res        1
bx           res        1
             global     ax, bx
             code
             addlw      3
             . . .
```

[<label> idata [<RAM address>]

This directive declares the beginning of a section of initialized data. If *<label>* is not specified, the section is named "*.data*". The starting address is initialized to the specified address or will be assigned at link time if no address is specified. No code can be generated in this section. The **res, db,** and **dw** directives may be used to reserve space for variables. The **res** directive will generate an initial value of zero. The **db** directive will initialize successive bytes of RAM. The **dw** directive will initialize successive bytes of RAM, one word at a time, in *low-byte/high-byte* order. The following example illustrates the use of this directive:

```
             idata
i            dw         0
j            dw         0
t_cnt        dw         20
flags        db         0
prompt       db         "hello there!"
```

[<label>] org <expr>

This directive sets the program origin for subsequent code at the address defined in *<expr>*. If *<label>* is specified, it will be given the value of the *<expr>*. If no **org** is specified, code generation will begin at address zero. Some examples of this directive follow:

```
reset        org 0x00
                . . .          ; reset vector code goes have
             goto start
             org 0x100
start           . . .          ; code of our program
```

processor <processor_type>

This directive sets the processor type. For example,

processor p18F8720

[<label>] udata [<RAM address>]

This directive declares the beginning of a section of uninitialized data. If *<label>* is not specified, the section is named "**.udata**". The starting address is initialized to the specified address or will be assigned at link time if no address is specified. No code can be generated in this section. The **res** directive should be used to reserve space or data. Here is an example that illustrates the use of this directive:

```
        udata
var1    res    1
var2    res    2
```

[<label>] udata_shr [<RAM address>]

This directive declares the beginning of a section of shared uninitialized data. If *<label>* is not specified, the section is named "**.udata_shr**". The starting address is initialized to the specified address or will be assigned at link time if no address is specified. This directive is used to declare variables that are allocated in RAM and are shared across all RAM banks. The **res** directive should be used to reserve space for data. The following examples illustrate the use of this directive:

```
temps   udata_shr
t1      res    1
t2      res    1
s1      res    2
s2      res    4
```

2.5 Representing the Program Logic

An embedded product designer must spend a significant amount of time on software development. It is important for the embedded product designer to understand software development issues.

Software development starts with the *problem definition.* The problem presented by the application must be fully understood before any program can be written. At the problem definition stage, the most critical thing is to get you, the programmer, and your end user to agree on what needs to be done. To achieve this, asking questions is very important. For complex and expensive applications, a formal, written definition of the problem is formulated and agreed on by all parties.

Once the problem is known, the programmer can begin to lay out an overall plan of how to solve the problem. The plan is also called an *algorithm.* Informally, an algorithm is any well-defined computational procedure that takes some value or a set of values as input and produces some value or set of values as output. An algorithm is thus a sequence of computational steps that transform input to output. An algorithm can also be viewed as a tool for solving a well-specified computational problem. The statement of the problem specifies in general terms the desired input/output relationship. The algorithm describes a specific computational procedure for achieving that input/output relationship.

An algorithm is expressed in *pseudocode* that is very much like C or Pascal. Pseudocode is distinguished from "real" code in that pseudocode employs whatever expressive method is most clear and concise to specify a given algorithm. Sometimes, the clearest method is English, so do not be surprised if you come across an English phrase or sentence embedded within a section of "real" code.

An algorithm provides not only the overall plan for solving the problem but also documentation to the software to be developed. In the rest of this book, all algorithms will be presented in the format as follows:

Step 1

. . .

Step 2

. . .

An earlier alternative for providing the overall plan for solving software problem is using flowcharts. A flowchart shows the way a program operates. It illustrates the logic flow of the program. Therefore, flowcharts can be a valuable aid in visualizing programs. Many people prefer using flowcharts in representing the program logic for this reason. Flowcharts are used not only in computer programming but in many other fields as well, such as business and construction planning.

The flowchart symbols used in this book are shown in Figure 2.1. The *terminal symbol* is used at the beginning and the end of each program. When it is used at the beginning of a program, the word *Start* is written inside it. When it is used at the end of a program, it contains the word *Stop*.

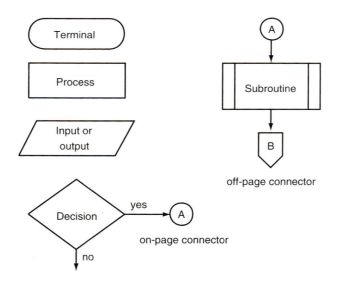

Figure 2.1 ■ Flowchart symbols used in this book

The *process box* indicates what must be done at this point in the program execution. The operation specified by the process box could be shifting the contents of one general-purpose register to a peripheral register, multiplying two numbers, decrementing a loop count, and so on.

The *input/output box* is used to represent data that are either read or displayed by the computer.

The *decision box* contains a question that can be answered either yes or no. A decision box has two exits, also marked yes or no. The computer will take one action if the answer is yes and will take a different action if the answer is no.

The *on-page connector* indicates that the flowchart continues elsewhere on the same page. The place where it is continued will have the same label as the on-page connector. The *off-page connector* indicates that the flowchart continues on another page. To determine where the flowchart continues, one needs to look at the following pages of the flowchart to find the matching off-page connector.

Normal flow on a flowchart is from top to bottom and from left to right. Any line that does not follow this normal flow should have an arrowhead on it.

When the program gets complicated, the flowchart that documents the logic flow of the program also becomes difficult to follow. This is the limitation of the flowchart. In this book, both the flowchart and the algorithm procedure are mixed to describe the solution to a problem.

After one is satisfied with the algorithm or the flowchart, one can convert it to source code in one of the assembly or high-level languages. Each statement in the algorithm (or each block in the flowchart) will be converted into one or multiple assembly instructions or high-level language statements. If an algorithmic step (or a block in the flowchart) requires many assembly instructions or high-level language statements to implement, then it might be beneficial to either (1) convert this step (or block) into a subroutine and just call the subroutine or (2) further divide the algorithmic step (or flowchart block) into smaller steps (or blocks) so that it can be coded with just a few assembly instructions or high-level language statements.

The next major step is *program testing*, which means testing for anomalies. Here one will first test for normal inputs that one always expects. If the result is as one expects, then one goes on to test the borderline inputs. Test for the maximum and minimum values of the input. When the program passes this test also, one continues to test for illegal input values. If the algorithm includes several branches, then enough values should be used to exercise all the possible branches to make sure that the program will operate correctly under all possible circumstances.

In the rest of this book, most of the examples are well defined. Therefore, our focus is on how to design the algorithm that solves the specified problem and also convert the algorithm into source code.

2.6 A Template for Writing Assembly Programs

When testing a user program, it should be considered as the only program executed by the computer. To achieve that, it should be written in a way that can be executed immediately out of reset. The following format will allow us to do that:

```
        org     0x0000
        goto    start
        org     0x08
        . . .              ;high-priority interrupt service routine
        org     0x18
        . . .              ;low-priority interrupt service routine
start   . . .
        . . .              ;your program
        end
```

The PIC18 MCU reserves a small block of memory locations to hold the reset handling routine and high-priority and low-priority interrupt service routines. The reset, the high-priority interrupt, and the low-priority interrupt service routines start at 0x000, 0x0008, and 0x0018, respectively. The user program should start somewhere after 0x0018.

In an application, the reset handling routine will be responsible for initializing the MCU hardware and performing any necessary housekeeping functions. The reset signal will provide default values to many key registers and allow the MCU to operate. At this moment, the user will take advantage of this by simply using the **goto** instruction to jump to the starting point of the user program. By doing this, the user program can be tested.

Since interrupt handling has not been covered yet, we will be satisfied by making both the high- and the low-priority interrupt service routines dummy routines that do nothing but simply return. This can be achieved by placing the **retfie** instruction in the default location. Therefore, the following template will be used to test the user program until interrupts are discussed:

```
        org     0x0000
        goto    start       ;reset handling routine
        org     0x08
        retfie              ;high-priority interrupt service routine
        org     0x18
        retfie              ;low-priority interrupt service routine
start   . . .
        . . .               ;your program
        end
```

2.7 Case Issue

The PIC18 instructions can be written in uppercase or lowercase. However, Microchip MPASM cross assembler is case sensitive. Microchip provides a free integrated development environment MPLAB IDE to all of its users. The MPLAB IDE provides an *include* file for every MCU made by Microchip. Each of these include files provides the definitions of all special-function registers for the specific MCU. All function registers and their individual bits are defined in uppercase. Since one will include one of these MCU include files in his or her assembly program, using uppercase for special-function register names becomes necessary. The convention adopted in this book is to use lowercase for instruction and directive mnemonics but uppercase for all function registers and their bits.

2.8 Writing Programs to Perform Arithmetic Computations

The PIC18 MCU has instructions for performing 8-bit addition, subtraction, and multiplication operations. Operations that deal with operands longer than eight bits can be synthesized by using a sequence of appropriate instructions. The PIC18 MCU provides no instruction for division, and hence this operation must also be synthesized by an appropriate sequence of instructions. The algorithm for implementing division operation will be discussed in Chapter 4.

In this section, smaller programs that perform simple computations will be used to demonstrate how a program is written.

2.8.1 Perform Addition Operations

As discussed in Chapter 1, the PIC18 MCU has two ADD instructions with two operands and one ADD instruction with three operands. These ADD instructions are designed to perform 8-bit additions. The execution result of the ADD instruction will affect all flag bits of the STATUS register.

The three-operand ADD instruction will be needed in performing multibyte ADD operations. For an 8-bit MCU, a multibyte addition is also called a *multiprecision* addition. A multiprecision addition must be performed from the least significant byte toward the most significant byte, just like numbers are added from the least significant digit toward the most significant digit.

When dealing with multibyte numbers, there is an issue regarding how the number is stored in memory. If the least significant byte of the number is stored at the lowest address, then the byte order is called *little-endian*. Otherwise, the byte order is called *big-endian*. This text will follow the little-endian byte order in order to be compatible with the MPLAB IDE software from Microchip. MPLAB IDE will be used throughout this text.

Example 2.4
▼

Write a program that adds the three numbers stored in data registers at 0x20, 0x30, and 0x40 and places the sum in data register at 0x50.

Solution: The algorithm for adding three numbers is as follows:

Step 1
Load the number stored at 0x20 into the WREG register.

Step 2
Add the number stored at 0x30 and the number in the WREG register and leave the sum in the WREG register.

Step 3
Add the number stored at 0x40 and the number in the WREG register and leave the sum in the WREG register.

Step 4
Store the contents of the WREG register in the memory location at 0x50.
The program that implements this algorithm is as follows:

```
        #include <p18F8720.inc>
        org       0x00
        goto      start
        org       0x08
        retfie
        org       0x18
        retfie
start   movf      0x20,W,A      ;copy the contents of 0x20 to WREG
        addwf     0x30,W,A      ;add the value in 0x30 to that of WREG
        addwf     0x40,W,A      ;add the value in 0x40 to that of WREG
        movwf     0x50,A        ;save the sum in memory location 0x50
        end
```

▲

Example 2.5
▼

Write a program that adds the 24-bit integers stored at 0x10 . . . 0x12 and 0x13 . . . 0x15, respectively, and stores the sum at 0x20 . . . 0x22.

Solution: The addition starts from the least significant byte (at the lowest address for little-endian byte order). One of the operand must be loaded into the WREG register before addition can be performed. The program is as follows:

```
        #include <p18F8720.inc>
        org      0x00
        goto     start
        org      0x08
        retfie
        org      0x18
        retfie
start   movf     0x10,W,A    ;copy the value of location at 0x10 to WREG
        addwf    0x13,W,A    ;add & leave the sum in WREG
        movwf    0x20,A      ;save the sum at memory location 0x20
        movf     0x11,W,A    ;copy the value of location at 0x11 to WREG
        addwfc   0x14,W,A    ;add with carry & leave the sum in WREG
        movwf    0x21,A      ;save the sum at memory location 0x21
        movf     0x12,W,A    ;copy the value of location at 0x12 to WREG
        addwfc   0x15,W,A    ;add with carry & leave the sum in WREG
        movwf    0x22,A      ;save the sum at memory location 0x22
        end
```

▲

2.8.2 Perform Subtraction Operations

The PIC18 MCU has two two-operand and two three-operand SUBTRACT instructions. SUBTRACT instructions will also affect all the flag bits of the STATUS register. Like other MCUs, the PIC18 MCU executes a SUBTRACT instruction by performing two's complement addition. When the subtrahend is larger than the minuend, a borrow is needed, and the PIC18 MCU flags this situation by clearing the C flag of the STATUS register.

Three-operand subtraction instructions are provided mainly to support the implementation of multibyte subtraction. A multibyte subtraction is also called a *multiprecision subtraction*. A multiprecision subtraction must be performed from the least significant byte toward the most significant byte.

Example 2.6

▼

Write a program to subtract 5 from memory locations 0x10 to 0x13.

Solution: The algorithm for this problem is as follows:

Step 1
Place 5 in the WREG register.

Step 2
Subtract WREG from the memory location 0x10 and leave the difference in the memory location 0x10.

Step 3
Subtract WREG from the memory location 0x11 and leave the difference in the memory location 0x11.

Step 4
Subtract WREG from the memory location 0x12 and leave the difference in the memory location 0x12.

Step 5

Subtract WREG from the memory location 0x13 and leave the difference in the memory location 0x13.

The assembly program that implements this algorithm is as follows:

```
        #include <p18F8720.inc>
        org      0x00
        goto     start
        org      0x08
        retfie
        org      0x18
        retfie
start   movlw    0x05            ;place the value 5 in WREG
        subwf    0x10,F,A        ;subtract 5 from memory location 0x10
        subwf    0x11,F,A        ;subtract 5 from memory location 0x11
        subwf    0x12,F,A        ;subtract 5 from memory location 0x12
        subwf    0x13,F,A        ;subtract 5 from memory location 0x13
        end
```

Example 2.7

Write a program that subtracts the number stored at 0x20 . . . 0x23 from the number stored at 0x10 . . . 0x13 and leaves the difference at 0x30 . . . 0x33.

Solution: The logic flow of this problem is shown in Figure 2.2.

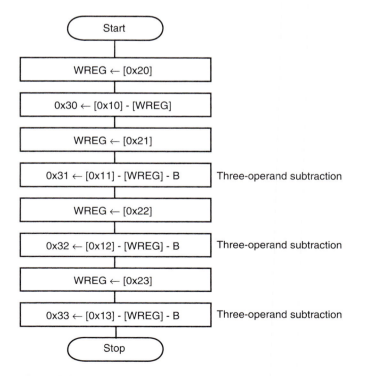

Figure 2.2 ■ Logic flow of Example 2.7

The program that implements the logic illustrated in Figure 2.2 is as follows:

```
          #include <p18F8720.inc>
          org      0x00
          goto     start
          org      0x08
          retfie
          org      0x18
          retfie
start     movf     0x20, W, A
          subwf    0x10, W, A      ;subtract the least significant byte
          movwf    0x30, A
          movf     0x21, W, A
          subwfb   0x11, W, A      ;subtract the second to least significant byte
          movwf    0x31, A
          movf     0x22, W, A
          subwfb   0x12, W, A      ;subtract the second to most significant byte
          movwf    0x32, A
          movf     0x23, W, A
          subwfb   0x13, W, A      ;subtract the most significant byte
          movwf    0x33, A
          end
```

2.8.3 Binary Coded Decimal Addition

All computers perform arithmetic using binary arithmetic. However, input and output equipment generally uses decimal numbers because we are used to decimal numbers. Computers can work on decimal numbers as long as they are encoded properly. The most common way to encode decimal numbers is to use four bits to encode each decimal digit. For example, 1234 is encoded as 0001 0010 0011 0100. This representation is called *binary-coded decimal* (BCD). If BCD format is used, it must be preserved during the arithmetic processing.

The BCD representation simplifies input/output conversion but complicates the internal computation. The use of the BCD representation must be carefully justified.

The PIC18 MCU performs all arithmetic in binary format. The following instruction sequence appears to cause the PIC18 MCU to add the decimal numbers 31 and 47 and store the sum at the memory location 0x50:

```
          movlw    0x31
          addlw    0x47
          movwf    0x50, A
```

This instruction sequence performs the following addition:

```
   h 3 1
 + h 4 7
   h 7 8
```

When the PIC18 MCU executes this instruction sequence, it adds the numbers according to the rules of binary addition and produces the sum h'78', which is a correct decimal number. However, a problem occurs when the PIC18 MCU adds two BCD digits that yield a sum larger than 9:

```
  h 2 4      h 3 6      h 2 9
+ h 6 7    + h 4 7    + h 4 7
  h 8 B      h 7 D      h 7 0
```

The first two additions are obviously incorrect because the results have illegal characters. The third example does not contain any illegal character. However, the correct result should be 76 instead of 70. There is a carry from the lower digit to the upper digit.

In summary, a sum in BCD is incorrect if the sum is greater than 9 or if there is a carry from the lower digit to the upper digit. Incorrect BCD sums can be adjusted by performing the following operations:

1. Add 0x6 to every sum digit greater than 9.

2. Add 0x6 to every sum digit that had a carry of 1 to the next higher digit.

These problems are corrected as follows:

```
  h 2 4      h 3 6      h 2 9
+ h 6 7    + h 4 7    + h 4 7
  h 8 B      h 7 D      h 7 0
+ h   6    + h   6    + h   6
  h 9 1      h 8 3      h 7 6
```

The bit 1 of the STATUS register is the digit carry (DC) flag that indicates if there is a carry from bit 3 to bit 4 of the addition result. The decimal adjust WREG (**daw**) instruction adjust the 8-bit value in the WREG register resulting from the earlier addition of two variables (each in packed BCD format) and produces a correctly packed BCD result. Multibyte decimal addition is also possible by using the DAW instruction.

Example 2.8
▼

Write an instruction sequence that adds the decimal number stored in 0x23 and 0x24 together and stores the sum in 0x25. The result must also be in BCD format.

Solution: The instruction is as follows:

```
movf    0x23,W,A
addwf   0x24,W,A
daw
movwf   0x25,A
```

▲

Example 2.9
▼

Write an instruction sequence that adds the decimal numbers stored at 0x10 . . . 0x13 and 0x14 . . . 0x17 and stores the sum in ox20 . . . 0x23. All operands are in the access bank.

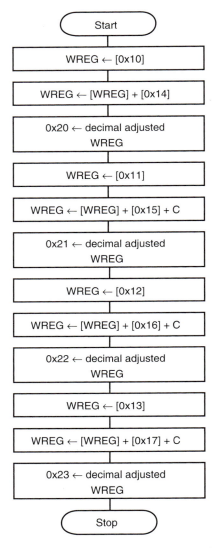

Figure 2.3 ■ Logic flow of Example 2.9

Solution: In order to make sure that the sum is also in decimal format, decimal adjustment must be done after the addition of each byte pair. The logic flow of this problem is shown in Figure 2.3. The program is as follows:

```
#include <p18F8720.inc>
org      0x00
goto     start
org      0x08
retfie
org      0x18
retfie
```

```
start    movf      0x10,W,A      ;WREG ← [0x10]
         addwf     0x14,W,A      ;WREG ← [0x10] + [0x14]
         daw                     ;decimal adjust WREG
         movwf     0x20,A        ;save the least significant sum digit
         movf      0x11,W,A      ;WREG ← [0x11]
         addwfc    0x15,W,A
         daw
         movwf     0x21,A
         movf      0x12,W,A      ;WREG ← [0x12]
         addwfc    0x16,W,A
         daw
         movwf     0x22,A
         movf      0x13,W,A      ;WREG ← [0x13]
         addwfc    0x17,W,A
         daw
         movwf     0x23,A        ;save the most significant sum digit
         end
```

▲

2.8.4 Multiplication

The PIC18 MCU provides two unsigned multiply instructions. The **mullw k** instruction multiplies an 8-bit literal with the WREG register and places the 16-bit product in the register pair PRODH:PRODL. The upper byte of the product is placed in the PRODH register, whereas the lower byte of the product is placed in the PRODL register. The **mulwf f,a** instruction multiplies the contents of the WREG register with that of the specified file register and leaves the 16-bit product in the register pair PRODH:PRODL. The upper byte of the product is placed in the PRODH register, whereas the lower byte of the product is placed in the PRODL register.

Example 2.10

▼

Write an instruction sequence to multiply two 8-bit numbers stored in data memory locations 0x10 and 0x11, respectively, and place the product in data memory locations 0x20 and 0x21.

Solution: The instruction sequence is as follows:

```
movf   0x10, W,A
mulwf  0x11,A
movff  PRODH, 0x21
movff  PRODL, 0x20
```

The unsigned multiply instructions can also be used to perform multiprecision multiplications. In a multiprecision multiplication, the multiplier and the multiplicand must be broken down into 8-bit chunks, and multiple 8-bit by 8-bit multiplications must be performed. Assume that we want to multiply a 16-bit hex number **P** by another 16-bit hex number **Q**. To illustrate the procedure, we will break P and Q down as follows:

$P = P_H P_L$
$Q = Q_H Q_L$

where P_H and Q_H are the upper eight bits of P and Q, respectively, and P_L and Q_L are the lower eight bits. Four 8-bit by 8-bit multiplications are performed, and then the partial products are added together as shown in Figure 2.4.

▲

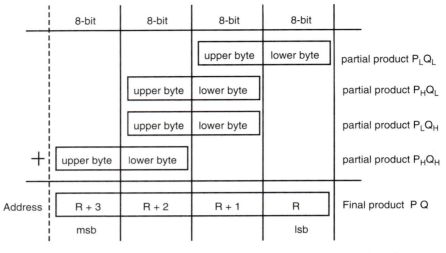

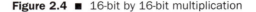

Note: msb stands for most significant byte and lsb stands for least significant byte

Figure 2.4 ■ 16-bit by 16-bit multiplication

Example 2.11

▼

Write a program to multiply two 16-bit unsigned integers assuming that the multiplier and multiplicand are stored in data memory locations M1 . . . M1 + 1 and N1 . . . N1 + 1, respectively. Store the product in data memory locations PR . . . PR + 3. The multiplier, the multiplicand, and the product are located in the access bank.

Solution: The algorithm for the unsigned 16-bit multiplication is as follows:

Step 1
Compute the partial product $M1_L N1_L$ and save it in locations PR and PR + 1.

Step 2
Compute the partial product $M1_H N1_H$ and save it in locations PR + 2 and PR + 3.

Step 3
Compute the partial product $M1_H N1_L$ and add it to memory locations PR + 1 and PR + 2. The C flag may be set to 1 after this addition.

Step 4
Add the C flag to memory location PR + 3.

Step 5
Compute the partial product $M1_L N1_H$ and add it to memory locations PR + 1 and PR + 2. The C flag may be set to 1 after this addition.

Step 6
Add the C flag to memory location PR + 3.

The assembly program that implements this algorithm is as follows:

```
        #include <p18F8720.inc>
n1_h    equ     0x37            ; upper byte of the first number
n1_1    equ     0x23            ; lower byte of the first number
m1_h    equ     0x66            ; upper byte of the second number
m1_1    equ     0x45            ; lower byte of the second number
M1      set     0x00            ; multiplicand
N1      set     0x02            ; multiplier
PR      set     0x06            ; product
        org     0x00
        goto    start
        org     0x08
        retfie
        org     0x18
        retfie
start   movlw   m1_h            ; set up test numbers
        movwf   M1 + 1,A        ; "
        movlw   m1_1            ; "
        movwf   M1,A            ; "
        movlw   n1_h            ; "
        movwf   N1+1,A          ; "
        movlw   n1_1            ; "
        movwf   N1,A            ; "
        movf    M1+1,W,A
        mulwf   N1+1,A          ; compute $M1_H \times N1_H$
        movff   PRODL, PR+2
        movff   PRODH,PR+3
        movf    M1,W,A          ; compute $M1_L \times N1_L$
        mulwf   N1,A
        movff   PRODL, PR
        movff   PRODH,PR+1
        movf    M1,W,A
        mulwf   N1+1,A          ; compute $M1_L \times N1_H$
        movf    PRODL,W,A       ; add $M1_L \times N1_H$ to PR
        addwf   PR+1,F,A        ; "
        movf    PRODH,W,A       ; "
        addwfc  PR+2,F,A        ; "
        movlw   0               ; "
        addwfc  PR+3,F,A        ; add carry
        movf    M1+1,W,A
        mulwf   N1,A            ; compute $M1_H \times N1_L$
        movf    PRODL,W,A       ; add $M1_H \times N1_L$ to PR
        addwf   PR+1,F,A        ; "
        movf    PRODH,W,A       ; "
        addwfc  PR+2,F,A        ; "
        movlw   0               ; "
        addwfc  PR+3,F,A        ; add carry
        nop
        end
```

Multiplication of other lengths (such as 32-bit by 32-bit or 24-bit by 16-bit) can be performed using an extension of the same method.

▲

2.9 Program Loops

One of the most powerful features of a computer is its ability to perform the same operation repeatedly without making any error. In order to tell the computer to perform the same operation repeatedly, program loops must be written.

A loop may be executed for a finite number of times or forever. A *finite loop* is a sequence of instructions that will be executed for a finite number of times, while an *endless loop* is a sequence of instructions that will be repeated forever.

2.9.1 Program Loop Constructs

There are four major looping methods:

1. **Do statement S forever.** This is an infinite loop in which the statement S will be executed forever. In some applications, the user may add the statement "If C then exit" to get out of the infinite loop. An infinite loop is illustrated in Figure 2.5.

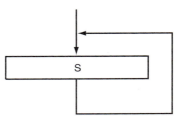

Figure 2.5 ■ An infinite loop

An infinite loop requires the use of *"goto target"* or *"bra target"* as the last instruction of the loop for the PIC18 MCU, where **target** is the label of the start of the loop.

2. **For i = n1 to n2 do S** or **For i = n2 downto n1 do S.** In this construct, the variable i is used as the loop counter that keeps track of the remaining times that the statements S is to be executed. The loop counter can be incremented (the first case) or decremented (the second case). The statement S is executed **n2 − n1 + 1** times. The value of n2 is assumed to be larger than n1. If there is concern that the relationship n1 ≤ n2 may not hold, then it must be checked at the beginning of the loop. Four steps are required to implement a *For loop*:

Step 1
Initialize the loop counter.

Step 2
Compare the loop counter with the limit to see if it is within bounds. If it is, then perform the specified operations. Otherwise, exit the loop.

Step 3
Increment (or decrement) the loop counter.

Step 4
Go to Step 2.

A **For-loop** is illustrated in Figure 2.6.

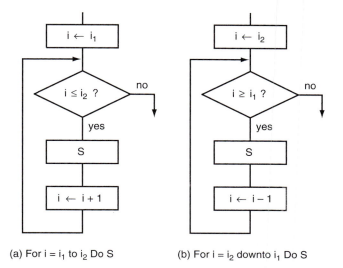

(a) For i = i₁ to i₂ Do S   (b) For i = i₂ downto i₁ Do S

Figure 2.6 ■ A **For-loop** looping construct

3. **While C Do S.** In this looping construct, the condition C is tested at the start of the loop. If the condition C is true, then the statement S will be executed. Otherwise, the statement S will not be executed. The **While C Do S** looping construct is illustrated in Figure 2.7. The implementation of a **while loop** consists of four steps:

Step 1
Initialize the logical expression C.

Step 2
Evaluate the logical expression C.

Step 3
Perform the specified operations if the logical expression C evaluates to true. Update the logical expression C and go to Step 2. The expression C may be updated by external events, such as interrupt.

Step 4
Exit the **while loop.**

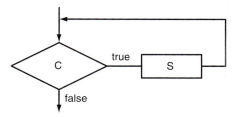

Figure 2.7 ■ The **While ... Do** looping construct

4. **Repeat S until C.** The statement S is executed, and then the logical expression C is evaluated. If C is true, then the statement S will be executed again. Otherwise, the next statement will be executed, and the loop is ended. The action of this looping construct is illustrated in Figure 2.8. The statement S will be executed at least once. This looping construct consists of three steps:

Step 1
Initialize the logical expression C.

Step 2
Execute the statement S.

Step 3
If the logical expression C is true, then go to Step 2. Otherwise, exit the loop.

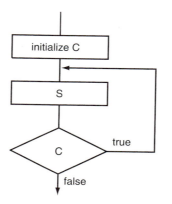

Figure 2.8 ■ The **Repeat ... Until** looping construct

2.9.2 Changing the Program Counter

A normal program flow is one in which the CPU executes instructions sequentially starting from lower addresses toward higher addresses. The implementation of a program loop requires the capability of changing the direction of a normal program flow. The PIC18 MCU supplies a group of instructions (shown in Table 2.9) that may change the normal program flow.

In the normal program flow, the program counter value is incremented by 2 or 4 (for two-word instructions). The PIC18 program counter is 21 bits wide. The low byte is called the PCL register. The high byte is called the PCH register. This register contains the PC<15:8> bits and is not directly readable or writable. Updates to the PCH register may be performed through the PCLATH register. The upper byte is called the PCU register. This register contains the PC<20:16> bits and is not directly readable or writable. Updates to the PCU register may be performed through the PCLATU register.

Figure 2.9 shows the interaction of the PCU, PCH, and PCL registers with the PCLATU and PCLATH registers.

Mnemonics, operands	Description	16-bit instruction word	Status affected
BC n	Branch if carry	1110 0010 nnnn nnnn	None
BN n	Branch if negative	1110 0110 nnnn nnnn	None
BNC n	Branch if no carry	1110 0011 nnnn nnnn	None
BNN n	Branch if not negative	1110 0111 nnnn nnnn	None
BNOV n	Branch if not overflow	1110 0101 nnnn nnnn	None
BNZ n	Branch if not zero	1110 0001 nnnn nnnn	None
BOV n	Branch if overflow	1110 0100 nnnn nnnn	None
BRA n	Branch unconditionally	1110 0nnn nnnn nnnn	None
BZ n	Branch if zero	1110 0000 nnnn nnnn	None
CALL n,s	Call subroutine	1110 110s kkkk kkkk 1111 kkkk kkkk kkkk	None
GOTO n	Go to address	1110 1111 kkkk kkkk 1111 kkkk kkkk kkkk	None
RCALL n	Relative call	1101 1nnn nnnn nnnn	None
RETLW k	Return with literal in WREG	0000 1100 kkkk kkkk	None
RETURN s	Return from subroutine	0000 0000 0001 001s	None

Table 2.9 ■ PIC18 instructions that change program flow

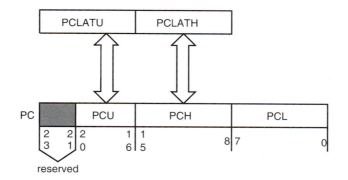

Figure 2.9 ■ Program counter structure

The low byte of the PC register is mapped in the data memory. PCL is readable and writable just as is any other data register. PCU and PCH are the upper and high bytes of the PC, respectively, and are not directly addressable. Registers PCLATU<4:0> and PCLATH<7:0> are used as holding latches for PCU and PCH and are mapped into data memory. Any time the PCL is read, the contents of PCH and PCU are transferred to PCLATH and PCLATU, respectively. Any time PCL is written into, the contents of PCLATH and PCLATU are transferred to PCH and PCU, respectively. The resultant effect is a branch. This is shown in Figure 2.10.

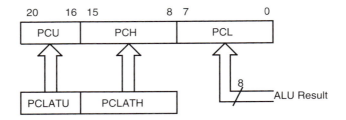

Figure 2.10 ■ Action taken when an instruction uses PCL as the destination

The PC addresses bytes rather than words in the program memory. Because the PIC18 MCU must access the instructions in program memory on an even byte boundary, the least significant bit of the PC register is forced to 0, and the PC register increments by two for each instruction. The least significant bit of PCL is readable but not writable. Any write to the least significant bit of PCL is ignored.

Finite loops require the use of one of the conditional branch instructions. The PIC18 MCU can make a forward or a backward branch. When a branch instruction is being executed, the 8-bit signed value contained in the branch instruction is added to the current PC. When the signed value is negative, the branch is backward. Otherwise, the branch is forward. The branch distance (in the unit of word) is measured from the byte immediately after the branch instruction as shown in Figure 2.11. The branch distance is also called *branch offset*. Since the branch offset is 8-bit, the range of branch is between −128 and +127 words.

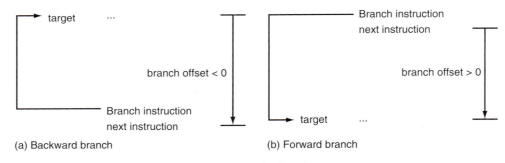

(a) Backward branch   (b) Forward branch

Figure 2.11 ■ Sign of branch offset and branch direction

The PIC18 CPU makes the branch decision using the condition flags of the STATUS register. Using the conditional branch instruction as the reference point, the instruction

    bn   -10

will branch backward nine words if the N flag is set to 1.

    bc   10

will branch forward 11 words if the C flag is set to 1.

Usually, counting the number of words to branch is not very convenient. Therefore, most assemblers allow the user to use the label of the target instruction to replace the branch offset. For example, the **bn –10** can be written as

```
is_minus   . . .

           . . .
      bn       is_minus
```

Using the label of the target instruction to replace the branch offset has another advantage: the user does not need to recalculate the branch offset if one or more instructions are added or deleted between the branch instruction and the target instruction.

The following two instructions are often used to increment or decrement the loop counter and hence update the condition flags:

```
incf f,d,a      ; increment file register f
decf f,d,a      ; decrement file register f
```

In addition to conditional branch instructions, the PIC18 MCU can also use the **goto** instruction to implement program loops. This method will require one to use another instruction that performs a compare, decrement, increment, or bit test operation to set up the condition for making a branch decision. These instructions are listed in Table 2.10.

Mnemonics, operands	Description	16-bit instruction word	Status affected
CPFSEQ f,a	Compare f with WREG, skip =	0110 001a ffff ffff	None
CPFSGT f,a	Compare f with WREG, skip >	0110 010a ffff ffff	None
CPFSLT f,a	Compare f with WREG, skip <	0110 000a ffff ffff	None
DECFSZ f,d,a	Decrement f, skip if 0	0010 11da ffff ffff	None
DCFSNZ f,d,a	Decrement f, skip if not 0	0100 11da ffff ffff	None
INCFSZ f,d,a	Increment f, skip if 0	0011 11da ffff ffff	None
INFSNZ f,d,a	Increment f, skip if not 0	0100 10da ffff ffff	None
TSTFSZ f,a	Test f, skip if 0	0110 011a ffff ffff	None
BTFSC f,b,a	Bit test f, skip if clear	1011 bbba ffff ffff	None
BTFSS f,b,a	Bit test f, skip if set	1010 bbba ffff ffff	None
goto n	goto address n (2 words)	1110 1111 kkkk kkkk 1111 kkkk kkkk kkkk	None

Table 2.10 ■ Non-branch Instructions that can be used to implement conditional branch

Suppose the loop counter is referred to as **i_cnt** and that the loop limit is placed in WREG. Then the following instruction sequence can be used to decide whether the loop should be continued:

```
i_loop   . . .

         . . .                  ; i_cnt is incremented in the loop
         cpfseq    i_cnt,A      ; compare i_cnt with WREG and skip if equal
         goto      i_loop       ; executed when i_cnt ≠ loop limit
```

Suppose that a program loop is to be executed **n** times. Then the following instruction sequence can do just that:

```
n           equ     20              ;n has the value of 20
lp_cnt      set     0x10            ; assign file register 0x10 to lp_cnt
            . . .
            movlw   n
            movwf   lp_cnt          ; prepare to repeat the loop for n times
loop        . . .                   ; program loop
            . . .                   ; "
            decfsz  lp_cnt,F,A      ; decrement lp_cnt and skip if equal to 0
            goto    loop            ; executed if lp_cnt ≠ 0
```

If the loop label is within 128 words from the branch point, then one can also use the one-word **bra loop** instruction to replace the **goto loop** instruction. The previously mentioned loop can also be implemented using the **bnz loop** instruction as follows:

```
lp_cnt      set     0x10            ; use file register 0x10 as lp_cnt
            . . .
            movlw   n
            movwf   lp_cnt          ; prepare to repeat the loop for n times
loop        . . .                   ; program loop
            . . .                   ; "
            decf    lp_cnt,F,A      ; decrement lp_cnt
            bnz     loop            ; executed if lp_cnt ≠ 0
```

The **btfsc f,b,a** and **btfss f,b,a** instructions are often used to implement a loop that waits for a certain flag bit to be cleared or set during an I/O operation. For example, the following instructions will be executed repeatedly until the ADIF bit (bit 6) of the PIR1 register is set to 1:

```
again       btfss   PIR1, ADIF,A    ; wait until ADIF bit is set to 1
            bra     again
```

The following instruction sequence will be executed repeatedly until the DONE bit (bit 2) of the ADCON0 register is cleared:

```
wait_loop   btfsc   ADCON0,DONE,A   ; wait until the DONE bit is cleared
            bra     wait_loop
```

Example 2.12
▼

Write a program to compute $1 + 2 + 3 + \ldots + n$ and save the sum at 0x00 and 0x01 assuming that the value of n is in a range such that the sum can be stored in two bytes.

Solution: The logic flow for computing the desired sum is shown in Figure 2.12. This flowchart implements the **For i = i$_1$ to i$_2$ Do S** loop construct.

The following program implements the algorithm illustrated in Figure 2.12:

```
            #include <p18F8720.inc>
n           equ     D'50'
sum_hi      set     0x01            ;high byte of sum
sum_lo      set     0x00            ;low byte of sum
i           set     0x02            ;loop index i
            org     0x00            ;reset vector
```

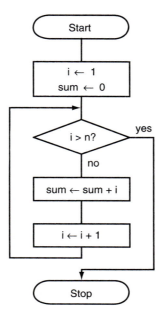

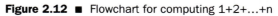

Figure 2.12 ■ Flowchart for computing 1+2+...+n

```
                goto        start
                org         0x08
                retfie
                org         0x18
                retfie
start           clrf        sum_hi,A            ; initialize sum to 0
                clrf        sum_lo,A            ; "
                clrf        i,A                 ; initialize i to 0
                incf        i,F,A               ; i starts from 1
sum_lp          movlw       n                   ; place n in WREG
                cpfsgt      i,A                 ; compare i with n and skip if i > n
                bra         add_lp              ; perform addition when i ≤ 50
                bra         exit_sum            ; it is done when i > 50
add_lp          movf        i,W,A               ; place i in WREG
                addwf       sum_lo,F,A          ; add i to sum_lo
                movlw       0
                addwfc      sum_hi,F,A          ; add carry to sum_hi
                incf        i,F,A               ; increment loop index i by 1
                bra         sum_lp
exit_sum        nop
                end
```

Example 2.13

Write a program to find the largest element stored in the array that is stored in data memory locations from 0x10 to 0x5F.

Solution: We use the indirect addressing mode to step through the given data array. The algorithm to find the largest element of the array is as follows:

Step 1
Set the value of the data memory location at 0x10 as the current temporary *array max.*

Step 2
Compare the next data memory location with the current temporary array max. If the new memory location is larger, then replace the current array max with the value of the current data memory location.

Step 3
Repeat the same comparison until all the data memory locations have been checked.

The flowchart of this algorithm is shown in Figure 2.13.

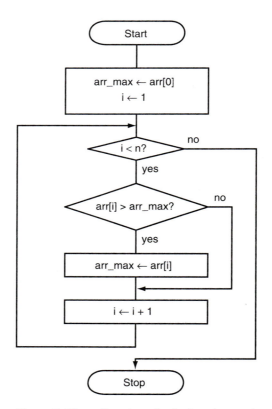

Figure 2.13 ■ Flowchart for finding the maximum array element

We use the **While C Do S** looping construct to implement the program loop. The condition to be tested is *i* < *n.* The PIC18 assembly program that implements the algorithm shown in Figure 2.13 is as follows:

```
arr_max    equ     0x00
i          equ     0x01
n          equ     D'80'              ; the array count
           #include <p18F8720.inc>
           org     0x00
           goto    start
           org     0x08
           retfie
           org     0x18
           retfie
start      movf    0x10,W,A           ; set arr[0] as the initial array max
           movwf   arr_max,A          ; "
           lfsr    FSR0,0x11          ; place 0x11 (address of arr[1]) in FSR0
           clrf    i,A                ; initialize loop count i to 0
again      movlw   n - 1              ; establish the number of comparisons to be made
; the next instruction implements the condition C(i = n)
           cpfslt  i,A                ; skip if i < n - 1
           goto    done               ; all comparisons have been done
; the following 7 instructions update the array max
           movf    POSTINC0,W         ; place arr[i] in WREG and increment array pointer
           cpfsgt  arr_max,A          ; is arr_max > arr[i]?
           goto    replace            ; no
           goto    next_i             ; yes
replace    movwf   arr_max,A          ; update the array max
next_i     incf    i,F,A
           goto    again
done       nop
           end
```

▲

2.10 Reading and Writing Data in Program Memory

The PIC18 program memory is 16-bit, whereas the data memory is 8-bit. In order to read and write program memory, the PIC18 MCU provides two instructions that allow the processor to move bytes between the program memory and the data memory:

- Table read (TBLRD)
- Table write (TBLWT)

Because of the mismatch of bus size between the program memory and data memory, the PIC18 MCU moves data between these two memory spaces through an 8-bit register (TABLAT). Figure 2.14 shows the operation of a table read with program memory and data memory.

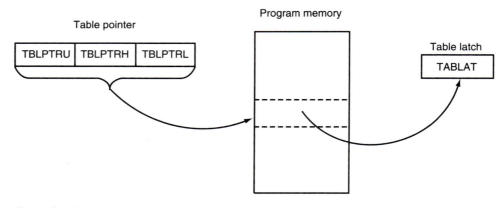

Figure 2.14 ■ Table read operation (redrawn with permission of Microchip)

Table-write operations store data from the data memory space into holding registers in program memory. Figure 2.15 shows the operation of a table write with program memory and data memory.

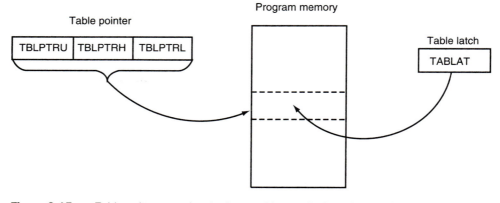

Figure 2.15 ■ Table write operation (redrawn with permission of microchip)

The on-chip program memory is either EPROM or flash memory. The erasure operation must be performed before an EPROM or flash memory location can be correctly programmed. The erasure and write operations for EPROM or flash memory take much longer time than the SRAM. This issue will be discussed in Chapter 14.

Eight instructions are provided for reading from and writing into the table in the program memory. These instructions are shown in Table 2.11.

Mnemonic, operator	Description	16-bit instruction word	Status affected
TBLRD*	Table read	0000 0000 0000 1000	none
TBLRD*+	Table read with post-increment	0000 0000 0000 1001	none
TBLRD*-	Table read with post-decrement	0000 0000 0000 1010	none
TBLRD+*	Table read with pre-increment	0000 0000 0000 1011	none
TBLWT*	Table write	0000 0000 0000 1100	none
TBLWT*+	Table write with post-increment	0000 0000 0000 1101	none
TBLWT*-	Table write with post-decrement	0000 0000 0000 1110	none
TBLWT+*	Table write with pre-increment	0000 0000 0000 1111	none

Table 2.11 ■ PIC18 MCU table read and write instructions

The table pointer (TBLPTR) addresses a byte within the program memory. The TBLPTR is comprised of three special-function registers (table pointer upper byte, high byte, and low byte). These three registers together form a 22-bit-wide pointer. The low-order 21 bits allow the device to address up to 2 MB of program memory space. The 22nd bit allows read-only access to the device ID, the user ID, and the configuration bits. The table pointer is used by the TBLRD and TBLWT instructions. Depending on the versions of these two instructions (shown in Table 2.11), the table pointer may be postdecremented (decremented after it is used), preincremented (incremented before it is used), or postincremented (incremented after it is used).

Whenever a table-read instruction is executed, a byte will be transferred from program memory to the table latch (TABLAT). All PIC18 members have a certain number of holding registers to hold data to be written into the program memory. Holding registers must be filled up before the program-memory-write operation can be started. The write operation is complicated by the EPROM and flash memory technology. The table-write operation to on-chip program memory (EPROM and flash memory) will be discussed in Chapter 14.

Example 2.14

▼

Write an instruction sequence to read a byte from program memory location at 0x60 into TABLAT.

Solution: The first step to read the byte in program memory is to set up the table pointer. The following instruction sequence will read the byte from the program memory:

```
clrf    TBLPTRU,A   ; set TBLPTR to point to data memory at
clrf    TBLPTRH,A   ; 0x60
movlw   0x60        ; "
movwf   TBLPTRL,A   ; "
tblrd*              ; read the byte into TABLAT
```

In assembly language programming, the programmer often uses a label to refer to an array. The MPASM assembler allows the user to use symbolic names to extract the values of the upper, the high, and the low bytes of a symbol:

- **upper** name refers to the upper part of **name.**
- **high** name refers to the middle part of **name.**
- **low** name refers to the low part of **name.**

Suppose that the symbol **arr_x** is the name of an array. Then the following instruction sequence places the address represented by **arr_x** in the table pointer:

```
movlw    upper arr_x
movwf    TBLPTRU,A      ; set up the upper part of the table pointer
movlw    high arr_x
movwf    TBLPTRH,A      ; set up the middle part of the table pointer
movlw    low arr_x
movwf    TBLPTRL,A      ; set up the lower part of the table pointer
```

2.11 Logic Instructions

The PIC18 MCU provides a group of instructions (shown in Table 2.12) that perform logical operations. These instructions allow the user to perform AND, OR, exclusive-OR, and complementing on 8-bit numbers.

Mnemonic, operator	Description	16-bit instruction word				Status affected
ANDWF f,d,a	AND WREG with f	0001	01da	ffff	ffff	Z,N
COMF f,d,a,	Complement f	0001	11da	ffff	ffff	Z,N
IORWF f,d,a	Inclusive OR WREG with f	0001	00da	ffff	ffff	Z,N
NEGF f,a	Negate f	0110	110a	ffff	ffff	all
XORWF f,d,a	Exclusive OR WREG with f	0001	10da	ffff	ffff	Z,N
ANDLW k	AND literal with WREG	0000	1011	kkkk	kkkk	Z,N
IOLW k	Inclusive OR literal with WREG	0000	1001	kkkk	kkkk	Z,N
XORLW k	Exclusive OR literal with WREG	0000	1010	kkkk	kkkk	Z,N

Table 12.12 ■ PIC18 MCU logic instructions

Logical operations are useful for looking for array elements with certain properties (e.g., divisible by power of 2) and manipulating I/O pin values (e.g., set certain pins to high, clear a few pins, toggle a few signals, and so on).

Example 2.15
▼

Write an instruction sequence to do the following:

(a)   Set bits 7, 6, and 0 of the PORTA register to high
(b)   Clear bits 4, 2, and 1 of the PORTB register to low
(c)   Toggle bits 7, 5, 3, and 1 of the PORTC register

Solution: These requirements can be achieved as follows:

(a) movlw   B'11000001'
      iorwf   PORTA, F, A
(b) movlw   B'11101001'
      andwf   PORTB, F, A
(c) movlw   B'10101010'
      xorwf   PORTC

▲

Example 2.16

▼

Write a program to find out the number of elements in an array of 8-bit elements that are a multiple of 8. The array is in the program memory.

Solution: A number is a multiple of 8 if its least significant three bits are 000. This can be tested by ANDing the array element with B'00000111'. If the result of this operation is zero, then the element is a multiple of 8. The algorithm is shown in the flowchart in Figure 2.16. This algorithm uses the **Repeat S until C** looping construct. The program is as follows:

```
                #include <p18F8720.inc>
ilimit    equ       0x20            ; loop index limit
count     set       0x00
ii        set       0x01            ; loop index
mask      equ       0x07            ; used to masked upper five bits
          org       0x00
          goto      start
          org       0x08            ; high-priority interrupt service routine
          retfie
          org       0x18            ; low-priority interrupt service routine
          retfie
start     clrf      count,A
          movlw     ilimit
          movwf     ii              ; initialize ii to ilimit
          movlw     upper array
          movwf     TBLPTRU,A
          movlw     high array
          movwf     TBLPTRH,A
          movlw     low array
          movwf     TBLPTRL,A
          movlw     mask
i_loop    tblrd*+                   ; read an array element into TABLAT
          andwf     TABLAT,F,A
          bnz       next            ; branch if not a multiple of 8
          incf      count, F,A      ; is a multiple of 8
next      decfsz    ii,F,A          ; decrement loop count
          bra       i_loop
          nop
array     db        0x00,0x01,0x30,0x03,0x04,0x05,0x06,0x07,0x08,0x09
          db        0x0A,0x0B,0x0C,0x0D,0x0E,0x0F,0x10,0x11,0x12,0x13
          db        0x14,0x15,0x16,0x17,0x18,0x19,0x1A,0x1B,0x1C,0x1D
          db        0x1E,0x1F
          end
```

▲

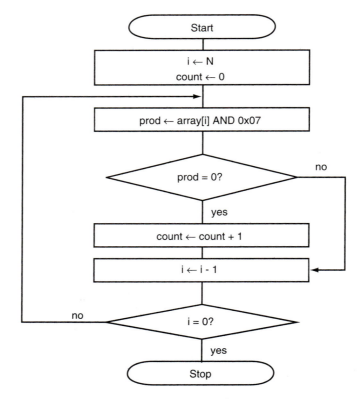

Figure 2.16 ■ Flowchart for Example 2.16

2.12 Using Program Loop to Create Time Delays

A time delay can be created by repeating an appropriate instruction sequence for certain number of times.

Example 2.17

Write a program loop to create a time delay of 0.5 ms. This program loop is to be run on a P18F8680 demo board clocked by a 40-MHz crystal oscillator.

Solution: Because each instruction cycle consists of four oscillator cycles, one instruction cycle lasts for 100 ns. The following instruction sequence will take 2 µs to execute:

```
loop_cnt  equ    0x00
again     nop
          nop
          nop
          nop
          nop
          nop
          nop
          nop
          nop
          nop
          nop
          nop
          nop
          nop
          nop
          nop
          dcfsnz    loop_cnt,F,A        ; decrement and skip the next instruction
bra       again
```

This instruction sequence can be shortened by using the following macro:

```
dup_nop   macro    kk           ; this macro will duplicate the nop instruction
          variable  i           ; kk times
i = 0
          while    i < kk
          nop
i += 1
          endw
          endm
```

The *nop* instruction performs no operation. Each of these instructions except *bra again* takes one instruction cycle to execute. The *bra again* instruction takes two instruction cycles to complete. Therefore, the previous instruction sequence takes 20 instruction cycles (or 2 µs) to execute.

To create a delay of 0.5 ms, the previous instruction sequence must be executed 250 times. This can be achieved by placing 250 in the **loop_cnt** register. The following program loop will create a time delay of 0.5 ms:

```
          movlw    D'250'
          movwf    loop_cnt, A
again     dup_nop  D'17'          ; 17 instruction cycle
          decfsz   loop_cnt,F,A   ; 1 instruction cycle (2 when [loop_cnt] = 0)
          bra      again          ;2 instruction cycle
```

This program tests the looping condition after 17 *nop* instructions have been executed, and hence it implements the **repeat S until C** loop construct. Longer delays can be created by adding another layer of loop.

Example 2.18

Write a program loop to create a time delay of 100 ms. This program loop is to be run on a PIC18 demo board clocked by a 40-MHz crystal oscillator.

Solution: A 100-ms time delay can be created by repeating the program loop in Example 2.17 for 200 times. The program loop is as follows:

```
lp_cnt1    equ      0x21
lp_cnt2    equ      0x22
           movlw    D'200'
           movwf    lp_cnt1,A
loop1      movlw    D'250'
           movwf    lp_cnt2,A
loop2      dup_nop  D'17'        ; 17 instruction cycles
           decfsz   lp_cnt2,F,A  ; 1 instruction cycle (2 when [lp_cnt1] = 0)
           bra      loop2        ; 2 instruction cycles
           decfsz   lp_cnt1,F,A
           bra      loop1
```

2.13 Rotate Instructions

The PIC18 MCU provides four rotate instructions. These four instructions are listed in Table 2.13.

Mnemonic, operator	Description	16-bit instruction word	Status affected
RLCF f, d, a	Rotate left f through carry	0011 01da ffff ffff	C, Z,N
RLNCF f, d, a	Rotate left f (no carry)	0100 11da ffff ffff	Z,N
RRCF f, d, a	Rotate right f through carry	0011 00da ffff ffff	C, Z,N
RRNCF f, d ,a	Rotate right f (no carry)	0100 00da ffff ffff	Z, N

Table 2.13 ■ PIC18 MCU rotate instructions

Rotate instructions can be used to manipulate bit fields and multiply or divide a number by a power of 2.

The operation performed by the **rlcf f,d,a** instruction is illustrated in Figure 2.17. The result of this instruction may be placed in the WREG register (d = 0) or the specified **f** register (d = 1).

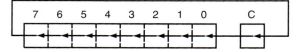

Figure 2.17 ■ Operation performed by the **rlcf f,d,a** instruction

The operation performed by the **rlncf f,d,a** instruction is illustrated in Figure 2.18. The result of this instruction may be placed in the WREG register (d = 0) or the specified **f** register (d = 1).

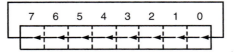

Figure 2.18 ■ Operation performed by the **rlncf f,d,a** instruction

The operation performed by the **rrcf f,d,a** instruction is illustrated in Figure 2.19. The result of this instruction may be placed in the WREG register (d = 0) or the specified **f** register (d = 1).

Figure 2.19 ■ Operation performed by the **rrcf f,d,a** instruction

The operation performed by the **rrncf f,d,a** instruction is illustrated in Figure 2.20. The result of this instruction may be placed in the WREG register (d = 0) or the specified **f** register (d = 1).

Figure 2.20 ■ Operation performed by the **rrncf f,d,a** instruction

Example 2.19
▼

Compute the new values of the data register 0x10 and the C flag after the execution of the **rlcf 0x10,F,A** instruction. Assume that the original value in data memory at 0x10 is 0xA9 and that the C flag is 0.

Solution: The operation of this instruction is shown in Figure 2.21.

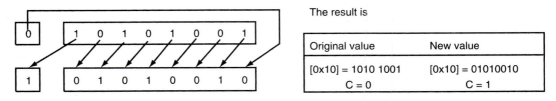

The result is

Original value	New value
[0x10] = 1010 1001	[0x10] = 01010010
C = 0	C = 1

Figure 2.21 ■ Operation of the RLCF 0x10,F,A instruction

▲

Example 2.20

Compute the new values of the data register 0x10 and the C flag after the execution of the rrcf **0x10,F,A** instruction. Assume that the original value in data memory at 0x10 is 0xC7 and that the C flag is 1.

Solution: The operation of this instruction is shown in Figure 2.22.

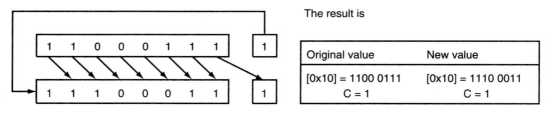

Figure 2.22 ■ Operation of the **rrcf 0x10,F,A** instruction

Example 2.21

Compute the new values of the data memory location 0x10 after the execution of the **rrncf 0x10,F,A** instruction and the **rlncf 0x10,F,A** instruction, respectively. Assume that the data memory location 0x10 originally contains the value of 0x6E.

Solution: The operation performed by the **rrncf 0x10,F,A** instruction and its result are shown in Figure 2.23.

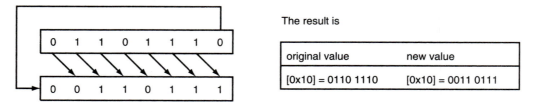

Figure 2.23 ■ Operation performed by the **rrncf 0x10,F,A** instruction

The operation performed by the **rlncf 0x10,F,A** instruction and its result are shown in Figure 2.24.

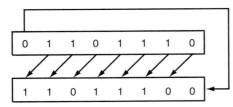

The result is

Before	After
[0x10] = 0110 1110	[0x10] = 1101 1100

Figure 2.24 ■ Operation performed by the **rlncf 0x10,F,A** instruction

2.14 Using Rotate Instructions to Perform Multiplications and Divisions

The operation of multiplying by the power of 2 can be implemented by shifting the operand to the left an appropriate number of positions, whereas dividing by the power of 2 can be implemented by shifting the operand to the right a certain number of positions.

Since the PIC18 MCU does not provide any shifting instructions, the shift operation must be implemented by using one of the rotate-through-carry instructions. The carry flag must be cleared before the rotate instruction is executed. As shown in Table 2.14, the PIC18 provides three instructions for manipulating an individual bit of a register. The **b** field specifies the bit position to be operated on.

Mnemonic, operator	Description	16-bit instruction word	Status affected
BCF f, b, a	Bit clear f	1001 bbba ffff ffff	none
BSF, f, b, a	Bit set f	1000 bbba ffff ffff	none
BTG f, b, a	Bit toggle f	0111 bbba ffff ffff	none

Table 2.14 ■ PIC18 bit manipulation instructions

Example 2.22

▼

Write an instruction sequence to multiply the three-byte number located at 0x00 to 0x02 by 8.

Solution: Multiplying by 8 can be implemented by shifting to the left three places. The left-shifting operation should be performed from the least significant byte toward the most significant byte. The following instruction sequence will achieve the goal:

```
        movlw    0x03           ; set loop count to 3
loop    bcf      STATUS, C, A   ; clear the C flag
        rlcf     0x00, F, A     ; shift left one place
        rlcf     0x01, F, A     ; "
        rlcf     0x02, F, A     ; "
        decfsz   WREG, W, A     ; have we shifted left three places yet?
        goto     loop           ; not yet, continue
```

▲

Example 2.23

▼

Write an instruction sequence to divide the three-byte number stored at 0x10 to 0x12 by 16.

Solution: Dividing by 16 can be implemented by shifting the number to the right four positions. The right-shifting operation should be performed from the most significant byte toward the least significant byte. The following instruction sequence will achieve the goal:

```
       movlw     0x04              ; set loop count to 4
loop   bcf       STATUS, C, A      ; shift the number to the right 1 place
       rrcf      0x12, F, A        ; "
       rrcf      0x11, F, A        ; "
       rrcf      0x10, F, A        ; "
       decfsz    WREG, W, A        ; have we shifted right four places yet?
       bra       loop              ; not yet, continue
```

▲

2.15 Summary

An assembly program consists of three types of statements: assembler directives, assembly language instructions, and comments. An assembler directive tells the assembler how to process subsequent assembly language instructions. Directives also provide a way for defining program constants and reserving space for dynamic variables. A statement of an assembly program consists of four fields: label, operation code, operands, and comment.

Although the PIC18 MCU can perform only 8-bit arithmetic operations, numbers that are longer than eight bits can still be added, subtracted, or multiplied by performing multiprecision arithmetic. Examples are used to demonstrate multiprecision addition, subtraction, and multiplication operations.

The multiprecision addition can be implemented with the **addwfc f,d,a** instruction, and multiprecision subtraction can be implemented with the **subwfb f,d,a** instruction. To perform multiprecision multiplication, both the multiplier and the multiplicand must be broken down into 8-bit chunks. The next step is to generate partial products and align them properly before adding them together.

The PIC18 MCU does not provide any divide instruction, and hence a divide operation must be synthesized.

Performing repetitive operation is the strength of a computer. For a computer to perform repetitive operations, one must write program loops to tell the computer what instruction sequence to repeat. A program loop may be executed a finite or an infinite number of times. There are four looping constructs:

- **Do** statement **S** forever
- **For i** = i_1 **to** i_2 **Do** S or **For i** = i_2 **downto** i_1 **Do** S
- **While** C **Do** S
- **Repeat** S **until** C

The PIC18 MCU provides many flow-control and conditional branch instructions for implementing program loops. Instructions for initializing and updating loop indices and variables are also available.

Rotate instructions are useful for bit-field manipulations. They can also be used to implement multiplying and dividing a variable by a power of 2. All rotate instructions operate on 8-bit registers only. One can write a sequence of instructions to rotate or shift a number longer than 8 bits.

A PIC18 instruction takes either one or two instruction cycles to complete. By choosing an appropriate instruction sequence and repeating it for a certain number of times, a time delay can be created.

2.16 Exercises

E2.1   Identify the four fields of the following instructions:

(a)		addwf	0x10,W,A	;add register 0x10 to WREG
(b)	wait	btfss	STATUS,F,A	; skip the next instruction if the C flag is 1
(c)		decfsz	cnt, F, A	; decrement *cnt* and skip if it is decremented to 0

E2.2   Find the valid and invalid labels in the following instructions and explain why an invalid label is invalid.

column 1
    ↓

a. sum_hi	equ	0x20	
b. low_t	incf	WREG, W, A	; increment WREG by 1
c. abc:	movwf	0x30, A	
d. 5plus3	clrf	0x33, A	
c. _may	decf	0x35, F, A	
f. ?less	iorwf	0x1A, F, A	
g. two_three	goto	less	

E2.3 Use an assembler directive to define a string "Please make a choice (1/2):" in program memory.

E2.4 Use assembler directives to define a table of all uppercase letters. Place this table in program memory starting from location 0x2000. Assign one byte to one letter.

E2.5 Use assembler directives to assign the symbols *sum, lp_cnt, height,* and *weight* to data memory locations at 0x00, 0x01, 0x02, 0x03, respectively.

E2.6 Write an instruction sequence to decrement the contents of data memory locations 0x10, 0x11, and 0x12 by 5, 3, and 1, respectively.

E2.7 Write an instruction sequence to add the 3-byte numbers stored in memory locations 0x11–0x13 and 0x14–0x16 and save the sum in memory locations 0x20–0x22.

E2.8 Write an instruction sequence to subtract the 4-byte number stored in memory locations 0x10–0x13 from the 4-byte number stored in memory locations 0x00–0x03 and store the difference in memory locations 0x20–0x23.

E2.9 Write an instruction sequence to shift the 4-byte number stored in memory locations 0x20–0x23 to the right arithmetically four places and leave the result in the same location.

E2.10 Write a program to shift the 64-bit number stored in data memory locations 0x10–0x17 to the left four places.

E2.11 Write an instruction sequence to multiply the 24-bit unsigned numbers stored in data memory locations 0x10–0x12 and 0x13–0x15 and store the product in data memory locations 0x20–0x25.

E2.12 Write an instruction sequence to multiply the unsigned 32-bit numbers stored in data memory locations 0x00–0x03 and 0x04–0x07 and leave the product in data memory memory locations 0x10–0x17.

E2.13 Write a program to compute the average of an array of 32 unsigned 8-bit integers stored in the program memory. Leave the array average in WREG. (Hint, the array contains 32 8-bit numbers. Therefore, the array average can be computed by using shift operation instead of division.)

E2.14 Write an instruction sequence that can extract the bit 6 to bit 2 of the WREG register and store the resultant value in the lowest five bits of the data register 0x10.

E2.15 Write an instruction sequence to create a time delay of 1 second.

E2.16 Write an assembly program to count the number of odd elements in an array of **n** 16-bit integers. The array is stored in program memory starting from the label **arr_x.**

E2.17 Write a PIC18 assembly program to count the number of elements in an array that are greater than 20. The array consists of **n** 8-bit numbers and is stored in program memory starting from the label **arr_y.**

E2.18 Write an assembly program to find the smallest element of an array of **n** 8-bit elements. The array is stored in program memory starting with the label **arr_z.**

E2.19 The sign of a signed number is the most significant bit of that number. A signed number is negative when its most significant bit is 1. Otherwise, it is positive. Write a program to count the number of elements that are positive in an array of **n** 8-bit integers. The array is stored in bank 1 of data memory starting from 0x00.

E2.20 Determine the number of times the following loop will be executed:

```
         #include <p18F8720.inc>
         movlw   0x80
loop     bcf     STATUS, C, A    ; clear the carry flag
         rrcf    WREG, W, A
         addwf   WREG, W, A      ; add WREG to itself
         btfsc   WREG, 7, A      ; test bit 7
         goto    loop
         . . .
```

E2.21 What will be the value of the carry flag after the execution of each of the following instructions? Assume that the WREG register contains 0x79 and that the carry flag is 0 before the execution of each instruction.

(a) addlw 0x30
(b) addlw 0xA4
(c) sublw 0x95
(d) sublw 0x40

E2.22 Write a program to compute the average of the squares of 32 8-bit numbers stored in the access bank from data memory location 0x00 to 0x1F. Save the average in the data memory locations 0x21–0x22.

E2.23 Suppose the contents of the WREG register and the C flag are 0x95 and 1, respectively. What will be the contents of the WREG register and the C flag after the execution of each of the following instructions?

(a) rrcf WREG, W, A
(b) rrncf WREG, W, A
(c) rlcf WREG, W, A
(d) rlncf WREG, W, A

E2.24 Write a program to swap the first element of the array with the last element of the array, the second element with the second-to-last element, and so on. Assume that the array has 20 8-bit elements and is stored in data memory. The starting address of this array is 0x10 in the access bank.

3

PIC18 Development Tools

3.1 Objectives

After completing this chapter, you should be able to

- Explain the function of software tools
- Explain the basic function of hardware tools
- Use MPLAB® IDE to enter programs and build the executable code
- Use MPLAB ICD2 to perform software debugging for the PIC18 microcontroller
- Use MPLAB simulator in perform software debugging for the PIC18 microcontroller

3.2 Development Tools

Development tools for microprocessors/microcontrollers can be classified into two categories: software and hardware. Software tools include programmers' editors, assemblers, compilers, simulators, debuggers, communication programs, and integrated development environment. Hardware tools include demo boards, logic analyzers, emulators, oscilloscopes, and logic probes.

Discussing all these tools is beyond the scope of this book. However, hardware and software tools made by Microchip and a few vendors for the PIC18 microcontrollers are reviewed in this book.

3.3 Software Tools

Software tools include text editors, cross assemblers, cross compilers, simulators, source-level debuggers, and integrated development environment.

3.3.1 Text Editors

The text editor is a program that allows us to enter and edit programs and text files. Editors range from very primitive to very sophisticated. A simple editor like the Notepad bundled with Windows 98/2000/XP provides simple editing functions in four different categories: *file, edit, search*, and *option*.

The Wordpad program bundled with the Windows 98/2000/XP is another simple editor that you can use to enter your program.

The Microchip MPLAB IDE has an embedded text editor for the user to enter programs. Most of the basic text editing functions are available in the MPLAB editor.

For professional programmers, neither Notepad nor Wordpad is adequate because of their primitive functions. A programmer's editor provides additional features, such as automatic keyword completion, keyword highlighting, syntax checking, and parentheses matching. These functions can speed up the user's program entry speed dramatically. The PFE editor is a freeware programmer's editor available from the Internet and can be downloaded from www.wintel.com or www.simtel.net.

3.3.2 Cross Assemblers and Cross Compilers

Cross assemblers and cross compilers generate executable code to be placed in ROM, EPROM, EEPROM, or flash memory of a PIC18-based product. Currently, several vendors are providing cross assemblers and cross-C compilers for the PIC18 microcontroller. At the time of this writing, Microchip, CCS, IAR, and HI-TECH provide cross compilers for the PIC18 microcontroller.

3.3.3 Simulator

A simulator allows us to run a PIC18 program without having the actual hardware. The contents of registers, internal memory, external memory, and the program source code are displayed in separate windows. The user can set breakpoints to the program and examine the program execution results. The simulator allows the user to step through the program to locate program bugs. The MPLAB IDE development software from Microchip includes simulators for all PIC16/PICI7/PIC18 devices.

3.3.4 Source-Level Debugger

A source-level debugger is a program that runs on a PC (or a workstation) and allows you to find problems in your code at the high level (such as C) or assembly language level. A source-level debugger allows you to set breakpoints at statements either at the high or assembly language level. It can execute the program from the start of the program until the breakpoint and then display the values of program variables. A source-level debugger can also trace your program statement by statement. A source-level debugger may have the option to run your program on the demo board or simulator. Like a simulator, a source-level debugger can display the contents of program variables, registers, memory locations, and program code in separate windows. With a source-level debugger, all debugging activities are done at the source level.

A source-level debugger can also invoke a simulator to perform debugging activity. The C-Spy from IAR is a source-level debugger for the PIC18 microcontrollers that uses this approach. In addition to running the program on a demo board, the debugger of the MPLAB IDE can also use simulator to debug your program.

3.3.5 Integrated Development Environment

An integrated development environment (IDE) includes everything that you need to enter, assemble, compile, link, and debug your application program. It includes every software tool mentioned earlier. A tool is invoked by clicking on the icon of the corresponding tool. The MPLAB IDE is an IDE designed for all Microchip microcontrollers. It is free and can be downloaded from Microchip's Web site at www.microchip.com. The Embedded Workbench from IAR is also an IDE for the PIC18 microcontrollers.

3.4 Hardware Tools

Numerous hardware development tools are available for the PIC18 microcontrollers. Microchip provides the following hardware development tools to support the hardware development of PIC18-based product:

1. In-circuit emulators
2. Device programmers
3. Demo boards
4. In-circuit debugger (ICD2)

3.4.1 The Nature of Debugging Activities

When an embedded product is being designed, both the hardware and the software components are being developed in parallel. Therefore, software debugging needs to be done before the final hardware is completed. Software debugging in microcontroller-based product development can be classified into a software-only approach and a hardware-assisted approach. Using a demo board with a resident monitor or simulator to test the program is the software-only approach. In the hardware-assisted approach, either the logic analyzer or the emulator is used to trace the program execution. No matter what approach is used, one needs to run the application programs in order to find out if program execution results are what one expects. When the program execution results are not what one expects, one would examine the programs to find out if there are any obvious logical errors. Many errors can be identified in this manner. However, there are some errors that cannot be easily identified by examining the program.

Many microcontroller demo boards have an onboard monitor program. A good monitor allows the user to enter commands to display and modify the contents of registers and memory locations, set breakpoints, trace program execution, and download the user program onto the demo board for execution. By using appropriate monitor commands, the user will be able to determine whether his or her program executes correctly up to the breakpoint. Instruction tracing is also available from most monitors in case the user needs to pinpoint the exact instructions that cause the error. All peripheral functionality can be actually exercised using this approach. A source-level debugger can be written to communicate with the monitor on the demo board or invoke the simulator to execute the user program. Using this approach, you need not examine the contents of memory locations in order to find out if the program execution result is correct. Instead, values of program variables are available for examination during the process of program execution. In other words, debugging activity is carried out at the *source level.*

In the hardware-assisted approach, designers use an *in-circuit emulator* (ICE) to identify program errors. The ICE needs to reconstruct instructions being executed from the data flowing on the system bus of the target prototype (including address, data, and control signals). This approach may require certain built-in hardware support from the microcontroller used in the prototype in order to identify the boundary of instructions. The ICE also allows us to set breakpoints at those locations that are likely to have errors so that program execution result can be examined. This approach is especially useful when a demo board with a resident monitor is not available. The price of an ICE usually costs much more than a demo board with a resident monitor.

In the past decade, more and more microcontrollers provide hardware support for software debugging. The IEEE 1149.1 JTAG Boundary Scan Interface was initially proposed to standardize the interface for testing newly designed integrated circuits. Many companies (e.g., 8051-variant vendor Silicon Laboratory and Atmel) also utilize this interface to provide support for software debugging and programming the on-chip flash memory or one-time-programmable EPROM. Other companies (including Motorola and Microchip) provide a serial interface (called Background Debug Module) that utilizes two to four pins to support in-circuit programming and software debugging but do not provide boundary scan test capability. These interfaces allow nonintrusive inspection of memory and register contents, support breakpoints, and allow single stepping of user programs. Tool developers can take advantage of these capabilities to implement inexpensive source-level debugger. The ICD2 from Microchip is an in-circuit debugger that utilizes this interface.

3.4.2 ICE

All debugging activities for microcontroller-based products involve running the application program. Depending on the progress of the product development, some debugging activities must be performed before the hardware product is available. An ICE allows the user to debug his or her software before the final hardware is constructed. An ICE includes a target processor module to emulate the final hardware. It can reconstruct the instruction stream being executed on the fly.

Microchip manufactures the MPLAB ICE2000 ICE. As shown in Figure 3.1, the MPLAB ICE-2000 consists of four components:

1. *Emulator pod.* This unit communicates with a PC to receive the software program to be tested.

2. *Processor module.* This module executes the instructions downloaded into the emulator pod.

3. *Device adapter.* A device adaptor consists of IC sockets and additional circuitry to allow the MCU to be connected to the target hardware.

4. *Transition socket.* This socket and the device adapter plugs into the target hardware so that signals can be exercised to the target hardware.

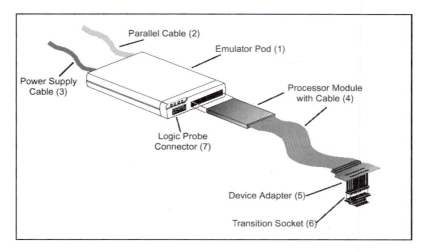

Figure 3.1 ■ MPLAB ICE2000 Emulator (reprint with permission of Microchip)

3.4.3 Device Programmer

The software program must be programmed into the on-chip ROM of the microcontroller before it can be tested. Microchip provides two programmers to meet the need of different users:

PICSTART® PLUS. PICSTART® Plus is a low-cost programmer for the PIC18 microcontrollers and other microcontrollers from Microchip. The photograph of this programmer is shown in Figure 3.2. A 40-pin socket is provided. By adding appropriate adapters, the PICSTART Plus can program PIC18 members with higher pin count. PICSTART Plus is connected to a PC through the serial port and is driven by the MPLAB IDE software.

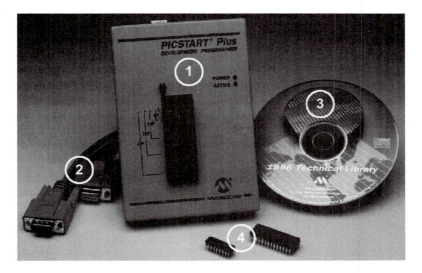

Figure 3.2 ■ PICSTART Plus programmer (reprint with permission of Microchip)

PRO MATE® II. The PRO MATE® II device programmer, shown in Figure 3.3, is another programmer made by Microchip that can program all microcontrollers from Microchip. Appropriate adapters are needed for different packages. Like the PICSTART Plus programmer, this programmer is also connected to the serial port of a PC and is controlled by the MPLAB IDE software.

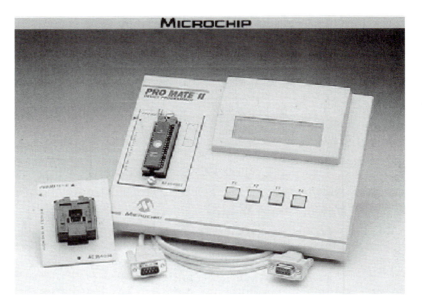

Figure 3.3 ■ PRO MATE II programmer (reprint with permission of Microchip)

In addition to programming microcontrollers, PRO MATE II can also program many EPROM and EEPROM devices.

3.4.4 In-Circuit-Debugger II

In-Circuit-Debugger II (ICD2) is designed to be a low-cost solution for debugging devices that support In-Circuit Serial Programming (ICSP) protocol. These devices provide on-chip flash memory to hold application programs. To enable in-circuit debugging, the ICD2 programs a small debug executive module into the target PIC® microcontroller device (i.e., the PIC18 microcontroller in your demo board). This debug executive module will reside at the end of the program memory. The debug executive works by communicating with special on-chip functions built inside the PIC18 microcontroller.

Utilizing the in-circuit debugging capability of the Flash PIC microcontroller and Microchip's ICSP protocol, the ICD2 also acts as a programmer. It operates under MPLAB IDE, connects to an application, and runs the actual microcontroller in the design.

The ICD2 module contains the logic for debugging, programming, and control. It is connected to either a PC's serial port via a nine-pin serial cable or a USB port via a USB cable and to the PIC18 demo board or target application using a six-wire modular cable. A photograph of ICD2 is shown in Figure 3.4.

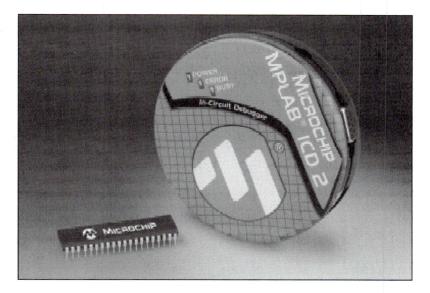

Figure 3.4 ■ Microchip ICD2 in-circuit debugger (reprint with permission of Microchip)

The module contains the software to provide serial communications to the PC and the target application or demo board and to program a supported PIC microcontroller device, all from the MPLAB IDE. The ICD2 module can provide power to the demo board/target application. The user can select V$_{DD}$ source on the Power tab in the ICD2 Settings dialog. If the target application draws over 200 mA, an additional power adapter must be connected to the demo board or target application. The target application also provides power to the ICD2 module only for the purpose of logic-level conversion.

The ICD2 interface cable must be plugged into a modular connector on the application circuit with the appropriate connections to the PIC microcontroller device. The interface cable carries the signals necessary to allow in-circuit debugging of the target application.

3.4.5 Demo Boards

Demo boards are useful for learning the microcontroller and testing the software before the final hardware is completed. As a learning tool, a well-designed demo board should allow the user to experiment with every peripheral function. Three demo boards will be reviewed in Section 3.8. Each of these demo boards can be used to test the programs in this book.

3.5 Using MPLAB IDE

MPLAB IDE is the center of Microchip's software development tool suite. It supports all the devices produced by Microchip that require software control. It consists of a text editor, a simulator, a cross assembler that supports all microcontrollers and digital signal processors, a simulator, and device drivers (for programmers, ICE, ICD, and ICD2) made by Microchip. A small number of third-party development tools (e.g., Hi-Tech C compiler) can also work with MPLAB IDE.

MPLAB IDE is a 32-bit window application that uses projects to manage software development tasks. A project may consist of a single or multiple files. This section will provide a step-by-step tutorial that sets up a project and gets you familiar with the debug capabilities of MPLAB IDE.

3.5.1 Getting Started with MPLAB IDE

MPLAB IDE provides the ability to do the following:

1. Create source code using the built-in editor.
2. Assemble, compile, and link source code using various language tools. An assembler, linker, and librarian come with MPLAB IDE, which also supports C compilers (MCC17 and MCC18) by Microchip. A limited number of third-party C compilers are also supported.
3. Debug the executable logic by watching program flow with the built-in simulator or in real time with the MPLAB ICE2000 emulator or MPLAB ICD2 in-circuit debugger.
4. Make timing measurements with the simulator or emulator.
5. View variables in Watch windows.
6. Program firmware (executable machine code) into devices with PICSTART® Plus or PRO MATE® II device programmers.
7. Find quick answer to questions from the MPLAB IDE online help.

To start the IDE, select Start>Programs>Microchip MPLAB IDE>MPLAB IDE from your monitor screen or simply double-click on the MPLAB IDE icon. A splash screen will display first, followed by the MPLAB IDE desktop as shown in Figure 3.5.

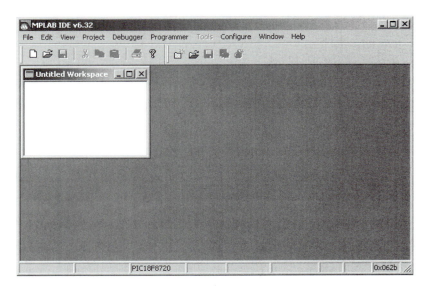

Figure 3.5 ■ MPLAB IDE desktop (reprint with permission of Microchip)

The major steps in software development using MPLAB IDE are as follows:

Step 1
Create a new project.

Step 2
Enter source files.

Step 3
Add source files into the project,

Step 4
Compile and build the executable code.

Step 5
Debug the project using the simulator, or ICD2, or ICE.

3.5.2 Creating a Simple Project

Before creating a new project, the user must configure the project and select the target device. After creating a new project, the user needs to set the language tool location and select the tool suite.

To configure the project, go to the *Configure> settings* menu and choose the Projects tab. Make sure the setup is as shown in Figure 3.6. Click on OK after making sure the setting is right.

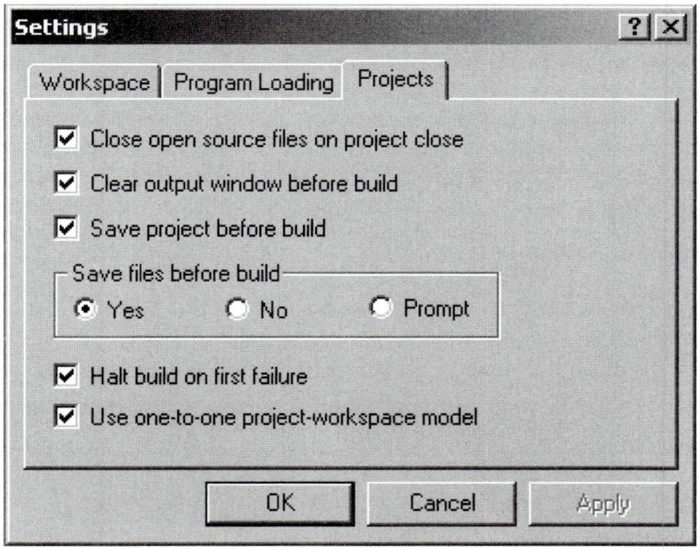

Figure 3.6 ■ Configure the project (reprint with permission of Microchip)

The next step is to make sure the right device is selected. For this tutorial, the PIC18F452 has been chosen. To select the target device, go to *Configure>Select Devices* . . . and make the selection as shown in Figure 3.7.

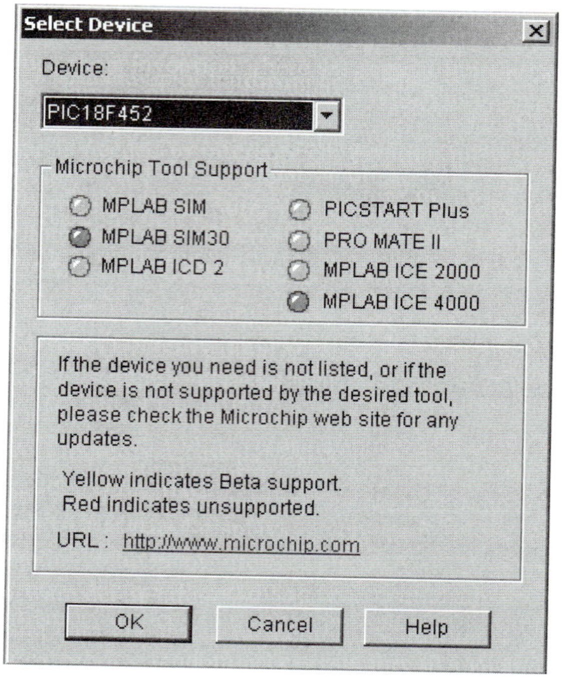

Figure 3.7 ■ Select device dialog (reprint with permission of Microchip)

After the correct target device has been selected, the user is ready to create a new project. When creating a new project, the user needs to decide where to place the project. This tutorial will assume that the user has decided to use the directory c:\pic18\ch03 to place his or her projects. Follow these steps to create a new project:

1. Select the *Project> New* menu. The New dialog will open as shown in Figure 3.8.
2. Enter the name of the new project (e,g., tutor1).

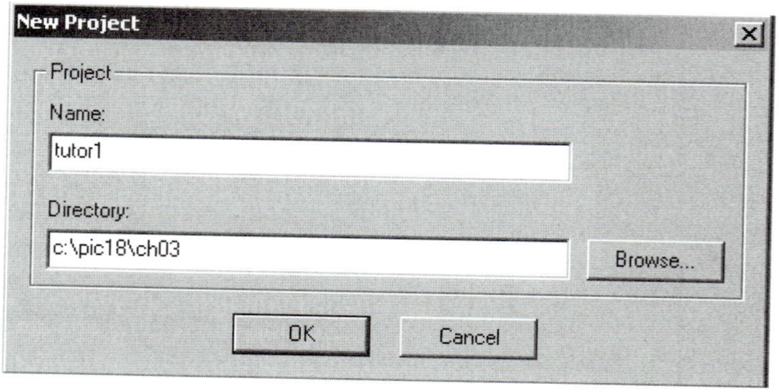

Figure 3.8 ■ New project dialog (reprint with permission of Microchip)

3. Use the Browse button to select the path to place the new project or type in the path.

4. When the Project name and Location are correct, click on OK.

After creating a new project, the user needs to make sure the language tool locations are set properly. To do this, select *Project> Set Language Tool Locations* to confirm the location of the Microchip Tool Suite. Click on MPASM Assembler (mpasmwin.exe). The full path to the MPASM Assembler executable should appear in the Location of Selected Tool text box as shown in Figure 3.9. If it is incorrect or empty, click on Browse to locate mpasmwin.exe. After the language tool location has been set up, the user can skip this step until he or she switches language tools.

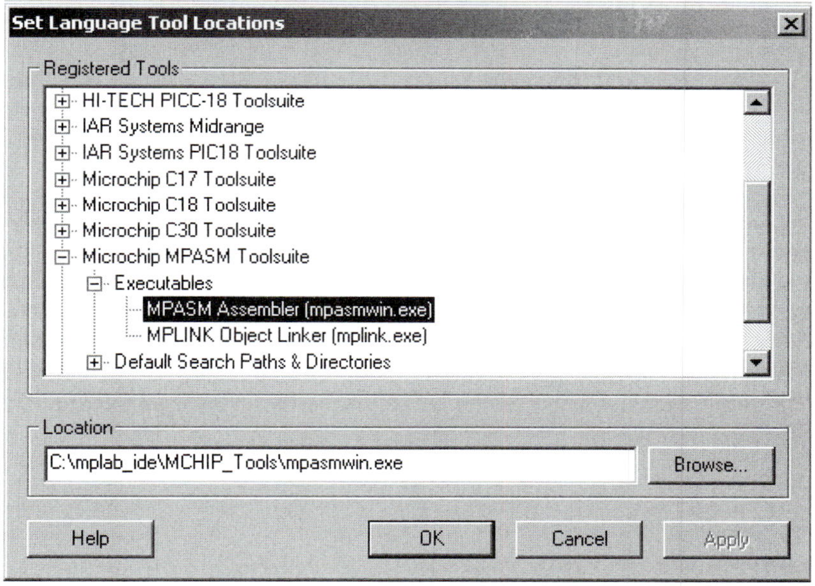

Figure 3.9 ■ Setting language tool location dialog (reprint with permission of Microchip)

Before starting entering your source code, one needs to set the language tool suite to be used in the new project. This allows the MPLAB IDE to tailor its operation more accurately with regard to context-sensitive editing and file extensions. Perform the following steps to select the language suite:

1. Select *Project> Set Language Toolsuite.*

2. For Active Toolsuite, select *Microchip MPASM Toolsuite* for this tutorial. The PICmicro® language tools will appear under Toolsuite Contents.

3. Click on OK.

The dialog for selecting language suite is shown in Figure 3.10.

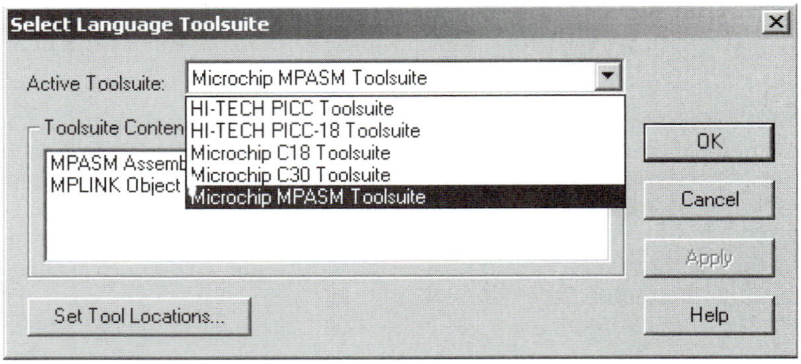

Figure 3.10 ■ Dialog for selecting language tool suite
(reprint with permission of Microchip)

3.5.3 Entering Source Code

After the new project has been configured properly, one can now enter the source code for the project.

Select *File> New*. A blank edit window will be opened in the workspace of MPLAB IDE. Enter the following program to the Edit window:

```
              title "Finding the Number of Elements That Are a Multiple of 4"
              #include <p18F452.inc>
ilimit     equ         0x20
count      set         0x00
loop_cnt   set         0x01
mask       equ         0x03            ; used to mask upper six bits
           org         0x00
           goto        start
           org         0x08            ; high-priority interrupt service routine
           retfie
           org         0x18            ; low-priority interrupt service routine
           retfie
start      clrf        count,A
           movlw       ilimit
           movwf       loop_cnt        ; initialize ii to N
           movlw       upper array     ; use table pointer to point to the array
           movwf       TBLPTRU, A      ; "
           movlw       high array      ; "
           movwf       TBLPTRH,A       ; "
           movlw       low array       ; "
           movwf       TBLPTRL,A       ; "
i_loop     movlw       mask
           tblrd*+                     ; read an array element into TABLAT
           andwf       TABLAT,F,A
           bnz         next            ; branch if not a multiple of 4
           incf        count,F,A       ; increment count if it is a multiple of 4
```

```
next        decfsz      loop_cnt,F,A    ; decrement loop count
            goto        i_loop
            nop
array       db          0x00, 0x01, 0x30, 0x03, 0x04, 0x05, 0x06, 0x07, 0x08, 0x09
            db          0x0A, 0x0B, 0x0C, 0x0D, 0x0E, 0x0F, 0x10, 0x11, 0x12, 0x13
            db          0x14, 0x15, 0x16, 0x17, 0x18, 0x19, 0x1A, 0x1B, 0x1C, 0x1D
            db          0x1E, 0x1F
            end
```

This program counts the number of elements in the given array that are divisible by 4. The lowest two bits of a number divisible by 4 are 00. This program starts with the first element and checks if it is divisible by 4. It increments the variable **count** by 1 if an element is divisible by 4 until the last element is checked. The array is located in program memory.

Once the source code has been entered, the user should select *File> Save* and save the file in the project directory as **tutor1.asm.** After the user has saved the code, the program is shown with identifying colors for readability. This context-sensitive colorization is customizable. For more information about the editor, see *Help> MPLAB Editor Help.*

3.5.4 Adding Source Files to the Project

To add a source file to the project, select *Project> add Files to Project.* A dialog window as shown in Figure 3.11 will appear on the screen. Select the file that you just created and saved (tutor1.asm) and click on Open. After this, the newly inserted file should appear in the project pane under Source Files (shown in Figure 3.12). If it appears under Unclassifiable, confirm that you have selected the correct language tool suite by selecting *Project> Set Language Toolsuite.* "Microchip Toolsuite" should appear as the Active Toolsuite. If not, select it and click on OK. The file name should now appear under Source Files.

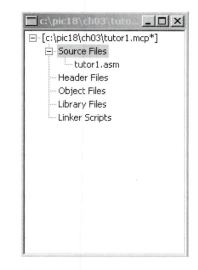

Figure 3.11 ■ Select input file dialog
(reprint with permission of Microchip)

Figure 3.12 ■ Project pane
(reprint with permission of Microchip)

Save the project by selecting *Project> Save.* You call also add files and save projects by using the right mouse button in the project pane. Experiment by clicking the right mouse button while the cursor is over Source Files. Note that the menu options are different.

3.5.5 Building the Project

After adding source files into the project, it is time to build the project. This step will compile the source code using the selected language tool suite.

To build the project, select *Project> Build All* or *Project> Make* (or press function key F10). This file (tutor.asm) should assemble successfully. If this project does not build correctly, check the following items and then build the project again:

- Check the spelling and format of the program in the editor window. If the assembler reported errors in the Output window, double-click on the error in the Output window. This will indicate the corresponding line in the source code with a green arrow in the gutter of the source code window.
- Check that the correct assembler (MPASM assembler) for the PICmicro device is being used.

On a successful build, the debug files (with file name extension .cod or .cof) generated by the language tool will be loaded. This file allows one to debug using the source code and view program variables symbolically in Watch windows.

For this one-file example, a project does not seem necessary. However, the real power of projects comes when many files are to be compiled/assembled and linked to form the final executable application. Projects keep track of all this for the user.

3.5.6 Debugging the Project

After building the project, one will want to check that it is functioning the way he or she intended. To do this, one will need to select a debug tool. There are five debug tools under the *Debugger>Select Tool* menu:

1. MPLAB ICD2
2. MPLAB ICE4000
3. MPLAB SIM
4. MPLAB ICE2000
5. MPLAB SIM30

Among these five debug tools, ICD2, SIM, and ICE2000 can be used to debug PIC18 microcontroller applications. The uses of SIM and ICD2 are discussed in the next two sections. The use of MPLAB ICE2000 or the ICE4000 is not discussed because of their higher cost.

3.6 Using the MPLAB SIM in Debugging PIC18 Applications

Program simulation involves the following major operations:

- Simulator setup
- Running the code
- Viewing variables

- Using Watch windows
- Setting breakpoints
- Tracing code

3.6.1 Setting Up the Simulator

The first step for simulating the program is to select MPLAB SIM as the debugging tool. This is done by selecting *Debugger>Select Tool> MPLAB SIM.* After selecting MPLAB SIM as the debugging tool, the status bar on the bottom of the MPLAB IDE window should change to "MPLAB SIM". Additional menu items should now appear in the Debugger menu. Additional toolbar elements would also appear on the MPLAB IDE window. This is shown in Figure 3.13.

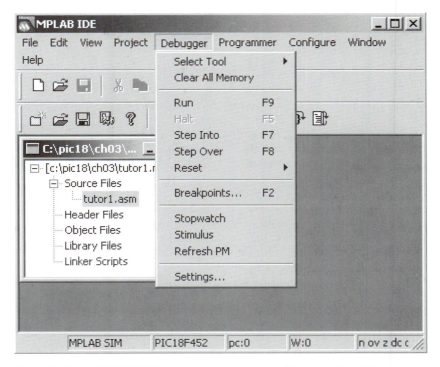

Figure 3.13 ■ MPLAB IDE window status change after selecting SIM.
Additional menu items also appear under Debugger menu.
(reprint with permission of Microchip)

3.6.2 Running Code under MPLAB SIM

Before running the program, one needs to set the program counter to the start of the program by selecting *Debugger> Reset* (or press function key F6). A green arrow should appear in the gutter of the source code window (in Figure 3.14), indicating the first source code line that will be executed.

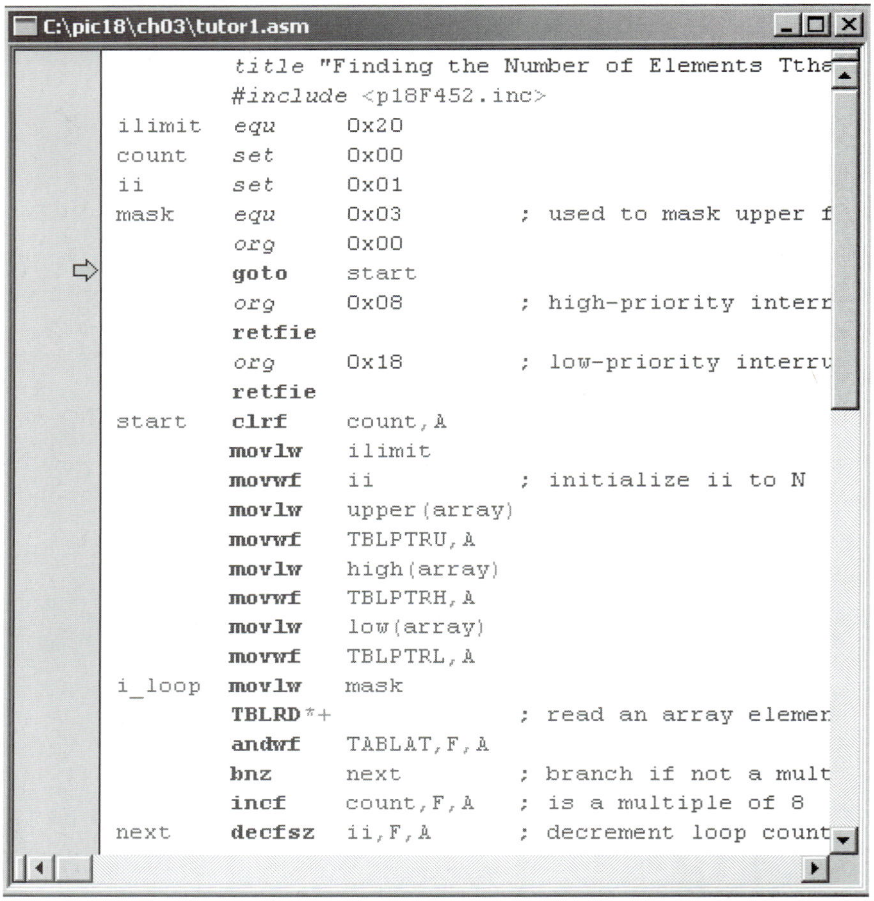

```
C:\pic18\ch03\tutor1.asm                                    _ □ ×
                    title "Finding the Number of Elements Ttha
                    #include <p18F452.inc>
        ilimit  equ     0x20
        count   set     0x00
        ii      set     0x01
        mask    equ     0x03            ; used to mask upper f
                org     0x00
    ⇨           goto    start
                org     0x08            ; high-priority interr
                retfie
                org     0x18            ; low-priority interru
                retfie
        start   clrf    count,A
                movlw   ilimit
                movwf   ii              ; initialize ii to N
                movlw   upper(array)
                movwf   TBLPTRU,A
                movlw   high(array)
                movwf   TBLPTRH,A
                movlw   low(array)
                movwf   TBLPTRL,A
        i_loop  movlw   mask
                TBLRD *+                ; read an array elemer
                andwf   TABLAT,F,A
                bnz     next            ; branch if not a mult
                incf    count,F,A       ; is a multiple of 8
        next    decfsz  ii,F,A          ; decrement loop count
```

Figure 3.14 ■ Source code window after Reset (reprint with permission of Microchip)

Select *Debugger>Run* (or press function key F9) to run the program. The message "Running. . ." will appear on the status bar.

Because the simulator was not told where to stop, it will keep running. To halt the program execution, select *Debugger > Halt* (or press function key F5). The line of code where the application halted will be indicated by the green arrow.

We can also single step through the application program by selecting *Debugger > Step Into* (or press function key F7). This will execute the currently indicated line of the code and move the arrow to the next line of code that will be executed.

3.6.3 Viewing Variables

One can see the values of variables at any time by putting the mouse cursor over their names anywhere in the source file. A small output window will pop up to show the current value. A screen shot for placing mouse over the variable **count** is shown in Figure 3.15.

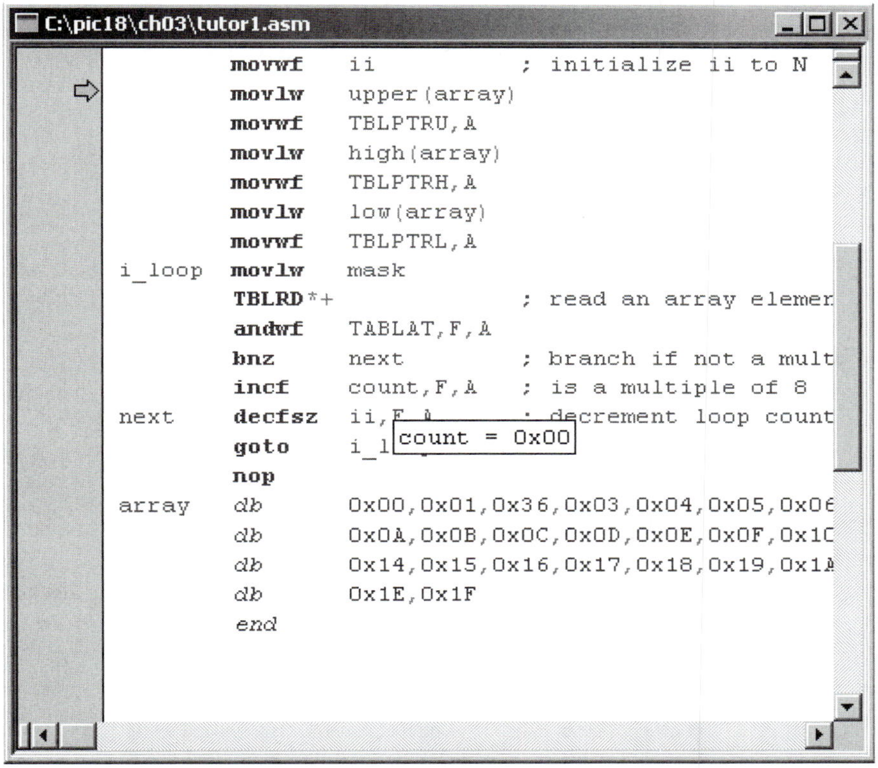

```
C:\pic18\ch03\tutor1.asm                                    _ □ ×

                movwf     ii              ; initialize ii to N
⇨               movlw     upper(array)
                movwf     TBLPTRU,A
                movlw     high(array)
                movwf     TBLPTRH,A
                movlw     low(array)
                movwf     TBLPTRL,A
        i_loop  movlw     mask
                TBLRD *+                  ; read an array elemen
                andwf     TABLAT,F,A
                bnz       next            ; branch if not a mult
                incf      count,F,A       ; is a multiple of 8
        next    decfsz    ii,F,A          ; decrement loop count
                goto      i_l count = 0x00
                nop
        array   db        0x00,0x01,0x36,0x03,0x04,0x05,0x06
                db        0x0A,0x0B,0x0C,0x0D,0x0E,0x0F,0x1C
                db        0x14,0x15,0x16,0x17,0x18,0x19,0x1A
                db        0x1E,0x1F
                end
```

Figure 3.15 ■ Mouse over variable "count" (reprint with permission of Microchip)

3.6.4 Using Watch Window to View Variables

Often one wants to watch certain key variables all the time. Rather than floating the mouse cursor over the name each time one wants to see the value, one can open a Watch window. The Watch window will remain on the screen and show the current variable values. Watch windows can be found under the View menu.

To open a Watch window, *View> Watch.* There are two ways to add a symbol into the Watch window:

1. Click the mouse in the Watch window and then type the name of the symbol and press the Enter key.
2. Select from the symbol selection box at the top of the window and then click on **Add Symbol** to add it to the Watch window list.

One can also add any special-function register (SFR) into the Watch window list. The procedure is identical to the procedure for adding symbols.

The only symbol of interest to us is **count.** Add it to the Watch window. The resultant Watch window is shown in Figure 3.16.

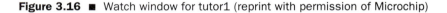

Figure 3.16 ■ Watch window for tutor1 (reprint with permission of Microchip)

Now we can watch the symbol value change as we step through the program:

1. Press function key F6 to reset your program.
2. Press function key F7 to step through the program one instruction at a time until you have stepped to the following program line:

   incf   count,F,A   ; increment count if it is a multiple of 4

3. Step one more time to see the value of **count** in the Watch window change from 0 to 1. The changed value appears in red color.
4. Continue to step through the program until you reach the instruction that follows **incf count,F,A.** You will see that the value of **count** increments if a value that is divisible by 4 is reached.
5. Continue to step through the program until all the array elements have been checked. The final value of **count** is 9 for this example.

3.6.5 Setting Breakpoints

Sometimes one wants to run the program to a specific location and then halt. This is accomplished by using breakpoints. To set a breakpoint, press the right mouse button on the code line that one wants to set a breakpoint. Suppose one wants to set a breakpoint at the **nop** instruction. Press the right mouse button, and a pop-up window as shown in Figure 3.17 will appear. It is correct that Figure 3.17 appeared overlapping.

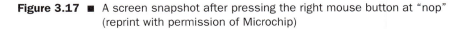

```
        goto     start
          o.  Set Breakpoint            high-priority interrupt service rou
          r   Breakpoints        ▶
          o.                             low-priority interrupt service rout
          r   Run to Cursor
 start    c.  Set PC at Cursor
          m
          m   Undo            Ctrl+Z
          m   Redo            Ctrl+Y     initialize ii to N
          m                              use table pointer to point to the a
          m   Cut             Ctrl+X       "
          m   Copy            Ctrl+C       "
          m   Paste           Ctrl+V       "
          m   Delete          Del          "
          m                                "
          m   Select All      Ctrl+A
 i_loop   m
          T   Find...         Ctrl+F     read an array element into TABLAT
          a   Find Next       F3
          b   Replace...      Ctrl+H     branch if not a multiple of 4
          i   Go To...        Ctrl+G     increment count if it is a multiple
 next     d                              decrement loop count
          g   Advanced        ▶
          nop Properties...
 array    db   0x00,0x01,0x30,0x03,0x04,0x05,0x06,0x07,0x08,0x09
          db   0x0A,0x0B,0x0C,0x0D,0x0E,0x0F,0x10,0x11,0x12,0x13
          db   0x14,0x15,0x16,0x17,0x18,0x19,0x1A,0x1B,0x1C,0x1D
          db   0x1E,0x1F
          end
```

Figure 3.17 ■ A screen snapshot after pressing the right mouse button at "nop" (reprint with permission of Microchip)

From the pop-up menu that appears, select *Set Breakpoint.* A "stop sign" should appear in the gutter next to the line (shown in Figure 3.18). Before running the program, press function key F6 to reset the program.

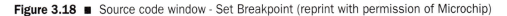

```
          andwf    TABLAT,F,A
          bnz      next          ; branch if not a multiple of 4
          incf     count,F,A     ; increment count if it is a multiple
 next     decfsz   ii,F,A        ; decrement loop count
          goto     i_loop
 Ⓑ        nop
 array    db   0x00,0x01,0x30,0x03,0x04,0x05,0x06,0x07,0x08,0x09
          db   0x0A,0x0B,0x0C,0x0D,0x0E,0x0F,0x10,0x11,0x12,0x13
          db   0x14,0x15,0x16,0x17,0x18,0x19,0x1A,0x1B,0x1C,0x1D
          db   0x1E,0x1F
          end
```

Figure 3.18 ■ Source code window - Set Breakpoint (reprint with permission of Microchip)

To run the program, select *Debugger> Run.* It should run briefly and then halt on the line at which the breakpoint was set. The screen should look like that in Figure 3.19 after the program halts.

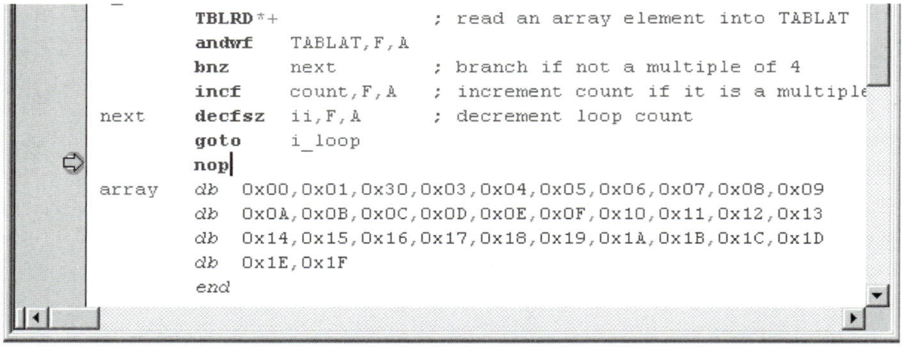
TBLRD*+ ; read an array element into TABLAT
andwf TABLAT,F,A
bnz next ; branch if not a multiple of 4
incf count,F,A ; increment count if it is a multiple
next decfsz ii,F,A ; decrement loop count
goto i_loop
nop
array db 0x00,0x01,0x30,0x03,0x04,0x05,0x06,0x07,0x08,0x09
db 0x0A,0x0B,0x0C,0x0D,0x0E,0x0F,0x10,0x11,0x12,0x13
db 0x14,0x15,0x16,0x17,0x18,0x19,0x1A,0x1B,0x1C,0x1D
db 0x1E,0x1F
end

Figure 3.19 ■ Source code window—Breakpoint Halt (reprint with permission of Microchip)

3.6.6 Tracing Code

There are times that the user will have difficulty identifying the program bugs. Tracing program becomes necessary under such situations. Although single stepping through the program would allow the user to identify the program bugs, it would be easier for the simulator to capture the execution trace of many instructions all at once while the user goes through them instruction by instruction.

The Simulator Trace can be used to record the execution of the user program. The user can enable the Simulator Trace by selecting *Debugger> Settings* and choosing the Pins/Trace tab.

As shown in Figure 3.20, there are two check boxes to control how the Simulator Trace collects data. When only the top box is checked, the simulator collects data when the simulator is in Run mode, it collects data until the user halts at a breakpoint or manually stops the simula-

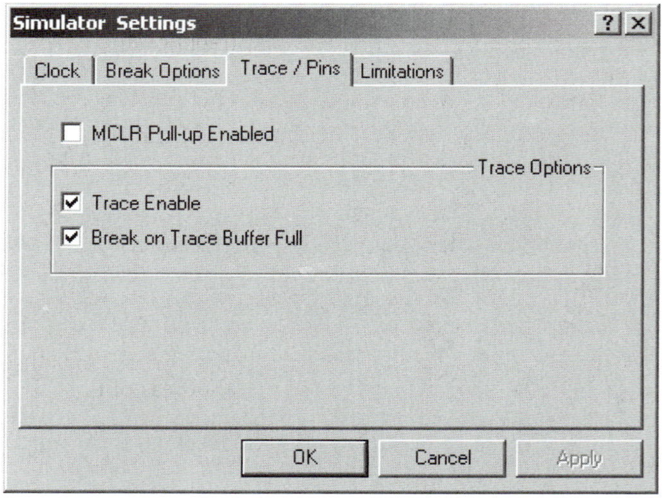
Simulator Settings
Clock
Break Options
Trace / Pins
Limitations
MCLR Pull-up Enabled
Trace Options
Trace Enable
Break on Trace Buffer Full
OK
Cancel
Apply

Figure 3.20 ■ Simulation Trace enable (reprint with permission of Microchip) tor. It will show the last 8192 cycles collected. This mode is useful if the user wants to see the record of instructions leading up to a breakpoint.

If the second button is also checked, the trace memory will collect 8192 cycles of data and then stop collecting and halt the user application at a breakpoint. This mode is useful for seeing the record of instructions after the user presses run.

After enabling Simulator Trace, make sure that the program is reset before running the program. Since tutor1.asm is a short program, the simulator will halt at the breakpoint. Select *View> Simulator Trace* to view the simulation trace. The trace display is shown in Figure 3.21. The trace display shows a time stamp at every cycle, and the data that were read or written into file registers will be captured and displayed.

Figure 3.21 ■ Simulation trace display (reprint with permission of Microchip)

There are 18 columns in the trace display. The meanings of these columns are as follows:

- Line. Decimal cycle number from start of trace session
- Addr. Program address of instruction
- Op. Numeric op code of instruction
- Label. Symbolic label of instruction, if known
- Instruction. Disassembled instruction
- SA. Source address, the register address of read operation
- SD. Source data, the data read from the register
- DA. Destination address, the register address of the destination
- DD. Destination data, the data written into the destination register
- Cycles. Time before the execution of an instruction, from reset
- *n* Probe (n = 7, 0). Irrelevant in simulator, used in MAPLAB ICE2000 emulator only (each n value is in one column)

If there is any dash in the row for these values, it means that the operation did not access any file register for this instruction. The column with the label of "Time" is the time stamp. This can be used to measure the execution time of routines. The time is calculated on the basis of the clock frequency entered in the *Debugger> Settings Clock* tab.

In line 15 of Figure 3.21, both the source register and the destination register are the same register (at 0x00, the count value). The value of count is 1 after the first array element (0x00) is checked because 0x00 is a multiple of 4. You can scroll down the Simulator Trace and find that the final value of count is 9.

If the user puts the cursor over the top row of the trace display where the column headings are listed and presses the right mouse button, a configuration dialog will pop up as shown in Figure 3.22. All the checked items will appear in the trace window. The user can uncheck columns to reduce clutter if the user is not interested in the data in those columns. The entries labeled as Probe 7, Probe 6, and so on are for the MPLAB ICE2000 emulator trace and are not relevant to the simulator. They should be unchecked.

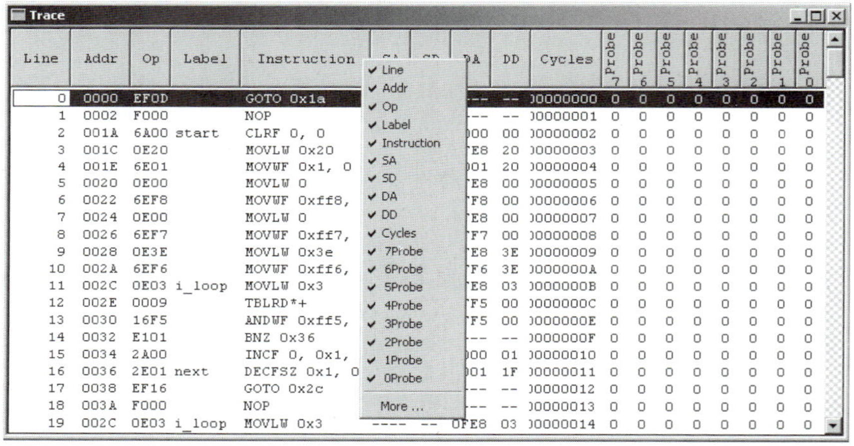

Figure 3.22 ■ Configure Simulation Trace (reprint with permission of Microchip)

3.6.7 Advanced Simulator Options

There are other characteristics of the simulator that can be configured from the MPLAB IDE dialogs. Normally, the default condition of the configuration bits has the Watch Dog Timer (WDT) enabled. This will cause the simulator to reset when the internal WDT times out.

Unless one wants to test the functioning of the WDT, one would be better off by disabling the WDT. To disable the WDT, one needs to bring up the Configuration Bits dialog by selecting *Configure> Configuration Bits.* The Configure Bits dialog is shown in Figure 3.23.

Click on the line that contains Watchdog Timer, and then a selection box will appear. Scroll down to select *Disabled* to prevent the WDT from causing the program to reset. The WDT selection box is shown in Figure 3.24. If the user uses ICD2 to perform software debugging, then *Low Voltage Program* should also be disabled, whereas Background Debug should be enabled.

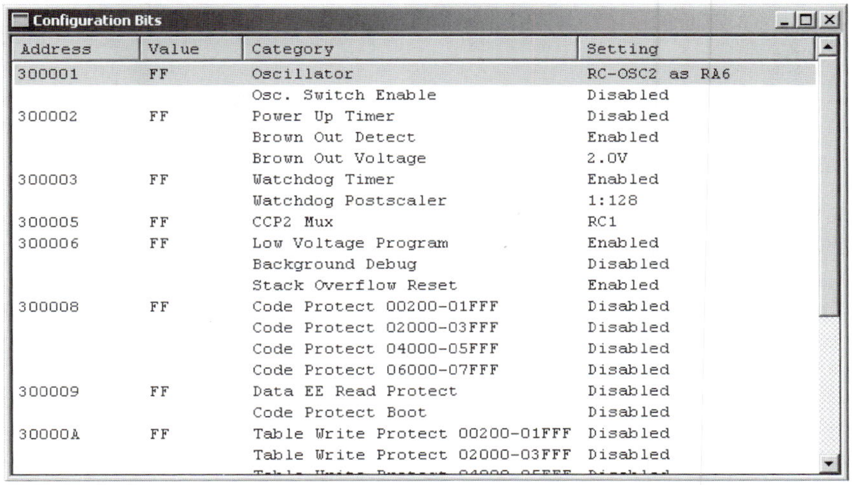

Figure 3.23 ■ Configuration Bits dialog (reprint with permission of Microchip)

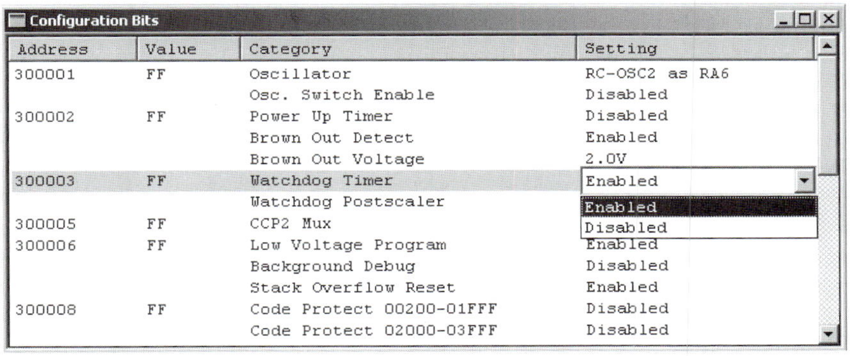

Figure 3.24 ■ Diable the WDT timer (reprint with permisson of Microchip)

3.7 Using the MPLAB ICD2

ICD2 can be connected to the PC via one of the COM ports or USB ports. The USB port connection is preferred because it provides much faster data transfer between the PC and the ICD2.

After building the project, the user can select ICD2 to debug his or her program. This choice is preferred when the hardware, such as a PIC18 demo board, is available. ICD2 can be selected by selecting *Debugger> Select Tool> MPLAB ICD2.*

Before using the ICD2 to debug the application, the user must make sure that his or her target hardware or demo board is connected to the ICD2 using a modular connector. The shape of the modular connector is shown in Figure 3.25.

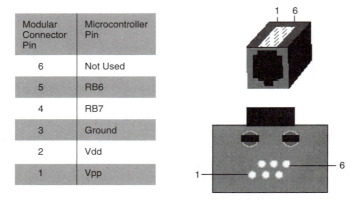

Modular Connector Pin	Microcontroller Pin
6	Not Used
5	RB6
4	RB7
3	Ground
2	Vdd
1	Vpp

Bottom View of Modular Connector
Pinout on Designer's Board

Figure 3.25 ■ MPLAB ICD2 modular connector (reprint with permission of Microchip)

3.7.1 ICD2 Settings

ICD2 needs to be set up properly before it can be used. Select *Debugger> Settings* (or *Programmer> Settings*) to configure ICD2. A setting window as shown in Figure 3.26 will appear on the screen. Figure 3.26 shows the status of the ICD programmer. The status of ICD2 is connected and it is automatically connected at startup.

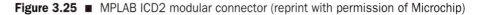

Figure 3.26 ■ ICD2 set up window (reprint with permission of Microchip)

The user must make sure that every setting is correct. There are three settings that need to be set properly: power, communication, and program. Click on the Power tab, and the screen will change to that in Figure 3.27. Notice that the Power Target Circuit From ICD2 setting is not selected. This is not desirable because ICD2 does not have enough power to drive the target hardware or demo board.

Click on the Communication tab shown in Figure 3.26, and the available settings will be made visible as shown in Figure 3.28. Choose USB if the PC has an available USB port. Otherwise, choose one of the COM ports. When selecting a COM port, the user will also need to set the baud rate. There are two baud rates to choose from: 19200 and 57600. Try the higher data rate. If it is not working properly, then switch to 19200.

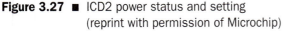

Figure 3.27 ■ ICD2 power status and setting
(reprint with permission of Microchip)

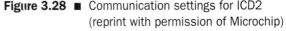

Figure 3.28 ■ Communication settings for ICD2
(reprint with permission of Microchip)

The last setting to be made is **Program.** Click on the Program tab, and its setting will be brought out as shown in Figure 3.29. The default settings are acceptable for this tutorial.

The remaining two tabs (Limitations and Versions) are for information purpose. The user should know the limitations of ICD2. Click on the Limitations tab, and its contents are shown in Figure 3.30.

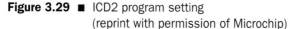

Figure 3.29 ■ ICD2 program setting
(reprint with permission of Microchip)

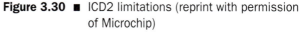

Figure 3.30 ■ ICD2 limitations (reprint with permission
of Microchip)

ICD2 will use certain program memory and special- function registers (SFRs) on the target microcontroller unit. By clicking on the Details tab, the user can look for device specific limitations.

After completing the settings of ICD2, the user is ready to debug his or her application on the target hardware with the help of ICD2. Most of the debug techniques applicable to MPLAB SIM are also applicable to ICD2.

Here we use a program that computes the sum of integers from 1 to n to illustrate the process. The program is as follows:

```
        #include <p18F452.inc>
n       equ     D'100'
sum_hi  set     0x01            ; high byte of sum
```

```
sum_lo    set     0x00          ; low byte of sum
i         set     0x02          ; loop index i
          org     0x00          ; reset vector
          goto    start
          org     0x08
          retfie
          org     0x18
          retfie
start     clrf    sum_hi,A      ; initialize sum to 0
          clrf    sum_lo,A      ; "
          clrf    i,A           ; initialize i to 0
          incf    i,F,A         ; i starts from 1
sum_lp    movlw   n
          cpfsgt  i,A           ; compare i with n and skip if greater than
          goto    add_lp        ; perform addition when i ≤ 50
          goto    done          ; it is done when i > 50
add_lp    movf    i,W,A         ; place i in WREG
          addwf   sum_lo,F,A    ; add i to sum_lo
          movlw   o
          addwfc  sum_hi,F,A    ; add carry to sum_hi
          incf    i,F,A         ; increment loop index i by 1
          goto    sum_lp
done      nop
          end
```

Perform the following steps to enter and build the project:

1. Follow the procedure described in Section 3.5.3 to enter the source code (call it tutor2.asm).

2. Follow the procedure described in Section 3.5.4 to add source code to the project (call the project tutor2).

3. Follow the procedure described in Section 3.5.5 to build the project.

After building the project, perform the following steps:

1. Configure the ICD2 debugger properly.

2. Before debugging the program using ICD2, the target device needs be programmed. Program the hex file (generated by the assembler) into the demo board by selecting the Program command under the Debugger menu.

3. Set up a watch window and add variables **sum_hi** and **sum_lo** into the Watch window.

4. Set a breakpoint at the statement of **done nop.**

5. Reset ICD2 by pressing function key F6.

6. Run the program (the program execution will be halted at the breakpoint).

The resultant screen on the MPLAB IDE window is shown in Figure 3.31. The Watch window displays the sum as a hex value 13BA (equivalent to decimal 5050) and is correct.

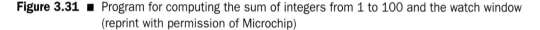

```
C:\books\pic18book\programs\ch03\tutor2.asm

                #include <P18F452.inc>
n       equ     D'100'
sum_hi  set     0x01        ; high byte of sum
sum_lo  set     0x00        ; low byte of sum
i       set     0x02        ; loop index i
        org     0x00        ; reset vector
        goto    start
        org     0x08
        retfie
        org     0x18
        retfie
start   clrf    sum_hi,A    ; initialize sum to 0
        clrf    sum_lo,A    ;    "
        clrf    i,A         ; initialize i to 0
        incf    i,F,A       ; i starts from 1
sum_lp  movlw   n
        cpfsgt  i,A         ; compare i with n and skip if grea
        goto    add_lp      ; perform addition when i   50
        goto    done        ; it is done when i > 50
add_lp  movf    i,W,A       ; place i in WREG
        addwf   sum_lo,F,A  ; add i to sum_lo
        movlw   0
        addwfc  sum_hi,F,A  ; add carry to sum_hi
        incf    i,F,A       ; increment loop index i by 1
        goto    sum_lp
done    nop
        end
```

```
Watch
Add SFR  ADCON0  ▼     Add Symbol ▼
Address   Symbol Name    Value
 0001       sum_hi          13
 0000       sum_lo          BA

Watch 1 | Watch 2 | Watch 3 | Watch 4
```

Figure 3.31 ■ Program for computing the sum of integers from 1 to 100 and the watch window (reprint with permission of Microchip)

If the program is not running correctly, then the user will need to use the techniques described in Sections 3.6.2 to 3.6.6 to debug the program. The user can combine the use of breakpoints, single stepping, and code tracing to identify the errors in the program.

3.8 Demo Boards from Shuan-Shizu Enterprise

Shuan-Shizu Enterprise has designed three PIC18 demo boards for learning the PIC18 microcontroller. The SSE452, the SSE8720, and the SSE8680 use the PIC18F452, the PIC18F8720, and the PIC18F8680, respectively, as their microcontrollers. Product information about these demo boards can be found at the Web site at www.evb.com.tw or by e-mail at Vincent-fan@umail.hinet.net. The features of these demo boards are described in the following sections.

3.8.1 SSE452 Demo Board

The main design feature of this demo board is that it allows the user to easily switch micro-controllers (see Figure 3.32). The SSE452 allows the user to specify the option of adding 28-pin and/or 40-pin ZIF sockets for the microcontroller. This allows the user to experiment with different 28-pin or 40-pin PIC18 microcontrollers.

Figure 3.32 ■ Photo of the SSE452 demo board (reprint with permission of Shuan-Shizu Enterprise)

In addition to the PIC18F452 on-chip peripheral functions, the SSE452 adds the following features:

1. One PCB suitable for any 28-pin or 40-pin PIC18 microcontroller
2. High-current sink/source: 25 mA
3. One RS-232 connector
4. Two debounced push-button switches (can be used as external interrupt sources)
5. One 8-bit DIP switch for digital input
6. One 4 × 4 keypad connector for interfacing with 16-key keypad
7. One rotary encoder with push button for optional input
8. One TC77 temperature sensor with SPI interface
9. One EEPROM (24LC04B) with I²C interface
10. One 2 × 20 bus expansion port to make signals available to end user
11. One potentiometer for exercising the A/D function
12. Optional devices: 2 × 20 character LCD, 48/28-pin ZIF socket
13. Digital signals with frequency from 1 Hz up to 8 MHz for exercising timer functions
14. ICD2 connector

3.8.2 SSE8720 Demo Board

This demo board uses the PIC18F8720 as its microcontroller and was designed for those users who need more I/O pins and more on-chip flash program memory (see Figure 3.33). The features of the SSE8720 are as follows:

1. Digital signals with frequency from 1 Hz up to 16 MHz for experimenting with timer functions
2. ICD2 connector for debugging
3. DB9 connector provides EIA232 interface to connect to USART1
4. Four debounced switches (connected to RB0/INT0, RB1/1NT1, RB2/INT2, and RB3/INT3) and one reset button
5. One potentiometer connected to RA0/AN0 pin for evaluating the A/D converter
6. One 8-bit DIP switch (Port F)
7. Eight LEDs
8. One 2 × 20 LCD
9. On-board 5-V regulator
10. One EEPROM with I²C interface
11. An SPI-compatible (four-wire) digital temperature sensor TC72
12. An 8-Ω speaker (driven through an NPN transistor) connected to RC2/CCP1 pin
13. One Dallas DS1306 SPI-compatible real-time clock chip
14. Two 2 × 20 connectors for accessing microcontroller signals

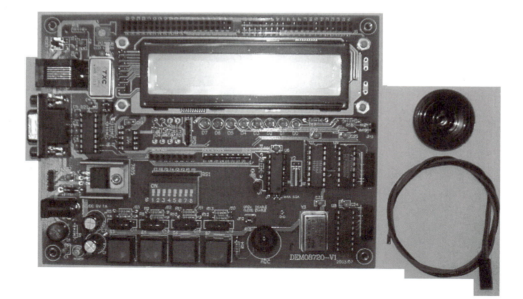

Figure 3.33 ■ SSE8720 demo board with speaker (reprint with permission of Shuan-Shizu Enterprise)

3.8.3 The SSE8680 Demo Board

The SSE8680 is designed for those users who are interested in experimenting with the CAN network (see Figure 3.34). Since this board uses the 80-pin PIC18F8680, it is also suitable for those applications that require many I/O pins. The CAN network is widely used in automotive and control applications. In addition to the on-chip peripheral functions of the PIC18F8680 microcontroller, the SSE8680 demo board has the following features:

1. Digital signals with frequency ranging from 1 Hz up to 16 MHz for experimenting with timer functions
2. ICD2 connector for debugging
3. DB9 connector provides EIA232 interface to connect to USART1
4. Four debounced switches (connected to RB0/INT0, RB1/INT1, RB2/INT2, and RB3/INT3) and one reset button (not connected to any pin)
5. One potentiometer connected to RA0/AN0 pin for evaluating the A/D converter
6. One 8-bit DIP switch (Port F)
7. Eight LEDs
8. One 2 × 20 LCD

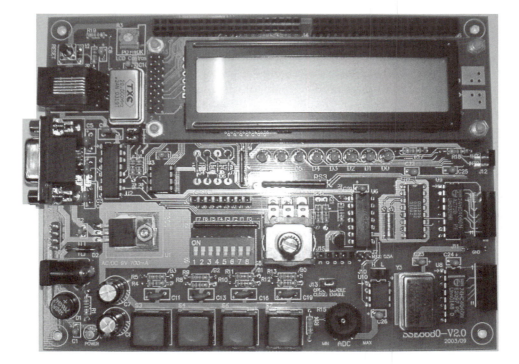

Figure 3.34 ■ SSE8680 demo board with LCD kit (reprint with permission of Shuan-Shizu Enterprise)

9. On-board 5-V regulator

10. One EEPROM with I²C interface

11. An SPI-compatible (four-wire) digital temperature sensor TC72

12. An 8-Ω speaker (driven through an NPN transistor) connected to RC2/CCP1 pin

13. One Dallas DS1306 SPI-compatible real-time clock chip

14. Two 2 × 20 connectors for accessing microcontroller signals

15. MCP2551 CAN transceiver

16. Rotary encoder

3.8.4 Debug Monitor

A debug monitor is being developed for these demo boards. The debug monitor will allow the user to download the hex file (created by MPLAB® IDE) onto the demo board without using the ICD2 in-circuit debugger. In addition, this monitor will allow the user to set breakpoints, single step the program, and display and modify the contents of register values. The monitor will also be able to display the program downloaded into the demo board. (It needs a disassembler to do this.). The beta version of the monitor is functioning. The stable version of the monitor should be available by the time this book is published.

Without a monitor, one needs the MPLAB IDE, a demo board, and an ICD2 to experiment with the hardware. With the monitor, one needs only the MPLAB IDE and a demo board in order to experiment with the hardware peripheral functions.

3.9 Summary

Hardware and software development tools are essential for learning the features of the microcontroller and developing microcontroller-based products. This chapter provides a brief review of most of the development tools from Microchip and also the demo boards made by Shuan-Shizu Enterprise.

Undoubtedly, the MPLAB® IDE is the most important software development tool from Microchip. This package consists of the following:

- Assemblers for all microcontrollers manufactured by Microchip
- MPLINK linker
- Simulators for all microcontrollers manufactured by Microchip
- Control programs for several hardware tools, such as MPLAB ICD2 debugger and other debugging hardware made by Microchip
- An IDE that combines all the software development tools and allows the user to perform development work from program entry until simulation without leaving the same environment

All microcontrollers provide certain features to support software debugging. Many microcontrollers utilize the JTAG interface to support software debugging in addition to performing chip-testing function. The PIC18 microcontroller provides the ICSP protocol to support software debugging and on-chip flash memory programming.

Tutorials on the use of MPLAB IDE, ICD2, and demo boards are provided at the end of this chapter.

3.10 Lab Exercises and Assignments

L3.1 Start the MPLAB IDE program and enter the following program as a text file with the file name progL1.asm:

```
            #include   <p18F452.inc>
            radix      dec
sum_hi      set        0x01
sum_lo      set        0x00
lp_cnt      set        0x02
kk          equ        50
            org        0x00
            goto       start
            org        0x08
            retfie
            org        0x18
            retfie
start       movlw      kk
            movwf      lp_cnt
            clrf       sum_hi,A
            clrf       sum _lo,A
            movlw      upper array
            movwf      TBLPTRU,A
            movlw      high array
            movwf      TBLPTRH,A
            movlw      low array
            movwf      TBLPTRL,A
loop        tblrd*+
            btfsc      TABLAT,0,A
            goto       next
            movf       TABLAT,W,A
            addwf      sum_lo,F,A
            clrf       WREG,A
            addwfc     sum_hi,F,A
next        decfsz     lp_cnt,F,A
            goto       loop
            nop
array       db         01,02,03,04,05,06,07,08,09,10
            db         11,12,13,14,15,16,17,18,19,20
            db         21,22,23,24,25,26,27,28,29,30
            db         31,32,33,34,35,36,37,38,39,40
            db         41,42,43,44,45,46,47,48,49,50
            end
```

Perform the following operations:

1. Create a new project called **progL1** in the same directory where the program **prog1L1.asm** is stored.
2. Configure MPLAB IDE properly.
3. Add the source code file progL1.asm into the project.

4. Build the project.
5. Select MPLAB SIM as the debug tool and perform appropriate configuration as described in Sections 3.6.1 to 3.6.7.
6. Set a breakpoint at the last instruction (**nop**).
7. Open a Watch window and enter symbols **sum_hi** and **sum_lo** into the window.
8. Reset the project by pressing the function key F6.
9. Run the program by pressing the function key F9.
10. Check the values of symbols **sum_hi** and **sum_lo**.
11. View the simulation trace to identify errors. You should not see any error if you type everything correctly.

Can you figure out what this program is doing?

L3.2 Enter the following program as a text file and name it progL2.asm:

```
        #include <p18F452.inc>
        radix      dec
ar_cnt  equ        30                    ; array count
lp_cnt  set        0x00                  ; loop count symbol
buffer  set        0x010
        org        0x00
        goto       start
        org        0x08
        retfie
        org        0x18
        retfie
start   movlw      upper array           ; set TBLPTR as the array pointer
        movwf      TBLPTRU,A             ; "
        movlw      high array            ; "
        movwf      TBLPTRH,A             ; "
        movlw      low array             ; "
        movwf      TBLPTRL,A             ; "
        lfsr       FSR0,buffer           ; use FSR0 as a pointer to buffer
        movlw      ar_cnt
        movwf      lp_cnt                ; set up loop count value
loop1   tblrd*+
        movff      TABLAT,POSTINC0       ; move from table latch to buffer in data memory
        decfsz     lp_cnt, F,A           ; decrement lp_cnt and skip if zero
        goto       loop1
        lfsr       FSR0,buffer           ; set FSR0 to point to the first array element
        lfsr       FSR1,buffer+ar_cnt-1  ; set FSR1 to point to the last array element
        movlw      ar_cnt/2
        movwf      lp_cnt                ; set loop count
loop2   movf       INDF0,W               ; copy array element
        movff      INDF1,POSTINC0        ; store and increment pointer FSR0
        movwf      POSTDEC1              ; store and decrement pointer FSR1
        decfsz     lp_cnt,F,A
        goto       loop2
        nop
```

```
forever    goto      forever
array      db        1,2,3,4,5,6,7,8,9,10
           db        11,12,13,14,15,16,17,18,19,20
           db        21,22,23,24,25,26,27,28,29,30
           end
```

Perform the following operations:

1. Create a new project called **progL2** in the same directory where the file progL2.asm is saved.
2. Add the file **progL2.asm** into the project.
3. Build the project.
4. Select ICD2 as your debug tool.
5. Make sure that ICD2 is configured properly.
6. Program the hex code into the microcontroller on your demo board. It takes a few seconds for the programming to be completed. You will see the message "Programming Target. . ." at the left bottom of the MPLAB IDE window. When programming is done, this message will disappear.
7. Open the Program Memory window by selecting *View>Program Memory*. Scroll the program memory so that program lines 44 to 60 can be seen on the window as shown in Figure L3.1. Look at the column with the title Opcode. The numbers from 1 to 30 can be seen starting from line 46. How does the Program Memory window store the data?

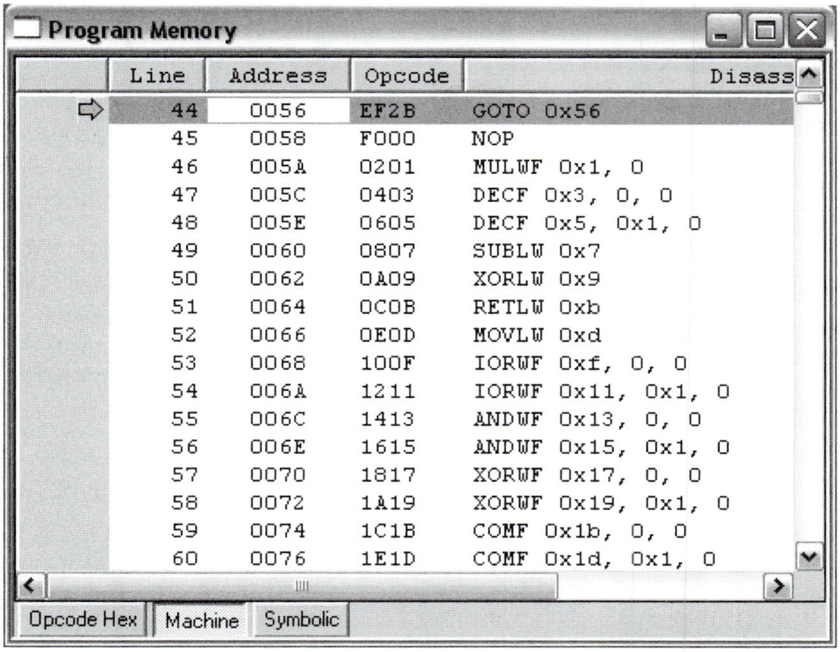

Line	Address	Opcode	Disass
44	0056	EF2B	GOTO 0x56
45	0058	F000	NOP
46	005A	0201	MULWF 0x1, 0
47	005C	0403	DECF 0x3, 0, 0
48	005E	0605	DECF 0x5, 0x1, 0
49	0060	0807	SUBLW 0x7
50	0062	0A09	XORLW 0x9
51	0064	0C0B	RETLW 0xb
52	0066	0E0D	MOVLW 0xd
53	0068	100F	IORWF 0xf, 0, 0
54	006A	1211	IORWF 0x11, 0x1, 0
55	006C	1413	ANDWF 0x13, 0, 0
56	006E	1615	ANDWF 0x15, 0x1, 0
57	0070	1817	XORWF 0x17, 0, 0
58	0072	1A19	XORWF 0x19, 0x1, 0
59	0074	1C1B	COMF 0x1b, 0, 0
60	0076	1E1D	COMF 0x1d, 0x1, 0

Opcode Hex | Machine | Symbolic

Figure L3.1 ■ Snap shot of program memory window for lab exercise L3.2 (reprint with permission of Microchip)

8. Open the File Register window. Resize and scroll so that the data memory locations 0x00 to 0x40 can be seen on the screen.

9. Set a breakpoint at line 30 and run the program until it halts at the breakpoint. The contents of the File Registers window should change to that in Figure L3.2.

10. Set a new breakpoint at line 42 (the last **nop** instruction) and rerun the program. The contents of the File Registers window should change to that in Figure L3.3.

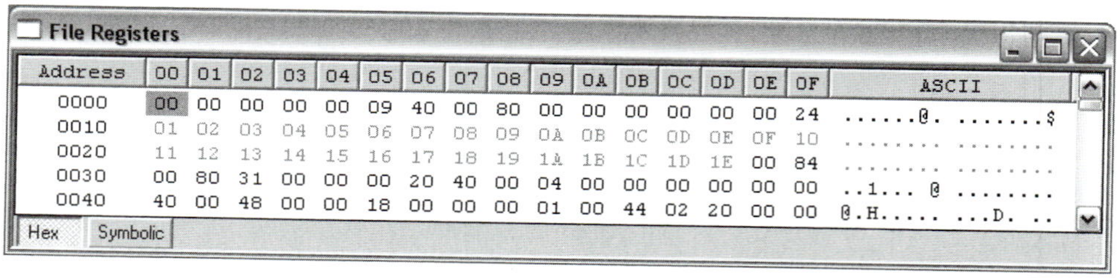

Address	00	01	02	03	04	05	06	07	08	09	0A	0B	0C	0D	0E	0F	ASCII
0000	00	00	00	00	00	09	40	00	80	00	00	00	00	00	00	24	......@. ....$
0010	01	02	03	04	05	06	07	08	09	0A	0B	0C	0D	0E	0F	10	....
0020	11	12	13	14	15	16	17	18	19	1A	1B	1C	1D	1E	00	84	....
0030	00	80	31	00	00	00	20	40	00	04	00	00	00	00	00	00	..1... @ ....
0040	40	00	48	00	00	18	00	00	00	01	00	44	02	20	00	00	@.H..... ...D. ..

Hex   Symbolic

Figure L3.2 ■ Contents of file registers at the first breakpoint (reprint with permission of Microchip)

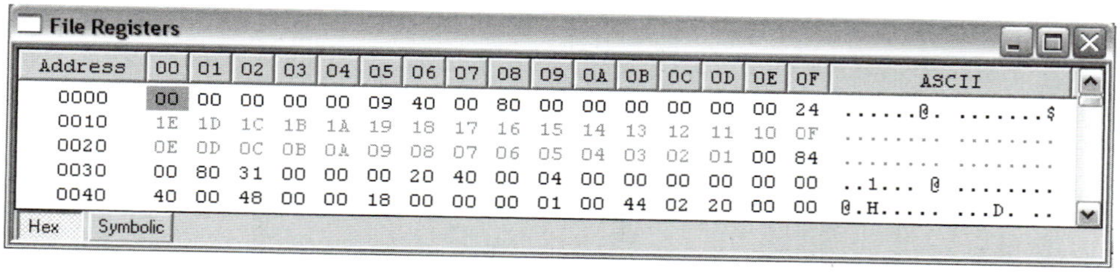

Address	00	01	02	03	04	05	06	07	08	09	0A	0B	0C	0D	0E	0F	ASCII
0000	00	00	00	00	00	09	40	00	80	00	00	00	00	00	00	24	......@. ....$
0010	1E	1D	1C	1B	1A	19	18	17	16	15	14	13	12	11	10	0F	....
0020	0E	0D	0C	0B	0A	09	08	07	06	05	04	03	02	01	00	84	....
0030	00	80	31	00	00	00	20	40	00	04	00	00	00	00	00	00	..1... @ ....
0040	40	00	48	00	00	18	00	00	00	01	00	44	02	20	00	00	@.H..... ...D. ..

Hex   Symbolic

Figure L3.3 ■ File register contents when program halts at the second breakpoint (reprint with permission of Microchip)

What operation is performed by this program?

L3.3 Write a program to compute the average of an array of 32 8-bit elements. The array is stored immediately after your program with the name of **array1.** Store the sum (two bytes) and average in data memory 0x00–0x01 and 0x02, respectively. Use the simulator to simulate the program. Hint: Divide-by-32 can be implemented by shifting to the right by five places.

L3.4 Write a program to count the number of elements that are greater than 30 in an array of n 8-bit elements using the **for i = n1 to n2 do** looping construct. Debug the program using the ICD2 and the demo board. Open a Watch window to view the result.

4

Advanced Assembly Programming

4.1 Objectives

After completing this chapter, you should be able to

- Perform signed arithmetic
- Perform logical and bit field operations
- Perform array and table manipulations
- Perform string operations
- Perform the conversion between integers and strings
- Compute square root using successive approximation method
- Write subroutines
- Delineate stack frames of subroutines
- Access variables in the stack frame
- Make subroutine calls

4.2 Introduction

Many quantities are unsigned and can be handled by unsigned arithmetic that you learned in Chapter 2. However, there are also many quantities that are signed (e.g., temperature, voltage, and profit). They can be positive and negative. Signed numbers are represented in two's complement format in a computer. Signed arithmetic operations are discussed in this chapter. Bit-field operations are also very common in microcontroller (MCU) applications. Such operations require the use of logical instructions (e.g., AND, OR, and XOR), rotate instructions, and data movement instructions. Bit-field operations are useful in input and output operations.

A program consists of *data structures* and *algorithms.* There are many data structures being used today. However, a full coverage of data structures is out of the scope of this textbook. This text deals only with arrays, stacks, and strings.

When a problem gets complicated, writing a program to solve the problem becomes difficult. A common approach is to use the divide-and-conquer strategy in which a large problem is divided into several smaller ones and each small problem solved by a subroutine.

The issues related to subroutine calls, including parameter passing, local variables allocation, and result returning, are discussed here. The instructions used in making subroutine calls and returning from the subroutine are also presented.

4.3 Signed Arithmetic

In a computer system, all negative numbers are represented in the two's complement format. One of the advantages in using the two's complement system is that it allows the hardware designer to use the adder to perform subtraction. When subtracting a number from the minuend, the PIC18 MCU actually adds the two's complement of the subtrahend to the minuend.

As long as the operands are represented in the two's complement format, additions and subtractions of signed numbers can be performed by executing the **add** and **subtract** instructions. This can be verified by using the MPLAB® simulator, ICD2 in-circuit debugger, and any PIC18 demo board.

To understand the following discussion, one must understand that the MCU is performing *modulus* arithmetic when performing an arithmetic operation. For example, an 8-bit MCU will drop (or discard) the part of the result that is equal to or larger than 2^8. A 16-bit MCU will drop the part of the result that is equal to or larger than 2^{16}. A 32-bit MCU will discard the part of the result that is equal to or larger than 2^{32}. Adding 2^8 to the result of the arithmetic operation performed by an 8-bit MCU has no effect on the result because this value (2^8) cannot be kept in the arithmetic hardware of an 8-bit MCU. In summary, in an n-bit MCU, the result of any arithmetic operation performed by an n-bit MCU is equal to the remainder of the initial result divided by 2^n (i.e., it is performing a modulus-2^n operation).

4.3.1 Signed 8-Bit Multiplication

The multiplication of signed numbers requires the programmer to consider the signs of the multiplier and the multiplicand. Let M and N represent the magnitudes of two numbers. There are four possible situations:

Case 1. Both operands are positive (**op1** = M, **op2** = N). The product of these two operands can be computed by using either the **mulwf f, A** or the **mullw k** instruction.

Case 2. The first operand is negative (**op1** = –M) whereas the second operand is positive (op2 = N). The first operand **op1** will be represented in two's complement of M (2^8 – **M**) in the PIC18. The product of –M and N can be rewritten as follows:

$$-M \times N = (2^8 - M) \times N = (2^8 \times N) - M \times N = \underbrace{2^{16} - M \times N}_{①} + \underbrace{2^8 \times N}_{②}$$

The value 2^{16} is added to this expression. Since in this case the PIC18 is performing a modulo-2^{16} arithmetic (PIC18 uses PRODH and PRODL to hold the product), adding the value of 2^{16} makes no difference to the result. Item 1 is the two's complement of the number –M × N and is the correct product. Part 2 is an extra term and should be subtracted from the product. This extra item can be eliminated by subtracting **op2** from the upper byte of the product of –M and N.

Case 3. The first operand is positive (**op1** = M), but the second operand is negative (**op2** = –N). Similar to Case 2, the product of M and –N can be rewritten as follows:

$$M \times (- N) = M \times (2^8 - N) = (2^8 \times M) - M \times N = \underbrace{2^{16} - M \times N}_{①} + \underbrace{2^8 \times M}_{②}$$

The value of 2^{16} is added to this expression and it makes no difference to the result for the same reason as in Case 2. Again, the first term is the correct product, which is represented as the two's complement of –M × N. The second term of this product is an extra term and can be eliminated by subtracting **op1** from the upper byte of the product of M and –N.

Case 4. Both operands are negative (**op1** = –M, **op2** = –N). The product of –M and –N can be rewritten as follows:

$$(-M) \times (-N) = (2^8 - M) \times (2^8 - N) = 2^{16} - 2^8 \times M - 2^8 \times N + M \times N$$
$$= M \times N + 2^{16} - 2^8 \times M + 2^{16} - 2^8 \times N$$
$$= \underbrace{M \times N}_{①} + \underbrace{2^8 \times (2^8 - M)}_{②} + \underbrace{2^8 \times (2^8 - N)}_{③}$$

The value of 2^{16} is added to this expression. It makes no difference to the result for the same reason as in Case 2. The first term is the product of –M and –N. The second term and the third term are extra terms and must be eliminated. The second term can be eliminated by subtracting **op1** from the upper byte of the product, whereas the third term can be eliminated by subtracting **op2** from the upper byte of the product.

By combining these four situations, we conclude that the signed 8-bit multiplication can be implemented by the following algorithm:

Step 1
Multiply two operands (i.e., compute **op1** × **op2**) disregarding their signs.

Step 2
If **op1** is negative, then subtract **op2** from the upper byte of the product.

Step 3
If **op2** is negative, then subtract **op1** from the upper byte of the product.

Example 4.1

▼

Write a program to compute the product of two 8-bit signed numbers represented by data memory locations **op1** and **op2** and leave the product in file registers PRODH and PRODL.

Solution:

```
        #include   <p18F8720.inc>
op1     set        0x00          ; the first operand
op2     set        0x01          ; the second operand
num1    equ        0x87          ; test number 1
num2    equ        0x98          ; test number 2
        org        0x00          ; reset vector
        goto       start
        org        0x08          ; high priority interrupt service routine
        retfie
        org        0x18          ; low priority interrupt service routine
        retfie
start   movlw      num1
        movwf      op1,A
        movlw      num2
        movwf      op2,A
        movf       op1,W,A
        mulwf      op2,A
        btfsc      op2,7,A        ; test the sign bit of the second operand
        subwf      PRODH,F,A      ; if negative, eliminated extra term
        movf       op2,W,A
        btfsc      op1,7,A        ; test the sign bit of the first operand
        subwf      PRODH,F,A      ; if negative, eliminated extra term
        nop
        end
```

▲

4.3.2 Signed 16-Bit Multiplication

Let P and Q be the magnitudes of two 16-bit numbers to be multiplied. There are four situations for the multiplication operation:

Case 1. Both operands are positive. (**op1** = P, **op2** = Q). The resultant product is P × Q. The method discussed in Section 2.7.4 should be used to carry out the multiplication.

Case 2. The first operand is negative (**op1** = –P), whereas the second operand is positive (**op2** = Q). The operand **op1** would be represented as 2^{16} – P. The product of **op1** and **op2** can be rewritten as follows:

$$- P \times Q = (2^{16} - P) \times Q = (2^{16} \times Q) - P \times Q = \underbrace{2^{32} - P \times Q}_{①} + \underbrace{2^{16} \times Q}_{②}$$

The number of 2^{32} is added to this expression. Adding 2^{32} makes no difference because the MCU is performing a modulo-2^{32} arithmetic in this case. The product of –P and Q is represented as the two's complement of P × Q, that is, 2^{32} – P × Q. The second item in this expression is an extra term and should be eliminated. This can be done by subtracting the second operand (**op2**) from the upper 16 bits of the product of –P and Q.

Case 3. The first operand is positive (**op1** = P), and the second operand is negative (**op2** = –Q). The second operand **op2** would be represented as $2^{16} - Q$. The product of **op1** and **op2** can be rewritten as follows:

$$P \times (-Q) = P \times (2^{16} - Q) = (2^{16} \times P) - P \times Q = \underbrace{2^{32} - P \times Q}_{①} + \underbrace{2^{16} \times P}_{②}$$

The value of 2^{32} is also added to the above expression. It makes no difference to the result for the same reason as in Case 2. The first term of this expression is the product, whereas the second item is an extra term and should be eliminated. This can be done by subtracting the first operand (**op1**) from the upper 16 bits of the product of P and –Q.

Case 4. Both operands are negative (**op1** = –P, **op2** = –Q). Since both operands are negative, they will be represented in two's complement forms. The first operand would be represented as $2^{16} - P$, whereas the second operand would be represented as $2^{16} - Q$. The product can be rewritten as follows:

$$(-P) \times (-Q) = (2^{16} - P) \times (2^{16} - Q) = 2^{32} - 2^{16} \times P - 2^{16} \times Q + P \times Q$$
$$= P \times Q + 2^{32} - 2^{16} \times P + 2^{32} - 2^{16} \times Q$$
$$= \underbrace{P \times Q}_{①} + \underbrace{2^{16} \times (2^{16} - P)}_{②} + \underbrace{2^{16} \times (2^{16} - Q)}_{③}$$

The value of 2^{32} is also added to this expression. It makes no difference to the result for the same reason as in Case 2. The first term in this expression is the product of –P and –Q. Both the second and the third terms are extra items and should be eliminated. The second term can be eliminated by subtracting the first operand (**op1**) from the upper 16 bits of the product of –P and –Q. The third term can be eliminated by subtracting the second operand (**op2**) from the upper 16 bits of the product.

By combining these four situations, we conclude that the signed 16-bit multiplication can be implemented by the following algorithm:

Step 1
Multiply two operands (i.e., compute **op1** × **op2**) disregarding their signs.

Step 2
If **op1** is negative, then subtract **op2** from the upper 16 bits of the product.

Step 3
If **op2** is negative, then subtract **op1** from the upper 16 bits of the product.

Example 4.2
▼

Write a program to compute the product of two 16-bit signed integers and leave the product in memory locations represented by **result** . . . **result+3**.

Solution: The program that implements the algorithm is as follows:

```
            #include     <p18F8720.inc>
arg1_hi     set          0x00            ; high byte of the first argument
arg1_lo     set          0x01            ; low byte of the first argument
arg2_hi     set          0x02            ; high byte of the second argument
arg2_lo     set          0x03            ; low byte of the second argument
result      set          0x04            ; location to hold the product
num1_hi     equ          0x83
num1_lo     equ          0x45
num2_hi     equ          0x81
```

```
num2_lo     equ       0x47
            org       0x00              ; reset vector
            goto      start
            org       0x08              ; high-priority interrupt service routine
            retfie
            org       0x18              ; low-priority interrupt service routine
            retfie
start       movlw     num1_hi           ; set up test number
            movwf     arg1_hi,A         ; "
            movlw     num1_lo           ; "
            movwf     arg1_lo,A         ; "
            movlw     num2_hi           ; "
            movwf     arg2_hi,A         ; "
            movlw     num2_lo           ; "
            movwf     arg2_lo           ; "
            movf      arg1_lo,W,A
            mulwf     arg2_lo           ; compute arg1_lo × arg2_lo
            movff     PRODL,result
            movff     PRODH,result+1
            movf      arg1_hi,W,A
            mulwf     arg2_hi,A         ; compute arg1_hi × arg2_hi
            movff     PRODL, result+2
            movff     PRODH,result+3
            movf      arg1_lo,W,A
            mulwf     arg2_hi,A         ; compute arg1_lo × arg2_hi
            movf      PRODL,W,A
            addwf     result+1,F,A
            movf      PRODH,W,A
            addwfc    result+2,F,A
            clrf      WREG,A
            addwfc    result+3,F,A      ; add carry to the most significant byte
            movf      arg1_hi,W,A
            mulwf     arg2_lo,A         ; compute arg1_hi × arg2_lo
            movf      PRODL,W,A
            addwf     result+1,F,A
            movf      PRODH,W,A
            addwfc    result+2,F,A
            clrf      WREG,A            ; add carry to the most significant byte
            addwfc    result+3,F,A      ; "
            btfss     arg2_hi,7,A       ; check the sign of arg2
            goto      sign_arg1
            movf      arg1_lo,W,A
            subwf     result+2,F,A      ; delete extra term
            movf      arg1_hi,W,A       ; "
            subwfb    result+3,F,A      ; "
sign_arg1   btfss     arg1_hi,7,A       ; check the sign of arg1
            goto      more              ; continue to perform other operation
            movf      arg2_lo,W,A
            subwf     result+2,F,A      ; delete extra term
            movf      arg2_hi,W,A       ; "
            subwfb    result+3,F,A      ; "
more        nop
            end
```

4.4 Unsigned Divide Operation

The PIC18 MCU does not provide any divide instruction. Therefore, a divide operation must be synthesized by other instructions. A simple but popular divide algorithm in use today is the *repeated subtraction method*. This method performs unsigned divide operation. The hardware required for implementing the repeated subtraction method is shown in Figure 4.1.

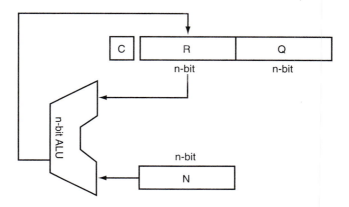

Figure 4.1 ■ Division hardware

Before performing the repeated subtraction operation, one needs to load 0, the dividend, and the divisor into registers R, Q, and N, respectively. The carry flag is used to indicate whether the subtraction result is negative. The ALU can perform n-bit unsigned addition and subtraction operations. The repeated subtraction method consists of n steps. Each division step consists of three parts:

Step 1
Shift the register pair (R, Q) one place to the left.

Step 2
Subtract the contents of N from R and put the result back to R if the result is positive.

Step 3
If the result of Step 2 is negative, then set the least significant bit of Q to 0. Otherwise, set the least significant bit of Q to 1.

Example 4.3

Write a program to divide an 8-bit number into another 8-bit number.

Solution: The following program is a direct translation of the paper-and-pencil division algorithm:

```
                #include <p18F8720.inc>
lp_cnt    set       0x00
rem       set       0x01        ; register to hold remainder
quo       set       0x02        ; register to hold quotient
dsr       set       0x03        ; register to hold divisor
dvd       set       0x04        ; register to hold dividend
dd        equ       0xf5        ; value used as dividend
```

```
dr          equ        0x11              ; value used as divisor
            org        0x00
            goto       start
            org        0x08
            retfie
            org        0x18
            retfie
start       movlw      dd
            movwf      quo,A             ; initialize Q register in Figure 4.1.
            movlw      dr
            movwf      dsr,A             ; initialize N register in Figure 4.1
            clrf       rem,A             ; initialize R register in Figure 4.1
            movlw      0x08
            movwf      lp_cnt            ; initialize loop count to 8
loop        bcf        STATUS,C,A        ; clear the C flag
            rlcf       quo,F,A           ; rotate (R, Q) pair to the left one place
            rlcf       rem,F,A           ; "
            movf       dsr, W,A
            subwf      rem,W,A           ; subtract and leave the difference in WREG
            btfss      STATUS,C,A        ; skip if carry is 1
            goto       negative
            bsf        quo,0,A           ; set the least significant bit of Q1 to 1
            mowvf      rem,A             ; place the difference in rem
            goto       next
negative    bcf        quo,0,A           ; set the quotient bit to 0
next        decfsz     lp_cnt,F,A        ; decrement the loop count and skip if zero
            goto       loop
            nop
            end
```

▲

Example 4.4

▼

Write a program to divide an unsigned 16-bit number into another unsigned 16-bit number.

Solution: The assembly program for the 16-bit unsigned division is as follows:

```
            #include <p18F8720.inc>
lp_cnt      set        0x00              ; loop count
temp        set        0x01              ; temporary storage
dsr         set        0x04              ; divisor
quo         set        0x06              ; quotient
rem         set        0x08              ; remainder
dd_h        equ        0x68              ; high byte of dividend test number
dd_l        equ        0x20              ; low byte of "
dr_h        equ        0x01              ; high byte of divisor test number
dr_l        equ        0x48              ; low byte of divisor test number
            org        0x00
            goto       start
            org        0x08
            retfie
```

```
                 org           0x18
                 retfie
start            movlw         dd_h            ; initialize Q register in Figure 4.1
                 movwf         quo+1,A         ; "
                 movlw         dd_l            ; "
                 movwf         quo,A           ; "
                 movlw         dr_h            ; initialize N register in Figure 4.1
                 movwf         dsr+1,A         ; "
                 movlw         dr_l            ; "
                 movwf         dsr,A           ; "
                 clrf          rem,A           ; initialize R register in Figure 4.1 to 0
                 clrf          rem+1,A         ; "
                 movlw         D'16'
                 movwf         lp_cnt,A        ; initialize loop count to 16
loop             bcf           STATUS,C,A      ; clear the C flag
                 rlcf          quo,F,A         ; rotate (R, Q) pair to the left one place
                 rlcf          quo+1,F,A       ; "
                 rlcf          rem,F,A         ; "
                 rlcf          rem+1,F,A       ; "
                 movf          dsr,W,A
                 subwf         rem,W,A
                 movwf         temp,A          ; save the low byte of the difference
                 movf          dsr+1,W,A
                 subwfb        rem+1,W,A
                 btfss         STATUS,C        ; skip if carry is 1
                 goto          less
                 bsf           quo,o,A         ; set the quotient bit to 1
                 movwf         rem+1,A         ; place the difference in R register
                 movff         temp,rem        ; "
                 goto          next
less             bcf           quo,0,A         ; set the quotient bit to 0
next             decfsz        lp_cnt,F,A      ; decrement the loop count and skip is zero
                 goto          loop
                 nop
                 end
```

Unsigned division program for numbers in other lengths (e.g., 32-bit by 32-bit) can be written in the same way and hence is left for you as an exercise. ▲

4.5 Signed Divide Operation

The one complication for the signed division is that we must also set the sign of the remainder. The following equation must always hold for division:

Dividend = Quotient × Divisor + Remainder

Our common sense requires that the magnitude of the quotient be the same as long as the magnitudes of the dividends are the same and the magnitudes of the divisors are the same. We can determine the sign of the remainder on the basis of this principle. To illustrate, let's use $(\pm 35) \div (\pm 6)$ as an example. The first situation is simple:

35 ÷ 6: Quotient = +5, Remainder = +5

If we change the sign of the dividend, the quotient must be changed as well:

–35 ÷ 6: Quotient = –5

Rewriting our basic formula to find the remainder,

$$\text{Remainder} = \text{Dividend} - \text{Quotient} \times \text{Divisor}$$
$$= -35 - (-5 \_ 6) = -35 + 30 = -5$$

If we change the sign of the divisor and keep the sign of dividend unchanged,

35 ÷ (–6): Quotient = –5
Remainder = 35 – (–5 × –6) = 35 – 30 = 5

If we change the signs of both the dividend and the divisor,

–35 ÷ –6: Quotient = 5
Remainder = –35 – (–5 × –6) = –35 + 30 = –5

From this discussion, we conclude that the correctly signed division algorithm negates the quotient if the signs of the operands are opposite and makes the sign of the nonzero remainder match the dividend.s

Example 4.5

Write a PIC18 program that performs the 8-bit signed divide operation. This program will leave the quotient and remainder in the data registers represented by **quo** and **rem**, respectively.

Solution: The following program implements the 8-bit signed divide operation described in this section:

```
              #include <p18F8720.inc>
sign      set        0x00
dvd       set        0x01              ; dividend
dsr       set        0x02              ; divisor
quo       set        0x03              ; quotient
rem       set        0x04              ; remainder
lp_cnt    set        0x05              ; loop count
dd        equ        0x82              ; testing number for dividend
dr        equ        0xf5              ; testing number for divisor
          org        0x00
          goto       start
          org        0x08
          retfie
          org        0x18
          retfie
start     bcf        sign,2,A          ; initialize the sign of quotient to positive
          bcf        sign,1,A          ; initialize the sign of dividend to positive
          bcf        sign,0,A          ; initialize the sign of divisor to positive
          movlw      dd
          movwf      dvd,A
          movlw      dr
          movwf      dsr,A
          btfss      dvd,7,A           ; check the sign of dividend
          goto       second
          btg        sign,2            ; change the sign of quotient
```

```
                   bsf          sign,1             ; record the sign bit of the dividend
                   negf         dvd,A              ; compute the magnitude of dividend
       second      btfss        dsr,7,A            ; check the sign of the divisor
                   goto         do_it
                   btg          sign,2,A           ; change the sign of quotient
                   bsf          sign,0,A           ; set the sign of the divisor
                   negf         dsr,A              ; compute the magnitude of divisor
       do_it       movf         dvd,W,A
                   movwf        quo,A
                   clrf         rem,A              ; initialize R register in Figure 4.1
                   movlw        0x08
                   movwf        lp_cnt,A           ; initialize loop count to 8
       loop        bcf          STATUS,C,A         ; clear the C flag
                   rlcf         quo,F,A            ; rotate (R, Q) pair to the left one place
                   rlcf         rem,F,A            ; "
                   movf         dsr,W,A
                   subwf        rem,W,A            ; subtract and leave the difference in WREG
                   btfss        STATUS,C,A         ; skip if carry is 1
                   goto         negative
                   bsf          quo,0,A            ; set the least significant bit of Q1 to 1
                   movwf        rem,A              ; place the difference in R1
                   goto         next
       negative    bcf          quo,0,A            ; set the quotient bit to 0
       next        decfsz       lp_cnt,F,A         ; decrement the loop count and skip if zero
                   goto         loop
                   btfss        sign,2,A           ; skip if sign of quotient is negative
                   goto         check_re           ; "
                   negf         quo,A
       check_re    btfss        sign,1,A           ; skip if dividend is negative
                   goto         ok_skip            ; "
                   negf         rem,A
       ok_skip     nop
                   end
```

Example 4.6

Write a program to divide a signed 16-bit number into another 16-bit signed integer.

Solution: The following program will perform the signed 16-bit divide operation:

```
                #include <p18F8720.inc>
    sign        set          0x00       ; keep track of the signs of dividend and divisor
    dvd         set          0x02       ; dividend
    dsr         set          0x04       ; divisor
    quo         set          0x06       ; quotient
    rem         set          0x08       ; remainder
    lp_cnt      set          0x0A       ; loop count
    temp        set          0x0B       ; temporary storage
    dd_h        equ          0xD9       ; testing number for dividend
```

```
dd_l        equ         0xB8            ; "
dr_h        equ         0xFF            ; testing number for divisor
dr_l        equ         0x80            ; "
            org         0x00
            goto        start
            org         0x08
            retfie
            org         0x18
            retfie
start       bcf         sign,2,A        ; initialize the sign of quotient to positive
            bcf         sign,1,A        ; initialize the sign of dividend to positive
            bcf         sign,0,A        ; initialize the sign of divisor to positive
            movlw       dd_l            ; set up dividend
            movwf       dvd,A           ; "
            movlw       dd_h            ; "
            movwf       dvd+1,A         ; "
            movlw       dr_l            ; set up divisor
            movwf       dsr,A           ; "
            movlw       dr_h            ; "
            movwf       dsr+1,A         ; "
            btfss       dvd+1,7,A       ; check the sign of dividend
            goto        second
            btg         sign,2          ; change the sign of quotient
            bsf         sign,1          ; record the sign bit of the dividend
            comf        dvd,F,A         ; compute the magnitude of dividend
            comf        dvd+1,F,A       ; "
            incf        dvd,F,A         ; "
            movlw       0x00            ; "
            addwfc      dvd+1,F,A       ; "
second      btfss       dsr+1,7,A       ; check the sign of the divisor
            goto        do_it
            btg         sign,2,A        ; change the sign of quotient
            bsf         sign,0,A        ; set the sign of the divisor
            comf        dsr,F,A         ; compute the magnitude of divisor
            comf        dsr+1,F,A       ; "
            incf        dsr,F,A         ; "
            movlw       0x00            ; "
            addwfc      dsr+1,F,A       ; "
do_it       movff       dvd,quo         ; place dividend in Q register in Figure 4.1
            movff       dvd+1,quo+1
            clrf        rem,A           ; initialize R register in Figure 4.1
            clrf        rem+1,A         ; "
            movlw       D'16'
            movwf       lp_cnt,A        ; initialize loop count to 8
loop        bcf         STATUS,C,A      ; clear the C flag
            rlcf        quo,F,A         ; rotate (R, Q) pair to the left one place
            rlcf        quo+1,F,A       ; "
            rlcf        rem,F,A         ; "
            rlcf        rem+1,F,A       ; "
            movf        dsr,W,A         ; compute R-N and places the difference
```

```
            subwf     rem,W,A          ; in WREG and temp
            movwf     temp,A           ; "
            movf      dsr+1,W,A        ; "
            subwfb    rem+1,W,A        ; "
            btfss     STATUS,C,A       ; skip if carry is 1
            goto      negative
            bsf       quo,0,A          ; set the least significant bit of Q to 1
            movwf     rem+1,A          ; place the difference in R in Figure 4.1
            movff     temp,rem         ; "
            goto      next
negative    bcf       quo,0,A          ; set the quotient bit to 0
next        decfsz    lp_cnt,F,A       ; decrement the loop count and skip if zero
            goto      loop
            btfss     sign,2,A         ; skip if sign of quotient is negative
            bra       check_re         ; "
            comf      quo,F,A          ; complement the quotient
            comf      quo+1,F,A        ; "
            incf      quo,F,A          ; "
            movlw     0x00             ; "
            addwfc    quo+1,F,A
check_re    btfss     sign,1,A         ; skip if dividend is negative
            bra       ok_skip          ; "
            comf      rem,F,A          ; complement the remainder
            comf      rem+1,F,A        ; "
            incf      rem,F,A          ; "
            movlw     0x00             ; "
            addwfc    rem+1,F,A        ; "
ok_skip     nop
            end
```

Signed division program for numbers in other lengths (e.g., 32-bit by 32-bit division) can be written in the same manner and hence are left for you as an exercise. ▲

4.6 The Stack

A stack is a first-in-last-out (or last-in-first-out) data structure. To implement a stack, two things are needed:

1. A stack pointer that points to the top (or the byte immediately above the top) of the stack
2. A block of RAM of adequate size

The PIC18 MCU has no register designated as the stack pointer. However, the user can use one of the FSR registers as the stack pointer and use one or more banks of the data memory to implement the data stack. The stack implemented this way is called a *software stack.*

A stack can grow from a low address toward higher addresses or from a high address toward lower addresses. This text follows the convention used by the Microchip C18 compiler:

1. Use the FSR1 register as the stack pointer and set it to point to the next available byte on the stack as shown in Figure 4.2.
2. Grow the stack from a low address toward higher addresses.

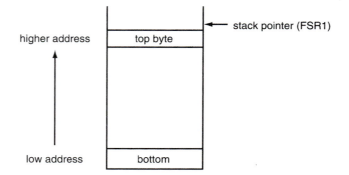

Figure 4.2 ■ The PIC18 Software Stack

A software stack is used mainly in saving program contexts and providing space for local variables during a subroutine call. The contents of some CPU registers such as WREG and STATUS are the program contexts, whereas local variables are those variables that exist only when a subroutine is entered (being called). The appropriate size for the stack is dependent on the application. An application that makes a lot of recursive subroutine calls will need a larger software stack. Unless the application requires, one data memory bank is used for the software bank in this text. Any unused data memory bank can be used for the software stack. This text uses the highest data memory bank for the software stack. For example, the bank that covers the address range from 0x500 to 0x5FF is used for the PIC18F452 and PIC18F458, whereas the bank that covers the address range from 0xE00 to 0xEFF is used for the PIC18F8720.

4.6.1 Stack Operations

The stack pointer needs to be initialized before the stack can be accessed. The user can initialize the stack pointer by using the **LFSR** instruction. For example, the instruction

    lfsr   FSR1,0x500

will initialize the stack pointer for the PIC18F452 and PIC18F458. The instruction

    lfsr   FSR1,0xE00

will initialize the stack pointer for the PIC18F8720.

The most common stack operations are the *push* and *pop* (also called *pull*) operations. A push operation places the contents of a register into the stack location pointed to by the stack pointer and then increments the stack pointer by one. A pop operation decrements the stack pointer by one and then copies the contents of the memory location pointed to by the stack pointer to the specified register.

The following instruction will push the WREG register onto the stack:

    movwf   POSTINC1

The following instruction will push the STATUS register onto the stack:

    movff   STATUS,POSTINC1

The following instruction sequence will pull the top byte of the stack onto the STATUS register:

    movff   POSTDEC1,STATUS   ; get the stack pointer decremented
    movff   INDF1,STATUS   ; pop the stack onto STATUS

To make the push and pop operations more understandable, the user can create macros for these operations:

- **pushr**—macro to push a file register onto the stack
- **popr**—macro to pop a byte off the stack

The definitions of these two macros are as follows:

```
pushr      macro         arg                  ; macro to push the WREG register
           movff         arg,POSTINC1
           endm
popr       macro         arg                  ; arg is a file register
           movff         POSTDEC1,arg         ; decrement the stack pointer
           movff         INDF1,arg            ; pop off a byte from the stack onto arg
           endm
```

A third macro that pushes a value into the stack can be created by calling the **pushr** macro:

```
push_dat   macro         dat                  ; this macro push the value dat onto the stack
           movlw         dat
           pushr         WREG
           endm
```

Example 4.7

▼

Utilize the macro defined earlier to push the WREG and STATUS register onto the stack and pop the top two bytes of the stack onto the data memory locations at 0x00 and 0x01.

Solution: The required operations can be performed as follows:

1. Pushing WREG: pushr   WREG
2. Pushing STATUS: pushr   STATUS
3. Popping onto 0x00:   popr   0x00
4. Popping onto 0x01:   popr   0x01

▲

4.7 Subroutines

A *subroutine* is a sequence of instructions that can be called from many different places in a program. Parameters can be passed to the subroutine so that it can perform the same operation on different variables. There are two main reasons for creating subroutines:

1. The problem at hand is too complicated to be solved by a single program. It would be easier to use the divide-and-conquer strategy to divide the problem into smaller ones and create a subroutine to solve each smaller problem. By doing this, a complicated problem will be more manageable and can easily be solved.

2. There are several places in a program that need to perform the same operation on possibly different values. By creating a single subroutine to perform the same operation on different values, the program can be dramatically shortened.

When dealing with a complicated problem, one normally begins with a simple main program whose steps clearly outline the logical flow of the algorithm and then assigns the execution details to subroutines. Subroutines may also call other subroutines. Programs written in this manner are called *structured programs*. Writing programs in this manner is called *structured programming*. The structure of a complicated program may look like that in Figure 4.3.

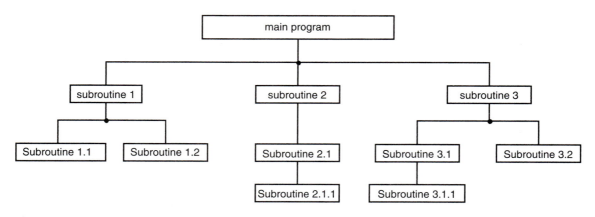

Figure 4.3 ■ A structured program

The PIC18 MCU provides the following mechanisms to support subroutine calls:

1. The **rcall n** and **call k[,s]** instructions for making the subroutine call and the **return** instruction for returning from the subroutine
2. A **return address stack** for saving and restoring return addresses
3. A **return stack pointer** that points to the top of the return address stack
4. The **push** and **pull** instructions for pushing values onto the return address stack and pulling values from the return address stack
5. A one-layer-deep **fast register stack** for fast saving and restoring the STATUS, WREG, and BSR registers during interrupts and subroutine calls

4.7.1 Instructions for Supporting Subroutine Calls

The PIC18 MCU provides two instructions for making subroutine calls. The subroutine being called can be up to 1 K words away from the **rcall n** instruction. The return address (PC + 2) will be pushed onto the return address stack before the subroutine is entered. The subroutine being called is normally referred to by its name. For example,

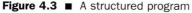

calls the subroutine **find_max.** The subroutine **find_max** must be no more than 1 K words away from the **rcall** instruction.

When the subroutine is farther than 1 K words, the **call k[,s]** instruction should be used to call the subroutine. This instruction can call a subroutine in the entire 2-MB memory range. The **call** instruction has an option to push the STATUS, WREG, and BSR registers onto their respective shadow registers. This option is enabled by setting the **s** flag to 1. For example,

    call   sq_root,1

will call the subroutine **sq_root** and save the BSR, STATUS, and WREG registers in the fast register stack (also called *shadow registers*). Because this is a 32-bit instruction, the saved return address is the address of the **call** instruction plus 4.

The last instruction of a subroutine must be a **return [s]** instruction. When this instruction is executed, the return address stack is popped, and the top of the stack is loaded into the program counter. Program control is thus transferred back to the caller of the subroutine. If the **s** flag is 1, then the BSR, STATUS, and WREG registers are reloaded from the fast register stack.

4.7.2 Return Address Stack

The return address stack allows any combination of up to 31 subroutine calls and interrupts to occur. The program counter is pushed onto the stack when a **push, call**, or **rcall** instruction is executed or an interrupt is acknowledged. On a **return, retlw,** or **retfie** instruction, the PC value is pulled from the return address stack. The **retlw** instruction is explained later.

The return address stack operates as a 31-word-by-21-bit stack memory and a 5-bit stack pointer (STKPTR). The stack space is neither part of the program nor part of the data memory space. The stack pointer is readable and writable, and the data on the top of the return address stack is readable and writable through SFR registers. Status bits (in the STKPTR register) indicate whether the stack pointer is at or beyond the 31 levels provided.

4.7.3 Top-of-Stack Access

The top of the return address stack is readable and writable through three register locations: TOSU, TOSH, and TOSL. These three registers allow access to the contents of the stack location indicated by the STKPTR register. The current PC value can be pushed into this stack by using the **push** instruction. The user can pop the top of the return address stack into the program counter by invoking the **pull** instruction.

4.7.4 Return Address Stack Pointer (STKPTR)

The contents of the STKPTR register are shown in Figure 4.4. In addition to the 5-bit pointer, the STKPTR also contains the STKFUL and STKUNF flags to indicate whether the return address stack has been pushed or pulled too many times.

7	6	5	4	3	2	1	0
STKFUL	STKUNF	--	SP4	SP3	SP2	SP1	SP0

STKFUL: stack full flag bit
    0 = stack has not become full or overflowed
    1 = stack has become full
STKUNF: stack underflow flag bit
    0 = stack underflow did not occur
    1 = stack underflow occurred
SP4:SP0: stack pointer location bits

Figure 4.4 ■ The STKPTR register (at 0xFFC) (redraw with permission of Microchip)

4.7.5 Fast Register Stack

A subroutine call usually requires the saving and restoring of certain CPU registers, such as the BSR, STATUS, and WREG registers. The PIC18 MCU provides a **fast return** option to speed up this process. The fast register stack is provided for the BSR, STATUS, and WREG registers and is only one layer deep. Stated in another way, each of the BSR, STATUS, and WREG registers has a shadow register in the fast register stack. The fast register stack is not readable nor writable.

To use the fast register stack, the **s** flag in the **call** instruction and the corresponding **return** instruction must be set to 1, In assembly language, this is represented by the symbol **FAST.** For example, the following instruction sequence utilizes the fast register stack:

```
        call      sub_x, FAST    ; save BSR, STATUS, and WREG in fast
        . . .                    ; register stack
        . . .
sub_x   . . .
        . . .
        return    FAST           ; restore values saved in fast register stack
```

4.7.6 Table Lookup instruction

A lookup table can be implemented by using the **call** and **retlw** instructions together. The following code segment will perform the desired table lookup:

```
        movlw     i              ; prepare to lookup the i^th entry
        call      look_up
        . . .
        . . .
look_up addwf     PCL
        retlw     k0
        retlw     k1
        . . .
        retlw     kn             ; end of table
```

4.8 Issues Related to Subroutine Calls

The subroutine being called is referred to as *callee*, whereas the program that makes the call is referred to as *caller*. A subroutine is *entered* when it is executed. There are three issues involved in the subroutine call:

- *Parameter passing.* The user usually wants the subroutine to perform a certain operation on the values passed to it. There are several parameter passing methods:

  1. *Use general-purpose CPU registers.* In this method, parameters are placed in registers before the subroutine is called.

  2. *Use the stack.* In this method, parameters are pushed into the stack before the subroutine is called. The stack needs to be cleaned up after the subroutine completes the desired computation. This can be done by either the caller or the callee.

  3. *Use global memory.* Global memory is accessible to both the caller and the callee. As long as the caller places parameters in global memory before it calls the subroutine, the callee will be able to access them. For the PIC18 MCU, this method is the same as the first one because general-purpose registers are identical to the global memory.

- *Result returning.* The result of a computation performed by the subroutine can be returned to the caller using three methods:

1. *Use general-purpose CPU registers.* The subroutine places the computation result in one or more general-purpose CPU registers before returning to the caller.

2. *Use the stack.* The caller makes a hole in the stack before calling the subroutine. The callee places the computation result in the hole before returning to the caller.

3. *Use global memory.* The callee places the result in global memory, and the caller will be able to use them. For the PIC18 MCU, this method is identical to the first one because general-purpose CPU registers are identical to the global memory.

- *Allocation of local variables.* In addition to the parameters passed to it, the subroutine may need memory space to hold temporary variables and results. These temporary variables are called *local variables* because they exist only when the subroutine is entered. Local variables are always allocated in the stack so that they are not accessible to any other program units. For the PIC18 MCU, local variables can be allocated by adding a number to the stack pointer (FSR1). Before the subroutine returns, the space allocated for local variables must be *deallocated.* Deallcoation is the reverse operation of allocation, and it can be done by subtracting a number (same as the number added to the stack pointer earlier) from the stack pointer.

When writing a subroutine, the programmer should keep in mind that the subroutine is to be invoked by many different programs and should try to write it in a general way. This will make the subroutine more useful.

Example 4.8

▼

Write a macro to allocate **n** bytes for local variables and a macro to deallocate **n** bytes from the stack by following the convention described in Section 4.6 where n is an 8-bit number.

Solution: The macro that allocates **n** bytes in the stack for local variables is as follows:

```
alloc_stk      macro     n              ; this macro allocates n bytes in stack
               movlw     n
               addwf     FSR1L,F,A
               movlw     0x00
               addwfc    FSR1H,F,A
               endm
```

The macro that deallocates **n** bytes from the stack is as follows:

```
dealloc_stk    macro     n              ; this macro deallocates n bytes from stack
               movlw     n
               subwf     FSR1L,F,A
               movlw     0
               subwfb    FSR1H,F,A      ; subtract borrow from high byte of FSR1
               endm
```

▲

4.8.1 The Stack Frame

The stack is used heavily during a subroutine call. The caller may pass parameters to the callee, and the callee may need to save registers and allocate local variables in the stack. The area in the stack that holds incoming parameters, saved registers, and local variables is referred to as the *stack frame*. Some microprocessors have a dedicated register for managing the stack frame—the register is referred to as the *frame pointer*. For the PIC18 MCU, the FSR2 register can be used as the frame pointer (The C18 compiler uses the FSR2 register as the frame pointer.) The stack frame for the PIC18 MCU is shown in Figure 4.5.

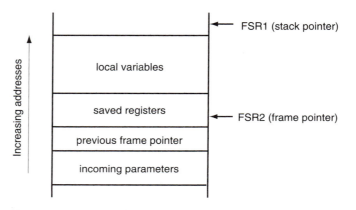

Figure 4.5 ■ PIC18 Stack frame

The frame pointer is mainly used to reference the location on the stack that separates the stack-based arguments and the stack-based local variables. Incoming parameters are located at negative offsets from the frame pointer whereas the local variables are located at positive offsets from the frame pointer.

Immediately on entry to the subroutine, the callee pushes the value of FSR2 onto the stack and copies the value of FSR1 to FSR2, thereby saving the caller's frame pointer and initializing the frame pointer for the current subroutine. The callee may save some registers other than those saved in the fast register stack and then add the size of the local variables for the callee to the stack pointer (FSR1). References to stack-based incoming parameters and stack-based local variables are resolved according to offsets from the frame pointer.

The reason for having a dedicated frame pointer is due to the fact that the stack pointer may change during the lifetime of a subroutine. Once the stack pointer changes value, there can be problem in accessing the variables stored in the stack frame. The frame pointer is added to point to a fixed location in the stack and can avoid this problem.

4.8.2 Accessing Locations in the Stack Frame

As stated earlier, a local variable can be accessed by adding a positive value (placed in WREG) to the frame pointer to form an address and then performing an indirect access, whereas an incoming parameter can be accessed by adding a negative value (also placed in WREG) to the frame pointer to form an address and then performing an indirect access.

For illustration purpose, assume that we have a stack frame as shown in Figure 4.6. The offsets from FSR2 for local variables **i, j, k, sum, in_order,** and **loop_cnt** are 0, 1, 2, 3, 4, and 5 respectively. The offsets from FSR2 for incoming parameters n and m are –2 and –3, respectively.

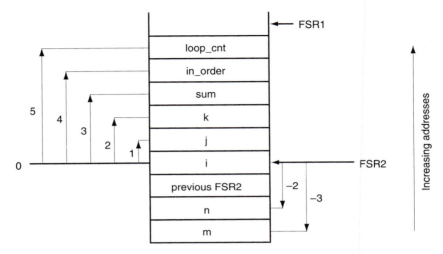

Figure 4.6 ■ Stack frame and variable offsets

The following instruction sequence will load the local variable **sum** into the WREG register:

```
movlw      0x3
movf       PLUSW2,W,A
```

The following instruction sequence will load the incoming variable **m** into the WREG register:

```
movlw      -3            ; place -3 in WREG
movf       PLUSW2,W,A    ; load m into WREG
```

To make one's program more readable, symbolic names should be used to access local variables and incoming parameters. This can be achieved as follows:

```
loop_cnt   equ         5         ; offset of loop_cnt from frame pointer
in_order   equ         4         ; offset of in_order from frame pointer
sum        equ         3         ; offset of sum from frame pointer
k          equ         2         ; offset of k from frame pointer
j          equ         1         ; offset of j from frame pointer
i          equ         0         ; offset of i from frame pointer
n          equ         -2        ; offset of n from frame pointer
m          equ         -3        ; offset of m from frame pointer
; loading sum into WREG
           movlw       sum
           movf        PLUSW2,W,A
; loading m into WREG
           movlw       m
           movf        PLUSW2,W,A
```

The following example illustrates the use of all the stack macros defined earlier and the accessing of local variables in the stack frame.

Example 4.9

▲

Write a subroutine that finds the maximum element of an array of 8-bit elements in program memory. Pass the array count and starting address in the software stack.

Solution: The algorithm of this problem has been explained in Figure 2.13. The stack frame of this subroutine is shown in Figure 4.7, which shows that the subroutine **find_amax** will do the following:

1. Save the current frame pointer in the stack
2. Save the table pointer in the stack
3. Allocate two bytes for local variables—one byte for loop count **lp_cnt** and one byte for temporary array max **ar_max.**

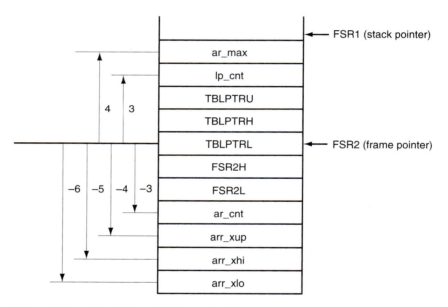

Figure 4.7 ■ Stack frame for Example 4.9

The testing program and the subroutine are as follows:

```
            #include        <p18F8720.inc>
            radix           dec
pushr       macro           arg              ; macro to push the arg register into stack
            movff           arg,POSTINC1
            endm
popr        macro           arg              ; macro to pop the arg register
            movff           POSTDEC1,arg     ; decrement the stack pointer
            movff           INDF1,arg        ; pop off a byte from the stack onto arg
            endm
alloc_stk   macro           n                ; this macro allocates n bytes in the software
            movlw           n                ; stack
            addwf           FSR1L,F,A
```

```
                         movlw          0x00
                         addwfc         FSR1H,F,A
                         endm
dealloc_stk              macro          n                      ; this macro deallocate n bytes from
                         movlw          n                      ; the software stack
                         subwf          FSR1L,F,A
                         movlw          0x00
                         subwfb         FSR1H,F,A
                         endm
acnt                     equ            D'32'                  ; array count
lp_cnt                   equ            3                      ; stack frame offset for lp_cnt
ar_max                   equ            4                      ; stack frame offset for array max
ar_cnt                   equ            -3                     ; stack frame offset for array count
arr_xup                  equ            -4                     ; stack frame offset for array base
arr_xhi                  equ            -5                     ; "
arr_xlo                  equ            -6                     ; "
array_max                set            0x00                   ; register to hold array max
                         org            0x00                   ; reset vector
                         goto           start
                         org            0x08
                         retfie
                         org            0x18
                         retfie
start                    lfsr           FSR1,0xE00             ; set up software stack pointer
                         movlw          low arr_x              ; pass array base in the stack
                         pushr          WREG                   ; "
                         movlw          high arr_x             ; "
                         pushr          WREG                   ; "
                         movlw          upper arr_x            ; "
                         pushr          WREG                   ; "
                         movlw          acnt                   ; pass array count in stack
                         pushr          WREG                   ; "
                         call           find_amax,FAST         ; call the subroutine
                         movff          PRODL,array_max        ; save the result
                         dealloc_stk    4                      ; clean up the allocated stack space
                         nop
here                     bra            here
; ***********************************************************************
; The following subroutine finds the maximum element of an array of 8-bit elements
; and return the result in the PRODL register. Both the array base and array count
; are passed in the software stack.
; ***********************************************************************
find_amax                pushr          FSR2L                  ; save callers frame pointer
                         pushr          FSR2H                  ; "
                         movff          FSR1L,FSR2L            ; set up new stack frame
                         movff          FSR1H,FSR2H            ; pointer
                         pushr          TBLPTRL                ; save table pointer in stack
                         pushr          TBLPTRH                ; "
                         pushr          TBLPTRU                ; "
                         alloc_stk      2                      ; allocate two bytes for local variables
```

```
                    movlw      ar_cnt
                    movff      PLUSW2,PRODL
                    decf       PRODL,F,A
                    movlw      lp_cnt
                    movff      PRODL,PLUSW2         ; initialize lp_cnt to arcnt-1
                    movlw      arr_xlo              ; place array pointer in TBLPTR
                    movff      PLUSW2,TBLPTRL       ; "
                    movlw      arr_xhi              ; "
                    movff      PLUSW2,TBLPTRH       ; "
                    movlw      arr_xup              ; "
                    movff      PLUSW2,TBLPTRU       ; "
                    tblrd*+                         ; assign arr_x[0] as the initial array max
                    movlw      ar_max               ; "
                    movff      TABLAT,PLUSW2        ; "
cmp_lp              movlw      lp_cnt
                    tstfsz     PLUSW2
                    goto       next
                    goto       done
next                movlw      lp_cnt
                    decf       PLUSW2,F             ; decrement the loop index
                    tblrd*+                         ; read in the next array element
                    movlw      ar_max
                    movf       PLUSW2,W,A
                    cpfsgt     TABLAT
                    goto       cmp_lp
                    movlw      ar_max
                    movff      TABLAT,PLUSW2        ; update the current array max
                    goto       cmp_lp
done                movlw      ar_max
                    movff      PLUSW2,PRODL
                    dealloc_stk 2                   ; deallocate local variables
                    popr       TBLPTRU
                    popr       TBLPTRH
                    popr       TBLPTRL
                    popr       FSR2H
                    popr       FSR2L
                    return     FAST
; define an array of 32 8-bit elements. The even bytes will be at lower addresses
arr_x               db         1,2,3,4,5,6,7,8,9,10,11,12,13,14,15,16
                    db         17,18,19,20,21,22,23,24,25,26,27,28,29,30,31,32
                    end
```

Searching an array or a file is a common operation in computer applications. When the array or file is not sorted, the search program will need to search the whole array. However, if the array has been sorted, more efficient search algorithms can be applied. The *binary search* algorithm is one such algorithm. Assuming that the array is sorted in ascending order, the binary search algorithm will divide the array into three sections:

- The upper half
- The middle element
- The lower half

To make the algorithm more specific, three integers are used to index the array:

- **max**—upper bound of array indices for search
- **mean**—middle index for array search
- **min**—lower bound of array indices for search

A search key will be compared to the middle element of the search range. There are three possibilities for the search outcome:

1. The key is equal to the middle element of the search range. In this case, the search operation is over.
2. The key is greater than the middle element of the search range. In this case, the search operation should be continued in the upper half of the search range.
3. The key is smaller than the middle element of the search range. In this case, the search operation should be continued in the lower half of the search range.

The exact algorithmic steps are as follows:

Step 1
Use *max* and *min* as the upper and lower indices of the array for searching and initialize *max* and *min* to **n – 1** and **0,** respectively.

Step 2
If *max* < *min*, then stop. No element matches the key.

Step 3
Let *mean* = (min + max)/2.

Step 4
If key = array_y[mean], then key is found; stop.

Step 5
If key < array_y[mean], then set **max** to **mean –1** and go to Step 2.

Step 6
If key > array_y[mean], then set *min* to **mean +1** and go to Step 2.

▲

Example 4.10

▼

Write a subroutine to implement the binary search algorithm. The starting address of the array, array count, and search key are passed in the stack. A value of 1 will be returned in PRODL if the key is found in the array. Otherwise, a value of 0 will be returned.

Solution: The stack frame for calling the binary search subroutine is shown in Figure 4.8. To make the search easy, each array element is stored in one word. By doing this, the address of the element **arr_x[mean]** is equal to **arr_x[0] + 2 × mean** (the PIC18 MPASM follows little-endian storage format). The binary search subroutine and its testing program are as follows:

```
                #include <p18F8790.inc>
                radix              dec                   ; set default radix to decimal
; ——
; include stack macros pushr, push_dat, popr, alloc_stk, and dealloc_stk here
; ——
acnt            equ                D'32'                 ; array count
local_var       equ                4                     ; number of bytes for local variables
```

```
        skey        equ             D'7'                ; key for search
        min         equ             3                   ; offset for lower bound of search range index
        mean        equ             4                   ; offset for middle element index
        max         equ             5                   ; offset for upper bound of search range index
        result      equ             6                   ; offset for search result
        key         equ             -3                  ; offset for search key
        ar_cnt      equ             -4                  ; stack frame offset for array count
        arr_xup     equ             -5                  ; stack frame offset for array base
        arr_xhi     equ             -6                  ; "
        arr_xlo     equ             -7                  ; "
        find_it     set             0x00                ; register to hold array max
                    org             0x00                ; reset vector
                    goto            start
                    org             0x08
                    retfie
                    org             0x18
                    retfie
        start       lfsr            FSR1,0xE00          ; set up software stack pointer
                    movlw           low arr_x           ; pass array base in the stack
                    pushr           WREG                ; "
                    movlw           high arr_x          ; "
                    pushr           WREG                ; "
                    movlw           upper arr_x         ; "
                    pushr           WREG                ; "
                    movlw           acnt                ; pass array count in stack
                    pushr           WREG                ; "
                    movlw           skey                ; pass key for search
                    pushr           WREG
                    call            bin_search, FAST    ; call the subroutine
                    movff           PRODL,find_it       ; save the result
                    dealloc_stk     5                   ; clean up the allocated stack space
                    nop
        here        bra             here
        bin_search  pushr           FSR2L               ; save callers frame pointer
                    pushr           FSR2H               ; "
                    movff           FSR1L, FSR2L        ; set up frame pointer
                    movff           FSR1H,FSR2H         ; "
                    pushr           TBLPTRL             ; save table pointer in stack
                    pushr           TBLPTRH             ; "
                    pushr           TBLPTRU             ; "
                    alloc_stk       local_var           ; allocate space for local variables
                    movlw           ar_cnt              ; max <= ar_cnt - 1
                    movff           PLUSW2,PRODL        ; "
                    decf            PRODL,F,A           ; "
                    movlw           max                 ; "
                    movff           PRODL, PLUSW2       ; "
                    movlw           min                 ; min <= 0
                    clrf            PLUSW2,A            ; "
                    movlw           result
                    clrf            PLUSW2              ; set the search result to not found
```

```
bloop          movlw               max
               movff               PLUSW2,PRODL        ; place max index in PRODL
               movlw               min
               movff               PLUSW2,WREG         ; place min index in W
               cpfslt              PRODL               ; is max < min?
               goto                do_it               ; max >= min and continue
               goto                done                ; no match is found in the array
do_it          addwf               PRODL,F,A           ; compute (min+max), this may set C flag
               bcf                 STATUS,C            ; compute (min+max)/2 and place it
               rrcf                PRODL,F,A           ; in PRODL
               movlw               mean
               movff               PRODL, PLUSW2       ; place mean index in the stack frame slot
; compare array[mean] with the search key
               movlw               arr_xlo             ; place array base in TBLPTR
               movff               PLUSW2,TBLPTRL      ; "
               movlw               arr_xhi
               movff               PLUSW2,TBLPTRH      ; "
               movlw               arr_xup
               movff               PLUSW2,TBLPTRU      ; "
; add (mean×2) to TBLPTR to compute the address of arr_x[mean]
               bcf                 STATUS,C,A
               rlcf                PRODL,F,A           ; multiply mean by 2
               movlw               0
               addwfc              TBLPTRH,F,A         ; add the msb of (2*mean) to TBLPTRH
               addwfc              TBLPTRU,F,A         ; "
               movf                PRODL,W,A           ; add lower 8 bits of 2 x mean to
               addwf               TBLPTRL,F,A         ; TBLPTR
               movlw               0                   ; "
               addwfc              TBLPTRH,F,A         ; "
               addwfc              TBLPTRU,F,A         ; "
               tblrd*                                  ; read array[mean] into TABLAT
; compare key with array[mean]
               movlw               key
               movf                PLUSW2,W,A          ; place key in WREG
               cpfseq              TABLAT,A
               goto                not_equal
               movlw               result
               incf                PLUSW2,F            ; set search result to "found"
               goto                done
not_equal      cpfslt              TABLAT,A
               goto                go_low
               movlw               mean
               movff               PLUSW2,PRODL
               incf                PRODL,F,A
               movlw               min
               movff               PRODL,PLUSW2        ; set min to mean+1 to
               goto                bloop               ; search upper half
go_low         movlw               mean
               movff               PLUSW2,PRODL
```

```
          decf          PRODL,F,A
          movlw         max
          movff         PRODL,PLUSW2      ; set max to mean - 1 and
          goto          bloop             ; search lower half
done      movlw         result
          movff         PLUSW2,PRODL      ; place the result in PRODL before return
          dealloc_stk   local_var         ; deallocate local variables
          popr          TBLPTRU
          popr          TBLPTRH
          popr          TBLPTRL
          popr          FSR2H
          popr          FSR2L
          return        FAST
; define an array of 32 8-bit elements for testing purpose. The element arr_x[i] is
; located at arr_x[0] + 2 * i due to the fact that PIC18 assembler follows
; Little-Endian storage format.
arr_x     dw            1,2,3,4,5,6,7,8,9,10
          dw            11,12,13,14,15,16,17,18,19,20
          dw            21,22,23,24,25,26,27,28,29,30
          dw            31,32
          end
```

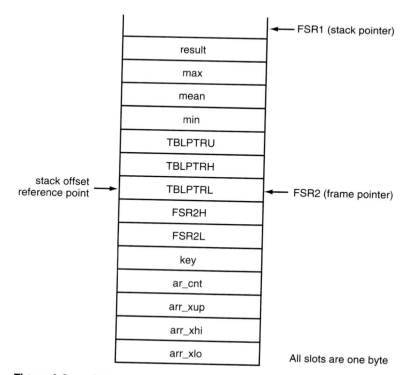

Figure 4.8 ■ Stack frame of example 4.10

4.9 String Processing

A string is a sequence of characters terminated by a NULL (ASCII code 0) or other character such as EOT (ASCII code 4). The NULL character is used to terminate a string in this book. Common operations applied to strings include string concatenation, character and word counting, string matching, and string insertion and deletion.

Strings are also needed in input and output operations. Before a number (in binary format in memory) is output, it must be converted to ASCII code because most output devices accept only ASCII code. A number can be output in BCD or hex format.

Assuming that the user wants to output the unsigned binary number 0010 1010 0011 0101 (equivalent to decimal value 10805), the computer must convert it to the ASCII code sequence 0x31 0x30 0x38 0x30 0x35 so that the user can quickly recognize its value.

To convert a binary number into its equivalent decimal digit string, the repeated divide-by-10 operation is performed. The hex number 0x30 is then added to the remainder to obtain its corresponding ASCII code. Since division operation is needed, the unsigned 16-bit division program will be converted to a subroutine so that it can be used in the binary-to-decimal-string conversion process.

Example 4.11
▼

Convert the unsigned 16-bit division program into a subroutine. The caller of this subroutine will push the dividend and divisor into the stack and make a hole in the stack for the callee to return the remainder and the quotient.

Solution: The stack frame of the subroutine is shown in Figure 4.9.

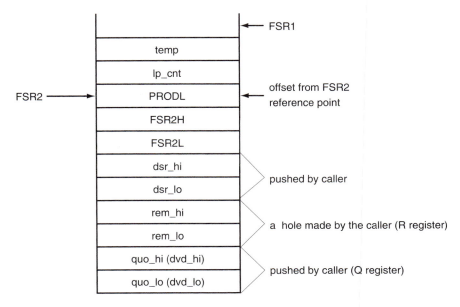

Figure 4.9 ■ Stack frame of example 4.11

As shown in Figure 4.9, the caller of the divu16 subroutine will perform the following operations before calling the div16u subroutine:

- Push the dividend into the stack
- Make a hole of two bytes to hold the remainder
- Push the divisor into the stack

The div16u routine will perform the following operations before executing the divide algorithm:

- Save the previous frame pointer in the stack
- Set up the current frame pointer
- Save the PRODL register in the stack
- Allocate space for local variables

The div16u subroutine and its testing program are as follows:

```
                        #include <p18F8720.inc>
; -
; include macros pushr, popr, alloc_stk, and dealloc_stk here
; -
        loc_var     equ     2               ; local variable size
        lp_cnt      equ     1               ; loop count
        temp        equ     2               ; temporary storage
        quo_hi      equ     -7              ; offset for quotient and dividend
        quo_lo      equ     -8              ; "
        rem_hi      equ     -5              ; offset for remainder
        rem_lo      equ     -6              ; "
        dsr_hi      equ     -3              ; offset for divisor
        dsr_lo      equ     -4              ; "
        dd_h        equ     0xEC            ; high byte of dividend test number
        dd_l        equ     0x46            ; low byte of "
        dr_h        equ     0x00            ; high byte of divisor test number
        dr_l        equ     0x87            ; low byte of "
        quo         set     0x00            ; memory space to hold the quotient
        rem         set     0x02            ; memory space to hold the remainder
                    org     0x00
                    goto    start
                    org     0x08
                    retfie
                    org     0x18
                    retfie
start               lfsr    FSR1,0xE00      ; set up software stack pointer
                    movlw   dd_l            ; pass dividend in stack
                    pushr   WREG            ; "
                    movlw   dd_h            ; "
                    pushr   WREG            ; "
                    alloc_stk  2           ; make a hole for remainder
                    movlw   dr_l            ; pass divisor in stack
                    pushr   WREG            ; "
                    movlw   dr_h            ; "
                    pushr   WREG            ; "
                    call    div16u,FAST     ; call 16-bit unsigned divide subroutine
```

```
                    dealloc_stk    2                     ; deallocate stack space
                    popr           rem+1                 ; retrieve the quotient (hi byte)
                    popr           rem                   ; " (lo byte)
                    popr           quo+1                 ; retrieve the quotient (hi byte)
                    popr           quo                   ; " (lo byte)
                    nop
forever             bra            forever
div16u              pushr          FSR2L                 ; save the previous frame pointer
                    pushr          FSR2H                 ; "
                    movff          FSR1L,FSR2L           ; set up frame pointer
                    movff          FSR1H,FSR2H           ; "
                    pushr          PRODL                 ; save PRODL in stack
                    alloc_stk      loc_var               ; allocate space for local variables
                    movlw          rem_hi                ; clear R register to 0
                    clrf           PLUSW2                ; "
                    movlw          rem_lo                ; "
                    clrf           PLUSW2                ; "
                    movlw          D'16'                 ; initialize loop count to 16
                    movwf          PRODL                 ; "
                    movlw          lp_cnt                ; "
                    movff          PRODL, PLUSW2         ; "
loop                bcf            STATUS,C,A            ; clear the C flag
                    movlw          quo_lo                ; rotate (R, Q) pair to the left one place
                    rlcf           PLUSW2,F              ; "
                    movlw          quo_hi                ; "
                    rlcf           PLUSW2,F              ; "
                    movlw          rem_lo                ; "
                    rlcf           PLUSW2,F              ; "
                    movlw          rem_hi                ; "
                    rlcf           PLUSW2,F              ; "
                    movlw          rem_lo                ; get the low byte of the remainder in
                    movff          PLUSW2,PRODL          ; PRODL
                    movlw          dsr_lo                ; get the low byte of the divisor in WREG
                    movf           PLUSW2,W              ; "
                    subwf          PRODL,F,A
                    movlw          temp
                    movff          PRODL,PLUSW2          ; save the low byte of difference at temp
                    movlw          rem_hi
                    movff          PLUSW2,PRODL
                    movlw          dsr_hi
                    movf           PLUSW2,W
                    subwfb         PRODL, F              ; subtract the high byte
                    btfss          STATUS,C              ; skip if carry is 1
                    goto           less
                    movlw          quo_lo
                    bsf            PLUSW2,0              ; set the quotient bit to 1
                    movlw          rem_hi
                    movff          PRODL, PLUSW2         ; place the difference in the R register
                    movlw          temp                 ; "
                    movff          PLUSW2,PRODL          ; "
```

```
              movlw    rem_lo            ; "
              movff    PRODL, PLUSW2     ; "
              goto     next
less          movlw    quo_lo
              bcf      PLUSW2,0          ; set the quotient bit to 0
next          movlw    lp_cnt
              decfsz   PLUSW2,F          ; decrement the loop count and skip if zero
              goto     loop
dealloc_stk   loc_var                    ; deallocate local variables
              popr     PRODL
              popr     FSR2H
              popr     FSR2L
              return   FAST
              end
```

▲

Example 4.12

▼

Write a subroutine to convert a 16-bit signed number to an ASCII string that represents its equivalent decimal value.

Solution: Let **p, rem, quo,** and **sign** represent the integer to be converted, the remainder after the divide-by-10 operation, the quotient after the divide-by-10 operation, and the sign of **p,** respectively. The algorithm for converting a binary number into a decimal string is as follows:

Step 1
If **p** = 0, then store $30 and $00 in the buffer and stop.

Step 2
If **p** < 0, then set sign to 1. Otherwise, set sign to 0.

Step 3
Divide **p** by 10 and leave the remainder and the quotient in **rem** and **quo,** respectively.

Step 4
Add $30 to the remainder and save it in the next available space in the buffer.

Step 5
If **quo** ≠ 0 then **p←quo** and go to Step 3.

Step 6
Push the characters in the buffer into the stack.

Step 7
If – is 1, then store the ASCII code of the minus sign as the first character in the buffer.

Step 8
Pop out all the decimal digits from the stack into the buffer.

The repeated divide-by-10 operation will derive the one's digit first, then the ten's digit, and so on. Since the corresponding decimal digits are generated from one's digit toward the most significant digit, they will be saved in the buffer in the reverse order. Hence, the algorithm needs to reverse it in Step 7 and Step 8.

The stack frame for this subroutine is shown in Figure 4.10. This function has only one local variable: **sign.**

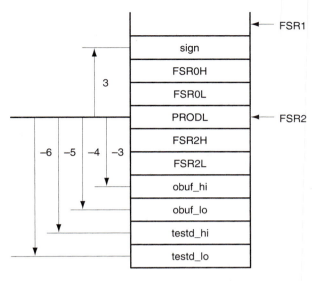

Figure 4.10 ■ Stack frame for example 4.12

The assembly that performs the conversion is as follows:

```
            #include <p18F8720.inc>
; –
; include macro definitions of pushr, pop, push_dat, alloc_stk, and dealloc_stk here
; –
tstd_hi     equ         0x88            ; test data high byte
tstd_lo     equ         0x89            ; test data low byte
sign        equ         3               ; stack offset for sign in bin2dec
testd_lo    equ         -6              ; stack offset for test data in bin2dec
testd_hi    equ         -5              ; "
obuf_lo     equ         -4              ; stack offset for conversion buffer
obuf_hi     equ         -3              ; "
minus       equ         0x2D            ; ASCII code of minus sign
loc_varc    equ         1               ; local variable size of bin2dec routine
obuf        set         0x0100
            org         0x00
            goto        start
            org         0x08            ; high-priority interrupt vector
            retfie
            org         0x18            ; low-priority interrupt vector
            retfie
start       lfsr        FSR1,0xE00      ; set up stack pointer
            push_dat    tstd_lo         ; pass test data in stack
            push_dat    tstd_hi         ; "
            push_dat    low(obuf)       ; "
            push_dat    high(obuf)      ; "
            call        bin2dec,FAST    ; call bin2dec subroutine
            dealloc_stk 4               ; clean up stack space
con_loop    nop
            bra         con_loop
; ***************************************************************
; This routine converts a 16-bit binary number into a decimal number
; represented by an ASCII string. The number to be converted is
```

```
; passed in stack.
; ************************************************************
bin2dec    pushr            FSR2L              ; save previous frame pointer
           pushr            FSR2H              ; "
           movff            FSR1L,FSR2L        ; set up new frame pointer
           movff            FSR1H,FSR2H        ; "
           pushr            PRODL
           pushr            FSR0L              ; save FSR0 in stack
           pushr            FSR0H              ; "
           alloc_stk        loc_varc
; set FSR0 as the pointer to the buffer that holds the conversion result
           movlw            obuf_lo
           movff            PLUSW2,FSR0L
           movlw            obuf_hi
           movff            PLUSW2,FSR0H
; initialize the sign of the number to be converted to positive
           movlw            sign               ; "
           clrf             PLUSW2             ; "
; test to find out if the number is negative. If yes, compute its two's complement.
           movlw            testd_hi
           btfss            PLUSW2,7           ; compute the magnitude when negative
           goto             tst_zero           ; no need to complement when positive
           comf             PLUSW2,F           ; complement the upper byte
           clrf             PRODL,A            ; add carry to testd_hi (carry may be 0)
           movlw            testd_lo           ; compute the two's complement of data
           negf             PLUSW2             ; to be converted
           movlw            testd_hi           ; "
           movf             PLUSW2, W          ; "
           addwfc           PRODL, F           ; "
           movlw            testd_hi           ; "
           movff            PRODL,PLUSW2       ; "
; change sign to 1 to indicate minus
           movlw            sign               ; "
           incf             PLUSW2, F          ; "
           bra              normal
; check if the number to be tested is zero
tst_zero   movlw            testd_hi           ; test high byte
           tstfsz           PLUSW2             ; "
           bra              normal
           movlw            testd_lo           ; test low byte
           tstfsz           PLUSW2             ; "
           bra              normal
           movlw            0x30
           movwf            POSTINC0
           clrf             INDF0              ; terminate the buffer with NULL character
           goto             done
; normal repeated divide-by-10 loop starts here
normal     movlw            testd_lo           ; pass the dividend
           movf             PLUSW2,W           ; "
           pushr            WREG               ; "
           movlw            testd_hi           ; "
```

```
                movf            PLUSW2,W            ; "
                pushr           WREG                ; "
div_lp          alloc_stk       2                   ; make room for remainder
                push_dat        0x0A                ; push 10 into the stack as divisor
                push_dat        0                   ; "
                call            div16u              ; call subroutine to perform division
                dealloc_stk     3                   ; deallocate the stack space (3 bytes remain)
                popr            WREG                ; pop off low byte of remainder
                addlw           0x30                ; convert the remainder into ASCII code
                movwf           POSTINC0            ; and save it in buffer
; is quotient equal to 0?
                movlw           -1
                tstfsz          PLUSW1
                goto            div_lp
                movlw           -2
                tstfsz          PLUSW1
                goto            div_lp
                clrf            INDF0               ; terminate the buffer with NULL
                dealloc_stk     2                   ; clean up the stack
; set FSR0 to point to the start of the buffer to reverse the string
                movlw           obuf_lo             ; "
                movff           PLUSW2,FSR0L        ; "
                movlw           obuf_hi             ; "
                movff           PLUSW2,FSR0H        ; "
                push_dat        0                   ; push a NULL character onto the stack
push_lp         movf            POSTINC0,W
                bz              to_pop              ; is this the NULL character
                pushr           WREG
                goto            push_lp
; reverse the converted string
to_pop          movlw           obuf_lo             ; set FSR0 to point to the start of buffer
                movff           PLUSW2, FSR0L       ; "
                movlw           obuf_hi             ; "
                movff           PLUSW2,FSR0H        ; "
                movlw           sign
                movf            PLUSW2,W            ; is the converted data negative?
                bz              pop_loop            ; "
; add a minus sign if the number is negative
                movlw           minus
                movwf           POSTINC0
pop_loop        popr            WREG                ; reverse string loop
                movwf           POSTINC0
                tstfsz          WREG,A              ; reach the end of string?
                goto            pop_loop
done            popr            WREG                ; get rid of sign
                popr            FSR0H
                popr            FSR0L
                popr            PRODL
                popr            FSR2H
                popr            FSR2L
                return          FAST
```

To convert a decimal digit string into a binary number, we will need a subroutine to perform 16-bit by 16-bit multiplication or even 32-bit by 32-bit multiplication, The following example will convert the 16-bit by 16-bit unsigned multiplication program into a subroutine.

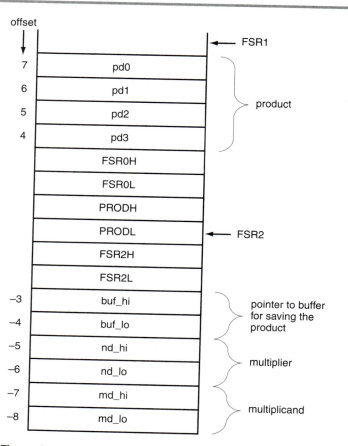

Figure 4.11 ■ Stack frame for Example 4.13

Example 4.13

Write a subroutine that computes the product of two unsigned 16-bit integers. The numbers to be multiplied and the pointer to the buffer for storing the product are passed in the stack.

Solution: The subroutine will save all the used registers in the stack and allocates four bytes for local variables to hold the product. The stack frame is shown in Figure 4.11.

The function **mul_16U** and its testing instruction sequence are as follows:

```
                #include <p18F8720.inc>
; –
; include macro definitions of pushr, popr, push_dat, alloc_stk, and dealloc_stk here
; –
M_hi            equ         0x53            ; test num1 high byte
M_lo            equ         0x29            ; test num1 low byte
```

```
N_hi          equ       0x84              ; test num2 high byte
N_lo          equ       0x37              ; test num2 low byte
ptr_hi        equ       0x01              ; address of buffer to hold product
ptr_lo        equ       0x00              ; "
loc_varm      equ       4                 ; number of bytes used for local variable
pdo           equ       7                 ; offset of pdo from frame pointer
pd1           equ       6                 ; offset of pd1 from frame pointer
pd2           equ       5                 ; offset of pd2 from frame pointer
pd3           equ       4                 ; offset of pd3 from frame pointer
md_lo         equ       -8                ; offset of md_lo from frame pointer
md_hi         equ       -7                ; offset of md_hi from frame pointer
nd_lo         equ       -6                ; offset of nd_lo from frame pointer
nd_hi         equ       -5                ; offset of nd_hi from frame pointer
buf_lo        equ       -4                ; offset of buf_lo from frame pointer
buf_hi        equ       -3                ; offset of buf_hi from frame pointer
              org       0x00
              goto      start
              org       0x08              ; high-priority interrupt vector
              retfie
              org       0x18              ; low-priority interrupt vector
              retfie
start         lfsr      FSR1,0xE00        ; set up stack pointer
              push_dat  M_lo              ; pass test data in stack
              push_dat  M_hi              ; "
              push_dat  N_lo              ; "
              push_dat  N_hi              ; "
              push_dat  ptr_lo            ; pass buffer pointer
              push_dat  ptr_hi            ; "
              call      mul_16U,FAST      ; call bin2dec subroutine
              dealloc_ stk  6             ; clean up stack space
              nop
tst_loop      bra       tst_loop
;*****************************************************************
;
; This routine performs a 16-bit unsigned multiplication. Both the multiplicand
; and multiplier and address of the buffer to hold the product are passed in stack.
;*****************************************************************
;
mul_16U       pushr     FSR2L             ; save previous frame pointer
              pushr     FSR2H             ; "
              movff     FSR1L, FSR2L      ; set up new frame pointer
              movff     FSR1H, FSR2H      ; "
              pushr     PRODL             ; save PRODL in stack
              pushr     PRODH             ; save PRODH in stack
              pushr     FSR0L             ; save FSR0 in stack
              pushr     FSR0H             ; "
              alloc_stk loc_varm
; compute md_lo × nd_lo and place in pd1..pd0
              movlw     md_lo
              movff     PLUSW2,PRODL      ; place md_lo in PRD0L
              movlw     nd_lo
              movf      PLUSW2,W          ; place nd_lo in WREG
              mulwf     PRODL             ; compute nd_lo * md_lo
```

```
                movlw           pdo
                movff           PRODL,PLUSW2
                movlw           pd1
                movff           PRODH,PLUSW2
; compute md_hi × nd_hi and place the product in pd3..pd2
                movlw           md_hi
                movff           PLUSW2,PRODL      ; place md_hi in PRDOL
                movlw           nd_hi
                movf            PLUSW2,W          ; place nd_hi in WREG
                mulwf           PRODL             ; compute nd_hi * md_hi
                movlw           pd2
                movff           PRODL,PLUSW2
                movlw           pd3
                movff           PRODH,PLUSW2
;compute md_lo × nd_hi
                movlw           md_lo
                movff           PLUSW2,PRODL
                movlw           nd_hi
                movf            PLUSW2,W
                mulwf           PRODL
                movlw           pd1               ; add to pd1
                movf            PLUSW2,W          ; "
                addwf           PRODL,F           ; "
                movlw           pd1               ; "
                movff           PRODL,PLUSW2      ; "
                movlw           pd2
                movf            PLUSW2,W
                addwfc          PRODH,F
                movlw           pd2
                movff           PRODH,PLUSW2
                clrf            PRODL
                movlw           pd3
                movf            PLUSW2,W
                addwfc          PRODL, F
                movlw           pd3
                movff           PRODL,PLUSW2
; compute md_hi × nd_lo
                movlw           md_hi
                movff           PLUSW2,PRODL
                movlw           nd_lo
                movf            PLUSW2,W
                mulwf           PRODL
                movlw           pd1               ; add to pd1
                movf            PLUSW2,W          ; "
                addwf           PRODL,F           ; "
                movlw           pd1               ; "
                movff           PRODL,PLUSW2      ; "
                movlw           pd2               ; add to pd2
                movf            PLUSW2,W          ; "
                addwfc          PRODH,F           ; "
                movlw           pd2               ; "
                movff           PRODH,PLUSW2      ; "
                clrf            PRODL             ; add carry to most significant byte
                movlw           pd3               ; "
```

```
                movf         PLUSW2,W        ; "
                addwfc       PRODL, F        ; "
                movlw        pd3             ; "
                movff        PRODL,PLUSW2    ; "
; use FSR0 as a pointer to the buffer that holds the product
                movlw        buf_lo
                movff        PLUSW2,FSR0L
                movlw        buf_hi
                movff        PLUSW2,FSR0H
; save product in the buffer
                popr         WREG
                movwf        POSTINC0
                popr         WREG
                movwf        POSTINC0
                popr         WREG
                movwf        POSTINC0
                popr         WREG
                movwf        POSTINC0
                popr         FSR0H
                popr         FSR0L
                popr         PRODH
                popr         PRODL
                popr         FSR2H
                popr         FSR2L
                return       FAST
                end
```

An application may require the user to enter a decimal number from the keyboard. The number is represented in a decimal digit ASCII string. For example, when the user enters the decimal number 1234, it will be represented in the ASCII code sequence 0x31 0x32 0x33 0x34. This string must be converted to the binary number that it represents before the number can be manipulated. Let the variables **sign, error,** and **number** represent the sign of the number, whether error occurs, and the equivalent binary number that the ASCII string represents, respectively. Suppose **in_ptr** is the pointer to the input ASCII string; then the algorithm to convert an ASCII string into a binary number is as follows:

Step 1
sign ← 0
error ← 0
number ← 0

Step 2
If the character pointed to by in_ptr is the minus sign, then
sign ← 1
in_ptr ← in_ptr + 1

Step 3
If the character pointed to by in_ptr is the NULL character, then go to Step 4; otherwise, if the character is not a BCD digit (i.e., m[in_ptr] > \$39 or m[in_ptr] < \$30), then
error ← 1;
go to Step 4;
otherwise,
number ← number × 10 + m[in_ptr] – \$30;
in_ptr ← in_ptr + 1;
go to Step 3

Step 4
If sign = 1 and error = 0, then
number ← two's complement of number;
otherwise, stop
The following example converts a decimal digit ASCII string into a binary number.

Example 4.14

Write a subroutine that converts a decimal digit string into the binary number that it represents. The string is stored in the data memory, and its starting address is passed to this function in the FSR0 register. The string may represent a positive or a negative number and is no longer than seven bytes. Thus, the converted binary number can be accommodated in two bytes. The converted result will be returned in PRODL and PRODH. Do not check for illegal characters.

Solution: This program needs to call the **mul_16U** subroutine to carry out the multiplication required in the previously mentioned algorithm. The **mul_16U** subroutine requires the caller to pass a pointer to the buffer that holds the product. A buffer will be reserved in the stack for this purpose. The frame pointer FSR2 is also the pointer to this buffer. The stack frame of this subroutine is shown in Figure 4.12. This subroutine will check the sign of the decimal digit string. If the first character is the minus sign, the converted number will be negated.

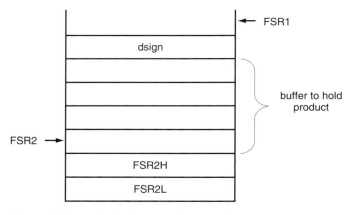

Figure 4.12 ■ Stack frame for Example 4.14

The assembly program is as follows:

```
#include <p18F8720.inc>
; --
; include macro definitions pushr, push_dat, popr, alloc_stk, and dealloc_stk here
; --
loc_vard        equ         5               ; size of local variables for d2b
dsign           equ         4               ; offset of dsign from FSR2
dec_strg        set         0x10            ; address of the buffer for the decimal string
                org         0x00
                goto        start
                org         0x08            ; high-priority interrupt vector
                retfie
                org         0x18            ; low-priority interrupt vector
```

```
                  retfie
start             lfsr        FSR1,0xE00         ; set up stack pointer
                  movlw       high dec_strg      ; use FSR0 as the string pointer
                  movwf       FSR0H              ; "
                  movlw       low dec_strg       ; "
                  movwf       FSR0L              ; "
                  movlw       0x2D               ; set up a test string
                  movwf       POSTINC0           ; "
                  movlw       0x39               ; "
                  movwf       POSTINC0           ; "
                  movlw       0x38               ; "
                  movwf       POSTINC0           ; "
                  movlw       0x38               ; "
                  movwf       POSTINC0           ; "
                  movlw       0x37               ; "
                  movwf       POSTINC0           ; "
                  clrf        INDF0              ; terminate the string with a NULL character
                  movlw       high dec_strg      ; re-establish the string pointer
                  movwf       FSR0H              ; "
                  movlw       low dec_strg       ; "
                  movwf       FSR0L              ; "
                  call        dec2bin            ; call the decimal to binary conversion function
forever           goto        forever
dec2bin           pushr       FSR2L              ; save the current frame pointer
                  pushr       FSR2H              ; "
                  movff       FSR1L, FSR2L       ; set up new frame pointer
                  movff       FSR1H,FSR2H        ; "
                  alloc_stk   loc_vard           ; allocate 5 bytes for local variables
                  movlw       0
                  clrf        PLUSW2             clear the buffer that holds the product
                  incf        WREG,W             ; "
                  clrf        PLUSW2             ; "
                  incf        WREG,W             ; "
                  clrf        PLUSW2             ; "
                  incf        WREG,W             ; "
                  clrf        PLUSW2             ; "
                  movlw       dsign              ; initialize sign to positive
                  clrf        PLUSW2             ; "
                  movff       INDF0,PRODL        ; get a character from string
                  movlw       0x2D               ; place minus sign in WREG
                  cpfseq      PRODL              ; is the first character a minus sign?
                  goto        con_lp
                  movlw       dsign
                  incf        PLUSW2             ; set sign to minus
                  movf        POSTINC0,W         ; move pointer to next character
con_lp            movff       POSTINC0,PRODL
                  tstfsz      PRODL,A            ; reach the NULL character?
                  goto        normal             ; not yet
                  goto        done               ; yes, reach NULL character
normal            movlw       0x30
                  subwf       PRODL,F,A          ; convert to digit value
```

```
          movlw      0                    ; push multiplicand
          movf       PLUSW2,W             ; "
          pushr      WREG                 ; "
          movlw      1                    ; "
          movf       PLUSW2,W             ; "
          pushr      WREG                 ; "
          push_dat   0x0A                 ; push multiplier 10 into stack
          push_dat   0x00                 ; "
          pushr      FSR2L                ; push buffer pointer
          pushr      FSR2H                ; "
          call       mul_16U              ; multiply the current result by 10
          dealloc_stk 6                   ; clean up stack space
          movf       PRODL,W
          addwf      INDF2,F              ; add the converted digit to the product
          clrf       PRODL,A              ; "
          movlw      1
          movf       PLUSW2,W             ; place the upper byte of the product in WREG
          addwfc     PRODL,F              ; "
          movlw      1                    ; "
          movff      PRODL,PLUSW2         ; "
          goto       con_lp
; check the sign of the converted binary number before return
done      dealloc_stk 3                   ; move down the stack pointer
          popr       PRODH                ; place upper byte of product in PRODH
          popr       PRODL                ; place lower byte of product in PRODL
          movlw      dsign                ; check the sign
          tstfsz     PLUSW2               ; "
          goto       negate
          goto       getback
negate    negf       PRODL                ; find 2's complement of the result
          comf       PRODH                ; "
          movlw      0                    ; "
          addwfc     PRODH,F              ; "
getback   popr       FSR2H
          popr       FSR2L
          return
; ——
; include MUL16U and its associated **equ** directives here.
; ——
          end
```

This program does not check for illegal characters in the string. The checking of illegal characters in the string is straightforward and hence is left for you as an exercise.

▲

4.10 More Examples on Subroutine Calls

Several commonly used subroutines are developed in this section. The first one is a sorting subroutine. Sorting is important for those applications that require searching the same set of data repeatedly. A sorted data set allows more efficient search methods to be applied. The *bubble sort* is a simple, inefficient, but widely known sorting method. Many other more efficient sorting methods require the use of recursive subroutine calls that fall outside the scope of this book.

The basic idea underlying the bubble sort is to go through the array or file sequentially several iterations with each iteration placing one element in its right position. Each iteration consists of comparing each element in the array or file with its successor (**x[i]** with **x[i+1]**) and interchanging the two elements if they are not in proper order (either ascending or descending).

For an array with **n** elements, **n – 1** comparisons are performed during the first iteration. As more and more iterations are performed, more and more elements would be moved to their right positions. Fewer comparisons are needed. In the worst case, **n – 1** iterations are needed, and only one comparison is made during the last iteration.

The bubble sort can be made more efficient by keeping track of whether swap operations have been performed. If no swap is made in an iteration, then the array is already in sorted order, and the process should be terminated. The logic flow of the bubble sort algorithm is illustrated in Figure 4.13.

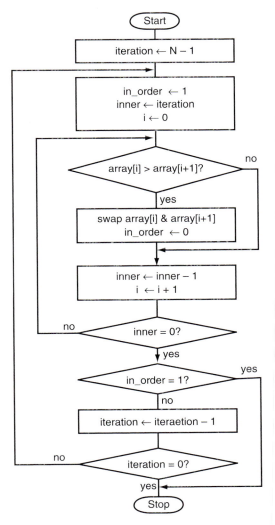

Figure 4.13 ■ Logic flow of bubble sort

Example 4.15

▼

Write a subroutine to implement the bubble sort algorithm and a sequence of instructions along with a set of test data for testing this subroutine. Use an array in data memory that consists of **n** 8-bit unsigned integers for testing purpose.

Solution: The subroutine has the following local variables:

- **in_order**—flag to indicate whether the array is in order after an iteration
- **inner**—number of comparisons remained to be performed in an iteration
- **iteration**—number of iterations remained to be performed

The caller of this subroutine will pass the array base address and the array count in the stack. The stack frame of this subroutine is shown in Figure 4.14. In many PIC18-based systems, program memory is in ROM rather than in RAM. This example will declare test data in the program memory and use a program loop to copy it to data memory (in SRAM) so that the bubble sort subroutine can be tested.

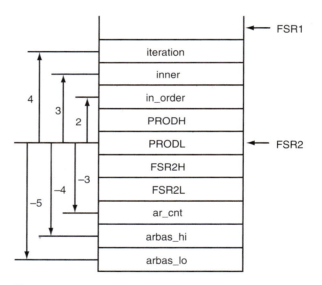

Figure 4.14 ■ Stack frrame for Example 4.16

The bubble sort subroutine and its testing instruction sequence are as follows:

```
                #include <p18F8720.inc>
;——
; include macro definitions pushr, push_dat, popr, alloc_stk, and dealloc_stk here
;——
NN          equ     D'30'           ; array count
loc_varb    equ     3               ; number of bytes of local variables
arbas       equ     0x0100          ; array base address
lp_cnt      set     0               ; use data memory location 0 for loop count
in_order    equ     2               ; offset of local variable from frame pointer
```

```
inner          equ        3               ; "
iteration      equ        4               ; "
ar_cnt         equ        -3              ; "
arbas_hi       equ        -4              ; "
arbas_lo       equ        -5              ; "
               org        0x00
               goto       start
               org        0x08            ; high-priority interrupt vector
               retfie
               org        0x18            ; low-priority interrupt vector
               retfie
start          lfsr       FSR1,0xE00      ; set up stack pointer
               movlw      high arbas      ; use FSR0 as the pointer to the array
               movwf      FSR0H,A         ; to be sorted
               movlw      low arbas       ; "
               movwf      FSR0L,A         ; "
               movlw      upper xarr      ; set up table pointer to point to
               movwf      TBLPTRU,A       ; the array in program memory
               movlw      high xarr       ; "
               movwf      TBLPTRH,A       ; "
               movlw      low xarr        ; "
               movwf      TBLPTRL,A       ; "
               movlw      NN
               movwf      lp_cnt
;  *******************************************************************************
; copy an array from program memory to data memory to test the bubble sort subroutine.
;  *******************************************************************************
copy_lp        tblrd*+                    ; read a byte into the table latch
               movff      TABLAT,POSTINC0 ; copy one byte
               decfsz     lp_cnt,F,A
               bra        copy_lp
               movlw      low arbas       ; push array base address into stack
               pushr      WREG            ; "
               movlw      high arbas      ; "
               pushr      WREG            ; "
               push_dat   NN              ; push array count
               call       bubble,FAST     ; call the decimal to binary conversion function
forever        goto       forever
;  *****************************************************************
; This function implement bubble sort. The array base address and array
; count are passed to this function via the stack.
;  *****************************************************************
bubble         pushr      FSR2L           ; save the current frame pointer
               pushr      FSR2H           ; "
               movff      FSR1L,FSR2L     ; set up new frame pointer
               movff      FSR1H,FSR2H     ; "
               pushr      PRODL           ; save PRODL in stack
               pushr      PRODH           ; save PRODH in stack
               alloc_stk  loc_varb
```

```
              movlw      ar_cnt
              movff      PLUSW2,PRODL
              decf       PRODL,F,A
              movlw      iteration        ; set iteration to array count – 1
              movff      PRODL,PLUSW2     ; "
ploop         movlw      in_order
              clrf       PLUSW2           ; set in_order flag to true (0)
              movlw      iteration
              movff      PLUSW2,PRODL
              movlw      inner
              movff      PRODL,PLUSW2     ; initialize inner loop count
              movlw      arbas_lo         ; use FSR0 as the array pointer
              movff      PLUSW2,FSR0L     ; "
              movlw      arbas_hi         ; "
              movff      PLUSW2,FSR0H     ; "
cloop         movff      INDF0,PRODL      ; place arr[i] in PRODL
              movlw      1
              movf       PLUSW0,W         ; place arr[i+1] in WREG
              cpfsgt     PRODL,A
              goto       looptst
              movwf      PRODH            ; swap arr[i] with arr[i+1]
              movlw      1                ; "
              movff      PRODL,PLUSW0     ; "
              movff      PRODH,INDF0      ; "
              movlw      in_order         ; set the in_order flag to false (1)
              bsf        PLUSW2,0         ; "
looptst       movf       POSTINC0,W       ; increment array pointer
              movlw      inner            ; decrement inner loop count and skip if it
              decfsz     PLUSW2,F         ; has been decremented to zero
              goto       cloop
              movlw      in_order
              tstfsz     PLUSW2           ; is the array in order?
              goto       nexti            ; not yet
              goto       done             ; yes, return.
nexti         movlw      iteration
              decfsz     PLUSW2,F
              goto       ploop
done          dealloc_stk loc_varb
              popr       PRODH
              popr       PRODL
              popr       FSR2H
              popr       FSR2L
              return     FAST
xarr          db         0x56,0x1F,0x01,0x08,0x11,0x47,0x21,0x20,0x30,0x07
              db         0x19,0x18,0x17,0x16,0x15,0x14,0x13,0x12,0x29,0x28
              db         0x27,0x26,0x25,0x24,0x23,0x22,0x06,0x05,0x04,0x03
              end
```

4.10.1 Square Root Computation

The computation of the square root of a number is complicated, and hence no commercial 8- or 16-bit MCU provides an instruction for computing the square root. The square root can be computed by using the *successive approximation method.* This method has been widely used to perform analog-to-digital conversion. Most of the 8-bit or 16-bit MCUs use this method to implement their A/D converters.

The square root of a *2n-bit* number would be an *n-bit* number. Let SAR be an n-bit register and Q be a 2n-bit number that we want to find its square root. The algorithm for finding the square root of Q is illustrated in Figure 4.15.

The algorithm illustrated in Figure 4.15 will always compute an approximate integer square root smaller than or equal to the actual one. The better approximate square root could be the value in SAR or [SAR]+1. By comparing Q – [SAR*SAR] and ([SAR] + 1)^2 – Q, the better approximate square root can be found. The following example incorporates this consideration.

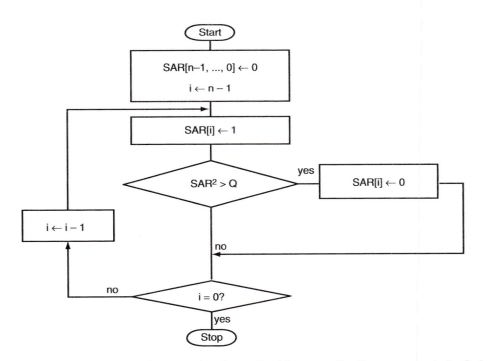

Figure 4.15 ■ Successive approximation method for computing the square root of a 2n-bit number Q

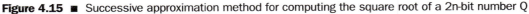

Example 4.16

Write an assembly routine to implement the successive approximation method for computing the square root of a 16-bit integer. The 16-bit integer will be pushed into the stack. The square root is returned in PRODL.

Solution: The stack frame of the subroutine is shown in Figure 4.16. The subroutine **find_sqr** and its calling instruction sequence are as follows:

```
                        #include <p18F8720.inc>
; ──
; include macro definitions pushr, push_dat, popr, alloc_stk, and dealloc_stk here
; ──
        testd_hi    equ         0x41            ; data to be tested
        testd_lo    equ         0x2E            ; "
        loc_vars    equ         3               ; number of bytes for local variable
        Q_hi        equ         -3              ; offsets of Q from frame pointer
        Q_lo        equ         -4              ; "
        sar         equ         0               ; "
        mask        equ         1               ; "
        lp_cnt      equ         2               ; "
        sq_root     set         0x00            ; memory location to save square root
                    org         0x00            ; reset vector
                    goto        start
                    org         0x08            ; high priority interrupt vector
                    retfie
                    org         0x18            ; low priority interrupt vector
                    retfie
start               lfsr        FSR1,0xE00      ; set stack pointer to 0xE00
                    push_dat    testd_lo        ; pass test data in stack
                    push_dat    testd_hi        ; "
                    call        find_sqr        ; call the square root routine
                    movff       PRODL,sq_root   ; save the square root in memory
                    dealloc_stk 2
forever             goto        forever
find_sqr            pushr       FSR2L           ; save caller's frame pointer
                    pushr       FSR2H           ; "
                    movff       FSR1L,FSR2L     ; set up new frame pointer
                    movff       FSR1H,FSR2H     ; "
                    alloc_stk   loc_vars        ; allocate local variables
```

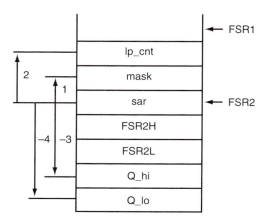

Figure 4.16 ■ Stack frame for example 4.17

```
                movlw       8                       ; initialize loop count to 8
                movwf       PRODL,A                 ; "
                movlw       lp_cnt                  ; "
                movff       PRODL,PLUSW2            ; "
                movlw       0x80                    ; start with the mask 0x80
                movwf       PRODL                   ; "
                movlw       mask                    ; "
                movff       PRODL,PLUSW2            ; "
                clrf        INDF2                   ; initialize SAR to 0
loop            movlw       mask
                movf        PLUSW2,W                ; place mask in WREG
                iorwf       INDF2,W                 ; set the ith bit of SAR
                mulwf       WREG,A                  ; compute SAR*SAR
                movlw       Q_lo                    ; compute SAR*SAR – Q
                movf        PLUSW2,W                ; "
                subwf       PRODL,F                 ; "
                movlw       Q_hi                    ; "
                movf        PLUSW2,W                ; "
                subwfb      PRODH,F                 ; "
                btfsc       STATUS,C                ; skip if Q > SAR*SAR
                goto        nextbit
                movlw       mask                    ; the guess is right, so
                movf        PLUSW2,W                ; set bit i of SAR
                iorwf       INDF2,F                 ; "
nextbit         movlw       lp_cnt                  ; decrement the loop count
                decf        PLUSW2,F                ; "
                bz          done                    ; it is done if lp_cnt = 0
                bcf         STATUS,C,A              ; clear the C flag
                movlw       mask
                rrcf        PLUSW2,F                ; shift the mask to the right
                goto        loop
;*****************************************************************************
; Before return, find out if SAR*SAR or (SAR+1)**2 is closer to Q and
; return SAR or SAR+1 accordingly.
;*****************************************************************************
done            movf        INDF2,W                 ; get the estimated square root
                mulwf       WREG,A                  ; compute SAR*SAR
                movlw       Q_lo                    ; compute SAR*SAR – Q (is < 0)
                movf        PLUSW2,W                ; "
                subwf       PRODL,F,A               ; "
                movlw       Q_hi                    ; "
                movf        PLUSW2,W                ; "
                subwfb      PRODH,F,A               ; "
                comf        PRODH,F,A               ; compute the magnitude of
                negf        PRODL,A                 ; |SAR*SAR – Q|
                movlw       0                       ; "
                addwfc      PRODH,F,A               ; "
                movlw       lp_cnt                  ; copy |SAR*SAR – Q| into lp_cnt
                movff       PRODH,PLUSW2            ; and mask
                movlw       mask                    ; "
                movff       PRODL,PLUSW2            ; "
```

```
          movf        INDF2,W          ; compute SAR+1
          incf        WREG,W,A         ; "
          mulwf       WREG,A           ; compute (SAR+1)**2
          movlw       Q_lo             ; compute (SAR+1)**2 – Q
          movf        PLUSW2,W         ; and leave the difference in
          subwf       PRODL,F,A        ; PRODH:PRODL
          movlw       Q_hi             ; "
          movf        PLUSW2,W         ; "
          subwfb      PRODH,F,A        ; "
          movlw       mask             ; compare [Q-(SAR*SAR)] with
          movf        PLUSW2,W         ; (SAR+1)**2 – Q
          subwf       PRODL,F,A        ; "
          movlw       lp_cnt           ; "
          movf        PLUSW2,W         ; "
          subwfb      PRODH,F,A        ; "
          btfsc       STATUS,C         ; "
          goto        sel_SAR          ; "
          incf        INDF2,F          ; increment SAR by 1
sel_SAR   dealloc_stk 2               ; remove lp_cnt and mask from stack
          popr        PRODL            ; place SAR in PRODL
          popr        FSR2H
          popr        FRS2L
          return
          end
```

The subroutine for finding the square root of a 32-bit number is similar and hence is left for you as an exercise.

▲

4.10.2 Finding Prime Numbers

Finding prime numbers would require the MCU to perform a lot of divide operations. One could use the square root of the number under test to set the limit of number of divide operations that need to be performed.

Example 4.17

▼

Write a subroutine to test whether a 16-bit unsigned integer is a prime number. The integer to be tested is pushed into the stack, and the test result is returned in the WREG register. If the number is a prime number, the subroutine returns a 1. Otherwise, a 0 is returned.

Solution: The most efficient algorithm for testing whether a number is a prime number is to divide the given number by all the prime numbers smaller or equal to the square root of the given number. If the given number cannot be divided by any of these testing numbers, then it is a prime number. Since we don't have a list of prime numbers at hand, we will be satisfied by dividing the given number from 2 to the square root of the given number. As long as a number cannot be divided by all integers from 2 to its integral square root, it is a prime number.

The stack frame for the **prime_test** subroutine is shown in Figure 4.17. This subroutine will use the local variable **i** to test divide the given number. This subroutine calls the **find_sqr** subroutine to compute the square root of the given number and use it as the upper limit of **i**. This upper limit is stored in the stack slot labeled **ilimit**. The stack slots **num_hi** and **num_lo** are

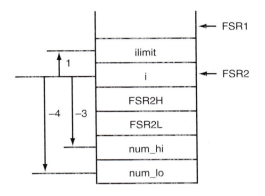

Figure 4.17 ■ Stack frame for example 4.18

used to hold the upper byte and the lower byte of the number to be tested. The subroutine **div16u** is invoked to carry out the division.

The prime_test subroutine and its testing instruction sequence are as follows:

```
                #include <p18F8720.inc>
; ──
; include macro definitions pushr, push_dat, popr, alloc_stk, and dealloc_stk here
; ──
loc_varp        equ         2              ; local variable size for prime_test
ilimit          equ         1              ; offset of ilimit from FSR2
num_hi          equ         -3             ; offset of test number high byte from FSR2
num_lo          equ         -4             ; offset of test number low byte from FSR2
                org         0x00
                goto        start
                org         0x08
                retfie
                org         0x18
                retfie
start           lfsr        FSR1,0xE00     ; set up stack pointer
                push_dat    pdat_lo
                push_dat    pdat_hi
                call        prime_tst
                nop
                dealloc_stk 2
forever         bra         forever
prime_tst       pushr       FSR2L
                pushr       FSR2H
                movff       FSR1L, FSR2L   ; set up frame pointer
                movff       FSR1H,FSR2H    ; "
                alloc_stk   loc_varp       ; allocate local variables
                movlw       num_lo         ; get a copy of the low byte
                movf        PLUSW2,W       ; of the incoming number
                pushr       WREG
                movlw       num_hi         ; get a copy of the high byte
                movf        PLUSW2,W       ; "
```

```
                    pushr          WREG
                    call           find_sqr           ; compute the upper limit for prime test
                    dealloc_stk    2                  ; clean up the pushed data from stack
                    movlw          ilimit             ; set up the upper limit for prime
                    movff          PRODL, PLUSW2      ; test
                    movlw          2
                    movwf          INDF2              ; initialize i to 2
loop                movlw          ilimit             ; place the ilimit in WREG
                    movf           PLUSW2,W           ; "
                    cpfsgt         INDF2              ; is i > ilimit?
                    goto           inloop
                    goto           isprime            ; it is prime if i > ilimit
inloop              movlw          num_lo             ; push the number to be test for prime
                    movf           PLUSW2,W           ; onto the stack
                    pushr          WREG               ; "
                    movlw          num_hi             ; "
                    movf           PLUSW2,W           ; "
                    pushr          WREG               ; "
                    alloc_stk      2                  ; allocate two bytes for R register
                    movf           INDF2,W            ; get the value of i
                    pushr          WREG               ; push i into the stack
                    push_dat       0                  ; push 0 as the upper byte of, divisor
                    call           div16u
                    dealloc_stk    3                  ; remove the pushed top three bytes
                    movlw          -1                 ; the low byte of the remainder is one
                    tstfsz         PLUSW1             ; below the byte pointed to by FSR1
                    goto           nexti              ; remainder is not zero, test next i
                    goto           not_prime
nexti               incf           INDF2,F
                    dealloc_stk    3                  ; deallocate the pushed remaining 3 bytes
                    goto           loop
not_prime           dealloc_stk    2+loc_varp         ; deallocate the pushed remaining 3 bytes
                                                      ; and two bytes of local variables
                    popr           FSR2H
                    popr           FSR2L
                    movlw          0                  ; return a 0 to indicate non_prime
                    return
isprime             dealloc_stk    loc_varp           ; deallocate local variables
                    popr           FSR2H
                    popr           FSR2L
                    movlw          1
                    return
; ——
; include subroutine find_sqr and its associated equ directives here
; ——
; ——
; include subroutine div16u and its associated equ directives here
; ——
                    end
```

4.11 Summary

Many applications require the software to perform signed and unsigned multiprecision multiplication and division. These operations must be implemented as programs. To be reusable in other applications, these programs should be written as subroutines.

A subroutine is a sequence of instructions that can be called from many different places in a program. There are three issues involved in the subroutine call: parameter passing, result returning, and local variables allocation. For the PIC18 MCU, parameters can be passed in the stack or global memory, and the result can also be returned in the stack or global memory.

To solve a complicated problem, software developers should follow the structured programming methodology. In the structured programming approach, the software developer begins with a simple main program whose steps clearly outline the logical flow of the algorithm and then assigns the execution details to subroutines. Subroutines may also call other subroutines. The structured programming methodology makes a complicated problem more manageable.

The PIC18 MCU provides the **call n** and **rcall n** instructions for making the subroutine call. The subroutine will use the **return** instruction to transfer the program control back to the caller.

The **stack** data structure is important in making subroutine calls. The PIC18 MCU has a 31-word by 21-bit return address stack for saving returning addresses. The return address stack can be accessed by executing the **push** and **pop** instructions. The PIC18 MCU does not have a stack for passing incoming parameters and returning results. However, the user can use one of the FSR registers as the stack pointer and implement the data stack in data memory. In this text, we use the FSR1 register as the data stack pointer. During the lifetime of a subroutine, the stack pointer might change, which makes the access of variables in the data stack tricky. To solve this problem, the frame pointer is added. The frame pointer points to a fixed location in the data stack and is never changed during the lifetime of a subroutine. The FSR2 register is used as the frame pointer in this text.

The data stack grows from a low address toward higher addresses. The data structure that consists of incoming parameters, frame pointer, saved registers, and local variables is called a stack frame. The frame pointer points to the memory slot immediately above the saved frame pointer. Therefore, incoming parameters have negative offsets relative to the frame pointer, whereas local variables have positive offsets relative to the frame pointer.

To write a subroutine, the user must lay out the exact location of each local variable and each incoming parameter. The offsets (relative to the byte pointed to by the frame pointer) of them are defined by using the equate directive. Using the equate directive allows the local variables and incoming parameters to be added or moved around without changing your subroutine. The stack frame for each subroutine should be clearly laid out before the subroutine is written. The subroutine examples in this chapter follow this convention. Following this convention will make the subroutine more reusable.

Most of the subroutine examples in this chapter pass parameters in the stack. They return the result in registers when only one or two bytes are needed. Otherwise, they return the result in the stack. In this situation, the caller of the subroutine will need to allocate space in the stack before making the subroutine call.

4.12 Exercises

E4.1 Write a subroutine that computes the product of two signed 8-bit numbers. The incoming parameters are passed in the data stack, and the result is also returned in the stack.

E4.2 Write a subroutine to compute the product of two 16-bit signed numbers. Pass the multiplier and the multiplicand in the stack and also return the product in the stack.

E4.3 What will be the contents of the top 10 bytes of the stack after the execution of the following instructions? Enter xx in the stack slot that has an unknown value.

```
          org        0x20
          lfsr       FSR1,0xE00
          lfsr       FSR2, 0xE00
          push_dat   0x55
          push_dat   0x20
          call       sub_abc
          . . .
sub_abc   pushr      FSR2L
          pushr      FSR2H
          pushr      PRODL
          pushr      PRODH
          alloc_stk  4
```

E4.4 What will be the contents of the top 10 bytes of the stack after the execution of the following instructions? Enter xx in the stack slot that has an unknown value.

```
          org        0x20
          lfsr       FSR1,0xE00
          lfsr       FSR2,0xE00
          movlw      0x30
          pushr      WREG
          push_dat   0x20
          push_dat   0x39
          call       find_lcm
          . . .
find_lcm  pushr      FSR2L
          pushr      FSR2H
          pushr      PRODL
          alloc_stk  4
```

E4.5 Write a subroutine to create a delay of a multiple of 1 ms assuming that the frequency of the crystal oscillator is 32 MHz. The number of milliseconds to be delayed is passed in the PRODL register.

E4.6 Write a subroutine that will divide a 32-bit unsigned number into another 32-bit unsigned number. Pass the divisor and dividend in the stack and also return the remainder and quotient in the stack.

E4.7 Write a subroutine that will divide a 16-bit signed number into another 16-bit signed number. Pass the divisor and dividend in the stack and also return the remainder and quotient in the stack.

E4.8 Write a subroutine that will divide a 32-bit signed number into another 32-bit signed number. Pass the divisor and dividend in the stack and also return the remainder and quotient in the stack.

E4.9 Write a subroutine that will convert all the lowercase letters in a string into uppercase without changing other nonalphabetic characters. The starting address of the string is passed in FSR0, and the string is stored in data memory.

E4.10 Write a subroutine to convert all the uppercase letters in a string into lowercase without changing other nonalphabetic characters. The starting address of the string is passed in FSR0, and the string is stored in data memory.

E4.11 Write a subroutine that will return a 16-bit random number to the caller in the PRODL and PRODH register pair. This subroutine would return different number each time it is called.

E4.12 Write a subroutine that will compute the greatest common divisor (GCD) of two 16-bit integers passed in the stack and return the GCD in the PRODL and PRODH register pair.

E4.13 Write a subroutine that will convert a 32-bit signed number to an ASCII string that represents its equivalent decimal value. The number to be converted and the starting address of the buffer to hold the resultant string is passed in the stack.

E4.14 Write a subroutine that converts the temperature in Fahrenheit scale to Celsius scale or vise versa. The temperature value to be converted is passed in the PRODH register, whereas the intended conversion (from °F to °C (0) or °C to °F (1)) is passed in the PRODL register. The converted result is returned in the PRODL register.

E4.15 Write a subroutine to convert a 16-bit binary number into four hex digits (in ASCII code). The number to be converted is passed in the stack, and the converted result is also returned in the stack.

E4.16 Write a subroutine that will compute the least common multiple (LCM) of two 16-bit integers passed in the stack and return the LCM in the stack as a 4-byte number.

E4.17 Write a subroutine to count the number of characters and words contained in a string terminated by a NULL character. The string is stored in the program memory, and the starting address (three bytes) of the string is passed to this subroutine in TBLPTR. This routine returns the character count and word count in PRODL and PRODH, respectively.

E4.18 Write a subroutine to compute the square root of a 24-bit integer and also write a program to test it. The 24-bit number is passed in the stack, and the square root is returned in the PRODH:PRODL register pair. The upper four bits of PRODH should be all 0s.

E4.19 Write a subroutine to compute the square root of a 32-bit integer and also write a program to test it. The 32-bit number is passed in the stack, and the square root is returned in the PRODH:PRODL register pair.

E4.20 Write a subroutine to find the prime numbers between **n1** and **n2** and store the result in a buffer specified by the caller of this routine. Both **n1** and **n2** are 16-bit integers and are passed to this routine in the stack. The pointer to the data memory buffer to hold the prime numbers is also passed in the stack. The caller will push **n1, n2,** and the pointer to the buffer onto the stack in that order. Write a program to test this subroutine.

E4.21 Write a subroutine to test if a 32-bit integer is prime. The number to be tested is passed in the stack. If the number-under-test is prime, then this subroutine returns a 1 in the WREG register. Otherwise, it returns a 0.

E4.22 Modify the subroutine in Example 4.14 so that it checks invalid character in the input string and returns an indicator in WREG to the caller.

4.13 Lab Exercises and Assignments

L4.1 Write a subroutine to count the number of characters and words contained in a string. The string is stored in the program memory, and its starting address is passed in TBLPTR. The subroutine will return the character count and word count in PRODL and PRODH, respectively. Use the MPLAB IDE to enter, compile, and simulate the program.

L4.2 Write a subroutine to search for a word from a string. The starting addresses of the word to be looked for and the string are passed in the stack. The string is stored in the program memory, whereas the word to be looked for is stored in data memory. Write a test program to prepare the word and the string and then call this subroutine. The algorithm for searching a word is shown in Figure L4.1. The variables used in Figure L4.1 are the following:

str_ptr	string pointer
wd_ptr	word pointer
match_flag	flag to indicate search result
m[str_ptr]	the character pointed to by **str_ptr**
m[wd_ptr]	the character pointed to by **wd_str**

L4.3   Write a subroutine to implement **selection sort** and write a program to test it. The selection sort method works like this:

1. Suppose we have an array of integer values. Search through the array, find the largest value, and exchange it with the value stored in the first array element position. Next, find the second-largest value in the array and exchange it with the value stored in the second array location. Repeat the same process until the end of the array is reached.

2. Store the array in data memory. The caller of this subroutine passes the starting address of the array and the array count in the software stack.

Before testing the subroutine, one needs to prepare an array in data memory. This can be done by declaring an array in program memory and copying it into the data memory in the caller.

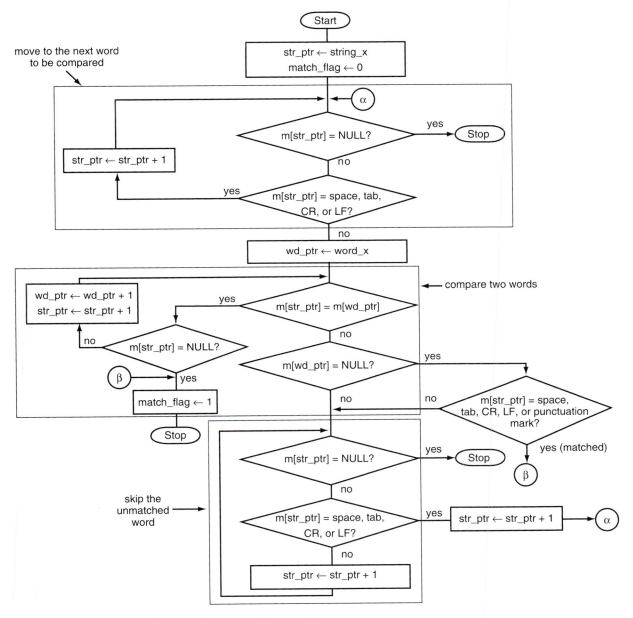

Figure L4.1 ■ Algorithm for searching a word from a string

5

A Tutorial to the C Language
and the Use of the C Compiler

5.1 Objectives

After completing this chapter, you should be able to

- Explain the overall structure of a C language program

- Use appropriate operators to perform desired operations in C language

- Understand the basic data types and expressions of C language

- Write program loops in C language

- Write functions and make function calls in C language

- Use arrays and pointers for data manipulation

- Perform basic I/O operations in C language

- Explain the application project build process

- Explain the functions of a linker

- Understand the linker file

- Create a new project that contains C programs

- Set up project options

5.2 Introduction to C

This tutorial is not intended to provide a complete coverage of the C language. Instead, it provides only a summary of those C language constructs that are used in this book. It should be adequate to deal with the basic PIC18 interface programming.

C language is gradually replacing assembly language in many embedded applications because it has several advantages over assembly language. The most important one is that it allows us to work on the program logic at a level higher than the assembly language; thus, programming productivity is greatly improved.

A C program, whatever its size is, consists of functions and variables. A function contains statements that specify the operations to be performed. The types of statements in a function could be *declaration, assignment, function call, control,* and *null.* A variable stores a value to be used during the computation. The **main()** function is required in every C program and is the one to which control is passed when the program is executed. A simple C program is as follows:

```
1.  #include <stdio.h>        /*includes information about standard library*/
2.  /* this is where program execution begins */
3.  main (void)               /*defines a function named main that receives */
                              /*no argument values*/
4.  {                         /*statements of main are enclosed in braces*/
5.  int a, b, c;              /*defines three variables of type int*/
6.  a = 3;                    /*assigns 3 to variable a*/
7.  b = 5;                    /*assigns 5 to variable b*/
8.  c = a + b;                /*adds a and b together and assigns it to c*/
9.  printf(" a + b = %d \n", c); /*calls library function printf to print the result*/
10. return 0;                 /*returns 0 to the caller of main*/
11. }                         /*the end of main function*/
```

The first line of the program,

```
#include <stdio.h>
```

causes the file **stdio.h** to be included in your program. This line appears at the beginning of many C programs. The header file **stdio.h** contains the prototype declarations of all I/O routines that can be called by the user program and constant declarations that can be used by the user program. The C language requires a function prototype be declared before that function can be called if a function is not defined when it is called. The inclusion of the file **stdio.h** allows the function **printf(. . .)** to be invoked in the program.

The second line is a comment. A comment explains what will be performed and will be ignored by the compiler. A comment in C language starts with /* and ends with */. Everything in between is ignored. Comments provide documentation to our code and enhance readability. Comments affect only the size of text file and do not increase the size of executable code. Many commercial C compilers also allow us to use two slashes (//) for commenting out a single line. The third line **main()** is where program execution begins. The opening brace on the fourth line marks the start of the **main()** function's code. Every C program must have one and only one **main()** function. Program execution is also ended with the **main** function. The fifth line declares 3 integer variables **a, b,** and **c.** In C, all variables must be declared before they can be used.

The sixth line assigns 3 to the variable **a.** The seventh line assigns 5 to the variable **b.** The eighth line computes the sum of variables **a** and **b** and assigns it to the variable **c.** Assignment statements are major components in the C program.

The ninth line calls the library function **printf** to print the string **a** + **b** = followed by the value of **c** and move the cursor to the beginning of the next line. The tenth line returns a 0 to the caller of **main**(). The closing brace in the eleventh line ends the **main**() function.

5.3 Types, Operators, and Expressions

Variables and constants are the basic objects manipulated in a program. Variables must be declared before they can be used. A variable declaration must include the name and the type of the variable and may optionally provide its initial value. A variable name may start with a letter (**A** through **Z** or **a** through **z**) or underscore character followed by zero or more letters, digits, or underscore characters. Variable names cannot contain arithmetic signs, dots, apostrophes, C keywords, or special symbols, such as @, #, or ?. Adding the underscore character (_) may sometimes improve the readability of long variables. Do not begin variable names with underscore, however, since library routines often use such names. Upper- and lowercase letters are distinct.

5.3.1 Data Types

There are only a few basic data types in C: **void, char, int, float,** and **double.** A variable of type void represents nothing. The type **void** is used most commonly with functions. A variable of type **char** can hold a single byte of data. A variable of type **int** is an integer, which is normally the natural size for a particular machine. For most commercial PIC18 C compilers, type integer has the length of 16 bits. The type **float** refers to a 32-bit, single-precision, floating-point number. The type **double** represents a 64-bit double-precision, floating-point number. In addition, there are a number of qualifiers that can be applied to these basic types. **Short** and **long** apply to integers. These two qualifiers will modify the lengths of integers. The modifier **short** does not change the length of an integer. The modifier **long** doubles a 16-bit integer to 32-bit. Both the type **float** and type **double** are 32 bits in length for the PIC18 C compiler.

5.3.2 Variable Declarations

All variables must be declared before their use. A declaration specifies a type and contains a list of one or more variables of that type, as in

```
int i, j, k;
char cx, cy;
```

A variable may also be initialized when it is declared, as in

```
int i = 0;
char echo = 'y'; /* the ASCII code of letter y is assigned to variable echo. */
```

5.3.3 Constants

There are four kinds of constants: integers, characters, floating-point numbers, and strings. A character constant is an integer, written as one character within single quotes, such as 'x'. A character constant is represented by the ASCII code of the character. A string constant is a sequence of zero or more characters surrounded by double quotes, as in

"PIC 18F8680 is a microcontroller made by Microchip"

or

" "   /* an empty string */

Each individual character in the string is represented by its ASCII code.

An integer constant such as 3241 is an **int.** A long constant is written with a terminal 1 (ell) or L, as in 44332211L. The following constant characters are predefined in C language:

\a alert (bell) character	\\	backslash
\b backspace	\?	question mark
\f formfeed	\'	single quote
\n newline	\"	double quote
\r carriage return	\ooo	octal number
\t horizontal tab	\xhh	hexadecimal number
\v vertical tab		

As in assembly language, a number in C can be specified in different bases. The method to specify the base of a number is to add a prefix to the number. The prefixes for different bases are the following

base	prefix	example	
decimal	none	1357	
octal	0	04723	; preceded by a zero
hexadecimal	0x	0x2A	

5.3.4 Arithmetic Operators

There are seven arithmetic operators:

+	add and unary plus
–	subtract and unary minus
*	multiply
/	divide
%	modulus (or remainder)
++	increment
—	decrement

The expression

a % b

produces the remainder when **a** is divided by **b.** The % operator cannot be applied to float or double. The ++ operator adds 1 to the operand, and the – – operator subtracts 1 from the operand. The / operator truncates the quotient to integer when both operands are integers.

5.3.5 Bitwise Operators

C provides six operators for bit manipulations; these may be applied only to integral operands, that is, **char, short, int,** and **long,** whether signed or unsigned:

&	AND
|	OR
^	XOR
~	NOT
>>	right shift
<<	left shift

The **&** operator is often used to clear one or more bits to zero. For example, the statement

PORTC = PORTC & 0xBD; /* PORTC is 8 bits */

clears bits 6 and 1 of PORTC to 0.

The | operator is often used to set one or more bits to 1. For example, the statement

```
PORTB = PORTB | 0x40;    /* PORTB is 8 bits */
```

sets the bit 6 of PORTB to 1.

The **XOR** operator can be used to toggle a bit. For example, the statement

```
abc = abc ^ 0xF0;          /* abc is of type char */
```

toggles the upper four bits of the variable **abc.**

The >> operator shifts the involved operand to the right for the specified number of places. For example,

```
xyz = abc >> 3;
```

shifts the variable **abc** to the right three places and assigns it to the variable **xyz.**

The << operator shifts the involved operand to the left for the specified number of places. For example,

```
xyz = xyz << 4;
```

shifts the variable *xyz* to the left four places.

The assignment operator = is often combined with the operator. For example,

```
PORTD = PORTD & 0xBD;
```

can be rewritten as

```
PORTD &= 0xBD;
```

The statement

```
PORTB = PORTB | 0x40;
```

can be rewritten as

```
PORTB |= 0x40;
```

5.3.6 Relational and Logical Operators

Relational operators are used in expressions to compare the values of two operands. If the result of the comparison is true, then the value of the expression is 1. Otherwise, the value of the expression is 0. Here are the relational and logical operators:

==	equal to (two "=" characters)
!=	not equal to
>	greater than
>=	greater than or equal to
<	less than
<=	less than or equal to
&&	and
||	or
!	not (one's complement)

Here are some examples of relational and logical operators:

```
if (!(ADCTL & 0x80))
     statement₁;          /* if bit 7 is 0, then execute statement₁ */
if (i > 0 && i < 10)
     statement₂;          /* if 0 < i < 10 then execute statement₂ */
if (a1 == a2)
     statement₃           /* if a1 equals a2 then execute statement₃ */
```

5.3.7 Precedence of Operators

Precedence refers to the order in which operators are processed. The C language maintains a precedence for all operators. The precedence for all operators is shown in Table 5.1. Operators at the same level are evaluated from left to right. A few examples that illustrate the precedence of operators are listed in Table 5.2.

Precedence	Operator	Associativity
Highest	( ) [ ] → .	left to right
	! ~ ++ -- + - * & (type) sizeof	right to left
	* / %	left to right
	+ -	left to right
	<< >>	left to right
	<<= >>=	left to right
	== !=	left to right
	&	left to right
	^	left to right
			left to right
	&&	left to right
	||	left to right
	?:	right to left
	= += -= *= /= %= &= ^= |= <= >>=	right to left
Lowest	'	left to right

Table 5.1 ■ Table of precedence of operators

Expression	Result	Note
15 - 2 * 7	1	* has higher precedence than +
(13 - 4) * 5	45	
(0x20 | 0x01)!= 0x01	1	
0x20 | 0x01 != 0x01	0x20	!= has higher precedence than |
1 << 3 + 1	16	+ has higher precedence than <<
(1 << 3) + 1	9	

Table 5.2 ■ Examples of operator precedence

5.4 Control Flow

The control-flow statements specify the order in which computations are performed. In C language, the semicolon is a statement terminator. Braces { and } are used to group declarations and statements together into a *compound statement*, or *block*, so that they are syntactically equivalent to a single statement. A compound statement is not terminated by a semicolon.

5.4.1 If Statement

The **if** statement is a conditional statement. The statement associated with the **if** statement is executed based on the outcome of a condition. If the condition evaluates to nonzero, the statement is executed. Otherwise, it is skipped.

The syntax of the **if** statement is as follows:

```
if (expression)
    statement;
```

Here is an example of the **if** statement:

```
if (a > b)
    sum += 2;
```

The value of sum will be incremented by 2 if the viable **a** is greater than the variable **b.**

5.4.2 If-Else Statement

The **if-else** *statement* handles conditions where a program requires one statement to be executed if a condition is nonzero and a different statement to be executed if the condition is zero.

The syntax of an **if-else** statement is

```
if (expression)
    statement₁
else
    statement₂
```

The expression is evaluated; if it is true (nonzero), statement$_1$ is executed. If it is false, **statement$_2$** is executed. Here is an example of the **if-else** statement:

```
if (a ! = 0)
    r = b;
else
    r = c;
```

The **if-else** statement can be replaced by the **?:** operator. The statement

```
r = (a ! = 0) ? b: c;
```

is equivalent to the previous **if** statement.

5.4.3 Multiway Conditional Statement

A multiway decision can be expressed as a cascaded series of **if-else** statements. Such series looks like this:

```
if (expression₁)
    statement₁
else if (expression₂)
    statement₂
else if (expression₃)
    statement₃
    . . .
else
    statementₙ
```

Here is an example of a three-way decision:

```
if (abc > 0) return 5;
else if (abc == 0) return 0;
else return –5;
```

5.4.4 Switch Statement

The **switch** statement is a multiway decision based on the value of a control expression. The syntax of the **switch** statement is

```
switch (expression) {
        case const_expr₁:
                statement₁;
                break;
        case const_expr₂:
                statement₂;
                break;
        . . .
        default:
                statementₙ;
}
```

As an example, consider the following program fragment:

```
switch (i) {
        case 1: printf("*");
                break;
        case 2: printf("**");
                break;
        case 3: printf("***");
                break;
        case 4: printf("****");
                break;
        case 5: printf("*****");
        default:
                printf("\n");
}
```

The number of * characters printed is equal to the value of **i**. The **break** keyword forces the program flow to drop out of the switch statement so that only the statements under the corresponding *case label* are executed. If any break statement is missing, then all the statements from that case label until the next break statement within the same switch statement will be executed.

5.4.5 For-Loop Statement

The syntax of a **for-loop** statement is

```
for (expr1; expr2; expr3)
        statement;
```

where, **expr1** and **expr3** are assignments or function calls and **expr2** is a relational expression. For example, the following **for** loop computes the sum of the squares of integers from 1 to 9:

```
sum = 0;
for (i = 1; i < 10; i++)
        sum = sum + i * i;
```

The following **for** loop prints out the first 10 odd integers:

```
for (i = 1; i < 20; i++)
        if (i % 2) printf("%d ", i);
```

5.4.6 While Statement

While an expression is nonzero, the **while** loop repeats a statement or block of code. The value of the expression is checked prior to each execution of the statement. The syntax of a **while** statement is

```
while (expression)
      statement;
```

The expression is evaluated. If it is nonzero (true), **statement** is executed and the expression is reevaluated. This cycle continues until the expression becomes zero (false), at which point execution resumes after **statement.** The **statement** may be a *null* statement. A null statement does nothing and is represented by a semicolon.

Consider the following program fragment:

```
int_cnt = 5;
while (int_cnt);
```

The CPU will do nothing before the variable **int_cnt** is decremented to zero. In microcontroller applications, the decrement of **int_cnt** is often done by external events, such as interrupts.

5.4.7 Do-While Statement

The **while** and **for** loops test the termination condition at the beginning. By contrast, the **do-while** statement tests the termination condition at the end of the statement; the body of the statement is executed at least once. The syntax of the statement is

```
do
      statement
while (expression);
```

The following **do-while** statement displays the integers 9 down to 1:

```
int digit = 9;
do
      printf("%d ", digit--);
while (digit >= 1);
```

5.4.8 GOTO Statement

Execution of a **goto** statement causes control to be transferred directly to the labeled statement. This statement must be located in the same function as the **goto** statement. The use of **goto** statement interrupts the normal sequential flow of a program and thus makes it harder to follow and decipher. For this reason, the use of **goto** statements is not considered good programming style, and it is recommended that you do not use them in your program.

The syntax of the **goto** statement is

```
goto label
```

An example of the use of **goto** statement is as follows:

```
if (x > 100)
      goto fatal_error;
 . . .
fatal_error:
    printf("Variable x is out of bound!\n");
```

5.5 Input and Output

Input and output facilities are not part of C language itself. However, input and output are fairly important in application. The ANSI standard defines a set of library functions that must be included so that they can exist in compatible form on any system where C exists.

Some of these functions deal with file input and output. Others deal with text input and output. In this section, we look at the following four input and output functions:

1. **int *getchar* ()**. This function returns a character when it is called. The following program fragment returns a character and assigns it to the variable **xch:**

   ```
   char xch;
   xch = getchar ();
   ```

2. **int *putchar* (int).** This function outputs a character on the standard output device. The following statement outputs the letter a from the standard output device:

   ```
   putchar ('a');
   ```

3. **int *puts* (const char *s).** This function outputs the string pointed to by **s** on the standard output device. The following statement outputs the string **Learning microcontroller is fun!** from the standard output device:

   ```
   puts ("Learning microcontroller is fun! \n");
   ```

4. **int *printf* (*formatting string*, arg$_1$, arg$_2$, . . ., arg$_n$).** This function converts, formats, and prints its arguments on the standard output under control of *formatting string,* arg$_1$, arg$_2$, . . ., arg$_n$ are arguments that represent the individual output data items. The arguments can be written as constants, single variable or array names, or more complex expressions. The formatting string is composed of individual groups of characters, with one character group associated with each output data item. The character group corresponding to a data item must start with %. In its simplest form, an individual character group will consist of the percent sign followed by a *conversion character* indicating the type of the corresponding data item.

   Multiple character groups can be contiguous or separated by other characters, including white-space characters. These other characters are simply transferred directly to the output device where they are displayed. A subset of the more frequently used conversion characters are listed in Table 5.3.

Conversion character	Meaning
c	data item is displayed as a single character
d	data item is displayed as a signed decimal number
e	data item is displayed as a floating-point value with an exponent
f	data item is displayed as a floating-point value without an exponent
g	data item is displayed as a floating-point value using either e-type or f-type conversion, depending on value; trailing zeros, trailing decimal point will not be displayed
i	data item is displayed as a signed decimal integer
o	data item is displayed as an octal integer, without a leading zero
s	data item is displayed as a string
u	data item is displayed as an unsigned decimal integer
x	data item is displayed as a hexadecimal integer, without the leading 0s

Table 5.3 ■ Commonly used conversion characters for data output

Between the % character and the conversion character, there may be, in order, the following:

- A minus sign, which specifies left adjustment of the converted argument.
- A number that specifies the minimum field width. The converted argument will be printed in a field at least this wide. If necessary, it will be padded on the left (or right, if left adjustment is called for) to make up the field width.
- A period, which separates the field width from precision.
- A number, the precision, that specifies the maximum number of characters to be printed from a string, or the number of digits after the decimal point of a floating-point value, or the minimum number of digits for an integer.
- An **h** if the integer is to be printed as a short or an l (letter ell) if as a long.

Several valid **printf** calls are as follows:

printf ("this is a challenging course!\n");	/* outputs only a string */
printf ("%d %d %d", x1, x2, x3);	/* outputs variables **x1, x2, x3** using minimal number of digits with one space separating each value */
printf("Today's temperature is %4.1d \n", temp);	/* display the string **Today's temperature is** followed by the value of **temp.** Display one fractional digit and use at least four digits for the value. */

The **printf** () function is not supported by the Microchip PIC18 C compiler.

5.6 Functions and Program Structure

Every C program consists of one or more functions. If a program consists of multiple functions, their definitions cannot be embedded within another. The same function can be called from several different places within a program. Generally, a function will process information passed to it from the calling portion of the program and return a single value. Information is passed to the function via special identifiers called *arguments* (also called *parameters*) and returned via the **return** statement. Some functions, however, accept information but do not return anything (e.g., the library function **printf**).

The syntax of a function definition is as follows:

```
return_type function_name (declarations of arguments)
{
        declarations and statements
}
```

The declaration of an argument in the function definition consists of two parts: the *type* and the *name* of the variable. The return type of a function is *void* if it does not return any value to the caller. An example of a function that converts a lowercase letter to an uppercase letter is as follows:

```
char lower2upper (char cx)
{
        if (cx > = 'a' && cx <= 'z') return (cx – ('a' – 'A'));
        else return cx;
}
```

A character is represented by its ASCII code. A letter is in lowercase if its ASCII code is between 97 (0x61) and 122 (0x7A). To convert a letter from lowercase to uppercase, subtract its ASCII code by the difference of the ASCII codes of letters **a** and **A**.

To call a function, simply write down the name of the function and replace the argument declarations by actual arguments or values and terminate it with a semicolon.

Example 5.1

▼

Write a function that finds the largest element of an integer array.

Solution: The parameters passed to this function include the array count and the starting address of the array. The function is as follows:

```
int arr_max(int ar_cnt, int arr[])
{
    int i, temp;
    temp = arr[0];//set the first array element as the current max
    for (i = 1; i < ar_cnt; i++) {
        if (arr[i] > temp)
                temp = arr[i];
    }
    return temp;
}
```

▲

Example 5.2

▼

Write a function to compute the square root of a 32-bit number using the successive approximation method.

Solution: The function is as follows:

```
unsigned sq_root (unsigned long int xz)
{
    unsigned int sar, guess_mask, rest_mask;
    unsigned int i;
    sar = 0;       /*successive approximation register is initialized to 0 */
    guess_mask = 0x8000;      /*this mask is used to guess the ith bit to be 1 */
    i = 16;
    do {
        rest_mask = ~guess_mask; /*rest_mask is used to cancel the incorrect guess */
        sar |= guess_mask; /*guess the ith bit to be 1 */
        if (((unsigned long)sar * (unsigned long)sar) > xz)
                sar &= rest_mask; //change the bit to 0
        guess_mask >> 1;
        i--;
    } while (i > 0);
    if ((xy - sar * sar) < ((sar + 1)*(sar + 1) - xy))
            return sar;
    else return (sar + 1);
}
```

A function cannot be called before it has been defined. This dilemma is solved by using the function prototype statement. The syntax for a function prototype statement is as follows:

return_type function_name (declarations of arguments);

The statement

char test_prime (int a);

is an example of the function prototype declaration.
The following examples illustrate the use of prototype declaration:

Example 5.3

Write a function to test whether an integer is a prime number.

Solution: The integer 1 is not a prime number. A number is prime if it is indivisible by any integer between 2 and its square root:

```
unsigned sq_root (unsigned long int xz);
/* this function returns a 1 if ka is prime. Otherwise, it returns a 0. */
char test_prime (unsigned long int ka)
{
    unsigned int i, limit;
    if (ka == 1) return 0;
    else if (k2 == 2) return 1;
    limit = sq_root (ka);        /* find the square root of ka */
    for (i = 2; i <= limit; i++)
        if ((a % i) == 0) return 0;
    return 1;
}
```

Example 5.4

Write a program to find out the number of prime numbers between 1000 and 10000.

Solution: We can find the number of prime numbers between 1000 and 10000 by calling the function in Example 5.3 to find out if a number is prime.

```
#include <stdio.h>
char test_prime (unsigned int a);        /* prototype declaration of test_prime () */
unsigned sq_root (unsigned long int xz); /* prototype of sq_root () */
main ( )
{
    unsigned int i, prime_count;
    prime_count = 0;
    for (i = 1000; i <= 10000; i++) {
        if (test_prime(i))
            prime_count ++;
    }
    printf("\n The total prime numbers between 1000 and 10000 is %d\n", prime_count);
}
/* include the functions sq_root () and test_prime () here */
```

5.7 Pointers, Arrays, Structures, and Unions

5.7.1 Pointers and Addresses

A *pointer* is a variable that holds the address of a variable. Pointers are used frequently in C, as they have a number of useful applications. For example, pointers can be used to pass information back and forth between a function and its reference (calling) point. In particular, pointers provide a way to return multiple data items from a function via function arguments. Pointers also permit references to other functions to be specified as arguments to a given function. This has the effect of passing functions as arguments to the given function.

Pointers are also closely associated with arrays and therefore provide an alternate way to access individual array elements. The syntax for declaring a pointer type is

```
type_name *pointer_name;
```

For example,

```
int *ax;
```

declares that the variable **ax** is a pointer to an integer,

```
char *cp;
```

declares that the variable **cp** is a pointer to a character.

To access the value pointed to by a pointer, use the *dereferencing* operator *. For example,

```
int a, *b;        /* b is a pointer to int */
. . .
a = *b;
```

assigns the value pointed to by **b** to variable **a.**

We can assign the address of a variable to a pointer by using the unary operator **&.** The following example shows how to declare a pointer and how to use **&** and *:

```
int x, y;
int *ip;          //ip is a pointer to an integer
ip = &x;          //assigns the address of the variable x to ip
y = *ip;          //y gets the value of x
```

Many beginners are confused by the pointer concept. The main difference between an ordinary variable and a pointer variable is the following:

When specifying the name of an ordinary variable, one gets the value of that variable. However, when specifying the name of a pointer variable, one gets the address of the memory location where the value of the variable is stored.

One can perform arithmetic of the pointer variables. Adding 1 to a pointer variable will actually add to the pointer variable the size of the object pointed to by the pointer variable. For example, if the variable **iptr** is a pointer to a long integer, then **iptr + 1** will actually add 4 to the current value of **iptr.**

Example 5.5

▼

Write the bubble sort function to sort an array of integers.

Solution: The algorithm for bubble sort: was described in Chapter 4. Here is the C language version of the function:

```
void swap (int *, int *);
void bubble (int a[], int n)    /*n is the array count*/
{
    int i,j;
    for (i = 0; i < n - 1; i++)
        for (j = 0; j < n - (i + 1); j++)
            if (a[j] > a[j+1])
                swap (&a[j], &a[j+1]);
}
void swap (int *px, int *py)
{
    int temp;
    temp = *px;
    *px = *py;
    *py = temp;
}
```

5.7.2 Arrays

Many applications require the processing of multiple data items that have common characteristics (e.g., a set of numerical data, represented by $x_1, x_2, . . ., x_n$). In such situations, it is more convenient to place data items into an *array*, where they will all share the same name. The individual data items can be characters, integers, floating-point numbers, and so on. They must all be of the same type and the same storage class.

Each array element is referred to by specifying the array name followed by one or more *subscripts*, with each subscript enclosed in brackets. Each subscript must be expressed as a nonnegative integer. Thus, the elements of an n-element array x are $x[0], x[1], . . ., x[n - 1]$. The number of subscripts determines the dimensionality of the array. For example, x[i] refers to an element of an one-dimensional array. Similarly, y[i][j] refers to an element of a two-dimensional array. Higher dimensional arrays can be formed by adding additional subscripts in the same manner. However, higher dimensional arrays are not used very often in 8- and 16-bit microcontroller applications.

In general, a one-dimensional array can be expressed as

data-type array_name[expression];

A two-dimensional array is defined as

data-type array_name[expr1][expr2];

An array can be initialized when it is defined. This is a technique used in table lookup, which can speed up the computation process. For example, a data acquisition system that utilizes an analog-to-digital converter can use table lookup technique to speed up the conversion (from the digital value back to the original physical quantity) process.

5.7.3 Pointers and Arrays

In C, there is a strong relationship between pointers and arrays. Any operation that can be achieved by array subscripting can also be done with pointers. The pointer version will in general be faster but somewhat harder to understand. For example,

int ax[20];

defines an array **ax** of 20 integral numbers.

The notation ax[i] refers to the *i*-th element of the array. If **ip** is a pointer to an integer, declared as

```
int *ip;
```

then the assignment

```
ip = &ax[0];
```

makes **ip** contain the address of ax[0]. Now the statement

```
x = *ip;
```

will copy the contents of ax[0] into x.

If **ip** points to ax[0], then ip + 1 points to ax[1], ip + i points to ax[i], and so on.

5.7.4 Passing Arrays to a Function

An array name can be used as an argument to a function, thus permitting the entire array to be passed to the function. To pass an array to a function, the array name must appear by itself, without brackets or subscripts, as an actual argument within the function call. When declaring a one-dimensional array as a formal argument, the array name is written with a pair of empty square brackets. The size of the array is not specified within the formal argument declaration. If the array is two-dimensional, then there should be two pairs of empty brackets following the array name.

The following program outline illustrates the passing of an array from the main portion of the program to a function:

```
int average (int n, int arr[]);
void main ( )
{
    int n, avg;              /* variable declaration */
    int arr[50];            /* array definition */
    . . .
    avg = average(n, arr);  /* function call */
    . . .
}
int average (int k, int brr[])   /* function definition */
{
    . . .
}
```

Within **main,** we see a call to the function **average.** This function call contains two actual arguments—the integer variable **n** and the one-dimensional, integer array **arr,** Note that **arr** appears as an ordinary variable within the function call.

In the first line of the function definition, we see two formal arguments **k** and **brr.** The formal argument declarations establish **k** as an integer variable and **brr** as a one-dimensional integer array. Note that the size of **brr** is not defined in the function definition.

As formal parameters in a function definition,

```
int brr[];
```

and

```
int *brr;
```

are equivalent.

5.7.5 Initializing Arrays

C allows initialization of arrays. Standard data type arrays may be initialized in a straight-forward manner. The syntax for initializing an array is as follows:

array_declarator = { value-list }

The following statement shows a five-element integer array initialization:

int i[5] = {10, 20, 30, 40, 50};

The element **i[0]** has the value of 10, and the element **i[4]** has the value of 50.

A string (character array) can be initialized in two ways. One method is to make a list of each individual character:

char strgx[5] = {'w', 'x', 'y', 'z', 0};

The second method is to use a string constant:

char myname [16] = "microcontroller";

A null character is automatically appended at the end of microcontroller. When initializing an entire array, the array size (which is one more than the actual length) must be included:

char prompt [24] = "Please enter an integer:";

5.7.6 Structures

A structure is a group of related variables that can be accessed through a common name. Each item within a structure has its own data type, which can be different from the other data items. The syntax of a structure declaration is as follows:

```
struct struct_name {          /* struct_name is optional */
       type1 member1;
       type2 member2;
. . .
};
```

The **struct_name** is optional and, if it exists, defines a *structure tag*. A **struct** declaration defines a type. The right brace that terminates the list of members may be followed by a list of variables, just as for any basic type. The following example is for a card catalog in a library:

```
struct catalog_tag {
       char author [40];
       char title [40];
       char pub [40];
       unsigned int date;
       unsigned char rev;
} card;
```

where the variable **card** is of type **catalog_tag.**

A structure definition that is not followed by a list of variables reserves no storage; it merely describes a template or the shape of a structure. If the declaration is tagged (i.e., it has a name), however, the tag can be used later in definitions of instances of the structure. For example, suppose we have the following structure declaration:

```
struct point {
       int x;
       int y;
};
```

We can then define a variable **pt** of type **point** as follows:

struct point pt;

A member of a particular structure is referred to in an expression by a construction of the form

structure-name.member

or

structure-pointer → member

The structure member operator "." connects the structure name and the member name. As an example, the square of the distance of a point to the origin can be computed as follows:

long integer sq_distance;

. . .

sq_distance = pt.x * pt.x + pt.y * pt.y;

Structure can be nested. One representation of a circle consists of the center and radius as shown in Figure 5.1.

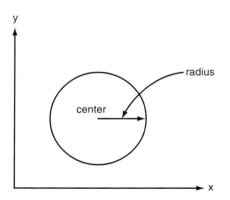

Figure 5.1 ■ A circle

This circle can be defined as follows:

struct circle {
        struct point center;
        unsigned int radius;
};

5.7.7 Unions

A *union* is a variable that may hold (at different times) objects of different types and sizes, with the compiler keeping track of size and alignment requirements. Unions provide a way to manipulate different kinds of data in a single area of storage without embedding any machine-dependent information in the program.

The syntax of the union is as follows:

```
union union_name {
        type-name₁ element₁;
        type-name₂ element₂;
        . . .
        type nameₙ elementₙ;
};
```

The field **union_name** is optional is also called a *union tag.* You can declare a union variable at the same time when you declare a union type. The union variable name should be placed after the right brace "}".

Suppose we want to represent the current temperature using both the integer and the string; we can use the following declaration:

```
union u_tag {
        int i;
        char c[4];
} temp;
```

Four characters must be allocated to accommodate the larger of the two types. The integer type is good for internal computation, whereas the string type is suitable for output. Of course, some conversion may be needed before we make certain kinds of interpretations. Using this method, we can interpret the variable **temp** as an integer or a string, depending on our purpose.

Syntactically, members of a union are accessed as

union-name.member

or

union-pointer → member

just as for structures.

5.8 Miscellaneous Items

5.8.1 Automatic/External/Static/Volatile

A variable defined inside a function is an *internal variable* of that function. These variables are called *automatic* because they come into existence when the function is entered and disappear when it is left. Internal variables are equivalent to local variables in assembly language. *External variables* are defined outside of any function and are thus potentially available to many functions. Because external variables are globally accessible, they provide an alternative to function arguments and return values for communicating data between functions. Any function may access an external variable by referring to it by name, if the name has been declared somehow. External variables are also useful when two functions must share some data, yet neither calls the other.

The use of **static** with a local variable declaration inside a block or a function causes the variable to maintain its value between entrances to the block or function. Internal static variables are local to a particular function just as automatic variables are, but unlike automatic variables, they remain in existence rather than coming and going each time the function is activated. When a variable is declared static outside of all functions, its scope is limited to the file that contains the definition. A function can also be declared as static. When a function is declared as static, it becomes invisible outside the file that defines the function.

A *volatile variable* has a value that can be changed by something other than the user code. A typical example is an input port or a timer register. These variables must be declared as volatile so that the compiler makes no assumptions on their values while performing optimizations. The keyword *volatile* prevents the compiler from removing apparently redundant references through the pointer.

5.8.2 Scope Rules

The functions and external variables that make up a C program need not all be compiled at the same time; the source text of the program may be kept in several files, and previously compiled routines may be loaded from libraries.

The scope of a name is the part of the program within which the name can be used. For a variable declared at the beginning of a function, the scope is the function in which the name is declared. Local (internal) variables of the same name in different functions are unrelated.

The scope of an external variable or a function lasts from the point at which it is declared to the end of the file being compiled. Consider the following program segment:

```
. . .
void f1 (. . .)
{
        . . .
}
int a, b, c;
void f2 (. . .)
{
        . . .
}
```

Variables **a, b,** and **c** are accessible to function **f2** but not to **f1.**

When a C program is split into several files, it is convenient to put all global variables into one file so that they can be accessed by functions in different files. Functions residing in different files that need to access global variables must declare them as external variables. In addition, we can place the prototypes of certain functions in one file so that they can be called by functions in other files.

The following example is a skeletal outline of a two-file C program that makes use of external variables:

```
In file1:
extern int xy;
extern long arr[];
main ( )
{
        . . .
}
void foo (int abc) { . . . }
long soo (void) { . . . }
In file2:
int xy;
long arr[100];
```

Free C compiler is available for PC platform if the user installs Linux or Free BSD Unix as the machine's operating system. By making the PC dual-bootable to Linux or Free BSD Unix, one can use the Gnu C compiler to practice C language programming.

5.9 Introduction to Project Build Process of the Microchip PIC18 Compiler

The process for building a project that consists of multiple files using the Microchip MPLAB® IDE is shown in Figure 5.2. As shown in the figure, a project may consist of a mix of assembly and C language program files. There are four major steps in the process:

1. *Source code entering and editing.* In this step, the user will use the text editor provided by the MPLAB IDE to enter the source code and save the source code in one or multiple files.

2. *Object code generation.* In this step, MPLAB IDE invokes the cross assembler and/or cross compiler to assemble and/or compile source code into object code. In Figure 5.2, **mpasm.exe** is a cross assembler, whereas **mcc18.exe** is a cross compiler. As shown in the figure, the cross compiler invokes a C language preprocessor (**cpp18.exe**) to make constants and macros substitutions and directives processing.

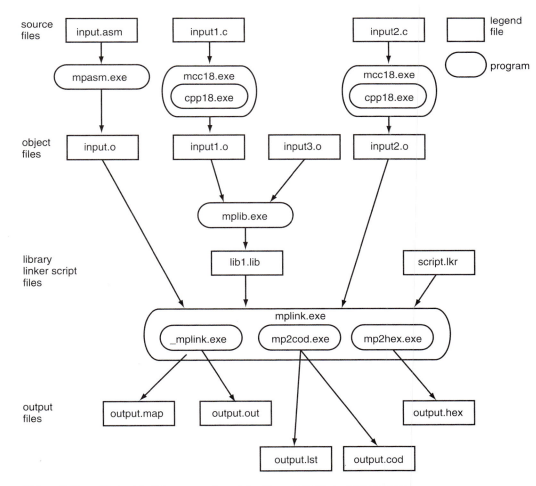

Figure 5.2 ■ Project build process (reprint with permission of Microchip)

3. *Library files creation and maintenance.* This step is optional. The user may optionally put related reusable function modules in a single library file, which will make the reuse of code easier. The linking process will also become easier. The program **mplib.exe** is responsible for the creation and maintenance of library functions.

4. *Program linking and executable code generation.* The linker (**mplink.exe**) is invoked in this step. When a project consists of multiple files, the linker is indispensable. A function in a file may do the following:

   - Reference variables contained in other files
   - Call functions contained in other files
   - Call library functions

   These issues are resolved by the linker. As shown in Figure 5.2, several files are created in this step. The file **output.hex** represents the executable code in hex format so that it can be programmed into the target device. The file **output.cod** provides information needed in debugging process. The file **output.lst** contains source code side by side with final binary code and line numbers so that users may examine the code generated. The file **output.out** is an intermediate file used by the linker to generate cod file, hex file, and listing files. The file **output.map** shows the memory layout after linking. It indicates used and unused memory regions. The programs that generate these files are also indicated in Figure 5.2.

5.10 The MPLINK Linker

The MPLINK® linker (hereafter the linker) is used with the Microchip MPASM® relocatable assembler and the Microchip MPLAB C18 C compilers to generate executable (hex) code. The MPLINK linker performs many functions:

- *Locating code and data.* The MPLINK linker accepts relocatable source files as input. Using the linker script, it decides where the code will be placed in the program memory and where variables will be placed in data RAM.

- *Resolving addresses.* The cross assembler and cross compiler generate relocation entries in the object file with respect to external references in a source file. After the linker locates code and data, it uses this relocation information to update all external references with the actual addresses.

- *Generating an executable.* The linker produces an executable file (with **.exe** as the suffix) that can be programmed into a PIC18 device or loaded into an emulator or simulator to be executed.

- *Configuring stack size and location.* The linker allows MPLAB **mcc18** to set aside RAM space for dynamic stack usage.

- *Identifying address conflicts.* The linker checks to ensure that program data do not get assigned to space that has already been assigned or reserved.

- *Providing symbolic debug information.* The linker outputs a file that the MPLAB IDE uses to track address labels, variable locations, and line/file information for source-level debugging.

A linker file (with a suffix .lkr to the file name) is usually needed to provide instruction to the linker regarding how program and data memory can be allocated. A linker file contains scripts that specify the following:

- Program and data memory for the target part
- Stack size and location (for MPLAB MCC18)
- Logical sections used in source code to place code and data

Linker script directives form the command language that controls the linker's behavior. There are four basic categories of linker script directives:

1. *Command line directives.* There are four directives in this category:

   LIBPATH: Library and object files that do not have a path are searched for using the library/object search path. Additional search paths can be specified using this directive. The following directive appends additional search directories to the library/object search path:

   LIBPATH 'libpath'

   where, **libpath** is a semicolon-delimited list of directories. A period without other modifiers represents the current project directory.

   LKRPATH: Linker command files that do not have a path are searched for using the linker command file search path. The following directive appends additional search directories to the linker command file search path:

   LKRPATH 'lkrpath'

   where **lkrpath** is a semicolon-delimited list of directories.
   FILES: This directive specifies additional files to be linked.
   INCLUDE: This directive specifies additional linker command files to be included.

2. *Memory region definition directives.* These directives define ROM and RAM regions.

   A ROM directive starts with the keyword **CODEPAGE.** The **CODEPAGE** directive is used for program code, initialized data values, constant data values, and external memory for PIC18 devices. It has the following format:

   CODEPAGE NAME=memName START=addr END=addr [PROTECTED]
   [FILL=fillvalue]

   where **memName** is any ASCII string used to identify a **CODEPAGE, addr** is a decimal or hexadecimal number specifying an address, and **fillValue** is a value that fills any unused portion of a memory block. If this value is in decimal notation, it is assumed to be a 16-bit quantity. If it is in hexadecimal notation (e.g., 0x2346), it may be any length divisible by full words (16 bits). The optional keyword **PROTECTED** indicates a region of memory that can be used only by program code that specifically requests it.

   A RAM directive can be used to define a databank, a shared bank, or an access bank. The formats for these directives are as follows:

   Banked registers:

   DATABANK NAME=memName START=addr END=addr [PROTECTED]

   Unbanked registers:

   SHAREBANK NAME=memName START=addr END=addr [PROTECTED]

   Access registers:

   ACCESSBANK NAME=memName START=addr END=addr [PROTECTED]

   where **memName** is any ASCII string used to identify an area in RAM and **addr** is a decimal or hexadecimal number specifying an address. The optional keyword **PROTECTED** indicates a region of memory that can be assigned only to variables that are specifically identified in the source code.

3. *Logical section definition directives.* Logical sections are used to specify which of the defined memory regions should be used for a portion of source code. To use logical sections, define the section in the linker script file with the **SECTION** directive and then reference that name in the source file using the built-in mechanism of that language (e.g., **#pragma** section for MPLAB C18).

The section directive defines a section using the following syntax:

SECTION NAME='secName' {ROM='memName' │ RAM='memName'}

where **secName** is an ASCII string used to identify a **SECTION** and **memName** is a previously defined **ACCESSBANK, SHAREBANK, DATABANK,** or **CODEPAGE.**

The **ROM** attribute must always refer to program memory previously defined using a **CODEPAGE** directive. The **RAM** attribute must always refer to data memory previously defined with an **ACCESSBANK, DATABANK,** or **SHAREBANK** directive.

4. *Stack definition directive.* The MPLAB C18 requires a software stack to be set up. This command specifies the size and the data bank used to implement the software stack. The syntax of this command is as follows:

STACK SIZE='allocSize' [RAM='memName']

where **allocSize** is the size in bytes of the stack and **memName** is the name of a memory bank previously declared using a **ACCESSBANK, DATABANK,** or **SHAREBANK** statement.

Examples of linker scripts are shown in Figures 5.3a and 5.3b. The MPLAB C18 compiler provides two sample linker files for each PIC18 microcontroller: one for use with source-level debugger and the other for use without debugger. For example, the linker file **18f8720.lkr** is to be used when debugger is not involved. The linker file **18f8720i.lkr** is to be used for source-level debugging. These two files are shown in Figures 5.3a and 5.3b.

Figure 5.3a shows the following:

- The linker command "LIBPATH." adds the project (current) directory as one of the search paths for the library functions.
- The linker command "FILES c018i.o" adds the startup code **c018i** into the project.
- The linker command "FILES clib.lib" adds the general and math library functions into the project.
- The linker command "FILES p18f8720.lib" adds peripheral library functions into the project.
- The linker command "CODEPAGE NAME=vectors . . ." informs the linker that the section named **vectors** is assigned to the address range from 0x00 to 0x29. No other section can be assigned to this address range.
- The linker command "CODEPAGE NAME=page . . ." informs the linker that the section named **page** is assigned to the address range from 0x2A to 0x1FFFFF. This is the page where your executable code should be placed or programmed into.
- The linker command "CODEPAGE NAME=idlocs . . ." informs the linker that the section named **idlocs** is assigned to the address range from 0x200000 to 0x200007. No other section can be assigned to this address range. The user can store checksum or other code identification numbers in this area.

```
/ / Sample linker command file for 18F8720
/ / $Id: 18f8720.lkr,v 1.4 2002/08/22 20:53:50 sealep Exp $
LIBPATH  .
FILES c018i.o
FILES clib.lib
FILES p18f8720.lib
```

CODEPAGE	NAME=vectors	START=0x0	END=0x29	PROTECTED
CODEPAGE	NAME=page	START=0x2A	END=0x1FFFFF	
CODEPAGE	NAME=idlocs	START=0x200000	END=0x200007	PROTECTED
CODEPAGE	NAME=config	START=0x300000	END=0x30000D	PROTECTED
CODEPAGE	NAME=devid	START=0x3FFFFE	END=0x3FFFFF	PROTECTED
CODEPAGE	NAME=eedata	START=0xF00000	END=0xF000FF	PROTECTED
ACCESSBANK	NAME=accessram	START=0x0	END=0x5F	
DATABANK	NAME=gpr0	START=0x60	END=0xFF	
DATABANK	NAME=gpr1	START=0x100	END=0x1FF	
DATABANK	NAME=gpr2	START=0x200	END=0x2FF	
DATABANK	NAME=gpr3	START=0x300	END=0x3FF	
DATABANK	NAME=gpr4	START=0x400	END=0x4FF	
DATABANK	NAME=gpr5	START=0x500	END=0x5FF	
DATABANK	NAME=gpr6	START=0x600	END=0x6FF	
DATABANK	NAME=gpr7	START=0x700	END=0x7FF	
DATABANK	NAME=gpr8	START=0x800	END=0x8FF	
DATABANK	NAME=gpr9	START=0x900	END=0x9FF	
DATABANK	NAME=gpr10	START=0xA00	END=0xAFF	
DATABANK	NAME=gpr11	START=0xB00	END=0xBFF	
DATABANK	NAME=gpr12	START=0xC00	END=0xCFF	
DATABANK	NAME=gpr13	START=0xD00	END=0xDFF	
DATABANK	NAME=gpr14	START=0xE00	END=0xEFF	
ACCESSBANK	NAME=accesssfr	START=0xF60	END=0xFFF	PROTECTED
SECTION	NAME=CONFIG	ROM=config		
STACK SIZE=0x100	RAM=gpr14			

Figure 5.3a ■ P18F8720 linker file without debugging support (reprint with permission of Microchip)

- The linker command "CODEPAGE NAME=config . . ." informs the linker that the section named **config** is assigned to the address range from 0x300000 to 0x30000D. The configuration information for the MCU is stored in this section. No other section can be assigned to this address range (i.e., **PROTECTED**).
- The linker command "CODEPAGE NAME=devid . . ." informs the linker that the section named **devid** is assigned to the address range from 0x3FFFFE to 0x3FFFFF.

```
// Sample linker command file for 18F8720i for MPLAB ICD2
// $Id: 18f8720i.lkr,v 1.3 2002/11/07 23:23:51 sealep Exp $
LIBPATH  .
FILES c018i.o
FILES clib.lib
FILES p18f8720.lib
```

CODEPAGE	NAME=vectors	START=0x0	END=0x29	PROTECTED
CODEPAGE	NAME=page	START=0x2A	END=0x1FDBF	
CODEPAGE	NAME=debug	START=0x1FDC0	END=0x1FFFF	PROTECTED
CODEPAGE	NAME=idlocs	START=0x200000	END=0x200007	PROTECTED
CODEPAGE	NAME=config	START=0x300000	END=0x30000D	PROTECTED
CODEPAGE	NAME=devid	START=0x3FFFFE	END=0x3FFFFF	PROTECTED
CODEPAGE	NAME=eedata	START=0xF00000	END=0xF000FF	PROTECTED
ACCESSBANK	NAME=accessram	START=0x0	END=0x5F	
DATABANK	NAME=gpr0	START=0x60	END=0xFF	
DATABANK	NAME=gpr1	START=0x100	END=0x1FF	
DATABANK	NAME=gpr2	START=0x200	END=0x2FF	
DATABANK	NAME=gpr3	START=0x300	END=0x3FF	
DATABANK	NAME=gpr4	START=0x400	END=0x4FF	
DATABANK	NAME=gpr5	START=0x500	END=0x5FF	
DATABANK	NAME=gpr6	START=0x600	END=0x6FF	
DATABANK	NAME=gpr7	START=0x700	END=0x7FF	
DATABANK	NAME=gpr8	START=0x800	END=0x8FF	
DATABANK	NAME=gpr9	START=0x900	END=0x9FF	
DATABANK	NAME=gpr10	START=0xA00	END=0xAFF	
DATABANK	NAME=gpr11	START=0xB00	END=0xBFF	
DATABANK	NAME=gpr12	START=0xC00	END=0xCFF	
DATABANK	NAME=gpr13	START=0xD00	END=0xDFF	
DATABANK	NAME=gpr14	START=0xE00	END=0xEF3	
DATABANK	NAME=dbgspr	START=0xEF4	END=0xEFF	PROTECTED
ACCESSBANK	NAME=accesssfr	START=0xF60	END=0xFFF	PROTECTED
SECTION	NAME=CONFIG	ROM=config		
STACK SIZE=0x100	RAM=gpr13			

Figure 5.3b ■ P18F8720 linker file with debugging support (reprint with permission of Microchip)

No other section can be assigned to this address arrange. Information that identifies the MCU part number is stored in these two bytes.

- The linker command "CODEPAGE NAME=eedata . . ." informs the linker that the section named **eedata** is assigned to the address range from 0xF00000 to 0xF000FF. No other section can be assigned to this address range.
- The linker command "ACCESSBANK NAME=accessram . . ." informs the linker that the section named **accessram** (RAM access bank) is assigned to the address range 0x00 to 0x5F.
- The following 15 linker commands define 15 data banks that are assigned with the names **gpr0 . . . gpr14.**
- The linker command "ACCESSBANK NAME=accesssfr . . ." defines the access bank that covers the part of special function registers.
- The linker command "SECTION NAME=CONFIG ROM=config" associates the section named **CONFIG** with the code page **config** in program memory.
- The last linker command "STACK SIZE=0x100 RAM=gpr14" informs the linker that the databank **gpr14** is used as the stack and its size is 256 bytes.

The linker file in Figure 5.3b is quite similar except for a few things:

- The end address of the CODE page is changed from 0x1FFFFF to 0x1FDBF. This implies that the ICD2 currently supports only on-chip flash program memory debugging.
- The last DATABANK gpr14 is reduced to 244 bytes instead of 256 bytes.
- The data RAM from 0xEF4 to 0xEFF are reserved for the debugger and is named **dbgspr.**
- The data bank **gpr13** rather than **gpr14** is used as the stack.

Figure 5.3 shows two possible linker files recommended by Microchip. Users should make modifications to these files to meet the requirements of their applications at hand.

5.11 A Tutorial on Using the MCC18 Compiler

Compared to the project with a single assembly file, a few extra steps are needed to build a project that contains C programs. The following C program will be used to illustrate the project build process:

```
int counter;
void main (void)
{
        counter = 0;
        TRISB = 0;                  /* configure PORTB for output */
        while (counter <= 15)
        {
                PORTB = counter;    /* display the value of 'counter' on the LEDs */
                counter++;
        }
}
```

5.11.1 Creating a New Project

Before creating a new project, the user must configure the project and select the target device. After creating a new project, the user needs to select the tool suite and set the language tool location.

To configure the project, go to the *Configure> Settings* menu and choose the Projects tab. Make sure that the setup is as shown in Figure 3.8. Click on OK after making sure the setting is right. You also need to select the proper target device. The device selection dialog is shown in Figure 3.9.

After the right target device has been selected, the user is ready to create a new project. When creating a new project, one needs to decide where to place the project. This tutorial will use the directory c:\pic18\ch05 to place the project. Follow these steps to create a new project.

1. Select the *Project> New* menu. The New dialog will open as shown in Figure 5.4.

2. Enter a name for your new project (e.g., **ctutor1**).

3. Use the Browse button to select the path you chose to place the new project or type in the path.

When the project name and location are correctly specified, click on OK.

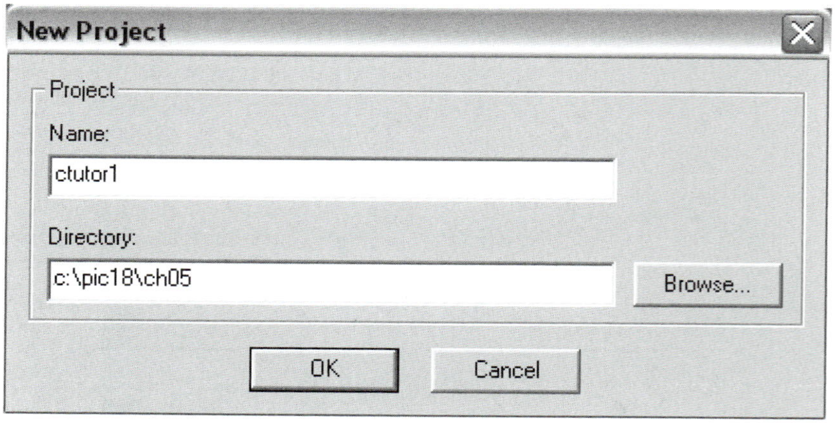

Figure 5.4 ■ Creating a new project (reprint with permission of Microchip)

After creating a new project, one needs to make sure the right language tool suite is selected. To do this, select *Project> Select Language Toolsuite.* Select Microchip C18 Toolsuite as the active toolsuite. The resultant screen is shown in Figure 5.5. Click on Set Tool Locations on the lower left to make sure that the tool suite location is correct.

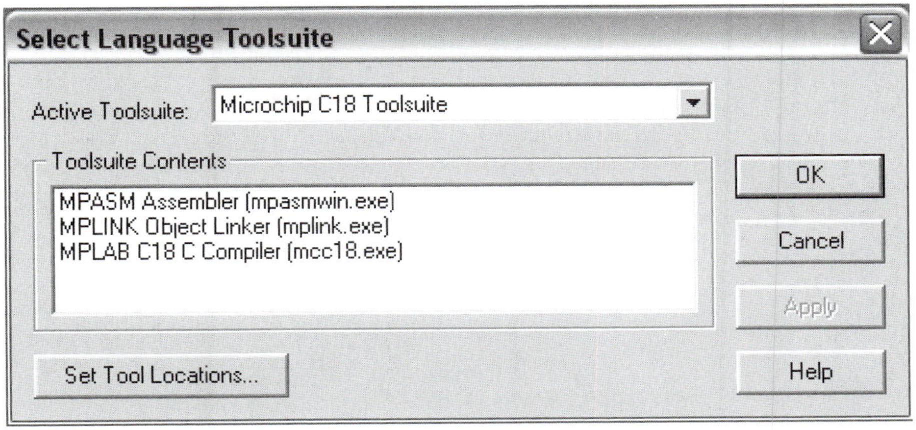

Figure 5.5 ■ Screen for selecting C language toolsuite (reprint with permission of Microchip)

The full path to the **mcc18** C compiler executable should appear in the Location of Selected Tool text box as shown in Figure 5.6. If it is incorrect or empty, click on Browse to locate **mcc18.exe.** After the language tool location has been set up, one can skip this step until switching language tools.

After the language tool suite is set up properly, one can enter the C program by following the same process as shown in Chapter 3, Section 3.5.3.

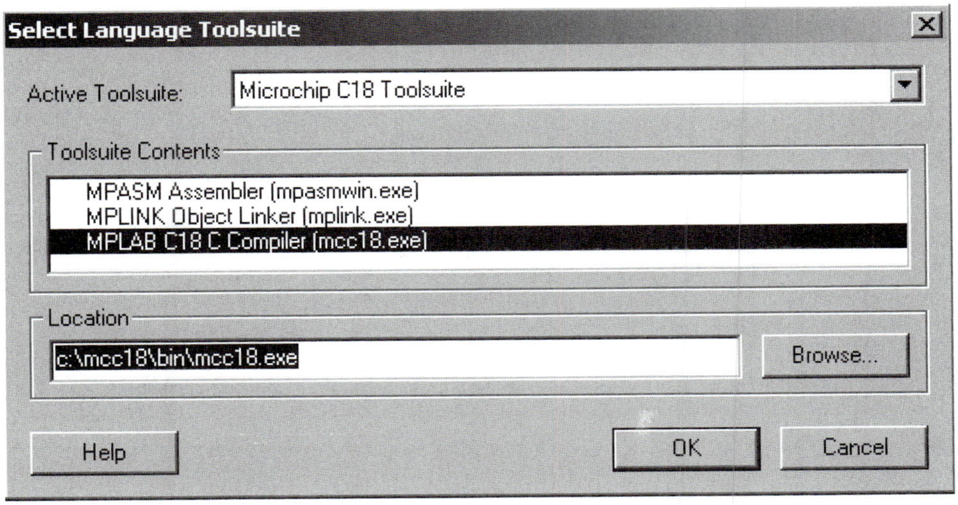

Figure 5.6 ■ Setting the location of the MPLAB C18 compiler (mcc18.exe) (reprint with permission of Microchip)

5.11.2 Setting Build Options

Before building the project, one needs to set the build options. This is done by selecting *Project>Build Options.* This will bring up the Build Options dialog. Click on General, and the general options will be brought up. The last three options similar to those shown in Figure 5.7 need to be filled in. The default settings for MPASM, MPLAB C18, and MPLINK are acceptable and need not be changed.

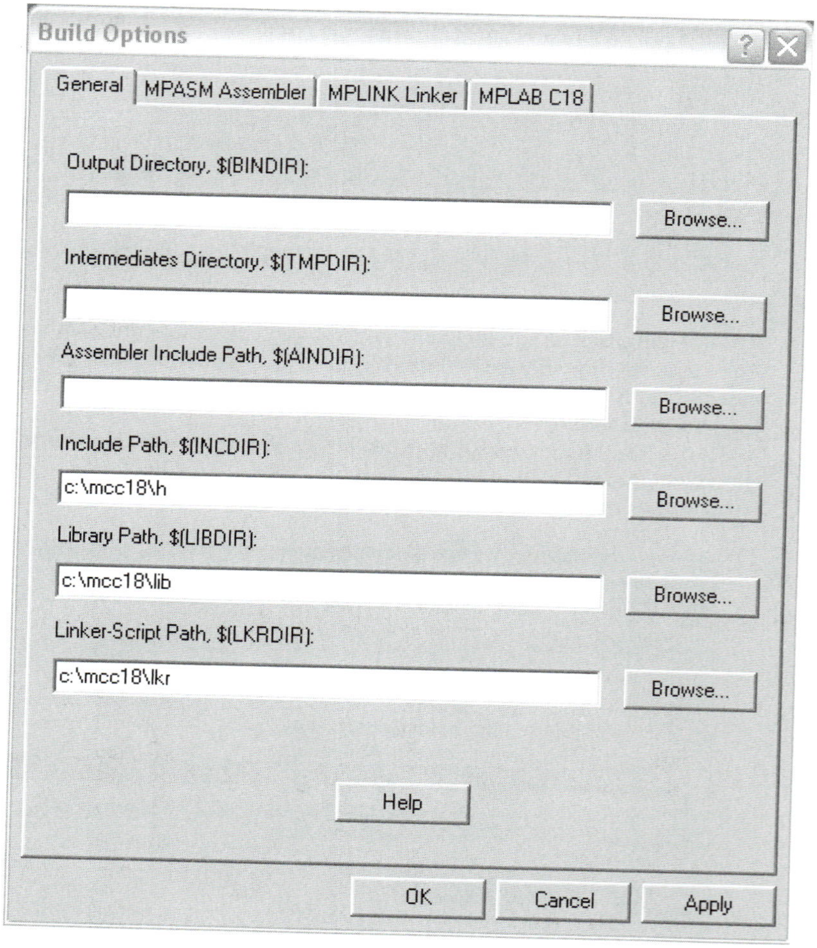

Figure 5.7 ■ Build options dialog (reprint with permission of Microchip)

5.11.3 Add Source Files to Project

To add a source file to the project, select *Project>Add Files to Project.* A dialog window as shown in Figure 5.8 will appear on the screen. Select the file that you just created and saved (**example1.c**) and click on Open. After this, the newly inserted file should appear in the project pane under *Source Files.*

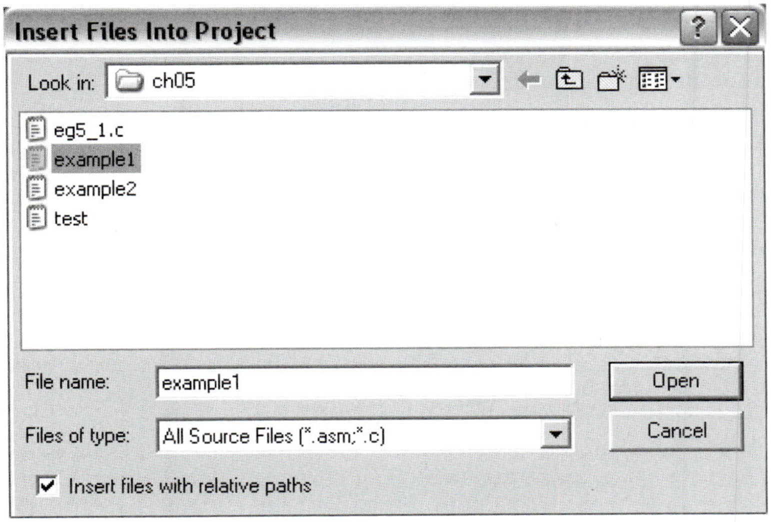

Figure 5.8 ■ Insert C source code into project (reprint with permission of Microchip)

One also needs to add the linker file into this project. Use the same method and browse the MCC18 C compiler installation directory (under **c:\mcc18\lkr**) and add the appropriate linker file (select **p18f452i.lkr** for this tutorial) into this project. The dialog is shown in Figure 5.9. Select different linker file if the demo board contains different MCU.

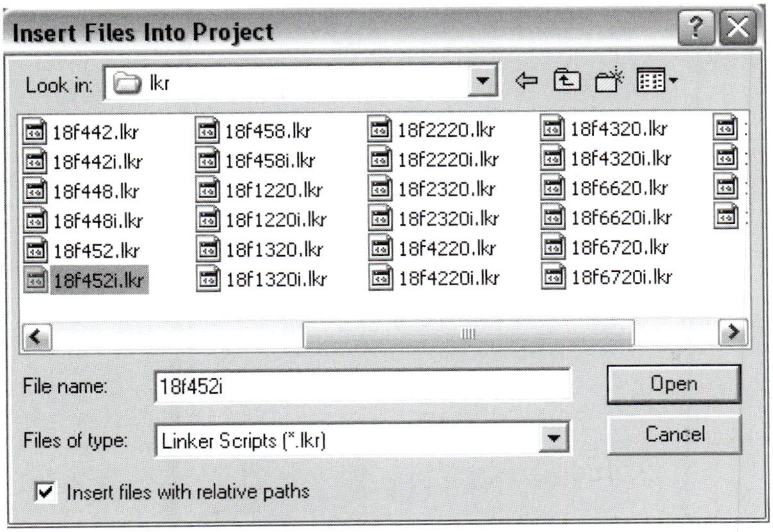

Figure 5.9 ■ Adding linker script file into the project (reprint with permission of Microchip)

5.11.4 Building the Project

To build the project, select *Project>Build All* or *Project>Make* (or press function key F10). This file should be compiled successfully.

5.11.5 Choosing Debug Tool

The C program in this tutorial increments a counter and outputs it to the LEDs driven by the lower four port B pins. It would be easier to see the result by running this program on the demo board and use ICD2 to debug the program. Select ICD2 as the debugging tool by selecting *Debugger>Select Tool>MPLAB ICD2*.

5.11.6 Build the Project

Press function key F10 to build the project. This step should run correctly. After the project is built successfully, program the hex file into the target device on the demo board.

5.11.7 Checking the ICD2 Settings

Before programming the target device, one needs to make sure the settings of ICD2 are correct. Select *Debugger>Settings* to bring up the dialog. Make sure the settings on Power, Program, and Communication are correct by following the procedures described in Chapter 3, Section 3.7.1.

One also needs to make sure the configuration bits are set up correctly. Bring up the dialog by selecting *Configure>Configuration Bits*. The settings should be as shown in Figure 5.10.

Configuration Bits

Address	Value	Category	Setting
300001	FF	Oscillator	RC-OSC2 as RA6
		Osc. Switch Enable	Disabled
300002	FF	Power Up Timer	Disabled
		Brown Out Detect	Enabled
		Brown Out Voltage	2.0V
300003	FE	Watchdog Timer	Disabled
		Watchdog Postscaler	1:128
300005	FF	CCP2 Mux	RC1
300006	7B	Low Voltage Program	Disabled
		Background Debug	Enabled
		Stack Overflow Reset	Enabled
300008	FF	Code Protect 00200-01FFF	Disabled
		Code Protect 02000-03FFF	Disabled
		Code Protect 04000-05FFF	Disabled
		Code Protect 06000-07FFF	Disabled
300009	FF	Data EE Read Protect	Disabled
		Code Protect Boot	Disabled
30000A	FF	Table Write Protect 00200-01FFF	Disabled
		Table Write Protect 02000-03FFF	Disabled
		Table Write Protect 04000-05FFF	Disabled
		Table Write Protect 06000-07FFF	Disabled
30000B	FF	Data EE Write Protect	Disabled
		Table Write Protect Boot	Disabled
		Config. Write Protect	Disabled
30000C	FF	Table Read Protect 00200-01FFF	Disabled
		Table Read Protect 02000-03FFF	Disabled
		Table Read Protect 04000-05FFF	Disabled
		Table Read Protect 06000-07FFF	Disabled
30000D	FF	Table Read Protect Boot	Disabled

Figure 5.10 ■ Configuration bits settings for MPLAB ICD2 (reprint with permission of Microchip)

5.11.8 Program the Target Device

Program the target device by selecting *Debugger>Program,* and the status bar of the MPLAB IDE window will change to that as shown in Figure 5.11.

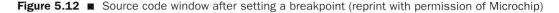

| Programming Target... | MPLAB ICD 2 | PIC18F452 | pc:0 | n ov z dc c | 0x4536 | Ln 12, Col 1 | INS | WR |

Figure 5.11 ■ Status bar of the MPLAB IDE window when target device programming is proceeded (reprint with permission of Microchip)

5.11.9 Running the Program

Before running the program, set a breakpoint at the following statement:

```
PORTB = counter; /* display value of 'counter' on the LEDs */
```

Do this by clicking the left mouse button on this statement, pressing the right button, and selecting the first item (Set Breakpoint) on the pop-up window. The source code window will change to that as shown in Figure 5.12.

```
C:\pic18\ch05\example1.c

int counter;
void main (void)
{
    counter = 0;
    TRISB = 0;                /* configure PORTB for output */
    while (counter <= 15)
    {
        PORTB = counter;    /* display value of 'counter' on the LEDs */
        counter++;
    }
}
```

Figure 5.12 ■ Source code window after setting a breakpoint (reprint with permission of Microchip)

One can also (optionally) set a **Watch window** on the MPLAB IDE window. The only variable to watch for in this tutorial is **counter.** Follow the procedure described in Chapter 3, Section 3.6.4, to set a Watch window.

Before running the program, reset the target device by pressing the function key F6. Run the program by pressing the function key F9. The program should stop at the breakpoint, the value of the variable **counter** in the Watch window should be 0, and no LED would be lighted. Press the function key F9 to rerun the program. The program should again stop at the same statement, the value of counter should be incremented by 1, and one LED would be lighted. Repeat this process for a few more times.

5.12 C Language Implementation in the MCC18 Compiler

The MCC18 compiler is not completely ANSI C compliant. It deviates from the standard wherever the standard conflicts with efficient support for the PIC18 devices. Certain extensions are added to better support the PIC18 devices. These specifics are detailed in the following sections.

5.12.1 Data Types and Limits

The MPLAB C18 compiler supports the standard integer types. The ranges of the standard integer types are listed in Table 5.4. In addition to the traditional integer types, MPLAB C18 also supports a 24-bit integer type **short long int** (or **long short int**) in both signed and unsigned varieties.

type	size	minimum	maximum
char[1,2]	8 bits	-128	127
signed char	8 bits	-128	127
unsigned char	8 bits	0	255
int	16 bits	-32768	32767
unsigned int	16 bits	0	65535
short	16 bits	32768	32767
unsigned short	16 bits	0	65535
short long	24 bits	-8,388,608	8,388,607
unsigned short long	24 bits	0	16,777,215
long	32 bits	-2,147,483,648	2,147,483,647
unsigned long	32 bits	0	4,294,967,295

Note 1. A plain **char** is signed by default.
      2. A plain **char** may be unsigned by default via the –k command line option.

Table 5.4 ■ Integer data type sizes and limits

The MPLAB C18 compiler also supports floating-point data types. Floating-point numbers are declared as **double** or **float** data types. The range of the magnitudes of the floating-point numbers is

$$2^{-126} \sim 2^{-128} \times (2 - 2^{-15})$$

or

$$1.17549435 \times 10^{-38} \sim 6.80564693 \times 10^{38}$$

The MPLAB C18 compiler uses a floating-point representation different from the IEEE 754 format. The IEEE 754 format can be converted to MPLAB C18 format by rotating to the left one place, whereas a right rotation will convert from the MPLAB C18 format to the IEEE 754 format. Shown in Table 5.5 is the comparison of these two formats.

MPLAB C18 stores data in little-endian format. Bytes at lower addresses have lower significance.

Standard	Exponent byte	Byte 0	Byte1	Byte2
IEEE 754	$sd_0d_1d_2d_3d_4d_5d_6$	$d_7ddd\ dddd$	dddd dddd	dddd dddd
MPLAB C18	$d_0d_1d_2d_3d_4d_5d_6d_7$	sddd dddd	dddd dddd	dddd dddd

Table 5.5 ■ MPLAB C18 floating pointformatvs. IEEE754

5.12.2 Storage Classes

MPLAB C18 supports the standard storage classes including **auto, extern, register, static,** and **typedef.**

Function parameters can have storage class **auto** or **static.** An **auto** parameter is pushed onto the *software stack* (defined in Chapter 4). A static parameter is allocated globally, enabling direct access for generally smaller code. The default storage class for function parameters is **auto.** This can be changed by configuring the build options of C18 compiler. The user can bring up the Build Options dialog by selecting *Project>Build Options>Project*. When the Build Options dialog appears, click on the MPLAB C18 button and then click on the default storage class selection window, which will bring up three options for you to choose from: Auto, Static, and Overlay. The dialog is shown in Figure 5.13. The storage class **overlay** is explained shortly.

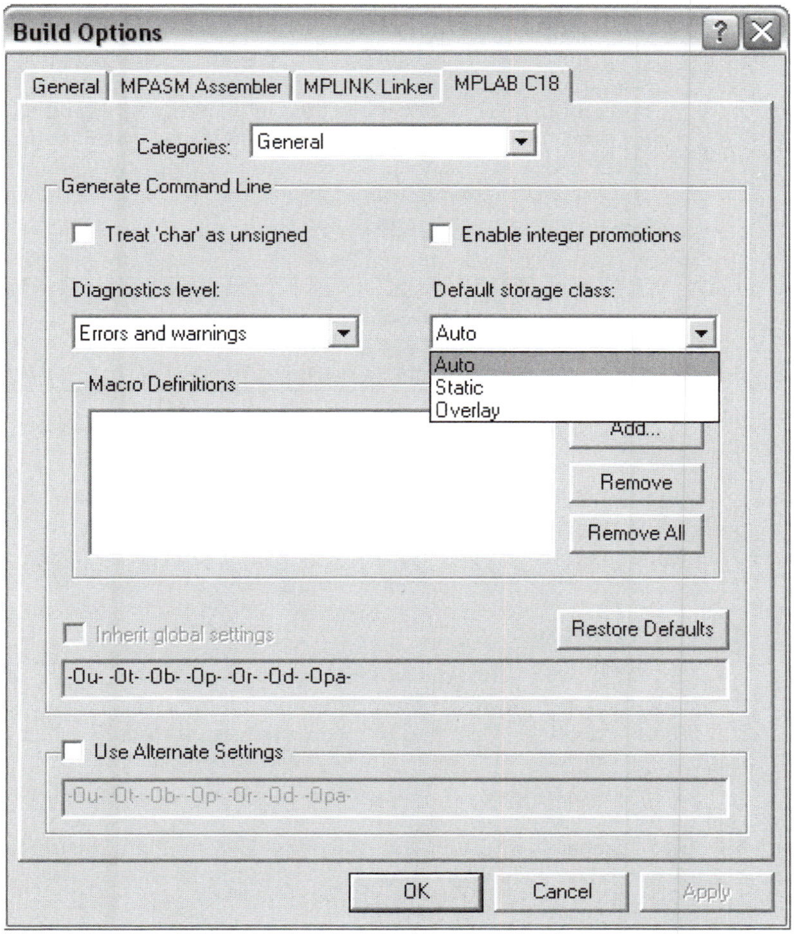

Figure 5.13 ■ Select the default storage class for MPLAB C18 compiler (reprint with permission of Microchip)

5.12.3 Storage Qualifiers

In addition to the standard storage qualifiers (**const, volatile**), the MPLAB C18 compiler introduces storage qualifiers of **far, near, rom,** and **ram.** Table 5.6 shows the location of an object based on the storage qualifiers specified when it was defined. The default storage qualifiers for an object defined without explicit storage qualifiers are **far ram.**

Near/far data memory objects. The qualifier **far** is used to denote that a variable that is located in data memory resides in a memory bank such that a bank switching instruction is required prior to accessing this variable. The qualifier **near** is used to denote that a variable is located in **access bank.**

Near/Far program memory objects. The qualifier **far** is used to denote that a variable that is located in program memory can be found anywhere in program memory or, if a pointer, that it can access up to and beyond 64 KB of program memory space. The qualifier **near** is used to denote that a variable located in the program memory is found at an address less than 64 KB or, if a pointer, that it can access only up to 64 KB of program memory space.

Rom/ram qualifiers. Because the program memory and data memory of the PIC18 are separate, MPLAB C18 requires extensions to distinguish between data located in program memory and data located in data memory. The ANSI/ISO C standard allows for code and data to be in separate address spaces, but this is not sufficient to locate data in the code space as well. To this purpose, MPLAB C18 introduces the **rom** and **ram** qualifiers. The qualifier **rom** denotes that the object is located in the program memory, whereas the qualifier **ram** denotes that the object is located in the data memory.

	rom	ram
far	Anywhere in program memory	Anywhere in data memory (default)
near	In program memory with address less than 64K	In access memory

Table 5.6 ■ Location of object based on storage qualifiers

Pointers can point to either the data memory (**ram** pointer) or the program memory (**rom** pointer). Pointers are assumed to be **ram** pointers unless declared as **rom.** The size of a pointer is dependent on the type of the pointer. Data memory and **near** program pointers are 16 bits, whereas **far** program memory pointers are 24 bits.

5.12.4 Overlay

The MPLAB C18 introduces a storage class of **overlay,** which allows more than one variable to occupy the same physical memory location or more than one section to occupy the same physical memory section. The overlay mechanism is often used when physical memory resource is limited.

The overlay storage class may be applied to local variables (but not formal parameters, function definitions, or global variables). The overlay storage class will allocate the associated symbols into a function-specific, static overlay section. Such a variable will be allocated statically but initialized on each function entry. For example, in

```
void func_x(void)
{
        overlay int x1 = 10;
        . . .
}
```

The variable **x1** will be initialized to 10 on each function entry, although its storage will be statically allocated. If no initializer is present, then its value on function entry is undefined. The MPLINK linker will attempt to overlay local variables specified as **overlay** from functions that are guaranteed not to be active simultaneously.

5.13 ANSI/ISO Divergences

There are some divergences between the MPLAB C18's implementation of C language and the ANSI/ISO standard. These divergences are described in this section.

5.13.1 Integer Promotions

ISO mandates that all arithmetic be performed at **int** precision or greater. By default, MPLAB C18 will perform arithmetic at the size of the largest operand, even if both operands are smaller than an integer. The ISO-mandated behavior can be instated via the **–Oi** command line option or instated in the MPLAB IDE Build Options as shown in Figure 5.14. This is done by clicking on the box to the left of Use Alternate Settings and entering the **–Oi** option below it.

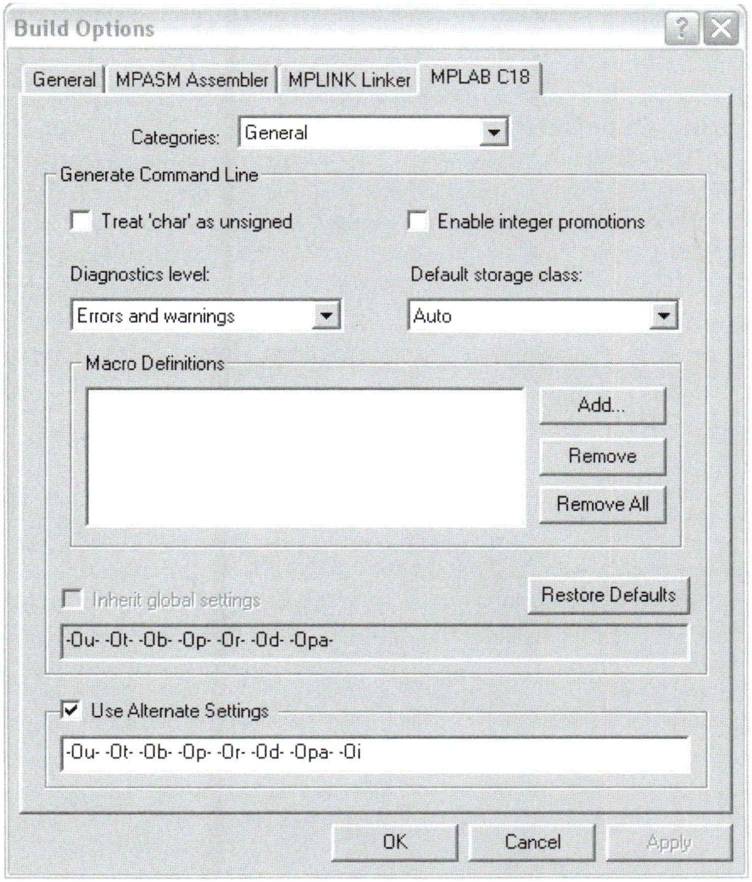

Figure 5.14 ■ Reinstate ISO integer promotion rule dialog (reprint with permission of Microchip)

For example, in the following code sequence

```
unsigned char j, k;
unsigned i;
j = 0x79;
k = 0x87;
i = j + k;
```

the ISO standard requires that 0x100 be assigned to **i,** but MPLAB C18 will assign 0 to **i.** This divergence also applies to constant literals. The chosen type for constant literals is the first one from the appropriate group that can represent the value of the constant without overflow. For example,

```
#define   X     0x20       /* X will be considered a char unless –Oi is specified */
#define   Y     0x5        /* Y will be considered a char unless –Oi is specified */
#define   Z     (X) * (Y)
unsigned i;
i = Z;                     /* ISO requires i == 0x100, but in C18 i == 0 */
```

5.13.2 Numeric Constants

MPLAB C18 supports the standard prefixes for specifying hex (0x) and octal (0) values and adds support for specifying binary values using the **0b** prefix. For example, the value 127 may be denoted as 0b01111111.

5.13.3 String Constants

Strings are often stored in program memory. MPLAB C18 automatically places all string constants in program memory. The section that contains all constant strings is called **.stringtable.** Because data memory and program memory are in different memory spaces, the C18 has four different variants for each standard function that deals with strings.

When using the MPLAB C18 compiler, a string table in program memory can be declared using one of the statements in the following:

```
rom const table1 [][20] = {"string 1", "string 2", . . ., "string 10"};
rom const char *rom table2[] = {"string 1", "string 2", . . ., "string 10"};
```

The declaration of **table1** declares an array of 10 strings that are each 20 characters long and takes 100 words of program memory. **table2** is declared as an array of pointers to program memory. The **rom** qualifier after the * character places the array of pointers in program memory as well. All the strings in **table2** are nine bytes long, and the array has 10 elements. Therefore, **table2** takes (10*9 + 10*2)/2 = 55 words of program memory. Access to **table2** will be less efficient than accesses to **table1** because of the additional level of indirection required by the pointer.

An important consequence of the separate address spaces for the MPLAB C18 is that pointers to data in program memory and pointers to data in data memory are not compatible. Two pointer types are not compatible unless they point to objects of compatible types and the objects they point to are located in the same address space.

A function to copy a string from program to data memory could be written as follows:

```
void str2ram (static char *dest, static char rom *src)
{
       while ((*dest++ = *src++) != '\0');
}
```

5.13.4 Anonymous Structures

The MPLAB C18 compiler supports anonymous structures inside unions. An anonymous structure has the form

```
struct { member-list };
```

An anonymous structure defines an unnamed object. The names of the members of an anonymous structure must be distinct from other names in the scope in which the structure is declared. The members are used directly in that scope without the usual access syntax. For example,

```
union spic {
    struct {
        int a1;
        int b1;
    };
    char cx;
} tpic;
. . .
tpic.a1 = tpic.cx; /* 'a1' is a member of the anonymous structure 'tpic' */
```

A structure for which objects or pointers are declared is not an anonymous structure. For example,

```
union spic {
    struct {
        int b1;
        int b2;
    }fx, *ptr;
    char cx;
} tpic;
. . .
tpic.b1 = tpic.cx; /* error */
tpic.ptr -> b1 = tpic.cx;/* OK */
```

The assignment to **tpic.b1** is illegal because the member name is not associated with any particular object.

5.14 Inline Assembly

The MPLAB C18 compiler provides an internal assembler using a syntax similar to that of the MPASM assembler. The block of inline assembly code must begin with **_asm** and end with **_endasm.** The syntax within the block is

```
[label:] [<instruction> [arg1 [, arg2[, arg3]]]]
```

The internal assembler differs from the stand-alone assembler as follows:

- No directive support
- Comments must be C or C++ notation
- Full text mnemonics must be used for table reads/writes, that is,

TBLRD	(not TBLRD*)
TBLRDPOSTDEC	(not TBLRD*–)
TBLRDPOSTINC	(not TBLRD*+)
TBLRDPREINC	(not TBLRD+*)

TBLWT	(not TBLWT*)
TBLWTPOSTDEC	(not TBLWT*−)
TBLWTPOSTINC	(not TBLWT*+)
TBLWTPREINC	(not TBLWT+*)

- No defaults for instruction operands—all operands must be fully specified
- Default radix is decimal
- Label must include colon

For example,

```
-asm
      clrf       count,0
loop:
      movlw      0x20             // check loop count
      cpfseq     count,0          //
      goto       doit
      goto       done
doit:
      movf       sum_lo,0,0       //move sum_lo to WREG
      addwf      count,0,0        //add count to sum_lo
      movwf      sum_lo,0         //update sum_lo
      movlw      0                //add carry to high byte of sum
      addwf      sum_hi,1,0       // "
      incf       count,1,0
      goto       loop
done:
      nop
-endasm
```

It is generally recommended to limit the use of inline assembly to a minimum. Any functions containing inline assembly will not be optimized by the compiler. If one needs to write a large fragment of assembly code, it is recommended that it be done with the stand-alone assembler and link the module to the C modules using the MPLINK linker.

5.15 Bit Field Manipulation in C Language

One can use the bitwise logical operators to manipulate bit fields. For example, the following statement will set pin 7 and pin 0 of the PORTD to 1s:

```
PORTD |= 0x81;
```

The following statement will clear bits 7 to 4 of the TRISD register to 0:

```
TRISD &= 0x0F;
```

The following statement will toggle all PORTD pins:

```
PORTD ^= 0xFF;      //perform an exclusive-OR operation
```

We can summarize as follows:

- To clear a few bits of a register (or variable) to 0s, one needs to AND the register (or variable) with a mask that has 0s in those bit positions to be cleared to 0s.
- To set a few bits of a register (or variable) to 1s, one needs to OR the register (or variable) with a mask that has 1s in those bit positions to be set to 1s.

- To toggle a few bits of a register (or variable), one needs to XOR (exclusive-OR) the register with a mask that has 1s in those bit positions to be toggled.

MPLAB C18 provides an alternative for the user to access bit fields. This alternative is made possible through the processor-specific header files and processor-specific definition files.

5.15.1 Processor-Specific Header Files

The processor-specific header file is a C file that contains external declarations for the special-function registers, which are defined in the register definition file. For example, in the PIC18F8720 processor-specific header file, port B is declared as

```
extern volatile near unsigned char PORTB;
```

and as

```
extern volatile near union {
    struct {
        unsigned RB0:1;
        unsigned RB1:1;
        unsigned RB2:1;
        unsigned RB3:1;
        unsigned RB4:1;
        unsigned RB5:1;
        unsigned RB6:1;
        unsigned RB7:1;
    };
    struct {
        unsigned INT0:1;
        unsigned INT1:1;
        unsigned INT2:1;
        unsigned INT3:1;
        unsigned KB10:1;
        unsigned KB11:1;
        unsigned KB12:1;
        unsigned KB13:1;
    };
    struct {
        unsigned :3;
        unsigned CCP2:1;
        unsigned :1;
        unsigned PGM:1;
        unsigned PGC:1;
        unsigned PGD:1;
    };
} PORTBbits;
```

The first declaration specifies that PORTB is a byte (**unsigned char**). The **extern** modifier is needed since the variables are declared in the register definition file. The **volatile** modifier tells the compiler that it cannot assume that port B retains values assigned to it. The **near** modifier specifies that the port is located in access RAM.

The second declaration specifies that PORTBbits is a union of bit-addressable anonymous structures. Since individual bits in a special function register may have more than one function (and hence more than one name), there are multiple structure definitions inside the union, all referring to the same register. Respective bits in all structure definitions refer to the same bit in the register. Where a bit has only one function for its position, it is simply padded in other structures. For example, bits 7, 6, 5, and 3 on port B are padded in the third structure because they have only two names, whereas bits 4, 2, 1 and 0 have three names and are specified in each of the structures.

Any of the following statements can be written to use the PORTB special-function register:

```
PORTB           = 0x29;     /* assign the value 0x29 to the port */
PORTBbits.INT3  = 1;        /* sets the INT3 pin to high */
PORTBbits.CCP2  = 1;        /* sets the CCP2 pin to high, same pin as INT3 */
```

Many of the I/O interfacing applications require the generation of clock edge. A clock edge can be rising or falling. A falling edge on port B pin RB0 can be created as follows:

In assembly language,

```
bsf PORTB,RB0,A    ; pull high first
bcf PORTB,RB0,A    ; pull low later
```

In C language,

```
PORTBbits.RB0 = 1;  // pull high first
PORTBbits.RB0 = 0;  // pull low later
```

In summary, one can access a bit of a register in C language as follows:

1. Appending "bits" to the register name
2. Appending a period to the symbol resulted in Step 1
3. Specifying the bit name after the period

In addition to register declarations, the processor-specific header file defines inline assembly macros. These macros represent certain PIC18 instructions that an application may need to execute from C code. Although these instructions could be included as inline assembly instructions, they are provided as C macros as a convenience for the programmer. These assembly macros are listed in Table 5.7.

Instruction macro	Action
Nop ()	Executes a no operation (NOP)
ClrWdt ()	Clears the watchdog timer(CLRWDT)
Sleep ()	Executes a SLEEP instruction
Reset ()	Executes a RESET instruction
Rlcf (var, dest, access)	Rotate var to the left through the carry bit
Rlncf(var, dest, access)	Rotate var to the left without going through the carry bit
Rrcf (var, dest, access)	Rotate var to the right through the carry bit
Rrncf (var, dest, access)	Rotate var to the right without going through the carry bit
Swap (var, dest, access)	Swaps the upper and lower nibbles of var

Note 1. var must be an 8-bit quantity (i.e., char) and not located on the stack.
2. If dest is 0, the result is stored in WREG, and if dest is 1, the result is located in var. If access is 0, the access bank will be selected, overriding the BSR value. If access is 1, then the bank will be selected as per the BSR value.
3. Each of the macros affects MPLAB C18's ability to perform optimization on the functions using these macros.

Table 5.7 ■ PIC 18 instructions provided as C macros (reprint with permission of Microchip)

In order to use the processor-specific header file, choose the header file that pertains to the device being used (e.g., we use p18F452.h in the tutorial C program) in the application. The processor-specific header files are located in the c:\mcc18\h directory, where c:\mcc18 is the directory where the compiler is installed.

5.15.2 Processor-Specific Register Definition Files

The processor-specific register definition file is an assembly file that contains definitions for all the special-function registers on a given device. The processor-specific register definitions file, when compiled, will become an object file that will need to be linked with the application (e.g., p18F8720.asm compiles to p18F8720.o). This object file, together with others, is contained in **clib.lib.**

The source code for the processor-specific register definitions file is found in the **c:\mcc18\ src\proc** directory, and compiled object code is found in the **c:\mcc18\lib** directory.

For example, port A is defined in the PIC18F452 processor-specific register definitions file as

```
SFR_UNBANKED0 UDATA_ACS H'F80'
            PORTA
            PORTAbits   RES 1   ; 0xF80
```

The first line specifies the file register bank where port A is located and the starting address for that bank. Port A has two labels, **PORTAbits** and **PORTA,** both referring to the same location (in this case 0xF80).

5.16 The #pragma Statement

A compiler writer can add implementation-dependent options to the C language standard by using the #**pragma** statement. This statement is often used to declare data or code sections, declare section attributes, locate code or data, declare interrupt vectors and interrupt service routines, and many other implementation-dependent features. Here we discuss the use of #**pragma** statement in declaring sections and defer the discussion of the #**pragma** statement used in interrupts to Chapter 6.

5.16.1 #pragma Section Type

The MPLAB C18 compiler uses sections to manage the memory. A section is a portion of an application located at a specific address of memory. Sections can contain code or data. A section can be located in either program or data memory. There are two types of sections for each type of memory:

- Program memory

  code—contains executable instructions

  romdata—contains variables and constants

- Data memory

  udata—contains statically allocated uninitialized user variables

  idata—contains statically allocated initialized user variables

Sections can be absolute, assigned, or unassigned. An absolute section is one that is given an explicit address via the location clause of the section declaration **pragma.** An assigned section is one that is ascribed to a specific section via the **section** directive of the linker script. An unassigned section is one that is neither absolute nor assigned.

The syntax for declaring sections is as follows:

```
#pragma udata [attribute-list] [section-name [location]]
#pragma idata [attribute-list] [section-name [location]]
#pragma romdata [overlay] [section-name] [location]
#pragma code [overlay] [section-name] [location]
```

where **attribute-list** can be access, overlay, or both; **section name** must be a valid C identifier; and **location** is the starting address of the section and must be a valid integer. The access attribute tells the compiler to locate the specified section to the access bank (defined as ACCESSBANK in the linker script file). The overlay attribute permits other sections to be located at the same physical address to conserve memory. The overlay attribute can be used in conjunction with the access attribute.

Following a **#pragma code** directive, all generated code will be assigned to the specified code section until another **#pragma code** directive is encountered. An absolute code section is declared by specifying the starting address of the section in the **#pragma code** directive. For example,

```
#pragma code util_code = 0x3000
```

will locate the code section **util_code** at program memory address 0x3000.

Data can be placed either in data or in program memory with the MPLAB C18 compiler. Data that is placed in on-chip program memory can be read but not written without additional user-supplied code because on-chip program memory is in either EPROM or flash memory. Data placed in external program memory can generally be either read or written without additional user-supplied code.

For example, the following statement declares a section for statically allocated uninitialized data (**udata**) at absolute address 0x200:

```
#pragma udata my_array section = 0x200
```

The **rom** keyword informs the compiler that a variable should be placed in program memory. The compiler will allocate this variable in the current **romdata** type section. For example,

```
#pragma romdata const_table
const rom char A2Dtable[50] = {0, 0, 1, 2, . . ., 49};
```

5.17 Mixing C and Assembly

Occasionally, one may have the need to call an assembly function from the C program or call a C function from the assembly program. This will be an easy task as long as one knows how function parameters are passed and how results are returned.

5.17.1 Calling Conventions

The MPLAB C18 compiler uses both the software stack and global memory (in data memory) to allocate local variables, pass parameters, and return results. The software stack definition and manipulation were discussed in Chapter 4. As shown in Figures 5.3a and 5.3b (the last lines of the linker files), 256 bytes are reserved for the software stack by default. When necessary, one can increase the size of the software stack (by changing the value in the linker file).

As described in Chapter 4, the software stack grows from a low address toward higher addresses. The stack pointer (FSR1) always points to the next available stack location. MPLAB C18 uses FSR2 as the frame pointer, providing quick access to local variables and parameters. When a function is invoked, its stack-based arguments are pushed onto the stack in right-to-left order. The leftmost (first) function argument is on the top of the software stack on entry into the function. Figure 5.15 shows the software stack immediately prior to a function call.

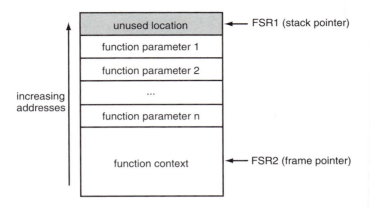

Figure 5.15 ■ Example of software stack immediately prior to a function call

The frame pointer references the location on the stack that separates the stack-based arguments from the stack-based local variables. Stack-based arguments are located at negative offsets from the frame pointer, and stack-based local variables are located at positive offsets from the frame pointer. Immediately on entry into a C function, the called function pushes the value of FSR2 onto the stack and copies the value FSR1 into FSR2, thereby saving the context of the calling function and initializing the frame pointer of the current function. Then the total size of stack-based local variables for the function is added to the stack pointer, allocating stack space for those variables. When necessary the current function may save certain registers before allocating local variables. Figure 5.16 shows a software stack following a call to a C function. This figure shows that the current function does not save any registers in the software stack.

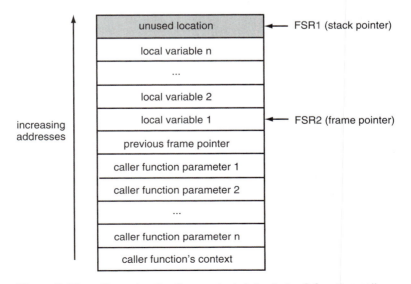

Figure 5.16 ■ Example of software stack following a C function call

A function parameter or a local variable defined with the keyword **static** will be allocated in the global memory. In general, stack-based local variables and function parameters require more code to access than **static** local variables and function parameters. Functions that use stack-based variables are more flexible in that they can be reentrant and/or recursive.

5.17.2 Return Values

The location of the return value is dependent on the size of the return value. Table 5.8 details the location or the return value based on its size.

Return value size	Return value location
8 bits	WREG
16 bits	PRODH: PRODL
24 bits	(AARGB2 + 2) (AARGB2 + 1) AARGB3
> 32 bits	on the stack, and FSRO points to the return value

Table 5.8 ■ MCC18 return values (reprint with permission of Microchip)

The variables AARGB2 and AARGB3 are defined in the file **aarg.asm** under the directory c:\mcc18\src\math\18Cxx. Both are global variables. Their exact addresses are determined during the link time.

5.17.3 Calling C Routines from an Assembly Program

C functions are global unless defined as **static.** To call a C function, the C function name must be declared as an **extern** symbol in the assembly program. A CALL or RCALL instruction must be used to make the function call.

Before calling a C function, the assembly program must push **auto** parameters onto the software stack from right (last function parameter) to left (first function parameter). For multibyte data, the last byte is pushed onto the software stack first.

Example 5.6
▼

Given the following prototype of a C function,

```
char xyz(auto char a, auto char b);
```

write an instruction sequence to call the function **xyz** with a = 0x20 and b = 0x30.

Solution: Since **b** is the rightmost parameter, it must be pushed onto the software stack first. After that, the parameter **a** is then pushed onto the software stack.

The following instruction sequence can be used to call the function **xyz** and save the returned value at memory location **result:**

```
extern      xyz          ; defined in C module
. . .
push_dat    0x30         ; push b onto the stack
push_dat    0x20         ; push a onto the stack
call        xyz
movwf       result       ; result is returned in WREG
```

▲

Example 5.7

▼

Given the following C function prototype,

    int **hwh** (int xx, int yy);

write an instruction sequence to call the function **hwh** with xx = 0x1234 and yy = 0x5678.

Solution: The parameter **yy** should be pushed onto the stack followed by **xx.** The following instruction sequence can be used to call the function **hwh:**

```
extern      hwh                ; defined in C module
. . .
push_dat    0x78               ; push the low byte of yy
push_dat    0x56               ; push the high byte of yy
push_dat    0x34               ; push the low byte of xx
push_dat    0x12               ; push the high byte of xx
call        hwh
movff       PRODL,result       ; save the low byte of result
movff       PRODH,result+1     ;save the high byte of result
```

Static parameters are allocated globally, enabling direct access. The naming convention for static parameters is **__function_name:n,** where **function_name** is replaced by the name of the function and **n** is the parameter, with numbering starting from 0. For example, given the following prototype for a C function,

    char xyz(static char x, static char y);

the value for y is accessed by using **__xyz:1,** and the value of **x** is accessed by using **__xyz:0.** However, because ":" is not a valid character in MPASM labels, accessing **static** parameters in assembly functions is not supported.

▲

5.17.4 Calling Assembly Functions from a C Program

The following conditions must be met before a C function can call an assembly function:

- The function label must be declared as **global** in the assembly module.
- The function must be declared as **extern** in the C module.
- The function must maintain the MPLAB C18 compiler's runtime model (e.g., return values must be returned in the locations specified in Table 5.8).
- The function is called from the C routine using the standard C function notation.

For example, given the following function in assembly language,

```
            udata_acs
dly_cnt     res     1
            code
delay1      setf    dly_cnt     ; initialize dly_cnt to 255
loop        nop
            nop
            nop
            decf    dly_cnt, F
            bnz     loop
            return
            global  delay1      ; export so linker can see it
            end
```

to call the assembly function **delay1** from a C source file, an external prototype for the assembly function must be added, and the function is then called using a standard C function statement:

```
extern void delay1 (void);
void main (void)
(
        . . .
        delay1 ();
        . . .
}
```

This assembly function creates a fix delay. Suppose the time delay (the only parameter) to be created is passed in the software stack. Then the previous assembly code is modified to

```
            #include <p18Fxxx.inc>
            code
delay2      movlw      -2
            nop
            nop
            decf       PLUSW2, F       ; the function parameter is 2 bytes below FSR2
            bnz        delay2
            return
global      delay2     ; export so linker can see it
            end
```

The following C statements will call the **delay2** function with 0xC0 as the parameter:

```
extern void delay2 (unsigned char);
void main (void)
{
        . . .
        delay2(0xC0);
        . . .
}
```

5.17.5 Referencing C Variables in an Assembly Function

It is possible to access C variables from an assembly program. To do this, the following must be done:

- ■ The C variable must have a global scope in the C source file.
- ■ The C variable must be declared as an **extern** symbol in the assembly file.

Suppose a C file consists of the following C statements:

```
unsigned int xyz;
void main (void)
{
        . . .
}
```

The following assembly statements will assign 0x2002 to the C variable **xyz:**

```
            extern     xyz       ; xyz is defined in C module
            code
abc_func    movlw      0x02
            movwf      xyz       ; set the low byte to 0x02
            movlw      0x20
```

```
        movwf      xyz+1        ; set the high byte to 0x20
        . . .
        return
        global     abc_func     ; export so linker can see it
        end
```

5.17.6 Referencing Assembly Variables in a C Function

It is also possible to reference variables declared in an assembly file from a C program. To do this, the following must be done:

- The variable must be declared as **global** in the assembly module.
- The variable must be declared as **extern** in the C module.

Suppose the following assembly statements are contained in an assembly program file:

```
asm_data    udata
abc         res     2        ; abc is a 2-byte variable
            . . .
            global  abc      ; export so linker can see it
            end
```

The following C statements can assign the value 0x2002 to the assembly variable **abc**:

```
extern unsigned int abc;
void main (void)
{
        . . .
        abc = 0x2002;
        . . .

}
```

5.17.7 Startup Code

The MPLAB C18 startup begins at the **reset** vector (at address 0). The reset handler jumps to a function that initializes FSR1 and FSR2 to reference the software stack, optionally calls a function to initialize **idata** sections (data memory initialized data) from program memory, and loops on a call to the application's **main** () function.

There are three versions of the startup code (c018.o, c018i.0, and c018iz.o) that can be chosen from to be linked with the user application. Whether the startup code initializes the **idata** section is determined by which startup code module is linked with the application. The c018i.o module performs the initialization, while the c018.o module does not. The default linker scripts provided by the MPLAB C18 link with c018i.o. The startup code modules are located under the directory c:\mcc18\lib.

The ANSI standard requires that all objects with static storage duration that are not initialized explicitly be set to zero. With both the c018.o and c018i.o startup code modules, this requirement is not met. A third startup module, c018iz.o, is provided to meet this requirement. If this startup code is linked with the application, then, in addition to initializing **idata** sections, all objects with static storage duration that are not initialized explicitly are set to zero.

After the startup code sets up the stack and optionally copies initialized data, it calls the **main()** function of the C program. There are no arguments passed to **main()**. MPLAB C18 transfers control to **main()** via a looped call, that is,

```
loop:
main();
goto loop;
```

If one has the need to customize the startup code, one can edit the c018.c, c018i.c, or c018iz.c to add any desirable customized startup code. After editing, one need to recompile it to generate their object codes to be linked later. These startup C programs are located at c:\mcc18\src\startup.

5.18 MPLAB C18 Library Functions

A library is a collection of functions grouped for reference and ease of linking. The MPLAB C18 libraries are included in the **lib** subdirectory of the installation. They can be linked directly into an application using the MPLINK linker.

C18 libraries can be divided to two groups: processor-independent libraries and processor-specific libraries.

5.18.1 Processor-Specific Libraries

The processor-specific library files contain definitions that may vary across individual members of the PIC18 family. This includes all the peripheral routines and the special-function register definitions. The peripheral routines that are provided include both those designed to use the hardware peripherals and those that implement a peripheral interface using general-purpose I/O lines. The processor-specific libraries are named

pprocessor.lib

For example, the library file for the PIC18F8720 is named p18f8720.lib. The functions included in this library are explored in subsequent chapters.

5.18.2 Processor-Independent Libraries

The processor-independent library consists of general functions and math functions. These functions are supported across all members in the PIC18 family and are contained in the **clib.lib** library.

MPLAB C18 compiler implements mathematical operations that are not directly supported by the hardware by calling appropriate mathematical library functions. Examples are 16-bit by 16-bit multiplication, 32-bit by 32-bit multiplication, division operation in any length, and floating-point operations. Whenever one's C program performs any mathematical operations that are not directly supported by the hardware, one needs to include the **clib.lib** library. This has been done in the linker file shown in Figure 5.3.

The MPLAB C18 general library supports the following categories of routines:

- Character classification functions
- Data conversion functions
- Delay functions
- Memory and string manipulation functions

Functions in these categories are discussed in the following sections.

5.18.3 Character Classification Functions

Character classification functions provided by the MPLAB C18 compiler are consistent with the ANSI 1989 standard C library functions of the same name. The functions and prototype declarations of these routines are listed in Tables 5.9a and 5.9b.

One needs to include the header file **ctype.h** in one's C program in order to use these routines.

name	Description
isalnum	Determine if a character is alphanumeric
isalpha	Determine if a character is alphabetic
iscntrl	Determine if a character is a control character
isdigit	Determine if a character is a decimal digit
isgraph	Determine if a character is a graphical character
islower	Determine if a character is a lower case alphabetical character
isprint	Determine if a character is a printable character
ispunct	Determine if a character is punctuation character
isspace	Determine if a character is a white space character
isupper	Determine if a character is an upper case alphabetical character
isxdgit	Determine if a character is a hex digit

Table 5.9a ■ Names and functions of character classification C routines in MPLAB C18 compiler (reprint with permission of Microchip)

name	Prototype declaration	
isalnum	unsigned char	**isalnum** (unsigned char **ch**);
isalpha	unsigned char	**isalpha** (unsigned char **ch**);
iscntrl	unsigned char	**iscntrl** (unsigned char **ch**);
isdigit	unsigned char	**isdigit** (unsigned char **ch**);
isgraph	unsigned char	**isgraph** (unsigned char **ch**);
islower	unsigned char	**islower** (unsigned char **ch**);
isprint	unsigned char	**isprint** (unsigned char **ch**);
ispunct	unsigned char	**ispunct** (unsigned char **ch**);
isspace	unsigned char	**isspace** (unsigned char **ch**);
isupper	unsigned char	**isupper** (unsigned char **ch**);
isxdigit	unsigned char	**isxdigit** (unsigned char **ch**);

Table 5.9b ■ Names and prototype declarations of character classification C routines in MPLAB C18 compiler

5.18.4 Data Conversion Library Functions

Most of these functions are useful for I/O operations. The need for data conversion between internal and external number representations was explained in Chapter 4, Section 4.9. MPLAB C18 provides a set of data conversion functions to be called by user applications. The functions and prototype declarations of these routines are listed in Tables 5.10a and 5.10b.

One needs to include the header file **stdlib.h** in one's C program in order to use these functions.

name	Description
atob	Convert a string to an 8-bit signed byte
atof	Convert a string into a floating point value
atoi	Convert a string to a 16-bit signed integer
atol	Convert a string into a long integer representation
btoa	Convert an 8-bit signed byte into a string
itoa	Convert a 16-bit signed integer to a string
ltoa	Convert a signed long integer to a string
rand	Generate a pseudo-random integer
srand	Set the starting seed for the pseudo-random number generator
tolower	Convert a character to a lower-case alphabetical ASCII character
toupper	Convert a character to an upper-case alphabetical ASCII character
ultoa	Convert an unsigned long integer to a string

Table 5.10a ■ Names and functions of character conversion C routines in MPLAB C18 compiler (reprint with permission of Microchip)

name	Prototype declaration
atob	signed char **atob** (constchar ***s**);
atof	double **atof** (const char ***s**);
atoi	int **atoi** (constchar ***s**);
atol	long **atol** (const char ***s**);
btoa	char ***btoa** (signed char **value**, char ***string**);
itoa	char ***itoa** (int **value**, char ***string**);
ltoa	char ***ltoa** (long **value**, char ***string**);
rand	int **rand** (void);
srand	void **srand** (unsigned int **seed**);
tolower	char **tolower** (char **ch**);
toupper	char **toupper** (cha **ch**);
ultoa	char ***ultoa** (unsigned long **value**, char ***string**);

Table 5.10b ■ Names and prototype declarations of character conversion C routines in MPLABC 18 compiler

5.18.5 Memory and String Manipulation Functions

Most of these functions are consistent with the ANSI standard C library functions of the same name. The functions and prototype declarations of these functions are listed in Tables 5.11a and 5.11b.

One needs to include the header file **string.h** in one's C program in order to use these routines.

name	Description
memchr	Search for a value in a specified memory region
memcmp	Compare the contents of two arrays (data ram and data ram)
memcmppgm	Compare the contents of two arrays (data ram and program memory)
memcmppgm2ram	Compare the contents of two arrays (program memory and data ram)
memcmpram2pgm	Compare the contents of two arrays (data ram and program memory)
memcpy	Copy a buffer from data memory to data memory.
memcpypgm2ram	Copy a buffer from program memory to data memory.
memmove	Copy a buffer from data memory to data memory.
memmovepgm2ram	Copy a buffer from program memory to data memory.
memset	Initialize an array with a single repeated value
strcat	Append a copy of the source string to the end of the destination string.
strcatpgm2ram	Append a copy of the source string to the end of the destination string.
strchr	Locate the first occurrence of a value in a string
strcmp	Compare two strings in data memory.
strcmppgm2ram	Compare a string in data memory with a string in program memory.
strcpy	Copy a string from data memory to data memory.
strcpypgm2ram	Copy a string from program memory to data memory.
strcspn	Calculate the number of consecutive characters at the beginning of a string that are not contained in a set of characters.
strlen	Determine the length of a string.
strlwr	Convert all upper-case characters in a string to lower-case.
strncat	Append a specified number of characters from the source string to the end of the destination string.
strncatpgm2ram	Append a specified number of characters from the source string to the end of the destination string.
strncmp	Compare two strings, up to the specified number of characters.
strncpy	Copy characters from the source string to the destination string up to the specified number of characters.
strncpypgm2ram	Copy characters from the source string to the destination string up to the specified number of characters.
strpbrk	Search a string for the first occurrence of a character from a set of characters.
strrchr	Locate the last occurrence of a specified character in a string.
strspn	Calculate the number of consecutive characters at the beginning of a string that are contained in a set of characters.
strstr	Locate the first occurrence of a string inside another string
strtok	Break a string into substrings, or tokens, by inserting null characters in place of specified delimiters
strupr	Convert all lower-case characters in a string to upper case.

Table 5.11a ■ Names and functions of memory and string manipulation C routines in MPLAB C18 compiler (reprint with permission of Microchip).

Name	Description
memchr	void *memchr (const void ***mem,** unsigned char **c,** size_t **n**);
memcmp	signed char memcmp (const void ***buf1,** const void ***buf2,** size_t **memsize**);
memcmppgm	signed char memcmppgm (const rom ***buf1,** const rom ***buf2,** sizerom_t **memsize**);
memcmppgm2ram	signed char memcmppgm2ram (const void ***buf1,** const rom ***buf2,** sizeram_t **memsize**);
memcmpram2pgm	signed char memcmpram2pgm (const rom ***buf1,** const void ***buf2,** sizeram_t **memsize**);
memcpy	void *memcpy (void ***dest,** const void ***src,** size_t **memsize**);
memcpypgm2ram	void *memcpypgm2ram (void ***dest,** const rom ***src,** size_t **memsize**);
memmove	void *memmove (void ***dest,** const void ***src,** sized_t **memsize**);
memmovepgm2ram	void *memmovepgm2ram (void ***dest,** const rom void ***src,** sizeram_t **memsize**);
memset	void *memset, (void ***dest,** unsigned char **value,** size_t **memsize**);
strcat	char *strcat (char ***dest,** const char ***src**);
strcatpgm2ram	char *strcatpgrm2ram (char ***dest,** const rom char ***src**);
strchr	char *strchr (const char ***str,** const char **c**);
strcmp	signed char strcmp (const char ***str1,** const char ***str2**);
strcmppgm2ram	signed char strcmppgm2ram (const char ***str1,** const rom char ***str2**);
strcpy	char *strcpy (char ***dest,** const char ***src**);
strcpypgm2ram	char *strcpypgm2ram (char ***dest,** const rom char ***src**);
strcspn	size_t *strcspn (const char ***str1,** const char *str2);
strlen	size_t strlen (const char ***str**);
strlwr	char *strlwr (char ***str**);
strncat	char *strncat (char ***dest,** const char ***src,** size_t **n**);
strncatpgm2ram	char *strncatpgm2ram (char ***dest,** const **rom** char ***src,** sizeram_t n);
strncmp	signed char strncmp (const char ***str1,** const char ***str2,** size_t **n**);
strncpy	char *strncpy (char ***dest,** const char ***src,** size_t **n**);
strncpypgm2ram	char *strncpypgm2ram (char ***dest,** const rom char ***src,** sizeram_t n);
strpbrk	char *strpbrk (const char ***str1,** const char ***str2**);
strrchr	char *strrchr (const char ***str,** const char **c**);
strspn	size_t *strspn (const char ***str1,** const char ***str2**);
strstr	char *strstr (const char ***str,** const char ***substr**);
strtok	char *strtok (char ***str,** const char ***delim**);
strupr	char *strupc (char ***str**);

Table 5.11b ■ Names and functions of memory and string manipulation C routines in MPLAB C18 compiler

5.18.6 Delay Functions

MPLAB C18 provides a set of delay functions that will create a delay equal to a multiple of instruction cycles. The functions and prototype declarations of these delay functions are listed in Tables 5.12a and 5.12b.

One needs to include the header file **delays.h** in one's C program in order to use these delay functions.

name	Description
Delay 1TCY	Delay one instruction cycle.
Delay 10TCYx	Delay in multiples of 10 instruction cycles.
Delay 100TCYx	Delay in multiples of 100 instruction cycles.
Delay 1KTCYx	Delay in multiples of 1000 instruction cycles.
Delay 10KTCYx	Delay in multiples of 10,000 instruction cycles.

name	Description
Delay1TCY	void Delay1TCY (void)
Delay10TCYx	void Delay10TCYx (unsigned char unit);
Delay100TCYx	void Delay100TCYx (unsigned char unit);
Delay1KTCYx	void Delay1KTCYx (unsigned char unit);
Delay10KTCYx	void Delay10KTCYx (unsigned char unit);

Table 5.12a ■ Names and functions of C delay routines in MPLAB C18 comiler (reprint with permission of Microchip)

Table 5.12b ■ Names and prototype declarations of C delay routines in MPLAB C18 compiler (reprint with permission of Microchip)

Example 5.8

▼

Assuming that there is a demo board (e.g., SSE8720) using a 16-MHz crystal oscillator to generate its clock signal and its port D is driving eight LEDs, write a program to flash these LEDs with the following patterns forever:

1. Turn on all LEDs for half a second and turn them off for half a second.
2. Repeat Step 1 three more times.
3. Turn on one LED at a time from left to right with each LED turned on for half a second.
4. Repeat Step 3 three more times.
5. Turn on one LED at a time from right to left with each LED turned on for half a second.
6. Repeat Step 5 three more times.
7. Turn on LEDs driven by pins RD7 and RD0 on and off four times. The on time and off time are both half a second.
8. Turn on LEDs driven by pins RD6 and RD1 on and off four times. The on time and off time are both half a second.
9. Turn on LEDs driven by pins RD5 and RD2 on and off four times. The on time and off time are both half a second.
10. Turn on LEDs driven by pins RD4 and RD3 on and off four times. The on time and off time are both half a second.
11. Turn on LEDs driven by pins RD4 and RD3 on and off four times. The on time and off time are both half a second.
12. Turn on LEDs driven by pins RD5 and RD2 on and off four times. The on time and off time are both half a second.
13. Turn on LEDs driven by pins RD6 and RD1 on and off four times. The on time and off time are both half a second.
14. Turn on LEDs driven by pins RD7 and RD0 on and off four times. The on time and off time are both half a second.

Solution: To turn on the LEDs on the SSE8720 (or SSE8680), one needs to output a 0 to the port pin that drives the LED. By outputting different values to PORTD pins, different LEDs can be lighted. For example, outputting the value 0x7E will turn on the LEDs driven by pins RD7 and RD0. By placing all the LED patterns in a table, reading one pattern at a time and outputting it to port D pins, waiting for half a second, the desired LED lighting sequence can be achieved. The instruction cycle time for the 16-MHz crystal oscillator is 250 ns. The C program that implements this idea is as follows:

```c
#include <delays.h>
#include <p18F8720.h>
unsigned rom char led_tab [] = {0x00,0xFF,0x00,0xFF,0x00,0xFF,0x00,0xFF,
                                0x7F,0xBF,0xDF,0xEF,0xF7,0xFB,0xFD,0xFE,
                                0x7F,0xBF,0xDF,0xEF,0xF7,0xFB,0xFD,0xFE,
                                0x7F,0xBF,0xDF,0xEF,0xF7,0xFB,0xFD,0xFE,
                                0x7F,0xBF,0xDF,0xEF,0xF7,0xFB,0xFD,0xFE,
                                0xFE,0xFD,0xFB,0xF7,0xEF,0xDF,0xBF,0x7F,
                                0xFE,0xFD,0xFB,0xF7,0xEF,0xDF,0xBF,0x7F,
                                0xFE,0xFD,0xFB,0xF7,0xEF,0xDF,0xBF,0x7F,
                                0xFE,0xFD,0xFB,0xF7,0xEF,0xDF,0xBF,0x7F,
                                0x7E,0xFF,0x7E,0xFF,0x7E,0xFF,0x7E,0xFF,
                                0xBD,0xFF,0xBD,0xFF,0xBD,0xFF,0xBD,0xFF,
                                0xDB,0xFF,0xDB,0xFF,0xDB,0xFF,0xDB,0xFF,
                                0xE7,0xFF,0xE7,0xFF,0xE7,0xFF,0xE7,0xFF,
                                0xE7,0xFF,0xE7,0xFF,0xE7,0xFF,0xE7,0xFF,
                                0xDB,0xFF,0xDB,0xFF,0xDB,0xFF,0xDB,0xFF,
                                0xBD,0xFF,0xBD,0xFF,0xBD,0xFF,0xBD,0xFF,
                                0x7E,0xFF,0x7E,0xFF,0x7E,0xFF,0x7E,0xFF};
void main (void)
{
        unsigned char i;
        TRISD = 0x00; /* configure PORTB for output */
        while (1) {
                for (i = 0; i < 136; i++) {
                        PORTD = led_tab[i]; /* output a new LED pattern */
                        Delay 10KTCYx(200)); /* stay for about half a second */
                }
        }
}
```

5.19 Using the HI-TECH C18 Compiler

Commercial C compilers for the PIC18 are available from several software companies. Among them, the HI-TECH C18 compiler can work with MPLAB and support source-level debugging at the time of this writing, HI-TECH C18 compiler has been used by many users. A tutorial on the use of this compiler is provided in Appendix C.

5.20 Summary

A C program consists of one or more functions and variables. The **main** () function is required in every C program. It is the entry point of a C program. A function contains statements that specify the operations to be performed. The types of statements in a function could be declaration, assignment, function call, control, and null.

A variable stores a value to be used during the computation. A variable must be declared before it can be used. The declaration of a variable consists of the name and the type of the variable. There are four basic data types in C: **int, char, float,** and **double.** Several qualifiers can be added to the variable declarations. They are short, long, signed, and unsigned.

Constants are often needed in forming a statement. There are four types of constants: integers, characters, floating-point numbers, and strings.

There are seven arithmetic operators: +, −, *, /, %, ++, --. There are six bitwise operators: &, |, ^, ~, >>, and <<. Bitwise operators can be applied only to integers. Relational operators are used in control statements. They are ==, !=, >, >= <, <=, &&, ||, and !.

The control-flow statements specify the order in which computations are performed. Control-flow statements include if-else statement, multiway conditional statement, switch statement, for-loop statement, while-statement, and do-while statement.

Every C program consists of one or more functions. If a program consists of multiple functions, their definitions cannot be embedded within another. The same function can be called from several different places within a program. Generally, a function will process information passed to it from the calling portion of the program and return a single value. Information is passed to a function via special identifiers called arguments (also called parameters) and returned via the return statement. Some functions, however, accept information but do not return anything (e.g., the library function **printf**).

A pointer is a variable that holds the address of another variable. Pointers can be used to pass information back and forth between a function and its reference (calling) point. In particular, pointers provide a way to return multiple data items from a function via function arguments. Pointers also permit references to other functions to be specified as arguments to a given function. Two operators are related with pointers: * and **&.** The * operator returns the value of the variable pointed to by the pointer. The **&** operator returns the address of a variable.

Data items that have common characteristics are placed in an array. An array may be one-dimensional or multidimensional. The dimension of an array is specified by the number of square bracket pairs ([]) following the array name. An array name can be used as an argument to a function, thus permitting the entire array to be passed to the function. To pass an array to a function, the array name must appear by itself, without brackets or subscripts. An alternate way to pass arrays to a function is to use pointers.

A variable defined inside a function is an internal variable of that function. External variables are defined outside of any function and are thus potentially available to many functions. The scope of a name is the part of the program within which the name can be used. The scope of an external variable or a function lasts from the point at which it is declared to the end of the file being compiled.

The process of project development consists of four major steps:

1. Source code entering and editing
2. Object code generation

3. Library code creation and maintenance

4. Program linking and executable code generation

This chapter provided a step-by-step tutorial on how to use the MPLAB C18 compiler in program development and debugging.

The MPLAB C18 compiler is not completely ANSI C compliant. It deviates from the standard wherever the standard conflicts with efficient support for the PIC18 devices. Certain extensions are added to better support the PIC18 devices. This chapter discussed these deviations and extensions in detail.

The MPLAB C18 compiler provides many library routines to facilitate the development of PIC18 applications. These library functions are divided into processor-dependent and processor-independent categories. Most of these functions are compliant with ANSI C standard. The library functions in the processor-independent category were discussed in this Chapter, whereas those functions in the processor-dependent category are discussed in later chapters.

5.21 Exercises

E5.1 Write a program to generate a table of powers of integers. For each integer compute its powers from 1 to 6. Print the powers of integers from 2 to 10.

E5.2 Write a function that can convert all uppercase letters in a string to lowercase. The starting address of the string is passed to this function.

E5.3 Let n_0 be a given positive integer. For i = 0, 1, 2, . . . define

$n_{i+1} = n_i / 2$ if n_i is even; $3n + 1$ if n_i is odd;

The sequence stops whenever n_i has the value 1. Numbers that are generated this way are called *hailstones.* Write a program that generates some hailstones. The function

```
void hailstones (int n)
{
. . .
}
```

should be used to compute and print the sequence generated by n. The output of your program might look as follows:

```
Hailstones generated by 77:
77    232   116   58    29    88
44    22    11    34    17    52
26    13    40    20    10    5
16    8     4     2     1
Number of hailstones generated: 23
```

Write a main program to call the hailstones function and use two different values to test it.

E5.4 Write a C function that implements the binary search algorithm.

E5.5 Write a function **setbits (int x, char p, char n)** that returns **x** with the **n** bits that begin at position **p** inverted (i.e., 1 changed to 0 and vice versa).

E5.6 Write the function **itoa (char *cptr, long int kk)** that accepts a 32-bit binary number **(kk)** and converts into the ASCII string that represents its value. The resultant string is to be stored in a buffer pointed by **cptr**. The string must be terminated by a NULL character.

5.22 Lab Exercises and Assignments

L5.1   Create a new project and call it eg5_L1 and add the following C program in the project:

```
#include <P18F452.h>
unsigned char root1,root2;
unsigned char sq_root_16(unsigned int xx);
void main(void)
{
        root1 = sq_root_16(3600);
        root2 = sq_root_16(9600);
        Nop();
}
unsigned char sq_root_16(unsigned int xx)
{
        unsigned char mask, root, test, i;
        mask = 0x80;
        root = 0x00;
        for(i = 0; i < 8; i++) {
                test = mask | root;
                if (test*test < xx)
                        root |= mask;
                mask = mask >> 1;
        }
        if ((xx - (root*root)) < ((root+1)*(root+1)- xx))
                return root;
        else
                return (root+1);
}
```

This program implements the successive-approximation method to compute the square root of a 16-bit unsigned integer.

Perform the following steps:

Step 1

Set up build options as shown in Figure 5.6,

Step 2

Add the linker file **p18f452i.lkr** into the project.

Step 3

Select **SIM** simulator as the debugging tool (do not select ICD2).

Step 4

Build the project.

Step 5

Set a breakpoint at the statement **Nop();**

Step 6

Open a Watch window and enter the symbols **root1** and **root2** into the Watch window

Step 7

Run the program. What are the values of root1 and root2 in the Watch window? You have 0xFF and 0xFF for root1 and root2. Do you? This is obviously wrong! Do you know why?

Perform the following steps:

Step 8
Move the following statement to immediately before the "void main (void)" statement:

unsigned char mask, root, test, i;

Step 9
Rebuild the project.

Step 10
Add variables **mask, root,** and **test** into the Watch window.

Step 11
Set a breakpoint at the first statement in the **sq_root_16** function.

Step 12
Run the program, and the program should stop at the new breakpoint.

Step 13
Step through the function **sq_root_16** by pressing the function key F7 and pay attention to the change of the variable values in the Watch window. What do you notice? You probably notice that the following if statement is always true:

if (test*test < xx)
        root |=mask;

Why is that?
Do you remember the discussion in Section 5.6.1? Follow the description in that section and rebuild the project.

Step 14
Reenable the breakpoint at the statement **Nop().** Rerun the program. Now you should get the correct values for root1 (0x3C) and root2 (0x62).

L5.2   Write a C program to find two prime numbers that are closest to but larger than 10000 and also two prime numbers that are closest to but smaller than 10000.

L5.3   Write a C program to flash the LEDs on your demo board with the following pattern (use port B in SSE452 and use port D in SSE8720 or SSE8680):

1. Turn on all LEDs for a quarter second and turn them off for a quarter second.
2. Repeat Step 1 three more times.
3. Turn on one LED (at odd position) at a time from left to right with each LED turned on for half a second.
4. Repeat Step 3 three more times.
5. Turn on one LED (at even position) at a time from right to left with each LED turned on for a quarter second.
6. Repeat Step 5 three more times.
7. Turn on LEDs driven by pins RB4 and RB3 on and off twice. The on time and off time are both a quarter second.
8. Turn on LEDs driven by pins RB5 and RB2 on and off twice. The on time and off time are both half a second.
9. Turn on LEDs driven by pins RB6 and RB1 on and off twice. The on time and off time are both half a second.
10. Turn on LEDs driven by pins RB7 and RB0 on and off twice. The on time and off time are both a quarter second.

6

Interrupts, Resets, and Configuration

6.1 Objectives

After completing this chapter, you should be able to

- Explain interrupts and resets
- Describe the handling procedures for interrupts and resets
- Enable and disable maskable interrupts
- Set the PIC18 interrupt priority to high or low
- Write interrupt service routines
- Write interrupt-driven applications

6.2 Basics of Interrupts

Interrupt is a mechanism provided by a microprocessor or a computer system to synchronize I/O operations, handle error conditions and emergency events, coordinate the use of shared resources, and so on. Without the interrupt mechanism, many of these operations will become either impossible or very difficult to implement.

6.2.1 What Is an Interrupt?

An interrupt is an event that requires the CPU to stop normal program execution and perform some service related to the event. An interrupt can be generated internally (inside the chip) or externally (outside the chip). An external interrupt is generated when the external circuitry asserts an interrupt signal to the CPU. An internal interrupt can be generated by the hardware circuitry inside the chip or caused by software errors. In some microcontrollers (e.g., the PIC18), timers, I/O interface functions, and the CPU are incorporated on the same chip, and these subsystems can generate interrupts to the CPU. Abnormal situations that occur during program execution, such as illegal opcodes, overflow, divided-by-zero, and underflow, are called *software interrupts*. The terms *traps* and *exceptions* are also used to refer to software interrupts by some companies.

A good analogy for interrupt is how you act when you are sitting in front of a desk to read this book and the phone rings. You probably will act like this:

1. Remember the page number or place a bookmark on the page that you are reading, close the book, and put it aside.
2. Pick up the phone and say, "Hello, this is so and so."
3. Listen to the voice over the phone to find out who is calling or ask who is calling if the voice is not familiar.
4. Talk to that person.
5. Hang up the phone when you finish talking.
6. Open the book and turn to the page where you placed the bookmark and resume reading this book.

The phone call example spells out a few things that are similar to how the microprocessor handles the interrupt:

1. As a student, you spend most of your time in studying. Answering a phone call happens only occasionally. Similarly, the microprocessor is executing application programs most of the time. Interrupts will only force the microprocessor to stop executing the application program briefly and take some necessary actions.
2. Before picking up the phone, you finish reading the sentence and then place a bookmark to remind you of the page number that you are reading so that you can resume reading after finishing the conversation over the phone. Most microprocessors will finish the instruction they are executing and save the address of the next instruction in memory (usually in the stack) so that they can resume the program execution later.
3. You find out the person who called you by listening to the voice over the phone, or you ask questions so that you can decide what to say. Similarly, the microprocessor needs to identify the cause of the interrupt before it can take appropriate actions. This is built into the microprocessor hardware.
4. After identifying the person who called you, you start the phone conversation with that person on some appropriate subjects. Similarly, the microprocessor will take some actions appropriate to the interrupt source.

5. When finishing the phone conversation, you hang up the phone, open the book to the page where you placed the bookmark, and resume reading. Similarly, after taking some actions appropriate to the interrupt, the microprocessor will jump back to the next instruction when the interrupt occurred and resume program execution. This can be achieved easily because the address of the instruction to be resumed was saved in memory (stack). Most microprocessors do this by executing a *return-from-interrupt instruction.*

6.2.2 Why Interrupts?

Interrupts are useful in many applications, such as the following:

- *Coordinating I/O activities and preventing the CPU from being tied up during the data transfer process.* The CPU needs to know if the I/O device is ready before it can proceed. Without the interrupt capability, the CPU will need to check the status of the I/O device periodically. The interrupt mechanism is often used by the I/O device to inform the CPU that it is ready for data transfer. CPU time can thus be utilized more efficiently because of the interrupt mechanism. Interrupt-driven I/O operations are explained in more detail in later chapters.

- *Performing time-critical applications.* Many emergent events, such as power failure and process control, require the CPU to take action immediately. The interrupt mechanism provides a way to force the CPU to divert from normal program execution and take immediate actions.

- *Providing a graceful way to exit from application when a software error occurs.* The service routine for a software interrupt may also output some useful information about the error so that it can be corrected.

- *Reminding the CPU to perform routine tasks.* There are many microprocessor applications that require the CPU to perform routine work, such as the following:

   Keeping track of time of day. Without the timer interrupt, the CPU will need to use program loops in order to update the current time. The CPU cannot do anything else without a timer interrupt in this application. The periodic timer interrupts prevent the CPU from being tied up.

   Periodic data acquisition. Some applications are designed to acquire data periodically.

   Task switching in a multitasking operating system. In a modern computer system, multiple application programs are resident in the main memory, and the CPU time is divided into many short slots (one slot may be from 10 to 20 ms). A multitasking operating system assigns a program to be executed for one time slot. At the end of a time slot or when a program is waiting for the completion of an I/O operation, the operating system takes over and assigns another program for execution. This technique is called *multitasking.* Multitasking can dramatically improve the CPU utilization and is implemented by using periodic timer interrupts.

6.2.3 Interrupt Maskability

Depending on the situation and application, some interrupts may not be desired or needed and should be prevented from interrupting the CPU. Most microprocessors and microcontrollers have the option of ignoring these interrupts. These types of interrupts are called *maskable interrupts.* There are other types of interrupts that the CPU cannot ignore and must take immediate actions for; these are *nonmaskable interrupts.* A program can request the CPU to service or ignore a maskable interrupt by setting or clearing an *enable bit.* When an interrupt is enabled, the CPU will respond to its happening. When an interrupt is disabled, the CPU will ignore it. An interrupt is said to be *pending* when it is active but not yet serviced by the CPU. A pending interrupt may or may not be serviced by the CPU, depending on whether it is enabled.

6.2.4 Interrupt Priority

If a computer is supporting multiple interrupt sources, then it is possible that several interrupts would be pending at the same time. The CPU has to decide which interrupt should receive service first in this situation. The solution is to prioritize all interrupt sources. An interrupt with higher priority always receives service before interrupts at lower priorities. Many microcontrollers (e.g., the 68HC11) prioritize interrupts in hardware For those microcontrollers that do not prioritize interrupts in hardware, the software can be written to handle certain interrupts before others. By doing this, interrupts are essentially prioritized. For most microprocessors and microcontrollers, interrupt priorities are not programmable.

6.2.5 Interrupt Service

The CPU provides service to an interrupt by executing a program called an *interrupt service routine*. After providing service to an interrupt, the CPU must resume normal program execution. How can the CPU stop the execution of a program and resume it later? It achieves this by saving the program counter and the CPU status information before executing the interrupt service routine and then restoring the saved program counter and CPU status before exiting the interrupt service routine. The complete interrupt service cycle involves the following:

1. Saving the program counter value.
2. Saving the CPU status (including the CPU status register and some other registers) in the stack. This step is optional for some microcontrollers and microprocessors.
3. Identifying the source of the interrupt.
4. Resolving the starting address of the corresponding interrupt service routine.
5. Executing the interrupt service routine.
6. Restoring the CPU status from the stack.
7. Restoring the program counter from the stack.
8. Resuming the interrupted program.

For all maskable hardware interrupts, the microprocessor starts to provide service when it completes the execution of the current instruction (the instruction being executed when the interrupt occurred). For some nonmaskable interrupts, the CPU may start the service without completing the current instruction. Many software interrupts are caused by an error in instruction execution that prevents the instruction from being completed. The service to this type of interrupt is simply to output an error message and abort the program.

6.2.6 Interrupt Vector

To provide service to an interrupt, the processor must know the starting address of the service routine. The starting address of the interrupt service routine is called *interrupt vector*. The interrupt vector can be determined by one of the following methods:

1. *Predefined.* In this method, the starting address of the service routine is predefined when the microcontroller is designed. The processor would jump to certain predefined location to execute the service routine. All Microchip microcontrollers and 8051 variants use this approach.

2. *Fetch the vector from a predefined memory location.* For many microprocessors and microcontrollers, the interrupt vector of each interrupt source is stored at a predefined location. The block of memory locations where all interrupt vectors are stored is referred to as the *interrupt vector table.* All Motorola microcontrollers use this approach.

3. *Execute an interrupt acknowledge cycle to fetch a vector number in order to locate the interrupt vector.* During the interrupt acknowledge cycle, the microprocessor performs a read bus cycle, and the external I/O device that requested the interrupt places a number on the data bus to identify itself. This number is called an *interrupt vector number.* The address of the memory location where the interrupt vector is stored is usually the sum of a multiple (2 and 4 are most common) of the vector number and the starting address of the interrupt vector table. The CPU needs to perform a read cycle in order to obtain the interrupt vector number. This method is used by Intel Pentium microprocessor and many older microprocessors. However, it is the least efficient method and hence is not used by any microcontroller.

It is a common practice for multiple interrupt sources to share the same interrupt vector. However, each interrupt source would have a flag associated with it to identify itself. The source of the interrupt can be identified by examining those flags. The PIC18 MCU has only two interrupt vectors to handle all interrupt sources. These interrupts are divided into high and low priority levels. Within each interrupt priority level, the service routine needs to check the flag bits of the appropriate interrupt request register in order to identify the cause of the pending interrupt.

6.2.7 Interrupt Programming

There are three steps in interrupt programming:

Step 1.
Write the service routine
An interrupt service routine is similar to a subroutine—the only difference is the last instruction. An interrupt service routine uses the return-from-interrupt (**retfie**) instruction instead of the **return** instruction to return to the interrupted program. In principle, the interrupt service routine should be as short as possible. An interrupt service routine might simply increment or decrement a variable and return. For example, the following instruction sequence could serve as the service routine for the PIC18 INT0 pin interrupts:

```
cnt     set     0x10
        .
        .
        org     0x08      ; interrupt vector
        incf    cnt,F,A
        retfie
```

The interrupt service routine may or may not return to the interrupted program, depending on the cause of the interrupt. It makes no sense to return to the interrupted program if the interrupt is caused by a software error such as divided-by-zero or overflow because the program is unlikely to produce correct results under these circumstances. In such situations, the service routine would return to the monitor program or the operating system instead.

Step 2.
Initialize the interrupt vector table (i.e., place the starting address of each interrupt service routine in the table)
This can be done by using the assembler directive **ORG** (or its equivalent):

```
ORG     0xkk
dw      ISR_1
dw      ISR_2
        .
        .
        .
```

where **ISR_i** is the starting address of the service routine for interrupt source **i**. This step is not needed for the PIC18 MCUs because their interrupt vector locations are fixed.

Step 3.
Enable interrupts to be serviced
Appropriate interrupt enable bits must be set in order to enable interrupt to occur.

6.2.8 Interrupt Overhead

The interrupt mechanism involves some overhead. The overhead is due to the saving and restoring of the machine state and the execution time required by the interrupt service routine. The default interrupt overhead of the PIC18 MCU is relatively light because the PIC18 saves and restores only the program counter, the WREG, the STATUS, and the BSR registers. The saving and restoring of these registers take only two clock cycles each. However, several special-function registers may need to be saved by the interrupt service routine in order to make sure that the application program can run correctly. Other microcontrollers may save and restore many more registers during an interrupt. For example, the Motorola 68HC11/12 MCUs save and restore six registers (nine bytes) automatically when an interrupt is serviced.

6.3 Resets

The initial values of some CPU registers, flip-flops, and control registers in the I/O interface chips must be established before the computer can operate properly. The reset mechanism is provided for establishing the initial condition of a computer.

There are at least two types of reset in each microprocessor: the *power-on reset* and the *manual reset*. A power-on reset allows the microprocessor to establish the initial values of registers and flip-flops and to initialize all I/O interface chips when power to the microprocessor is turned on. A manual reset without turning off the power allows the computer to get out of most error conditions (if the hardware has not failed) and reestablish the initial conditions. The computer *reboots* itself after a reset.

The reset service routine has a fixed starting address and is stored in the read-only memory of all microprocessors. At the end of the service routine, program control should be returned to the monitor program or the operating system. An easy way to do that (for the PIC18 MCU) is to place the upper, the middle, and the low bytes of the starting address of monitor in PCLATU, PCLATH, and PCL, respectively.

Like nonmaskable interrupts, resets are unmaskable. However, the microcontroller does not save any register on a reset.

6.4 The PIC18 Interrupts

The PIC18 MCU has the following interrupt sources:
- Four edge-triggered INT pin (INT0 . . . INT3) interrupts.
- Port B pins change (any one of upper four port B pins) interrupts.
- On-chip peripheral function interrupts. Because PIC18 members do not implement the same number of peripheral functions, they do not have the same number of peripheral interrupts.

6.4.1 PIC18 Interrupt Priority

The PIC18 MCU allows the user to have the option to divide all interrupts into two categories: high-priority group and low-priority group. When the application requires certain interrupts to receive closer attention, the user can place them in the high-priority group. If the application does not need to differentiate the importance of interrupts, the user can choose not to enable priority scheme.

6.4.2 Registers Related to Interrupts

A PIC18 member may use up to 13 registers to control interrupt operation. These registers are the following:
- RCON
- INTCON
- INTCON2
- INTCON3
- PIR1, PIR2, and PIR3
- PIE1, PIE2, and PIE3
- IPR1, IPR2, and IPR3

Each interrupt source has three bits to control its operation. These bits are the following:
- A flag bit—indicating whether an interrupt event has occurred.
- An enable bit—enabling or disabling the interrupt source.
- A priority bit—selecting high priority or low priority. This bit has effect only when the priority scheme is enabled.

6.4.3 RCON Register

The RCON register has a bit (IPEN) to enable interrupt priority scheme. The other bits are used to indicate the cause of reset. The contents of the RCON register are shown in Figure 6.1.

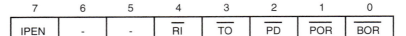

7	6	5	4	3	2	1	0
IPEN	-	-	$\overline{RI}$	$\overline{TO}$	$\overline{PD}$	$\overline{POR}$	$\overline{BOR}$

IPEN: Interrupt priority enable bit

   0: Disable priority levels on interrupts

   1: Enable priority levels on interrupts

$\overline{RI}$: RESET instruction flag bit

   0: The reset instruction was executed causing a device reset

   1: The reset instruction was not executed

$\overline{TO}$: Watchdog timeout flag bit

   0: A watchdog timeout occurred

   1: After power-up, CLRWDT instruction, or SLEEP instruction

$\overline{PD}$: Power-down detection flag bit

   0: By execution of the SLEEP instruction

   1: After power up or by the CLRWDT instruction

$\overline{POR}$: Power-on reset status bit

   0: A power-on reset has occurred

   1: A power-on reset has not occurred

$\overline{BOR}$: Brown-out reset status bit (PIC18CX01 does not have this bit)

   0: A brown-out reset has occurred

   1: A brown-out reset has not occurred

Figure 6.1 ■ The RCON Register (reprint with permission of Microchip)

6.4.4 Interrupt Control Registers (INTCON, INTCON2, INTCON3)

These three registers contain enable, priority, and flag bits for external INT pins, port B pin change, and Timer 0 (TMR0) overflow interrupts. The contents of these three registers are shown in Figures 6.2a, 6.2b, and 6.2c. Careful readers may have discovered that the INT0 pin interrupt does not have a priority bit for the user to select its interrupt priority. In fact, INT0 interrupt is always at the high priority because its request appears in both the high-priority and the low-priority logic circuit.

It is important to know that when the priority scheme is enabled, the user needs to set the GIEH bit (bit 7) in order to enable the low-priority interrupts. Setting the GIEL bit without setting the GIEH bit will not enable any low-priority interrupts.

7	6	5	4	3	2	1	0
GIE/GIEH	PEIE/GIEL	TMR0IE	INT0IE	RBIE	TMR0IF	INT0IF	RBIF

GIE/GIEH: Global interrupt enable bit
  when IPEN (RCON<7>)= 0
  0: disables all interrupts
  1: enables all interrupts
  when IPEN = 1
  0: disables all interrupts
  1: enables all high priority interrupts

PEIE/GIEL: Peripheral interrupt enable bit
  when IPEN = 0:
  0: disables all peripheral interrupts
  1: enables all peripheral interrupts
  when IPEN = 1
  0: disables all low priority interrupts
  1: enables all low priority interrupts

TMR0IE: TMR0 overflow interrupt enable bit
  0: disables TMR0 overflow interrupt
  1: enables TMR0 overflow interrupt

INT0IE: INT0 pin interrupt enable
  0: disables INT0 pin interrupt
  1: enables INT0 pin interrupt

RBIE: PORTB port change interrupt enable bit
  0: disables PORTB port change interrupt
  1: enables PORTB port change interrupt

TMR0IF: TMR0 overflow interrupt flag bit
  0: TMR0 has not overflowed
  1: TMR0 has overflowed

INT0IF: INT0 pin interrupt flag bit
  0: the INT0 pin interrupt did not occur
  1: the INT0 pin interrupt has occurred

PORTB port change interrupt flag bit
  0: none of the RB7:RB4 pins have changed state
  1: at least one of the RB7:RB4 pins change state

Figure 6.2a ■ The INTCON register (reprint with permission of Microchip)

7	6	5	4	3	2	1	0
R̄B̄P̄Ū	INTEDG0	INTEDG1	INTEDG2	INTEDG3	TMR0IP	INT3IP	RBIP

RBPU: PORTB pull-up enable bit

    0: all PORTB pull-ups are enabled

    1: all PORTB pull-ups are disabled

INTEDG0..INTEDG3: INT0..INT3 interrupt pins edge select

    0: interrupt on falling edge

    1: interrupt on rising edge

TMR0IP: TMR0 overflow interrupt priority bit

    0: low priority

    1: high priority

INT3IP: INT3 interrupt priority bit (not available in P18FXX8 & P18CX01)

    0: low priority

    1: high priority

RBIP: PORTB change interrupt priority bit

    0: low priority

    1: high priority

Note. 1. PIC18FXX8 does not have INTEDG2 & INTEDG3)

       2. PIC18C601/801 does not have INTEDG3

Figure 6.2b ■ The INTCON2 register (reprint with permission of Microchip)

7	6	5	4	3	2	1	0
INT2IP	INT1IP	INT3IE	INT2IE	INT1IE	INT3IF	INT2IF	INT1IF

INT2IP..INT1IP: INT2..INT1 interrupt priority bit

    0: low priority

    1: high priority

INT3IE..INT1IE: INT3..INT1 interrupt enable bit

    0: disable interrupt

    1: enable interrupt

INT3IF..INT1IF: INT3..INT1 interrupt flag bit

    0: interrupt did not occur

    1: interrupt occurred

Note. 1. PIC18FXX2 , PIC18CXX2, PIC18CXX8, and PIC18FXX8 do not

       have INT2 and INT3 enable and flag bits

      2. PIC18C601/801 does not have INT3 enable and flag bits

Figure 6.2c ■ The INTCON3 register (reprint with permission of Microchip)

6.4.5 PIR1 . . . PIR3 Registers

The PIR registers contain the individual flag bits for the peripheral interrupts. Because not all PIC18 members implement the same peripheral functions, some bits may not be present in some PIC18 members. The flag bits allow the interrupt service routine to identify the cause of the interrupt. The contents of these registers am shown in Figures 6.3a to 6.3d.

The PIC18C601/801 does not have the upper four bits of the PIR2 register, whereas the PIC18CXX2/PIC18FXX2 does not have the upper three bits of this register. The PIC18C601/801 and the PIC18CXX2/PIC18FXX2 do not have the PIR3 register. The PIR3 registers are used as the interrupt flag bits for the CAN controller for the PIC18 devices that have a CAN controller.

7	6	5	4	3	2	1	0
PSPIF[1]	ADIF	RC1IF	TX1IF	SSPIF	CCP1IF	TMR2IF	TMR1IF

PSPIF: Parallel slave port Read/Write interrupt flag bit [1]

    0: no read or write has occurred

    1: a read or write operation has taken place (must be cleared in software)

ADIF: A/D converter interrupt flag bit

    0: the A/D conversion is not completed

    1: an A/D conversion completed (must be cleared in software)

RCIF: USART receive interrupt flag bit

    0 = the USART receive buffer is empty

    1 = the USART receive buffer is full (cleared by reading RCREG)

TXIF: USART transmit interrupt flag bit

    0 = the USART transmit buffer is full

    1 = the USART transmit buffer is empty

SSPIF: Synchronous serial port interrupt flag bit

    0 = waiting to transmit/receive

    1 = the transmission/reception is complete (must be cleared in software)

CCP1IF: CCP1 interrupt flag bit

Capture mode

    0 = no TMR1 or TMR3 register capture occurred

    1 = a TMR1 or TMR3 register capture occurred (must be cleared in software

Compare mode

    0 = no TMR1 or TMR3 register compare match occurred

    1 = a TMR1 or TMR3 register compare match occurred (must be cleared in :

PWM mode

    Unused in this mode

TMR2IF: TMR2 to PR2 match interrupt flag bit

    0 = No TMR2 to PR2 match occurred

    1 = TMR2 to PR2 match occurred (must be cleared in software)

TMR1IF: TMR1 overflow interrupt flag bit

    0 = TMR1 register did not overflow

    1 = TMR1 register overflowed (must be cleared in software)

Note 1. Enabled only in Microcontroller mode for the PIC18F8X20 devices

    2. PIC18CX01 device does not have PSPIF flag bit

Figure 6.3a ■ The PIC18 PIR1 register (reprint with permission of Microchip)

7	6	5	4	3	2	1	0
-	CMIF	-	EEIF	BCLIF	LVDIF	TMR3IF	CCP2IF

CMIF: Comparator interrupt flag bit

    0: the comparator input has not changed

    1: the comparator input has changed (must clear in software)

EEIF: Data EEPROM/FLASH write operation interrupt flag bit

    0: the write operation is not complete

    1: the write operation is complete (must cleared in software)

BCLIF: Bus collision interrupt flag bit

    0: no bus collision

    1: a bus collision occurred while SSP module was transmission (I2C mode)

       (must be cleared in software)

LVDIF: low voltage detect interrupt flag bit

    0: the device voltage is above the low voltage detect trip point

    1: a low voltage condition occurred (must be cleared in software)

TMR3IF: TMR3 overflow interrupt flag bit

    0: TMR3 register did not overflow

    1: TMR3 register overflowed

CCP2IF: CCP2 interrupt flag bit

    Capture mode

    0: no TMR1 or TMR3 register capture occurred

    1: a TMR1 or TMR3 register capture occurred (must be cleared in software)

    Compare mode

    0: no TMR1 or TMR3 register compare match occurred

    1: TMR1 or TMR3 register compare match occurred (must be cleared in software)

    PWM mode:

    unused in this mode

Figure 6.3b ■ The PIC18 PIR2 register (reprint with permission of Microchip)

7	6	5	4	3	2	1	0
-	-	RC2IF	TX2IF	TMR4IF	CCP5IF	CCP4IF	CCP3IF

RC2IF: USART2 receive interrupt flag bit

    0 = the USART receive buffer is empty

    1 = the USART receive buffer is full (cleared when RCREG is read)

TX2IF: USART2 transmit interrupt flag bit

    0 = the write operation is not complete

    1 = the USART transmit buffer is empty (cleared when TXREG is written)

TMR4IF: TMR4 overflow interrupt flag bit

    0 = TMR4 did not overflow

    1 = TMR4 register overflowed (must be cleared in software)

CCPxIF: CCPx interrupt flag bit (x = 3, 4, 5)

    Capture mode:

    0 = no TMR1 or TMR3 register capture occurred

    1 = a TMR1 or TMR3 register capture occurred (must be cleared in software)

    Compare mode:

    0 = no TMR1 or TMR3 register compare match occurred

    1 = a TMR1 or TMR3 register compare match occurred (must cleared in software)

    PWM mode: (not used)

Figure 6.3c ■ The PIR3 register (PIC18FXX20)(reprint with permission of Microchip)

7	6	5	4	3	2	1	0
IRXIF	WAKIF	ERRIF	TXB2IF/ TXBnIF	TXB1IF	TXB0IF	RXB1IF/ RXBnIF	RXB0IF

IRXIF: Invalid message received interrupt flag bit

   0: an invalid message has not occurred on the CAN bus

   1: an invalid message has occurred on the CAN bus

WAKIF: Bus activity wakeup interrupt flag bit

   0: activity on the CAN bus has not occurred

   1: activity on the CAN bus has occurred

ERRIF: CAN bus error interrupt flag big

   0: an error has not occurred in the CAN module

   1: an error has occurred in the CAN module (multiple sources)

When CAN is in mode 0

TXB2IF..TXB0IF: Transmit buffer 2..0 interrupt flag bit

   0: transmit buffer 2..0 has not completed transmission of a message

   1: transmit buffer 2..0 has completed transmission of a message and may be reloaded

When CAN is in mode 1 or mode 2

TXBnIF: Any transmit buffer interrupt flag bit

   0: No message was transmitted

   1: One or more transmit buffer has completed transmission of a message and may
      reloaded

TXB1IF and TXB0IF are forced to 0 in mode 1 and mode 2

When CAN is in mode 0

RXB1IF..RXB0IF: Receive buffer 1..0 interrupt flag bit

   0: receive buffer 1 (or 0) has not received a new message

   1: receive buffer 1 (or 0) has received a new message

When CAN is in mode1 or mode 2

RXBnIF: CAN receive buffer interupt flag big

   0: No receive buffer has received a new message

   1: One or more receive buffer has received a new message

RXB0IF is forced to 0 when in mode 1 or mode 2

Figure 6.3d ■ The PIC18 PIR3 register (PIC18FXX8 or other devices with CAN) (reprint with permission of Microchip)

6.4.6 PIE1 . . . PIE3 Registers

The PIE registers contain the individual enable bits for the peripheral interrupts. When the IPEN bit (RCON<7>) is 0, the PEIE bit must be set to enable any of these peripheral interrupts. The contents of these three registers are shown in Figures 6.4a to 6.4d.

The 40-pin or smaller devices do not have the PIE3 register. The PIC18Fxx20 uses this register to enable or disable the second USART, TMR4, and three input capture interrupts. The PIC18 devices with a CAN controller use this register to enable/disable interrupts related to the operation of the CAN controller.

7	6	5	4	3	2	1	0
PSPIE[1]	ADIE	RC1IE	TX1IE	SSPIE	CCP1IE	TMR2IE	TMR1IE

PSPIE: Parallel slave port Read/Write interrupt enable bit [1]

    0: disables the PSP read/write interrupt

    1: enables the PSP read/write interrupt

ADIE: A/D converter interrupt enable bit

    0: disables the A/D interrupt

    1: enables the A/D interrupt

RC1IE: USART1 receive interrupt enable bit

    0 = disable the USART1 receive interrupt

    1 = enable the USART1 receive interrupt

TX1IE: USART1 transmit interrupt enable bit

    0 = disable the USART1 transmit interrupt

    1 = enable the USART1 transmit interrupt

SSPIE: Synchronous serial port interrupt enable bit

    0 = disables the MSSP interrupt

    1 = enables the MSSP interrupt

CCP1IE: CCP1 interrupt enable bit

    0 = disables the CCP1 interrupt

    1 = enables the CCP1 interrupt

TMR2IE: TMR2 to PR2 match interrupt enable bit

    0 = disables the TMR2 to PR2 match interrupt

    1 = enables the TMR2 to PR2 match interrupt

TMR1IE: TMR1 overflow interrupt enable bit

    0 = disables TMR1 register overflow interrupt

    1 = enables TMR1 register overflow interrupt

Note 1. Enabled only in Microcontroller mode for the PIC18F8X20 devices

    2. The PIC18C601/801 does not have PSPIE flag bit

Figure 6.4a ■ The PIE1 register (reprint with permission of Microchip)

7	6	5	4	3	2	1	0
-	CMIE	-	EEIE	BCLIE	LVDIE	TMR3IE	CCP2IE

CMIE: Comparator interrupt enable bit

    0: disables the comparator interrupt

    1: enables the comparator interrupt

EEIE: Data EEPROM/FLASH write operation interrupt enable bit

    0: disables the write operation interrupt

    1: enables the write operation interrupt

BCLIE: Bus collision interrupt enable bit

    0: disables the bus collision interrupt

    1: enables the bus collision interrupt

LVDIE: low voltage detect interrupt enable bit

    0: disables the device voltage detect interrupt

    1: enables the device voltage detect interrupt

TMR3IE: TMR3 overflow interrupt enable bit

    0: disables the TMR3 overflow interrupt

    1: enables the TMR3 overflow interrupt

CCP2IE: CCP2 interrupt enable bit

    0: disables the CCP2 interrupt

    1: enables the CCP2 interrupt

Note 1. The PIC18C601/801 has only the lower four bits of this register.

      2. The PIC18FXX2 does not have the CMIF bit

Figure 6.4b ■ The PIE2 register (reprint with permission of Microchip)

7	6	5	4	3	2	1	0
-	-	RC2IE	TX2IE	TMR4IE	CCP5IE	CCP4IE	CCP3IE

RC2IE: USART2 receive interrupt enable bit

    0: the comparator input has not changed

    1: the comparator input has changed (must clear in software)

TX2IE: USART2 transmit interrupt enable bit

    0: disables the USART2 transmit enable bit

    1: enables the USART2 transmit enable bit

TMR41E: TMR4 overflow interrupt enable bit

    0: disables the TMR4 overflow interrupt

    1: enables the TMR4 overflow interrupt

CCPxIE: CCPx interrupt enable bit (x = 3, 4, or 5)

    0: disables the CCPx interrupt

    1: enables the CCPx interrupt

Figure 6.4c ■ The PIE3 register (PIC18FXX20) (reprint with permission of Microchip)

7	6	5	4	3	2	1	0
IRXIE	WAKIE	ERRIE	TXB2IE/ TXBnIE	TXB1IE	TXB0IE	RXB1IE/ RXBnIE	RXB0IE

IRXIE: Invalid message received interrupt enable bit

  0: disables the invalid CAN message received interrupt

  1: enables the invalid CAN message received interrupt

WAKIE: Bus activity wakeup interrupt enable bit

  0: disables the bus activity wakeup interrupt

  1: enables the bus activity wakeup interrupt

ERRIE: CAN bus error interrupt enable big

  0: disables the CAN bus error interrupt

  1: enables the CAN bus error interrupt

When CAN is in mode 0:

TXB2IE..TXB0IE: Transmit buffer 2..0 interrupt enable bit

  0: disables transmit buffer 2..0 interrupt

  1: enables transmit buffer 2..0 interrupt

When CAN is in mode 1 or mode 2:

TXBnIE: CAN transmit buffer interrupt enable bit

  0: disable all transmit buffer interrupts

  1: enable transmit buffer interrupt; individual interrupt is enabled by TXBIE abd BIE0

When CAN is in mode 0:

TXB1IE and TXB0IE are forced to 0 in mode 1 and mode 2.

RXB1IE..RXB0IE: Receive buffer 1..0 interrupt enable bit

  0: disables receive buffer 1 (or 0) interrupt

  1: enables receive buffer 1 (or 0) interrupt

When CAN is in mode 1 or mode 2:

RXBnIE: CAN receive buffer interrupt enable bit

  0: disable all receive buffer interrupts

  1: enable receive buffer interrupts; individual interrupts is enabled by BIE0

RXB0IE is forced to 0 when in mode 1 and mode 2.

Figure 6.4d ■ Contents of the PIC18 PIE3 register (PIC18FXX8 and other devices with CAN) (reprint with permission

6.4.7 Interrupt Priority Registers (IPR1 . . . IPR3)

The IPR registers contain the individual priority bits for the peripheral interrupts. These registers have effect only when the interrupt priority enable (IPEN) bit is set. By enabling interrupt priority and setting the associated priority bit, the user can place any peripheral interrupt at high or low priority. The contents of these registers are shown in Figure 6.5.

7	6	5	4	3	2	1	0
PSPIP[1]	ADIEP	RC1IP	TX1IP	SSPIP	CCP1IP	TMR2IP	TMR1IP

PSPIP: Parallel slave port Read/Write interrupt priority bit [1]
    0: low priority
    1: high priority
ADIP: A/D converter interrupt priority bit
    0: low priority
    1: high priority
RCIP: USART receive interrupt priority bit
    0: low priority
    1: high priority
TXIP: USART transmit interrupt priority bit
    0: low priority
    1: high priority
SSPIP: Synchronous serial port interrupt priority bit
    0: low priority
    1: high priority
CCP1IP: CCP1 interrupt priority bit
    0: low priority
    1: high priority
TMR2IP: TMR2 to PR2 match interrupt priority bit
    0: low priority
    1: high priority
TMR1IP: TMR1 overflow interrupt priority bit
    0: low priority
    1: high priority

Note 1. Enabled only in Microcontroller mode for the PIC18F8X20 devices
    2. The PIC18C601/801 does not have the PSPIP priority bit

Figure 6.5a ■ The IPR1 register (reprint with permission of Microchip)

7	6	5	4	3	2	1	0
-	CMIP	-	EEIP	BCLIP	LVDIP	TMR3IP	CCP2IP

CMIP: Comparator interrupt priority bit
  0: low priority
  1: high priority
EEIP: Data EEPROM/FLASH write operation interrupt priority bit
  0: low priority
  1: high priority
BCLIP: Bus collision interrupt priority bit
  0: low priority
  1: high priority
LVDIP: low voltage detect interrupt priority bit
  0: low priority
  1: high priority
TMR3IP: TMR3 overflow interrupt priority bit
  0: low priority
  1: high priority
CCP2IP: CCP2 interrupt priority bit
  0: low priority
  1: high priority
Note 1. The PIC18C601/801 has only the lower four bits of this register.
     2. The PIC18FXX2 does not have the CMIP bit

Figure 6.5b ■ The IPR2 register (reprint with permission of Microchip)

7	6	5	4	3	2	1	0
-	-	RC2IP	TX2IP	TMR4IP	CCP5IP	CCP4IP	CCP3IP

RC2IP: USART2 receive interrupt priority bit
  0: low priority
  1: high priority
TX2IP: USART2 transmit interrupt priority bit
  0: low priority
  1: high priority
TMR4IP: TMR4 overflow interrupt priority bit
  0: low priority
  1: high priority
CCPxIP: CCPx interrupt priority bit (CCP modules 3, 4, and 5)
  0: low priority
  1: high priority

Figure 6.5c ■ The IPR3 register (PIC18FXX20) (reprint with permission of Microchip)

7	6	5	4	3	2	1	0
IRXIP	WAKIP	ERRIP	TXB2IP/ TXBnIP	TXB1IP	TXB0IP	RXB1IP/ RXBnIP	RXB0IP

IRXIP: Invalid message received interrupt priority bit

  0: low priority

  1: high priority

WAKIP: Bus activity wakeup interrupt enable bit

  0: low priority

  1: high priority

ERRIP: CAN bus error interrupt enable big

  0: low priority

  1: high priority

TXB2IP..TXB0IP: Transmit buffer 2..0 interrupt enable bit

  In mode 0

  0: low priority

  1: high priority

  In mode 1 and mode 2

TXBnIP is used as the priority bit for all CAN transmit buffers

  0: low priority

  1: high priority

  Bit 3 and 2 are forced to 0 in CAN mode 1 and mode 2.

RXB1IP..RXB0IP: Receive buffer 1..0 interrupt enable bit

  In mode 0

  0: low priority

  1: high priority

  In mode 1 and 2

RXBnIP is used to enable the priority of all receive buffer

  0: low priority

  1: high priority

  Bit 0 is forced to 0 in CAN mode 1 and mode 2.

Figure 6.5d ■ The IPR3 register (devices with CAN) (reprint with permission of Microchip)

6.5 PIC18 Interrupt Operation

The PIC18 MCU can enable or disable the priority scheme. Once enabled, the priority scheme divides all interrupt sources into high-priority and low-priority groups. They operate differently from the nonpriority scheme. Every interrupt source has an associated priority bit. The priority bits of interrupt sources in the core group are contained in one of the interrupt control registers (INTCON, INTCON2, and INTCON3), whereas the priority bits of peripheral interrupts are contained in one of the IPR registers (IPR1 . . . IPR3).

When the priority scheme is enabled, all high-priority interrupts are under the control of a two-level interrupt enabling mechanism, whereas all low-priority interrupts are under the control of a three-level interrupt enabling scheme. If the priority scheme is disabled, then all peripheral interrupts are under the control of a three-level enabling scheme. Interrupts in the core group are under the control of a two-level enabling scheme. The following interrupt sources are in the core group:

- INT pin (INT0 . . . INT3) interrupts. Devices with 44, 40, and 28 pins may have only INT0 and INT1 pins (devices with CAN) or INT0 through INT2 pins (devices without CAN).
- TMR0 overflow interrupt.
- PORTB input pin (upper four pins only) change interrupts.

6.5.1 PIC18 Interrupt without Setting Priority

When bit 7 (IPEN bit) of the RCON register is cleared, the PIC18 MCU disables the interrupt priority and works in the compatible mode (compatible with the PIC16 MCU). All interrupt sources share the common interrupt vector at 0x000008.

To enable an interrupt that is under the control of the two-level enabling mechanism, one needs to set the GIE bit of the INTCON register and its associated interrupt bit. For example, one needs to set both the GIE bit and the T0IE bit (INTCON<5>) in order to enable the TMR0 overflow interrupt.

To enable an interrupt that is under the control of the three-level enabling mechanism, one needs to set the GIE bit, the PEIE bit, and the interrupt enable bit associated with the interrupt source. For example, one needs to set the following three bits in order to enable the A/D conversion-complete interrupt:

1. The GIE bit (INTCON<7>)
2. The PEIE bit (INTCON<6>)
3. The ADIE bit (PIE1<6>)

As long as one of these three enable bits is not set, the A/D conversion-complete interrupt will be disabled. If the PEIE bit is cleared, then none of the peripheral interrupts will occur.

When an interrupt is responded to, the GIE bit is automatically *cleared* to disable any further interrupt, the return address is pushed onto the return address stack, and the program counter is loaded with the interrupt vector. Most interrupt flags must be cleared by the software. This must be done in the interrupt service routine to avoid repetitive interrupts from the same source.

The last instruction of an interrupt service routine must be the **retfie** instruction. The execution of this instruction will cause the top word of the hardware stack to be popped into the program counter and the GIE bit to be set. After this, program control is returned to the interrupted program, and interrupt is reenabled. This instruction has the option to restore the WREG, the STATUS, and the BSR registers from the fast register stack if the application requires it.

6.5.2 PIC18 Interrupt with Priority Enabled

This mechanism is useful when certain interrupts need to be serviced promptly. As shown in Figure 6.2a, the GIE bit (bit 7) is used as the global high-priority interrupt enable bit (GIEH), whereas the PEIE bit (bit 6) is used as the global low-priority interrupt enable (GIEL) bit when the priority scheme is enabled. The GIEH bit must also be set in order to enable low-priority interrupts.

All unmasked interrupts are enabled by setting the GIEH bit, and all unmasked low-priority interrupts are enabled by setting the GIEL bit. By default, all interrupts are placed in high-priority level after reset. High-priority interrupts can interrupt a low-priority interrupt under service. A high-priority interrupt will be enabled when both the GIEH bit and its associated enable bit are set to 1. To enable a low-priority interrupt, the user needs to set the GIEH bit, the GIEL bit, and its associated enable bit.

To place an interrupt source at the low-priority level, clear its corresponding priority bit in one of the IPR registers. All low-priority interrupts share the same interrupt vector at 0x000018, whereas all high-priority interrupts share the same interrupt vector at 0x000008.

The pending interrupts in the high-priority group will always be serviced before the pending interrupts in the low-priority group. When the **retfie** instruction is executed and the priority scheme is functioning, then either the GIEH or the GIEL bit will be set to reenable the interrupt. Interrupt flags must be cleared in the software to avoid recursive interrupts. In addition, it is a good practice to clear the interrupt flag bit before enabling any interrupt.

6.5.3 INT Pin Interrupts

All INT pin interrupts (INT0 . . . INT3) are edge triggered. The edge-select bits are contained in the INTCON2 register. When an edge-select bit is set to 1, the corresponding INT pin interrupt is requested on the rising edge. Otherwise, the INT pin interrupt is requested on the falling edge.

6.5.4 Port B Pins Input Change

An input change on pins RB7 . . . RB4 sets the flag bit RBIF (INTCON<0>). The interrupt can be enabled/disabled by setting/clearing the RBIE bit (INTCON<3>).

All port B pins are general I/O pins. They can be configured as input or output. Only pins configured as inputs can cause this interrupt to occur. Any of these input changes will cause the RBIF (INTCON<0>) bit to set. Reading the PORTB port will clear the RBIF flag bit. The user can also use the **bcf** instruction to clear this flag bit.

The interrupt-on-change feature is mainly used for wakeup on key depression operation and operations where PORTB is used only for the interrupt-on-change feature.

6.5.5 TMR0 Overflow Interrupt

The Timer0 module consists of a 16-bit timer/counter (TMR0) and can operate in 8-bit or 16-bit mode. The high byte of TMR0 is the TMR0H register, whereas the low byte is the TMR0L register. Whenever the register TMR0 counts from 0xFF to 0x00 in the 8-bit mode or from 0xFFFF to 0x0000 in the 16-bit mode, an interrupt may be generated. The clock source to this timer/counter can be either the T0CKI pin or the internal instruction clock. The control bits for this module are in the T0CON register. We defer the discussion of this interrupt until Chapter 8.

6.5.6 Peripheral Interrupts

The PIC18 has the following peripheral interrupts:

- Parallel slave port interrupt (not available in devices with 28 pins or less)
- Analog-to-digital (A/D) conversion complete interrupt
- USART1 receive interrupt
- USART1 transmit interrupt
- Master synchronous serial port (SPI) interrupt
- CCP1 interrupt
- TMR2 to PR2 match interrupt
- TMR1 overflow interrupt
- Comparator interrupt
- Data EEPROM/Flash write operation interrupt
- Bus collision interrupt

- Low-voltage detect interrupt
- TMR3 overflow interrupt
- CCP2 interrupt
- USART2 receive interrupt (not all PIC18 devices have this interrupt)
- USART2 transmit interrupt (not all PIC18 devices have this interrupt)
- TMR4 overflow interrupt (not all PIC18 devices have this interrupt)
- CCP5 interrupt (not all PIC18 devices have this interrupt)
- CCP4 interrupt (not all PIC18 devices have this interrupt)
- CCP3 interrupt (not all PIC18 devices have this interrupt)
- Invalid message on CAN bus interrupt (available only in devices with CAN controller)
- Activity on CAN bus interrupt (available only in devices with CAN controller)
- CAN bus error interrupt (available only in devices with CAN controller)
- CAN transmission buffer 2 interrupt (available only in devices with CAN controller)
- CAN transmission buffer 1 interrupt (available only in devices with CAN controller)
- CAN transmission buffer 0 interrupt (available only in devices with CAN controller)
- CAN receive buffer 1 interrupt (available only in devices with CAN controller)
- CAN receive buffer 0 interrupt (available only in devices with CAN controller)

The functioning of these interrupt sources is discussed in appropriate chapters.

6.6 PIC18 Interrupt Programming

There are two steps in the PIC18 interrupt programming:

Step 1
Write the interrupt service routine and place it in the predefined program memory location.

Step 2
Set the appropriate interrupt enable bits to enable the desired interrupt.

6.6.1 Interrupt Programming in Assembly language

Interrupt programming in assembly language is straightforward. The following two examples show that. Example 1 illustrates how to deal with an INT pin interrupt. Example 2 demonstrates how to handle a low-priority timer 1 overflow interrupt. Both examples utilize LEDs to indicate the occurrence of interrupts.

Example 6.1
▼

Suppose you are given a circuit as shown in Figure 6.6. Write a main program and an INT0 interrupt service routine in assembly language. The main program initializes a counter to 0, enables the INT0 interrupt, and then stays in a **while-loop** to wait forever. The INT0 interrupt service routine simply increments the counter by 1 and outputs it to the LEDs. Whenever the count is incremented to 15, the service routine resets it to 0. Choose appropriate component values so that the PIC18 receives an INT0 interrupt roughly every second. The frequency of the V_{OUT} signal is given in Figure. 6.6. (SSE452, SSE8680, and SSE8720 demo boards have a 1-Hz clock source.)

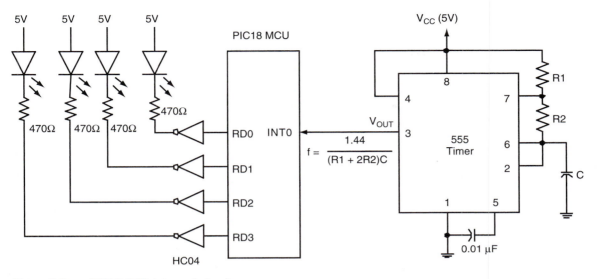

Figure 6.6 ■ PIC18 INT0 interrupt circuit

Solution: Set R1, R2, and C to be 6.2K Ω, 150K Ω, and 4.7μF, respectively. The frequency of the pin 3 output will be about 1 Hz. Since the output of pin 3 is a periodic square waveform and there is only one rising edge and one falling edge per period, there is no need to specify the active edge for interrupting the PIC18 MCU on the INT0 pin. The main program and the INT0 interrupt service routine are as follows:

```
        #include   <p18F8720.inc>
count   set        0x00          ; count value to be output to LEDs
        org        0x00
        goto       main
        org        0x08          ; INT interrupt vector
        goto       int0_ISR
        retfie
        org        0x18
        retfie                   ; pseudo interrupt service routine
main    clrf       count,A       ; initialize the count to 0
        movlw      0xF0          ; configure port D lower 4 pins
        movwf      TRISD         ; for output
        movff      count,PORTD   ; output count to port D
        bsf        RCON,IPEN,A   ; enable priority interrupt
        bcf        INTCON,INTOIF ; clear INT0 flag
        movlw      0x90          ; enable INT0 interrupt
        movwf      INTCON,A      ; "
forever goto       forever
;*******************************
; INT0 pin interrupt service routine
;*******************************
```

```
int0_ISR     bcf           INTCON,INT0IF           ; clear the INT0 interrupt flag bit
             movlw         0x0F
             cpfseq        count, A                ; was count equal to 15 already?
             goto          add_1
             clrf          count,A                 ; reset count to 0 after reaching 15
             goto          update
add_1        incf          count,F,A
update       movff         count,PORTD             ; output the count to LEDs
             retfie        fast
end
```

Most PIC18 devices have four timers. The timer function is a good vehicle for experiment-ing interrupt programming. The user can configure one of the timers (e.g., timer 1) as follows:

- Low-priority interrupt
- Use instruction clock as the clock signal
- Set prescale factor properly so that TMR1 overflows twice per second

▲

Example 6.2

▼

Suppose port D pins RD3 . . . RD0 are driving four LEDs. Write an assembly program that chooses the instruction clock as the clock source for timer 1 and set the prescale factor so that it overflows twice every second. The program will increment a counter by one in every timer 1 overflow interrupt and output the count value to LEDs. Program the timer 1 interrupt to the low-priority group. Assume that the crystal oscillator used in the demo board is 4 MHz.

Solution: The program is as follows:

```
             #include >p18F452.inc>
led_cnt      set           0x00
             org           0x00
             goto          start
             org           0x08
             retfie
             org           0x18
             goto          tmr1_ISR                ;jump to timer 1 interrupt service routine
             retfie
start        movlw         0xF0                    ; configure port D upper four pins for input
             movwf         TRISD,A                 ; and port B lower four pins for output
             clrf          led_cnt                 ; initialize the count to 0
             movlw         0xB1                    ; enables TMR1, 16-bit mode, set prescaler
             movwf         T1CON,A                 ; value to 8, choose FOSC/4, enable TMR1
             bcf           PIR1,TMR1IF,A           ; clear TMR1 flag bit
             bsf           PIE1,TMR1IE,A           ; enable TMR1 overflow interrupt
             bsf           RCON,IPEN,A             ; enable priority interrupts
             bcf           IPR1,TMR1IP,A           ; set TMR1 to low priority
             bsf           INTCON,GIEH,A           ; enable high priority interrupt
             bsf           INTCON,GIEL,A           ; enable low priority interrupt
             movlw         0xDB                    ; let Timer 1 overflow in
             movwf         TMR1L                   ; half a second
```

```
                    movlw       0x0B                  ; "
                    movwf       TMR1H                 ; "
forever             goto        forever
                    nop
; ********************************
;
; Timer 1 low-priority interrupt service routine
; ********************************
;
tmr1_ISR            bcf         PIR1,TMR1IF,A         ; clear TMR1 overflow flag
                    movlw       0xDB
                    movwf       TMR1L
                    movlw       0x0B
                    movwf       TMR1H
                    movlw       0x0F
                    cpfseq      led_cnt,A             ; is count value equal to 15?
                    goto        add1
                    clrf        led_cnt,A             ; reset count value to 0
                    goto        update
add1                incf        led_cnt,A
update              movff       led_cnt,PORTD         ; update the LED values
                    retfie
                    end
```

When running this program, you can try clearing the GIEH bit to see if LEDs are still changing. When programming the microcontroller on the demo board, the user can disable background debug mode. This is done by choosing the ICD2 as a programmer in MPLAB IDE and selecting no debugging tool under the Debugger menu. After disabling the background debug mode, the demo board will start executing the user program after power on reset (the ICD 2 must be unplugged from the demo board).

6.6.2 Interrupt Programming in C Language

In C language, the user needs to declare the function that handles the interrupt to be an interrupt service routine by using the **pragma** directive. Since there are high- and low-priority interrupts, the user needs to make the following declarations:

```
#pragma interruptlow function_name     ; low priority interrupt service routine
#pragma interrupt function_name        ; high priority interrupt service routine
```

The MCC18 C compiler does not automatically place an interrupt service routine at the interrupt vector. Usually, a **goto** instruction is placed at the interrupt vector for transferring control to the proper interrupt service routine.

A template for writing interrupt service routine in C language is as follows:

```
#include <p18Fxxx.h>
void low_ISR(void);
void high_ISR(void);
#pragma code high_vector = 0x08          // force the following statement to start at 0x08
void high_interrupt (void)
{
        _asm
        goto high_ISR;
        _endasm
```

```
        }
        #pragma code                                //return to the default code section
        #pragma interrupt high_ISR
        void high_ISR (void)
        {
                . . .                               //handle high-priority interrupts
        }
        #pragma code low_vector = 0x18              //force the following statements to start at
        void low_interrupt (void)                   //0x18
        {
                _asm
                goto low_ISR;
                _endasm
        }
        #pragma code                                //return to the default code section
        #pragma interruptlow low_ISR
        void low_ISR (void)
        {
                . . .                               //handle low-priority interrupts
        }
```

Example 6.3

▼

Write a C language version of the program in Example 6.2.

Solution: The C language version of the program is as follows:

```
#include                 <p18F452.h>
#include                 <timers.h>
#define                  NUMBER_OF_LEDS          4
void timer_isr (void);
static unsigned char s_count = 0;

#pragma code low_vector = 0x18
void low_interrupt (void)
{
        _asm goto timer_isr _endasm
}

#pragma code
#pragma interruptlow timer_isr save = PROD        //inform C compiler to save PROD
void timer_isr (void)
        PIR1bits.TMR1IF = 0;                       //clear TMR1IF flag
        s_count = (++s_count) % 16;                //set the count range to 0..15
        TMR1 = 0xBDB;
        PORTD = s_count;                           //update LEDs
}
void main (void)
{
```

```
                TRISD = 0xF0;              //configure lower 4 Port D pins for output
                PORTD = 0;                 //initialize LEDs to 0
                T1CON = 0xB1;              //enable TMR1 and its interrupt
                RCONbits.IPEN = 1;         //enable interrupt priority scheme
                IPR1bits.TMR1IP = 0;       //configure TMR1 interrupt to low priority
                PIR1bits.TMR1IF = 0;       //clear the TMR1 interrupt flag bit
                INTCONbits.GIEL = 1;       //enable low interrupts
                INTCONbits.GIEH = 1;       //enable global interrupts
                PIE1bits.TMR1IE = 1;       //enable TMR1 overflow interrupt
                TMR1 = 0xBDB;
                while (1);                 //do nothing infinite loop
        }
```

The purpose of this example is to illustrate interrupt programming in C language, Since timer functions have not been discussed yet, some details might not be clear.

This example considers only the TMR1 overflow interrupt. If an application enables multiple interrupts, then the common interrupt service routine needs to check other interrupt flag bits to identify the cause of interrupt and take actions accordingly.

6.6.3 Context Saving during Interrupts

When an interrupt occurs, the PIC18 MCU automatically saves WREG, BSR, and STATUS in the fast register stack. The user can optionally use the **retfie fast** instruction to restore these three registers when returning from the interrupt.

If both low- and high-priority interrupts are enabled, the registers saved in the fast register stack cannot be used reliably for low-priority interrupts. If a high-priority interrupt occurs while the MCU is servicing a low-priority interrupt, the register values saved by the low-priority interrupt will be overwritten. The low-priority interrupt must save the key registers in the software stack in order not to lose their values. This is usually done immediately after the low-priority interrupt service routine is entered. Since the low-priority interrupt service routines need to save WREG, BSR, and STATUS in the software stack, the **retfie fast** instruction shall only be used by the high-priority interrupt service routine. The user can invoke the stack macros defined in Chapter 4 to save and restore the key registers in the low-priority interrupt service routine.

In C language, the user can add a **save** clause to the **#pragma** statement to inform the C compiler to generate appropriate instructions for saving additional registers (in software stack). The syntax is as follows:

```
#pragma interrupt high_ISR save        = reg1, .., regn
#pragma interruptlow low_ISR save      = reg1, .., regn
```

You can also save a whole section of data when necessary. The syntax is as follows:

```
#pragma interrupt high_ISR save        = section("section name")
#pragma interruptlow low_ISR save      = section("section name")
```

For example, the following statement tells the C compiler to generate code to save the section of **.tmpdata** when the interrupt service routine is entered:

```
pragma interrupt high_ISR save         =section(".tmpdata")
```

This statement can eliminate many unexpected errors when the interrupt service routine needs to call a subroutine.

6.7 The PIC18 Resets

The main function of reset is to establish or reestablish appropriate values for key registers so that the MCU can start or restart properly. The PIC18 MCU differentiates between various kinds of reset:

- Power-on reset (POR)
- $\overline{MCLR}$ pin reset during normal operation
- $\overline{MCLR}$ pin reset during SLEEP
- Watchdog timer (WDT) reset (during normal operation)
- Programmable brown-out reset (BOR)
- RESET instruction
- Stack full reset
- Stack underflow reset

The block diagram of the PIC18 reset circuit is shown in Figure 6.7. The $\overline{MCLR}$ signal is the manual reset signal which will set the S and R inputs of the SR latch to 1 and 0, respectively. Therefore, the **Chip_Reset** signal is asserted and resets the microcontroller.

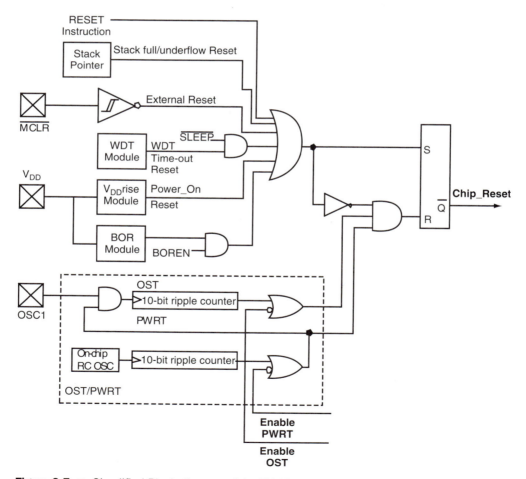

Figure 6.7 ■ Simplified Block diagram of the PIC18 on-clip reset circuit (redraw with permission of Microchip)

Although most registers are not affected by a reset, a few registers will be forced to a "RESET state" on power-on reset, $\overline{\text{MCLR}}$ pin reset, WDT reset, brown-out reset, $\overline{\text{MCLR}}$ reset during sleep, and by the RESET instruction. The cause of reset can be determined by looking at the RCON register (shown in Figure 6.1). Register values after reset can be found in Appendix A.

After the CPU leaves the reset state, the CPU will start program execution from program memory location 0x0000. This address is called *reset vector*.

6.7.1 Power-On Reset

A power-on reset (POR) pulse is generated when V_{DD} rise is detected. To take advantage of the POR circuitry, tie the $\overline{\text{MCLR}}$ pin through a 1-KΩ to 10-KΩ resistor to V_{DD}. This will eliminate external RC components usually needed to create a power-on reset delay.

6.7.2 Power-Up Timer

The power-up timer (PWRT) provides a fixed nominal time-out only on power-up. This timer operates on an internal RC oscillator. The chip is kept in RESET state as long as the PWRT is active. The PWRT creates a time delay that allows V_{DD} to rise to an acceptable level. A configuration bit is provided to enable/disable the PWRT timer. MPLAB IDE allows the user to deal with this bit before programming the target device. The power-up time delay will vary from chip to chip because of V_{DD}, temperature, and process variation.

6.7.3 Oscillator Start-Up Timer

All crystal oscillators require certain amount of time to stabilize after power is first turned on. The oscillator start-up timer provides 1024 oscillator cycles of delay (from OSC1 input) after the PWRT delay is over. This ensures that the crystal oscillator or resonator has started and stabilized. The oscillator start-up timer (OST) time-out is invoked only on power-on reset or wakeup from sleep mode.

6.7.4 Brown-Out Reset

"Brown-out" refers to the situation of power level going low temporarily. The voltage level on the V_{DD} input may fluctuate occasionally. When V_{DD} drops below a trip point (BV$_{DD}$, typically 4.2 V for a 5-V power supply), the microcontroller can no longer operate correctly. At this point, the CPU should reset itself until the V_{DD} input returns to its proper level. The brown-out reset (BOR) circuit provides this reset until V_{DD} goes above BV$_{DD}$. If the power-up timer is enabled, it will be invoked after V_{DD} rises above BV$_{DD}$; it will then keep the chip in RESET for an additional delay (power-up timer period). PIC18 provides a configuration bit (BOREN) for enabling and disabling the brown-out reset circuitry.

6.7.5 The Reset Instruction

This instruction will reset all registers and flags that are affected by a $\overline{\text{MCLR}}$ reset. One can use this instruction to force the MCU to enter a known state before the target application is launched.

6.8 Summary

Interrupt is a mechanism provided by a microprocessor or computer system to synchronize I/O operations, handle error conditions and emergency events, and coordinate the use of shared resources. Interrupts that can be ignored by the processor are called maskable interrupts. Interrupts that cannot be ignored by the processor are called nonmaskable interrupts. The processor provides service to the interrupt by executing a program called interrupt service routine, The last instruction of the interrupt service routine is the **retfie** instruction for the PIC18 MCU. When there are multiple active interrupts, the processor uses the priority assigned to each interrupt source to choose one of the pending interrupt requests to service. The active interrupt with the highest priority will receive service first. The PIC18 MCU implements a two-level interrupt priority scheme. When this scheme is enabled, any interrupt source can be placed in either group (high or low priority).

The term *interrupt vector* refers to the starting address of an interrupt service routine. Some microprocessors store all interrupt vectors in a table, whereas others (all Microchip microcontrollers) have predefined interrupt vectors. There is no need to store interrupt vectors in a table for these processors. The PIC18 MCU allocates eight words for high-priority interrupt service routine. If the interrupt service routine is longer than eight words, the common practice is to use the **goto** instruction to jump to the actual interrupt service routine.

The PIC18 MCU mixes the use of two- and three-level interrupt enabling mechanisms. When the priority scheme is enabled for the high-priority group, the user needs to set both the global high-priority interrupt enable (GIEH) and individual interrupt enable bits to enable any interrupt in this group. For the low-priority group, the user needs to set the GIEH, the global low-priority interrupt enable (GIEL), and individual interrupt enable bits in order to enable an interrupt in this group.

The PIC18 MCU divides all interrupt sources into two groups when the priority scheme is disabled:

- *Core group.* This group consists of four INT pins, timer 0 overflow, and port B pin change interrupts. This group is controlled by a two-level enabling mechanism. To enable an interrupt in this group, one needs to set both the global interrupt enable (GIE) and the individual interrupt enable bits.
- *Peripheral group.* All other interrupts belong to this group. To enable any interrupt in this group, the user needs to set three bits: GIE, PEIE, and the individual interrupt enable bits.

There are two steps in programming the PIC18 MCU interrupt: (1) write the interrupt service routine and (2) enable interrupt by setting appropriate interrupt enable bit(s).

The main function of reset is to establish or reestablish appropriate values for key registers so that the microcontroller can start or restart properly. The PIC18 MCU has eight different reset sources:

- Power-on reset
- $\overline{\text{MCLR}}$ reset during normal operation
- $\overline{\text{MCLR}}$ reset during SLEEP
- Watchdog timer reset (during normal operation)
- Programmable brown-out reset
- RESET instruction
- Stack full reset
- Stack underflow reset

6.9 Exercises

E6.1 What is the name given to a routine that is executed in response to an interrupt?

E6.2 What are the advantages of using interrupts to handle data input and output?

E6.3 What is the last instruction of most interrupt service routines? What does this instruction do? Is there any option for this instruction?

E6.4 What registers will be saved in the fast register stack when an interrupt occurs?

E6.5 Is the PIC18 interrupt priority programmable?

E6.6 Write an instruction sequence to set the INT1 pin interrupt to low priority and enable its interrupt.

E6.7 Write an instruction sequence to enable the timer 0 overflow interrupt and set it to high priority.

E6.8 Write an instruction sequence to enable the A/D interrupt and set it to high priority level.

E6.9 Write an instruction sequence to enable the USART1 transmit interrupt and set it to low priority level.

E6.10 Select appropriate values for R1, R2, and C in Figure 6.6 so that the INT interrupt is generated every 5 seconds instead of every 1 second.

E6.11 For the circuit in Figure 6.6, write a main program and an INT0 interrupt service routine in C language. The main program initializes a counter to 0, enables the INT0 interrupt, and then stays in a **while-loop** to wait forever. The INT0 interrupt service routine simply increments the counter by 1 and outputs it to the LEDs. Whenever the count reaches 15, the service routine resets it to 0. Choose appropriate component values so that the PIC18 receives an INT0 interrupt roughly every second.

E6.12 Give a C statement to declare **low_ISR** as the low-priority interrupt service routine and inform the C compiler to generate an instruction sequence to save the PRODL and PRODH registers in the software stack.

E6.13 Describe how you would avoid corrupting the STATUS, WREG, and BSR registers when both the low- and the high-priority interrupts are enabled. Describe how you can achieve this in your low- or high-priority interrupt service routine.

6.10 Lab Exercises and Assignments

L6.1 Enter the program shown in Example 6.1 in MPLAB IDE, assemble it, program the PIC18 MCU on your demo board using the ICD2, and run the program. Watch the change of the LEDs.

L6.2 Enter the program shown in Example 6.3 in MPLAB IDE, compile it, program the PIC18 MCU on your demo board using the ICD2, and run the program. Watch the change of the LEDs.

L6.3 Repeat L6.1 but replace the 555 timer circuit with the output of a debounced switch. Experiment with the circuit by pressing the key switch and see how LEDs change.

7

Parallel Ports

7.1 Objectives

After completing this chapter, you should be able to

- Define I/O addressing methods
- Explain the data transfer synchronization methods between the I/O interface chip and the I/O device
- Explain input and output handshake protocols
- Input data from simple switches
- Input data from keypads and keyboards
- Output data to LED and LCD displays
- Interface with a D/A converter to generate waveforms
- Explain the use of the slave parallel port

7.2 Introduction

Embedded products are often designed to allow the user to provide inputs (or commands) to specify the operation to be performed and to receive results related to the operation. This is made possible by I/O devices, also called *peripheral* devices. Key switches, keypads, keyboards, magnetic and optical scanners, and sensors are among the most popular input devices. LEDs, seven-segment displays, LCDs, printers, and CRT displays are among the most popular output devices.

The speed and electrical characteristics of I/O devices are quite different from those of the microprocessor, and hence it is not feasible to connect them directly to the microprocessor. Usually, interface chips are added to resolve the differences between the microprocessor and I/O devices.

The major function of the interface chip is to synchronize data transfer between the CPU and I/O devices. When interfacing to the I/O devices, resistors may be needed to limit the current flow, whereas buffer chips (or transistors) may be needed to increase the current flow required by the I/O devices.

An interface chip consists of control registers, data registers, status registers, data direction registers, and control circuitry. Control registers allow the user to set up parameters for the desired I/O operation. Status registers provide information about the progress and status of the I/O operation. The data direction register controls the direction of data transfer on I/O pins. Data registers are used to hold data to be sent to output devices and data latched from input devices. The combination of a group of I/O pins (usually eight pins or less) along with its associated control, status, data, and data direction registers are referred to as an *I/O port.*

In an input operation, the CPU reads data from the data register. In an output operation, the CPU places data on the data register, which will hold the data until it is fetched by the output device.

An interface chip has data pins that are connected to the microprocessor data bus and I/O port pins that are connected to the I/O device, as illustrated in Figure 7.1. In a sophisticated embedded

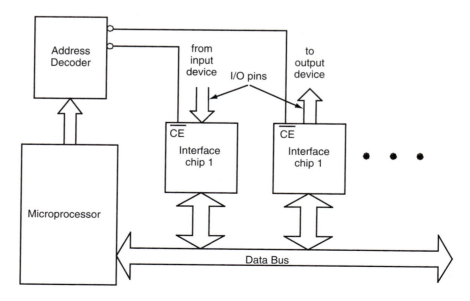

Figure 7.1 ■ Interface chip, I/O devices, and microprocessor

product, many I/O devices are attached to the data bus via the interface chip. However, only one device is allowed to drive data to the data bus at a time. Otherwise, data *bus contention* can occur. Bus contention is a situation in which two or more devices attempt to drive different voltages to the same data bus and cause severe damage. The address encoder in Figure 7.1 ensures that one and only one device is allowed to drive the data bus at a time. An interface chip is allowed to access the data bus only when its chip enable ($\overline{CE}$) input signal is asserted (at low voltage). Otherwise, the interface chip is electrically isolated from the data bus (the data pins of the interface chip are in high-impedance state).

Data transfer between the I/O devices can be proceeded bit by bit (serially) or in multiple bits (in parallel) at a time. In general, people use the serial method for slower I/O devices and use the parallel method for high-speed I/O devices. This chapter covers only parallel data transfer.

7.3 I/O Addressing

During an I/O operation, the CPU needs to access the registers of the interface chip to carry out the desired I/O operation. Since most interface chips have multiple registers, an address is needed to select each individual register. There are two design decisions to be made for microprocessor designers:

1. *Address space sharing.* Should I/O devices share the same memory spaces with memory devices such as SRAMs, EPROM, EEPROM, and flash memory, or should they have a dedicated memory space? Some early microprocessors, such as Intel 8085 and Zilog Z80, provided dedicated memory space for I/O devices, whereas other microprocessors, such as Motorola 6800, used the same memory space for I/O devices and memory components.

2. *Instruction set and addressing modes sharing.* Should dedicated I/O instructions and addressing modes be provided for I/O operations, or should all I/O operations and memory references use the same instruction set and addressing modes? Again, early microprocessors, such as Intel 8085 and Zilog Z80, have dedicated instructions and addressing modes for I/O operations, whereas Motorola 6800 and others used the same instruction set and addressing modes to perform I/O operations and memory references.

7.4 I/O Synchronization

The role of an interface chip is shown in Figure 7.2. The microprocessor interacts with the interface chip rather than the I/O devices. The electronics in the I/O devices converts electrical signals into mechanical actions or vice versa. As described in Chapter 1, the functions of an interface chip and the CPU are incorporated onto the same chip in a microcontroller.

When the CPU needs data from an input device, it reads from the interface chip (or interface circuit for the microcontroller) rather than directly from the input device. There must be a mechanism to make sure that the CPU does not read the same data more than once. When the CPU needs to output data, it sends data to the interface chip. Since the output device may be much slower than the CPU, the previous output operation may not have been completed when the CPU intends to output new data. There must be a mechanism to inform the CPU whether the interface chip is ready for new data. Since the interface chip needs to deal with the CPU and the I/O device, there are two aspects in I/O synchronization: the synchronization between the CPU and the interface chip and the synchronization between the interface chip and the I/O device.

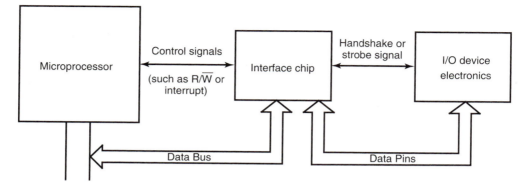

Figure 7.2 ■ The role of an interface chip

7.4.1 Synchronizing the CPU with the Interface Chip

To obtain valid data from an input device, the microprocessor must make sure that the interface chip has latched new data from the input device. There are two methods to achieve that:

1. *The polling method.* The interface chip uses a status flag to indicate whether it has valid data (stored in a data register) for the microprocessor. By checking the flag bit, the microprocessor can determine whether the interface chip has received new data from the input device. The microprocessor can check the status flag continuously or periodically (e.g., once every millisecond). This approach may tie up the CPU or incur very high CPU overhead.

2. *Interrupt-driven method.* In this method, the interface chip asserts an interrupt signal to the microprocessor whenever it has received new data from the input device. The interrupt service routine is then executed, and the microprocessor reads the input data.

To output data to the output device, the microprocessor must make sure that the output device is not busy. There are also two methods to do that:

1. *The polling method.* The interface device uses a status flag to indicate whether the data in its data register has been sent to the output device and therefore can accept new data. By checking the status flag continuously or periodically (e.g., every 100 μs), the microprocessor can determine whether the interface chip can accept new data. This method may tie up the CPU or incur very high CPU overhead.

2. *Interrupt-driven method.* The interface chip asserts an interrupt signal to the microprocessor whenever its data register is empty and can accept new data. The microprocessor then executes the interrupt service routine and sends the data to the interface chip.

None of the PIC18 parallel ports supports either method directly in the hardware. However, if the user adds an external parallel interface chip, such as an Intel i8255, to the PIC18 microcontroller, then both methods can be used.

7.4.2 Synchronizing the Interface Chip with I/O Devices

The interface chip is responsible for making sure that data are properly transferred to and from the I/O devices. The following methods have been used to synchronize data transfer between the interface chip and I/O devices:

1. *No method.* No signal transaction is used for data transfer synchronization. For input, the interface chip returns the voltage levels on the input pins to the microprocessor. For output, the interface chip makes the data written by the microprocessor directly available on output pins. This method is useful in situations in which the timing of data is unimportant. It can be used to test the voltage level of a signal, set the voltage of an output pin to high or low, or drive LEDs. All PIC18 I/O ports perform I/O operations without using any special synchronization method.

2. *The strobe method.* This method uses a strobe signal to indicate whether data is stable on input or output pins. During an input operation, the input device asserts a strobe signal whenever data is stable on the input port. The interface chip latches data into the data register using the strobe signal. For output, the interface chip first places data on the output port pins. When data becomes stable, the interface chip asserts a strobe signal to inform the output device to latch data on the output port pins. None of the PIC18 parallel ports supports this synchronization method. The PIC18 port D is an exception in which it can be configured as a parallel slave port to allow other microcontrollers (or microprocessors) to read data from or write data into it. In this configuration, PORTD is not used to deal with an I/O device.

3. *The handshake method.* The previous method cannot guarantee correct data transfer between an interface chip and an I/O device when the timing of data is critical. For example, it takes much longer time for a dot-matrix printer to print a character than it does to send a character to the printer. Therefore, data should not be sent to the printer if it is still printing. The solution is to use a handshake protocol between the interface chip and the printer electronics. There are two handshake methods: *interlocked mode* handshake and *pulse-mode* handshake. Whichever handshake protocol is used, two handshake signals are needed—one (call it H1) is asserted by the interface chip and the other (call it H2) is driven by the I/O device.

The signal transaction of the *input handshake protocol* is illustrated in Figure 7.3 and takes place in three steps:

Step 1
The interface chip asserts (or pulses) H1 to indicate its intent to acquire new data.

Step 2
The input device places valid data on the data pins and asserts (or pulses) H2.

Step 3
The interface chip latches the data and deasserts H1. After some delay, the input device also deasserts H2.

The complete process will be repeated if the interface chip wants to input more data.

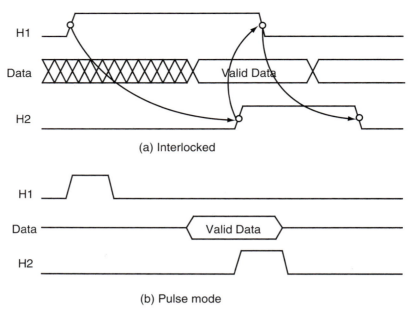

(a) Interlocked

(b) Pulse mode

Figure 7.3 ■ Input Handshakes

The signal transaction of the *output handshake protocol* is shown in Figure 7.4. It also takes place in three steps:

Step 1
The interface chip places data on the data pins and asserts (or pulses) H1 to indicate that it has data to be output.

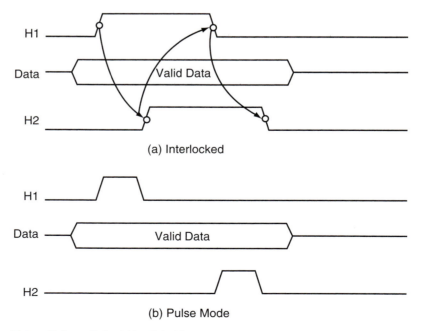

(a) Interlocked

(b) Pulse Mode

Figure 7.4 ■ Output Handshaking

Step 2

The output device latches the data and asserts (or pulses) H2 to acknowledge the receipt of data.

Step 3

The interface chip deasserts H1 following the assertion of H2. The output device then deasserts H2. The whole process will be repeated if the microprocessor has more data to be output.

Handshake protocol is common in older interface chips and I/O devices. However, the microcontrollers (including the PIC18) designed in the past 10 years rarely provide this mechanism. This is probably due to the shift of applications of the parallel I/O ports.

7.5 Overview of the PIC18 Parallel Ports

A PIC18 microcontroller may have as many as 10 I/O ports. The pins of an I/O port are often multiplexed with one or more peripheral functions (e.g., timer or serial peripheral interface) on the device. In general, when a peripheral function is enabled, that pin may not be used as a general-purpose I/O pin. General-purpose I/O pins can be considered the simplest of peripherals. They allow the microcontroller to monitor and control other devices.

The following generalities hold:

- 18-pin and 20-pin devices (e.g., PIC18F1220 and 1320) have two parallel ports: A and B.
- 28-pin devices (e.g., PIC18F242, 252, 2220, and 2320) have three parallel ports: A, B, and C.
- 40-pin (DIP package) and 44-pin (PLCC package) devices (e.g., PIC18F452, 458, 4320, and 4220) have five parallel ports: A, B, C, D, and E.
- 64-pin (TQFP package) and 68-pin (PLCC package) devices (e.g., PIC18F6620 and 6720) have seven ports: A, B, C, D, E, F, and G.
- Most 80-pin (TQFP package) and 84-pin (PLCC package) devices (e.g., PIC18F8620 and 8720) have nine ports: A, B, C, D, E, F, G, H, and J. Some 80-pin (TQFP package) and 84-pin (PLCC package) devices (e.g., the PIC18C858) have 10 ports: A, B, C, D, E, F, G, H, J, and K.

I/O ports do not have the same number of pins. The number of pins available in each I/O port is listed in Table 7.1. Port J of the PIC18C858 has only four pins.

Port Name	No. of Pins	Pin Name
A	7	RA6..RA0
B	8	RB7..RB0
C	8	RC7..RC0
D	8	RD7..RD0
E	8	RE7..RE0
F	8	RF7..RF0
G	5	RG4..RG0
H	8	RH7..RH0
J*	8	RJ7..RJ0
K	4	RK3..RK0

Note. The Port J of the PIC18C858 has only four pins

Table 7.1 ■ Number of pins available in each parallel port

Each I/O port has three registers for its operation. These registers are the following:

- TRIS register (data direction register)
- PORT register (reads the voltage levels on the pins of the device)
- LAT register (output latch)

The register names for each port can be derived by adding the port name to these general register names. For example, port A has TRISA, PORTA, and LATA registers.

Figure 7.5 shows a typical I/O port. This does not take into account peripheral functions that may be multiplexed onto the I/O pin. Reading the PORT register reads the status of the pins, whereas writing to it will write to the port latch. All write operations (such as BSF and BCF instructions) are read-modify-write operations. Therefore, a write to a port involves three steps:

1. Reading the pin levels
2. Modifying the read value
3. Writing the modified value to the port latch

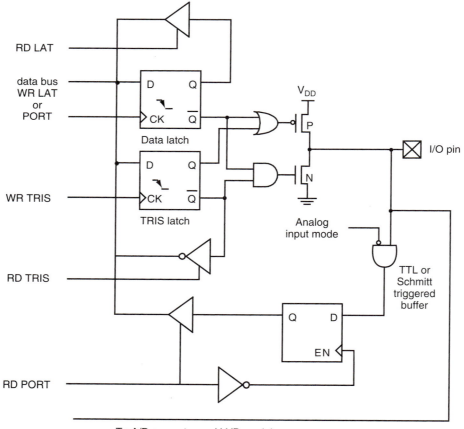

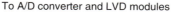
To A/D converter and LVD modules

Figure 7.5 ■ Block diagram of a typical I/O port pin (reprint with permission of Microchip)

To configure a port pin for input, set the associated bit in the TRIS register to 1. To configure a port pin for output, clear the associated bit in the TRIS register to 0.

Example 7.1
▼

Write an instruction sequence to output the hex value 0x26 to port C.

Solution: To output a value to an I/O port, one has to configure the I/O port for output. The following instruction sequence will configure port C for output and also output the value 0x26 to it:

```
clrf     TRISC,A      ; configure the port C for output
movlw    0x26         ; output the value 0x26 to the port C
movwf    PORTC,A      ; "
```
▲

Example 7.2
▼

Write an instruction sequence to read the current value of port D into the WREG register.

Solution: To read a value from an I/O port, one has to configure the I/O port for input. The following instruction sequence will configure the port D for input and read a value from it:

```
setf     TRISD,A      ; configure the port D for input
movf     PORTD,W,A    ; read the current value of the port D into WREG
```
▲

Example 7.3
▼

Write an instruction sequence to configure the lower four pins of port D for input and the upper four pins for output.

Solution: The following instruction sequence will configure port D as desired:

```
movlw    0x0F
movwf    TRISD,A
```

Most I/O ports also serve alternate functions. Their alternate functions are discussed in later chapters.
▲

7.5.1 Port A

Port A is a 7-bit-wide, bidirectional port. The RA4 pin is multiplexed with the timer 0 module clock input to become the RA4/T0CKI pin. The RA4/T0CKI pin is a Schmitt trigger input and an open-drain output. All other RA port pins have TTL input levels and full CMOS output drivers.

The RA6 pin is enabled only as a general I/O pin. The other port A pins are multiplexed with analog inputs and the analog-to-digital converter reference voltage inputs V_{REF+} and V_{REF-}. The function of each RA pin is shown in Figure 7.6.

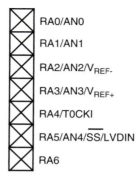

Figure 7.6 ■ Port A pins and their alternate functions

7.5.2 Port B

Port B is an 8-bit-wide, bidirectional port. Each of the port B pins has a weak internal pull-up. These pull-ups can be turned on by clearing the RBPU bit (bit 7) of the INTCON2 register. The weak pull-up is automatically turned off when the port pin is configured as an output. The pull-ups are disabled on a power-on reset. The function of each RB pin is shown in Figure 7.7. The RB3 pin can be used as the CCP2 pin when the CCP2MX configuration bit is enabled. The programming of the PIC18 configuration registers is discussed in Chapter 15.

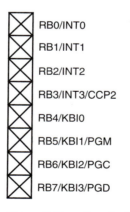

Figure 7.7 ■ Port B pins and their alternate functions

The lower four port B pins (RB3-RB0) are the external interrupt pins, INT3 through INT0. The other four pins (RB7-RB4) have an interrupt-on-change feature. Only pins configured as inputs can cause this interrupt to occur. The input pins (of RB7-RB4) are compared with the old value latched on the last read of port B. The "mismatch" outputs of RB7-RB4 are OR'ed together to generate the RB port change interrupt flag bit RBIF (INTCON<0>).

For the PIC18F8X20 devices, RB3 can be configured as the alternate peripheral pin for the CCP2 module (a timer capture module). This is available only when the device is configured in microprocessor, microprocessor with boot block, or extended microcontroller operation modes.

The RB5 pin can be used as the low-voltage programming (LVP) programming pin. When the LVP configuration bit is programmed, this pin loses the I/O function and becomes a programming test function.

7.5.3 Port C

Port C is an 8-bit-wide, bidirectional port. As shown in Figure 7.8, port C is multiplexed with several peripheral functions. All port C pins have Schmitt trigger input buffers. When the CCP2MX configuration bit is disabled, the RC1 pin can be used as the CCP2 pin.

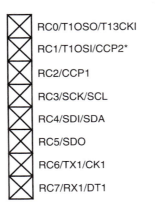

RC0/T1OSO/T13CKI

RC1/T1OSI/CCP2*

RC2/CCP1

RC3/SCK/SCL

RC4/SDI/SDA

RC5/SDO

RC6/TX1/CK1

RC7/RX1/DT1

Figure 7.8 ■ Port B pins and their alternate functions

7.5.4 Port D

As shown in Figure 7.9, port D is also an 8-bit-wide, bidirectional port. All port D pins have Schmitt trigger buffers. Every port D pin can be configured as input or output. Port D is multiplexed with the system bus (lowest eight address and data pins) as the external memory interface. I/O port functions are available only when the system bus is disabled. Port D can also be configured as an 8-bit-wide parallel slave port (PSP). In this mode, the input buffers are TTL compatible. The function of parallel slave port is explained in Section 7.11.

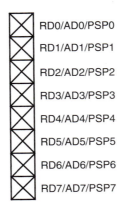

RD0/AD0/PSP0

RD1/AD1/PSP1

RD2/AD2/PSP2

RD3/AD3/PSP3

RD4/AD4/PSP4

RD5/AD5/PSP5

RD6/AD6/PSP6

RD7/AD7/PSP7

Figure 7.9 ■ Port D pins and their alternate functions

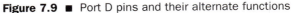

7.5.5 Port E and Port F

Both port E and port F are 8-bit-wide, bidirectional port with Schmitt trigger input buffers. All port E and F pins serve multiple functions. Their functions are shown in Figure 7.10 and 7.11, respectively. The RE7 pin can be configured as the alternate peripheral pin for CCP2 when the device is operating in the microcontroller mode. This is done by clearing the configuration bit CCP2MX in the CONFIG3H configuration register.

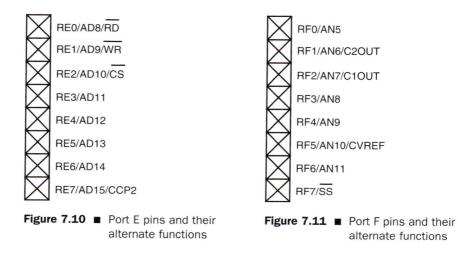

Figure 7.10 ■ Port E pins and their alternate functions

Figure 7.11 ■ Port F pins and their alternate functions

7.5.6 Port G

Port G is a 5-bit, bidirectional I/O port with Schmitt trigger input buffers. As shown in Figure 7.12, all port G pins serve multiple functions.

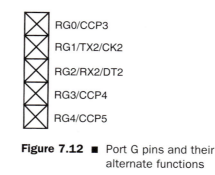

Figure 7.12 ■ Port G pins and their alternate functions

7.5.7 Port H and Port J

Port H and Port J are 8-bit-wide, bidirectional ports with Schmitt trigger input buffers. As shown in Figure 7.13, the upper four pins of port H are multiplexed with analog input AN15-AN12, whereas the lower four pins are multiplexed with system address pins A19-A16.

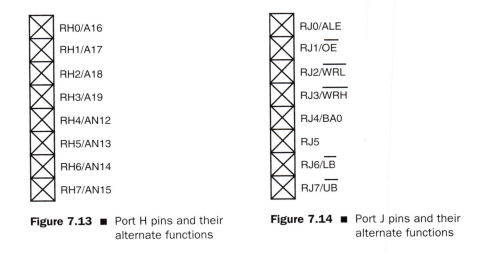

Figure 7.13 ■ Port H pins and their alternate functions

Figure 7.14 ■ Port J pins and their alternate functions

7.5.8 Port K

Port K is a 4-bit-wide, bidirectional I/O port with Schmitt trigger input buffers available only in the PIC18C858 device. Port K pins do not have alternate functions. (The PIC18C858 datasheet stated that port K has eight pins. However, the PIC18C858 pin diagram shows that port K has only four pins.)

7.6 Interfacing with Simple Output Devices

Many embedded systems only require simple input and output devices, such as switches, light-emitting devices (LEDs), keypads, and seven-segment displays.

7.6.1 Interfacing with LEDs

Simple as an LED, it is often used to indicate if power is on, if operation is normal, and so on. An LED can illuminate when it is forward biased and has sufficient current flowing through it. The current required to light an LED may range from a few to more than 10 mA. The voltage drop across the LED when it is forward biased can range from about 1.6 V to more than 2.2 V.

When configured for output, a PIC18 pin can drive an LED directly. A resistor (300 Ω to 500 Ω) is required to limit the current flow. A typical circuit connection is shown in Figure 7.15.

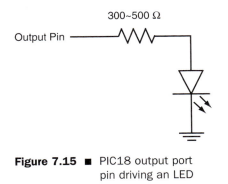

Figure 7.15 ■ PIC18 output port pin driving an LED

Example 5.3 demonstrates how to use a parallel port to perform LED flashing. Example 6.6 provides another method of driving LEDs using a parallel port. The circuit shown in Figure 6.6 will not draw much current from the parallel port and hence should be used when the microcontroller is heavily loaded.

7.6.2 Interfacing with Seven-Segment Displays

Seven-segment displays are often used when the embedded product needs to display only a few digits. Seven-segment displays are used mainly to display decimal digits and a subset of letters.

Although a PIC18 device has enough current to drive a seven-segment display, it is not advisable to do so when a PIC18-based embedded product needs to drive many other I/O devices. In Figure 7.16, port D drives a common-cathode seven-segment display through the buffer chip 74HC244.

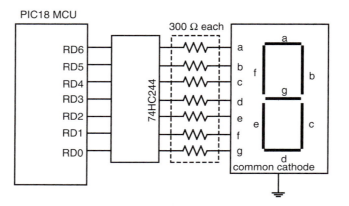

Figure 7.16 ■ Driving a single seven-segment display

The V_{OH} (output high voltage) value of the 74HC244 is about 5 V. Adding a 300-Ω resistor will set the display segment current to about 10 mA, which should be sufficient to light an LED segment. The light patterns corresponding to decimal digits are shown in Table 7.2. The numbers in Table 7.2 require that segment **a**, . . ., **g** be connected from the most significant pin to the least significant pin of the output port.

Decimal digit	Segments a	b	c	d	e	f	g	Corresponding Hex Number
0	1	1	1	1	1	1	0	0x7E
1	0	1	1	0	0	0	0	0x30
2	1	1	0	1	1	0	1	0x6D
3	1	1	1	1	0	0	1	0x79
4	0	1	1	0	0	1	1	0x33
5	1	0	1	1	0	1	1	0x5B
6	1	0	1	1	1	1	1	0x5F
7	1	1	1	0	0	0	0	0x70
8	1	1	1	1	1	1	1	0x7F
9	1	1	1	1	0	1	1	0x7B

Table 7.2 ■ Decimal to seven-segment decoder

When an application needs to display multiple BCD digits, the time-multiplexing technique is often used. An example of circuit that displays six BCD digits is shown in Figure 7.17. In Figure 7.17, the common cathode of a seven-segment display is connected to the collector of an NPN transistor. When a port B pin voltage is high, the connected NPN transistor will be driven into the saturation region. The common cathode of the display will then be pulled down to low (about 0.1 V), allowing the display to be lighted. By turning the six NPN transistors on and off in turn many times in 1 second, multiple digits can be displayed. A 2N2222 transistor can sink from 100 mA to 300 mA of current. The maximum current that flows into the common cathode is about 70 mA (7 × 10 mA) and hence can be sunk by a 2N2222 transistor. The resistor R should be selected to allow the 2N2222 transistor to be driven into the saturation region. A value of several hundred to 1000 Ω will work.

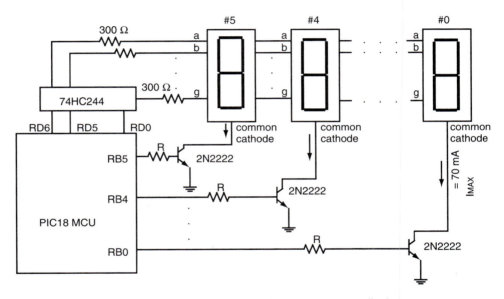

Figure7.17 ■ PortB and Port D together drive six seven-segment displays

Example 7.4

▼

Write an instruction sequence to display 5 on the seven-segment display #5 in Figure 7.17.

Solution: To display the value 5 on the seven-segment display #5, one needs to do the following:

1. Configure both port B and port D for output
2. Drive the RB5 pin to high
3. Drive the RB4 . . . RB0 pins to low
4. Output the hex value 0x5B to port D

The instruction is as follows:

```
clrf      TRISB,A     ; configure Port B for output
clrf      TRISD,A     ; configure Port D for output
movlw     0x5B
```

```
movwf      PORTD      ; output the segment pattern of 5
movlw      0x20
movwf      PORTB      ; enable display #5 to light
```

In C language, this can be achieved by the following statements:

```
TRISB   = 0x00;     /* configure Port B for output */
TRISD   = 0x00;     /* configure Port D for output */
PORTD   = 0x5B;     /* output the segment pattern of 5 */
PORTB   = 0x20;     /* enable display #5 to light */
```

The circuit in Figure 7.17 can display six digits simultaneously by using the time-multiplexing technique, in which each seven-segment display is lighted in turn briefly and then turned off. When one display is lighted, all other displays are turned off. Within 1 second, each seven-segment display is lighted and then turned off many times. Because of the *persistence of vision*, the six displays will appear to be lighted simultaneously.

Example 7.5

Write a program to display 654321 on the six seven-segment displays shown in Figure 7.17 assuming that the PIC18 is running with a 40-MHz crystal oscillator.

Solution: The digits 6, 5, 4, 3, 2, and 1 will be displayed on display #5, . . ., and #0, respectively. The values to be output to port B and port D to display one digit at a time are shown in Table 7.3.

Seven-segment display	Displayed BCD digit	Port D	Port B
#5	6	0x5F	0x20
#4	5	0x5B	0x10
#3	4	0x33	0x08
#2	3	0x79	0x04
#1	2	0x6D	0x02
#0	1	0x30	0x01

Table 7.3 ■ Table of display patterns for Example 7.5

This table can be created (in program memory) by the following assembler directives:

```
display_tab    db      0x5F,0x20
               db      0x5B,0x10
               db      0x33,0x08
               db      0x79,0x04
               db      0x6D,0x02
               db      0x30,0x01
```

The program logic of this example is shown in Figure 7.18.

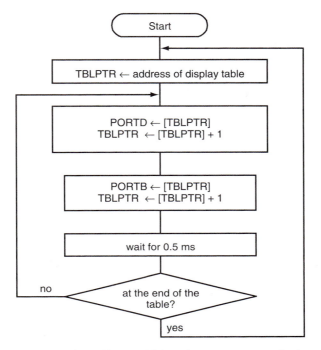

Figure 7.18 ■ Time-multiplexed seven-segment display algorithm

The assembly program that implements this algorithm is as follows:

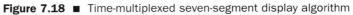

```
            #include <p18F8680.inc>
dup_nop     macro       kk          ; this macro will duplicate the "nop" instruction
            variable    i           ; kk times
i = 0
            while       i < kk
            nop
i += 1
            endw
            endm
lp_cnt      set         0x00        ; use data memory 0 as loop count
lp_cnt1     set         0x01        ; another loop count
            org         0x00
            goto        start
            org         0x08
            retfie
            org         0x18
            retfie
start       clrf        TRISB,A     ; configure Port B for output
            clrf        TRISD,A     ; configure Port D for output
forever     movlw       0x06        ; there are six digits to be displayed
            movwf       lp_cnt      ; "
```

```
                movlw          upper disp_tab      ; set up the display table pointer
                movwf          TBLPTRU,A           ; "
                movlw          high disp_tab       ; "
                movwf          TBLPTRH,A           ; "
                movlw          low disp_tab        ; "
                movwf          TBLPTRL,A           ; "
loop            tblrd*+
                movff          TABLAT,PORTD        ; output the seven-segment pattern
                tblrd*+
                movff          TABLAT,PORTB        ; turn on one display
                call           wait_hms            ; wait for half a millisecond
                decfsz         lp_cnt,F,A          ; decrement the loop count
                goto           loop                ; read the next pattern
                goto           forever             ; go to the start of the pattern table
; ************************************************************************
; The subroutine wait_hms creates a delay of 0.5 ms for a demo board running
; at 40 MHz crystal oscillator.
; ************************************************************************
wait_hms        movlw          D'250'
                movwf          lp_cnt1, A
again           dup_nop        D'17'               ; 17 instruction cycles
                decfsz         lp_cnt1,F,A         ; 1 instruction cycle (2 when [lp_cnt1] = 0)
                goto           again               ; 2 instruction cycles
                return
disp_tab        db             0x5F,0x20
                db             0x5B,0x10
                db             0x33,0x08
                db             0x79,0x04
                db             0x6D,0x02
                db             0x30,0x01
                end
```

The C language version of this program is as follows:

```c
#include <delays.h>
#include <p18F8680.h>
char display[6][2] = {{0x5F,0x20}, {0x5B,0x10}, {0x33,0x08}, {0x79,0x04}, {0x6D,0x02}, {0x30,0x01}};
void main ()
{
        int i;
        TRISB = 0x00;                       /* configure Port B for output */
        TRISD = 0x00;                       /* configure Port D for output */
        while (1) {
            for (i = 0; i < 6; i++) {
                PORTD = display[i][0];      /* output display pattern */
                PORTB = display[i][1];      /* turn on one display */
                Delay1KTCYx(5);             /* wait for 0.5 ms */
            }
        }
}
```

7.6.3 Liquid Crystal Display

Although seven-segment displays are easy to use, they are bulky and quite limited in the set of characters that they can display. When more than a few letters and digits are to be displayed, seven-segment displays become inadequate. Liquid crystal displays (LCDs) come in handy when the application requires the display of many characters.

An LCD has the following advantages:

- High contrast
- Low power consumption
- Small footprint
- Ability to display both characters and graphics

The basic construction of an LCD is shown in Figure 7.19. The most common type of LCDs allows light to pass through when activated. A segment is activated when a low-frequency bipolar signal in the range of 30 Hz to 1000 Hz is applied to it. The polarity of the voltage must alternate, or else the LCD will not be able to change very quickly.

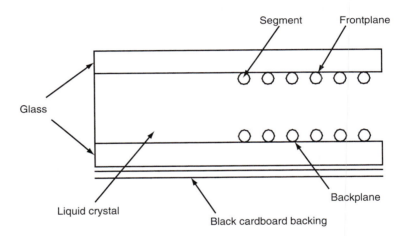

Figure 7.19 ■ A liquid crystal display (LCD)

When a voltage is applied across the segment, an electrostatic field is set up that aligns the crystals in the liquid. This alignment allows the light to pass through the segment. If no voltage is applied across a segment, the crystals appear to be opaque because they are randomly aligned. Random alignment is assured by the AC excitation voltage applied to each segment. In a digital watch, the segment appear darker when they are activated because light passes through the segment to a black cardboard backing that absorbs all light. The area surrounding the activated segment appears brighter in color because the randomly aligned crystals reflect much of the light. In a backlit computer display, the segment appears to grow brighter because of a light placed behind the display; the light is allowed to pass through the segment when it is activated.

In recent years, the price of LCD displays has dropped to such an acceptable level that most PC vendors bundle LCD displays instead of CRT displays with their PC systems. Notebook computers used LCDs as displays right from the beginning. Because of the price reduction of LCDs, the prices of notebook computers have also dropped sharply, and more and more computer users have switched from desktop to notebook computers. Although LCDs can display graphics and characters, only character-based LCDs are discussed in this text.

LCDs are often sold in a module that consists of the LCD and its controller. The Hitachi HD44780 is one of the most popular LCD display controllers in use today. The following section examines the operation and programming of this controller.

7.7 The HD44780 LCD Controller

The block diagram of an LCD kit that incorporates the HD44780 controller is shown in Figure 7.20. The pin assignment shown in Table 7.4 is the industry standard for character-based LCD modules with a maximum of 80 characters. The pin assignment shown in Table 7.5 is the industry standard for character-based LCD modules with more than 80 characters.

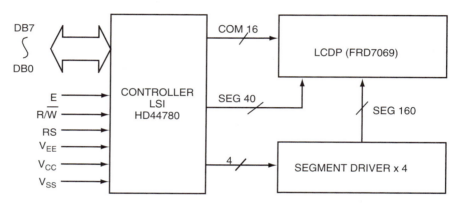

Figure 7.20 ■ Block diagram of a HD44780-based LCD kit

Pin No.	Symbol	I/O	Function
1	VSS	-	Power supply (GND)
2	VCC	-	Power supply (+5V)
3	VEE	-	Contrast adjust
4	RS	I	0 = instruction input, 1 = data input
5	R/$\overline{W}$	I	0 = write to LCD, 1 = read from LCD
6	E	I	enable signal
7	DB0	I/O	data bus line 0
8	DB1	I/O	data bus line 1
9	DB2	I/O	data bus line 2
10	DB3	I/O	data bus line 3
11	DB4	I/O	data bus line 4
12	DB5	I/O	data bus line 5
13	DB6	I/O	data bus line 6
14	DB7	I/O	data bus line 7

Table 7.4 ■ Pin assignment for displays with no more than 80 characters

Pin No.	Symbol	I/O	Function
1	DB7	I/O	data bus line 7
2	DB6	I/O	data bus line 6
3	DB5	I/O	data bus line 5
4	DB4	I/O	data bus line 4
5	DB3	I/O	data bus line 3
6	DB2	I/O	data bus line 2
7	DB1	I/O	data bus line 1
8	DB0	I/O	data bus line 0
9	E1	I	enable signal row 0 & 1
10	R/$\overline{W}$	I	0 = write to LCD, 1 = read from LCD
11	RS	I	0 = instruction input, 1 = data input
12	VEE	-	Contrast adjust
13	VSS	-	Power supply (GND)
14	VCC	-	Power supply (+5V)
15	E2	I	Enable signal row 2 & 3
16	N.C	-	

Table 7.5 ■ Pin assignment for displays with more than 80 characters

The DB7-DB0 pins are used to exchange data with the microcontroller. The E pin is an enable signal to the kit. The R/$\overline{\text{W}}$ signal determines the direction of data transfer. The RS signal selects the register to be accessed. When the RS signal is high, the *data register* is selected. Otherwise, the *instruction register* is selected. The V_{EE} pin is used to control the brightness of the display and is often connected to a potentiometer. The V_{EE} input should not be set to the maximum value (=V_{CC}) for an extended period of time to avoid burning the LCD.

An LCD module can be used as a memory-mapped device and be enabled by an address decoder. The E signal is normally connected to the address decoder output qualified by other control signals to meet the timing requirement. The R/$\overline{\text{W}}$ pin can be connected to the same pin of the microcontroller. The RS pin can be connected to the least significant bit of the address pin (A0) from the microcontroller. This approach is applicable only for those microcontrollers that support external memory. The LCD programming in this approach is generally easier and straightforward.

An LCD module can also be interfaced directly with an I/O port. In this configuration, the user will need to use I/O pins to control the signals E, R/$\overline{\text{W}}$, and RS. Programming will be slightly more cumbersome than the memory-mapped approach because of the need to tweak these three signals.

The HD44780 provides a set of instructions for the user to set up the LCD parameters. The LCD instruction set is shown in Table 7.6. The meanings of certain bits in the instructions are shown in Table 7.7.

Instruction	RS	R/$\overline{\text{W}}$	B7	B6	B5	B4	B3	B2	B1	B0	Description	Execution time
Clear display	0	0	0	0	0	0	0	0	0	1	Clear display and returns cursor to the home position (address 0).	1.64 ms
Cursor home	0	0	0	0	0	0	0	0	1	*	Returns cursor to home position (address 0). Also returns display being shifted to the original position. DDRAM contents remain unchanged.	1.64 ms
Entry mode set	0	0	0	0	0	0	0	1	I/D	S	Set cursor move direction (I/D), specifies to shift the display (S). These operations are performed during data read/write.	40 µs
Display on/off control	0	0	0	0	0	0	1	D	C	B	Sets on/off of all display (D), cursor on/off (C) and blink of cursor position character (B).	40 µs
Cursor / display shift	0	0	0	0	0	1	S/C	R/L	*	*	Sets cursor-move or display-(S/C), shift direction (R/L). DDRAM contents remains unchanged.	40 µs
Function set	0	0	0	0	1	DL	N	F	*	*	Sets interface data length (DL), number of display line (N) and character font (F).	40 µs
Set CGRAM address	0	0	0	1	CGRAM address						Sets the CGRAM address. CGRAM data is sent and received after this setting.	40 µs
Set DDRAM address	0	0	1	DDRAM address							Sets the DDRAM address. DDRAM data is sent and received after this setting.	40 µs
Read busy flag and address counter	0	1	BF	CGRAM/DDRAM address							Reads busy flag (BF) indicating internal operation is being performed and reads CGRAM or DDRAM address counter contents (depending on previous instruction).	0 µs
Write to CGRAM or DDRAM	1	0	write data								Writes data to CGRAM or DDRAM.	40 µs
Read from CGRAM or DDRAM	1	1	read data								Reads data from CGRAM or DDRAM.	40 µs

Table 7.6 ■ HD44780 instruction set

Bit name	Settings	
I/D	0 = decrement cursor position	1 = increment cursor position
S	0 = no display shift	1 = display shift
D	0 = display off	1 = display on
C	0 = cursor off	1 = cursor on
B	0 = cursor blink off	1 = cursor blink on
S/C	0 = move cursor	1 = shift display
R/L	0 = shift left	1 = shift right
DL	0 = 4-bit interface	1 = 8-bit interface
N	0 = 1/8 or 1/11 duty (1 line)	1 = 1/16 duty (2 lines)
F	0 = 5 × 7 dots	1 = 5 × 10 dots
BF	0 = can accept instruction	1 = internal operation in progress

Table 7.7 ■ LCD instruction bit names

The HD44780 can be configured to control one-, two-, and four-line LCDs. The mappings of character positions on the LCD screen and DDRAM addresses are not sequential and are shown in Table 7.8.

Display size	Visible	
	character positions	DDRAM address
1 * 8	00..07	0x00..0x07
1 * 16	00..15	0x00..0x0F
1 * 20	00..19	0x00..0x13
1 * 24	00..23	0x00..0x17
1 * 32	00..31	0x00..0x1F
1 * 40	00..39	0x00..0x27

Table 7.8a ■ DDRAM address usage for a 1-line LCD

Display size	Visible	
	character positions	DDRAM address
2 * 16	00..15	0x00..0x0F + 0x40..0x4F
2 * 20	00..19	0x00..0x13 + 0x40..0x53
2 * 24	00..23	0x00..0x17 + 0x40..0x57
2 * 32	00..31	0x00..0x1F + 0x40..0x5F
2 * 40	00..39	0x00..0x27 + 0x40..0x67

Table 7.8b ■ DDRAM address usage for a 2-line LCD

Display size	Visible	
	character positions	DDRAM aaddresses
4 * 16	00..15	0x00..0x0F + 0x40..0x4F + 0x14..0x23 + 0x54..0x63
4 * 20	00..19	0x00..0x13 + 0x40..0x53 + 0x14..0x27 + 0x54..0x67
4 * 40	00..39 on 1st controller and 00..39 on 2nd controller	0x00..0x27 + 0x40..0x67 on 1st controller and 0x00..0x27 + 0x40..0x67 on 2nd controller

Note: Two LCD controllers are needed to control LCD displays with 4 * 40 chanracters.

Table 7.8c ■ DDRAM address usage for a 4-line LCD

7.7.1 Display Data RAM

Display data RAM (DDRAM) stores display data represented in 8-bit character codes. Its extended capacity is 80×8 bits, or 80 characters. The area in DDRAM that is not used for display can be used as general data RAM. The relationships between DDRAM addresses and positions on the LCD are shown in Table 7.8.

7.7.2 Character Generator ROM

The character generator ROM (CGROM) generates 5×8 or 5×10 dot character patterns from 8-bit character codes. It can generate 208 5×8 dot character patterns and 32 5×10 dot character patterns.

7.7.3 Character Generator RAM

The user can rewrite character patterns into the character generator RAM (CGRAM) by program. For 5×8 fonts, eight character patterns can be written, and for 5×10 fonts, four character patterns can be written.

7.7.4 Registers

The HD44780 has two 8-bit registers, an instruction register (IR), and a data register (DR). The *IR register* stores instruction codes, such as **display clear** and **cursor shift,** and **address information** for DDRAM and CGRAM. The microcontroller writes commands into this register to set up the LCD operation parameters.

To write data into the DDRAM or CGRAM, the microcontroller writes data into the *DR register*. Data written into the DR register will be automatically written into DDRAM or CGRAM by an internal operation.

The DR register is also used for data storage when reading data from DDRAM or CGRAM. When address information is written into the IR register, data is read and then stored in the DR register from DDRAM or CGRAM by an internal operation. The microcontroller can then read the data from the DR register. After a read operation, data in DDRAM or CGRAM at the next address is sent to the DR register, and the microcontroller does not need to send another address. The IR and DR registers are distinguished by the RS signal. The IR register is selected when the RS input is low. The DR register is selected when the RS input is high. Register selection is illustrated in Table 7.9.

RS	R/$\overline{\text{W}}$	Operation
0	0	IR write as an internal operation (display clear, etc)
0	1	Read busy flag (DB7) and address counter (DB0 to DB6)
1	0	DR write as an internal operation (DR to DDRAM or CGRAM)
1	1	DR read as an internal operation (DDRAM or CGRAM to DR)

Table 7.9 ■ Register selection

The HD44780 has a *busy flag* (BF) to indicate whether the current internal operation is complete. When BF is 1, the HD44780 is still busy with an internal operation. When RS = 0 and R/$\overline{\text{W}}$ is 1, the BF is output to the DB7 pin. The microprocessor can read this pin to find out if the HD44780 is still busy.

The HD44780 uses a 7-bit *address counter* (AC) to keep track of the address of the next DDRAM or CGRAM location to be accessed. When an instruction is written into the IR register, the address information contained in the instruction is transferred to the AC register. The selection of DDRAM or CGRAM is determined by the instruction. After writing into (reading from) DDRAM or CGRAM, the content of the AC register is automatically incremented (decremented) by 1. The content of the AC register is output to the DB6-DB0 pins when the RS signal is low and the R/$\overline{\text{W}}$ signal is high.

7.7.5 Instruction Description

The functions of LCD instructions are discussed in this section.

The **clear display** instruction writes the space code 0×20 (character pattern for character code 0×20 must be a blank pattern) into all DDRAM locations. It then sets 0 into the address counter and returns the display to its original status if it was shifted. In other words, the display disappears, and the cursor or blinking goes to the left edge of the display. It also sets the I/D bit to 1 (increment mode) in entry mode.

The **return home** instruction sets DDRAM address 0 into the address counter and returns to its original status if it was shifted. The DDRAM contents do not change. The cursor or blinking goes to the first column of the first row on the display.

The I/D bit of the **entry mode set** instruction controls the incrementing (I/D = 1) or decrementing (I/D = 0) of the DDRAM address. The cursor or blinking will be moved to the right or left, depending on whether this bit is set to 1 or 0. The same applies to writing and reading of CGRAM.

The S bit of this instruction controls the shifting of the LCD display. The display shifts if S = 1. Otherwise, the display does not shift. If S = 1, it will seem as if the cursor does not move but the display does. The display does not shift when reading from DDRAM. Also, writing into or reading from CGRAM does not shift the display.

The **display on/off control** instruction has three bit parameters: D, C, and B. When the D bit is set to 1, the display is turned on. Otherwise, it is turned off. The cursor is turned on when the C bit is set to 1. The character indicated by the cursor will blink when the B bit is set to 1.

The **cursor or display shift** instruction shifts the cursor position to the right or left without writing or reading display data. The shifting is controlled by two bits as shown in Table 7.10. This function is used to correct or search the display. When the cursor has moved to the end of a line, it will be moved to the beginning of the next line.

S/C	R/L	Operation
0	0	Shifts the cursor position to the left. (AC is decremented by 1)
0	1	Shifts the cursor position to the right. (AC is incremented by 1)
1	0	Shifts the entire display to the left. The cursor follows the display shift.
1	1	Shifts the entire display to the right. The cursor follows the display shift.

Table 7.10 ■ Shift function

When the displayed data is shifted repeatedly, each line moves only horizontally. The second line of the display does not shift into the first row.

The contents of the address counter will not change if the only action performed is a display shift.

The **function set** instruction allows the user to set the interface data length, select the number of display lines, and select the character fonts. There are three bit variables in this instruction:

DL—Data is sent or received in 8-bit lengths (DB7–DB0) when DL is set to 1 and in 4-bit lengths (DB7–DB4) when DL is set to 0. When the pin count is at a premium for the application, the 4-bit data length should be chosen even though it is cumbersome to perform the programming.

N—This bit sets the number of display lines. When set to 0, one-line display is selected. When set to 1, two-line display is selected.

F—When set to 0, the 5×7 font is selected. When set to 1, the 5×10 font is selected.

The **CGRAM address** instruction contains the CGRAM address to be set into the address counter.

The **DDRAM address** instruction allows the user to set the address of the DDRAM (in the address counter).

The **read busy flag and address** instruction reads the busy flag (BF) and the address counter. The BF flag indicates whether the LCD controller is still executing the previously received instruction.

7.7.6 Interfacing the HD44780 to the PIC18 Microcontroller

For the PIC18 members that do not support external memory, the user must use I/O ports to interface with the LCD module. Possible circuit connections for the 8-bit- (used in SSE452, SSE8680, and SSE8720) and the 4-bit-wide data buses are shown in Figures 7.21a and 7.21b.

For the PIC18 members that support external memory, the user has the choice of using I/O ports to interface with the LCD module (as shown in Figures 7.21a and 7.21b) or treating the LCD as a memory device. This chapter treats the LCD only as an I/O device.

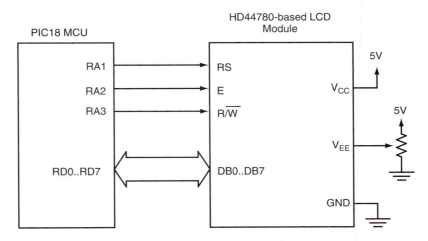

Figure 7.21a ■ LCD interface example (8-bit bus)

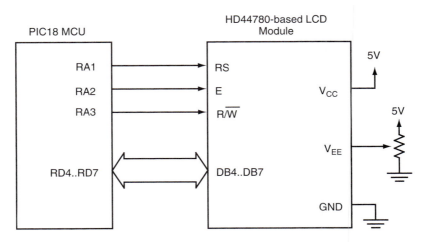

Figure 7.21b ■ LCD interface example (4-bit bus)

Certain timing parameters must be satisfied in order to achieve successful access to the LCD. The read and write timing diagrams are shown in Figure 7.22 and 7.23, respectively. The values of timing parameters depend on the frequency of the operation. HD44780-based LCDs can operate at either 1 MHz (cycle time of E signal) or 2 MHz. The values of timing parameters at these two frequencies are shown in Table 7.11 and 7.12, respectively.

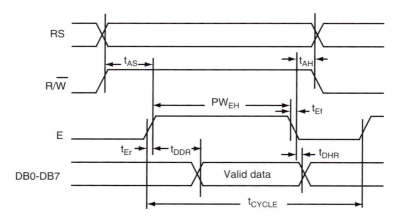

Figure 7.22 ■ HD44780 LCD controller read timing diagram

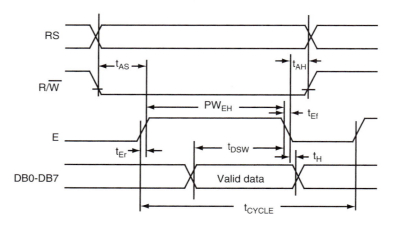

Figure 7.23 ■ HD44780 LCD controller write timing diagram

Example 7.6

There is a PIC18F452-based product that needs to interface with the LCD kit using the circuit shown in Figure 7.21a. The LCD kit has a 2 × 20 display and is to be operating at 2 MHz. Write a set of subroutines that perform the following functions, assuming that the PIC18F452 is running with a 20-MHz crystal oscillator:

1. Initialize the LCD
2. Write a byte (passed in WREG) into the LCD

Symbol	Meaning	Min	Typ	Max.	Unit
t_{CYCLE}	Enable cycle time	1000	-	-	ns
PW_{EH}	Enable pulse width (high level)	450	-	-	ns
t_{Er}, t_{Ef}	Enable rise and decay time	-	-	25	ns
t_{AS}	Address setup time, RS, R/$\overline{W}$, E	60	-	-	ns
t_{DDR}	Data delay time	-	-	360	ns
t_{DSW}	Data setup time	195	-	-	ns
t_H	Data hold time (write)	10	-	-	ns
t_{DHR}	Data hold time (read)	5	-	-	ns
t_{AH}	Address hold time	20	-	-	ns

Table 7.11 ■ HD44780 bus timing parameters (1 MHz operation)

Symbol	Meaning	Min	Typ	Max.	Unit
t_{CYCLE}	Enable cycle time	500	-	-	ns
PW_{EH}	Enable pulse width (high level)	230	-	-	ns
t_{Er}, t_{Ef}	Enable rise and decay time	-	-	20	ns
t_{AS}	Address setup time, RS, R/$\overline{W}$, E	40	-	-	ns
t_{DDR}	Data delay time	-	-	160	ns
t_{DSW}	Data setup time	80	-	-	ns
t_H	Data hold time (write)	10	-	-	ns
t_{DHR}	Data hold time (read)	5	-	-	ns
t_{AH}	Address hold time	10	-	-	ns

Table 7.12 ■ HD44780 bus timing parameters (2 MHz operation)

3. Send a command to the LCD
4. Output a string to the LCD

Solution: Using a 20-MHz crystal oscillator, the CPU cycle time is 4/20 MHz = 200 ns. The common operations performed by the specified functions are as follows:

- Write a command (a byte) to the IR register
- Read a byte from the IR register
- Write a byte to the DR register

The procedure for writing a byte to the IR register is as follows:

Step 1
Pull the RS and the E signals to low.

Step 2
Pull the R/$\overline{W}$ signal to low.

Step 3
Pull the E signal to high.

Step 4
Output data to the output port attached to the LCD data bus. One needs to configure port D for output before writing data to the LCD kit.

Step 5
Pull the E signal to low.

The procedure to read a byte from the IR register is as follows:

Step 1
Pull the RS and the E signals to low.

Step 2
Pull the R/$\overline{W}$ signal to high.

Step 3
Pull the E signal to high.

Step 4
Read the value of the LCD data bus. One must configure port D for input before reading the LCD data.

Step 5
Pull the E signal to low.

The procedure for outputting a byte to the LCD data register is as follows:

Step 1
Pull the RS signal to high.

Step 2
Pull the R/$\overline{W}$ signal to low.

Step 3
Pull the E signal to high.

Step 4
Output data to the I/O port attached to the LCD data bus.

Step 5
Pull the E signal to low.

The following constant definitions will be used in the specified functions:

```
LCD_DATA      equ           PORTD           ; LCD data bus
#define        LCD_D_DIR     TRISD,A         ; LCD data port direction
#define        LCD_E         PORTA,1,A       ; LCD E clock
#define        LCD_E_DIR     TRISA,1,A
#define        LCD_RW        PORTA,2,A       ; LCD read/write line
#define        LCD_RW_DIR    TRISA,2,A
#define        LCD_RS        PORTA,3,A       ; LCD register select line
#define        LCD_RS_DIR    TRISA,3,A
#define        LCD_ctl_port  PORTA,A
```

The BF flag of the LCD can be read without waiting. However, the LCD write operation cannot be performed if the LCD is busy in completing its internal operation. The following routine can poll the LCD to make sure that the LCD has completed its internal operation:

```
LCD_rdy       setf          LCD_D_DIR       ; configure LCD data port for input
              bcf           LCD_RS          ; select IR register
              bsf           LCD_RW          ; setup to read busy flag
              bsf           LCD_E           ; pull LCD E-line to high
              movf          LCD_DATA,W,A    ; read busy flag and DDRAM address
              nop                           ; small delay to lengthen E pulse
              bcf           LCD_E           ; pull LCD E-line to low
              btfsc         WREG,7,A        ; is busy flag (BF) cleared
              goto          LCD_rdy
              clrf          LCD_D_DIR       ; configure data pins for output
              return
```

The user needs to send a command to the LCD whenever he or she wants to change the LCD configuration. The following routine will write a command (in WREG) into the IR register:

```
LCD_cmd      pushr        WREG           ; save WREG in the stack
             call         LCD_rdy        ; wait until LCD is ready
             bcf          LCD_RS         ; select IR register
             bcf          LCD_RW         ; Set write mode
             bsf          LCD_E          ; Setup to clock data
             popr         WREG           ; restore WREG
             movwf        LCD_DATA,A     ; send out the command in WREG
             nop                         ; small delay to lengthen E pulse
             bcf          LCD_E
             return
```

Before the LCD can display data, it must be configured properly. A common LCD configuration for a 4 × 20 LCD is as follows:

- Set the LCD to two-line display
- Turn on display, cursor, and blinking
- Shift cursor right
- Clear the display and return the cursor to home position

The following routine performs the LCD initialization:

```
LCD_init     clrf         LCD_ctl_port   ; make sure the LCD control port is low
             bcf          LCD_E_DIR      ; configure control lines
             bcf          LCD_RW_DIR     ; directions to output
             bcf          LCD_RS_DIR     ; "
             movlw        0x38           ; configure display to 2x40
             call         LCD_cmd        ; send command to LCD
             movlw        0x0F           ; turn on display and cursor
             call         LCD_cmd        ; "
             movlw        0x14           ; shift cursor right
             call         LCD_cmd        ; "
             movlw        0x01           ; clear cursor and return to home position
             call         LCD_cmd        ; "
             return
```

The second row of Table 7.8c shows the correspondence of the four rows of a 4 × 20 LCD display and the internal DDRAM locations. Because the mapping from the four rows of the LCD to the internal DDRAM locations is not contiguous, the user needs to use instructions to set the DDRAM to the first position of the next row when reaching the last character of a row. To do that, the user needs to find out the current DDRAM address. The following subroutine returns the DDRAM address in the PRODL register:

```
get_DDRAM_addr
             setf         LCD_D_DIR      ; configure LCD data port for input
             bcf          LCD_RS         ; select IR register
             bsf          LCD_RW         ; setup to read busy flag
             bsf          LCD_E          ; pull LCD E-line to high
             movf         LCD_DATA,W,A   ; read busy flag and DDRAM address
             nop                         ; small delay to lengthen E pulse
             bcf          LCD_E          ; pull LCD E-line to low
             movwf        PRODL,A
             clrf         LCD_D_DIR      ; configure LCD data port for ouput
             return
```

When outputting a character to the last column of a row, the character cannot be seen. One needs to resend it to the column 1 of the next row. The following subroutine will output the character in the WREG register to the LCD:

```
LCD_putch        pushr          WREG               ; save WREG
                 call           LCD_rdy            ; wait until LCD is not busy
                 popr           WREG               ; restore data in WREG
                 send2LCD                          ; write WREG content to LCD
                 pushr          WREG               ; save WREG data
                 call           get_DDRAM_addr     ; get the DDRAM address in PRODL
                 movlw          0x13
                 cpfseq         PRODL,A
                 goto           chk_0x53
; reach the end of first line
                 movlw          0xC0
                 call           LCD_cmd            ; set DDRAM address to 0x40
                 call           LCD_rdy
                 popr           WREG
                 send2LCD                          ; re-output the same character
                 return
chk_0x53         movlw          0x53
                 cpfseq         PRODL,A
                 goto           chk_0x27
; reach the end of the second line
                 movlw          0x94               ; set DDRAM address to 0x14 (start of
                 call           LCD_cmd            ; 3rd line)
                 call           LCD_rdy
                 popr           WREG
                 send2LCD
                 return
; reach the end of the third line
chk_0x27         movlw          0x27
                 cpfseq         PRODL,A
                 return
                 movlw          0xD4
                 call           LCD_cmd
                 call           LCD_rdy
                 popr           WREG
                 send2LCD
                 return
```

The **LCD_putch** subroutine invokes the following macro:

```
send2LCD         macro                             ; this macro writes WREG to LCD
                 bsf            LCD_RS             ; select DR register
                 bcf            LCD_RW             ; Set write mode
                 bsf            LCD_E              ; Setup to clock data
                 movwf          LCD_DATA,A         ; write into LCD DR
                 nop                               ; a short delay
                 bcf            LCD_E
                 endm
```

The routine for outputting a string in program memory (pointed to by the TBLPTR register) can be implemented by calling the **LCD_putch** routine repeatedly:

```
LCD_putstr   TBLRD*+                       ; read a character into TABLAT
             movf        TABLAT,W,A        ; copy to WREG
             tstfsz      WREG,A            ; test WREG and skip if zero
             goto        send_it
             return
send_it      call        LCD_putch
             goto        LCD_putstr
```

▲

Example 7.7

▼

Write a program to output the following messages to the LCD in two rows on the SSE8720 demo board:

PIC18 demo board

LCD display test

The SSE8720 demo board uses port H pins RH7, RH6, and RH5 to control the E, R/$\overline{\text{W}}$, and RS signals, respectively. Port E is used to drive data pins D7 to D0.

Solution: The test program is as follows:

```
             #include <P18F8720.inc>
LCD_DATA     equ         PORTE             ; LCD data bus
#define       LCD_D_DIR   TRISE,A           ; LCD data port direction
#define       LCD_E       PORTH,7,A         ; LCD E clock
#define       LCD_RW      PORTH,6,A         ; LCD read/write line
#define       LCD_RS      PORTH,5,A         ; LCD register select line
#define       lcd_cport   PORTH,A
#define       LCD_E_DIR   TRISH,7,A
#define       LCD_RW_DIR  TRISH,6,A
#define       LCD_RS_DIR  TRISH,5,A
pushr        macro       arg               ; macro to push the arg register into stack
             movff       arg,POSTINC1
             endm
popr         macro       arg               ; macro to pop the arg register from the stack
             movff       POSTDEC1,arg      ; decrement the stack pointer
             movff       INDF1,arg         ; pop off a byte from the stack onto arg
             endm
alloc_stk    macro       n                 ; this macro allocates n bytes in the software
             movlw       n                 ; stack
             addwf       FSR1L,F,A
             movlw       0x00
             addwfc      FSR1H,F,A
             endm
dealloc_stk  macro       n                 ; this macro deallocates n bytes from
             movlw       n                 ; the software stack
             subwf       FSR1L,F,A
             movlw       0x00
             subwfb      FSR1H,F,A
```

```
              endm
send2LCD      macro                                     ; this macro writes the WREG content into LCD
              bsf         LCD_RS                        ; select DR register
              bcf         LCD_RW                        ; Set write mode
              bsf         LCD_E                         ; Setup to clock data
              movwf       LCD_DATA,A                    ; write into LCD DR
              nop                                       ; a short delay
              nop
              bcf         LCD_E
              endm
              org         0x00
              goto        start
              org         0x08
              retfie
              org         0x18
              retfie
start         lfsr        FSR1,0xE00                    ; set up software stack pointer
              movlw       0x0E
              movwf       ADCON1,A
              call        init_lcd
              movlw       upper msg1                    ; set up pointer to msg1
              movwf       TBLPTRU,A                     ;
              movlw       high msg1                     ;
              movwf       TBLPTRH,A
              movlw       low msg1                      ;
              movwf       TBLPTRL,A
              movlw       0x80
              call        LCD_cmd                       ; set LCD cursor to column 1 of row 1
              call        LCD_putstr                    ; output "This is line one."
              movlw       upper msg2                    ;
              movwf       TBLPTRU,A
              movlw       high msg2                     ;
              movwf       TBLPTRH,A
              movwf       low msg2                      ;
              movwf       TBLPTRL,A
              movlw       0xC0
              call        LCD_cmd                       ; set LCD cursor to column 1 of row 2
              call        LCD_putstr                    ; output "This is line two."
forever       nop
              goto        forever
;************************************************
;
; This routine configure the LCD to 2x40 display, turn on display and
; cursor, shift cursor right, and clear display return to home
;************************************************
init_lcd      clrf        lcd_cport                     ; make sure the LCD control port is low
              bcf         LCD_E_DIR                     ; configure control lines
              bcf         LCD_RW_DIR                    ; directions to output
              bcf         LCD_RS_DIR                    ; "
              movlw       0x38                          ; configure display to 2 rows
              call        LCD_cmd                       ; send command to LCD
              movlw       0x0F                          ; turn on display and cursor
```

```
            call            LCD_cmd             ; "
            movlw           0x14                ; shift cursor right
            call            LCD_cmd             ; "
            movlw           0x01                ; clear cursor and return to home
            call            LCD_cmd             ; "
            return
;*************************************
; This routine sends a command to the LCD controller.
;*************************************
LCD_cmd     pushr           WREG                ; save WREG in the stack
            call            LCD_rdy             ; wait until LCD is ready
            bcf             LCD_RS              ; select IR register
            bcf             LCD_RW              ; Set write mode
            bsf             LCD_E               ; Setup to clock data
            popr            WREG                ; restore WREG
            movwf           LCD_DATA,A          ; send out the command in WREG
            nop                                 ; small delay to lengthen E pulse
            bcf             LCD_E
            return
;*************************************************************
; This routine polls the RF flag to make sure LCD is ready for new data or
; command. This routine also configure LCD data port for output
;*************************************************************
LCD_rdy     setf            LCD_D_DIR           ; configure LCD data port for input
            bcf             LCD_RS              ; select IR register
            bsf             LCD_RW              ; setup to read busy flag
            bsf             LCD_E               ; pull LCD E-line to high
            movf            LCD_DATA,W,A        ; read busy flag and DDRAM address
            nop                                 ; small delay to lengthen E pulse
            bcf             LCD_E               ; pull LCD E-line to low
            btfsc           WREG,7,A            ; is busy flag (BF) cleared
            goto            LCD_rdy
            clrf            LCD_D_DIR           ; configure data pins for output
            return
;*************************************
; This routine outputs the character in WREG to the LCD.
;*************************************
LCD_putch   pushr           WREG                ; save WREG
            call            LCD_rdy             ; wait until LCD is not busy
            popr            WREG                ; restore data in WREG
            send2LCD                            ; write WREG content to LCD
            pushr           WREG                ; save WREG data
            call            get_DDRAM_addr      ; get the DDRAM address in PRODL
            movlw           0x13
            cpfseq          PRODL,A
            goto            chk_0x53
; reach the end of first line
            movlw           0xC0
            call            LCD_cmd             ; set DDRAM address to 0x40 (start of 2nd line)
            call            LCD_rdy
            popr            WREG
```

```
                    send2LCD                              ; re-output the same character
                    return
chk_0x53            movlw           0x53
                    cpfseq          PRODL,A
                    goto            chk_0x27
; reach the end of the second line
                    movlw           0x94                  ; set DDRAM address to 0x14 (start of
                    call            LCD_cmd               ; 3rd line)
                    call            LCD_rdy
                    popr            WREG
                    send2LCD
                    return
; reach the end of the third line
chk_0x27            movlw           0x27
                    cpfseq          PRODL,A
                    return
                    movlw           0xD4
                    call            LCD_cmd
                    call            LCD_rdy
                    popr            WREG
                    send2LCD
                    return
; ************************************************
; This routine output a string to the LCD pointed to by TBLPTR.
; ************************************************
LCD_putstr          TBLRD*+                               ; read a character into TABLAT
                    movf            TABLAT,W,A            ; copy to WREG
                    tstfsz          WREG,A               ; test WREG and skip if zero
                    goto            send_it
                    return
send_it             call            LCD_putch
                    goto            LCD_putstr
; ******************************************************
; This routine reads the current DDRAM address and returns it in PRODL.
; ******************************************************
get_DDRAM_addr
                    setf            LCD_D_DIR            ; configure LCD data port for input
                    bcf             LCD_RS              ; select IR register
                    bsf             LCD_RW              ; setup to read busy flag
                    bsf             LCD_E               ; pull LCD E-line to high
                    movf            LCD_DATA,W,A        ; read busy flag and DDRAM address
                    nop                                 ; small delay to length E pulse
                    bcf             LCD_E               ; pull LCD E-line to low
                    movwf           PRODL,A
                    return
msg1                data            "PIC18 demo board."
msg2                data            "LCD display test."
                    end
```

Example 7.8

Write the C functions that perform the same LCD operations as specified in Example 7.7.

Solution: The C functions and their test program are as follows:

```c
#include <P18F8720.h>
#define     LCD_DATA        PORTE
#define     LCD_D_DIR       TRISE
#define     LCD_RS          PORTHbits.RH5
#define     LCD_RS_DIR      TRISHbits.TRISH5
#define     LCD_RW          PORTHbits.RH6
#define     LCD_RW_DIR      TRISHbits.TRISH6
#define     LCD_E           PORTHbits.RH7
#define     LCD_E_DIR       TRISHbits.TRISH7

void LCD_rdy(void);                /* These are function prototype definitions */
void LCD_cmd(char cx);
void LCD_init(void);
char get_DDRAM_addr(void);
void LCD_putch(char dx);
void LCD_putstr(rom char *ptr);
void send2LCD(char xy);

rom char *msg1 = "SSE8720 demo board.";
rom char *msg2 = "LCD display test.";
rom char *msg3 = "This is line three.";
rom char *msg4 = "This is line four.";

void main (void)
{
    ADCON1 = 0x0E;              /* configure PortA pins for digital */
    LCD_init();
    LCD_cmd(0x80);             /* set cursor to column 1 row 1 */
    LCD_putstr(msg1);
    LCD_cmd(0xC0);             /* set cursor to column 1 row 2 */
    LCD_putstr(msg2);
//  LCD_cmd(0x94);             /* set cursor to column 1 row 3 */
//  LCD_putstr(msg3);
//  LCD_cmd(0xD4);             /* set cursor to column 1 row 4 */
//  LCD_putstr(msg4);
    while (1);                 /* hold the program not to output to LCD */
}
/*****************************************/
/* the following function waits until the LCD is not busy. */
/*****************************************/
void LCD_rdy(void)
{
    char test;
    LCD_D_DIR       = 0xFF;      /* configure LCD data bus for input */
```

```
            test          = 0x80;
            while (test) {
                    LCD_RS        = 0;                      /* select IR register */
                    LCD_RW        = 1;                      /* Set read mode */
                    LCD_E         = 1;                      /* Setup to clock data */
                    test          = LCD_DATA;
                    Nop();
                    LCD_E         = 0;                      /* complete a read cycle */
                    test          &= 0x80;                  /* check bit 7 */
            }
            LCD_D_DIR     = 0x00;                           /* configure LCD data bus for output */
}
/**************************************/
/* The following function sends a command to the LCD. */
/**************************************/
void LCD_cmd(char cx)
{
    LCD_rdy();                                             /* wait until LCD is ready */
    LCD_RS        = 0;                                     /* select IR register */
    LCD_RW        = 0;                                     /* Set write mode */
    LCD_E         = 1;                                     /* Setup to clock data */
    Nop();
    LCD_DATA      = cx;                                    /* send out the command */
    Nop();                                                 /* small delay to lengthen E pulse */
    LCD_E         = 0;                                     /* complete an external write cycle */
}
/**************************************/
/* The following function initializes the LCD kit properly. */
/**************************************/
void LCD_init(void)
{
    PORTH         = 0;                                     /* make sure LCD control port is low */
    LCD_E_DIR     = 0;                                     /* configure LCD control port for output */
    LCD_RS_DIR    = 0;                                     /*        "        */
    LCD_RW_DIR    = 0;                                     /*        "        */
    LCD_cmd(0x38);                                         /* configure display to 2x40 */
    LCD_cmd(0x0F);                                         /* turn on display, cursor, and blinking */
    LCD_cmd(0x14);                                         /* shift cursor right */
    LCD_cmd(0x01);                                         /* clear display and move cursor to home */
}
/**************************************/
/* The following function obtains the LCD cursor address. */
/**************************************/
char get_DDRAM_addr (void)
{
    char temp;
    LCD_D_DIR     = 0xFF;                                  /* configure LCD data port for input */
    LCD_RS        = 0;                                     /* select IR register */
    LCD_RW        = 1;                                     /* setup to read busy flag */
```

```
    LCD_E           = 1;                    /* pull LCD E-line to high */
    temp            = LCD_DATA & 0x7F;      /* read DDRAM address */
    Nop();                                  /* small delay to length E pulse */
    LCD_E           = 0;                    /* pull LCD E-line to low */
    LCD_D_DIR       = 0x00;
    return temp;
}
/*********************************************************************/
/* The following function sends a character to the LCD kit. The character cannot be displayed at the */
/* end of a row, so we need to reoutput it to the first column of the next row.                       */
/*********************************************************************/
void LCD_putch (char dx)
{
    char addr;
    LCD_rdy();                              /* wait until LCD internal operation is complete */
    send2LCD(dx);
    LCD_rdy();                              /* wait until LCD internal operation is complete */
    addr = get_DDRAM_addr();
    if (addr == 0x13) {
        LCD_cmd(0xC0);
        LCD_rdy();
        send2LCD(dx);                       /* output it to the column 1 of the next row */
    }
    else if(addr == 0x53) {
        LCD_cmd(0x94);
        LCD_rdy();
        send2LCD(dx);                       /* output it to the column 1 of the next row */
    }
    else if(addr == 0x27){
        LCD_cmd(0xD4);
        LCD_rdy();
        send2LCD(dx);                       /* output it to the column 1 of the next row */
    }
}
/*************************************/
/* The following function outputs a string to the LCD.  */
/*************************************/
void LCD_putstr(rom char *ptr)
{
    while (*ptr) {
        LCD_putch(*ptr);
        ptr++;
    }
}
/*************************************/
/* The following function outputs a string to the LCD.  */
/*************************************/
void send2LCD(char xy)
{
    LCD_RS          = 1;
```

```
        LCD_RW          = 0;
        LCD_E           = 1;
        LCD_DATA        = xy;
        Nop();
        Nop();
        LCD_E           = 0;
}
```

7.8 Interfacing with DIP Switches

A set of DIP switches usually consists of four or eight switches. An eight-input DIP package can be connected to any input port with eight pins, such as PORTB, PORTC, or PORTD. As shown in Figure 7.24, a set of DIP switches is connected to port B. One needs to enable the port B pull-up in order to make the circuit to work. When a switch is open, the associated port B pin is pulled up to high internally and will be read as 1. When a switch is closed, the associated pin will be pulled to low and read as 0.

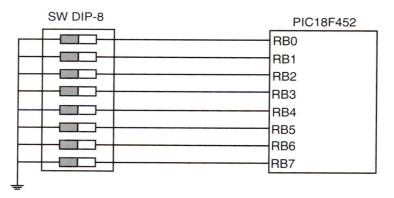

Figure 7.24 ■ Connecting a set of eight DIP switches to port B of the PIC18F452

Example 7.9

Write an instruction sequence to read a byte from the DIP switches into the WREG register.

Solution: Solution: One needs to configure port B for input and then read from the port pins:

```
bsf       INTCON2,RBPU,A      ; enable Port B pins pullup
movlw     0xFF
movwf     TRISB,A             ; configure Port B for input
movf      PORTB,W,A           ; copy port B data to WREG
```

7.9 Interfacing with a Keypad

The keypad is another commonly used input device. Like a keyboard, a keypad is arranged as an array of switches that can be mechanical, membrane, capacitors, or Hall-Effect in construction. In mechanical switches, two metal contacts are brought together to complete an electrical circuit. In membrane switches, a plastic or rubber membrane presses one conductor onto another; this type of switch can be made very thin. Capacitive switches internally comprise two plates of a parallel plate capacitor; pressing the key cap effectively increases the capacitance between the two plates. Special circuitry is needed to detect this change in capacitance. In Hall-Effect key switches, the motion of the magnetic flux lines of a permanent magnet perpendicular to a crystal is detected as voltage appearing between the two faces of the crystal—it is this voltage that *registers* a switch closure.

Mechanical keypads are most popular because of their low cost and strength in construction. However, mechanical switches have a common problem called *contact bounce.* Instead of producing a single, clean pulse output, pressing a mechanical switch generates a series of pulses because the switch contacts do not come to rest immediately. This phenomenon is illustrated in Figure 7.25.

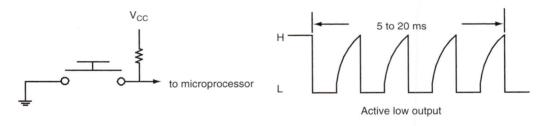

Figure 7.25 ■ Key switch contact bounce

When the key is not pressed, the voltage output to the computer is 5 V. In order to detect which key has been pressed, the microprocessor needs to scan every key switch of the keypad. A human being cannot press and release a key switch faster than 20 ms. During this interval, the microprocessor can scan the same key switch closure tens or even hundreds of thousand times, interpreting each low signal as a new input when in fact only one input should be sent.

Because of the contact bounce and the disparity in speed between the microprocessor and the human key pressing, a *debouncing* process is needed. A keypad input program can be divided into three stages:

1. *Keypad scanning* to find out which key was pressed
2. *Key switch debouncing* to make sure a key is indeed pressed
3. *Table lookup* to find the ASCII code of the key that was pressed

7.9.1 Keypad Scanning

Keypad scanning is usually performed row by row, column by column. A 16-key keypad can easily be interfaced with any available I/O port. Figure 7.26 shows a 16-key keypad organized into four rows with each row consisting of four switches.

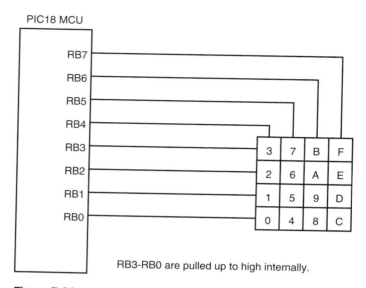

Figure 7.26 ■ Sixteen-key keypad connected to PIC18 MCU

For the keypad input application, the upper four pins (RB7 . . . RB4) of port B should be configured for output, whereas the lower four pins (RB3 . . . RB0) of port B should be configured for input.

The rows and columns of a keypad are simply conductors. In Figure 7.26, port B pins RB3 . . . RB0 are pulled up to high by pull-up resistors. Whenever a key switch is pressed, the corresponding rows and columns are shorted together. In order to distinguish the row being scanned and those not being scanned, the row being scanned is driven low, whereas the other rows are driven high. The row selection of the 16-key keypad is shown in Table 7.13.

PB7	PB6	PB5	PB4	Selected keys
1	1	1	0	0, 1, 2, and 3
1	1	0	1	4, 5, 6, and 7
1	0	1	1	8, 9, A, and B
0	1	1	1	C, D, E, and F

Table 7.13 ■ Sixteen-key keypad row selections

7.9.2 Keypad Debouncing

Contact bounce is due to the dynamics of a closing contact. The signal falls and rises a few times within a period of about 5 ms as a contact bounces. Since a human being cannot press and release a switch in less than 20 ms, a debouncer will recognize that the switch is closed after the voltage is low for about 10 ms and will recognize that the switch is open after the voltage is high for about 10 ms.

Both hardware and software solutions to the key bounce problem are available. Hardware solutions to contact bounce include an analog circuit that uses a resistor and a capacitor to smooth the voltage and two digital solutions that use set-reset latches or CMOS buffers and double-throw switches. Dedicated scanner chips that perform keypad scanning and debouncing are also available. National Semiconductor 74C922 and 74C923 are two examples.

The following are hardware debouncing techniques:

- *Set-reset latches.* A key switch can be debounced by using the set-reset latch shown in Figure 7.27a. Before being pressed, the key is touching the set input, and the Q voltage is high. When pressed, the key moves toward the reset position. When the key touches the reset position, the Q voltage will go low. When the key is bouncing and touching neither the set nor the reset input, both set and reset inputs are pulled low by the pull-down resistors. Since both set and reset are low, the Q voltage will remain low, and the key will be recognized as pressed.

- *Noninverting CMOS buffer with high input impedance.* The CMOS buffer output is identical to its input. When the switch is pressed, the input of the buffer chip 4050 is grounded, and hence V_{OUT} is forced to low. When the key switch is bouncing (not touching the input), the resistor R keeps the output voltage low. This is due to the high input impedance of 4050, which causes a negligible voltage drop on the feedback resistor. Thus, the output is debounced. This solution is shown in Figure 7.27b.

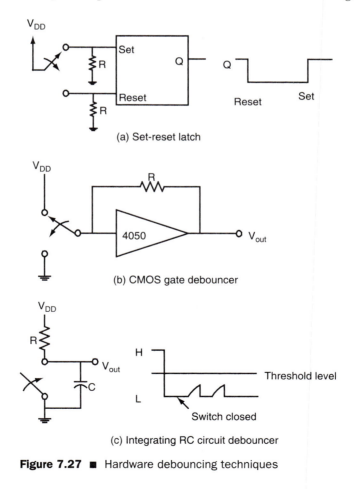

(a) Set-reset latch

(b) CMOS gate debouncer

(c) Integrating RC circuit debouncer

Figure 7.27 ■ Hardware debouncing techniques

- *Integrated debouncers.* The RC constant of the integrator determines the rate at which the capacitor charges up toward the supply voltage once the ground connection via the switch has been removed. As long as the capacitor voltage does not exceed the logic zero threshold value, the V_{OUT} signal will continue to be recognized as a logic zero. This solution is shown in Figure 7.27c.

A simple software debouncing method is wait-and-see; that is, wait for 10 ms and reexamine the same key to see if it is still pressed. A key is considered pressed if its output voltage is still low 10 ms after it is detected low. If the output voltage is high, the program will consider that the signal was noise or that the key is still bouncing. In either case, the program will continue to perform the scanning.

A more complicated method could be for the software to implement an integrated counter. For example, the program could initialize a counter to zero when it detects that a key was pressed for the first time. The program then rechecks the same key once every one millisecond within a 10-ms period. Whenever it detects the key to be pressed, it increments the counter by 1. Otherwise, it decrements the counter by 1. At the end of the 10-ms period, the program will consider the key pressed if the counter is larger than 5 (you can choose a different threshold value). For most applications, the simple wait-and-see method is adequate.

For an application that needs a keyboard, the easiest way to find out the ASCII code of the pressed key is to perform table lookup. However, table lookup is not necessary for the keypad because ASCII code lookup can be embedded in the program that performs scanning and debouncing.

Example 7.10

▼

Write a program to perform keypad scanning and debouncing and to return the ASCII code of the pressed key to the caller.

Solution: The following assembly routine will perform scanning and debouncing and return the ASCII code of the pressed key in WREG to the caller:

```
keypd_dir    equ      TRISB          ; keypad port direction register
keypad       equ      PORTB          ; keypad port
; *****************************************************
;
; The following routine performs keypad scanning, debouncing, and returns
; a character to the caller.
; *****************************************************
;
get_key      movlw    0x0F           ; configure the upper four pins of keypad
             movwf    keypd_dir      ; port for output, others for input
scan_r0      movlw    0xFF           ; prepare to scan row 0 (driven by RB4)
             iorwf    keypad,F       ;    "
             bcf      keypad,4       ;    "
scan_k0      btfss    keypad,0       ; check key 0
             goto     db_key0
scan_k1      btfss    keypad,1       ; check key 1
             goto     db_key1
scan_k2      btfss    keypad,2       ; check key 2
             goto     db_key2
scan_k3      btfss    keypad,3       ; check key 3
             goto     db_key3
```

```
scan_r1     movlw      0xFF          ; prepare to scan row 1 (driven by RB5)
            iorwf      keypad,F      ;      "
            bcf        keypad,5      ;      "
scan_k4     btfss      keypad,0      ; check key 4
            goto       db_key4
scan_k5     btfss      keypad,1      ; check key 5
            goto       db_key5
scan_k6     btfss      keypad,2      ; check key 6
            goto       db_key6
scan_k7     btfss      keypad,3      ; check key 7
            goto       db_key7
scan_r2     movlw      0xFF          ; prepare to scan row 2 (driven by RB6)
            iorwf      keypad,F      ;      "
            bcf        keypad,6      ;      "
scan_k8     btfss      keypad,0      ; check key 8
            goto       db_key8
scan_k9     btfss      keypad,1      ; check key 9
            goto       db_key9
scan_kA     btfss      keypad,2      ; check key A
            goto       db_keyA
scan_kB     btfss      keypad,3      ; check key B
            goto       db_keyB
scan_r3     movlw      0xFF          ; prepare to scan row 3 (driven by RB7)
            iorwf      keypad,F      ;      "
            bcf        keypad,7      ;      "
scan_kC     btfss      keypad,0      ; check key C
            goto       db_keyC
scan_kD     btfss      keypad,1      ; check key D
            goto       db_keyD
scan_kE     btfss      keypad,2      ; check key E
            goto       db_keyE
scan_kF     btfss      keypad,3      ; check key F
            goto       db_keyF
            goto       scan_r0
db_key0     call       wait10ms      ; wait for 10 ms
            btfsc      keypad,0      ; check key 0 again
            goto       scan_k1       ; key 0 not pressed, check key 1
            movlw      0x30          ; return the ASCII code of digit 0
            return
db_key1     call       wait10ms
            btfsc      keypad,1
            goto       scan_k2
            movlw      0x31
            return
db_key2     call       wait10ms
            btfsc      keypad,2
            goto       scan_k3
            movlw      0x32
            return
```

```
db_key3     call        wait10ms
            btfsc       keypad,3
            goto        scan_r1
            movlw       0x33
            return
db_key4     call        wait10ms
            btfsc       keypad,0
            goto        scan_k5
            movlw       0x34
            return
db_key5     call        wait10ms
            btfsc       keypad,1
            goto        scan_k6
            movlw       0x35
            return
db_key6     call        wait10ms
            btfsc       keypad,2
            goto        scan_k7
            movlw       0x36
            return
db_key7     call        wait10ms
            btfsc       keypad,3
            goto        scan_r2
            movlw       0x37
            return
db_key8     call        wait10ms
            btfsc       keypad,0
            goto        scan_k9
            movlw       0x38
            return
db_key9     call        wait10ms
            btfsc       keypad,1
            goto        scan_kA
            movlw       0x39
            return
db_keyA     call        wait10ms
            btfsc       keypad,2
            goto        scan_kB
            movlw       0x41
            return
db_keyB     call        wait10ms
            btfsc       keypad,3
            goto        scan_r3
            movlw       0x42
            return
db_keyC     call        wait10ms
            btfsc       keypad,0
            goto        scan_kD
```

```
                     movlw        0x43
                     return
db_keyD      call         wait10ms
                     btfsc        keypad,1
                     goto         scan_kE
                     movlw        0x44
                     return
db_keyE      call         wait10ms
                     btfsc        keypad,2
                     goto         scan_kF
                     movlw        0x45
                     return
db_keyF      call         wait10ms
                     btfsc        keypad,3
                     goto         scan_r0
                     movlw        0x46
                     return
; ****************************************************************************
; The routine wait10ms creates a 10 ms time delay using timer function with 32 MHz crystal oscillator.
; ****************************************************************************
delay10ms    bcf          INTCON,TMR0IF        ; clear TMR0IF flag
                     bcf          INTCON,TMR0IE        ; disable TMR0 interrupt
                     movlw        0x02                 ; set TMR0 prescaler to 8
                     movwf        T0CON                ;       "
                     movlw        0xDB                 ; load 55,536 into TMR0 so that it
                     movwf        TMR0H                ; overflows in 10 ms
                     movlw        0xF0                 ;       "
                     movwf        TMR0L                ;       "
                     bsf          T0CON,TMR0ON         ; enable TMR0 to count
dlyloop1     btfss        INTCON,TMR0IF        ; wait until 10 ms is over
                     bra          dlyloop1
                     return
                     end
```

The C language version of the subroutine is as follows:

```c
#include <p18F452.h>
void wait_10ms(void);
char get_key(void );
void main(void)
{
     char xc;
     xc = get_key();
}
char get_key(void)
{
     TRISB = 0x0F;                        /* configure RB7..RB4 for output RB3..RB0 for input */
     while (1) {
     PORTB |= 0xFF;                       /* set RB4 to low to scan the first row */
     PORTBbits.RB4 = 0;
```

```
if (!(PORTBbits.RB0)) {              /* is key 0 pressed? */
        wait_10ms( );
        if (!(PORTBbits.RB0))
                return 0x30;         /* return the ASCII code of 0 */
}
if (!(PORTBbits.RB1)) {              /* is key 1 pressed? */
        wait_10ms( );
        if (!(PORTBbits.RB1))
                return 0x31;         /* return the ASCII code of 1 */
}
if (!(PORTBbits.RB2)) {              /* is key 2 pressed? */
        wait_10ms();
        if (!(PORTBbits.RB2))
                return 0x32;         /* return the ASCII code of 2 */
}
if (!(PORTBbits.RB3)) {              /* is key 3 pressed? */
        wait_10ms();
        if (!(PORTBbits.RB3))
                return 0x33;         /* return the ASCII code of 3 */
}
PORTB |= 0xFF;                       /* set RB5 to low to scan the second row */
PORTBbits.RB5 = 0;
if (!(PORTBbits.RB0)) {
        wait_10ms();
        if (!(PORTBbits.RB0))
                return 0x34;         /* return the ASCII code of 4 */
}
if (!(PORTBbits.RB1)) {
        wait_10ms();
        if (!(PORTBbits.RB1))
                return 0x35;         /* return the ASCII code of 5 */
}
if (!(PORTBbits.RB2)) {
        wait_10ms();
        if (!(PORTBbits.RB2))
                return 0x36;         /* return the ASCII code of 6 */
}
if (!(PORTBbits.RB3)) {
        wait_10ms();
        if (!(PORTBbits.RB3))
        return 0x37;                 /* return the ASCII code of 7 */
}
PORTB |= 0xFF;                       /* set RB6 to low to scan the third row */
PORTBbits.RB6 = 0;
if (!(PORTBbits.RB0)) {
        wait_10ms();
        if (!(PORTBbits.RB0))
                return 0x38;         /* return the ASCII code of 8 */
}
if (!(PORTBbits.RB1)) {
```

```
                        wait_10ms();
                        if (!(PORTBbits.RB1))
                                return 0x39;        /* return the ASCII code of 9 */
            }
            if (!(PORTBbits.RB2)) {
                        wait_10ms();
                        if (!(PORTBbits.RB2))
                                return 0x41;        /* return the ASCII code of A */
            }
            if (!(PORTBbits.RB3)) {
                        wait_10ms();
                        if (!(PORTBbits.RB3))
                                return 0x42;        /* return the ASCII code of B */
            }
            PORTB |= 0xFF;                            /* set RB7 to low to scan the fourth row */
            PORTBbits.RB7 = 0;
                        if (!(PORTBbits.RB0)) {
                        wait_10ms();
                        if (!(PORTBbits.RB0))
                                return 0x43;        /* return the ASCII code of C */
            }
            if (!(PORTBbits.RB1)) {
                        wait_10ms();
                        if (!(PORTBbits.RB1))
                                return 0x44;        /* return the ASCII code of D */
            }
            if (!(PORTBbits.RB2)) {
                        wait_10ms();
                        if (!(PORTBbits.RB2))
                                return 0x45;        /* return the ASCII code of E */
            }
            if (!(PORTBbits.RB3)) {
                        wait_10ms();
                        if (!(PORTBbits.RB3))
                                return 0x46;        /* return the ASCII code of F */
            }
        }
    }
}
void wait_10ms(void)
{
    INTCONbits.INTOIE = 0;
    INTCONbits.INTOIF = 0;
    TOCON = 0x03;                            /* enable timer 0 to 16-bit mode with 1:16 prescaler */
    TMROH = 0xB1;
    TMROL = 0xEB;
    TOCONbits.TMROON = 1;
    while(!INTCONbits.INTOIF);
}
```

7.10 Interfacing with a D/A Converter

A digital-to-analog converter (DAC) converts a digital code into an analog signal (can be voltage or current). A DAC has many applications. Examples include digital gain and offset adjustment, programmable voltage and current sources, programmable attenuators, digital audio, digital video, closed-loop positioning, and robotics.

Although there are a few microcontrollers (e.g., some 8051 variants from Cygnal and Analog Devices) that incorporate DAC(s) on the chip, most microcontrollers still use off-chip DAC to perform the digital-to-analog conversion function. The PIC18 microcontroller is no exception. A DAC may use a serial or parallel interface to obtain digital code from the microprocessor or microcontroller.

7.10.1 The MAX5102 DAC

The MAX5102 is a dual-channel 8-bit DAC from MAXIM that has a parallel interface with the microprocessor or microcontroller. The MAX5102 converts an 8-bit digital value into an analog voltage. The pin configuration and block diagram of the MAX5102 are shown in Figures 7.28 and 7.29. The pin functions are as follows:

- OUTA, OUTB (analog voltage output).
- D7-D0 (data inputs): Digital code to be converted is sent to MAX5102 over these eight pins.
- REF (reference voltage input): This signal controls the magnitude of OUTA and OUTB.
- $\overline{WR}$ (write): When this signal is low, new digital code can be written into the MAX5102.
- A0 (DAC address select bit): DAC address select bit.
- V_{DD} (positive supply voltage): The range of V_{DD} is from 5 V to 15 V.
- GND (ground).
- SHDN (shutdown): When set to high, this signal shuts down the chip to save power. This signal should be grounded for normal operation.

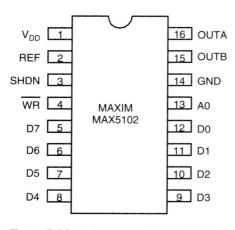

Figure 7.28 ■ MAX5102 Pin configuration

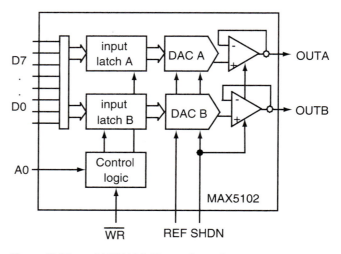

Figure 7.29 ■ MAX5102 Pin configuration

The MAX5102 uses a matrix decoding architecture for the DACs. The external reference voltage is divided down by a resistor string placed in a matrix fashion. Row and column decoders select the appropriate tab from the resistor string to provide the needed analog voltages. The resistor network converts the 8-bit digital input into an equivalent analog output voltage in proportion to the applied reference voltage input. The resistor string presents a code-independent input impedance to the reference and guarantees a monotonic output.

The MAX5102 has a shutdown mode that reduces current consumption to 1 nA. A high voltage at the SHDN pin shuts down the DACs and the output amplifiers. In shutdown mode, the output amplifiers enter a high-impedance state. When bringing the device out of shutdown, the user should allow at least 13 μs for the output to stabilize.

The MAX5102 provides a code-independent input impedance (R_{REF}) on the REF input. Input impedance is typically 460 KΩ in parallel with 15 pF, and the reference input voltage range is 0 to V_{DD}. The reference input accepts positive DC signals as well as AC signals with peak values between 0 and V_{DD}. The voltage at REF pin sets the full-scale output voltage for the DAC. The output voltage (V_{OUT}) for any DAC is represented by a digitally programmable voltage source as follows:

$$V_{OUT} = (NB \bullet V_{REF})/256$$

where NB is the numeric value of the DAC binary input code.

The address line A0 selects the DAC that receives data from D7 . . . D0 as shown in Table 7.14. When the $\overline{WR}$ input is low, the addressed DAC's input latch is transparent. Data is latched when the $\overline{WR}$ input is high. The DAC outputs (OUTA, OUTB) represent the data held in the 8-bit input latches. To avoid glitches in the MAX5102, the user must ensure that data is valid before the $\overline{WR}$ signal goes low.

$\overline{WR}$	A0	Latch state
H	X	Input data latched
L	L	DAC A input latch transparent
L	H	DAC B input latch transparent

Table 7.14 ■ MAX5102 addressing table

The reference source resistance (R_S) must be considerably less than the reference input resistance (R_{REF}). To keep within 1 LSB error in an 8-bit system, R_S must be less than $R_{REF}/256$. The user must therefore maintain a value of $R_S < 1$ KΩ to ensure 8-bit accuracy. If V_{REF} is DC only, bypass REF to GND with a 0.1-μF capacitor.

When sending data to the MAX5102, the control signals A0 and $\overline{WR}$ and the data D7-D0 must satisfy the timing requirement specified in Figure 7.30 and Table 7.15.

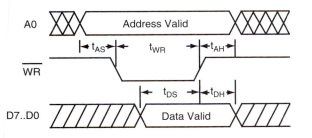

Figure 7.30 ■ MAX5102 Timing diagram

Symbol	Parameter	Min	Units
t_{AS}	Address to $\overline{WR}$ setup	5	ns
t_{AH}	Address to $\overline{WR}$ hold	0	ns
t_{DS}	Data to $\overline{WR}$ setup	25	ns
t_{DH}	Data to $\overline{WR}$ hold	0	ns
t_{WR}	$\overline{WR}$ pulse width	20	ns

Table 7.15 ■ MAX5102 Digital timing

7.10.2 Interfacing the MAX5102 with the PIC18 Microcontroller

Interfacing the MAX5102 with a microcontroller can be very simple. A possible connection is shown in Figure 7.31. An unused 8-bit I/D port can be used to send data to be converted to the MAX5102. Two I/O pins should be used to control the $\overline{WR}$ and A0 inputs to the MAX5102.

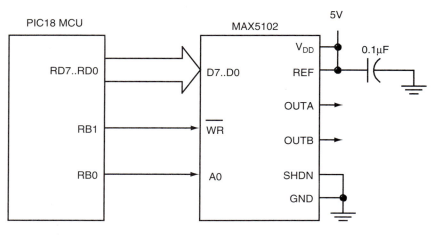

Figure 7.31 ■ Circuit connection between the MAX5102 AND PIC18 MCU

Example 7.11
▼

Write a program to generate a 1-KHz square wave from OUTA pin, assuming that the instruction cycle time is 200 ns (corresponding to a 20-MHz crystal oscillator).

Solution: The procedure for generating a 1-KHz square waveform from the OUTA pin is as follows:

Step 1
Configure the pins RD7-RD0, RB1-RB0 for output.

Step 2
Clear the RB0 pin to low to select OUTA channel.

Step 3
Pull the RB1 pin to low.

Step 4
Output 255 to port D.

Step 5
Pull the RB1 pin to high.

Step 6
Wait for 0.5 ms.

Step 7
Pull the RB1 pin to low.

Step 8
Output 0 to port D.

Step 9
Pull the RB1 pin to high.

Step 10
Wait for 0.5 ms.

Step 11
Go to Step 3.

The assembly program that implements this algorithm is as follows:

```
                    #include <p18F452.inc>
DAC_data            equ         PORTD               ; DAC data bus
DAC_DIR             equ         TRISD               ; DAC data bus direction
DAC_ctl             equ         PORTB               ; DAC control port
DAC_ctl_DIR         equ         TRISB               ; DAC control port direction
DAC_A0              equ         RB0
DAC_WR              equ         RB1
loop_cnt            equ         PRODL
dup_nop             macro       kk                  ; this macro duplicates the nop instruction
                    variable    i                   ;kk times
1 = 0
                    while       i < kk
                    nop
i += 1
                    endw
                    endm
                    org         0x00
                    goto        start
                    org         0x8
                    retfie
                    org         0x18
                    retfie
start               clrf        DAC_DIR             ; configure Port D for output
                    movlw       0xFC
                    andwf       DAC_ctl_DIR,F       ; configure RB1 & RB0 for output
                    bcf         DAC_ctl,DAC_A0      ; pull A0 to low to select OUTA channel
forever             bcf         DAC_ctl,DAC_WR      ; pull WR to low
                    setf        DAC_data            ; output 255 to DAC
                    bsf         DAC_ctl,DAC_WR      ; pull WR to high
                    rcall       wait500us           ; wait for 0.5 ms
                    bcf         DAC_ctl,DAC_WR      ; pull WR to low
                    clrf        DAC_data            ; output 0 to DAC
                    bsf         DAC_ctl,DAC_WR      ; pull WR to high
                    rcall       wait500us
                    goto        forever
;************************************************
; The following routine use program loop to create 0.5 ms delay.
;************************************************
wait500us           movlw       D'125'              ; 125 × 20 × 0.2 us = 500 us = 0.5 ms
                    movwf       loop_cnt
again               dup_nop     D'17'               ; 17 instruction cycles
                    decfsz      loop_cnt,F          ; 1 instruction cycle (2 when [loop_cnt] = 0)
                    goto        again               ; 2 instruction cycle
                    end
```

The C language version of the program is as follows:

```
#include    <p18F452.h>
#include    <delays.h>
#define     DAC_data       PORTD              /* DAC data port */
#define     DAC_dat_dir    TRISD              /* DAC data bus direction register */
#define     DAC_ctl_dir    TRISB              /* DAC control pins direction */
#define     DAC_WR         PORTBbits.RB1      /* DAC write control pin */
#define     DAC_A0         PORTBbits.RB0      /* DAC A0 pin */
#define     output         0x00
#define     high           1
#define     low            0
void main (void)
{
      DAC_dat_dir = output;                   /* configure DAC data bus direction to output */
      DAC_ctl_dir &= 0xFC;                    /* configure WR & A0 pins for output */
      DAC_A0 = 0;                             /* select OUTA channel */
      while (1) {
            DAC_WR = low;
            DAC_data = 255;                   /* output a high voltage */
            DAC_WR = high;
            Delay100TCYx(25);                 /* wait for 0.5 ms */
            DAC_WR = low;
            DAC_data = 0;                     /* output a low voltage */
            DAC_WR = high;
            Delay100TCYx(25);                 /* wait for 0.5 ms */
      }
}
```

Other periodic waveforms such as triangular and sinusoidal waveforms can be created easily by using the MAX5102 DAC. They are left as exercise problems for you.

▲

7.11 Parallel Slave Port

PORTD can be configured as an 8-bit wide *parallel slave port* (PSP; also called a *microprocessor port*) by setting the PSPMODE bit of the TRISE (or PSPCON for the PIC18F8X20 and the PIC18F6X20) register (bit 4). After PORTD is configured as a PSP, other microprocessors or microcontrollers can read from and write into this port. This allows the PIC18 microcontroller and other microprocessors to exchange messages.

Three PORTE pins are used to control the access of this PSP:

- RE0/$\overline{\text{RD}}$: This pin is active low and is used as the read control signal.
- RE1/$\overline{\text{WR}}$: This pin is active low and is used as the write control signal.
- RE2/$\overline{\text{CS}}$: This pin is active low and is used as the chip-select signal.

These three pins must be configured as digital inputs. To achieve this, the user must do the following:

- Set the lowest three bits of the TRISE register to 1s.
- Set the lowest three bits of the ADCON1 register to 1s.
- Set the PSPMODE bit (bit 4) of the PSPCON (or TRISE for 40-pin devices) register to 1 to allow these PORTE pins to be used for slave port control signals. The contents of the PSPCON register are shown in Figure 7.32. The upper four bits of the PSPCON register are available in the TRISE register in devices without a PSPCON register.

7	6	5	4	3	2	1	0
IBF	OBF	IBOV	PSPMODE	—	—	—	—

IBF: Input buffer full status bit
  0 = No word has been received.
  1 = A word has been received and is waiting to be read by the CPU.
OBF: Output buffer full status bit
  0 = The output buffer has been read.
  1 = The output buffer still holds a previously written word.
IBOV: Input buffer overflow detect bit.
  0 = No overflow occurred.
  1 = A write occurred when a previously input word has not been read.
PSPMODE: Parallel slave port mode select bit
  0 = general purpose I/O mode
  1 = parallel slave port mode

Figure 7.32 ■ PSPCON register (reprint with permission of Microchip)

The timing waveform for the PSP write operation is shown in Figure 7.33. A four-phase clocking scheme is used in the design of the PIC18 microcontroller. These four phases are referred to as Q1, Q2, Q3, and Q4. The PIC18 microcontroller uses the *instruction cycle clock* signal to synchronize its internal operations. One instruction cycle consists of four crystal oscillator cycles, which correspond to Q1, Q2, Q3, and Q4 in Figure 7.33.

A write to the PSP occurs when both the $\overline{CS}$ and the $\overline{WR}$ lines are first detected low and takes two instruction cycles to complete. When either the $\overline{CS}$ or the $\overline{WR}$ signal becomes high, the input buffer full status flag IBF is set on the Q4 clock cycle. The interrupt flag bit, PSPIF (in PIR1), is also set on the same Q4 clock cycle. The IBF flag bit is inhibited from being cleared for an additional instruction cycle. The *input buffer overflow* status bit (IBOV) is set if a second write to the PSP is attempted when the previous byte has not been read out of the buffer.

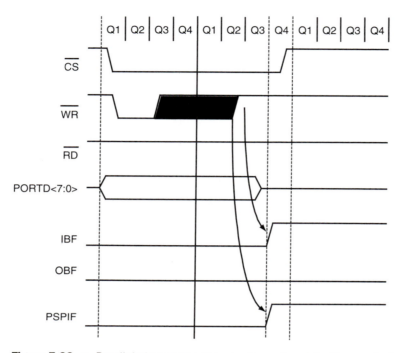

Figure 7.33 ■ Parallel slave port write waveforms (reprint with permission of Microchip)

The timing diagram of a PSP read operation is shown in Figure 7.34. A read of the PSP from the external system occurs when both the $\overline{CS}$ and the $\overline{RD}$ lines are first detected low. This cycle is completed in two instruction cycles. The output buffer full status flag (OBF) is cleared immediately, indicating that the PORTD latch was read by the external bus. When either the $\overline{CS}$ or the $\overline{RD}$ signal becomes high, the interrupt flag bit, PSPIF, is set on the Q4 clock cycle during the second instruction cycle. The OBF flag remains low until data is written to PORTD by the user program.

When not in the PSP mode, the IBF and OBF bits are held clear. The IBOV flag must be cleared in software if it was previously set. An interrupt is generated when a read or a write operation is completed. The PSPIF flag bit must be cleared by the user software, and the interrupt can be disabled by clearing the PSPIE bit.

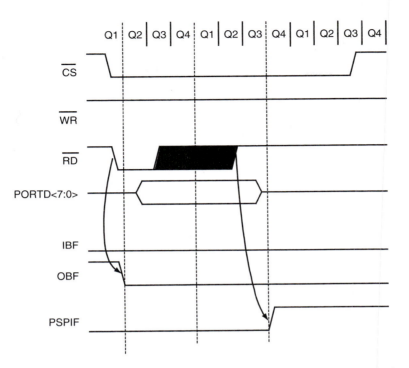

Figure 7.34 ■ Parallel slave port read waveforms (reprint with permission of Microchip

Example 7.12

▼

Write an instruction sequence to enable PORTD to be used as the PSP port.

Solution: The following instruction sequence will configure PORTD as desired:

```
        movlw    0x07
        movwf    ADCON1,A          ; configure PE2..PE0 pins for digital I/O
        movwf    TRISE,A           ; configure PE2..PE0 pins for input
        bsf      PSPCON,PSPMODE,A  ; enable PSP mode (for P18F8X20)
;       bsf      TRISE,PSPMODE,A   ; 40-pin or 44-pin (in PLCC) devices
```

The following C statements will perform the same configuration:

```
    ADCON1    = 0x07;              /* configure PE2-PE0 for digital I/O */
    TRISE     = 0x07;              /* configure PE2-PE0 for input */
    PSPCONbits.PSPMODE = 1;         /* enable PSP mode */
//    TRISEbits.PSPMODE = 1;
```

The PSP allows the PIC18 device to exchange data with other microcontrollers over a short distance. To use this port, the application should be written to follow certain protocols that coordinate the operation of two devices. It is usually a good idea to enable the PSP interrupt so that the PIC18 device will not become bogged down by polling the PSPIF flag in order to find out if a byte has been read from or written into the PSP.

▲

7.12 Summary

Embedded systems require the use of I/O devices to exchange data with a computer. Examples of I/O devices are switches, keypads, keyboards, LEDs, seven-segment displays, LCDs, printers, D/A converters, A/D converters, bar code scanners, magnetic scanners, and so on. Because the electrical characteristics and speeds of I/O devices are significantly different from those of a microprocessor, I/O devices are rarely connected directly to the microprocessor. Interface chips are used to resolve those differences between the microprocessor and peripheral devices.

The major function of an interface chip is to synchronize the data transfer between the processor and an I/O device. An interface chip consists of control registers, status registers, data registers, data direction registers, and control circuitry. There are two aspects in I/O transfer synchronization: the synchronization between the CPU and the interface chip and the synchronization between the interface chip and the I/O device. There are two methods to synchronize the CPU and the interface chip: polling and interrupt. There are three methods for synchronizing the interface chip and the I/O device: no method, strobe, and handshake.

Some microprocessors have dedicated instructions and addressing modes for performing I/O operations, whereas other microprocessors use the same set of instructions and addressing modes to access memory and I/O devices. In some microprocessors, memory components and I/O devices share the same memory space, whereas in others, I/O devices and memory components have different memory spaces.

A PIC18 microcontroller may have as few as 18 or as many as 84 I/O pins. Most of these pins serve multiple functions. In addition to being used for general-purpose I/O, these pins can be used as system address and data signals and to support peripheral functions, such as serial communications, A/D conversion, timer inputs, or generating waveforms.

At the time of this writing, the PIC18 family has devices with 18 to 84 pins. In the near future, devices with fewer I/O pins (e.g., eight) will also be introduced. Each I/O port has a port data register and a data direction register. All I/O ports can be configured for input or output. Writing a 1 to a bit in the data direction register will configure the corresponding port pin for input. Writing a 0 to a bit in the data direction register will configure the corresponding pin for output.

The interfacing and programming of DIP switches, keypad, LEDs, seven-segment displays, LCDs, and DAC were illustrated in this chapter.

Port D can be configured as a parallel slave port. When enabled, the parallel slave port allows the PIC18 to exchange message with another PIC18 or any other microprocessor. The parallel slave port can be enabled by setting the PSPIE bit in the TRISE (or the PSPCON register of the PIC18F8X20) register.

7.13 Exercises

E7.1 What is an I/O port?

E7.2 What is I/O polling?

E7.3 What is I/O handshaking? Are PIC18 devices supporting I/O handshaking directly?

E7.4 Write an instruction sequence to output the value 0x39 to port D.

E7.5 Write an instruction sequence to read a byte from port H to the WREG register.

E7.6 Assuming that port D pins RD4 . . . RD0 are connected to green, yellow, red, blue, and purple LEDs, write a program to light the green, yellow, red, blue, and purple LEDs in turn for 0.1, 0.2, 0.3, 0.4, and 0.5 seconds forever. Use both assembly and C language.

E7.7 Assuming that port D pins RD5 . . . RD0 are driving green, yellow, red, green, yellow, and red LEDs, write a program to simulate the traffic light controller. Assign RD5 . . . RD3 to drive east–west and RD2 . . . RD0 to drive north–south direction lights. The traffic light patterns and durations for traffic heading east–west and north–south directions are given in Table E7.7. Write the program in both assembly and C language to control the light patterns and connect the circuit to demonstrate the changes of lights.

East-west			North-south			
green	**yellow**	**red**	**green**	**yellow**	**red**	**Duration**
1	0	0	0	0	1	25
0	1	0	0	0	1	5
0	0	1	1	0	0	20
0	0	1	0	1	0	4

Table E7.7 ■ Traffic light pattern and duration

E7.8 Write a program in C and assembly languages to display the following rows in the LCD connected to the PIC18 demo board (display the message from the first column of each row):

Name: Peter Jennings

Job: anchorman

Ranking: first

Resident: New York

E7.9 Write a program to generate a sawtooth waveform from the DAC circuit shown in Figure 7.31. What is the period of the sawtooth waveform?

E7.10 Write a program to generate a sine wave from the DAC circuit shown in Figure 7.31. What is the period of this waveform?

E7.11 Write a program to generate the waveform shown in Figure E7.11, assuming that V_{REF} is 5 V.

E7.12 Write a program to generate a triangular waveform from the circuit shown in Figure 7.31. What is the period of this waveform?

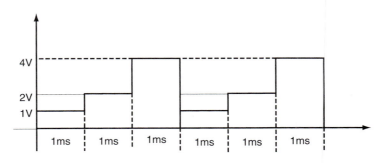

Figure E7.11 ■ A waveform to be generated

E7.13 Write a program in assembly and C language to display the time of day in the circuit shown in Figure 7.17. The program will read the time of day from a DIP switch connected to port F when the program starts to execute.

E7.14 Write a program to display decimal digits on the seven-segment displays shown in Figure 7.18. The decimal digits start from 0 to 9. At the beginning, 0 is displayed for half a second on the leftmost display, then 1 follows 0 and both are displayed on the seven-segment displays for half a second. The complete patterns are as follows:

```
0
10
210
3210
43210
543210 ——— (1)
654321
765432
876543
987654
098765
109876
210987
321098
432109 ——— (10)
```

Repeat from 1 to 10 forever.

E7.15 Use the circuit shown in Figure 7.17 to display decimal digits 1, 2, 3, 4, 5, and 6 on each display from left to right. Each time, only one digit is displayed for 1 second. Repeat this operation forever.

7.14 Lab Exercises and Assignments

L7.1 *Two-digit counter and display.* Use RD6 . . . RD0 pins to drive the seven-segment pattern and use RB1 and RB0 to drive the digit select transistor similar to the circuit shown in Figure 7.17. Write a program to display 00 to 99 and then repeat on two seven-segment displays. Each value is displayed for a quarter second.

L7.2 *Time-of-day display on LCD.* Connect a 4 × 20 LCD (e.g., DMC-20434 from Optrex) to the PIC18 demo board and display the time in the format of **hh mm ss** on the center of second row of the LCD.

L7.3 *Interrupt-driven input.* Use a PIC18 demo board to connect an 8-DIP switch to an unused I/O port. Enable INT0 interrupt. Set up an appropriate value on the DIP switch. Press the INT0 switch to interrupt the microcontroller to read in the value. When the value is read in, display it on the LCD. The entered values are to be display in the following format:

> Value 1: ww
>
> Value 2: xx
>
> Value 3: yy
>
> Value 4: zz

Perform this experiment four times and then stop.

8

Timers and CCP Modules

8.1 Objectives

After completing this chapter, you should be able to

- Explain the overall timer structure of the PIC18 microcontroller
- Use timer function to create time delays
- Use timer function to measure the frequency of an unknown signal
- Use CCP in capture mode to measure the duration of a pulse or the period of a square wave
- Use CCP in capture mode to measure the duty cycle of a waveform or the phase difference between two waveforms having the same frequency
- Use CCP in capture mode to measure the frequency of an unknown signal
- Use CCP in compare mode to create time delays
- Use CCP in compare mode to trigger pin actions
- Use CCP in compare mode to generate digital waveforms or pulses
- Use CCP in compare mode to create music
- Use CCP in PWM mode to generate digital waveform with certain frequencies and duty cycles
- Use CCP and ECCP in motor control

8.2 Overview of PIC18 Timer Functions

In a digital system, *time* is represented by the *count* of a *timer*. There are many applications that are very difficult or even impossible to implement without the timer function:

- Event arrival time recording and comparison
- Periodic interrupt generation
- Pulse width and period measurement
- Frequency and duty cycle measurement of periodic signals
- Generation of waveforms with certain duty cycles and frequencies
- Time references
- Event counting

A PIC18 microcontroller may have four or five timers. These timers are Timer0, Timer1, Timer2, Timer3, and Timer4.

Timer0, Timer1, and Timer3 are 16-bit timers, whereas Timer2 and Timer4 are 8-bit timers. Whenever a 16-bit (8-bit) timer rolls over from 0xFFFF to 0x0000 (from 0xFF to 0x00), an interrupt will be requested if it is enabled. Both Timer2 and Timer4 use the instruction cycle clock as the clock source, whereas the other timers can also use external clock signal as their clock source.

CCP stands for *capture, compare,* and *pulse-width modulation* (PWM). A PIC18 microcontroller may have one, two, or five CCP modules. An 18-pin or 20-pin PIC18 device has only one CCP channel. The PIC18F8X2X and PIC18F6X2X devices have five CCP channels. Other PIC18 devices (at the time of this writing) have two CCP channels. Each CCP channel can be configured to perform capture, compare, or PWM functions. These three functions share the same signal pin and registers.

When configured as a capture function, the CCP module can be programmed to copy the contents of a timer into a capture register on every falling edge, every rising edge, every fourth rising edge, or every 16th rising edge. This capability can be utilized to measure the pulse width, period, duty cycle, or frequency. The capture function can also be used for timing reference.

When configured as a compare function, the CCP module compares the contents of the 16-bit CCPRx (x = 1 . . . 5) register with the TMR1 (or TMR3) register in every clock cycle. When these two registers match, the CCP module can optionally drive the associated pin to high or low or simply toggled. The compare function can be used to create a time delay, generate a few pulses, or generate periodic waveforms.

When configured as a PWM function, the CCP module can be programmed to generate a waveform with a certain frequency and duty cycle. This function is often used in motor control and light dimming applications.

8.3 Timers

A PIC18 microcontroller may have four or five timers. The PIC18F8720, 18F8621, 18F8620, 18F8525, 18F8520, 18F6720, 18F6620, 18F6621, 18F6525, and 18F6520 have five timers. Timer4 is available only in these devices. A detailed discussion of these timers is given in this section.

8.3.1 Timer0

Timer0 can be configured as an 8-bit or a 16-bit timer or counter. The user can choose the internal instruction cycle clock or the external T0CKI signal as the clock source of Timer0. The user can choose to divide the clock signal by a prescaler before it is connected to the clock input to Timer0. The block diagrams of Timer0 in 8-bit and 16-bit mode are shown in Figures 8.1a and 8.1b.

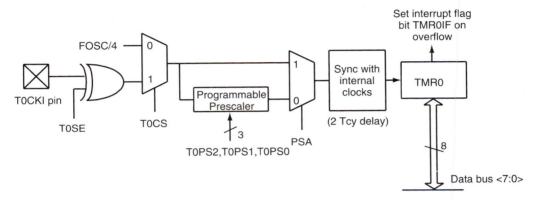

Figure 8.1a ■ Timer0 block diagram in 8-bit mode (redraw with permission of Microchip)

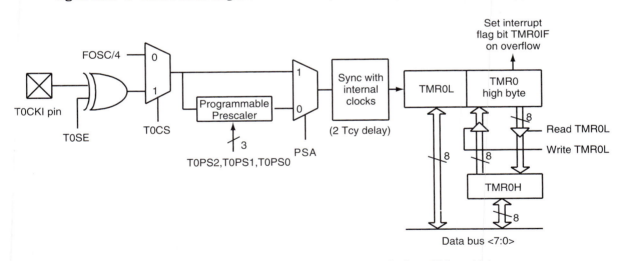

Figure 8.1b ■ Timer0 block diagram in 16-bit mode (redraw with permission of Microchip)

The T0CON register is a readable and writable register that controls the operation parameters of Timer0, including the prescaler selection. The contents of the T0CON register are shown in Figure 8.2.

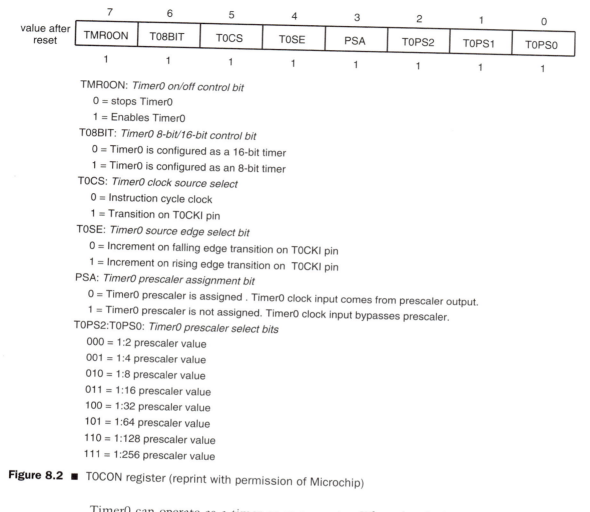

Figure 8.2 ■ T0CON register (reprint with permission of Microchip)

Timer0 can operate as a timer or as a counter. When the clock source is the instruction cycle clock, Timer0 operates as a timer. When the clock source comes from the T0CKI pin, Timer0 operates as a counter. By setting the T0CS bit of the T0CON register to 0, the internal instruction cycle clock is used as the clock source to Timer0. Otherwise, the signal from the T0CKI pin will be selected as the clock source, and Timer0 will operate as a counter. When the T0CKI signal is selected, the user has the choice of using either the rising or the falling edge of the T0CKI signal to increment the counter register pair TMR0L and the TMR0 high byte.

As shown in Figure 8.1b, the 8-bit register TMR0H acts as a buffer between the TMR0 high byte and the internal data bus. When the CPU reads the TMR0L register, the content of the TMR0 high byte is transferred to TMR0H. When the MCU reads the high byte of Timer0, it reads from TMR0H. Since the PIC18 microcontroller is 8-bit, it can read only eight bits at a time. Because of this feature, the 16-bit count value will be the value when the CPU reads the TMR0L register. Otherwise, the user may read an invalid value because of the rollover between successive reads of the high and low bytes. This feature is not available in the PIC16 and PIC17 devices.

Example 8.1

Suppose Timer0 does not have the buffer register TMR0H. What would be the value read from Timer0 with the following instruction sequence, assuming that the count value is 0x35FE when the first byte is read?

```
movff     TMR0L,PRODL
movff     TMR0H,PRODH
```

Solution: The **movff** instruction takes two clock cycles to execute. When the second instruction is executed, Timer0 has rolled over to 0x3600. Therefore, the value transferred to PRODH will be 0x36. Since the value transferred to PRODL is 0xFE, the value copied to the PRODH..PRODL pair becomes 0x36FE and is larger than the valid value by 256.

This problem cannot be solved by reversing the order of reading. The reason for this problem is left for you as an exercise problem.

Creating time delays is an important application for Timer0. By setting the prescaler properly, a wide range of delays can be created. The following example illustrates the procedure.

Example 8.2

Write a subroutine to create a time delay that is equal to 100 ms times the contents of the PRODL register, assuming that the crystal oscillator is running at 32 MHz.

Solution: The instruction clock frequency is 8 MHz and has a period of 125 ns. The 100-ms time delay can be created as follows:

1. Place the value 15535 into the TMR0 high byte and the TMR0L register so that Timer0 will overflow in 50000 clock cycles.

2. Choose internal instruction cycle clock as the clock source and set the prescaler to 16 so that Timer0 will roll over in 100 ms.

3. Enable Timer0.

4. Wait until Timer0 overflows.

The subroutine that implements this procedure is as follows:

```
delay     movlw     0x83           ; enable TMR0, select internal clock,
          movwf     T0CON          ; set prescaler to 16
loopd     movlw     0x3C           ; load 15535 into TMR0 so that it will
          movwf     TMR0H          ; roll over in 50000 clock cycles
          movlw     0xAF           ; "
          movwf     TMR0L          ; "
          bcf       INTCON,TMR0IF  ; clear the TMR0IF flag
wait      btfss     INTCON,TMR0IF  ;
          bra       wait           ; wait until 100 ms is over
          decfsz    PRODL,F
          bra       loopd
          return
```

In C language, this function can be written as follows:

```
void delay (char cx)
{
    int i;
    T0CON = 0x83;           /* enable TMR0, select instruction clock, prescaler set to 16 */
    for (i = 0; i < cx; i++) {
        TMR0 = 15535;       /* load 15535 into TMR0 so that it rolls */
                            /* over in 50000 clock cycles */
        INTCONbits.TMR0IF = 0;
        while(!(INTCONbits.TMR0IF)); /* wait until TMR0 rolls over */
    }
    return;
}
```

8.3.2 Timer1

Timer1 is a 16-bit timer/counter, depending on the clock source. The 16-bit timer/counter (TMR1H/TMR1L) is readable and writable. An interrupt will be requested whenever Timer1 rolls over from 0xFFFF to 0x0000. The block diagram of Timer1 is shown in Figure 8.3. Timer1 can be reset when the CCP module is configured in compare mode to generate a *special event trigger*. An A/D conversion will also be started by this reset.

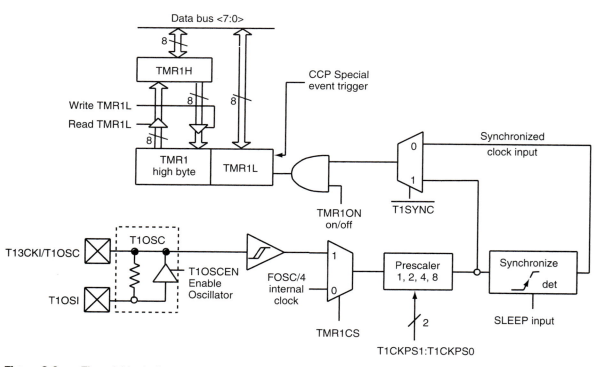

Figure 8.3 ■ Timer1 block diagram: 16-bit mode (redraw with permission of Microchip)

Timer1 operation is controlled by the T1CON register. The contents of the T1CON register are shown in Figure 8.4.

	7	6	5	4	3	2	1	0
value after reset	RD16	--	T1CKPS1	T1CKPS0	T1OSCEN	$\overline{\text{T1SYNC}}$	TMR1CS	TMR1ON
	0	0	0	0	0	0	0	0

RD16: 16-bit read/write mode enable bit

   0 = Enables read/write of Timer1 in two 8-bit operations

   1 = Enable read/write of Timer1 in 16-bit operation

T1CKPS1:T1CKPS0: Timer1 input clock prescale select bits

   00 = 1:1 prescale value

   01 = 1:2 prescale value

   10 = 1:4 prescale value

   11 = 1:8 prescale value

T1OSCEN: Timer1 oscillator enable bit

   0 = Timer1 oscillator is shut off

   1 = Timer1 oscillator is enabled

$\overline{\text{T1SYNC}}$: Timer1 external clock input synchronization select bit

When TMR1CS = 1

   0 = Synchronize external clock input

   1 = Do not synchronize external clock input

When TMR1CS = 0

   This bit is ignored.

TMR1CS: Timer1 clock source select bit

   0 = Instruction cycle clock (FOSC/4)

   1 = External clock from pin RC0/T1OSO/T13CKI

TMR1ON: Timer1 on bit

   0 = Stop Timer1

   1 = Enables Timer1

Figure 8.4 ■ T1CON contents (redraw with permission of Microchip)

When the clock source to Timer1 is the instruction cycle clock, it works as timer. When the external clock signal is selected, Timer1 can be a synchronous or an asynchronous counter, depending on the setting of the T1SYNC bit in the T1CON register. When this bit is clear, the external clock is synchronized with the instruction cycle clock.

When the TMRCS1 bit of the T1CON register is set, Timer1 increments on the rising edge of the external clock input (RC0/T1OSO/T1CKI pin) or the Timer1 oscillator, if enabled. When the Timer1 oscillator is enabled, the RC1/T1OSI and RC0/T1OSO/T1CKI pins become inputs. The TRISC$<1:0>$ value has no effect on the pin direction under this condition. The Timer1 oscillator is intended primarily for a 32-KHz crystal.

To take advantage of the feature of the special event trigger, Timer1 must be configured in compare mode. This event will reset Timer1 but will not set the interrupt flag TMR1IF.

Like Timer0, the high byte of Timer1 is double buffered. When reading TMR1L, the high byte of TMR1 will be loaded into the TMR1H register in Figure 8.3. This ensures that the values read from TMR1H and TMR1L belong together.

Like Timer0, Timer1 can also be used to create time delays. A more interesting application is *frequency measurement*. The method is to use one of the timers to create a 1-second time delay and another timer as a counter to count the incoming clock cycles within 1 second. The following example illustrates this idea.

Example 8.3

▼

Use Timer0 as a timer to create a 1-second delay and use Timer1 as a counter to count the rising (or falling) edges of an unknown signal (at the T0CKI pin) arriving in 1 second, which would measure the frequency of the unknown signal. Write a program to implement this idea, assuming that the PIC18 microcontroller is running with a 32-MHz crystal oscillator.

Solution: A 1-second delay can be created by loading 10 into the PRODL register and calling the subroutine in Example 8.2. Since Timer1 is only 16 bits, it may overflow many times in 1 second. If the unknown signal has the same frequency as the PIC18 microcontroller instruction cycle clock, it may overflow 122 times. Therefore, it is adequate to use one byte to keep track of the number of times that Timer1 overflows. Timer1 **overflow count** can be incremented by using interrupts.

Timer0 should be configured as follows:

- 16-bit mode
- Using instruction cycle clock as the clock source
- Set prescaler to 16
- Enabled to count

Timer1 should be configured as follows:

- 16-bit mode
- Prescaler value set to 1
- Disable oscillator
- Do not synchronize external clock input
- Select external T1CKI pin signal as clock source

The setting of interrupt is as follows:

- Enable priority interrupt
- Place Timer1 interrupt at high priority
- Enable only Timer1 rollover interrupt

The assembly program that implements the frequency measurement algorithm and settings is as follows:

```
           #include <p18F452.inc>
t1ov_cnt  set     0x00          ; Timer1 rollover interrupt count
freq      set     0x01          ; to save the contents of Timer1 at the end
          org     0x00
          goto    start
```

```
; high priority interrupt service routine
        org     0x08
        btfss   PIR1,TMR1IF     ; skip if Timer1 roll-over interrupt occurs
        retfie                  ; return if not Timer1 interrupt
        bcf     PIR1,TMR1IF     ; clear the interrupt flag
        incf    t1ov_cnt,F      ; increment Timer1 roll-over count
        retfie
; dummy low priority interrupt service routine
        org     0x18
        retfie
start   clrf    t1ov_cnt        ; initialize Timer1 overflow cnt to 0
        clrf    freq            ; initialize frequency to 0
        clrf    freq+1          ; "
        clrf    TMR1H           ; initialize Timer1 to 0
        clrf    TMR1L           ; "
        clrf    PIR1            ; clear all interrupt flags
        bsf     RCON,IPEN       ; enable priority interrupt
        movlw   0x01            ; set TMR1 interrupt to high priority
        movwf   IPR1            ; "
        movwf   PIE1            ; enable Timer1 roll-over interrupt
        movlw   0x87            ; enable Timer1, select external clock, set
        movwf   T1CON           ; prescaler to 1, disable crystal oscillator
        movlw   0xC0            ; enable global and peripheral interrupt
        movwf   INTCON          ; "
        movlw   0x0A
        movwf   PRODL           ; prepare to call delay to wait for 1 second
        call    delay           ; Timer1 overflow interrupt occur in this second
        movff   TMR1L,freq      ; save frequency low byte
        movff   TMR1H,freq+1    ; save frequency high byte
        bcf     INTCON,GIE      ; disable global interrupt
forever goto    forever
;***************************
; include the delay subroutine here.
;***************************
```

The C language version of the program is as follows:
```
#include  <p18F452.h>
unsigned  int t1ov_cnt;
unsigned  short long freq;
void high_ISR(void);
void low_ISR(void);
#pragma code high_vector = 0x08     // force the following statement to
void high_interrupt (void)          // start at 0x08
{
        _asm
        goto high_ISR
        _endasm
}
```

```
#pragma code low_vector = 0x18      //force the following statements to start at
void low_interrupt (void)           //0x18
{
        _asm
        goto      low_ISR
        _endasm
}
#pragma code                        //return to the default code section
#pragma interrupt high_ISR
void high_ISR (void)
{
        if(PIR1bits.TMR1IF){
           PIR1bits.TMR1IF = 0;
           t1ov_cnt ++;
           }
}
#pragma code
#pragma interrupt low_ISR
void low_ISR (void)
{
        _asm
        retfie 0
        _endasm
}
void delay (char cx);               /* prototype declaration */
void main (void)
{
        char t0_cnt;
        char temp;
        t1ov_cnt  = 0;
        freq      = 0;
        TMR1H     = 0;              /* force Timer1 to count from 0 */
        TMR1L     = 0;              /* " */
        PIR1      = 0;              /* clear Timer1 interrupt flag */
        RCONbits.IPEN = 1;         /* enable priority interrupt */
        IPR1      = 0x01;          /* set Timer1 interrupt to high priority */
        PIE1      = 0x01;          /* enable Timer1 roll-over interrupt */
        T1CON     = 0x83;          /* enable Timer1 with external clock, prescaler 1 */
        INTCON    = 0xC0;          /* enable global and peripheral interrupts */
        delay (10);                /* create one-second delay and wait for interrupt */
        INTCONbits.GIE = 0;        /* disable global interrupt */
        temp      = TMR1L;
        freq      = t1ov_cnt * 65536 + TMR1H * 256 + temp;
}
void delay (char cx)
{
        int i;
        T0CON = 0x83;              /* enable TMR0, select instruction clock, prescaler set to 16 */
```

```
for (i = 0; i < cx; i++) {
    TMR0H = 0x3C;                  /* load 15535 into TMR0 so that it rolls */
    TMR0L = 0xAF;                  /* over in 50000 clock cycles */
    INTCONbits.TMR0IF = 0;
    while(!(INTCONbits.TMR0IF)); /* wait until TMR0 rolls over */
    }
    return;
}
```

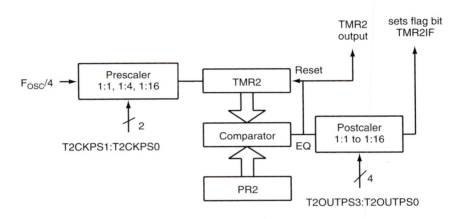

Figure 8.5 ■ Timer2 block diagram (redraw with permission of Microchip)

8.3.3 Timer2

Timer2 has an 8-bit timer (TMR2) and an 8-bit period register (PR2). Both registers are readable and writable. The block diagram of Timer2 is shown in Figure 8.5.

As shown in Figure 8.5, the clock source to TMR2 is $F_{OSC}/4$ divided by a programmable prescale factor. TMR2 is counting up and comparing with the value in the PR2 register in every clock cycle. When these two registers are equal (for one clock cycle only), the EQ signal will reset TMR2 to 0. The output of the comparator is divided by a programmable postscale factor. The equal signal (EQ) goes through this postscale circuit to generate a TMR2 interrupt. The output of TMR2 (before the postscaler) is fed to the synchronous serial port module, which may optionally use it to generate the shift clock.

The prescaler and postscaler counters are cleared when any of the following occurs:

- A write to the TMR2 register
- A write to the T2CON register
- Any device RESET

The operation parameters of Timer2 are configured by the T2CON register. The contents of T2CON are shown in Figure 8.6. Both the prescale and the postscale factors are programmed via this register.

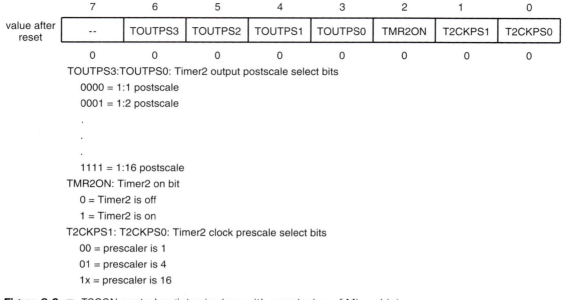

7	6	5	4	3	2	1	0
--	TOUTPS3	TOUTPS2	TOUTPS1	TOUTPS0	TMR2ON	T2CKPS1	T2CKPS0

value after reset

0	0	0	0	0	0	0	0

TOUTPS3:TOUTPS0: Timer2 output postscale select bits

0000 = 1:1 postscale

0001 = 1:2 postscale

.

.

.

1111 = 1:16 postscale

TMR2ON: Timer2 on bit

0 = Timer2 is off

1 = Timer2 is on

T2CKPS1: T2CKPS0: Timer2 clock prescale select bits

00 = prescaler is 1

01 = prescaler is 4

1x = prescaler is 16

Figure 8.6 ■ T2CON control register (redraw with permission of Microchip)

Because TMR2 is reset whenever it is equal to the PR2 register, it can be used to generate periodic interrupts. The next example illustrates this application.

Example 8.4

Assume that the PIC18F8680 is running with a 32-MHz crystal oscillator. Write an instruction sequence to generate periodic interrupts every 8 ms with high priority.

Solution: We need to compute the value to be written into the PR2 register to make this happen. Suppose we set the prescaler and postscaler to 16. Then the value to be written into PR2 register is

PR2 = number of counts in 8 ms = $8 \times 10^{-3} \times 32 \times 10^{6} \div 4 \div 16 \div 16 - 1 = 249$

The following instruction sequence will configure Timer2 to generate periodic interrupts every 8 ms:

```
        movlw    D'249'           ; load 249 into PR2 so that TMR2 counts up
        movwf    PR2,A            ; to 249 and reset
        bsf      RCON,IPEN,A      ; enable priority interrupt
        bsf      IPR1,TMR2IP,A    ; place TMR2 interrupt at high priority
        bcf      PIR1,TMR2IF,A    ;
        movlw    0xC0
        movwf    INTCON,A         ; enable global interrupt
        movlw    0x7E             ; enable TMR2, set prescaler to 16, set
        movwf    T2CON,A          ; postscaler to 16
        bsf      PIE1,TMR2IE,A    ; enable TMR2 overflow interrupt
```

8.3.4 Timer3

Timer3 consists of two readable and writable 8-bit registers (TMR3H and TMR3L). Timer3 can choose to use either the internal (instruction cycle clock) or external signal (T13CKI) as its clock source. The block diagram of Timer3 is quite similar to that of Timer3 and is shown in Figure 8.7.

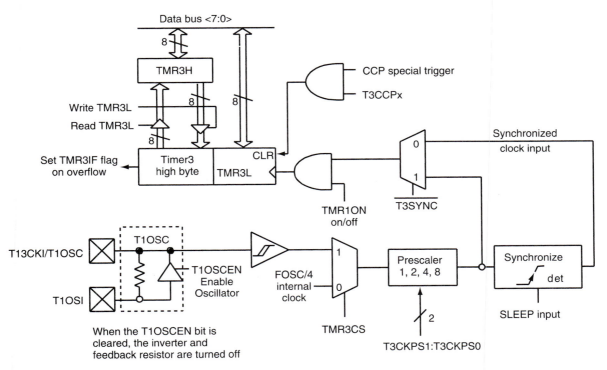

Figure 8.7 ■ Timer3 block diagram: 16-bit mode (redraw with permission of Microchip)

Timer3 can operate in three modes:

- Timer
- Synchronous counter
- Asynchronous counter

When the clock source is the instruction cycle clock, Timer3 operates as a timer. When the external clock source is selected, Timer3 operates as a counter. The external clock source could be the T13CKI signal or the crystal oscillator connected to the T1OSO and T1OSI pins. Timer3 also has an internal reset input. This RESET signal can be generated by the CCP module.

Timer3 may optionally generate an interrupt to the CPU when it rolls over from 0xFFFF to 0x0000. When rollover occurs, the TMR3IF flag (in PIR2 register) will be set to 1. Timer3 operation parameters are configured via the T3CON register. The contents of the T3CON register are shown in Figure 8.8.

7	6	5	4	3	2	1	0
RD16	T3CCP2	T3CKPS1	T3CKPS0	T3CCP1	T3SYNC	TMR3CS	TMR3ON
0	0	0	0	0	0	0	0

value after reset

RD16: 16-bit read/write mode enable bit

    0 = Enables read/write of Timer3 in two 8-bit operations

    1 = Enables read/write of Timer3 in 16-bit operation

T3CCP2:T3CCP1: Timer3 and Timer1 to CCPx enable bits

    00 = Timer1 and Timer2 are the clock sources for CCP1 through CCP5

    01 = Timer3 and Timer4 are the clock sources for CCP2 through CCP5;

        Timer1 and Timer2 are the clock sources for CCP1

    10 = Timer3 and Timer4 are the clock sources for CCP3 through CCP5;

        Timer1 and Timer2 are the clock sources for CCP1 and CCP2

    11 = Timer3 and Timer4 are the clock sources for CCP1 through CCP5

T3CKPS1:T3CKPS0: Timer3 input clock prescale select bits

    00 = 1:1 prescale value

    01 = 1:2 prescale value

    10 = 1:4 prescale value

    11 = 1:8 prescale value

T3SYNC: Timer3 external clock input synchronization select bit

When TMR3CS = 1

    0 = Synchronizes external clock input

    1 = Do not synchronize external clock input

When TMR3CS = 0

    This bit is ignored.

TMR3CS: Timer3 clock source select bit

    0 = Instruction cycle clock (FOSC/4)

    1 = External clock from pin RC0/T1OSO/T13CKI

TMR3ON: Timer3 on bit

    0 = Stops Timer3

    1 = Enables Timer3

Figure 8.8 ■ T3CON contents (redraw with permission of Microchip)

Reading the TMR3L register will transfer the high byte of Timer3 into the TMR3H register. Writing to the TMR3L register will transfer the value of TMR3H into the high byte of Timer3. This feature enables the 16-bit value written into and read (in two separate operations) from the Timer3 register to belong together.

8.3.5 Timer4

Timer4 is available only in the PIC18F8X20 and PIC18F6X20 devices. It is an 8-bit timer and has an 8-bit period register. The block diagram of Timer4 is shown in Figure 8.9.

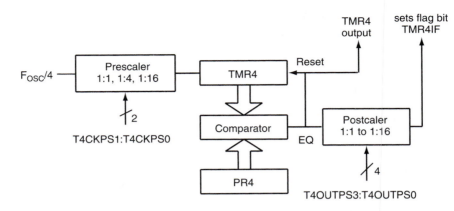

Figure 8.9 ■ Timer4 block diagram (redraw with permission of Microchip)

The output of Timer4 is used mainly as a PWM base timer for the CCP module (to be discussed later). The content of the T4CON register is identical to that of T2CON.

Timer4 increments from 0x00 until it matches the value of the PR4 register and then resets to 0x00 in the next increment cycle. On reset, the PR4 register is initialized to 0xFF. The input clock is divided by a programmable prescaler. The match output of TMR4 goes through a 4-bit postscaler to generate a TMR4 interrupt (when enabled).

The prescaler and postscaler counters are cleared when any of the following occurs:

- A write to the TMR4 register
- A write to the T4CON register
- Any device RESET (power-on reset, MCLR reset, watchdog timer reset, or brown-out reset)

8.3.6 C Library Functions for Timers

Microchip C18 compiler provides library functions for configuring, disabling, and reading from and writing into the timers. One needs to include the header file **timers.h** in order to use these library functions.

The library functions for disabling timers are the following:

void **CloseTimer0** (void);

void **CloseTimer1** (void);

void **CloseTimer2** (void);

void **CloseTimer3** (void);

void **CloseTimer4** (void);

The library functions for configuring timers are the following:

void **OpenTimer0** (unsigned char **config**);

void **OpenTimer1** (unsigned char **config**);

void **OpenTimer2** (unsigned char **config**);

void **OpenTimer3** (unsigned char **config**);

void **OpenTimer4** (unsigned char **config**);

The argument to the **OpenTimer0** function is a bit mask that is created by performing a bit-wise AND operation with a value from each of the categories listed here:

Enable Timer0 Interrupt

TIMER_INT_ON	enable interrupt
TIMER_INT_OFF	disable interrupt

Timer Width

T0_8BIT	8-bit mode
T0_16BIT	16-bit mode

Clock Source

T0_SOURCE_EXT	external clock source
T0_SOURCE_INT	internal clock source

External Clock Trigger

T0_EDGE_FALL	External clock on falling edge
T0_EDGE_RISE	External clock on rising edge

Prescale Value

T0_PS_1_n	1: n prescale (n = 1, 2, 4, 8, 16, 32, 64, 128, or 256)

The arguments to other timers are similar and are described in Appendix D. An example of a call to OpenTimer0 is as follows:

OpenTimer0 (TIMER_INT_ON & T0_8BIT & T0_SOURCE_INT & T0_PS_1_32);

The prototype definitions of library functions for reading timer values are the following:

unsigned int	**ReadTimer0** (void);
unsigned int	**ReadTimer1** (void);
unsigned char	**ReadTimer2** (void);
unsigned int	**ReadTimer3** (void);
unsigned char	**ReadTimer4** (void);

The use of these functions is straightforward. The following example reads the 16-bit value of Timer1:

unsigned int time_val;

time_val = ReadTimer1();

The prototype definitions of library functions for writing values into timers are the following:

void **WriteTimer0** (unsigned int **timer**);

void **WriteTimer1** (unsigned int **timer**);

void **WriteTimer2** (unsigned char **timer**);

void **WriteTimer3** (unsigned int **timer**);

void **WriteTimer4** (unsigned char **timer**);

where the parameter **timer** is the value to be written into the specified timer register. The following statement will write the value of 0x1535 into Timer0:

writeTimer0 (0x1535);

8.4 Capture/Compare/PWM Modules

A PIC18 device may have one or two or five capture/compare/PWM (CCP) modules. A CCP module contains a 16-bit capture register, a 16-bit compare register, or a PWM master/slave duty cycle register. A CCP module can be configured to operate in capture or compare or PWM mode. Each CCP module requires the use of timer resources. In capture or compare mode, the CCP module may need to use either Timer1 or Timer3 to operate. In PWM mode, either Timer2 or Timer4 may be needed.

The operations of all CCP modules are identical, with the exception of the special event trigger present on CCP1 and CCP2. The operation mode of the CCP module is configured by the CCPxCON (x = 1, 2, . . ., 5) register. The contents of these registers are shown in Figure 8.10.

	7	6	5	4	3	2	1	0
value after reset	--	--	DCxB1	DCxB0	CCPxM3	CCPxM2	CCPxM1	CCPxM0
	0	0	0	0	0	0	0	0

DCxB1:DCxB0: PWM duty cycle bit 1 and bit 0 for CCP module x

    capture mode:

        unused

    compare mode:

        unused

    PWM mode:

      These two bits are the lsbs (bit 1 and bit 0) of the 10-bit PWM duty cycle.

CCPxM3:CCPxM0: CCP module x mode select bits

    0000 = capture/compare/PWM disabled (resets CCPx module)

    0001 = reserved

    0010 = compare mode, toggle output on match (CCPxIF bit is set)

    0100 = capture mode, every falling edge

    0101 = capture mode, every rising edge

    0110 = capture mode, every 4th rising edge

    0111 = capture mode, every 16th rising edge

    1000 = compare mode, initialize CCP pin low, on compare match force CCP pin high

        (CCPxIF bit is set)

    1001 = compare mode, initialize CCP pin high, on compare match force CCP pin low

        (CCPxIF bit is set)

    1010 = compare mode, generate software interrupt on compare match (CCP pin

        unaffected, CCPxIF bit is set).

    1011 = compare mode, trigger special event (CCPxIF bit is set)

        For CCP1 and CCP2: Timer1 or Timer3 is reset on event

        For all other modules: CCPx pin is unaffected and is configured as an I/O port.

    11xx = PWM mode

Figure 8.10 ■ CCPxCON register (x = 1,..,5) (redraw with permission of Microchip)

8.4.1 CCP Module Configuration

Each capture/compare/PWM module is associated with a control register (CCPxCON) and a data register (CCPRx). The data register in turn is comprised of two 8-bit registers: CCPRxL (low byte) and CCPRxH (high byte).

The CCP modules utilize Timer1, Timer2, Timer3, or Timer4, depending on the mode selected. Timer1 and Timer3 are available to modules in capture or compare mode, while Timer2 and Timer4 are available for modules in PWM mode.

The assignment of a particular timer to a module is determined by the timer-to-CCP enable bits (bit 6 and bit 3) in the T3CON register. Depending on the selected configuration, up to four timers may be active at once, with modules in the same configuration sharing the same timer resources. The possible configurations are shown in Figure 8.11.

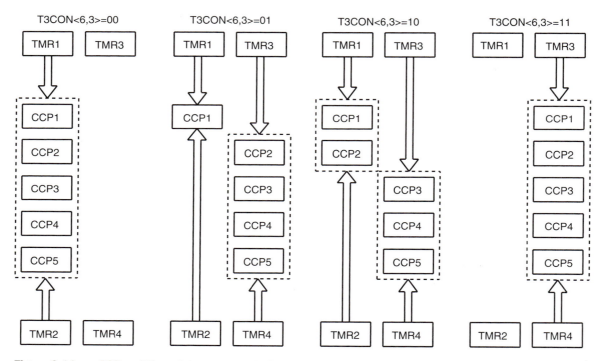

Figure 8.11 ■ CCP and Timer interconnect configurations (redraw with permission of Microchip)

In Figure 8.11,

when T3CON<6,3> = 00,

Timer1 is used for all capture and compare operations for all CCP modules. Timer2 is used for PWM operations for all CCP modules.

when T3CON<6,3> = 01,

Timer1 is used for the capture and compare operations for the CCP1 module, and Timer2 is used for the PWM operation for the CCP1 module. Timer3 is used for the capture and compare operations for the remaining four CCP modules, and Timer4 is used for the PWM operations for the remaining four CCP modules.

when T3CON<6,3> = 10,

Timer1 is used for the capture and compare operations for the CCP1 and CCP2 modules, whereas Timer2 is used for the PWM operations for the CCP1 and CCP2 modules. Timer3 is used for the capture and compare operations for the CCP3, CCP4, and CCP5 modules. Timer4 is used for the PWM operations for CCP3-CCP5.

when T3CON<6,3> = 11,

Timer3 is used for the capture and compare operations for all CCP modules, whereas Timer4 is used for the PWM operations for all CCP modules.

8.5 CCP in Capture Mode

Some applications need to know the arrival times of events. In a computer, *physical time* is represented by the count value in a counter, whereas the occurrence of an *event* is represented by a signal edge (can be a rising or a falling edge). The time when an event occurs can be recorded by latching the count value when a signal edge arrives, as illustrated in Figure 8.12. The capture mode of the PIC18 CCP module is designed for this type of application. As shown in Figure 8.13,

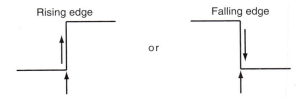

Figure 8.12 ■ Events represented by signal edges

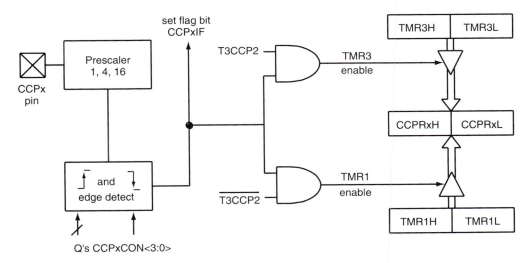

Figure 8.13 ■ Capture mode operation block diagram (redraw with permission of Microchip)

the 16-bit value of the TMR1 or the TMR3 register pair is captured into the register pair CCPRxH: CCPRxL when an event occurs on the CCPx pin. An event can be one of the following:

- Every falling edge
- Every rising edge
- Every 4th rising edge
- Every 16th rising edge

8.5.1 Capture Operation

When the capture is made, the interrupt request flag bit CCPxIF is set, which must be cleared in software. If another capture occurs before the value in the register CCPRx is read, the old value will be overwritten by the new captured value.

In capture mode, the CCPx pin must be configured as an input. The timers that are to be used with the capture operation (Timer1 and/or Timer3) must be running in the timer mode or synchronous counter mode. In the asynchronous counter mode, the capture operation may not work.

When the capture mode is changed, a false capture interrupt may be generated. The user should clear the CCPxIE bit prior to changing the capture mode to avoid false interrupts and should clear the flag bit CCPxIF following any such change in the operation mode.

There are four prescaler settings (one for falling edge capture and three for rising edge capture). Whenever the CCP module is turned off or the CCP module is not in the capture mode, the prescaler counter is cleared. Any reset will also clear the prescaler counter.

Switching from one capture prescaler to another may generate an interrupt. In addition, the prescaler counter will not be cleared during this switching. Therefore, the first capture may be from a nonzero prescaler. To prevent this, the user should turn off the CCP module before switching to a new capture prescaler.

8.5.2 Microchip C Library Functions for CCP in Capture Mode

Microchip MCC18 C compiler provides three functions (shown in Table 8.1) to support each CCP channel. The header file **capture.h** must be included in the user program to use these library functions.

Function	Description
CloseCapture*x*	Disable capture channel *x*
OpenCapture*x*	Configure capture channel *x*
ReadCapture*x*	Read a value from CCP channel *x*

Table 8.1 ■ MCC18 C library functions for CCP peripheral

The prototype declarations and parameter definitions are as follows:

void **CloseCapture1** (void);

void **CloseCapture2** (void);

void **CloseCapture3** (void);

void **CloseCapture4** (void);

void **CloseCapture5** (void);

void **OpenCapture1** (unsigned char **config**);

void **OpenCapture2** (unsigned char **config**);

void **OpenCapture3** (unsigned char **config**);

void **OpenCapture4** (unsigned char **config**);

void **OpenCapture5** (unsigned char **config**);

There are two values to be set for the parameter **config:** interrupt enabling and the edge to capture. There are two possible values for the CCP interrupt capture mode:

CAPTURE_INT_ON   : interrupt enabled

CAPTURE_INT_OFF   : interrupt disabled

There are four possible edges to be captured:

Cx_EVERY_FALL_EDGE   : capture on every falling edge

Cx_EVERY_RISE_EDGE   : capture on every rising edge

Cx_EVERY_4_RISE_EDGE   : capture on every 4th rising edge

Cx_EVERY_16_RISE_EDGE   : capture on every 16th rising edge

The following five functions read the values from the specified capture channel:

unsigned int **ReadCapture1** (void);

unsigned int **ReadCapture2** (void);

unsigned int **ReadCapture3** (void);

unsigned int **ReadCapture4** (void);

unsigned int **ReadCapture5** (void);

The following example configures CCP2 in capture mode that disables interrupt, capture on every 16th rising edge:

OpenCapture2 (CAPTURE_INT_OFF & C2_EVERY_16_RISE_EDGE);

8.5.3 Applications of Capture Mode

The capture mode of CCP module has many applications. Some examples are the following:

- *Event arrival time recording.* Some applications (e.g., a swimming competition) need to compare the arrival times of several different swimmers. The capture function is very suitable for this application. The number of events that can be compared is limited by the number of capture channels.

- *Period measurement.* To measure the period of an unknown signal, the capture function should be configured to capture the timer values corresponding to two consecutive rising or falling edges, as illustrated in Figure 8.14.

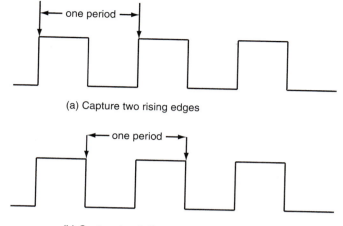

(a) Capture two rising edges

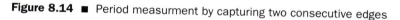

(b) Capture two falling edges

Figure 8.14 ■ Period measurment by capturing two consecutive edges

- *Pulse-width measurement.* To measure the width of a pulse, two adjacent rising and falling edges are captured, as shown in Figure 8.15.

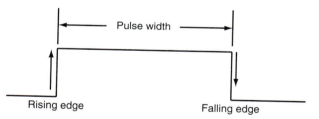

Figure 8.15 ■ Pulse-width measurement using input capture

- *Interrupt generation.* All capture inputs can serve as edge-sensitive interrupt sources. Once enabled, interrupt will be generated on the selected edge.
- *Event counting.* An event can be represented by a signal edge. A CCP channel in capture mode can be used in conjunction with a timer or another CCP channel in compare mode to count the number of events that occur during an interval. This application is similar to using a timer in counter mode.
- *Time reference.* In this application, a CCP channel in capture mode is used in conjunction with another CCP channel in compare mode. For example, if the user wants to activate a pulse at certain number of clock cycles after detecting an input event, the CCP channel in capture mode can be used to record the time at which the edge is detected. A number corresponding to the desired delay would be added to this captured value and stored in the CCPRx register of the CCPx channel in compare mode. This application is illustrated in Figure 8.16.

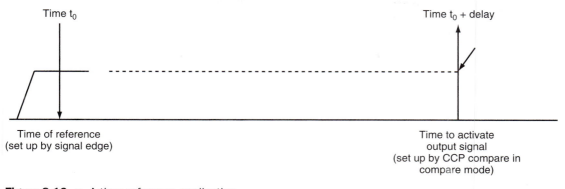

Figure 8.16 ■ A time reference application

- *Duty cycle measurement.* The duty cycle is the percentage of time that the signal is high within a period in a periodic digital signal. The measurement of duty cycle is illustrated in Figure 8.17.

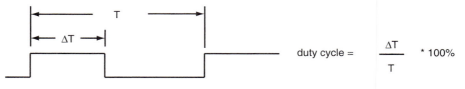

$$\text{duty cycle} = \frac{\Delta T}{T} * 100\%$$

Figure 8.17 ■ Definition of duty cycle

The unit used in most of the measurements is the number of clock cycles. When it is desirable, the unit should be converted to an appropriate unit such as seconds.

Example 8.5
▼

Period measurement. Use the CCP channel 1 in capture mode to measure the period of an unknown signal, assuming that the PIC18 microcontroller is running with a 16-MHz crystal oscillator. Use the number of clock cycles as the unit of period. The period of the unknown signal is shorter than 65536 clock cycles.

Solution: The user needs to capture either two consecutive rising or falling edges and calculate their difference in order to measure the period. The required settings are as follows:

- CCP1 (RC2) pin: Configured for input.
- Timer1: 16-bit operation, use instruction cycle clock as clock source, prescaler set to 1. Write the value 0x81 into the T1CON register.
- Timer3: Select Timer1 as the base timer for the CCP1 capture mode. Write the value 0x81 into the T3CON register.
- CCP1: Capture on every rising edge. Write the value 0x05 into the CCP1CON register.
- Disable the CCP1 interrupt. Clear the CCP1IE bit of the PIE1 register.

The assembly program that measures the period (placed in PRODH:PRODL) using this setting is as follows:

```
                #include <p18F8720.inc>
                org         0x00
                goto        start
                org         0x08
                retfie
                org         0x18
                retfie
start           bsf         TRISC,CCP1          ; configure CCP1 pin for input
                movlw       0x81                ; use Timer1 as the time base
                movwf       T3CON               ; of CCP1 capture
                bcf         PIE1,CCP1IE         ; disable CCP1 capture interrupt
                movlw       0x81                ; enable Timer1, prescaler set to 1,
                movwf       T1CON               ; 16-bit, use instruction cycle clock
                movlw       0x05                ; set CCP1 to capture on every rising edge
                movwf       CCP1CON             ; "
                bcf         PIR1,CCP1IF         ; clear the CCP1IF flag
edge1           btfss       PIR1,CCP1IF         ; wait for the first edge to arrive
                goto        edge1               ; "
                movff       CCPR1H,PRODH        ; save the first edge
                movff       CCPR1L,PRODL        ; "
                bcf         PIR1,CCP1IF         ; clear the CCP1IF flag
edge2           btfss       PIR1,CCP1IF         ; wait for the second edge to arrive
                goto        edge2               ; "
                clrf        CCP1CON             ; disable CCP1 capture
                movf        PRODL,W
                subwf       CCPR1L,W            ; subtract first edge from 2nd edge
                movwf       PRODL               ; and leave the period in PRODH:PRODL
                movf        PRODH,W             ; "
                subwfb      CCPR1H,W            ; "
                movwf       PRODH               ; "
forever         goto        forever             ;
                end
```

The C language version of the program is as follows:

```c
#include <p18F8720.h>
void main (void)
{
        unsigned int period;
        TRISCbits.TRISC2 = 1;               /* configure CCP1 pin for input */
        T3CON = 0x81;                       /* use Timer1 as the time base for CCP1 capture */
        PIE1bits.CCP1IE = 0;                /* disable CCP1 capture interrupt */
        PIR1bits.CCP1IF = 0;                /* clear the CCP1IF flag */
        T1CON = 0x81;                       /* enable 16-bit Timer1, prescaler set to 1 */
        CCP1CON = 0x05;                     /* capture on every rising edge */
        while (!(PIR1bits.CCP1IF));         /* wait for 1st rising edge */
        PIR1bits.CCP1IF = 0;
        period = CCPR1;                     /* save the first edge (CCPR1 is accessed as a 16-bit value) */
```

```
        while (!(PIR1bits.CCP1IF));     /* wait for the 2nd rising edge */
        CCP1CON = 0x00;                 /* disable CCP1 capture */
        period = CCPR1 - period;
}
```

The clock period of an unknown signal could be much longer than 2^{16} clock cycles. In this situation, the user will need to keep track of the number of times that the timer overflows. Each timer overflow adds 2^{16} clock cycles to the period. Let

ovcnt   = timer overflow count
diff   = the difference of two edges
edge1   = the captured time of the first edge
edge2   = the captured time of the second edge

The signal period can be calculated by the following equations:

Case 1: edge2 $\geq$ edge1
period = ovcnt $\times$ 2^{16} + diff
Case 2: edge1 > edge2
period = (ovcnt − 1) $\times$ 2^{16} + diff

In case 2, the timer overflows at least once even if the pulse width is shorter than $2^{16} - 1$ clock cycles. Therefore, we need to subtract 1 from the timer overflow count in order to get the correct result. The period is obtained by appending the difference of the two captured edges to the timer overflow count.

Example 8.6

Write a program to measure the period of a signal connected to the CCP1 (RC2) pin, assuming that the instruction clock is running at 5 MHz. Make the program more general so that it can also measure the period of a signal with very low frequency.

Solution: The logic flow of the program is illustrated in Figure 8.18. The assembly program that implements the algorithm shown in Figure 8.18 is as follows:

```
            #include <p18F8720.inc>
ov_cnt      set     0x00            ; timer overflow count
per_hi      set     0x01            ; high byte of edge difference
per_lo      set     0x02            ; low byte of edge difference
            org     0x00
            goto    start
            org     0x08
            goto    hi_pri_ISR      ; go to the high-priority service routine
            org     0x18
            retfie
start       clrf    ov_cnt          ; initialize overflow count by 1
            bcf     INTCON,GIE      ; disable all interrupts
            bsf     RCON,IPEN       ; enable priority interrupt
            bcf     PIR1,TMR1IF     ; clear the TMR1IF flag
            bsf     IPR1,TMR1IP     ; set Timer1 interrupt to high priority
            bsf     TRISC,CCP1      ; configure CCP1 pin for input
            movlw   0x81            ; use Timer1 as the time base
```

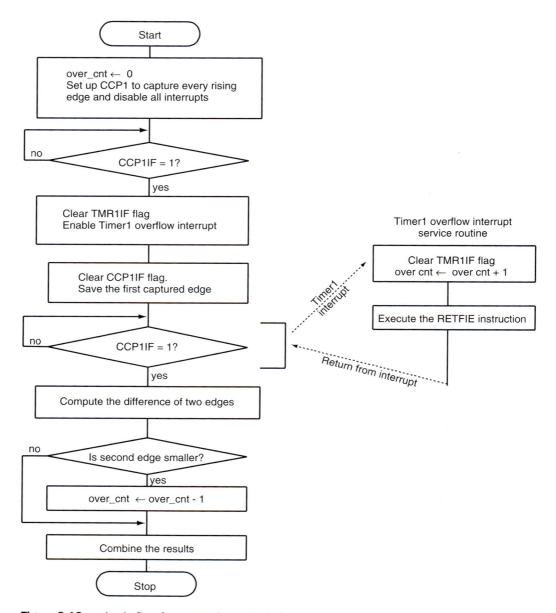

Figure 8.18 ■ Logic flow for measuring period of slow signals

```
movwf    T3CON          ; of CCP1 capture
bcf      PIE1,CCP1IE    ; disable CCP1 capture interrupt
movlw    0x81           ; enable Timer1, prescaler set to 1,
movwf    T1CON          ; 16-bit mode, use instruction cycle clock
movlw    0x05           ; set CCP1 to capture on every rising edge
movwf    CCP1CON        ; "
bcf      PIR1,CCP1IF    ; clear the CCP1IF flag
```

```
edge1      btfss      PIR1,CCP1IF       ; wait for the first edge to arrive
           goto       edge1             ; "
           movff      CCPR1H,per_hi     ; save the high byte of captured edge
           movff      CCPR1L,per_lo     ; save the low byte of captured edge
           bcf        PIR1,TMR1IF
           movlw      0xC0
           iorwf      INTCON,F          ; enable global interrupts
           bsf        PIE1,TMR1IE       ; enable Timer1 overflow interrupt
edge2      btfss      PIR1,CCP1IF       ; wait for the 2nd edge to arrive
           goto       edge2
           movf       per_lo,W
           subwf      CCPR1L,W
           movwf      per_lo            ; save the low byte of edge difference
           movf       per_hi,W
           subwfb     CCPR1H,W
           movwf      per_hi,A          ; save the high byte of edge difference
           btfsc      STATUS,C
           goto       normal
           decf       ov_cnt            ; 1st edge is larger, so decrement overflow count
normal     nop
forever    goto       forever
hi_pri_ISR btfss      PIR1,TMR1IF       ; high priority interrupt service routine
           retfie                       ; not Timer1 interrupt, so return
           incf       ov_cnt
           bcf        PIR1,TMR1IF       ; clear Timer1 overflow interrupt flag
           retfie
           end
```

Using one byte to keep track of the times that Timer1 has rolled over can measure a period of length up to $2^{24} \times 10^{-6} \times 0.2 = 3.355$ seconds. To measure an even slower signal, two bytes are needed to keep track of the Timer1 rollover count.

The C language version of the program is as follows:

```c
#include <p18F8720.h>
#include <timers.h>
#include <capture.h>
unsigned int ov_cnt, temp;
unsigned short long period;             /* 24-bit period value */
void high_ISR(void);
void low_ISR(void);
#pragma code high_vector = 0x08         // force the following statement to
void high_interrupt (void)              // start at 0x08
{
        _asm
        goto high_ISR
        _endasm
}
#pragma code low_vector = 0x18          //force the following statements to start at
void low_interrupt (void)               //0x18
{
        _asm
        goto       low_ISR
```

```
                _endasm
        }
        #pragma code                            //return to the default code section
        #pragma interrupt high_ISR
        void high_ISR (void)
        {
                if(PIR1bits.TMR1IF){
                        PIR1bits.TMR1IF = 0;
                        ov_cnt ++;
                }
        }
        #pragma code
        #pragma interrupt low_ISR
        void low_ISR (void)
        {
                _asm
                retfie 0
                _endasm
        }

        void main (void)
        {
                unsigned int temp1;
                ov_cnt = 0;
                INTCONbits.GIE = 0;             /* disable global interrupts */
                RCONbits.IPEN = 1;              /* enable priority interrupts */
                PIR1bits.TMR1IF = 0;
                IPR1bits.TMR1IP = 1;            /* promote Timer1 rollover interrupt to high priority */
                TRISCbits.TRISC2 = 1;          /* configure CCP1 pin for input */
                OpenTimer1 (TIMER_INT_ON & T1_16BIT_RW & T1_PS_1_1 &
                            T1_OSC1EN_OFF & T1_SYNC_EXT_OFF &
                            T1_SOURCE_INT);
                OpenTimer3 (TIMER_INT_OFF & T3_16BIT_RW & T3_PS_1_1 &
                            T3_SOURCE_INT & T3_PS_1_1 & T3_SYNC_EXT_ON & T12_SOURCE_CCP);
                            /* turn on Timer3 and appropriate parameters */
                OpenCapture1 (CAPTURE_INT_OFF & C1_EVERY_RISE_EDGE);
                PIE1bits.CCP1IE = 0;           /* disable CCP1 capture interrupt */
                PIR1bits.CCP1IF = 0;
                while(!(PIR1bits.CCP1IF));
                temp = ReadCapture1( );        /* save the first captured edge */
                PIR1bits.CCP1IF = 0;
                PIR1bits.TMR1IF = 0;
                INTCON |= 0xC0;                /* enable global interrupts */
                PIE1bits.TMR1IE = 1;           /* enable Timer1 rollover interrupt */
                while(!(PIR1bits.CCP1IF));
                CloseCapture1();               /* disable CCP1 capture */
                temp1 = ReadCapture1();
                if (temp1 < temp)
                        ov_cnt–;
                period = ov_cnt * 65536 + temp1 - temp;
        }
```

The algorithm for measuring the pulse width is similar to that for measuring the duty cycle and hence is left for you as an exercise problem.

When the signal frequency is very high, it is likely that the next edge will arrive even before the captured edge has been saved. In this situation, the user can choose to capture every 4th or even every 16th rising edge in order to measure the period of the unknown signal. The actual period is simply the difference of two captured edges divided by 4 or 16.

8.6 CCP in Compare Mode

In compare mode, the 16-bit CCPRx register value is constantly compared against either the TMR1 register pair value or the TMR3 register pair value. When a match occurs, one of the following actions may occur on the associated CCPx pin:

- Driven high
- Driven low
- Toggle output (high to low or low to high)
- Remains unchanged

8.6.1 Compare Mode Operation

The circuit related to the CCP compare mode operation is shown in Figure 8.19. In order to use the CCP in compare mode, the user must do the following:

1. Make a copy of the 16-bit timer value
2. Add to this copy a **delay count**
3. Store the sum in the CCPRxH:CCPRxL register pair

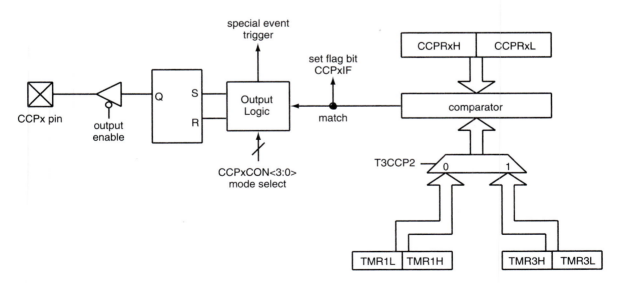

Figure 8.19 ■ Circuit for CCP in compare mode (redraw with permission of Microchip)

The CCPx pin must be configured for output and Timer1 and/or Timer3 must be running in timer mode or synchronized counter mode.

The CCP1 and the CCP2 modules in compare mode can also generate the special event trigger. The output of either the CCP1 or the CCP2 module resets the TMR1 or TMR3 register pair, depending on which timer resource is currently selected. This allows the CCPR1 register to effectively be a 16-bit programmable period register for Timer1 or Timer3. The CCP2 special event trigger will also start an A/D conversion if the A/D module is enabled.

8.6.2 Applications of CCP in Compare Mode

CCP in compare mode can be used to perform a variety of functions. Generation of a single pulse, a train of pulses, a periodic waveform with a certain duty cycle, and a specific time delay are among the most popular applications.

Example 8.7

Use CCP1 to generate a periodic waveform with 40% duty cycle and 1-KHz frequency, assuming that the instruction cycle clock frequency is 4 MHz.

Solution: The waveform of a 1-KHz digital signal with a 40% duty cycle is shown in Figure 8.20.

Figure 8.20 ■ 1KHz 40% duty cycle waveform

The logic flow of this problem is illustrated in Figure 8.21. Suppose we use Timer3 as the base timer and set the prescale factor of Timer3 to 1. The number of clock cycles that the signal is high and low will be 1600 and 2400, respectively.

The assembly program to implement the algorithm illustrated in Figure 8.21 is as follows:

```
            #include <p18F8720.inc>
hi_hi   equ     0x06            ; number (1600) of clock cycles that signal
hi_lo   equ     0x40            ; is high
lo_hi   equ     0x09            ; number (2400) of clock cycles that signal
lo_lo   equ     0x60            ; is low
        org     0x00
        goto    start
        org     0x08
        retfie
        org     0x18
        retfie
start   bcf     TRISC,CCP1      ; configure CCP1 pin for output
        movlw   0xC9            ; Configure Timer3 for 16-bit timer, prescaler 1,
        movwf   T3CON           ; use Timer3 for CCP1 time base, enable Timer3
        bcf     PIR1,CCP1IF     ; clear the CCP1IF flag
        movlw   0x09            ; CCP1 pin set high initially and
        movwf   CCP1CON         ; pull low on match
```

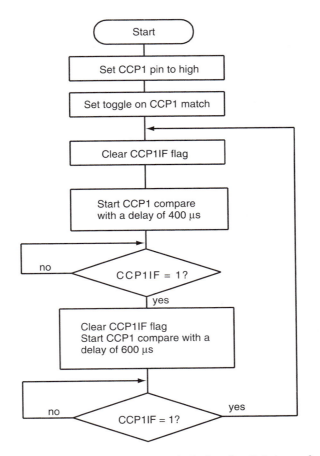

Figure 8.21 ■ The program logic flow for digital waveform generation

```
;
; CCPR1 <-- TMR3 + 1600; start a new compare operation
;
          movlw     hi_lo
          addwf     TMR3L,W
          movwf     CCPR1L          ; "
          movlw     hi_hi           ; "
          addwfc    TMR3H,W         ; "
          movwf     CCPR1H          ; save it in CCPR1 register pair
          bcf       PIR1,CCP1IF     ; clear the CCP1IF flag
hi_time   btfss     PIR1,CCP1IF
          bra       hi_time
          bcf       PIR1,CCP1IF
          movlw     0x08            ; CCP1 pin set low initially and
          movwf     CCP1CON         ; pull high on match
```

```
            ;
            ; CCPR1 <-- CCPR1 + 2400; start another compare operation
            ;
                    movlw       lo_lo
                    addwf       CCPR1L,F        ;
                    movlw       lo_hi
                    addwfc      CCPR1H,F
            lo_time btfss       PIR1,CCP1IF
                    bra         lo_time
                    bcf         PIR1,CCP1IF
                    movlw       0x09            ; CCP1 pin set high initially
                    movwf       CCP1CON         ; and pull low on match
                    movlw       hi_lo
                    addwf       CCPR1L,F
                    movlw       hi_hi
                    addwfc      CCPR1H,F
                    bra         hi_time
                    end
```

The C language version of the program is as follows:

```c
#include <p18F8720.h>
void main (void)
{
        TRISCbits.TRISC2 = 0;           /* configure CCP1 pin for output */
        T3CON = 0xC9;                   /* turn on TMR3 in 16-bit mode, prescaler is 1 */
        CCP1CON = 0x09;                 /* configure CCP1 pin to set high initially but pull low on match*/
        CCPR1 = TMR3 + 0x0640;          /* start CCP1 compare with delay equals 1600*/
        PIR1bits.CCP1IF = 0;            /* clear CCP1IF flag */
        while (1) {
                while (!(PIR1bits.CCP1IF));
                PIR1bits.CCP1IF = 0;
                CCP1CON = 0x08;
                CCPR1 += 0x0960;        /* start CCP1 compare with delay equals 2400*/
                while (!(PIR1bits.CCP1IF));
                PIR1bits.CCP1IF = 0;
                CCP1CON = 0x09;
                CCPR1 += 0x0640;        /* start CCP1 compare with delay equals 1600*/
        }
}
```

In this example, the CPU is tied up in generating the waveform. A better approach is to use interrupt. After starting the first pulse, the user program enables the CCP interrupt. The CCP interrupt service routine simply clears the interrupt flag and restarts the next compare operation. By using this approach, the CPU can still perform other useful operations. The following example illustrates this idea.

▲

Example 8.8

▼

Use the interrupt-driven approach to generate the waveform specified in Example 8.6.

Solution: This program will use a flag to select either 1600 (= 0) or 2400 (= 1) as the delay count for the compare operation. We will configure the CCP1 interrupt to high priority. The CCP1 interrupt service routine will check to make sure that the interrupt is caused by the compare match. If not, the service routine simply returns. If the interrupt is caused by the CCP1 compare match, the service routine will start a new compare operation with 1600 (when flag = 0) or 2400 (when flag = 1) as the delay. The assembly program is as follows:

```
#include <p18F8720.inc>
hi_hi       equ     0x06        ; number (1600) of clock cycles that signal
hi_lo       equ     0x40        ; is high
lo_hi       equ     0x09        ; number (2400) of clock cycles that signal
lo_lo       equ     0x60        ; is low
flag        set     0x00        ; select 1600 (=0) or 2400(=1) as delay
            org     0x00
            goto    start
            org     0x08
            goto    hi_ISR      ; go to the actual interrupt service routine
            org     0x18
            retfie
start       bcf     TRISC,CCP1  ; configure CCP1 pin for output
            movlw   0xC9        ; Enable Timer3 in 16-bit mode, prescaler 1,
            movwf   T3CON       ; use Timer3 as the base timer of CCP1
            bsf     PORTC,CCP1  ; pull the CCP1 pin to high
            movlw   0x09        ; configure CCP1 pin to set high initially and
            movwf   CCP1CON     ; pull low on match
;
; start a compare operation so that the CCP1 pin stay high for 400 us
            movlw   hi_lo       ; "
            addwf   TMR3L,W     ; "
            movwf   CCPR1L      ; "
            movlw   hi_hi       ; "
            addwfc  TMR3H,W     ; "
            movwf   CCPR1H      ; "
            bcf     PIR1,CCP1IF ; clear the CCP1IF flag
hi_LST      btfss   P1R1,CPP1IF
            bra     hi_lst
            bcf     P1R1,CCP1IF
            movlw   0x02        ; CCP1 pin
            movwf   CCP1CON     ; toggle on match
            movlw   lo_lo       ; start next
            addwf   CCPR1L,F    ; compare
            movlw   lo_hi       ; operation
            addwfc  CCPR1H,F    ; "
            bsf     IPR1,CCP1IP ; configure CCP1 interrupt to high priority
            movlw   0x00        ; clear the flag so that 1600 is the next delay
            movwf   flag        ; for CCP compare operation
            movlw   0xC0
            iorwf   INTCON,F    ; enable CCP1 interrupt
            bsf     PIE1,CCP1IE ; "
forever     nop                 ; wait for interrupt
            goto    forever     ; "
hi_ISR      btfss   PIR1,CCP1IF ; is the interrupt caused by CCP1?
            retfie
```

```
            bcf       PIR1,CCP1IF
            btfsc     flag,0              ; prepare to add 1600 if flag is 0
            goto      add_2400
            movlw     hi_lo              ; start a new compare operation
            addwf     CCPR1L,F           ; which will keep CCP1 pin high for
            movlw     hi_hi              ; 1600 clock cycles
            addwfc    CCPR1H,F           ; "
            btg       flag,0             ; toggle the flag
            retfie
add_2400    movlw     lo_lo              ; start a new compare operation which
            addwf     CCPR1L,F           ; will keep CCP1 pin low for 2400
            movlw     lo_hi              ; clock cycles
            addwfc    CCPR1H,F           ; "
            btg       flag,0             ; toggle the flag
            retfie
            end
```

The C language version of the program is straightforward and hence is left for you as an exercise problem.

The CCP in compare mode can be used to create a time delay. The following example illustrates this application.

▲

Example 8.9

▼

Assume that there is a PIC18 demo board (e.g., SSE8720) running with a 16-MHz crystal oscillator. Write a function that uses CCP1 in compare mode to create a time delay that is equal to 10 ms multiplied by the contents of PRODL.

Solution: The function uses a program loop to perform the number (specified in the PRODL register) of compare operations on CCP1. A 10-ms delay can be created by the following:

- Choosing instruction cycle clock as the clock source
- Setting the prescaler of Timer3 to 1
- Using 40000 as the delay count of the compare operation
- Using Timer1 as the base timer for CCP1 through CCP5

The assembly language version of the function is as follows:

```
;****************************************************************
; The following function creates a delay that is equal to the content of PRODL
; multiplied by 10 ms assuming that the PIC18 MCU uses a 16 MHz crystal oscillator.
;****************************************************************
dly_by10ms   movlw    0x81              ; enable Timer3 in 16-bit mode, 1:1 prescaler
             movwf    T3CON,A           ; use Timer1 as base times for CCP1
             movwf    T1CON,A           ; enable Timer1 in 16-bit mode with 1:1 prescaler
             movlw    0x0A              ; configure CCP1 to generate software
             movwf    CCP1CON,A         ; interrupt on compare match
             movf     TMR1L,W,A         ; to perform a CCP1 compare with
             addlw    0x40              ; 40000 cycles of delay
             movwf    CCPR1L,A          ; "
```

```
            movlw    0x9C          ; "
            addwfc   TMR1H,W,A     ; "
            movwf    CCPR1H,A      ; "
            bcf      PIR1,CCP1IF,A
loop        btfss    PIR1,CCP1IF,A ; wait until 40000 cycles are over
            bra      loop
            dcfsnz   PRODL,F,A     ; is loop count decremented to zero yet?
            return   0             ; delay is over, return
            bcf      PIR1,CCP1IF,A ; clear the CCP1IF flag
            movlw    0x40          ; start the next compare operation
            addwf    CCPR1L,F,A    ; with 40000 cycles delay
            movlw    0x9C          ; "
            addwfc   CCPR1H,F,A    ; "
            bra      loop
```

The C language version of the function is as follows:

```
/* ------------------------------------*/
/* The following function creates a delay equals to 10ms */
/* times kk.                              */
/* ------------------------------------*/
void dly_by10ms (unsigned char kk)
{
        CCP1CON         = 0x0A; /* configure CCPR1 to generate software interrupt */
        T3CON           = 0x81; /* enables Timer3 and select Timer1 as base timer for CCP1 */
        T1CON           = 0x81; /* enables Timer1 in 16-bit mode with 1:1 as the prescaler */
        CCPR1           = TMR1 + 40000; /* start compare operation with 40000 delay */
        PIR1bits.CCP1IF = 0;
        while (kk){
                while (!PIR1bits.CCP1IF);
                PIR1bits.CCP1IF = 0;
                kk--;
                CCPR1 += 40000;
        }
        return;
}
```

Using the CCP module (in compare mode) to make sound is easy. The basic idea is to create a digital waveform of appropriate frequency and use it to drive a speaker. A small speaker of 8-Ω resistance that consumes between 10 mW and 20 mW can be driven directly by the CCP output. The user can purchase such a speaker from Radio Shack or any electronic retailer. The next example illustrates how to use CCP1 in compare mode to generate a siren.

▲

Example 8.10
▼

Describe the circuit connection for siren generation and write a program that uses CCP1 in compare mode to generate a siren that oscillates between 440 Hz and 880 Hz.

Solution: A simple 8-Ω speaker has two terminals: one terminal is for signal input, whereas the other terminal is for ground connection. The circuit connection for siren generation is shown in Figure 8.22.

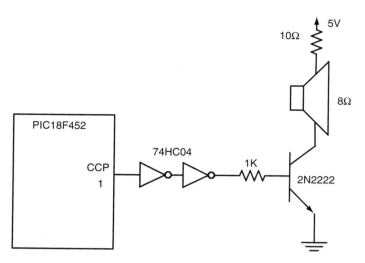

Figure 8.22 ■ Circuit connection for siren generation

The algorithm for generating the siren is as follows:

Step 1
Configure an appropriate CCP channel (CCP1 in this example) to operate in the compare mode and toggle output on match.

Step 2
Start a compare operation and enable its interrupt with a delay equal to half the period of the sound of the siren.

Step 3
Wait for certain amount of time (say, half of a second). During the waiting period, interrupts will be requested by the CCP compare match many times. The interrupt service routine simply clears the CCP flag and starts the next compare operation and then returns.

Step 4
At the end of the delay, choose a different delay time for the compare operation so that the siren sound with different frequency can be generated.

Step 5
Wait for the same amount of time as in Step 3. Again, interrupts caused by CCP compare match will be requested many times.

Step 6
Go to Step 2.

The assembly program that implements this algorithm for a PIC18F452 running with a 32-MHz crystal oscillator is as follows:

```
            #include <p18F452.inc>
hi_hi       equ        0x02        ; delay count to create 880 Hz sound
hi_lo       equ        0x38        ; "
lo_hi       equ        0x04        ; delay count to create 440 Hz sound
lo_lo       equ        0x70        ; "
            org        0x00
            goto       start
            org        0x08
```

```
                    goto        hi_ISR
                    org         0x18
                    retfie
      start         bcf         TRISC,CCP1        ; configure CCP1 pin for output
                    movlw       0x81              ; Enable Timer3 for 16-bit mode, use
                    movwf       T3CON             ; Timer1 as the base timer of CCP1
                    movlw       0xB1              ; enables Timer1 for 16-bit, prescaler set to 1:8
                    movwf       T1CON             ; "
                    movlw       0x02              ; CCP1 pin toggle on match
                    movwf       CCP1CON           ; "
                    bsf         RCON,IPEN         ; enable priority interrupt
                    bsf         IPR1,CCP1IP       ; configure CCP1 interrupt to high priority
                    movlw       hi_hi             ; load delay count for compare operation
                    movwf       PRODH             ; into PRODH:PRODL register pair
                    movlw       hi_lo             ; "
                    movwf       PRODL             ; "
                    movlw       0xC0
                    iorwf       INTCON,F          ; set GIE & PIE bits
                    movf        PRODL,W           ; start a new compare operation with
                    addwf       TMR1L,W           ; delay stored in PRODH:PRODL
                    movwf       CCPR1L            ; "
                    movf        PRODH,W           ; "
                    addwfc      TMR1H,W           ; "
                    movwf       CCPR1H            ; "
                    bcf         PIR1,CCP1IF       ; clear CCP1IF flag
                    bsf         PIE1,CCP1IE       ; enable CCP1 interrupt
      forever       call        delay_hsec        ; stay for half second in one frequency
                    movlw       lo_hi             ; switch to different frequency
                    movwf       PRODH,A           ; "
                    movlw       lo_lo             ; "
                    movwf       PRODL,A           ; "
                    call        delay_hsec        ; stay for half second in another frequency
                    movlw       hi_hi             ; switch to different frequency
                    movwf       PRODH,A           ; "
                    movlw       hi_lo             ; "
                    movwf       PRODL,A           ; "
                    goto        forever
      hi_ISR        bcf         PIR1,CCP1IF       ; clear the CCP1IF flag
                    movf        PRODL,W           ; start the next compare operation
                    addwf       CCPR1L,F          ; using the delay stored in PRODH:PRODL
                    movf        PRODH,W           ; "
                    addwfc      CCPR1H,F          ; "
                    retfie      fast
;*****************************************************************
; The following routine uses Timer 0 to create a delay of half a second.
; By setting the prescale factor to 64 and let Timer0 to count up from 3035 (0x0BDB),
; Timer0 will overflow in 62500 clock cycles (0.5 second).
;*****************************************************************
;
delay_hsec  movlw        0x0B
            movwf        TMR0H
```

```
        movlw        0xDB
        movwf        TMROL
        movlw        0x85          ; enable TMR0, select instruction clock,
        movwf        TOCON         ; prescaler set to 64
        bcf          INTCON,TMROIF
loopw   btfss        INTCON,TMROIF
        bra          loopw         ; wait for a half second
        return
        end
```

The C language version of the program is straightforward and hence is left for you as an exercise problem. The previous program can be modified to be run on a demo board with a crystal oscillator running at different frequencies.

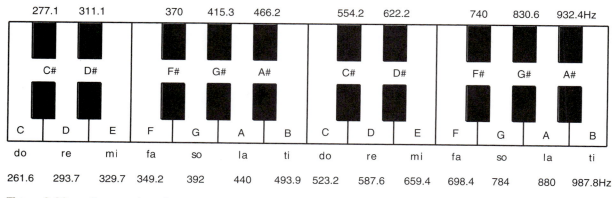

Figure 8.23 ■ Frequencies of music notes.

A siren can be considered as a song with only two notes. The frequencies of sounds over two octaves along with the black keys on a piano are shown in Figure 8.23. In general, a quarter note lasts for about 0.4 seconds (150 quarter notes per minute). The duration of other notes can be derived proportionally.

To play a song from the speaker, one places the frequencies and durations of all the notes in a music score in a table. For every note, the user program uses the CCP in compare mode to generate the digital waveform with the specified frequency and duration. The following example uses CCP1 in compare mode to play a song using a speaker. The frequencies of all the notes available on the piano are listed in Appendix E.

Example 8.11

For the circuit shown in Figure 8.22, write a C program to generate a simple song assuming that the demo board is running with a 4 MHz crystal oscillator.

Solution: The sample song to be played is a folk song. The following C program places in two tables (1) numbers to be added to the CCPR1 register to generate the waveform with the desired frequency for each note and (2) numbers that select the duration for each note:

```
#include     <P18F452.h>
#define      base        3125          /* counter count to create 0.1s delay */
```

```
#define        NOTES      38                /* 38 notes in the song to be played */
unsigned int half_cyc = 0;
unsigned rom int per_arr[38] = {
                           0x777, 0x470, 0x4FC, 0x470, 0x598, 0x777, 0x777, 0x3BC,
                           0x3F4, 0x3BC, 0x470, 0x470, 0x598, 0x353, 0x353, 0x353,
                           0x3BC, 0x470, 0x3BC, 0x3BC, 0x3F4, 0x470, 0x3F4, 0x3BC,
                           0x470, 0x598, 0x353, 0x2CC, 0x353, 0x3BC, 0x470, 0x3BC,
                           0x3BC, 0x3F4, 0x470, 0x3F4, 0x3BC, 0x470};

unsigned rom char wait[38] = {
                           3, 5, 3, 3, 5, 3, 3, 5,
                           3, 3, 5, 3, 3, 5, 3, 3,
                           5, 3, 3, 3, 2, 2, 3, 3,
                           6, 3, 5, 3, 3, 5, 3, 3,
                           3, 2, 2, 3, 3, 6};
void delay (unsigned char xc);
void high_ISR(void);
void low_ISR(void);
#pragma code high_vector = 0x08          // force the following statement to
void high_interrupt (void)               // start at 0x08
{
          _asm
          goto high_ISR
          _endasm
}
#pragma code low_vector = 0x18           //force the following statements to start at
void low_interrupt (void)                //0x18
{
          _asm
          goto        low_ISR
          _endasm
}
#pragma code                             //return to the default code section
#pragma interrupt high_ISR
void high_ISR (void)
{
          if(PIR1bits.CCP1IF){
                    PIR1bits.CCP1IF = 0;
                    CCPR1 + = half_cyc;
          }
}
#pragma code
#pragma interrupt low_ISR
void low_ISR (void)
{
          _asm
          retfie 0
          _endasm
}

void main (void)
```

```
{
        int i, j;
        TRISCbits.TRISC2 = 0; /* configure CCP1 pin for output */
        T3CON = 0x81; /* enables Timer3 in 16-bit mode, Timer1 for CCP1 time base */
        T1CON = 0x81; /* enable Timer1 in 16-bit mode */
        CCP1CON = 0x02; /* CCP1 compare mode, pin toggle on match */
        IPR1bits.CCP1IP = 1; /* set CCP1 interrupt to high priority */
        PIR1bits.CCP1IF = 0; /* clear CCP1IF flag */
        PIE1bits.CCP1IE = 1; /* enable CCP1 interrupt */
        INTCON |= 0xC0; /* enable high priority interrupt */
        for (j = 0; j < 3; j++) { /*play the song for times */
                i = 0;
                half_cyc = per_arr[0];
                CCPR1 = TMR1 + half_cyc;
                while (i < NOTES) {
                        half_cyc = per_arr[i]; /* get the cycle count for half period of the note */
                        delay (wait[i]);        /* stay for the duration of the note */
                        i++;
                }
                INTCON &= 0x3F; /* disable interrupt */
                delay (5);
                delay (6);
                INTCON |= 0xC0; /* re-enable interrupt */
        }
}
/* ─────────────────────────────────────────────────────────────*/
/* The following function runs on a PIC18 demo board running with a 4 MHz crystal  */
/* oscillator. The parameter xc specifies the amount of delay to be created        */
/* ─────────────────────────────────────────────────────────────*/
void delay (unsigned char xc)
{
        switch (xc){
        case 1:          /* create 0.1 second delay (sixteenth note) */
                TOCON = 0x84; /* enable TMR0 with prescaler set to 32 */
                TMR0 = 0xFFFF - base; /* set TMR0 to this value so it overflows in 0.1 second */
                INTCONbits.TMR0IF = 0;
                while (!INTCONbits.TMR0IF);
                break;
        case 2:          /* create 0.2 second delay (eighth note) */
                TOCON = 0x84; /* set prescaler to Timer0 to 32 */
                TMR0 = 0xFFFF - 2*base; /* set TMR0 to this value so it overflows in 0.2 second */
                INTCONbits.TMR0IF = 0;
                while (!INTCONbits.TMR0IF);
                break;
        case 3:          /* create 0.4 seconds delay (quarter note) */
                TOCON = 0x84; /* set prescaler to Timer0 to 32 */
                TMR0 = 0xFFFF - 4*base; /* set TMR0 to this value so it overflows in 0.4 second */
                INTCONbits.TMR0IF = 0;
                while (!INTCONbits.TMR0IF);
```

```
                     break;
        case 4:             /* create 0.6 s delay (3 eighths note) */
                     T0CON = 0x84; /* set prescaler to Timer0 to 32 */
                     TMR0 = 0xFFFF – 6*base; /* set TMR0 to this value so it overflows in 0.6 second */
                     INTCONbits.TMR0IF = 0;
                     while (!INTCONbits.TMR0IF);
                     break;
        case 5:             /* create 0.8 s delay (half note) */
                     T0CON = 0x84; /* set prescaler to Timer0 to 32 */
                     TMR0 = 0xFFFF – 8*base; /* set TMR0 to this value so it overflows in 0.8 second */
                     INTCONbits.TMR0IF = 0;
                     while (!INTCONbits.TMR0IF);
                     break;
        case 6:             /* create 1.2 second delay (3 quarter note) */
                     T0CON = 0x84; /* set prescaler to Timer0 to 32 */
                     TMR0 = 0xFFFF – 12*base; /* set TMR0 to this value so it overflows in 1.2 second */
                     INTCONbits.TMR0IF = 0;
                     while (!INTCONbits.TMR0IF);
                     break;
        case 7:             /* create 1.6 second delay (full note) */
                     T0CON = 0x84; /* set prescaler to Timer0 to 32 */
                     TMR0 = 0xFFFF – 16*base; /* set TMR0 to this value so it overflows in 1.6 second */
                     INTCONbits.TMR0IF = 0;
                     while (!INTCONbits.TMR0IF);
                     break;
        default:
                     break;
    }
}
```

Other songs can be created by changing the tables **per_arr[]** and **wait[]** to appropriate values.

8.7 CCP in PWM Mode

In PWM mode, the CCPx pin can output a 10-bit resolution periodic digital waveform with programmable period and duty cycle. To operate in PWM mode, a CCPx pin must be configured for output. A simplified block diagram of the CCP module in PWM mode is shown in Figure 8.24.

The duty cycle of the waveform to be generated is a 10-bit value of which the upper eight bits are stored in the CCPRxH register, whereas the lowest two bits are stored in bit 5 and bit 4 of the CCPxCON register. The duty cycle value is compared with TMRy (y = 2 or 4) cascaded with the 2-bit Q clocks in every instruction clock cycle. Whenever these two values are equal, the CCPx pin is pulled to low. The TMRy register is also compared with the PRy register in every clock cycle. Whenever these two registers are equal, the following three events occur on the next increment cycle:

- The CCPx pin is pulled high (exception: if PWM duty cycle is 0%, the CCPx pin will not be set).

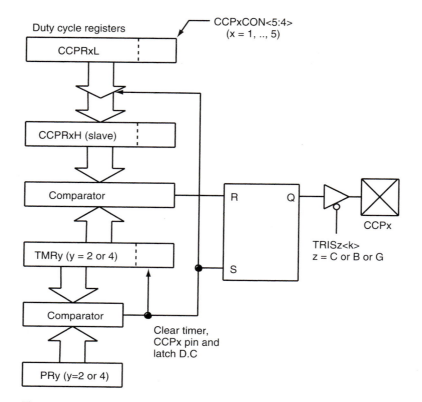

Figure 8.24 ■ Simplified PWM block diagram (redraw with permission of Microchip)

- TMRy register is cleared.
- The PWM duty cycle is latched from CCPRxL into CCPRxH.

The PWM period can be calculated using the following formula:

PWM period = [(PRy) + 1] × 4 × T_{OSC} × (TMRy prescale factor)

The PWM duty cycle can be calculated using the following formula:

PWM duty cycle = (CCPRxL:CCPxCON<5:4>) × T_{OSC} × (TMRy prescale factor)

The following steps should be taken when configuring the CCP module for PWM operation:

1. Set the PWM period by writing to the PRy (y = 2 or 4) register.
2. Set the PWM duty cycle by writing to the CCPRxL register and CCPxCON<5:4> bits.
3. Make the CCPx pin an output.
4. Set the TMRy prescale value and enable TMRy by writing to TyCON register.
5. Configure the CCPx module for PWM operation.

The user should be aware that the time unit used in period calculation is *instruction cycle clock*, whereas the time unit used in duty cycle calculation is *crystal oscillator cycle*. Using different time units can improve the accuracy of the waveform to be generated when multiplying the period value by the duty cycle is not an integer. The lowest two bits of the duty cycle can be used to represent the fractional part (0.0, 0.25, 0.5, or 0.75) of the duty cycle value. For example, if one wants to create a waveform with 62.5% duty cycle from CCP1 pin, then one can place 99 in the PR2 register and 62 in the CCPR1L and set bit 5 and bit 4 of the CCP1CON register to 10. However, since the user can place any value (from 0 to 255) in the CCPRxL register, the lowest two bits may not always represent the fractional part of the duty cycle.

Example 8.12

▼

Configure CCP1 in PWM mode to generate a digital waveform with 40% duty cycle and 10-KHz frequency assuming that the PIC18 microcontroller is running with a 32-MHz crystal oscillator.

Solution: A possible setting is as follows:

1. Use Timer2 as the base timer of CCP1 through CCP5 for PWM mode.
2. Enable Timer3 in 16-bit timer mode with 1:1 prescaler.
3. Set prescaler to Timer2 to 1:4.

The value to be written into PR2 is calculated as follows:

PR2 = $32 \times 10^6 \div 4 \div 4 \div 10^4 - 1 = 199$

The value to be written into the CCPR1L (and CCPR1H) register is

CCPR1L = $200 \times 40\% = 80$

The lowest two bits (bits 5 and 4) of the duty cycle value will be cleared to 0. The following instruction sequence will configure CCP1 in PWM mode as desired:

```
movlw    0xC7             ; set period value to 199
movwf    PR2,A            ; "
movlw    0x50             ; set duty cycle value to 80
movwf    CCPR1L,A         ; "
movwf    CCPR1H,A         ; "
bcf      TRISC,CCP1,A     ; configure CCP1 pin for output
movlw    0x81             ; enable Timer3 in 16-bit mode and use
movwf    T3CON,A          ; Timer2 as time base for PWM1 thru PWM5
clrf     TMR2,A           ; force TMR2 to count from 0
movlw    0x05             ; enable Timer2 and set its prescaler to 4
movwf    T2CON,A          ; "
movlw    0x0C             ; enable CCP1 PWM mode
movwf    CCP1CON,A        ; "
```

▲

Function	Description
ClosePWMx	Disable PWM channel x
OpenPWMx	Configure PWM channel x
SetDCPWMx	Write a new duty cycle value to PWM channel x
SetOutputPWMx	Sets the PWM output configuration bits for ECCP

Table 8.2 ■ Microchip C library functions for PWM

8.7.1 PWM C Library Functions

The Microchip PIC18 C compiler provides library functions (listed in Table 8.2) for configuring PWM parameters, disabling PWM, and setting duty cycles. The header file **pwm.h** must be included in order to use these functions.

The prototype declarations and parameter definitions are as follows:

    void **ClosePWM1** (void);

    void **ClosePWM2** (void);

    void **ClosePWM3** (void);

    void **ClosePWM4** (void);

    void **ClosePWM5** (void);

    void **OpenPWM1** (char **period**);

    void **OpenPWM2** (char **period**);

    void **OpenPWM3** (char **period**);

    void **OpenPWM4** (char **period**);

    void **OpenPWM5** (char **period**);

Each of the **OpenPWM** functions enables the specified CCP module for PWM mode and configures its period. The user can choose either Timer2 or Timer4 as the base timer of the PWM function. This can be done by configuring the T3CON register to select the desired base timer for PWM operation. The parameter **period** can be any value between 0x00 and 0xFF. This value determines the PWM frequency by using the following formula:

PWM period = [period + 1] $\times$ 4 $\times$ T$_{OSC}$ $\times$ TMRy (y = 2 or 4) prescaler

    void **SetDCPWM1** (unsigned int **dutycycle**);

    void **SetDCPWM2** (unsigned int **dutycycle**);

    void **SetDCPWM3** (unsigned int **dutycycle**);

    void **SetDCPWM4** (unsigned int **dutycycle**);

    void **SetDCPWM5** (unsigned int **dutycycle**);

These functions write a new duty cycle value to the specified PWM channel duty cycle registers. The value of **dutycycle** can be any 10-bit number:

void **SetOutputPWM1** (unsigned char **outputconfig**, unsigned char **outputmode**);

The value of **outputconfig** can be any one of the following values:

SINGLE_OUT	Single output
FULL_OUT_FWD	Full-bridge output forward
HALF_OUT	Half-bridge output
FULL_OUT_REV	Full-bridge reverse

The value of **outputmode** can be any one of the following values:

PWM_MODE_1	P1A and P1C active high,
	P1B and P1D active high
PWM_MODE_2	P1A and P1C active high,
	P1B and P1D active low
PWM_MODE_3	P1A and P1C active low,
	P1B and P1D active high

PWM_MODE_4   P1A and P1C active low,

P1B and P1D active low

This function can be used only with the PIC18 members that come with the enhanced CCP (ECCP) modules. ECCP is discussed in Section 8.8 and is placed in the complimentary CD.

Example 8.13

▼

Write a set of C statements to configure CCP4 to generate a digital waveform with 5-KHz frequency and 70% duty cycle, assuming that the PIC18F8720 is running with a 16-MHz crystal oscillator. Use Timer4 as the base timer.

Solution: To create a waveform with 5-KHz frequency, the user can set the timer prescaler to 8 and set the period value to 100 (99 to be written as the period value). The duty cycle value to be written is $100 \times 70\% \times 4 = 280$.

The following C statements will configure CCP4 to generate the desired waveform:

```
TRISGbits.TRISG3 = 0; /* configure CCP4 pin for output */
OpenTimer3 (TIMER_INT_OFF & T3_16BIT_RW & T3_SOURCE_INT & T3_PS_1_1 & T34_SOURCE_CCP &
        T3_OSC1EN_OFF);
OpenTimer4 (TIMER_INT_OFF & T4_PS_1_8 & T4_POST_1_1);
SetDCPWM4 (280);
OpenPWM4 (99);
```

▲

8.7.2 PWM Applications

There are some applications that respond to the average input voltage rather than the instantaneous input voltage. Lamp dimming and direct-current (DC) motor control are two examples. Both the brightness of a lamp and the speed of a DC motor are proportional to the average applied voltage rather than the instantaneous voltage value.

A PIC18 I/O pin can be used to switch the lightbulb through a PNP or an NPN transistor as long as the lightbulb power consumption is moderate. Two circuit examples are shown in Figure 8.25.

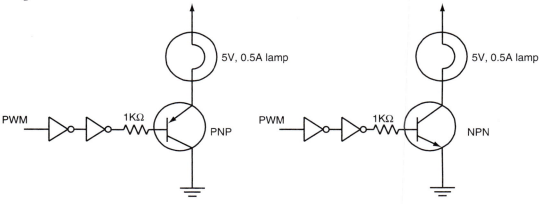

Figure 8.25 ■ Using PWM to control the brightness of a lightbulb

The collector current of the BJT transistor drives the lamp. A PIC18 I/O pin can sink or source up to 25 mA. The collector current is equal to h_{fe} times this current. The parameter of h_{fe} varies from around 15 to around 800. The typical value of h_{fe} is 100. However, one must keep in mind that the larger the collector current gets, the smaller the h_{fe} parameter becomes. The circuit shown in Figure 8.25 is not suitable for applications that require current larger than a few hundred milliamps. If an application requires larger current to drive, then it would be necessary to use some type of power MOSFETs. In Figure 8.25, a buffer is added to protect the microcontroller from being damaged by accidental overload of the transistor.

Example 8.14
▼

Assume that PWM1 is being used to control the brightness of a lamp. The circuit connection is shown in Figure 8.25. Write a program to dim the lamp to 10% brightness in 5 seconds. Assume that the PIC18 microcontroller is running with a 32-MHz crystal oscillator.

Solution: We will dim the lamp as follows:

1. Set duty cycle to 100% at the beginning. Load 99 and 400 as the initial period and duty cycle register values, respectively.

2. Dim the lamp by 10% in the first second by reducing the brightness in 10 steps. This can be achieved by decrementing the duty cycle value by 4 in each step (or decrementing the CCPR1L register by 1).

3. Dim the lamp down to 10% brightness in the next 4 seconds in 40 steps. This is achieved by decrementing the CCPR1L register by 2.

The assembly program that implements this algorithm is as follows:

```
          #include <p18F452.inc>
          org       0x00
          goto      start
          org       0x08
          retfie
          org       0x18
          retfie
start     bcf       TRISC,CCP1   ; configure CCP1 pin for output
          movlw     0x81         ; Use Timer2 as the base timer for PWM1
          movwf     T3CON        ; and enable Timer3 in 16-bit mode
          movlw     0x63         ; set 100 as the period of the digital
          movwf     PR2          ; waveform
          movlw     0x64         ; set 100 as the duty cycle
          movwf     CCPR1L       ; "
          movwf     CCPR1H       ; "
          movlw     0x05         ; enable Timer2 and set its prescaler to 4
          movwf     T2CON        ; "
          movlw     0x0C         ; enable PWM1 operation and set the lowest
          movwf     CCP1CON      ; two bits of duty cycle to 0
          movlw     0x0A         ; use PRODL as the loop count
          movwf     PRODL        ; "
loop_1s   call      delay        ; call "delay" to delay for 100 ms
          decf      CCPR1L,F     ; decrement the duty cycle value by 1
```

```
            decfsz      PRODL,F         ; check to see if loop index expired
            goto        loop_1s
            movlw       0x28            ; repeat the next loop 40 times
            movwf       PRODL           ; "
loop_4s     call        delay           ; call "delay" to delay for 100 ms
            decf        CCPR1L,F        ; decrement duty cycle value by 2
            decf        CCPR1L,F        ; "
            decfsz      PRODL,F         ; is loop index expired?
            goto        loop_4s
forever     nop
            goto        forever
;************************************************************
; The following function creates 100 ms delay for a 32 MHz crystal oscillator.
;************************************************************
delay       movlw       0x3C            ; load 15535 into TMR0 so that it will overflow
            movwf       TMR0H           ; in 50000 clock cycles
            movlw       0xAF            ; "
            movwf       TMR0L           ; "
            movlw       0x83            ; set prescaler of TMR0 to 16
            movwf       T0CON,A         ; "
            bcf         INTCON,TMR0IF   ; clear TMR0IF flag
wait_3      btfss       INTCON,TMR0IF   ; wait until the delay time is over
            goto        wait_3
            return
            end
```

The C language version of the program is as follows:

```c
#include <p18F452.h>
#include <pwm.h>
#include <timers.h>
void delay (void);
void main (void)
{
            int i;
            TRISCbits.TRISC2 = 0;       /* configure CCP1 pin for output */
            T3CON = 0x81;               /* use Timer2 as base timer for CCP1 */
            OpenTimer2 (TIMER_INT_OFF & T2_PS_1_4 & T2_POST_1_1);
            SetDCPWM1 (400);            /* set duty cycle to 100% */
            OpenPWM1 (99);              /* enable PWM1 with period equals 100 */
            for(i = 0; i < 10; i++) {
                    delay();
                    CCPR1L--;           /* decrement duty cycle value by 1 */
            }
            for(i = 0; i < 40; i++) {
                    delay();
                    CCPR1L -= 2;        /* decrement duty cycle value by 2 */
            }
}
void delay (void)
```

```
{
        TMR0 = 0x3CAF; /* load 15535 into TMR0 so that it overflows in 50000 clock cycles */
        OpenTimer0 (TIMER_INT_OFF & T0_16BIT & T0_SOURCE_INT & T0_PS_1_16);
                    /* enable Timer0 in 16-bit mode and set prescaler to 16 */
        INTCONbits.TMR0IF = 0; /* clear TMR0IF flag */
        while(!INTCONbits.TMR0IF); /* wait for 100 ms */
}
```

▲

8.7.3 DC Motor Control

A DC motor is devised to convert electrical power into mechanical power. In a DC motor, electrical energy is converted into mechanical energy through the interaction of two magnetic fields. One field is produced by a permanent magnet assembly (on the *stator*), and the other field is produced by an electrical current flowing in the motor winding (on the *rotor*). These two fields result in a torque that tends to rotate the rotor. As the rotor turns, the current in the windings is commutated to produce a continuous torque output. Instead of using permanent magnets to produce permanent magnetic fields, some DC motors use coils.

The rotation of magnetic field is achieved by switching current between coils within the motor. This action is called *commutation*. Most DC motors have built-in commutation in which mechanical brushes automatically commutate coils on the rotor as the motor rotates. Besides these *brush-type* DC motors, there is another DC motor type: *brushless*. A brushless DC motor relies on the external power drive to perform the commutation of stationary copper windings on the stator. This changing stator field makes the permanent magnet rotor rotate. A brushless permanent magnet motor is the highest-performing motor in terms of torque versus weight or efficiency. Brushless motors are usually the most expensive type of motor. Electronically commutated, brushless motors are widely used as drives for blowers and fans in electronics, telecommunications, and industrial equipment applications. There are a wide variety of different brushless motors for various applications. Some are designed to rotate at constant speed (such as those used in disk drives), and the speed of some can be controlled by varying the voltage applied to them.

DC motor speed is controlled by controlling its driving voltage. The higher the voltage, the higher the motor speed. In many applications, a simple voltage regulation would cause lots of power loss in the control circuit, so a PWM method is used in many DC motor-controlling applications. In the basic PWM method, the operating power to the motors is turned on and off to modulate the current to the motor. The ratio of on time to off time is what determines the speed of the motor.

Sometimes the rotation direction needs to be changed. In normal permanent magnet motors, the reversal of motor direction is implemented by changing the polarity of the operating power.

A PWM circuit can be implemented by using discrete components. However, this approach cannot provide the desired flexibility and controllability and is expensive. A better implementation method for PWM circuitry is to use the PWM functions available in many microcontrollers today. Most of the PIC18 devices have PWM functions.

The PIC18 microcontroller can interface with a DC motor through a driver, as shown in Figure 8.26. This circuit takes up only three I/O pins. The pin that controls the direction can be an ordinary I/O pin, but the pin that controls the speed must be a PWM pin. The pin that receives the feedback must be a CCP pin configured in capture mode.

Although some DC motors can operate at 5 V or less, the PIC18 microcontroller cannot supply the necessary current to drive a motor directly. The minimum current required by any practical DC motor is much larger than any microcontroller can supply. Depending on the size and rating of the motor, a suitable driver must be selected to take control signals from the PIC18 microcontroller and deliver the necessary voltage and current to the motor.

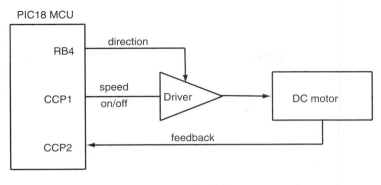

Figure 8.26 ■ A simplified circuit for DC motor control

DC MOTOR DRIVERS

Standard motor drivers are available in many current and voltage ratings. Examples are the L292 and L293 made by SGS Thompson Inc. The L293 has four channels and can deliver up to 1 A of current per channel with a supply of 36 V. It has a separate logic supply and takes a logic input (1 or 0) to enable or disable each channel. The L293D also includes clamping diodes needed to drain the *kickback* current generated from the inductive load during the motor reversal. The pin assignment and block diagram of the L293 are shown in Figure 8.27. There are two supply voltages: Vss and Vs. Vss is the logic supply voltage, which can be from 4.5 V to 36 V (normally 5.0 V). Vs is the analog supply voltage and can be as high as 36 V.

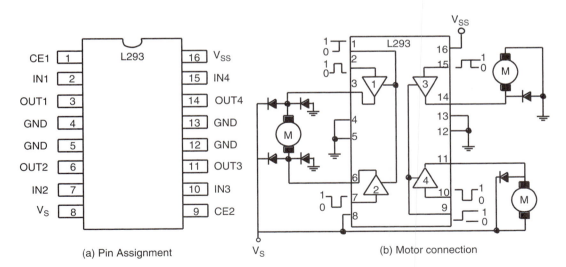

(a) Pin Assignment   (b) Motor connection

Figure 8.27 ■ Motor driver L293 pin assignment and motor connection

FEEDBACK

The DC motor controller needs information to adjust the voltage output to the motor driver circuit. The most important information is the speed of the motor that must be fed back from the motor by a sensing device. The sensing device may be an optical encoder, infrared detector,

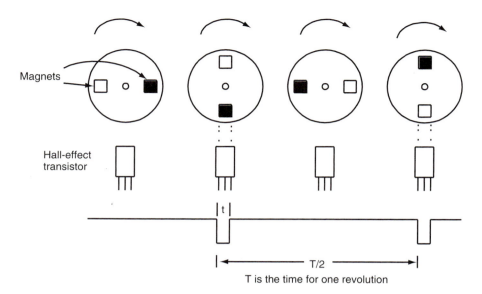

Figure 8.28 ■ The output waveform of the Hall-effect transistor

Hall-effect sensor, and so on. Whatever the means of sensing, the result is a signal that is fed back to the microcontroller. The microcontroller can use the feedback to determine the speed and position of the motor. Then it can make adjustments to increase or decrease the speed, reverse the direction, or stop the motor.

Assume that a Hall-effect transistor is mounted on the shaft (rotor) of a DC motor and that two magnets are mounted on the armature (stator). As shown in Figure 8.28, every time the Hall-effect transistor passes by the magnet, it generates a pulse. One can use the CCP module to capture the passing time of the pulse. The time between two captures is half a revolution. Thus, the motor speed can be calculated. By storing the value of the capture register each time, the controller can constantly measure and adjust the speed of the motor. A motor can therefore be run at a precise speed or synchronized with another event.

The schematic of a motor control system is illustrated in Figure 8.29. The PWM output from the CCP1 pin is connected to one end of the motor, whereas the RB4 pin is connected to the other end. The circuit is connected in such a way that the motor will rotate clockwise when the voltage of the RB4 pin is zero while the PWM output is positive. The direction of the motor rotation is illustrated in Figure 8.30. By applying appropriate voltages on the RB4 and CCP1 pins, the motor can rotate clockwise or counterclockwise or even stop. The CCP2 module in capture mode is used to capture the feedback from the Hall-effect transistor.

When a DC motor is first turned on, it cannot reach a steady speed immediately. A certain amount of startup time should be allowed for the motor to get to speed. A smaller motor usually can reach steady speed faster than a larger one. It is desirable for the motor speed to be a constant for many applications. However, when a load is applied to the motor, it will be slowed down. To keep the speed constant, the duty cycle of the voltage applied to the motor should be increased. When the load gets lighter, the motor will accelerate and run faster than desired. To slow down the motor, the duty cycle of the applied voltage should be reduced.

The response time will be slow if the change to the duty cycle is small. However, a large variation in the duty cycle tends to cause it to overreact and causes oscillation. There are control algorithms that you can find discussed in textbooks on control.

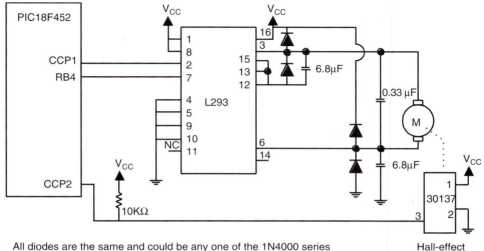

All diodes are the same and could be any one of the 1N4000 series

Hall-effect switch

Figure 8.29 ■ Schematic of a PIC18-based motor-control system

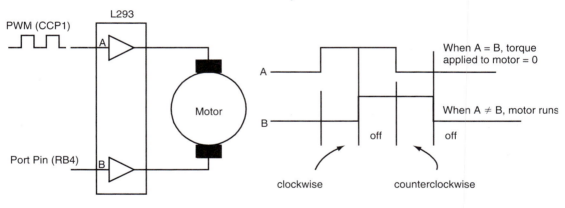

Figure 8.30 ■ The L293 motor driver

A DC motor cannot respond to the change of duty cycle instantaneously because of its inertia. A certain amount of time should be allowed for the motor to speed up or slow down before the effect of the change of duty cycle is measured.

ELECTRICAL BRAKING

Once a DC motor is running, it picks up speed. Turning off the voltage to the motor does not make it stop immediately because the momentum of the motor will keep it rotating. After the voltage is turned off, the momentum will gradually wear out because of friction. If the application does not require an abrupt stop, then the motor can be brought to a gradual stop by removing the driving voltage.

An abrupt stop may be required by certain applications in which the motor must run a few turns and stop quickly at a predetermined point. This can be achieved by electrical braking. Electrical braking is done by reversing the voltage applied to the motor. The length of time that the reversing voltage is applied must be precisely calculated to ensure a quick stop while not starting it in the reverse direction. A discussion of good motor braking algorithms is outside the scope of this textbook.

In a closed-loop system, the feedback can be used to determine where and when to start and stop braking and when to discontinue. In Figure 8.30, the motor can be braked by (1) reducing the PWM duty cycle to 0 and (2) setting the RB4 pin's output to high for an appropriate amount of time.

Example 8.15
▼

For the circuit shown in Figure 8.29, write a function in C language to measure the motor speed (in RPM), assuming that the PIC18 microcontroller is running with a 20-MHz crystal oscillator.

Solution: The motor speed can be measured by capturing two consecutive rising or falling edges. Let the difference of two captured edges and the clock frequency of Timer1 be **diff** and **f**, respectively. Then the motor speed (in RPM) is

$$\text{Speed} = 60 \times f \div (2 \times \text{diff})$$

The C function that measures the motor speed is as follows:

```
unsigned int motor_speed(void)
{
    unsigned int edge1, diff, rpm;
    long unsigned temp;
    T3CON = 0x81;                /* enables Timer3 in 16-bit mode and use Timer1 and Timer2 for CCP1 thru CCP2
                                   operations */
    OpenTimer1(TIMER_INT_OFF & T1_16BIT_RW & T1_SOURCE_INT & T1_PS_1_4);
                /* set Timer1 prescaler to 1:4 */
    PIR2bits.CCP2IF = 0;
    OpenCapture2(CAPTURE_INT_OFF & C2_EVERY_RISE_EDGE);
    while (!PIR2bits.CCP2IF);
    edge1 = CCPR2;              /* save the first rising edge */
    PIR2bits.CCP2IF = 0;
    while (!PIR2bits.CCP2IF);
    CloseCapture2();
    diff = CCPR2 – edge1;       /* compute the difference of two rising consecutive edges */
    temp = 1250000ul/(2 * diff);
    rpm = temp * 60;
    return rpm;
}
```

▲

Example 8.16
▼

Write a subroutine to perform the electrical braking.

Solution: Electrical braking is implemented by setting the duty cycle to 0% and setting the voltage on the RB4 pin to high for certain amount of time. The following subroutine will perform the desired electrical braking:

```
brake    bsf      PORTB,RB4,A    ; reverse the applied voltage to motor
         movlw    0x00           ; "
         movwf    CCPR1L,A       ; set PWM1 duty cycle to 0
         call     brake_time     ; wait for certain amount of time
         bcf      PORTB,RB4,A    ; stop braking
         return
```

The subroutine **brake_time** will wait for certain amount of time to allow for the motor to stop. It is similar to the delay routine that we learned earlier.

Instead of calling the **brake_time** function, the brake function can also keep monitoring the motor speed until it drops to a very low value (say, 5%) and then stop the PWM function. This method is also straightforward and hence is left for you as an exercise problem.

The motor braking can be invoked by using external interrupt. Using this approach, the brake function would be implemented as an interrupt service routine.

▲

8.8 Enhanced CCP Module

The PIC18F448/458, PIC18F1220/1320, and PIC18F4220/4320 have incorporated one enhanced CCP (ECCP) module. The PIC18F8621/8525 and PIC6621/6525 incorporate three ECCP modules. Many future PIC18 members will also have the enhanced ECCP module. The ECCP module is implemented as a standard CCP module with enhanced PWM capabilities with the intention to simplify the support of motor-control applications.

8.8.1 ECCP Pins

Depending on the operating mode, one ECCP may have up to four outputs. These outputs, designated P1A through P1D, are multiplexed with I/O pins on PORTB (18F1x20) or PORTC and PORTD (18F458) or PORTD (18F4x20). The pin assignments are summarized in Table 8.3. These four pins must be configured for output. Depending on the device, either RC2 or RD4 is used as the CCP1/P1A pin. The pin assignments for P1A–P1D, P2A–P2D, and P3A–P3D for other PIC18 devices can be found in appropriate datasheets.

ECCP mode	CCP1CON Configuration	RB3 or RC2 or RD4	RB2 or RD5	RB6 or RD6	RB7 or RD7
Compatible CCP	00xx11xx	CCP1	I/O pin	I/O pin	I/O pin
Dual PWM	10xx11xx	P1A	P1B	I/O pin	I/O pin
Quad PWM	x1xx11xx	P1A	P1B	P1C	P1D

Table 8.3 ■ Pin assignments for various ECCP modes

8.8.2 ECCP Registers[1]

The ECCP module differs from the CCP module with the addition of an enhanced PWM mode, which allows for two or four output channels, user-selectable polarity, deadband control, and automatic shutdown and restart. The control register for the CCP1 module is shown in Figure 8.31. The control registers for the ECCP2 and ECCP3 modules are identical and are called CCP2CON and CCP3CON, respectively.

---

[1]Section 8.8 can be skipped without affecting one's understanding of later chapters.

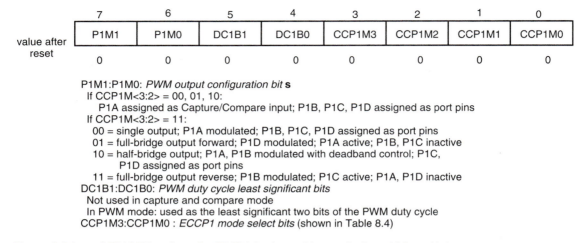

value after reset

7	6	5	4	3	2	1	0
P1M1	P1M0	DC1B1	DC1B0	CCP1M3	CCP1M2	CCP1M1	CCP1M0
0	0	0	0	0	0	0	0

P1M1:P1M0: *PWM output configuration bit* **s**
If CCP1M<3:2> = 00, 01, 10:
    P1A assigned as Capture/Compare input; P1B, P1C, P1D assigned as port pins
If CCP1M<3:2> = 11:
    00 = single output; P1A modulated; P1B, P1C, P1D assigned as port pins
    01 = full-bridge output forward; P1D modulated; P1A active; P1B, P1C inactive
    10 = half-bridge output; P1A, P1B modulated with deadband control; P1C,
        P1D assigned as port pins
    11 = full-bridge output reverse; P1B modulated; P1C active; P1A, P1D inactive
DC1B1:DC1B0: *PWM duty cycle least significant bits*
    Not used in capture and compare mode
    In PWM mode: used as the least significant two bits of the PWM duty cycle
CCP1M3:CCP1M0 : *ECCP1 mode select bits* (shown in Table 8.4)

Figure 8.31 ■ CCP1CON register for ECCP1 (redraw with permission of Microchip)

CCP1M3:CCP1M0	Description
0000	Capture/Compre/PWM off (resets ECCP module)
0001	Unused (reserved)
0010	Compare mode, toggle output on match (ECCP1IF bit is set)
0011	Unused
0100	Capture mode, every falling edge
0101	Capture mode, every rising edge
0110	Capture mode, every 4th rising edge
0111	Capture mode, every 16th rising edge
1000	Compare mode, set output on match (ECCP1IF bit is set)
1001	Compare mode, clear output on match (ECCP1IF bit is set)
1010	Compare mode, generate software interrupt on match (ECCP1IF bit is set, ECCP1 pin unaffected)
1011	Compare mode, trigger special event (ECCPIF bit is set; ECCP resets TMR1 or TMR2, and starts an A/D conversion, if the A/D module is enabled)
1100	PWM mode; P1A, P1C active high; P1B, P1D active high
1101	PWM mode; P1A, P1C active high; P1B, P1D active low
1110	PWM mode; P1A, P1C active low; P1B, P1D active high
1111	PWM mode; P1A, P1C active low; P1B, P1D active low

Table 8.4 ■ ECCP1A MODE select combinations (redraw with permission of Microchip)

In addition to the expanded functions of the CCP1CON register, the ECCP module has two registers associated with enhanced PWM operation and auto shutdown features:

- PWM1CON or ECCPxDEL (x = 1 . . . 3) for devices with 3 ECCP modules
- ECCPxAS (x = 1 . . . 3)

The contents of these registers are explained when the enhanced PWM mode is discussed in the next section. All other registers associated with the ECCP module are identical to those for the CCP1 module, including register and individual bit names.

8.8.3 Enhanced PWM Mode

The ECCP module can be configured to operate in four different modes:

1. Capture mode
2. Compare mode
3. Standard PWM mode (when PWM is configured in single-output mode)
4. Enhanced PWM mode

The first three modes are identical to their counterpart in a standard CCP module, whereas the fourth mode is designed to strengthen the PWM capability. The enhanced PWM mode provides additional PWM output options for a broader range of control applications.

A simplified block diagram of the enhanced PWM of the PIC18F4320 is shown in Figure 8.32. All control registers are double buffered and are loaded when the TMR2 register resets. The only exception is the ECCP1DEL register, which is loaded at either the duty cycle boundary or the boundary period (whichever comes first).

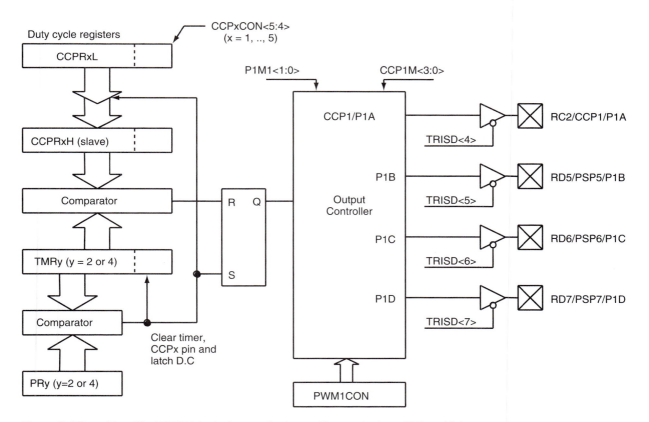

Figure 8.32 ■ Simplified PWM block diagram (redraw with permission of Microchip)

The P1M1:P1M0 bits in the CCP1CON register allows one of four configurations:

1. Single output
2. Half-bridge output
3. Full-bridge output, forward mode
4. Full-bridge output, reverse mode

The single output is the standard PWM mode discussed in Section 8.7. The general relationship of the outputs in all configurations is summarized in Figure 8.33.

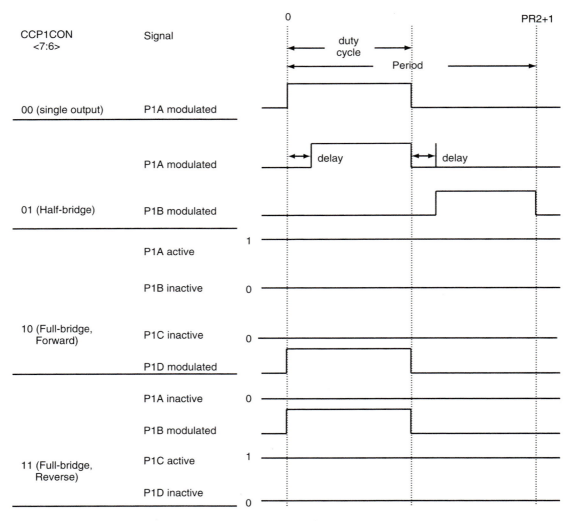

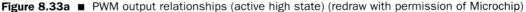

Figure 8.33a ∎ PWM output relationships (active high state) (redraw with permission of Microchip)

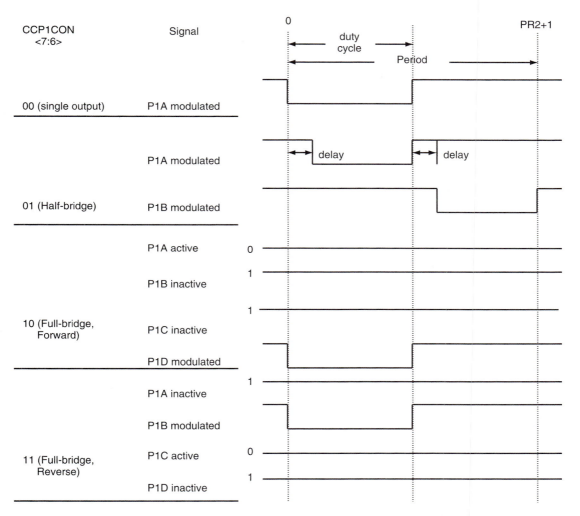

Figure 8.33b ■ PWM output relationships (active low state) (redraw with permission of Microchip)

HALF-BRIDGE MODE

In the half-bridge output mode, two pins are used as outputs to drive push-pull loads. The PWM output signal is output on the P1A pin, while the complementary PWM output signal is output on the P1B pin (shown in Figure 8.34). This mode can be used for half-bridge applications, as shown in Figure 8.35a, or for full-bridge applications, where four power switches are being modulated with two PWM signals (shown in Figure 8.35b).

In half-bridge output mode, the programmable deadband delay can be used to prevent shoot-through current in half-bridge power devices. In Figure 8.35a, when both FET transistors are turned on, there is a current flow directly from V_{DD} to the ground. This situation is called *shoot-through*. The values of bits PDC6:PDC0 set the number of instruction cycles before the output is driven active. If the value is greater than the duty cycle, the corresponding output remains inactive during the entire cycle.

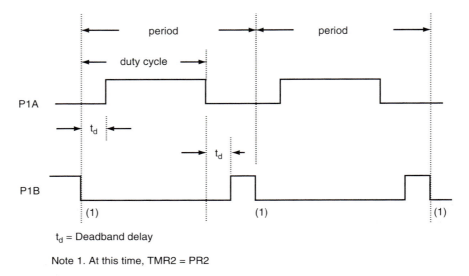

t_d = Deadband delay

Note 1. At this time, TMR2 = PR2

Figure 8.34 ■ Half-bridge PWM output

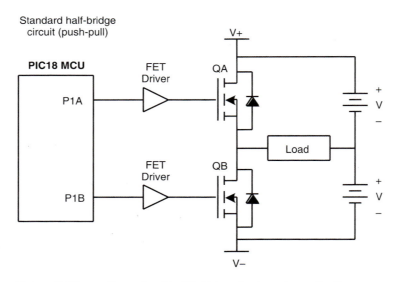

Figure 8.35a ■ Example of half-bridge output mode applications
(redraw with permission of Microchip)

Half-bridge output driving a
full-bridge circuit

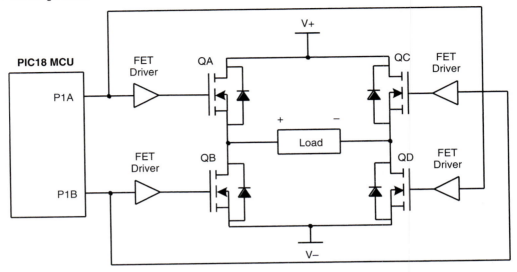

Figure 8.35b ■ Example of half-bridge output mode applications (redraw with permission of Microchip)

The FET driver in Figure 8.35 must be able to provide large amount of current. One such example is the Microchip motor driver chip TC4424, which can supply 3 A of current. The MAXIM MAX4427 is the same but with a lower current rating (1.5 A).

The circuit shown in Figure 8.35b is an H-bridge. The H-bridge circuit is often used in driving DC motors and stepper motors. An H-bridge allows the motor to be driven forward or backward at any speed, optionally using a completely independent power source (possible when the H-bridge is driven in full-bridge mode). In the normal operation, the DC motor is driven in forward direction. The H-bridge allows the user to reverse the motor driving direction briefly to perform electrical braking if such a need arises.

FULL-BRIDGE MODE

In full-bridge output mode, four pins are used as output; however, only two outputs are active at a time. As shown in Figure 8.36, there are two submodes under the full-bridge mode:

- Forward mode
- Reverse mode

In the forward mode, the P1A pin is continuously active, and the P1D pin is modulated. In the reverse mode, the P1C pin is continuously active, and the P1B pin is modulated.

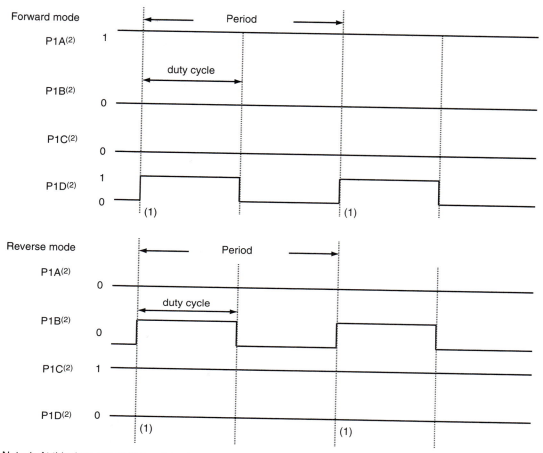

Note 1: At this time, the TMR2 register is equal to the PR2 register.
      2: Output signal is shown as active high.

Figure 8.36 ■ Full-bridge PWM output (redraw with permission of Microchip)

An example of full-bridge applications is illustrated in Figure 8.37. In this example, each of the four PWM outputs drives one FET driver, which in turn controls a switching FET. These switching FETs are all n-type.

In the forward mode, the transistor QA is turned on all the time, whereas the transistor QD is turned on only when the P1D signal is high. Both the QB and the QC transistors are turned off and hence will turn the motor in one direction. In the reverse mode, the transistor QC is turned on all the time, whereas QB is turned on when the P1B signal is high. Both the QA and QD transistors are turned off, and hence the DC motor will rotate in the reverse direction.

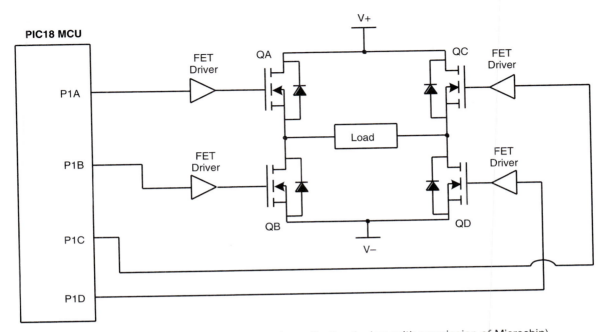

Figure 8.37 ■ Example of full-bridge output mode application (redraw with permission of Microchip) (Transistors QA through QD are all n-transistors)

In the full-bridge mode, the user can change forward/reverse direction by clearing/setting the P1M1 bit in the CCP1CON register. After this bit is changed, the module will assume the new direction on the next PWM cycle. Just before the end of the current PWM period, the modulated outputs (P1B and P1D) are placed in their inactive state, while the unmodulated outputs (P1A and P1C) are switched to drive in the opposite direction. This occurs in a timer interval of $4T_{OSC}$*(Timer2 prescale value) before the next PWM period begins. During the interval from the switch of the unmodulated outputs to the beginning of next period, the modulated outputs (P1B and P1D) remain inactive. This relationship is shown in Figure 8.38.

In the full-bridge mode, the ECCP does not provide any deadband delay. This is acceptable if the duty cycle is not close to 100% because shoot-through cannot happen. However, there is a situation where a deadband delay might be needed. This situation occurs when both of the following conditions are true:

1. The direction of the PWM output changes when the duty cycle of the output is at or near 100%.

2. The turn-off time of the power switch, including the power device and driver circuit, is greater than the turn-on time.

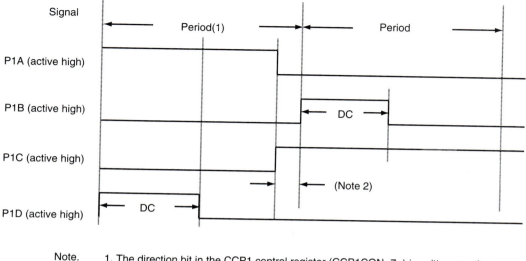

Note. 1. The direction bit in the CCP1 control register (CCP1CON<7>) is written any time during the PWM cycle.
2. When changing directions, the P1A and P1C signals switch before the end of the current PWM cycle at intervals of $4T_{OSC}$, $16T_{OSC}$, or $64T_{OSC}$, depending on the Timer2 prescaler value. The modulated P1B and P1D signals are inactive at this time.

Figure 8.38 ■ PWM direction change (redraw with permission of Microchip)

Figure 8.39 shows an example where the PWM direction changes from forward to reverse at a near 100% duty cycle. At time t1, the outputs P1A and P1D become inactive, while output P1C becomes active. In this example, since the turn-off time of the power devices is longer than the turn-on time, a shoot-through current may flow through power devices QC and QD for the duration of t. The same situation will occur to power devices QA and QB for PWM direction changes from reverse to forward.

When changing PWM direction at high duty cycle is required in an application, one of the following requirements must be met:

1. Reduce the PWM duty cycle for a PWM period before changing directions.

2. Use switch drivers that can drive the switches off faster than they can drive them on.

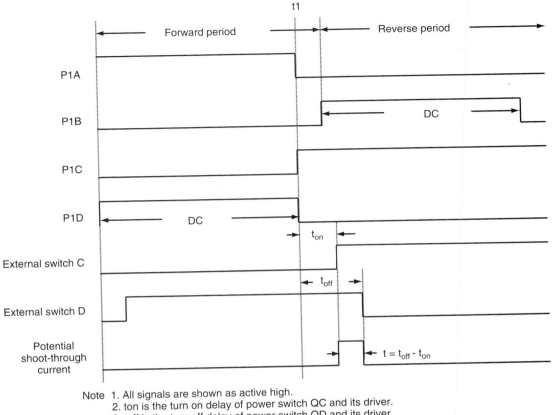

Note 1. All signals are shown as active high.
2. ton is the turn on delay of power switch QC and its driver.
3. toff is the turn off delay of power switch QD and its driver.

Figure 8.39 ■ PWM direction change at near 100% duty cycle (redraw with permission of Microchip)

PROGRAMMABLE DEADBAND DELAY

In half-bridge applications where all power switches are modulated at the PWM frequency at all times, the power switches normally require more time to turn off than to turn on. If both the upper and the lower power switches are switched at the same time, both switches may be on for a short period of time until one switch completely turns off. During this period, a current shoot-through occurs and can damage the driver circuit. To avoid this potentially destructive shoot-through current, turning on either of the power switches is normally delayed to allow the other switch to completely turn off.

In the half-bridge output mode, a programmable deadband delay is available to avoid shoot-through current from destroying the bridge power switches. The delay occurs at the signal transition from the nonactive state to the active state. The PWM1CON register (shown in Figure 8.40) sets the delay period in terms of microcontroller instruction cycles.

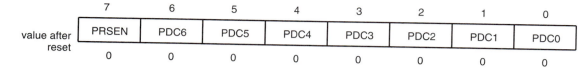

7	6	5	4	3	2	1	0
PRSEN	PDC6	PDC5	PDC4	PDC3	PDC2	PDC1	PDC0
0	0	0	0	0	0	0	0

value after reset

PRSEN: PWM restart enable bit
   0 = Upon auto shutdown, ECCPASE must be cleared in software to restart PWM.
   1 = Upon auto shutdown, the ECCPASE bit clears automatically once the
      shutdown event goes away; the PWM restarts automatically.
PDC<6:0>: PWM delay count bit
   Number of instruction cycles between the scheduled time when a PWM signal
   should transition active, and the actual time it transitions active.

Figure 8.40 ■ PWM1CON register (called ECCPxDEL for devices with 3 ECCPs) (redraw with permission of Microchip)

Enhanced PWM Auto Shutdown

When the ECCP is programmed for any of the enhanced PWM modes, the active output pins may be configured for auto shutdown. Auto shutdown immediately places the enhanced PWM output pins into a defined shutdown state when a shutdown event occurs.

A shutdown event can be caused by either of the two comparator modules or the INT0 pin or any combination of the three modules. The comparators may be used to monitor a voltage input proportional to a current being monitored in the bridge circuit. If the voltage exceeds a threshold, the comparator switches state and triggers a shutdown. The auto shutdown feature can be disabled by not selecting any auto shutdown sources. The auto shutdown sources are selected by programming the ECCPAS register shown in Figure 8.41. For devices with three ECCP modules, there are three ECCPAS registers: ECCP1AS, ECCP2AS, and ECCP3AS.

When a shutdown occurs, the output pins are asynchronously placed in their shutdown states, specified by the PSSAC1:PSSAC0 and PSSBD1:PSSBD0 bits of the ECCPAS register. Each pin pair (P1A/P1C and P1B/P1D) may be set to high, low, or tristated.

The ECCPASE bit is set by hardware when a shutdown event occurs. If automatic restarts are not enabled, the ECCPASE bit must be cleared by application software when the cause of the shutdown clears. If automatic restarts are enabled, the ECCPASE bit is automatically cleared when the cause of the auto shutdown has cleared.

If the ECCPASE bit is set when a PWM period begins, the PWM outputs remain in their shutdown state for that entire PWM period. When the ECCPASE bit is cleared, the PWM outputs will return to normal operation at the beginning of the next PWM period.

The auto shutdown feature can be configured to allow automatic restarts of the module following a shutdown event. This is enabled by setting the PRSEN bit of the PWM1CON register. In shutdown mode with PRSEN = 1 (shown in Figure 8.42), the ECC1PASE bit will remain set for as long as the cause of the shutdown continues. When the shutdown condition clears, the ECCP1ASE bit is cleared. If PRSEN = 0 (shown in Figure 8.43), once a shutdown condition occurs, the ECC1PASE bit will remain set until it is cleared by application software. Once ECCP1ASE is cleared, the enhanced PWM will resume at the beginning of the next PWM period.

Independent of the PRSEN bit setting, if the auto shutdown source is one of the comparators, the shutdown condition is a level. The ECCPASE bit cannot be cleared as long as the cause of the shutdown persists. The auto shutdown mode can be forced by writing a 1 to the ECCPASE bit.

8.8 ■ Enhanced CCP Module

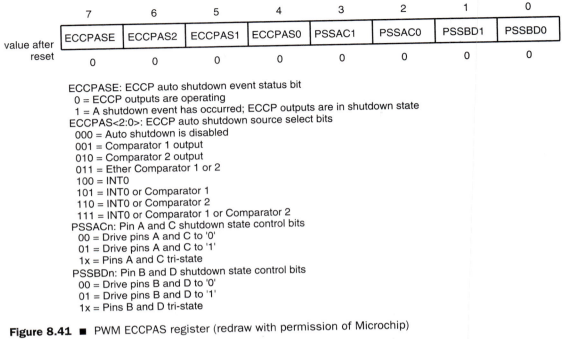

	7	6	5	4	3	2	1	0
	ECCPASE	ECCPAS2	ECCPAS1	ECCPAS0	PSSAC1	PSSAC0	PSSBD1	PSSBD0
value after reset	0	0	0	0	0	0	0	0

ECCPASE: ECCP auto shutdown event status bit
  0 = ECCP outputs are operating
  1 = A shutdown event has occurred; ECCP outputs are in shutdown state
ECCPAS<2:0>: ECCP auto shutdown source select bits
  000 = Auto shutdown is disabled
  001 = Comparator 1 output
  010 = Comparator 2 output
  011 = Ether Comparator 1 or 2
  100 = INT0
  101 = INT0 or Comparator 1
  110 = INT0 or Comparator 2
  111 = INT0 or Comparator 1 or Comparator 2
PSSACn: Pin A and C shutdown state control bits
  00 = Drive pins A and C to '0'
  01 = Drive pins A and C to '1'
  1x = Pins A and C tri-state
PSSBDn: Pin B and D shutdown state control bits
  00 = Drive pins B and D to '0'
  01 = Drive pins B and D to '1'
  1x = Pins B and D tri-state

Figure 8.41 ■ PWM ECCPAS register (redraw with permission of Microchip)

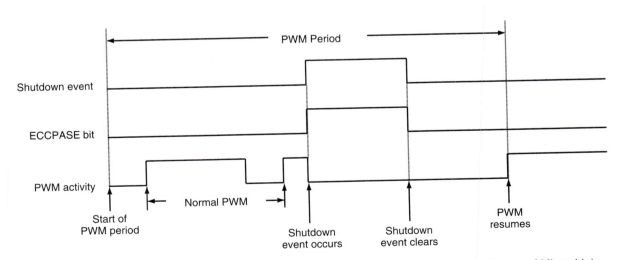

Figure 8.42 ■ PWM auto shutdown (PRSEN = 1, auto restart enabled) (redraw with permission of Microchip)

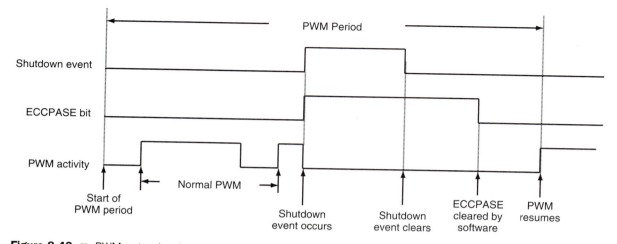

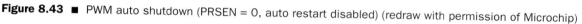

Figure 8.43 ■ PWM auto shutdown (PRSEN = 0, auto restart disabled) (redraw with permission of Microchip)

STARTUP CONSIDERATION

Prior to enabling the PWM outputs, the P1A, P1B, P1C, and P1D latches may not be in the proper states. Enabling the associated pins for output at the same time with the ECCP1 module being configured may cause damage to the power switch devices. The ECCP1 module must be enabled in the proper output mode with the associated pins (by setting associated bits in the TRISB or the TRISC or the TRISD register) configured as inputs. Once the ECCP1 module completes a full PWM cycle, the P1A, P1B, P1C, and P1D output latches are properly initialized. At this time, the data direction bits in the appropriate TRIS register can be cleared for outputs to start driving the power switch devices. The completion of a full PWM cycle is indicated by the TMR2IF bit being set as the second PWM period begins.

SETUP FOR PWM OPERATION

The following steps should be followed when configuring the ECCP1 module for PWM operation:

1. Configure the PWM pins P1A and P1B (and P1C and P1D, if used) for input.
2. Set the PWM period by loading the PR2 register.
3. Configure the ECCP module for the desired PWM mode and configuration by loading the CCP1CON register with the appropriate values:
    (a) Select one of the available output configurations and directions with the P1M1:P1M0 bits.
    (b) Select polarities of the PWM output signals with the CCP1M3:CCP1M0 bits.
4. Set the PWM duty cycle by loading the CCPR1L register and CCP1CON<5:4> bits.
5. For half-bridge output mode, set the deadband delay by loading PWM1CON<6:0> with the appropriate value.
6. If auto shutdown operation is required, load the ECCPAS register:
    (a) Select the shutdown sources of the PWM output pins by setting the ECCPS<2:0> bits.
    (b) Select the shutdown states of the PWM output pins by setting the PSSAC1:PSSAC0 and PSSBD1:PSSAD0 bits.

(c) Set the ECCPASE bit (ECCPAS<7>).

(d) Configure the comparators using the CMCON register (see datasheets of the PIC18F8x2x or the PIC18F6x2x devices).

(e) Configure the comparator inputs as analog inputs.

7. If auto restart operation is required, set the PRSEN bit (PWM1CON<7>).

8. Configure and start TMR2:

(a) Clear the TMR2 interrupt flag by clearing the TMR2IF bit.

(b) Set the TMR2 prescale value by loading the T2CKPS bits (T2CON<1:0>).

(c) Enable Timer2 by setting the TMR2ON bit.

9. Enable PWM outputs after a new PWM cycle has started:

(a) Wait until TMR2 overflows (TMR2IF bit is set).

(b) Enable the CCP1/P1A, P1B, P1C, and/or P1D pin outputs by clearing the respective TRISC and TRISD bits.

(c) Clear the ECCPASE bit (ECCPAS<7>).

Example 8.17

▼

Write an instruction sequence to generate a 5-KHz digital waveform with 60% duty cycle using the PIC18F4320 ECCP module so that it can be used in the half-bridge mode. Disable auto shutdown. To avoid a shoot-through problem, set the deadband delay to 16 instruction cycles. Assume that the PIC18F4320 is running with a 32-MHz crystal oscillator.

Solution: To generate a 5-KHz waveform, one can set the Timer2 prescaler to 16 and load the value of 99 into the PR2 register:

$$period = 32 \times 10^6 \div 16 \div 4 \div 5000 - 1 = 99$$

Then the value to be loaded into the CCPR1L register is $100 \times 60\% = 60$. The lowest two bits of the duty cycle register should be 00. The value to be written into the PWM1CON register for setting the deadband delay is 0x10.

The following instruction sequence will configure the ECCP1 module to generate a 60% duty cycle, 5-KHz waveform with the desired properties:

```
setf     TRISD,A          ; configure P1A..P1D for input
movlw    D'99'            ; set up period value
movwf    PR2,A            ; "
movlw    D'60'            ; set up duty cycle value
movwf    CCPR1L,A         ; "
movwf    CCPR1H,A         ; "
movlw    0x10             ; set up deadband delay to 16
movwf    PWM1CON,A        ; "
clrf     ECCPAS,A         ; disable auto shutdown
movlw    0x81             ; enable Timer3 to 16-bit mode and
movwf    T3CON,A          ; select Timer2 as the base timer for ECCP
movlw    0x06             ; enable Timer2 and set prescaler to 16
movwf    T2CON,A          ; "
movlw    0x8C             ; enable ECCP1 in half-bridge mode
movwf    CCP1CON,A        ; P1A thru P1D active high
bcf      PIR1,TMR2IF,A
```

```
wait1      btfss      PIR1,TMR2IF,A          ; wait until TMR2IF flag is set
           goto       wait1                  ; "
           movlw      0x1F                   ; configure P1A..P1D pins for output
           andwf      TRISD,F,A              ; "
           bcf        TRISC,RC2              ; "
           bcf        ECCPAS, ECCPASE,A      ; clear the auto shutdown flag
```

The following C statements can achieve the same setting:

```
TRISD = 0xFF;           /* configure P1A..P1D pins for input */
PR2 = 99;               /* set up period register */
CCPR1L = 60;            /* set up duty cycle */
CCPR1H = 60;            /* set up duty cycle */
PWM1CON = 0x10;         /* set up deadband delay */
ECCPAS = 0x00;          /* disable auto shutdown */
T3CON = 0x81;           /* enable Timer3 and select Timer2 as base timer */
T2CON = 0x06;           /* enable Timer2 and set prescaler to 16 */
SetOutputPWM1(HALF_OUT, PWM_MODE_1);
PIR1bits.TMR2IF = 0;
while (!PIR1bits.TMR2IF);
TRISD &= 0x1F;          /* configure P1A..P1D pins for output */
TRISCbits.TRISC2 = 0;
ECCPAS &= 0x7F;         /* clear ECCPASE flag */
```

▲

Example 8.18

▼

Write an instruction sequence to generate a 5-KHz digital waveform with an 80% duty cycle using the PIC18F4320 ECCP module so that it can be used in the full-bridge mode. Program shutdown state to '0'. Enable auto shutdown and auto restart. Assume that the PIC18F4320 is running with a 32-MHz crystal oscillator and that all PWM outputs are active high. Use comparator 2 output for shutdown detection.

Solution: We can use the same value to set up period register but change the duty cycle value to 80. The following instruction sequence will achieve the desired setting:

```
        setf     TRISD,A                ; configure P1A..P1D for input
        movlw    D'99'                  ; set up period value
        movwf    PR2,A                  ; "
        movlw    D'80'                  ; set up duty cycle value
        movwf    CCPR1L,A               ; "
        movwf    CCPR1H,A               ; "
        movlw    0x20                   ; use comparator 2 output for auto shutdown
        movwf    ECCPAS,A               ; enable and set shutdown state to '0'
        bsf      PWM1CON,PRSEN,A        ; enable auto restart
        movlw    0x81                   ; enable Timer3 to 16-bit mode and
        movwf    T3CON,A                ; select Timer2 as the base timer for ECCP
        movlw    0x06                   ; enable Timer2 and set prescaler to 16
        movwf    T2CON,A                ; "
        movlw    0x4C                   ; PWM mode, full-bridge output forward,
```

```
          movwf     CCP1CON,A          ; active high
          bcf       PIR1,TMR2IF,A
wait1     btfss     PIR1,TMR2IF,A      ; wait until TMR2IF flag is set
          goto      wait1              ; "
          movlw     0x1F               ; configure P1A..P1D pins for output
          andwf     TRISD,F,A          ; "
          bcf       TRISC,RC2          ; "
```

The following C statements can achieve the same setting:

```
TRISD       = 0xFF;       /* configure P1A..P1D pins for input */
PR2         = 99;         /* set up period register */
CCPR1L      = 80;         /* set up duty cycle */
CCPR1H      = 80;         /* " */
PWM1CON     = 0x80;       /* enable auto restart */
ECCPAS      = 0x20;       /* select comparator 2 output to enable auto shutdown */
T3CON       = 0x81;       /* enable Timer3 and select Timer2 as base timer */
T2CON       = 0x06;       /* enable Timer2 and set prescaler to 4 */
SetOutputPWM1(FULL_OUT_FORWARD, PWM_MODE_1);
PIR1bits.TMR2IF = 0;
while (!PIR1bits.TMR2IF);
TRISD & = 0x1F;           /* configure P1A..P1D pins for output */
TRISCbits.TRISC2 = 0;
ECCPAS & = 0x7F;          /* clear ECCPASE flag */
```

8.9 Summary

Timers and their associated components have been the center of many applications. Most microcontrollers introduced in the past few years have incorporated timer functions that include counting, capturing, comparing, and waveform-generating capabilities. They have been used in the following applications:

- Time delay creation
- Waveform generation, which may have applications in DC motor control, lamp dimming, siren and song generation, antilock brake, and so on
- Event arrival time capture
- Frequency measurement
- Pulse-width measurement
- Period measurement
- Duty cycle measurement
- Periodic interrupt generation

A PIC18 member may have four to five timers. Among them, Timer2 and Timer4 are 8-bit timers, whereas Timer0, Timer1, and Timer3 are 16-bit timers. Clock sources to these timers are selectable. Either the instruction cycle clock or an external signal can be selected as the clock source for these timers. A prescaler can be applied to the clock source before it is used to actually clock the timer module. This capability makes the time delay creation and waveform generation much easier to do.

A PIC18 microcontroller may have from two to five CCP modules: CCP1 through CCP5. These five CCP modules are identical in their functioning and configuration method. The CCP module has three operation modes:

- Capture
- Compare
- PWM

Either Timer1 and/or Timer3 can be used as the base timer in capture and compare modes. In the capture mode, the CCP module can be programmed to capture the falling or rising (every falling, every rising, every 4th rising, or every 16th rising) edge. When the selected edge is detected, the contents of the selected timer (Timer1 or Timer3) are captured in the CCPRxH and CCPRxL register pair. When a capture event occurs, the associated CCPxIF flag will be set to 1, and an interrupt may be requested if it is enabled.

In the compare mode, the contents of the CCPRxH and CCPRxL register pair are compared with those of the selected timer in every clock cycle. When they match, the associated CCP pin may be toggled, set to high, or pulled to low or stay the same. The associated CCPxIF flag will be set on a successful comparison. To use this mode, the user makes a copy of the selected timer, adds a delay to this copy, and then stores the sum in the CCPRxH:CCPRxL register pair. This CCP mode has been useful in creating time delays, generating digital waveforms, or creating software interrupts.

The base timer for CCP in PWM mode can be either Timer2 or Timer4 (if available). The main function of the PWM mode is waveform generation. The period of the waveform is specified by writing to the PR2 (or PR4) register, whereas the duty cycle is specified by writing into the CCPRxL register and bit 5 and bit 4 of the CCPxCON register. Whenever the base timer (Timer2 or Timer4) equals the PR2 (or PR4) register, three events occur:

- Base timer is cleared (TMR2 or TMR4).
- The CCPx pin is set to high unless the duty cycle is 0%.
- The PWM duty cycle is reloaded from CCPRxL into CCPRxH.

Whenever the value of the CCPRxL register equals the base timer, the CCPx pin is pulled to low. Most microcontrollers included the PWM function in order to support the motor control and many applications that require digital waveforms.

A few PIC18 members also implement the enhanced CCP function (ECCP). The ECCP module can be configured to perform exactly like an ordinary CCP module. However, the ECCP module is designed mainly to facilitate the DC motor control and applications that have similar requirements.

The ECCP module provides up to four output pins, including P1A through P1D. These four pins allow the ECCP to operate in four possible configurations:

- Single output
- Half-bridge output
- Full-bridge output, forward mode
- Full-bridge output, reverse mode

Both the half-bridge and the full-bridge outputs can be used in motor control or similar applications. The half-bridge configuration also allows the user to preset a deadband delay to avoid the shoot-through problem. In full-bridge mode, the user has the option to choose auto shutdown and auto restart the ECCP module, depending on whether the selected shutdown event exists. The cause for auto shutdown can be the INT0 pin level and/or the combinations of the two comparators outputs.

8.10 Exercises

E8.1 Suppose that the Timer0 counter does not have a buffer in its high byte. You are given the following instruction sequence and assume that the value of Timer0 is 0x48FE when the first instruction reads TMR0H. What value will be read into PRODH and PRODL?

```
movff    TMR0H,PRODH
movff    TMR0L,PRODL
```

E8.2 Assume that the PIC18F8720 is running with a 16-MHz crystal oscillator. Write an assembly subroutine to create a time delay that is equal to the value in PRODL multiplied by 100 ms. Use Timer0 to create the desired time delay.

E8.3 Assume that the PIC18F8720 is running with a 16-MHz crystal oscillator. Write an assembly instruction sequence and a sequence of C statements to configure Timer4 so that it generates periodic interrupts every 8 ms. The interrupt service routine is not needed.

E8.4 Give a value to be written into T3CON to set up the following parameters:

- Enable read/write of Timer3 in one 16-bit operation
- Select Timer3 and Timer4 as the base timers for CCP1 through CCP5
- Select external clock from pin T13CKI
- Synchronize external clock input

E8.5 For the assembly program given in Example 8.5, assume that the unknown signal frequency is about 1 MHz (not exact) and that the instruction cycle clock is 4 MHz. Which of the following options would allow the period be correctly measured?

1. Every rising edge
2. Every falling edge
3. Every 4th rising edge
4. Every 16th rising edge

E8.6 Modify the program in Example 8.6 so that it can measure the pulse width. Assume that the instruction cycle clock is 4 MHz. What is the shortest pulse width that can be measured with your program?

E8.7 Write a program to generate a 25-Hz digital waveform with a 50% duty cycle as long as the RB4 pin input is high, assuming that your PIC18 microcontroller operates with a 16-MHz crystal oscillator. Your program should consist of two parts:

1. **entry test**—As long as the RB4 signal is low, it stays at this loop.
2. **waveform generation body**—This part generates a pulse with 20 ms high time and 20 ms low time and at the end of a period tests the RB4 signal. If RB4 is still high, it generates the next pulse. Otherwise, it jumps to **entry test.**

This program simulates the behavior of an antilock brake system (ABS). The voltage level of the RB4 pin indicates whether the driver is pressing the brake pedal.

E8.8 Write a C language version of the siren generator that will work on the SSE452 demo board.

E8.9 Modify the program in Example 8.11 so that it can run on a demo board with a 32-MHz crystal oscillator.

E8.10 Modify the instruction sequence in Example 8.12 so that it can generate a 20-KHz digital waveform with a 70% duty cycle from the CCP2 pin.

E8.11 Modify the program in Example 8.11 so that it can play the song "Home, Sweet Home" and run on the SSE8720 demo board.

E8.12 Assume that you have a lamp controlled by a PIC18F8720 running with a 16-MHz crystal oscillator. Write a program that uses the PWM5 function to brighten the lamp from 0% brightness to 100% brightness in 5 seconds. Your program will use PWM5 to drive the lamp with the initial duty cycle set to 0%. After this setup, it will test the voltage level of the RB4 pin. If the RB4 pin remains low, the program simply loops. Otherwise, it starts to dim the lamp. Your program decreases the brightness to 0% at the end of 5 seconds in 10 steps.

E8.13 Write an assembly language version of the subroutine described in Example 8.15.

E8.14 Write a program that uses CCP2 in compare mode to trigger special event to generate periodic interrupt every 100 ms to be run on a PIC18 demo board with a 16-MHz crystal oscillator.

8.11 Lab Exercises and Assignments

L8.1 *Playing your favorite song.* Get an appropriate speaker connect to the CCP1 pin of your PIC18 demo board. Find a song with 50 notes or more. Avoid a song with too many identical consecutive notes. Modify the example program in this chapter. Compile (or assemble if a compiler is not available) and download the program onto the demo board and run the program. The circuit shown in Figure 8.25 can be used to drive speakers with higher power output.

L8.2 *Periodic interrupt generation using the CCP compare mode in special event trigger.* The special event trigger option of the CCP compare mode can be used to generate periodic interrupt. This feature can be used to generate time delay. Perform the following steps:

Step 1
Configure CCP1 in compare mode with the special event trigger option.

Step 2
Load the value 62500 (20-MHz crystal) or 50000 (16-MHz crystal) into the CCPR1H:CCPR1L register pair.

Step 3
Load 0 into TMR3H:TMR3L. Clear CCP1IF flag.

Step 4
Program Timer3 in 16-bit mode and select Timer3 as the base timer for CCP1. Set the prescaler of Timer3 to 8.

Step 5
Initialize a counter (call it **count_10**) to 10 and a byte counter **sec_cnt** to 0. Output the value of **sec_cnt** to the LEDs (eight) driven by an I/O port.

Step 6
Enable priority interrupt and set CCP1 interrupt at high priority. Enable CCP1 interrupt.

Step 7
Stay in this step forever.

Write a CCP1 interrupt service routine. This routine performs the following operations:

1. Clear CCP1IF flag.
2. Decrement **count_10** by 1.
3. If **count_10** is decremented to 0, then reset it to 10 and increment **sec_cnt** by 1 and output it to the LEDs.
4. Return from interrupt.

L8.3 *Lamp brightening and dimming.* Lamp dimming has been illustrated in Example 8.14. Lamp brightening is just the opposite operation of lamp dimming. Perform the following procedure to implement lamp dimming and brightening:

Step 1
Connect the lamp circuit properly.

Step 2
Use two flags to indicate whether the dimming or the brightening operation should be performed. One (DF) is for dimming, whereas the other (BF) is for brightening. Clear these two flags to zero at the beginning.

Step 3
Configure CCP1 in PWM mode and use it to drive the lamp circuit. Select Timer2 as the base timer for PWM1 and set the prescaler of Timer2 to an appropriate value. Set the initial duty cycle to 0%.

Step 4
Use two debounced switches (connect them to the INT0 and INT1 pins) as the user interface.

Step 5
Enable priority interrupt and place INT0 and INT1 in high priority. Write an interrupt service routine that checks the cause of interrupt and set the DF or BF flag accordingly. The service routine will set the DF flag if the interrupt source is INT0 and will set the BF flag if the interrupt source is INT1.

Step 6
Stay in a wait loop to check the DF and BF flags. Whenever the DF flag is 1, call the dimming routine and then clear the DF flag. Whenever the BF flag is 1, call the brightening routine and then clear the BF flag. This will ensure that one key press causes one operation.

L8.4 *Period and duty cycle measurement.* The algorithms for measuring period and duty cycle have been outlined in Section 8.5.3. Connect an unknown signal to the CCP1 pin. Configure the CCP1 module in capture mode. Write a program to measure the period and duty cycle of the unknown signal. Display the result in an LCD using the following format:

> Period = xxxxxx
>
> Duty cycle = yy%

9

Addressable Universal Synchronous Asynchronous Receiver Transceiver

9.1 Objectives

After completing this chapter, you should be able to

- Explain the four aspects of the EIA232 standard
- Explain the errors that could occur in data transmission
- Establish a null modem connection
- Explain the operation of a USART module
- Wire the USART pins to the EIA232 connector
- Program the USART module to perform data communications
- Use the USART module to add I/O ports to the PIC18 microcontroller by using the shift registers, such as the 74LS164 and the 74LS165

9.2 Overview of Serial Communication

Parallel I/O operation was discussed in Chapter 7. Parallel I/O is simple and fast. However, parallel I/O is not suitable for every application for the following reasons:

1. Parallel I/O uses many signal pins. In addition to eight data pins, other signal pins are needed for indicating data transfer direction and device handshaking. Many microcontrollers and microprocessors have a very limited number of pins. Using parallel I/O makes fewer pins available for other purposes.

2. It is usually not a problem for multiple data signals that belong to the same transfer to arrive at the destination simultaneously for a short distance. However, nobody can guarantee that parallel data can arrive at the destination simultaneously over a longer distance (e.g., 100 meters or longer).

3. Cable for parallel transfer is usually more expensive than serial cable.

4. Many applications do not need the high data rate provided by parallel data transfer.

Several serial data transfer protocols have been proposed to satisfy the data transfer requirements of different microcontroller applications:

1. USART (universal synchronous asynchronous receiver and transmitter)
2. SPI (synchronous peripheral interface, or Microwire)
3. I2C (interintegrated circuit)
4. CAN bus (controller area network bus)
5. LIN (local interconnect network)

The first four interface protocols are discussed in this text. LIN is not discussed mainly because it is not directly supported by any PIC18 microcontroller.

9.3 The EIA232 Standard

In 1960, the Electronic Industry Association (EIA) established the RS232 standard as a common interface for data communication equipment. The prefix "RS" stands for "recommended standard." At that time, data communications was thought to mean digital data communications between a mainframe computer and a remote terminal or between two terminals without a computer involved. These devices were linked together by telephone lines and consequently required a modem at each end for signal translation.

Over the past 40 years, the EIA232 standard has been revised several times. The latest revision, EIA232E, was published in July 1991. In this revision, the EIA replaced the prefix "RS" with "EIA" to identify the source of the standard. We refer to this standard as both RS232 and EIA232 throughout this chapter. In data communications, both computers and terminals are called *data terminal equipment* (DTE), whereas modems, bridges, and routers are called *data communication equipment* (DCE).

There are four aspects to the EIA232 standard:

1. *Electrical specifications*—Specify the voltage level, rise time and fall time of each signal, achievable data rate, and the distance of communication

2. *Functional specifications*—Specify the function of each signal

3. *Mechanical specifications*—Specify the number of pins and shape and dimensions of the connectors

4. *Procedural specifications*—Specify the sequence of events for transmitting data based on the functional specifications of the interface

9.3.1 Electrical Specifications

The following electrical specifications of the EIA232E are of interest to us:

1. *Data rates.* The EIA232 standard is applicable to data rates of up to 20000 bits per second (the usual upper limit is 19200 baud). Fixed baud rates are not set by the EIA 232 standard. However, the commonly used values are 300, 1200, 2400, 9600, and 19200 baud. Other accepted values that are not often used are 110 (mechanical teletype machines), 600, and 4800 baud.

2. *Signal state voltage assignments.* Voltages of –3 V to –25 V with respect to signal ground are considered logic "1" (the *mark* condition), whereas voltages of +3 V to +25 V are considered logic "0" (the *space* condition). The range of voltages between –3 V and +3 V is considered a transition region for which a signal state is not assigned.

3. *Signal transfer distance.* The signal should be able to transfer correctly within 15 meters. Greater distance can be achieved with good design.

9.3.2 EIA232E Functional Specifications

The EIA232E standard specifies 22 signals. A summary of these signals is given in Table 9.1. These signals can be divided into six categories:

1. *Signal ground and shield.*

2. *Primary communications channel.* This is used for data interchange and includes flow control signals.

3. *Secondary communications channel.* When implemented, this is used for control of the remote modem, requests for retransmission when errors occur, and governance over the setup of the primary channel.

4. *Modem status and control signals.* These signals indicate modem status and provide intermediate checkpoints as the telephone voice channel is established.

5. *Transmitter and receiver timing signals.* If a synchronous protocol is used, these signals provide timing information for the transmitter and receiver, which may operate at different baud rates.

6. *Channel test signals.* Before data is exchanged, the channel may be tested for its integrity and the baud rate automatically adjusted to the maximum rate that the channel could support.

Pin No.	Circuit	Description
1	-	Shield
2	BA	Transmitted data
3	BB	Received data
4	CA/CJ	Request to send/ready for receiving[1]
5	CB	Clear to send
6	CC	DCE ready
7	AB	Signal ground
8	CF	Received line signal detect
9	-	(reserved for testing)
10	-	(reserved for testing)
11	-	unassigned[3]
12	SCF/CI	Secondary received line signal detection/data rate selector (DCE source)[2]
13	SCB	Secondary clear to send
14	SBA	Secondary transmitted data
15	DB	Transmitter signal element timing (DCE source)
16	SBB	Secondary received data
17	DD	Receiver signal element timing
18	LL	Local loopback
19	SCA	Secondary request to send
20	CD	DTE ready
21	RL/CG	Remote loopback/signal quality detector
22	CE	Ring indicator
23	CH/CI	Data signal rate selector (DTE/DCE source)[2]
24	DA	Transmitter signal element timing (DTE source)
25	TM	Test mode

1. When hardware flow control is required, circuit CA may take on the functionality of circuit CJ. This is one change from the former EIA232.
2. For designs using interchange circuit SCF, interchange circuits CH and CI are assigned to pin 23. If SCF is not used, CI is assigned to pin 12.
3. Pin 11 is unassigned. It will not be assigned in future versions of EIA232. However, in international standard ISO 2110, this pin is assigned to select transmit frequency.

Table 9.1 ■ Functions of EIA232E signals

SIGNAL GROUND

Pins 7 and 1 and the shell are included in this category. Cables provide separate paths for each, but internal wiring often connects pin 1 and the cable shell/shield to signal ground on pin 7.

All signals are referenced to a common ground as defined by the voltage on pin 7. This conductor may or may not be connected to protective ground inside the DCE device.

PRIMARY COMMUNICATION CHANNEL

Pin 2 carries the transmit data (TxD) signal, which is active when data is transmitted from the DTE device to the DCE device. When no data is transmitted, the signal is held in the mark condition (logic "1." negative voltage).

Pin 3 carries the received data (RxD) signal, which is active when the DTE device receives data from the DCE device. When no data is received, the signal is held in the mark condition.

Pin 4 carries the request to send (RTS) signal, which will be asserted (logic "0," positive voltage) to prepare the DCE device for accepting transmitted data from the DTE device. Such preparation might include enabling the receive circuits or setting up the channel direction in half-duplex applications. When the DCE is ready, it acknowledges by asserting the clear to send signal.

Pin 5 carries the clear to send (CTS) signal, which will be asserted (logic "0") by the DCE device to inform the DTE device that transmission may begin. RTS and CTS are commonly used as handshaking signals to moderate the flow of data into the DCE device.

SECONDARY COMMUNICATION CHANNEL

Pin 14 carries the secondary transmitted data (STxD) signal. Pin 16 carries the secondary received data (SRxD) signal. Pin 19 carries the secondary request to send (SRTS) signal. Pin 13 carries the secondary clear to send (SCTS) signal.

These signals are equivalent to the corresponding signals in the primary communications channel. The baud rate, however, is typically much slower in the secondary channel for increased reliability.

MODEM STATUS AND CONTROL SIGNALS

This group includes the following signals:

Pin 6—DCE ready (DSR). When originating from a modem, this signal is asserted (logic "0") when all the following three conditions are satisfied:

1. The modem is connected to an active telephone line that is off-hook.

2. The modem is in data mode, not voice or dialing mode.

3. The modem has completed dialing or call setup functions and is generating an answer tone.

   If the line goes off-hook, a fault condition is detected or a voice connection is established, the DSR signal is deasserted (logic "1").

Pin 20—DTE ready (DTR). This signal is asserted (logic "0") by the DTE device when it wishes to open a communications channel. If the DCE device is a modem, the assertion of DTR prepares the modem to be connected to the telephone circuit and, once connected, maintains the connection. When DTR is deasserted, the modem is switched to on-hook to terminate the connection (same as placing the phone back to the telephone socket).

Pin 8—Received line signal detector (CD). Also called *carrier detect*, this signal is relevant when the DCE device is a modem. It is asserted (logic "0") by the modem when the telephone line is off-hook, a connection has been established, and an answer tone is being received from the remote modem. The signal is deasserted when no answer tone is being received or when the answer tone is of inadequate quality to meet the local modem's requirements.

Pin 12—Secondary received line signal detector (SCD). This signal is equivalent to the CD (pin 8) but refers to the secondary channel.

Pin 22—Ring indicator (RI). This signal is relevant when the DCE device is a modem and is asserted (logic "0") when a ringing signal is being received from the telephone line. The assertion time of this signal will approximately equal the duration of the ring signal, and it will be deasserted between rings or when no ringing is present.

Pin 23—Data signal rate selector. This signal may originate in either the DTE or the DCE devices (but not both) and is used to select one of two prearranged baud rates. The assertion condition (logic "0") selects the higher baud rate.

TRANSMITTER AND RECEIVER TIMING SIGNALS

This group consists of the following signals:

Pin 15—Transmitter signal element timing (TC). Also called *transmitter clock*, this signal is relevant only when the DCE device is a modem and is operating with a synchronous protocol. The modem generates this clock signal to control exactly the rate at which data is sent on TxD (pin 2) from the DTE device to the DCE device. The logic "1" to logic "0" (negative to positive transition) transition on this line causes a corresponding transition to the next data element on the TxD line. The modem generates this signal continuously, except when it is performing internal diagnostic functions.

Pin 17—Receiver signal element timing (RC). Also called *receiver clock*, this signal is similar to TC, except that it provides timing information for the DTE receiver.

Pin 24—Transmitter signal element timing (ETC). Also called *external transmitter clock*, with timing signals provided by the DTE device for use by a modem, this signal is used only when TC and RC (pins 15 and 17) are not in use. The logic "1" to logic "0" transition indicates the time center of the data element. Timing signals will be provided whenever the DTE is turned on regardless of other signal conditions.

CHANNEL TEST SIGNALS

This group consists of the following signals:

Pin 18—Local loopback (LL). This signal is generated by the DTE device and is used to place the modem into a test state. When LL is asserted (logic "0," positive voltage), the modem redirects its modulated output signal, which is normally fed into the telephone line, back into its receive circuitry. This enables data generated by the DTE to be echoed back through the local modem to check the condition of the modem circuitry. The modem asserts its test mode signal on pin 25 to acknowledge that it has been placed in LL condition.

Pin 21—Remote loopback (RL). This signal is generated by the DTE device and is used to place the remote modem into a test state. When RL is asserted (logic "0"), the remote modem redirects its received data back to its transmitted data input, thereby remodulating the received data and returning it to its source. When the DTE initiates such a test, transmitted data is passed through the local modem, the telephone line, the remote modem, and back to exercise the channel and confirm its integrity. The remote modem signals the local modem to assert test mode on pin 25 when the remote loopback test is under way.

Pin 25—Test Mode (TM). This signal is relevant only when the DCE device is a modem. When asserted (logic "0"), it indicates that the modem is in an LL or RL condition. Other internal self-test conditions may also cause TM to be asserted, depending on the modem and the network to which it is attached.

9.3.3 EIA232E Mechanical Specifications

The EIA232E uses a 25-pin D-type connector, as shown in Figure 9.1a. Since only a small subset of the 25 signals is actually used, a 9-pin connector (DB9) is used in most PCs. The signal assignment of DB9 is shown in Figure 9.1b. The DB9 is not part of the EIA232E standard.

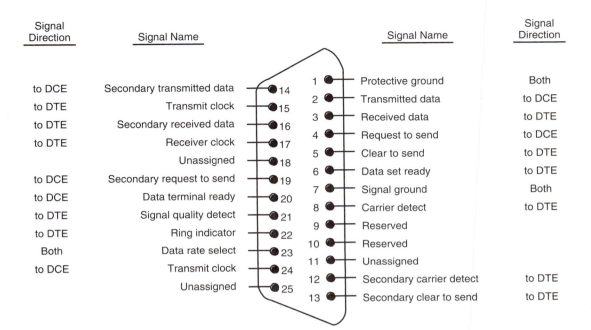

Signal Direction	Signal Name		Signal Name	Signal Direction
to DCE	Secondary transmitted data	14	1 Protective ground	Both
to DTE	Transmit clock	15	2 Transmitted data	to DCE
to DTE	Secondary received data	16	3 Received data	to DTE
to DTE	Receiver clock	17	4 Request to send	to DCE
	Unassigned	18	5 Clear to send	to DTE
to DCE	Secondary request to send	19	6 Data set ready	to DTE
to DCE	Data terminal ready	20	7 Signal ground	Both
to DTE	Signal quality detect	21	8 Carrier detect	to DTE
to DTE	Ring indicator	22	9 Reserved	
Both	Data rate select	23	10 Reserved	
to DCE	Transmit clock	24	11 Unassigned	
	Unassigned	25	12 Secondary carrier detect	to DTE
			13 Secondary clear to send	to DTE

Figure 9.1a ■ EIA232E DB25 connector and pin assignment

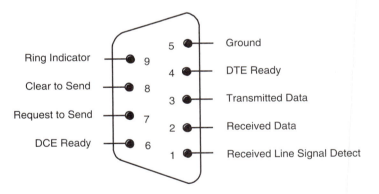

	5	Ground	
Ring Indicator	9		
	4	DTE Ready	
Clear to Send	8		
	3	Transmitted Data	
Request to Send	7		
	2	Received Data	
DCE Ready	6		
	1	Received Line Signal Detect	

Figure 9.1b ■ EIA232E DB9 connector and signal assignment

9.3.4 EIA232E Procedural Specifications

The sequence of events that occurs during data transition using the EIA232E is easier to understand by studying examples. Two examples are used to explain the procedure.

In the first example, two DTEs are connected with a point-to-point link using a modem. The modem requires only the following circuits to operate:

- Signal ground (AB)
- Transmitted data (BA)
- Received data (BB)

- Request to send (CA)
- Clear to send (CB)
- Data set ready (CC)
- Carrier detect (CF)

Before the DTE can transfer data, the DSR signal must be asserted to indicate that the modem is ready to operate. This signal should be asserted before the DTE attempts to make a request to send data. The DSR signal can simply be connected to the power supply of the DCE to indicate that it is switched on and ready to operate. When a DTE is ready to send data, it asserts the RTS signal. The modem responds, when ready, with CTS, indicating that data may be transmitted over circuit BA. If the arrangement is half duplex, then the RTS signal also inhibits the receive mode. The DTE sends data to the local modem bit serially. The local modem modulates the data into the carrier signal and transmits the resultant signal over the dedicated communication lines. Before sending out the modulated data, the local modem sends out a carrier signal to the remote modem so that the remote modem is ready to receive the data. The remote modem detects the carrier and asserts the CD signal. The assertion of the CD signal tells the remote DTE that the local modem is transmitting. The remote modem receives the modulated signal, demodulates it to recover the data, and sends it to the remote DTE over the received-data pin. The circuit connections are illustrated in Figure 9.2.

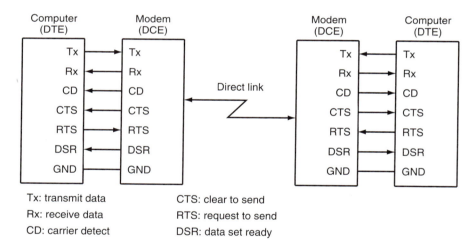

Tx: transmit data   CTS: clear to send
Rx: receive data   RTS: request to send
CD: carrier detect   DSR: data set ready

Figure 9.2 ■ Point-to-point asynchronous connection

The next example involves two computers exchanging data through a public telephone line. One of the computers (initiator) must dial the phone (automatically or manually) to establish the connection, just like the people talking over the phone. Two additional leads are required for this application:

- Data terminal ready (DTR)
- Ring indicator (RI)

The data transmission in this setting can be divided into three phases:

Establishing the connection. The following events occur in this phase:

1. The transmitting computer asserts the DTR signal to indicate to the local modem that it is ready to make a call.

2. The local modem opens the phone line and dials the destination number. The number can be stored in the modem or transmitted to the modem by the computer via the transmit data pin.

3. The remote modem detects a ring on the phone line and asserts the RI to inform the remote computer that a call has arrived.

4. The remote computer asserts the DTR signal to accept the call.

5. The remote modem answers the call by sending a carrier signal to the local modem via the phone line. It also asserts the DSR signal to inform the remote computer that it is ready for data transmission.

6. The local modem asserts both DSR and CD signals to indicate that the connection is established and it is ready for data communication.

7. For full-duplex data communication, the local modem also sends a carrier signal to the remote modem. The remote modem then asserts the CD signal.

Data transmission. The following events occur during this phase:

1. The local computer asserts the RTS signal when it is ready to send data.

2. The local modem responds by asserting the CTS signal.

3. The local computer sends data bits serially to the local modem over the TxD pin. The local modem then modulates its carrier signal to transmit the data to the remote modem.

4. The remote modem receives the modulated signal from the local modem, demodulates it to recover the data, and sends it to the remote computer over the received data pin.

Disconnection. Disconnection requires only two steps:

1. When the local computer has finished the data transmission, it drops the RTS signal.

2. The local modem then deasserts the CTS signal and drops the carrier (equivalent to hanging up the phone).

The circuit connection for this example is shown in Figure 9.3. A timing signal is not required in an asynchronous transmission.

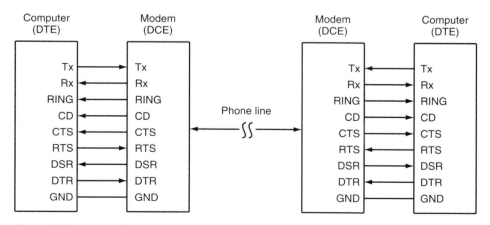

Figure 9.3 ■ Asynchronous connection over public phone line

9.3.5 Data Format

In asynchronous data transfer, data is transferred character by character. Each character is preceded by a start bit (a low), followed by seven to nine data bits, and terminated by one to two stop bits. The data format of a character is shown in Figure 9.4.

Figure 9.4 ■ The format of a character

As shown in Figure 9.4, the least significant bit is transmitted first, and the most significant bit is transmitted last. Stop bit(s) is (are) high. The start bit and stop bit identify the start and end of a character.

Since there is no clock information in the asynchronous format, the receiver uses a clock signal with a frequency that is a multiple (usually 16) of the data rate to sample the incoming data in order to detect the arrival of the start bit and determine the logical value of each data bit. A clock, with a frequency that is 16 times the data rate, can tolerate a difference of about 5% in the clocks at the transmitter and receiver.

The method for detecting the arrival of a start bit is similar among all microcontrollers: a majority detection circuit takes three samples (seventh, eighth, and ninth samples) on the RX/DT pin to determine if a high or a low level is present at the RX pin. Figure 9.5 shows the waveform for the sampling circuit. When the majority function output is low, the majority circuit determines that the start bit has arrived. The same method is used to determine the logic value of any data bit.

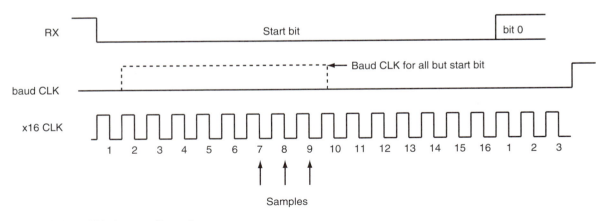

Figure 9.5 ■ RX pin sampling scheme

In older designs, a character can be terminated by one, one and a half, or two stop bits. In newer designs, only one stop bit is used. Using this format, it is possible to transfer character by character without any gaps.

Example 9.1

Sketch the output of the letter K when it is transmitted using the format of one start bit, eight data bits, and one stop bit.

Solution: Letters are represented in ASCII code. The ASCII code of letter K is $4B (= 01001011). The format of the output of letter K is shown in Figure 9.6.

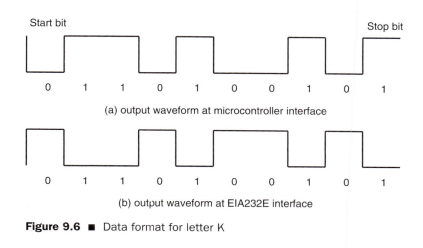

(a) output waveform at microcontroller interface

(b) output waveform at EIA232E interface

Figure 9.6 ■ Data format for letter K

9.3.6 Data Transmission Errors

The following errors may occur during the data transfer process using asynchronous serial transmission:

- *Framing error.* A framing error occurs when a received character is improperly framed by the start and stop bits; it is detected by the absence of the stop bit. This error indicates a synchronization problem, faulty transmission, or a break condition.
- *Receiver overrun.* One or more characters in the data stream were received but were not read from the buffer before subsequent characters were received.
- *Parity error.* A parity error occurs when an odd number of bits change value. It can be detected by a parity error-detecting circuit. Parity checking can also be implemented in software.

9.3.7 Null-Modem Connection

When two DTE devices are located side by side and use an EIA232E interface to exchange data, there is really no reason to use two modems to connect them. However, the EIA232E standard does not allow the direct connection of two DTEs. In order to make this scheme work, a *null modem* is needed. The null modem interconnects leads in such a way that fools both DTEs into thinking that they are connected to modems. The null-modem connection is shown in Figure 9.7.

Signal Name	DTE 1		DTE 2		Signal Name
	DB25 pin	DB9 pin	DB9 pin	DB25 pin	
FG (frame ground)	1	-	-	1	FG
TD (transmit data)	2	3	2	3	RD
RD (receive data)	3	2	3	2	TD
RTS (request to send)	4	7	8	5	CTS
CTS (clear to send)	5	8	7	4	RTS
SG (signal ground)	7	5	5	7	SG
DSR (data set ready)	6	6	4	20	DTR
CD (carrier detect)	8	1	4	20	DTR
DTR (data terminal ready)	20	4	1	8	CD
DTR (data terminal ready)	20	4	6	6	DSR

Figure 9.7 ■ Null Modem connection

In Figure 9.7, the transmitter timing and receiver timing signals are not needed in asynchronous data transmission. The RI signal is not needed, either, because the transmission is not made through a public phone line.

9.4 The PIC18 Serial Communication Interface

A serial communication interface can be called a USART or UART, depending on whether it supports both synchronous and asynchronous modes. A USART supports both asynchronous and synchronous modes, whereas a UART supports only asynchronous mode. USART is the acronym for "universal synchronous asynchronous receiver transmitter." The PIC18 device supports either one or two USART modules. If a PIC18 device has two USART modules, then these two modules are identical and can be configured independently. The USART module can be configured as a full-duplex asynchronous system that can communicate with peripheral devices, such as CRT terminals and personal computers, or as a half-duplex synchronous system that can communicate with peripheral devices, such as A/D or D/A integrated circuits and serial EEPROMs.

The pins of USART1 and USART2 are multiplexed with the functions of PORTC (RC6/TX1/CK1 and RC7/RX1/DT1) and PORTG (RG1/TX2/CK2 and RG2/RX2/DT2), respectively. The USART module can be configured to operate in three different modes:

- Asynchronous (full duplex)
- Synchronous—master (half duplex)
- Synchronous—slave (half duplex)

9.4.1 USART-Related Registers

Each USART module uses the following registers to control its operation:

- Transmit status register (TXSTA)
- Receive status register (RCSTA)
- Baud rate generate register (SPBRG)

- Transmit register (TXREG)
- Receive register (RCREG)

For those PIC18 devices with two USART modules, the suffix "1" or "2" is added to differentiate them. For example, TXSTA1 and TXSTA2 stand for the transmit status register for the USART1 and the USART2 module, respectively. No suffix should be added for those devices having only one USART module. The contents of the TXSTA register are shown in Figure 9.8. The TXSTA register enables/disables transmission, selects 8-bit or 9-bit transmission, selects asynchronous/synchronous mode, and also holds the ninth bit to be transferred.

	7	6	5	4	3	2	1	0
	CSRC	TX9	TXEN	SYNC	--	BRGH	TRMT	TX9D
Value after reset	0	0	0	0	0	0	1	0

CSRC: Clock Source Select bit
  Asynchronous mode: (don't care)
  Synchronous mode:
    0 = Slave mode (clock from external source)
    1 = Master mode (clock generated internally from BRG)
TX9: 9-bit Transmit Enable bit
    0 = selects 8-bit transmission
    1 = selects 9-bit transmission
TXEN: Transmit Enable Bit
    0 = Transmit disabled
    1 = Transmit enabled
SYNC: USART Mode Select Bit
    0 = Asynchronous mode
    1 = Synchronous mode
BRGH: High Baud Rate Select Bit
  Asynchronous mode:
    0 = low speed
    1 = high speed
  Synchronous mode (unused)
TRMT: Transmit Shift Register Status Bit
    0 = TSR full
    1 = TSR empty
TX9D: 9th bit of transmit data
    Can be Address/Data bit or a parity bit

Figure 9.8 ■ The TXSTA Register (redraw with permission of Microchip)

The RCSTA register performs the following functions:

- Enables/disables the USART module
- Selects 8-bit/9-bit reception
- Enables/disables single reception in synchronous mode
- Enables/disables continuous reception
- Enables/disables address detection
- Records framing and overrun error status
- Holds the received ninth bit

The contents of the RCSTA register are shown in Figure 9.9.

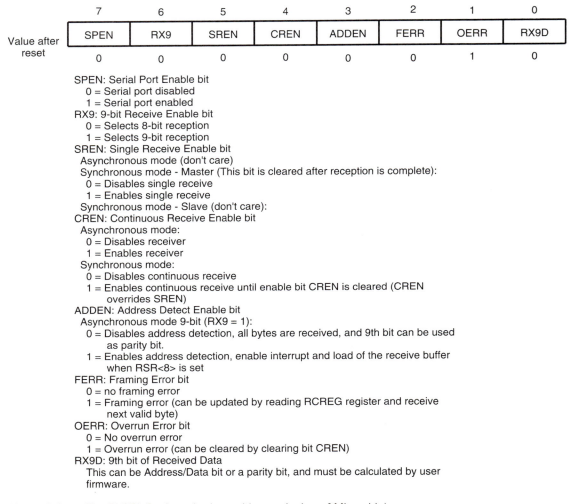

Value after reset

7	6	5	4	3	2	1	0
SPEN	RX9	SREN	CREN	ADDEN	FERR	OERR	RX9D
0	0	0	0	0	0	1	0

SPEN: Serial Port Enable bit
  0 = Serial port disabled
  1 = Serial port enabled
RX9: 9-bit Receive Enable bit
  0 = Selects 8-bit reception
  1 = Selects 9-bit reception
SREN: Single Receive Enable bit
  Asynchronous mode (don't care)
  Synchronous mode - Master (This bit is cleared after reception is complete):
  0 = Disables single receive
  1 = Enables single receive
  Synchronous mode - Slave (don't care):
CREN: Continuous Receive Enable bit
  Asynchronous mode:
  0 = Disables receiver
  1 = Enables receiver
  Synchronous mode:
  0 = Disables continuous receive
  1 = Enables continuous receive until enable bit CREN is cleared (CREN
      overrides SREN)
ADDEN: Address Detect Enable bit
  Asynchronous mode 9-bit (RX9 = 1):
  0 = Disables address detection, all bytes are received, and 9th bit can be used
      as parity bit.
  1 = Enables address detection, enable interrupt and load of the receive buffer
      when RSR<8> is set
FERR: Framing Error bit
  0 = no framing error
  1 = Framing error (can be updated by reading RCREG register and receive
      next valid byte)
OERR: Overrun Error bit
  0 = No overrun error
  1 = Overrun error (can be cleared by clearing bit CREN)
RX9D: 9th bit of Received Data
  This can be Address/Data bit or a parity bit, and must be calculated by user
  firmware.

Figure 9.9 ■ The RXSTA Register (redraw with permission of Microchip)

A framing error can be cleared, in effect, by reading the RCREG register. The error bit FERR will remain set until new data has been received and loaded into the RCREG register. Overrun can occur if the microcontroller does not read the receiver buffer for a while. Once an overrun error occurs, no new data will be received until the receive logic has been reset by clearing the receive enable bit, CREN, and enabling it again. A common symptom of an overrun error is that the USART stops receiving unexpectedly, often after the first two bytes.

The shift clock for the USART is generated by the *baud rate generator* (SPBRG). SPBRG supports both the asynchronous and the synchronous modes of the USART and is a dedicated 8-bit baud rate generator. In asynchronous mode, the BRGH bit of the TXSTA register is also involved in the control of baud rate. The value to be written into the SPBRG register is calculated using the formula given in Table 9.2.

Both the TXREG and the RCREG are 8-bit data registers.

SYNC bit	BRGH = 0 (low speed)	BRGH = 1 (high speed)
0	(Asynchronous) Baud rate = $F_{OSC}/(64 (X + 1))$	Baud Rate = $F_{OSC}/(16(X + 1))$
1	(Synchronous) Baud Rate = $F_{osc}/(4(X + 1))$	N/A

Note. X is the content of the SPBRG register
In asynchronous mode:
  When BRGH = 1, SPBRG = $(F_{OSC}/(16 \times \text{baud rate})) - 1$
  When BRGH = 0, SPBRG = $(F_{OSC}/(64 \times \text{baud rate})) - 1$
In synchronous mode:
  SPBRG = $(FOSC/(4 \times \text{baud rate})) - 1$

Table 9.2 ■ Formula for baud rate

Example 9.2

▼

Compute the value to be written into the SPBRG register to generate 9600 baud for asynchronous mode high-speed transmission, assuming that the frequency of the crystal oscillator is 20 MHz.

Solution: Apply the formula given in Table 9.2. The value (for BRGH = 1) to be written into the SPBRG register is

SPBRG = $20 \times 10^6 \div (16 \times 9600) - 1 = 130 - 1 = 129$

The actual baud rate is

$20,000,000 \div (16 \times 130) = 9615.4$

The resultant error rate is

$(9615.4 - 9600) \div 9600 \times 100\% = 0.16\%$

The same baud rate can also be achieved by using a low-speed (BRGH = 0) approach in which

SPBRG = $20,000,000 \div (64 \times 9600) - 1 = 31$

The actual baud rate is

$20000000 \div (64 \times 32) = 9765.6$

The resultant error rate is

$(9765.6 - 9600) \div 9600 \times 100\% = 1.7\%$

This approach has a higher error rate in its resultant baud rate. In most cases, both approaches give the same error rate. However, the user should compare the resultant error rates of both approaches before making a decision regarding which approach to use.

▲

9.4.2 USART Asynchronous Mode

In this mode, the USART uses the data format of one start bit, eight or nine data bits, and one stop bit for data transmission and reception. The least significant bit is transmitted and received first. The transmitter and receiver are independent but use the same data format and baud rate. The baud rate generator produces a clock with a frequency that equals either 16 or 64 times that of the bit shift rate, depending on the setting of the BRGH bit.

Asynchronous mode is stopped during the SLEEP mode. The USART transmitter block diagram is shown in Figure 9.10. The heart of the transmitter is the transmit (serial) shift register (TSR).

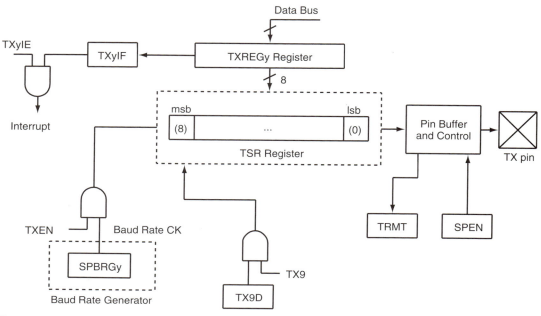

Note. y = 1 or 2

Figure 9.10 ■ USARTy Transmit block diagram (redraw with permission of Microchip)

Whenever user software has data to be transmitted, it writes into the TXREG register. If the TSR register is empty, the data in the TXREG register will be loaded into the TSR register. Once this occurs, the TXyIF (y = 1 or 2) bit will be set. An interrupt will be requested if it is enabled (if TXyIE = 1). If the TSR register is empty, then the TRMT bit is set to 1. This status does not have an associated interrupt flag.

When 9-bit format is selected, the ninth bit must be written into the TX9D bit of the TXSTA register before writing the lower eight bits of the data into the TXREG register. When a byte is written into the TXREG register, it will be loaded immediately into the TSR register and be shifted out.

Example 9.3

Write an instruction sequence to configure the USART1 to transmit data in asynchronous mode using an 8-bit data format, disable interrupt, and baud rate of 9600. Assume that the frequency of the crystal oscillator is 16 MHz.

Solution: The value to be written into the TXSTA1 register is 0x24. The SPEN bit of the RCSTA1 should be set to 1. The RC6/TX1 pin and RC7/RX1 pin should be configured for output and input, respectively. The following instruction sequence will configure the USART1 accordingly:

```
usart1_open    movlw    0x24        ; enable 8-bit transmission
               movwf    TXSTA1      ; "
               movlw    D'103'
               movwf    SPBRG1      ; set up baud rate to 9600
```

```
                    bsf        TRISC,RX          ; configure RX1 pin for input
                    bcf        TRISC,TX          ; configure TX1 pin for output
                    bsf        RCSTA1,SPEN       ; enable USART1
                    return
```

The following C statement will perform the same configuration:

```
TXSTA1          = 0x24;      /* enable 8-bit transmission */
SPBRG1          = 103;       /* set baud rate to 9600 */
TRISC           |= 0x80;     /* configure RC7/RX1 pin for input */
TRISC           &= 0xBF;     /* configure RC6/TX1 pin for output */
RCSTAbits.SPEN = 1;          /* enable USART1 */
```

▲

Example 9.4

▼

Write a subroutine to output the character in WREG to USART1 using the polling method.

Solution: The user should send the character to the transmitter only when the TX1IF flag is set to 1, which indicates that the TXREG1 register is empty. The following assembly subroutine implements this idea:

```
putc_usart1    btfss      PIR1,TXIF         ; wait until TXIF flag is set before output
               bra        putc_usart1
               movwf      TXREG1
               return
```

The C language version of the function is as follows:

```
void putc_usart1 (char xc);
{
    while (!PIR1bits.TX1IF);
    TXREG1 = xc;
}
```

▲

Example 9.5

▼

Write a subroutine to output a string (in program memory) pointed to by TBLPTR and terminated by a NULL character from USART1.

Solution: This subroutine will read one character from the string at a time and call **putc_usart1( )** to output it until the NULL character is reached:

```
; ****************************************************************
; The following subroutine outputs the string pointed to by TBLPTR.
; ****************************************************************
;
puts_usart1    tblrd*+                        ; read one character into TABLAT
               movf       TABLAT,W            ; place the character in WREG
               bz         done                ; is it a NULL character?
               call       putc_usart1         ; not NULL, output
               bra        puts_usart1         ; continue
done           return
```

The C language version of the function is as follows:

```
void puts_usart1 (unsigned rom char *cptr)
{
        while(*cptr)
        putc_usart1 (*cptr++);
}
```

The USART receiver block diagram is shown in Figure 9.11. The data is received on pin RC7/RX1/DT1 (USART1) or RG2/RX2/DT2 (USART2) and drives the data recovery block. The data recovery block is a high-speed shifter operating at 16 times the baud rate, whereas the main receive serial shifter operates at the bit rate or at F_{OSC}. This mode is typically used in EIA232 systems. When the 9-bit format is selected, the received ninth bit will be copied into the RX9D bit of the RCSTA register.

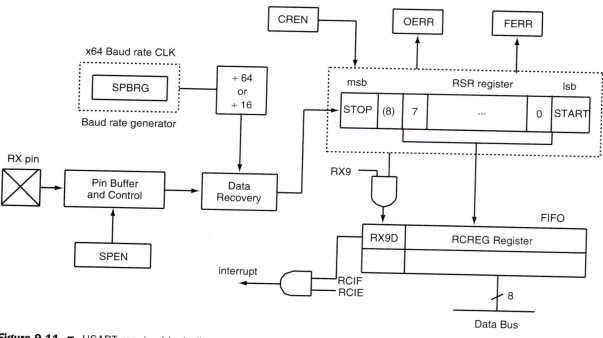

Figure 9.11 ■ USART receive block diagram (redraw with permission of Microchip)

Example 9.6

Write an instruction sequence to configure the USART1 to receive data in asynchronous mode using 8-bit data format, disable interrupt, and baud rate of 9600. Assume that the frequency of the crystal oscillator is 16 MHz.

Solution: The value to configure USART1 for reception with the specified configuration is 0x90. The instruction sequence to perform the desired configuration is as follows:

```
movlw      0x90          ; enable USART1 and receiver
movwf      RCSTA1
movlw      D'103'
movwf      SPBRG1        ; set up baud rate to 9600
bsf        TRISC,RX1     ; configure RX1 pin for input
bcf        TRISC,TX1     ; configure TX1 pin for input
```

The following C statements perform the same configuration:

```
RCSTA1     = 0x90;
SPBRG      = 103;
TRISC      |= 0x80;      ; configure RC7/RX pin for input
TRISC      &= 0xBF;      ; configure RC6/TX pin for output
```

Usually the configuration of the transmitter and receiver is done all at once instead of in two separate subroutines. This is left for you as an exercise problem.

▲

Example 9.7

▼

Write a subroutine to read a character from USART1 and return the character in WREG using the polling method. Ignore any errors.

Solution: A new character is received by USART1 if the RCIF flag of the PIR1 register is 1. If the RCIF flag is not 1, then the subroutine should wait until it is set to 1. The following subroutine checks the RCIF flag and reads the RCREG1 register:

```
getc_usart1    btfss    PIR1,RCIF       ; make sure a new character has been received
               bra      getc_usart1
               movf     RCREG1,W        ; read the character
               return
```

The C language version of the subroutine is as follows:

```
char   getc_usart1 (void)
{
       while (!PIR1bits.RCIF);
       return RCREG1;          /* also clears RCIF flag */
}
```

When any reception errors occur, the common practice is to clear the error flag and then wait for the next correctly received character. You can accommodate this into the **getc_usart1** subroutine in the previous code.

▲

Example 9.8

▼

Write a subroutine to read a string from USART1 and store the string in a buffer pointed to by FSR0.

Solution: The string from USART1 is terminated by the carriage return character. This subroutine is implemented by calling **getc_usart1** repeatedly until the CR character is received:

```
CR                         equ      0x0D
gets_usart1    call        getc_usart1
               movwf       INDF0                ; save the character in buffer
```

```
                    sublw       CR
                    bz          done
                    clrf        PREINC0,F        ; move the pointer
                    goto        gets_usart1
        done        clrf        INDF0            ; terminate the string with a NULL character
                    return
```

The C language version of the subroutine is as follows:

```
#define     CR      0x0D
void        gets_usart1 (char *ptr)
{
    char  xx;
    while (1)
    {
        xx = getc_usart1();     /* read a character */
        if (xx == CR) {         /* is it a carriage return? */
            *ptr = '\0';        /* terminate the string with a NULL */
            return;
        }
        ptr++ = xx;             /* store the received character in the buffer */
    }
}
```

Examples 9.3 to 9.8 did not consider 9-bit transmission and reception. If the application requires it, one should modify the previous subroutines to handle 9-bit transmission and reception.

In order to practice USART I/O, one will need to run a terminal program, such as the HyperTerminal provided in the Windows operating system. When the HyperTerminal program starts, a terminal window will be opened. Whatever data written to USART1 (by the user program) will appear on the HyperTerminal window, and whatever the user enters from the keyboard will be sent to the USART port connected to the PC communication port COM1 (or COM2, depending the setting in HyperTerminal).

▲

9.4.3 Flow Control of USART in Asynchronous Mode

The USART module will transmit data as fast as the baud rate allows. In some circumstances, the software that is responsible for reading the data from the RCREG register may not be able to do so as fast as the data is being received. In this case, there is a need for the PIC18 microcontroller to tell the transmitting device to suspend transmission of data temporarily.

Similarly, the PIC18 microcontroller may need to be told to suspend transmission temporarily. This is done by means of flow control. There are two common methods of flow control: XON/XOFF and hardware. Here, "X" stands for "transmission."

The XON/XOFF flow control can be implemented completely in software with no external hardware, but full-duplex communications is required. When incoming data needs to be suspended, an XOFF byte is transmitted back to the other device that is transmitting the data being received. To start the other device transmitting again, an XON byte is transmitted. XON and XOFF are standard ASCII control characters. This means that when sending raw data instead of ASCII text, care must be taken to ensure that XON and XOFF characters are not accidentally sent with the data. The ASCII codes of XON and XOFF are 0x11 and 0x13, respectively. The XON character is also called the device control 1 (DC1) character, whereas the XOFF character is also called the device control 3 (DC3) character.

Hardware flow control uses extra signals to control the flow of data. To implement hardware flow control on a PIC18 device, extra I/O pins must be used. Generally, an output pin is controlled by the receiving device to indicate that the transmitting device should suspend or resume transmissions. The transmitting device tests an input pin before a transmission to determine whether data can be sent.

9.4.4 C Library Functions for USART

The Microchip MCC18 compiler provides a list of extensive library functions for the USART modules. The library functions for the USART modules are listed in Table 9.3. The file **usart.h** should be included in your program before they can be called. The prototype declarations and functions of these functions are described next.

Function	Description
BusyUSART	Is the USART transmitting?
CloseUSART	Disable the USART
DataRdyUSART	Is data available in the USART read buffer?
getcUSART	Read a byte from USART
getsUSART	Read a string from USART
OpenUSART	Configure the USART
putcUSART	Write a byte to the USART
putsUSART	Write a string from data memory to the USART
putrsUSART	Write a string from program memory to the USART
ReadUSART	Read a byte from the USART
WriteUSART	Write a byte to the USART

Table 9.3a ■ Library functions for devices with only one USART

Function	Description
BusyxUSART	Is the USARTx transmitting?
ClosexUSART	Disable the USARTx
DataRdyxUSART	Is data available in the USARTx read buffer?
getcxUSART	Read a byte from USARTx
getsxUSART	Read a string from USARTx
OpenxUSART	Configure the USARTx
putcxUSART	Write a byte to the USARTx
putsxUSART	Write a string from data memory to the USARTx
putrsxUSART	Write a string from program memory to the USARTx
ReadxUSART	Read a byte from the USARTx
WritexUSART	Write a byte to the USARTx

Note. x = 1 or 2

Table 9.3b ■ Library functions for devices with multiple USARTs

The following functions return a "1" if the transmitter is currently busy. Otherwise, a "0" is returned. One of these functions should be used prior to commencing a new transmission:

char **BusyUSART** (void);	Used on devices with single USART
char **Busy1USART** (void);	Used on devices with two USARTs
char **Busy2USART** (void);	Used on devices with two USARTs

The following functions disable the transmitter and receiver and the interrupt of the specified USART channel:

void **CloseUSART** (void);	Used on devices with single USART
void **Close1USART** (void);	Used on devices with two USARTs
void **Close2USART** (void);	Used on devices with two USARTs

The following functions return a "1" if the RCIF flag is set. Otherwise, they return a "0":

char **DataRdyUSART** (void);	Used on devices with single USART
char **DataRdy1USART** (void);	Used on devices with two USARTs
char **DataRdy2USART** (void);	Used on devices with two USARTs

Any one of these functions read a byte from the receive buffer, including the ninth bit:

char **getcUSART** (void);	Used on devices with single USART
char **getc1USART** (void);	Used on devices with two USARTs
char **getc2USART** (void);	Used on devices with two USARTs
char **ReadUSART** (void);	Used on devices with single USART
char **Read1USART** (void);	Used on devices with two USARTs
char **Read2USART** (void);	Used on devices with two USARTs

The status bits and the ninth data bit are saved in a union with the following declaration:

```
union USART
{
        unsigned char val;
        struct
        {
                unsigned RX_NINE: 1;
                unsigned TX_NINE: 1;
                unsigned FRAME_ERROR: 1;
                unsigned OVERRUN_ERROR: 1;
                unsigned fill: 4;
        };
};
```

The ninth bit is read only if 9-bit mode is enabled. The status bits are always read.

The following functions read the specified number (**len**) of bytes from the specified USART module and saved in **buffer**:

void **getsUSART** (char *buffer, unsigned len);	Used on devices with single USART
char **gets1USART** (char *buffer, unsigned len);	Used on devices with two USART
char **gets2USART** (char *buffer, unsigned len);	Used on devices with two USART

The following functions write one character to the transmit buffer of the specified USART. If the 9-bit mode is enabled, the ninth bit is written from the field TX_NINE of the USART structure described above.

char **putcUSART** (char data);	Used on devices with single USART
char **putc1USART** (char data);	Used on devices with two USARTs
char **putc2USART** (char data);	Used on devices with two USARTs
char **WriteUSART** (char data);	Used on devices with single USART
char **Write1USART** (char data);	Used on devices with two USARTs
char **Write2USART** (char data);	Used on devices with two USARTs

The following functions output a NULL-terminated string in data memory pointed to by **data** to the specified USART module:

void **putsUSART** (char *data);	Used on devices with single USART
void **puts1USART** (char *data);	Used on devices with two USARTs
void **puts2USART** (char *data);	Used on devices with two USARTs

Any one of the following functions outputs a NULL-terminated string pointed to by **data** in program memory to the specified USART module:

void **putrsUSART** (const rom char *data);	Used on devices with single USART
void **putrs1USART** (const rom char *data);	Used on devices with two USARTs
void **putrs2USART** (const rom char *data);	Used on devices with two USARTs

The following functions configure the specified USART module to operate with the parameters and baud rate specified in the arguments **config** and **spbrg**:

void **OpenUSART** (unsigned char **config**, char **spbrg**);	Used on devices with single USART
void **Open1USART** (unsigned char **config**, char **spbrg**);	Used on devices with two USARTs
void **Open2USART** (unsigned char **config**, char **spbrg**);	Used on devices with two USARTs

The values of these parameters are defined in the file **usart.h** as follows:

Interrupt on Transmission:

USART_TX_INT_ON	Transmit interrupt ON
USART_TX_INT_OFF	Transmit interrupt OFF

Interrupt on Reception:

USART_RX_INT_ON	Receive interrupt ON
USART_RX_INT_OFF	Receive interrupt OFF

USART Mode:

USART_ASYNCH_MODE	Asynchronous mode
USART_SYNCH_MODE	Synchronous mode

Transmission Width:

USART_EIGHT_BIT	8-bit transmit/receive
USART_NINE_BIT	9-bit transmit/receive

Slave/Master Select:

USART_SYNC_SLAVE	Synchronous slave mode
USART_SYNC_MASTER	Synchronous master mode

Reception Mode:

USART_SINGLE_RX	Single reception
USART_CONT_RX	Continuous reception

Baud Rate: (applied to asynchronous mode only)

USART_BRGH_HIGH	High baud rate
USART_BRGH_LOW	Low baud rate

The parameter **spbrg** is the value to be written into the SPBRG register to determine the baud rate at which USART operates. The formulas for baud rate are the following:

Asynchronous mode, low speed:

$$F_{OSC}/(64 * (spbrg + 1))$$

Asynchronous mode, high speed:

$$F_{OSC}/(16 * (spbrg + 1))$$

Synchronous mode:

$$F_{OSC}/(4 * (spbrg + 1))$$

where F_{OSC} is the oscillator frequency.

The following statement will configure the USART1 to operate with the following features:

- Transmit interrupt turned off
- Receive interrupt turned off
- Asynchronous high-speed mode
- Continuous receive

Open1USART (USART_TX_INT_OFF & USART_RX_INT_OFF & USART_ASYNCH_
MODE & USART_EIGHT_BIT & USART_BRGH_HIGH, 103);

9.4.5 Interface Asynchronous Mode USART with EIA232

The USART in asynchronous mode is used mainly with the EIA232 interface. Because the USART module uses 0 V and 5 V to represent logic "0" and "1," respectively, it cannot be connected to the EIA232 interface circuit directly. A voltage translation circuit, which is called the **EIA232 transceiver,** is needed to translate the voltage levels of the USART signals (RX and TX) to and from those of the corresponding EIA232 signals.

EIA232 transceivers are available from many vendors. The LT1080/1081 from Linear Technology, the ST232 from SGS Thompson, the ICL232 from Intersil, the MAX232 from MAXIM, and the DS14C232 from National Semiconductor are EIA232 transceiver chips that can operate with a single 5-V power supply and generate EIA232-compatible outputs. These chips are also pin compatible with each other.

The MAX232A from MAXIM are discussed in this section. The pin assignment and the use of each pin are shown in Figure 9.12. Adding an EIA232 transceiver chip will then allow the PIC18 microcontroller to use the USART module to interface with the EIA232 interface circuit. The circuit shown in Figure 9.13 is used in SSE452, SSE8680, and SSE8720 demo boards. A null-modem wiring is followed to allow the use of a straight-through cable to communicate with a PC. Demo boards from Microchip also use the same wiring method but with the CTS and RTS pins floating.

If a demo board needs only to communicate with a PC, then the user needs only to connect the RC6/TX and RC7/RX pins with pin 2 and pin 3 of the DB9 connector via the MAX232A, respectively. Pin 5 of the DB9 connector needs to be grounded.

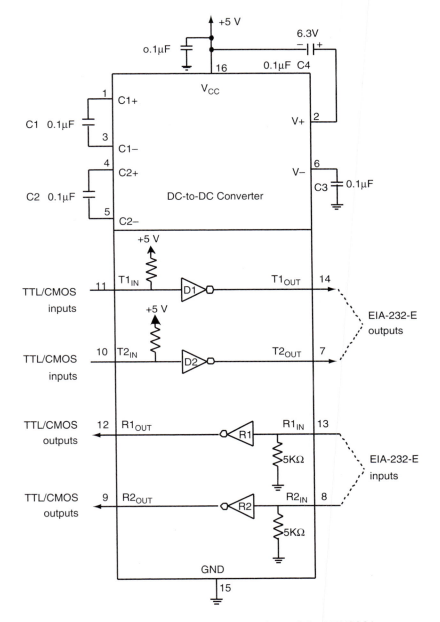

Figure 9.12 ■ Pin assignments and connections of the MAX232A

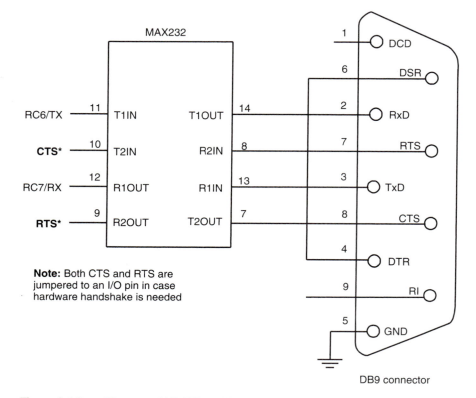

Figure 9.13 ■ Diagram of USART and EIA232 DB9 connector wiring in SSE452, SSE8680, and SSE8720 demo boards

9.4.6 USART Synchronous Master Mode

Synchronous master mode is entered by setting bits SYNC, SPEN, and CSRC. In synchronous master mode, data is transmitted in a half-duplex manner (i.e., transmission and reception do not occur at the same time). The clock signal is driven out of the CK pin in this mode.

SYNCHRONOUS MASTER TRANSMISSION

Before enabling synchronous transmission, one must set the TXEN bit, which will start the SPBRG circuit to generate the shift clock. However, the actual data transmission will not begin until data is written into the TXREG register and loaded into the TSR register. The first data bit will be shifted out on the rising edge of the CK clock and becomes stable on the falling edge of the same clock period. The transmission-timing diagram is shown in Figure 9.14.

SYNCHRONOUS MASTER RECEPTION

Once synchronous mode is selected, reception is enabled by setting either the enable bit SREN (single reception) or the enable bit CREN (continuous reception). Data is sampled on the falling edge of the CK clock. If the SREN bit is set, only a single byte is received. If the CREN bit is set, the reception is continuous until the CREN bit is cleared. If both bits are set, then the CREN bit takes precedence. The timing diagram of synchronous master reception (single reception) is shown in Figure 9.15.

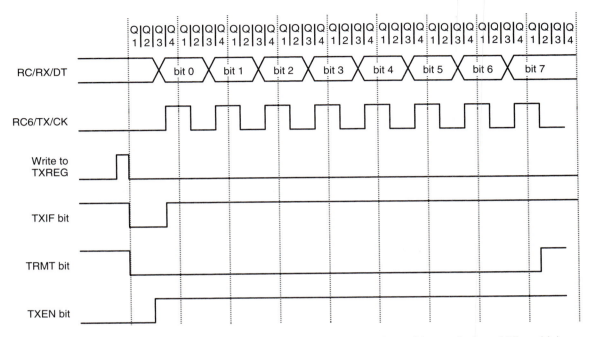

Figure 9.14 ■ USART transmission in synchronous master mode (redraw with permission of Microchip)

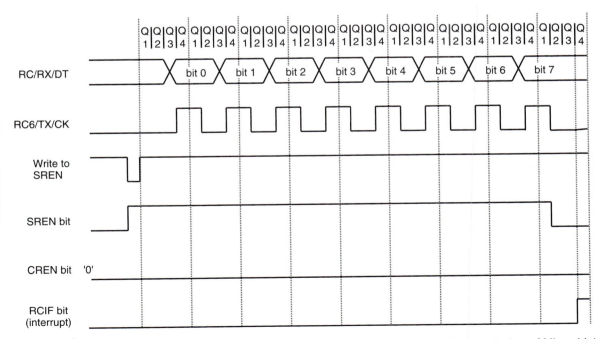

Figure 9.15 ■ USART reception in synchronous master mode (SREN=1) (redraw with permission of Microchip)

9.4.7 USART Synchronous Slave Mode

The USART can also be made to operate in slave mode. The slave mode differs from the master mode in the fact that the shift clock is provided externally at the TX/CK pin. The slave mode is entered by setting both the SYNC and the SPEN bits and clearing the CSRC bit.

SYNCHRONOUS SLAVE TRANSMISSION

The operation in this mode is almost identical to the master mode except that the TX/CK pin is an input and a clock input is expected. The procedure for setting up synchronous slave transmission is as follows:

1. Enable the synchronous slave transmission by setting both the SYNC and the SPEN bits and clearing the CSRC bit.
2. Clear the CREN bit.
3. Enable interrupts by setting the TXIE bit if interrupt is desired. If 9-bit transmission is to be used, then set the TX9 bit in the TXSTA register and copy the ninth bit into the TX9D bit of the TXSTA register.
4. Start transmission by loading data into the TXREG register.
5. Enable transmission by setting the TXEN bit.

SYNCHRONOUS SLAVE RECEPTION

The operation of this mode is almost identical to the master mode reception. The difference between the two modes of operation is the clock source. In the slave mode, the clock signal (CK) is an input rather than an output. The steps for setting up the synchronous slave reception are as follows:

1. Enable the synchronous slave mode by setting both the SYNC and the SPEN bits and clearing the CSRC bit.
2. Enable interrupts by setting the RCIE bit if interrupts are desirable. If 9-bit reception is to be used, set the RX9 bit of the RCSTA register. Set the CREN bit to enable reception.
3. When the RCIF bit is set, read the RCSTA to get the ninth bit (if 9-bit reception is enabled) and determine if any error has occurred. Read the received data from the RCREG register.
4. If any errors occurred, then clear the error by clearing the CREN bit.

9.4.8 Applications of USART Synchronous Mode

One of the applications for the synchronous mode of the USART is to expand the number of I/O ports. Since the USART module transmits data by least significant bit first, the user must pay attention to match the bit order. Many shift registers can be used to expand the I/O ports for the PIC18 devices. For example, the 74LS165 and 74HC589 are parallel-to-serial shift registers that can be used to add input ports, whereas the 74LS164 and 74HC595 are serial-to-parallel shift registers that can be used to add output ports to the PIC18 devices.

The pin assignment diagram and truth table for the 74LS165 are shown in Figure 9.16 and Table 9.4, respectively. According to the truth table, the 74LS165 will shift in data from the DS pin on the falling edge of either the CP1 or the CP2 clock signal while keeping the other clock signal low and the PL signal high. The whole shift register contents will be shifted to the right (from Q0 to Q7) during this shifting process. The highest shift frequency for the 74LS165 is 35 MHz.

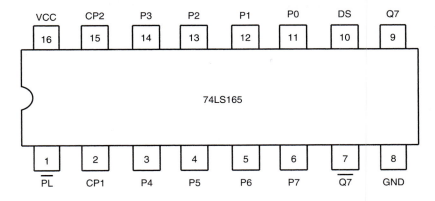

Figure 9.16 ■ 74LS165 pin assignment

	CP		**Contents**								
PL	**1**	**2**	**Q0**	**Q1**	**Q2**	**Q3**	**Q4**	**Q5**	**Q6**	**Q7**	**Response**
L	X	X	P0	P1	P2	P3	P4	P5	P6	P7	Parallel load
H	L	↑	DS	Q0	Q1	Q2	Q3	Q4	Q5	Q6	rightshift
H	H	↑	Q0	Q1	Q2	Q3	Q4	Q5	Q6	Q7	no change
H	↑	L	DS	Q0	Q1	Q2	Q3	Q4	Q5	Q6	rightshift
H	↑	H	Q0	Q1	Q2	Q3	Q4	Q5	Q6	Q7	no change

Table 9.4 ■ Truth table of 74LS165

Example 9.9

▼

Show the circuit connection for using one 74LS165 to add an input port to the PIC18F8720 using the USART2 in synchronous master mode. Write a subroutine to configure the USART2 properly and a subroutine to shift one byte of data, assuming that the PIC18F8720 is running with a 16-MHz crystal oscillator. Assume that the user uses eight switches to set the value to be input and use INT1 interrupt to inform the arrival of data.

Solution: The circuit connection between the PIC18F8720 and the 74LS165 is shown in Figure 9.17 where the following occur:

- PL is controlled by the port pin RB4.
- CP1 is grounded.
- CP2 is connected to RG1/CK2 pin.
- Q7 is connected to RG2/DT2.
- DS is grounded.

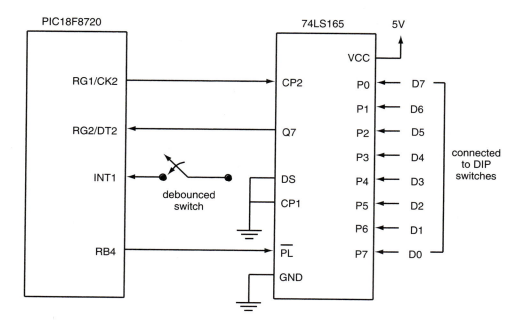

Figure 9.17 ■ Circuit connection between the PIC18F8720 and 74LS165

The USART2 module should be configured to operate with the following settings:
- Synchronous master mode for reception of 8-bit data
- TRISG<2> set to 1 (configure RG2/DT2 for input)
- TRISG<1> set to 0 (configure RG1/CK2 for output)
- Baud rate set to 1 Mbps
- Interrupt enabled

One possible interaction between the user and the 74LS165 is as follows:

1. The user sets up a new value to P0 through P7 pins using the DIP switches.
2. The user presses the debounced switch to request an interrupt from the INT1 pin.
3. The INT1 interrupt service routine makes sure that interrupt is from the INT1 pin and then enables the reception, waits until the RC2IF flag is set, and then reads in the data from the RCREG2 register.

The subroutine for configuring the USART2 module in synchronous mode and the interrupt service routine in assembly language are as follows:

```
           #include    <p18F8720.inc>
rcv_buf    set         0                  ; buffer to hold the received byte
           org         0x00
           goto        start
           org         0x08
           goto        hi_ISR
           org         0x18
           retfie
start      bcf         TRISG,RG1,A        ; configure CK2 pin for output
```

```
                bsf         TRISG,RG2,A        ; configure DT2 pin for input
                call        open_usart2,0
                call        open_INT1
forever         nop
                bra         forever
; ***************************************************
; The following function configures USART2 to operate in
; synchronous master mode with baud rate set to 10**6.
; ***************************************************
open_usart2
                movlw       0x03
                movwf       SPBRG2             ; set the shift rate to 1MHz
                movlw       0x90               ; set synchronous master mode
                movwf       TXSTA2             ; "
                movlw       0x80               ; enable USART2 and disable
                movwf       RCSTA2             ; reception
                return
; ***************************************************
; The following subroutine configures INT1 to high priority and
; then enable its interrupt.
; ***************************************************
open_INT1  bsf          RCON,IPEN          ; enable priority interrupt
                movlw       0x48               ; set INT1 interrupt to high priority,
                movwf       INTCON3            ; and then enable INT1 interrupt
                movlw       0xC0               ; enable global interrupt
                movwf       INTCON             ; "
                return
; ***************************************************
; The high priority interrupt service routine makes sure
; that the interrupt is caused by usart2 INT1IF and then enables
; reception by setting SPEN bit. Wait until a byte is received.
; ***************************************************
hi_ISR     btfss        INTCON3,INT1IF     ; is interrupt caused by INT1
                retfie                         ; interrupt is not caused by INT1
                bcf         INTCON3,INT1IF     ; clear INT1IF
                bcf         PORTB,RB4          ; load data into 74LS165
                nop                            ; "
                bsf         PORTB,RB4          ; disable new data into 74LS165
                bcf         PIR3,RC2IF         ;
                bsf         RCSTA2,SREN        ; enable single reception from USART2
wait            btfss       PIR3,RC2IF         ; wait for the arrival of a byte
                bra         wait
                movff       RCREG2,rcv_buf
                retfie
                end
```

The C language version of the program is straightforward and hence is left for you as an exercise problem.

The pin assignment diagram and the truth table for the 74LS164 are shown in Figure 9.18 and Table 9.5, respectively.

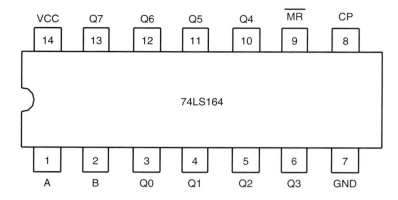

Figure 9.18 ■ 74LS164 pin assignment

Operating mode	Inputs				Outputs	
	$\overline{MR}$	CP	A	B	Q0	Q1–Q7
Reset (clear)	L	X	X	X	L	L–L
Shift	H	↑	L	L	L	Q0–Q6
	H	↑	L	H	L	Q0–Q6
	H	↑	H	L	L	Q0–Q6
	H	↑	H	H	H	Q0–Q6

Table 9.5 ■ Truth table of 74LS164

Example 9.10

▼

Show the circuit connection if you are to use one 74LS164 to add an output port to the PIC18F8720 using the USART2 module in synchronous master mode. Write a subroutine to shift out one byte of data, assuming that the PIC18F8720 is running with a 16-MHz crystal oscillator. The user may uses Q7-Q0 to drive LEDs.

Solution: The circuit connection between the PIC18F8720 and the 74LS164 is shown in Figure 9.19, where the following occur:

- B input is pulled up to high so that the input A can be used to receive data from the PIC18F8720.
- The $\overline{MR}$ pin is pulled up to high to disable reset.
- The CP pin is connected to the RG1/CK2 pin to receive the clock signal.
- The Q7 through Q0 pins can be used to drive LEDs or other output devices. Because the USART2 transmits data by least significant bit first, the user must use the data on Q0 through Q7 as the most significant to least significant bits.

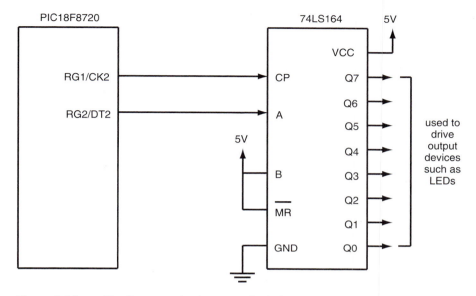

Figure 9.19 ■ Circuit connection between the PIC18F8720 and 74LS164

The following subroutine outputs the character in WREG to the USART2 using the polling method:

```
outc_usart2   bsf     RCSTA2,SPEN    ; enable USART port pins
poll2         btfss   PIR3,TX2IF     ; is the TX2IF flag set?
              bra     poll2
              movwf   TXREG2
              return
```

The USART synchronous mode can also be used to exchange data between two microcontrollers. Figure 9.20 illustrates the circuit connection. In Figure 9.20, one PIC18 microcontroller is configured as the master, whereas the other PIC18 microcontroller is configured as a slave.

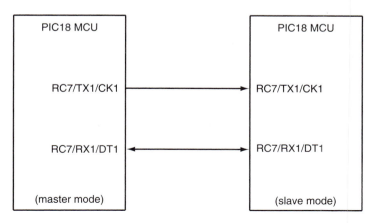

Figure 9.20 ■ Two PIC18s exchange data using USART synchronous mode

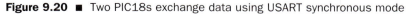

9.5 Enhanced USART

Newer PIC18 devices (e.g., PIC18F8680/8621/8625/8585/6680/6621/6585/6525) also implemented some enhancements to the original USART design. The enhanced USART adds the following features to the original USART module:

- Automatic baud rate detection and calibration
- Sync break reception
- 12-bit break character transmit

A baud rate control register (BAUDCONx, x = 1 or 2) is added for additional baud rate control. An additional high byte of the baud rate generator is added. These two baud rate generator registers can be configured in either 8-bit or 16-bit mode. These additional features are ideal for supporting the LIN bus systems that are becoming more widely used in Europe today. Since the LIN protocol is not discussed in this book, these features are not explored any further. The contents of the BAUDCONx register are shown in Figure 9.21.

An additional baud rate generator (SPBRGHx, x = 1 or 2) is added to provide the user with the option of choosing either the 16-bit or the 8-bit baud rate generation method.

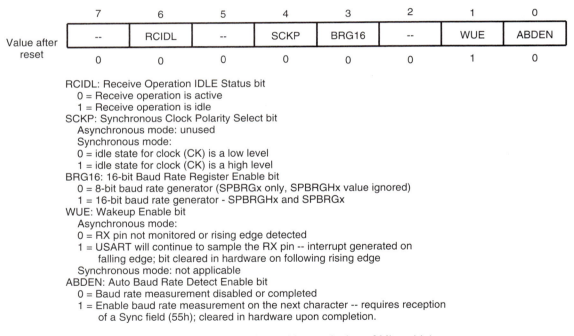

Value after reset

7	6	5	4	3	2	1	0
--	RCIDL	--	SCKP	BRG16	--	WUE	ABDEN
0	0	0	0	0	0	1	0

RCIDL: Receive Operation IDLE Status bit
   0 = Receive operation is active
   1 = Receive operation is idle
SCKP: Synchronous Clock Polarity Select bit
   Asynchronous mode: unused
   Synchronous mode:
   0 = idle state for clock (CK) is a low level
   1 = idle state for clock (CK) is a high level
BRG16: 16-bit Baud Rate Register Enable bit
   0 = 8-bit baud rate generator (SPBRGx only, SPBRGHx value ignored)
   1 = 16-bit baud rate generator - SPBRGHx and SPBRGx
WUE: Wakeup Enable bit
   Asynchronous mode:
   0 = RX pin not monitored or rising edge detected
   1 = USART will continue to sample the RX pin -- interrupt generated on
      falling edge; bit cleared in hardware on following rising edge
   Synchronous mode: not applicable
ABDEN: Auto Baud Rate Detect Enable bit
   0 = Baud rate measurement disabled or completed
   1 = Enable baud rate measurement on the next character -- requires reception
      of a Sync field (55h); cleared in hardware upon completion.

Figure 9.21 ■ Baud Rate Control Register (redraw with permission of Microchip)

9.6 Summary

The EIA232 is an asynchronous serial communication standard widely used in data communications. This standard was introduced in 1960 and has gone through several revisions. The latest revision was the E revision made in 1991.

The EIA232 standard has four aspects:

1. An electrical aspect that specifies the voltage levels, rise time and fall time of each signal, achievable data rate, and the distance of communication

2. A functional aspect that specifies the function of each signal

3. A mechanical aspect that specifies the number of pins and the shape and dimensions of the connectors

4. A procedural aspect that specifies the sequence of events for transmitting data based on the functional specifications of the interface

The EIA232 standard requires the use of modems to interconnect two DTEs. Computer and terminals are examples of DTEs. A modem is a type of DCE that can modulate the data onto a carrier signal before the transmission and can also recover the original signal from the modulated signals. In a short distance, two DTEs can be wired together without using modems by using the null-modem connection for the EIA232 interface. The EIA232 standard is often used to interface a demo board to a PC using the null-modem connection.

In the EIA232 standard, data is transferred character by character, and each character is transmitted using the following format:

- One start bit

- Eight or nine data bits

- One to two stop bits

Data communications may experience the following errors:

- Framing error

- Receiver overrun error

- Parity error

The PIC18 device may have one or two USART modules. The USART module can operate in asynchronous and synchronous modes. The asynchronous mode is intended to work with the EIA232 interface, whereas the synchronous mode can be used to add I/O ports to the PIC18 devices or perform data communications between multiple microcontrollers.

In the asynchronous mode, the PIC18 device use a clock signal with a frequency equal to either 16 or 64 times the baud rate to sample the incoming data signal to detect the arrival of a start bit and determine the logic value of a data bit. Three samples (seventh, eighth, and ninth of 16 samples) are taken, and the majority function of them determines whether the start bit has arrived. The logic value of a data bit is also determined by the same majority circuit.

The voltage levels used in the EIA232 standard are not compatible with the TTL levels. To interface with the EIA232 interface, the USART port of the PIC18 device needs a transceiver to perform the voltage level translation. Many semiconductor manufacturers produce the EIA232 transceiver chip. Most of them are pin compatible.

Shift registers are often used to add I/O ports to a microcontroller. The parallel-to-serial shift registers, such as 74LS165 and 74HC589, can be used to add input ports to the PIC18 microcontroller, whereas the serial-to-parallel shift registers 74LS164 and 74HC595 can be used to add output ports to the PIC18 devices. The USART in synchronous master mode can be used to carry out the data exchange between the PIC18 and these shift registers.

9.7 Exercises

E9.1 Sketch the output of the digit 9 and letter m when they are transmitted via the USART interface using the format of one start bit, eight data bits, and one stop bit.

E9.2 Write a subroutine to configure the USART1 and the USART2 to operate with the following settings:

- 19200 baud (f_{OSC} = 32 MHz)
- One start bit, eight data bits, one stop bit
- Both transmit and receive interrupts enabled
- Asynchronous mode
- Both transmit and receive enabled

E9.3 Write a subroutine to configure the USART1 and the USART2 to operate with the following settings:

- 9600 baud (f_{OSC} = 32 MHz)
- One start bit, nine data bits, one stop bit
- High baud rate, no address detect or interrupt
- Transmit and receive enabled
- Asynchronous mode

E9.4 Write a subroutine to configure the USART1 and the USART2 to operate with the following settings:

- Asynchronous mode
- One start bit, nine data bit, and one stop bit
- Address detect and interrupt enabled
- 19200 baud rate, high baud rate option
- Transmit and receive enabled

E9.5 Write a subroutine that will output the contents of the WREG register as two hex digits to the USART1. Use the asynchronous mode.

E9.6 Write a subroutine that will read two hex digits from the USART1 and echo them back to the USART1. Use the asynchronous mode.

E9.7 Write a subroutine to output the character in WREG to the USART1 using the synchronous slave mode.

E9.8 Write a subroutine to read a character from the USART1 and return the character in WREG using the synchronous slave mode.

E9.9 Multiple 74LS165s can be cascaded and driven by one USART. Draw a circuit to illustrate how to add two input ports to the PIC18 microcontroller using the USART synchronous mode. Write a subroutine to read 16-bit values from these two 74LS165s using the interrupt-driven method similar to that described in Example 9.9.

E9.10 Multiple 74LS164s (or 74HC164) can be cascaded and driven by one USART module. Draw a circuit to illustrate how to add two output ports to the PIC18 microcontroller using the USART1 synchronous mode. Write a subroutine to output a 16-bit values to these two 74LS164 using the method similar to that described in Example 9.10.

9.8 Lab Exercises and Assignments

L9.1 Use the circuit in Figure 9.19 to drive eight LEDs. Configure the USART1 to operate in synchronous master mode. Display the LEDs in the following manner:

1. Light all LEDs for a quarter second and off for a quarter second. Repeat this pattern four times.
2. Light one LED at a time for a quarter second—from the LED driven by Q7 to the LED driven by Q0. Repeat this pattern four times.
3. Light one LED at a time for a quarter second—from the LED driven by Q0 to the LED driven by Q7. Repeat this pattern four times.
4. Light the LEDs driven by Q7 and Q0 for a quarter second and then off for a quarter second. Repeat this pattern four times.
5. Light the LEDs driven by Q6 and Q1 for a quarter second and then off for a quarter second. Repeat this pattern four times.
6. Light the LEDs driven by Q5 and Q2 for a quarter second and then off for a quarter second. Repeat this pattern four times.
7. Light the LEDs driven by Q4 and Q3 for a quarter second and then off for a quarter second. Repeat this pattern four times.
8. Go to Step 1 and repeat forever.

L9.2 Output a message to the PC monitor display by following these steps:

Step 1
Run a terminal program such as HyperTerminal.

Step 2
Use an EIA232 cable to connect the USART1 port to the PC COM port.

Step 3
Write a short program to configure USART1 to operate in asynchronous mode with 8-bit data and output the following message on the HyperTerminal:

Today is Monday.
It is 1:20:50

L9.3 Write a program to read one character at a time from the USART port and output it to the I/O port connected to LEDs. Start a terminal program such as HyperTerminal and set the baud rate to be identical to that of the demo board. Type in one character at a time and see the change on the LED light. When testing this program, disconnect the ICD2 from the demo board.

10

Serial Peripheral Interface

10.1 Objectives

After completing this chapter, you should be able to

- Describe the operation of the SPI operation

- Interface an SPI master with one or multiple SPI slaves in general

- Use SPI to interface with one or multiple shift register 74HC595s

- Use SPI to interface with a digital temperature sensor TC72

- Use SPI to interface with one or multiple seven-segment display driver MAX7221s to display multiple digits

- Use SPI to interface with the alarm real-time clock DS1306 to keep track of time of day and set alarm times

10.2 Introduction

The master synchronous serial port (MSSP) module of the PIC18 microcontroller is a serial interface useful for communicating with other peripheral or microcontroller devices. These devices may be serial EEPROMs, shift registers, display drivers, A/D converters, D/A converters, digital temperature sensors, time-of-day chips, and so on. The MSSP port requires that both the transmitter and the receiver share a common clock signal. The common clock signal determines when to send and receive the data bits. Because of this common clock, the clock frequency used in data exchange need not be held constant or agreed on before the data exchange is started.

The MSSP can operate in one of two modes:

1. Serial peripheral interface (SPI)

2. Inter-integrated circuit (I^2C)

Both the SPI and I^2C are synchronous serial interfacing protocols. The SPI protocol was proposed by Motorola as a standard interfacing method to simplify the interfacing of peripheral devices to Motorola microcontrollers. Because of its ease of use and flexibility, SPI has become an industry standard. Microcontrollers from most companies provide this interface method. Many semiconductor manufacturers produce peripheral chips with the SPI interface to work with microcontrollers from many vendors.

I^2C is a two-wire bus developed and patented by Philips. In a system that incorporates the I^2C bus, peripheral devices are addressable, data transfers are acknowledged, and data transfers can be proceeded in three levels of data rates:

1. Normal mode: 100 Kbps

2. Fast mode: 400 Kbps

3. High-speed mode: 3.4 Mbps

I^2C is discussed in Chapter 11.

In a system that uses the SPI protocol, one device (must be an intelligent device, such as a microcontroller) serves as the *master,* and other devices are configured as *slaves.* The master is responsible for generating the clock signal to synchronize data transfer. Peripheral devices can only be slaves, whereas a microcontroller can be configured as the master or a slave.

In a system that implements the I^2C interface, peripheral devices are slaves and will be assigned with addresses. A microcontroller can be configured as a master or a slave. Unlike the SPI protocol, I^2C allows multiple masters to coexist.

The SPI and I^2C modes share the same signal pins, preventing these two protocols from being active at the same time. The following three pins are used by the MSSP module:

- Serial data out (SDO)—RC5/SDO
- Serial data in (SDI)—RC4/SDI/SDA
- Serial clock (SCK)—RC_3/SCK/SCL

Additionally, a fourth pin may be used when in a slave mode operation:

- Slave select ($\overline{SS}$)—RF7/$\overline{SS}$

The slave select signal is active low. In Motorola's literature, the signal SDO is called master-out-slave-in (MOSI) and hence is an output pin for a master device. The signal SDI is called master-in-slave-out (MISO) and is an input pin for a master device.

10.3 SPI Mode

The SPI mode allows eight bits of data to be synchronously transmitted and received simultaneously. In slave mode, all four MSSP signal pins are used. In master mode, the $\overline{SS}$ pin is not needed.

10.3.1 MSSP Registers

The MSSP module has four registers for SPI mode operation:

- MSSP control register 1 (SSPCON1)
- MSSP status register (SSPSTAT)
- Serial receive/transmit buffer (SSPBUF)
- MSSP shift register (SSPSR)—not directly accessible by the programmer

SSPCON1 and SSPSTAT are the control and status registers in SPI mode operation. The SSP-CON1 register is readable and writable. The lower six bits of the SSPSTAT register are read only. The upper two bits of the SSPSTAT are read/writable.

SSPSR is the shift register used for shifting data in or out. SSPBUF is the buffer register to which data bytes are written and read from. In receive operations, the SSPSR register and the SSPBUF register together create a double-buffered receiver. When the SSPSR register receives a complete byte, it is transferred to the SSPBUF register, and the SSPIF flag bit is set to 1. During transmission, the SSPBUF register is not double buffered. A write to the SSPBUF register will write to both the SSPBUF register and the SSPSR register.

The contents of the SSPSTAT register in the SPI mode are shown in Figure 10.1. The contents of the SSPCON1 register in the SPI mode are shown in Figure 10.2.

	7	6	5	4	3	2	1	0
value after reset	SMP	CKE	$D/\overline{A}$	P	S	$R/\overline{W}$	UA	BF
	0	0	0	0	0	0	0	0

SMP: Sample bit

SPI master mode:

  0 = Input data sampled at middle of data output time

  1 = Input sampled at the end of data output time

SPI slave mode:

  SMP must be cleared when SPI is used in slave mode

CKE: SPI clock edge select bit

When CKP = 0

  0 = Data transmitted on falling edge of SCK

  1 = Data transmitted on rising edge of SCK

When CKP = 1

  0 = Data transmitted on rising edge of SCK

  1 = Data transmitted on falling edge of SCK

$D/\overline{A}$: Data/Address bit

Used in I2C only

P: Stop bit

Used in I2C only

S: Start bit

Used in I2C only

$R/\overline{W}$: Read/Write bit information

Used in I2C only

UA: Update Address bit

Used in I2C only

BF: Buffer Full Status bit (Receive mode only)

  0 = Receive not complete, SSPBUF is empty

  1 = Receive complete, SSPBUF is full

Figure 10.1 ■ SSPSTAT Register **(SPI mode)** (redraw with permission of Microchip)

	7	6	5	4	3	2	1	0
value after reset	WCOL	SSPOV	SSPEN	CKP	SSPM3	SSPM2	SSPM1	SSPM0
	0	0	0	0	0	0	0	0

WCOL: Write collision detect bit (Transmit mode only)
  0 = No collision
  1 = The SSPBUF register is written while it is still transmitting the previous byte
SSPOV: Receive overflow indicator bit
  SPI slave mode:
  0 = No overflow
  1 = A new byte is received while the SSPBUF register is still holding the previous
      data. In case of overflow, the data in SSPSR is lost. Overflow can only occur
      in slave mode. The user must read the SSPBUF, even if only transmitting
      data, to avoid setting overflow (must cleared in software).
  SPI master mode: This bit is not set.
SSPEN: SPI enable bit
  0 = Disables serial port and configures SCK, SDO, SDI, and $\overline{SS}$ as input or output
  1 = Enables these pins as serial port pins
CKP: Clock polarity select bit
  0 = IDLE state for clock is low level
  1 = IDLE state for clock is high level
SSPM3:SSPM0: Synchronous serial port mode select bits
  0000 = SPI master mode, clock = $F_{OSC}/4$
  0001 = SPI master mode, clock = $F_{OSC}/16$
  0010 = SPI master mode, clock = $F_{OSC}/64$
  0011 = SPI master mode, clock = TMR2 output/2
  0101 = SPI slave mode, clock = SCK pin, $\overline{SS}$ pin control disabled, $\overline{SS}$ can be I/O
  0100 = SPI slave mode, clock = SCK pin, $\overline{SS}$ pin control enabled

Figure 10.2 ■ SSPCON1 Register **(SPI mode)** (redraw with permission of Microchip)

Only bit 7 and bit 6 of the SSPSTAT register are related to the SPI configuration. Bit 7 allows the user to select *input data sample time.* Input data sample time can be in the middle (SMP = 0) or the end (SMP = 1) of data output time. Bit 6, together with the CKP bit of the SSP-CON1 register, allows the user to select the clock edge for data shifting. Bit 0 is a flag bit that will be set to 1 when an SPI transfer is complete. The BF bit can be cleared by reading the SSP-BUF register. The SSPCON1 register allows the user to enable/disable the SPI module, select idle state for the SCK signal, and select the mode and data shift rate of the SPI module.

Because data is transmitted bit serially, it takes a few instruction cycles to shift out a byte. The user may attempt to write another byte into the SSPBUF register before the current byte has been shifted out. This situation is called the *write collision error* and is flagged in bit 7 of the SSPCON1 register.

SPI is double buffered during a reception. One byte can be held in the SSPBUF register when a byte is shifted in the SSPSR register. If the CPU is too busy to fetch the byte in the SSPBUF register and the second byte is shifted in, an SPI *reception overflow error* (SSPOV) occurs. The receive overflow error is flagged in bit 6 of the SSPCON1 register. The user can choose from four possible data rates in the SPI master mode by programming the lower four bits of the SSPCON1 register.

10.3.2 SPI Operation

A simplified block diagram of SPI connection between a master and a slave is shown in Figure 10.3.

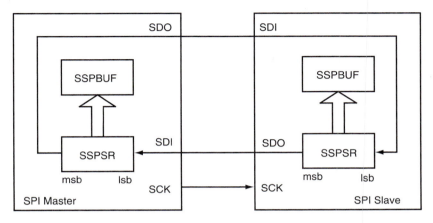

Figure 10.3 ■ Connection between an SPI master and an SPI slave

When making the connection, the SDI pin of one device should be connected to the SDO pin of another SPI device, whereas the SDO pin of one device should be connected to the SDI pin of another device.

The SCK signal is driven by the SPI master device. The SCK signal controls the data shifting in an SPI transfer.

To send data from the SPI master to an SPI slave, the master device writes a byte into the SSPBUF register. This byte will also be written into the SSPSR register. When a byte is written into the SSPBUF register, eight clock pulses will be triggered and sent out from the SCK pin, which will swap the contents of the master SSPSR register and that of the slave SSPSR register in Figure 10.3. At the end of the data shift operation, the SSPIF flag of the PIR1 register and the BF flag of the SSPSTAT register will be set.

To read data, the master device will perform the same operation, that is, write a byte into the SSPBUF register to trigger eight clock pulses to be sent out from the SCK pin. However, the value to be written into the SSPBUF does not matter for this purpose.

10.3.3 Configuration of MSSP for SPI Mode

In order to make sure that data transfer is carried out successfully, the following parameters must be set up properly:

DATA SHIFT RATE

In master mode, the SPI clock rate (bit rate) is user programmable to be one of the following:

- $F_{OSC}/4$ (or T_{CY})
- $F_{OSC}/16$ (or $4 \bullet T_{CY}$)
- $F_{OSC}/64$ (or $16 \bullet T_{CY}$)
- Timer2 output/2

The data shift rate is configured by programming the lowest four bits of the SSPCON1 register. The highest data rate for the 40-MHz crystal is 10 Mbps.

CLOCK EDGE FOR SHIFTING DATA

The user can set the CLK signal to be *idle high* (when SPI is not transmitting data) or *idle low*. This is done by setting or clearing the CKP bit of the SSPCON1 register. Depending on the requirement of the application, the user may choose to shift data on the rising or the falling edge of the CLK signal. The CKE bit of the SSPCON1 register allows the user to make the clock edge selection. The actual clock edge for shifting data is determined by the combination of the CKP and the CKE signals. The available clock formats are shown in Figures 10.4a, 10.4b, and 10.4c. The idle low clock state is used more often. In the slave mode, the $\overline{SS}$ signal must be asserted (low). Otherwise, no data shifting is possible.

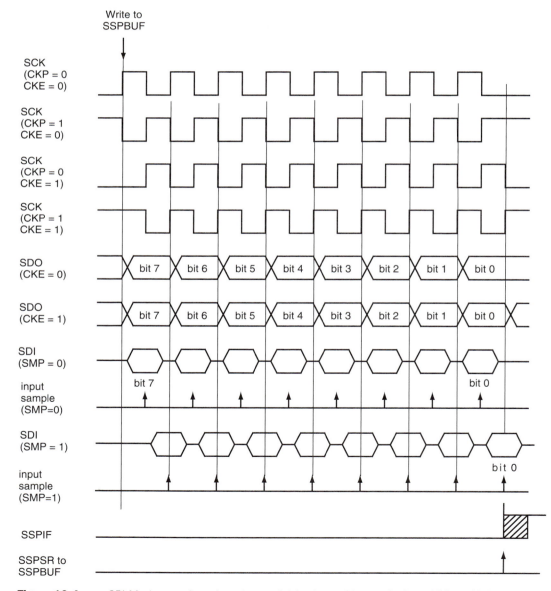

Figure 10.4a ■ SPI Mode waveform (master mode) (redraw with permission of Microchip)

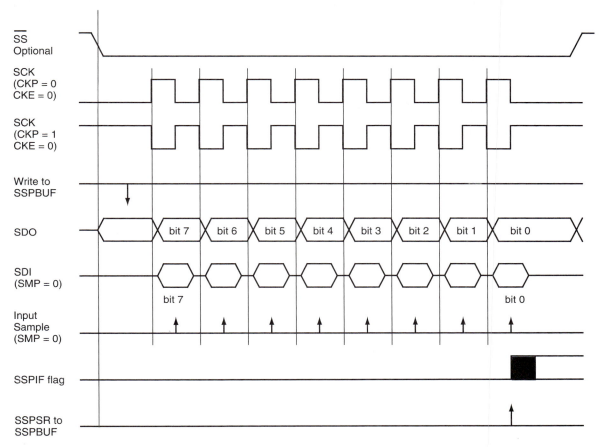

Figure 10.4b ■ SPI clock formats (slave mode with CKE = 0) (redraw with permission of Microchip)

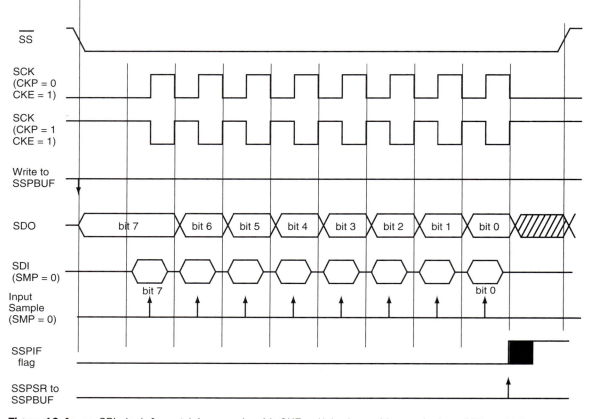

Figure 10.4c ■ SPI clock format (slave mode with CKE = 1) (redraw with permission of Microchip)

Example 10.1

Write an instruction sequence to configure the MSSP in the SPI master mode to shift data on the rising edge. The SCK signal must be idle low. The data rate should be set to at least 2 Mbps, assuming that f_{OSC} is 32 MHz.

Solution: The setting of the SSPCON1 register should be as follows:

- Bits 7 and 6 are status bits and should be cleared to 00.
- Bit 5 should be set to 1 to enable SPI port pins.
- Bit 4 should be set to 0 to select idle low.
- Bits 3 through 0 should be set to either 0000 or 0001 to set data rate to at least 2 Mbps and also select SPI master mode.

The setting of the SSPSTAT register should be as follows:

- Bit 7 should be set to 0 so that input data is sampled at the middle of data output time.
- Bit 6 should be set to 1 so that data can be transmitted on rising edge of SCK.
- Bits 5 through 0 are don't-cares and should be set to 0s.

Both the RC5/SDO and the RC3/SCK pin should be configured for output, whereas the RC4/SDI pin should be configured for input.

The following instruction sequence will configure the MSSP to SPI mode accordingly:

```
movlw   0x21          ; enable SPI master mode, select idle
movwf   SSPCON1,A     ; low and set data rate to 2 Mbps
movlw   0x40          ; select rising edge to shift out data and
movwf   SSPSTAT,A     ; sample input data in the middle of a bit time
bsf     TRISC,SDI     ; configure SDI pin for input
bcf     TRISC,SDO     ; configure SDO pin for output
bcf     TRISC,SCK     ; configure SCK pin for output
```

10.3.4 SPI Data Transfer

Because an SPI transfer takes quite a few clock cycles to complete, it is a good idea to wait until SPI transfer is complete before moving on to the next operation. When the SPI transfer is complete, the BF flag of the SSPSTAT register will be set to 1.

The following instruction sequence will send the contents of the WREG register to the SPI module and also clear the BF flag after the SPI transfer:

```
        movwf   SSPBUF,A      ; write the contents of WREG to SSPBUF
loop1   btfss   SSPSTAT,BF    ; wait until the SPI transfer is complete
        bra     loop1
        movf    SSPBUF,W      ; clear the BF flag
```

The following instruction sequence will read a byte into the WREG register from the SPI module:

```
        movwf   SSPBUF        ; trigger 8 SCK pulses to shift in data
loop2   btfss   SSPSTAT,BF    ; wait until SPI transfer is complete
        bra     loop2
        movf    SSPBUF,W      ; copy data from SSPBUF and also clear BF
```

10.3.5 SPI Circuit Connection

MSSP is often configured in the SPI master mode to interface with peripheral devices such as EEPROMs, A/D and D/A converters, display drivers, and many other special-function chips to the PIC18 microcontroller. The circuit connection for one master and one slave has been illustrated in Figure 10.3. The MSSP module in the SPI mode also allows the PIC18 microcontroller to interface with multiple peripheral devices. There could be many connection methods for a multiple slave environment. One possibility is shown in Figure 10.5.

For the connection method shown in Figure 10.5, the SDO pin of the PIC18 device is connected to the MOSI pin of all slave devices, whereas the SDI pin of the PIC18 device is connected to the MISO pin of all slave devices. The SCK pin is also connected to the SCK pin of all slaves. The PIC18 device uses a few unused I/O pins to control the $\overline{SS}$ input of slave devices. A slave device will be isolated from the connection whenever the $\overline{SS}$ input is high. Using this method, the PIC18 can send data to any selected device without affecting other slave devices. This method may not be appropriate when I/O pins are limited because it requires one pin for each slave device.

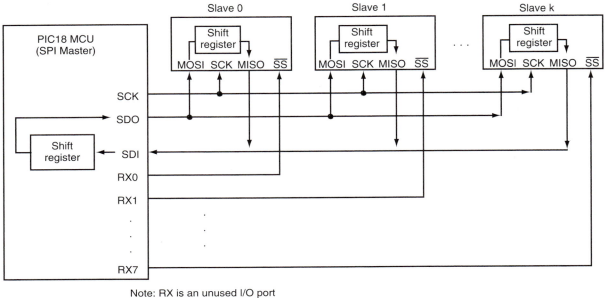

Note: RX is an unused I/O port

MOSI stands for master out, slave in

MISO stands for master in, slave out

Figure 10.5 ■ Single-master and multiple-slave device connection (method 1)

When the application requires the PIC18 microcontroller to send data to all peripheral devices connected with the SPI interface in all situations, then the connection method shown in Figure 10.6 can be used. In this method, the shift registers of all the slaves are cascaded into a longer shift register.

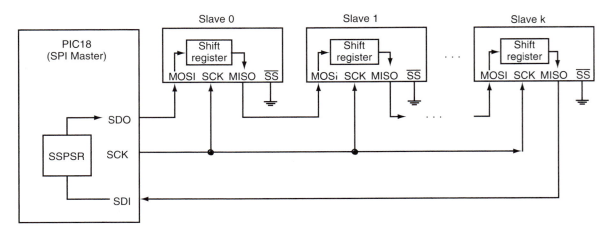

Figure 10.6 ■ Single-master and multiple-slave device connection (method 2)

10.3 SPI C Library Functions

Microchip C18 compiler provides library functions (listed in Table 10.1) to support SPI communications. The user must include the file **spi.h** in order to invoke these functions. The prototype declarations and arguments are explained as follows:

void **CloseSPI** (void);

unsigned char **DataRdySPI** (void);

A "0" is returned if there is no new data in the SSPBUF. Otherwise, a "1" is returned.

unsigned char **getcSPI** (void);

This function initiates an SPI bus cycle to acquire a byte of data.

unsigned char **ReadSPI** (void);

This function is identical to **getcSPI.**

void **getsSPI** (unsigned char *rdptr, unsigned char **length**);

This function reads **length** bytes from SPI device and save it in a buffer pointed to by **rdptr.**

void **OpenSPI** (unsigned char **sync_mode,** unsigned char **bus_mode,**

unsigned char **smp_phase**);

Function	Description
CloseSPI	Disable the SPI module used for SPI communications
DataRdySPI	Determine if a value is available from the SPI buffer.
getcSPI	Read a byte from the SPI bus.
getsSPI	Read a string from the SPI bus.
OpenSPI	Initialize the SSP module used for SPI communications.
putcSPI	Write a byte to the SPI bus.
putsSPI	Write a string to the SPI bus.
ReadSPI	Read a byte from the SPI bus.
WriteSPI	Write a byte to the SPI bus.

Table 10.1 ■ C18 SPI library functions

The argument **sync_mode** may have the following values:

SPI_FOSC_4	SPI master mode, clock = FOSC/4
SPI_FOSC_16	SPI master mode, clock = FOSC/16
SPI_FOSC_64	SPI master mode, clock = FOSC/64
SPI_FOSC_TMR2	SPI master mode, clock = TMR2 output/2
SLV_SSON	SPI slave mode, /SS pin control enabled
SLV_SSOFF	SPI slave mode, /SS pin control disabled

The argument **bus_mode** may have the following values:

MODE_00	Setting for SPI bus mode 0,0 (idle low, shift on rising edge)
MODE_01	Setting for SPI bus mode 0,1 (idle low, shift on falling edge)
MODE_10	Setting for SPI bus mode 1,0 (idle high, shift on falling edge)
MODE_11	Setting for SPI bus mode 1,1 (idle high, shift on rising edge)

The argument **smp_phase** may have the following values:

SMPEND   Input data sample at end of data out

SMPMID   Input data sample at middle of data out

This function also configures SDO and SCK pins for output and configures SDI pin for input.

unsigned char **putcSPI** (unsigned char **data_out**);

This function writes a single byte out and then checks for a bus collision. A "0" is returned if there is no collision. Otherwise, a "1" is returned.

unsigned char **writeSPI** (unsigned char **data_out**);

This function is identical to **putcSPI.**

void **putsSPI** (unsigned char *****wrptr**);

This function writes a NULL-terminated string to the SPI device.

10.5 Applications of SPI

MSSP in SPI mode can interface with many different peripheral devices. A few of them are discussed in the following sections.

10.6 Interfacing with the 74HC595

The block diagram of the 74HC595 is shown in Figure 10.7. The 74HC595 consists of an 8-bit shift register and an 8-bit D-type latch with tristate parallel outputs.

The functions of the pins in Figure 10.7 are as follows:

- *SI—serial data input.* The data on this pin is shifted into the 8-bit shift register.
- *SC—shift clock.* The rising edge of this input causes the data at the pin SI to be shifted into the 8-bit shift register.
- $\overline{Reset}$. A low on this pin clears the shift register portion of this device only. The 8-bit latch is not affected.
- *LC—latch clock.* The rising edge of this signal loads the contents of the shift register into the output latch.
- $\overline{SS}$—*output enable.* A low on this pin allows the data from the latch to be presented at the output pins Q_A . . . Q_H.
- *QA . . . QH.* Noninverted, tristate latch outputs.
- *SO—serial data output.* This is the output of the eighth stage of the 8-bit shift register. This output does not have tristate capability.

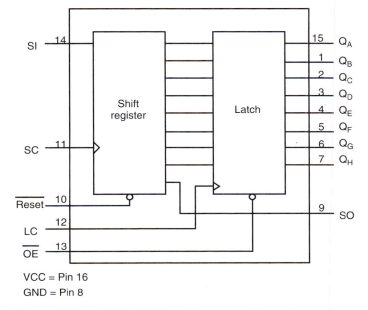

VCC = Pin 16
GND = Pin 8

Figure 10.7 ■ 74HC595 block diagram and pin assignment

10.6.1 Operation of the 74HC595

The 74HC595 is designed to shift in 8-bit data serially and then transfer it to the latch to be used as parallel data. The 74HC595 can be used to add output ports to the PIC18 microcontroller. Both the connection methods shown in Figures 10.5 and 10.6 can be used in appropriate applications.

Example 10.2

Describe how to use two 74HC595 to drive up to eight common-cathode seven-segment displays, assuming that the crystal oscillator frequency is 32 MHz.

Solution: Two 74HC595s can be cascaded using the method shown in Figure 10.6. One 74HC595 is used to hold the seven-segment pattern, whereas the other 74HC595 is used to carry digit-select signals. The circuit connection is shown in Figure 10.8.

Since there are only seven segments, the Q_H bit of the segment-control 74HC595 is not needed. The RB4 pin is used to control the LC input of the 74HC595. The time-multiplexing technique illustrated in Example 7.5 will be used to display multiple digits in Figure 10.8. To light the digit on display #7, the voltage at Q_H of the digit-select 74HC595 must be driven to high. To light the digit on display #6, the voltage at Q_G of the digit-select 74HC595 must be driven to high and so on. In order to turn on one digit at a time in turn, a periodic interrupt will be generated to switch the on and off of each digit. Timer2 will be configured to generate periodic interrupts to perform time-multiplexing operation.

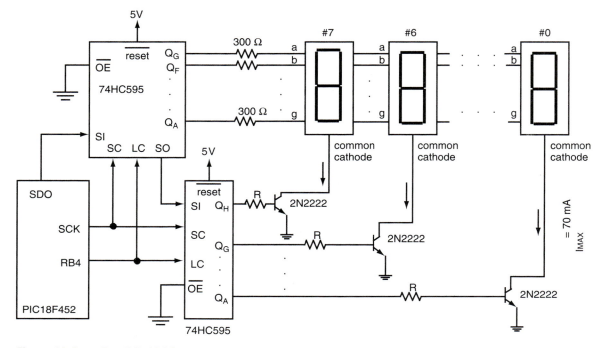

Figure 10.8 ■ Two 74HC595s together drive eight seven-segment displays

There are four parts in the program to be written:

Part 1. Display table initialization

1. Load display data table address into TBLPTR registers.

2. Initialize table entry count to 8.

Part 2. Timer2 setup

1. Configure Timer2 with prescaler set to 4 and postscaler set to 16.

2. Set the period of Timer2 counting to 124 by loading 124 into the PR register. This will generate an interrupt once every 1 ms.

Part 3. MSSP configuration

1. Configure MSSP in SPI mode to shift data at 8 MHz on the rising clock edge.

2. Send out the first digit on the display table to be displayed.

Part 4. Timer2 interrupt service routine

1. Decrement table entry count by 1. If the result is 0, then reload the table pointer to the start of the display table and reset the table entry count to 8.

2. Send digit pattern and digit select bytes to 74HC595s.

3. Transfer the values to the output latches of 74HC595s.

The following program will display 87654321 on display #7 through display #0:

```
                #include <p18F452.inc>
lp_cnt          set         0x00                    ; display loop count
                org         0x00
                goto        start
                org         0x08
                goto        hi_ISR
                org         0x18
                retfie
start           movlw       upper disp_tab          ; set up pointer to display
                movwf       TBLPTRU,A               ; pattern table
                movlw       high disp_tab           ; "
                movwf       TBLPTRH,A               ; "
                movlw       low disp_tab            ; "
                movwf       TBLPTRL,A               ; "
                movlw       0x08                    ; set loop count to 8
                movwf       lp_cnt,A                ; for hi_ISR
; configure Timer2 to generate periodic interrupt every 1 ms
                movlw       0x7D                    ; timer2 prescaler set to 4, postscaler
                movwf       T2CON,A                 ; set to 16, turn on Timer2
                movlw       D'124'                  ; Timer2 period set to 125
                movwf       PR2,A                   ; "
; configure SPI mode
                movlw       0xC0
                movwf       SSPSTAT,A               ; data transmitted on rising edge
                movlw       0x21                    ; select SPI master mode and shift
                movwf       SSPCON1,A               ; data at FOSC/16, enable SPI
                bsf         TRISC,SDI,A             ; configure SDI pin for input
                bcf         TRISC,SDO,A             ; configure SDO pin for output
                bcf         TRISC,SCK,A             ; configure SCK pin for output
                bcf         TRISB,RB4,A             ; configure RB4 pin for output
                tblrd*+                             ; read the digit select byte
                movff       TABLAT,SSPBUF           ; send to 74HC595
wait1           btfss       SSPSTAT,BF,A            ; is SPI transfer completed?
                bra         wait1
                movf        SSPBUF,W,A              ; clear the flag
                tblrd*+                             ; read the digit pattern value
                movff       TABLAT,SSPBUF           ; send to 74HC595
wait2           btfss       SSPSTAT,BF,A            ;
                bra         wait2
                movf        SSPBUF,W,A              ; clear the flag
                bcf         PORTB,RB4,A             ; transfer data to output latch
                bsf         PORTB,RB4,A             ; "
; configure interrupt
                bsf         RCON,IPEN,A             ; enable priority interrupt
                bcf         PIR1,TMR2IF,A           ; clear Timer2 interrupt flag
                movlw       0xC0
                movwf       INTCON,A                ; enable global interrupt
```

```
                    clrf          TMR2,A                  ; clear Timer2
                    bsf           PIE1,TMR2IE,A           ; enable Timer2 overflow interrupt
forever             nop
                    goto          forever                 ; wait forever
; ******************************************************
; The hi_ISR makes sure that the interrupt is caused by TMR2IF.
; If lp_cnt is decremented to zero, the ISR reloads lp_cnt and
; table pointer. Otherwise it sends the next digit to update displays.
; ******************************************************
hi_ISR              btfss         PIR1,TMR2IF,A           ; is interrupt caused by TMR2 overflow?
                    retfie                                ; return if interrupt not caused by TMR2IF
                    decf          lp_cnt,F,A
                    bz            reload
send_dat            tblrd*+                               ; read the digit select byte
                    movff         TABLAT,SSPBUF           ; send to 74HC595
wait3               btfss         SSPSTAT,BF,A            ; is SPI transfer completed?
                    bra           wait3
                    movf          SSPBUF,W,A              ; clear the flag
                    tblrd*+                               ; read the digit pattern value
                    movff         TABLAT,SSPBUF           ; send to 74HC595
wait4               btfss         SSPSTAT,BF,A            ;
                    bra           wait4
                    movf          SSPBUF,W,A              ; clear the flag
                    bcf           PORTB,RB4,A             ; transfer data to output latch
                    bsf           PORTB,RB4,A             ; "
                    retfie
reload              movlw         upper disp_tab          ; set up pointer to display
                    movwf         TBLPTRU,A               ; pattern table
                    movlw         high disp_tab           ; "
                    movwf         TBLPTRH,A               ; "
                    movlw         low disp_tab            ; "
                    movwf         TBLPTRL,A               ; "
                    movlw         0x08                    ; set loop count to
                    movwf         lp_cnt,A                ; 8
                    goto          send_dat
disp_tab            db            0x80,0x7F,0x40,0x70,0x20,0x5F,0x10,0x5B
                    db            0x08,0x33,0x04,0x79,0x02,0x6D,0x01,0x30
                    end
```

The C language version of the program is as follows:

```
include <p18F452.h>
int           icnt;
unsigned char disp_tab [8][2] = {{0x80,0x7F},{0x40,0x70},{0x20,0x5F}, {0x10,0x5B},{0x08,0x33},{0x04,0x79},
                                 {0x02,0x6D},{0x01,0x30}};
void high_ISR(void);
void low_ISR(void);
unsigned char   temp;
#pragma code high_vector = 0x08
void high_interrupt(void)
```

```
                {
                            _asm
                            goto              high_ISR
                            _endasm
                }
                #pragma code low_vector = 0x18
                void low_interrupt (void)
                {
                            _asm
                            retfie 0
                            _endasm
                }
                #pragma code
                #pragma interrupt high_ISR
                void high_ISR(void)
                {
                        if (!PIR1bits.TMR2IF) {
                            _asm
                            retfie 0
                            _endasm
                        }
                        PIR1bits.TMR2IF = 0;                    /* clear TMR2IF flag */
                        icnt ++;
                        if (icnt == 8)
                            icnt = 0;
                        SSPBUF = disp_tab[icnt][0];             /* send out digit select byte */
                        while(!SSPSTATbits.BF);                 /* wait until SPI transfer is complete */
                        temp = SSPBUF;                          /* clear the flag */
                        SSPBUF = disp_tab[icnt][1];             /* send out digit pattern */
                        while(!SSPSTATbits.BF);
                        temp = SSPBUF;                          /* clear the flag */
                        PORTBbits.RB4 = 0;                      /* create a rising edge on LC pin to */
                        PORTBbits.RB4 = 1;                      /* load data to output latch of 74HC595 */
                        _asm
                        retfie 0
                        _endasm
                }
                void main (void)
                {
                        /* set up MSSP */
                        SSPSTAT = 0xC0;                         /* use rising edge to shift data */
                        SSPCON1 = 0x21;                         /* enable SPI to shift data at 2 Mbps */
                        TRISCbits.TRISC5 = 0;                   /* configure SDO pin for output */
                        TRISCbits.TRISC3 = 0;                   /* configure SCK pin for output */
                        TRISCbits.TRISC4 = 1;                   /* configure SDI pin for input */
                        TRISBbits.TRISB4 = 0;                   /* configure RB4 pin for output */
                        /* configure interrupt priority and enabling */
                        RCON |= 0x80;                           /* enable priority interrupt */
```

```
                    INTCON = 0xC0;                  /* enable global interrupt */
                    PIR1bits.TMR2IF = 0;
                    PIE1bits.TMR2IE = 1;            /* enable Timer2 overflow interrupt */
                    PIR1bits.SSPIF = 0;             /* clear the SSPIF flag */
                    SSPBUF = disp_tab[0][0];        /* send out digit select byte */
                    while(!SSPSTATbits.BF);         /* wait until 8 bits have been shifted out */
                    temp = SSPBUF;                  /* clear the flag */
                    SSPBUF = disp_tab[0][1];        /* send out digit pattern */
                    while(!SSPSTATbits.BF);
                    temp = SSPBUF;                  /* clear the flag */
                    PORTBbits.RB4 = 0;              /* create a rising edge on LC pin to */
                    PORTBbits.RB4 = 1;              /* load data to output latch of 74HC595 */
                    /* configure Timer2 */
                    TMR2 = 0;                       /* Timer2 count from 0 */
                    T2CON = 0x7D;                   /* prescaler set to 4, postscaler set to 16 */
                    PR2 = 124;                      /* set Timer2 period to 125 */
                    icnt = 0;                       /* display table entry count set to 0 */
                    while (1) {};
           }
```

This example requires the use of eight discrete NPN transistors. The extra wiring required by these eight NPN transistors may not be desirable. A few semiconductor manufacturers produce seven-segment display driver chips that can drive from four to eight seven-segment displays. The MAX7221 from MAXIM is one such example.

▲

10.7 MAX7221 Seven-Segment Display Driver

The MAX7221 from MAXIM has an SPI interface and can drive up to eight seven-segment displays. In addition to driving seven-segment displays, the MAX7221 can also drive bar-graph displays, industrial controllers, panel meters, and LED matrix displays.

When driving seven-segment displays, the MAX7221 can address and update individual digit without rewriting the entire display. Only common-cathode seven-segment displays can be driven directly.

The user can choose to use the *code-B decode* method to represent digits 0 through 9 and characters -, E, H, L, P, and space. Other characters can be defined by using the *no-decode* method.

10.7.1 Signal Pins

The MAX7221 has 23 signal pins and is housed in a 24-pin DIP package. Its pin assignment is shown in Figure 10.9. The function of each pin is as follows:

- *DIN—serial data input.* Data is loaded into the internal 16-bit shift register on CLK's rising edge.

- *DIG0 . . . DIG7—eight-digit drive lines.* These lines sink current from the display common cathode. The MAX7221's digit drivers are in high impedance when they are turned off.

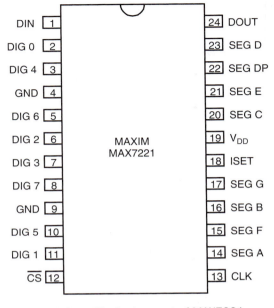

Figure 10.9 ■ Pin Assignment of MAX7221

- *GND. Ground.*
- *$\overline{CS}$—chip-select.* Serial data is loaded into the shift register while $\overline{CS}$ is low. The last 16 bits of serial data are latched on the rising edge of this signal.
- *CLK—serial-clock input.* Data is shifted into the internal shift register on the rising edge of CLK and shifted out of the DOUT pin on the falling edge of CLK. The maximum data shift rate is 10 MHz.
- *SEG A . . . SEG G, DP—seven-segment drives and decimal point drive.* These signals are in a high-impedance state when they are turned off.
- *ISET—current setting resistor.* This pin is connected to the V_{DD} through a resistor (R_{SET}) to set the peak segment current.
- *V_{DD}.* Positive power supply.
- *DOUT—serial-data out.* This pin is used to daisy-chain several MAX7221s and is never in a high-impedance state.

10.7.2 MAX7221 Functioning

The functional diagram of the MAX7221 is shown in Figure 10.10.

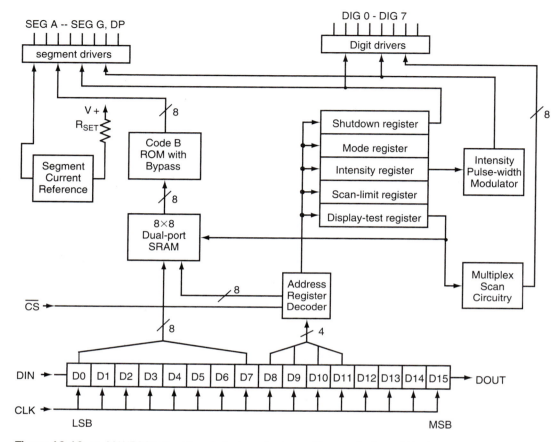

Figure 10.10 ■ MAX7221 Functional diagram (redraw with permission of Maxim)

SEGMENT DRIVERS

This block provides the current to drive the segment pattern outputs SEG A, . . ., SEG G, and DP.

CODE-B ROM WITH BYPASS

This ROM stores the patterns of decimal digits 0 through 9, -, E, H, L, and P. A 4-bit address presented by the dual-port SRAM is used to select the segment pattern from this ROM. The output of this ROM will be used to drive the segment outputs SEG A, . . ., SEG G, and DP. The intensity of the segment pattern is under the control of the output intensity register.

If the patterns provided by the code-B ROM cannot satisfy the need of the application, the user can bypass them and supply his or her own. The no-decode mode must be selected for this purpose. In no-decode mode, the user needs to store the desired patterns in the dual-port SRAM, and these patterns will be used to drive the segment outputs. The user can choose to bypass from one to all digits.

The segment patterns stored in the code-B ROM are shown in Table 10.2.

7-segment character	register data (output from SRAM)						On segment = 1						
	D7* D6-D4	D3	D2	D1	D0	DP*	A	B	C	D	E	F	G
0	x	0	0	0	0		1	1	1	1	1	1	0
1	x	0	0	0	1		0	1	1	0	0	0	0
2	x	0	0	1	0		1	1	0	1	1	0	1
3	x	0	0	1	1		1	1	1	1	0	0	1
4	x	0	1	0	0		0	1	1	0	0	1	1
5	x	0	1	0	1		1	0	1	1	0	1	1
6	x	0	1	1	0		1	0	1	1	1	1	1
7	x	0	1	1	1		1	1	1	0	0	0	0
8	x	1	0	0	0		1	1	1	1	1	1	1
9	x	1	0	0	1		1	1	1	1	0	1	1
-	x	1	0	1	0		0	0	0	0	0	0	1
E	x	1	0	1	1		1	0	0	1	1	1	1
H	x	1	1	0	0		0	1	1	0	1	1	1
L	x	1	1	0	1		0	0	0	1	1	1	0
P	x	1	1	1	0		1	1	0	0	1	1	1
blank	x	1	1	1	1		0	0	0	0	0	0	0

*The decimal point is set by bit D7 = 1

Table 10.2 ■ Code B pattern

DIGIT DRIVERS AND MULTIPLEX SCAN CIRCUITRY

The MAX7221 uses the time-multiplexing technique to display eight digits. When DIG i (i = 0 to 7) is asserted, the corresponding pattern is sent to the segment driver from either ROM or SRAM. Since only one digit output is asserted, only one digit will be lighted at a time. However, the multiplex scan circuitry will allow only one digit to be lighted for a short period of time (slightly over 1 ms) before it switches to display the next digit. In this way, all eight digits are lighted in turn many times in 1 second. All eight digits appear to be lighted simultaneously because of the effect of *persistence of vision.*

MODE REGISTER

The mode register sets code-B or no-decode operation for each digit. Each bit in the register corresponds to one digit. Logic high selects code-B decoding, while logic low bypasses the decoder. Examples of the decode mode control-register format are shown in Table 10.3.

When the code-B decode mode is used, the decoder looks only at the lower nibble of the data in the digit registers (D3-D0), disregarding bits D4 to D6. D7 is used to set the decimal point (DP) and is independent of the decoder. When D7 = 1, decimal point is turned on.

Decode mode	Register data								Hex Code
	D7	D6	D5	D4	D3	D2	D1	D0	
No decode for digits 7-0	0	0	0	0	0	0	0	0	00
Code B decode for odd digits No decode for even digits	1	0	1	0	1	0	1	0	AA
Code B decode for digits 3-0 No decode for digits 7-4	0	0	0	0	1	1	1	1	0F
Code B decode for all digits	1	1	1	1	1	1	1	1	FF

Table 10.3 ■ Decode-Mode register examples

INTENSITY CONTROL AND INTER-DIGIT BLANKING

MAX7221 allows the user to set the display brightness by setting the value of the R_{SET} resistance (connecting the V_{DD} and ISET pin) and setting the value of the *intensity register.*

Digital control of display brightness is provided by an internal pulse-width modulator, which is controlled by the lower nibble of the intensity register. The modulator scales the average segment current in 16 steps from a maximum of 15/16 down to 1/16 of the peak current set by R_{SET}. Table 10.4 lists the intensity register format.

duty cycle (min on)	D7	D6	D5	D4	D3	D2	D1	D0	Hex code
1/16	x	x	x	x	0	0	0	0	x0
2/16	x	x	x	x	0	0	0	1	x1
3/16	x	x	x	x	0	0	1	0	x2
4/16	x	x	x	x	0	0	1	1	x3
5/16	x	x	x	x	0	1	0	0	x4
6/16	x	x	x	x	0	1	0	1	x5
7/16	x	x	x	x	0	1	1	0	x6
8/16	x	x	x	x	0	1	1	1	x7
9/16	x	x	x	x	1	0	0	0	x8
10/16	x	x	x	x	1	0	0	1	x9
11/16	x	x	x	x	1	0	1	0	xA
12/16	x	x	x	x	1	0	1	1	xB
13/16	x	x	x	x	1	1	0	0	xC
14/16	x	x	x	x	1	1	0	1	xD
15/16	x	x	x	x	1	1	1	0	xE
15/16 (max on)	x	x	x	x	1	1	1	1	xF

Table 10.4 ■ Intensity register format

SCAN-LIMIT REGISTER

The number (from 1 to 8) of digits displayed is set by this register. Table 10.5 lists the scan-limit register format.

Scan Limit	Register data								Hex Code
	D7	D6	D5	D4	D3	D2	D1	D0	
Display digit 0 only	x	x	x	x	x	0	0	0	x0
Display digits 0 & 1	x	x	x	x	x	0	0	1	x1
Display digits 0 to 2	x	x	x	x	x	0	1	0	x2
Display digits 0 to 3	x	x	x	x	x	0	1	1	x3
Display digits 0 to 4	x	x	x	x	x	1	0	0	x4
Display digits 0 to 5	x	x	x	x	x	1	0	1	x5
Display digits 0 to 6	x	x	x	x	x	1	1	0	x6
Display digits 0 to 7	x	x	x	x	x	1	1	1	x7

Table 10.5 ■ Scan-Limit register format

SHUTDOWN MODE REGISTER

When the MAX7221 is in shutdown mode, the segment drivers are in a high-impedance state. Data in the digit and control registers remains unaltered. The shutdown mode can be used to save power or as an alarm to flash the display by successively entering and leaving shutdown mode. To enter the shutdown mode, clear bit 0 of the shutdown mode register to 0.

DISPLAY TEST REGISTER

The display test register operates in two modes: normal and display test. The display test mode turns all LEDs on by overriding but not altering all controls and digit registers (including the shutdown register). In display test mode, eight digits are scanned, and the duty cycle is 15/16. To enter display test mode, set bit 0 of the display test register to 1.

SERIAL SHIFT REGISTER

At the bottom of Figure 10.10 is a 16-bit serial shift register. The meaning of the content of this register is illustrated in Figure 10.11. When the $\overline{CS}$ signal is low, data will be shifted in from the DIN pin on the rising edge of the CLK signal. On the rising edge of the $\overline{CS}$ signal, the lower eight bits of the shift register will be transferred to a destination specified by the value represented by bit D11 through bit D8 of the shift register. The mapping of these four address bits is shown in Table 10.5. The upper four bits of this register are not used.

D15	D14	D13	D12	D11	D10	D9	D8	D7	D6	D5	D4	D3	D2	D1	D0
x	x	x	x	Address				MSB			Data				LSB

Figure 10.11 ■ Serial data format

In Table 10.6, digits 0 to 7 are registers in dual-port SRAM that are used to hold the decimal digits or the patterns (supplied by the user when no-decode mode is selected) to be displayed. Whenever the register Digit i (i = 0 . . . 7) is selected to supply segment pattern (directly in no-decode mode) or indirectly (in decode mode) to the segment drivers, the associated digit driver pin **DIG i** is also asserted.

Register	Register data D15–D12	D11	D10	D9	D8	Hex Code
No op	x	0	0	0	0	x0
Digit 0	x	0	0	0	1	x1
Digit 1	x	0	0	1	0	x2
Digit 2	x	0	0	1	1	x3
Digit 3	x	0	1	0	0	x4
Digit 4	x	0	1	0	1	x5
Digit 5	x	0	1	1	0	x6
Digit 6	x	0	1	1	1	x7
Digit 7	x	1	0	0	0	x8
Decode Mode	x	1	0	0	1	x9
Intensity	x	1	0	1	0	xA
Scan limit	x	1	0	1	1	xB
Shutdown	x	1	1	0	0	xC
Display test	x	1	1	1	1	xF

Table 10.6 ■ Register address map

The register name in the first row of the table is *no op*. This register is used when cascading multiple MAX7221s. To cascade multiple MAX7221s, connect all devices' $\overline{CS}$ inputs together and connect DOUT to DIN on adjacent devices. For example, if four MAX7221s are cascaded, then to write to the fourth chip without affecting other three, set the desired 16-bit word, followed by three no-op codes (hex X0XX, most significant bit first). When $\overline{CS}$ goes high, data is latched in all devices. The first three chips receive no-op commands, and the fourth receives the intended data.

10.7.3 Choosing the Value for R$_{SET}$

Appropriate values must be chosen for the R$_{SET}$ resistor in order to supply enough current to the LED segments. Table 10.7 shows us how to choose the R$_{SET}$ value for a given LED segment current (I$_{SEG}$) and the associated LED forward voltage drop (V$_{LED}$). Any other combinations of I$_{SEG}$ and V$_{LED}$ that are not in the table can be obtained by using interpolation. For example, supposing that V$_{LED}$ is 2.3 V when I$_{SEG}$ is 10 mA, the R$_{SET}$ value can be calculated as follows:

$$(2.3 - 2.0)/(2.5 - 2.0) = (63.7 - R_{SET})/(63.7 - 59.3)$$

The value of R$_{SET}$ is solved to be 61 KΩ, which is not a standard resistor. The closest standard resistor 62 KΩ will be used instead.

I$_{SEG}$ (mA)	V$_{LED}$ (V)				
	1.5	2.0	2.5	3.0	3.5
40	12.2 KΩ	11.8 KΩ	11.0 KΩ	10.6 KΩ	9.69 KΩ
30	17.8 KΩ	17.1 KΩ	15.8 KΩ	15.0 KΩ	14.0 KΩ
20	29.8 KΩ	28.0 KΩ	25.9 KΩ	24.5 KΩ	22.6 KΩ
10	66.7 KΩ	63.7 KΩ	59.3 KΩ	55.4 KΩ	51.2 KΩ

Table 10.7 ■ R$_{SET}$ vs. segment current and LED forward voltage drop

Example 10.3

▼

Use a MAX7221 and six seven-segment displays to display the value –42.5° C. Show the circuit connection and write a program to display the given value. Assume that enough brightness can be obtained when 10 mA of current are flowing through the LED and the V$_{LED}$ at this current is 2.3 V. Assume that the PIC18F8720 is running with a 16-MHz crystal.

Solution: The circuit connection is shown in Figure 10.12. The R$_{SET}$ value is chosen to be 62 KΩ to satisfy the V$_{LED}$ and I$_{SEG}$ requirements.

In this circuit, the user needs to send 11 16-bit values to the MAX7221 in order to complete the setup. The values to be written are as follows:

Scan-limit register—The given temperature is to be displayed on digits 0, 1, 2, 3, 4, and 5. The value to be written into the scan-limit register is 0x05. Since the address of the scan-limit register is 0x0B, the 16-bit value to be sent to MAX7221 is 0x0B05.

Decode mode register—Among the six characters (–42.5° C) to be displayed, the degree character and letter C cannot be decoded. The user can use decode mode for the leading four characters and use no-decode mode for the degree character and letter C. Therefore, the value to be written into the decode mode register is 0xFC. To achieve this, the 16-bit value 0x09FC should be sent to MAX7221.

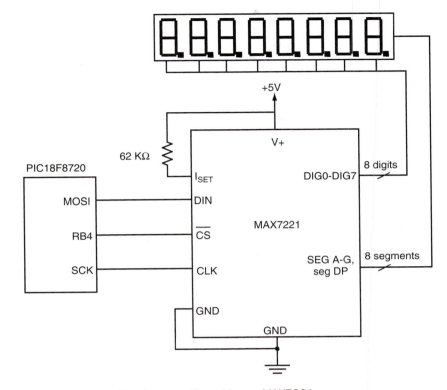

Figure 10.12 ■ Circuit connection with one MAX7221

Digit 0 register (to display letter C)—No-decode mode is used to display letter C. Segments a, d, e, and f should be turned on, whereas segments b, c, g and DP should be turned off. The address of this register is 1. The 16-bit value to be sent to the MAX7221 is 0x014E.

Digit 1 register (to display degree character)—The degree character is displayed in no-decode mode. The seven-segment pattern for a degree character is 0x63. The 16-bit value to be sent to MAX7221 is 0x0263.

Digit 2 register (to display the value 5)—This digit should be displayed in code-B decode mode. A decimal point is not associated with this digit. The address of this digit is 3. The 16-bit value to be sent to the MAX7221 is 0x0305.

Digit 3 register (to display the value 2)—This digit should be displayed using code-B decode mode. A decimal point is associated with this digit. The value to be stored in this register is 0x82. Therefore, the 16-bit value to be sent to the MAX7221 is 0x0482.

Digit 4 register (to display the value 4)—This digit should be displayed using code-B decode mode. No decimal point is associated with this digit. The value to be stored in this register is 0x04. The address of this register is 5. Therefore, the 16-bit value to be written into this register is 0x0504.

Digit 5 register (to display the minus sign)—This character should be displayed in code-B decode mode. No decimal point is associated with this character. The value to be stored in this register is 0x0A. The address of this register is 6. Therefore, the 16-bit value to be sent to the MAX7221 is 0x060A.

Intensity register—The user will want to display the value –42.5°C in full brightness. Therefore, the value to be stored in this register should be 0x0F. The address of this register is 0x0A. Therefore, the 16-bit value to be sent to the MAX7221 is 0x0A0F.

Shutdown register—The chip should perform in normal operation. Therefore, the value to be stored in this register is 0x01 (bit 0 should be set to 1). The address of this register is 0x0C. Therefore, the 16-bit value to be sent to the MAX7221 is 0x0C01.

Display test register—The chip should perform normal operation. Therefore, the value to be stored in this register should be 0x00, and the 16-bit value to be sent to the MAX7221 is 0x0F00.

The MSSP should be configured properly:

1. Pin directions: The SDO, SCK, and the RB4 pins should be configured for output.

2. SSPSTAT register: Input data should be sampled at the end of data output time, and output data is shifted out on the rising edge of SCK. Load 0xC0 into this register.

3. SSPCON1 register: Write the value 0x20 into this register to set SPI master mode and set the clock rate to 4 MHz ($f_{OSC}/4$). Set clock polarity to idle low.

The following subroutine will initialize the SPI module accordingly:

```
spi_init    movlw    0xC0
            movwf    SSPSTAT,A
            movlw    0x20
            movwf    SSPCON1,A
            bcf      TRISC,SDO,A    ; configure SDO pin for output
            bcf      TRISC,SCK,A    ; configure SCK pin for output
            bcf      TRISB,RB4,A    ; configure RB4 pin for output
            return
```

To configure the MAX7221, the user can use a table to hold the data (in program memory) to be transferred to the MAX7221 as follows:

```
disp_dat    db    0x0B,0x05    ; data for scan limit register
            db    0x09,0xFC    ; data for decode mode register
            db    0x0A,0x0F    ; data for intensity register
            db    0x0C,0x01    ; value for shutdown register
            db    0x0F,0x00    ; value for display test register
            db    0x01,0x4E    ; value for digit 0 register (letter C)
            db    0x02,0x63    ; value for digit 1 register (degree character)
            db    0x03,0x05    ; value for digit 2 register (value 5)
            db    0x04,0x82    ; value for digit 3 register (value 2 with DP)
            db    0x05,0x04    ; value for digit 4 register (value 4)
            db    0x06,0x0A    ; value for digit 5 register (minus sign)
```

The following subroutine will transfer the data in disp_dat to the MAX7221, and the value −42.5° C will be displayed on six seven-segment displays:

```
lp_cnt      equ     11
send2max    movlw   upper disp_dat      ; set TBLPTR to point to the start of disp_dat
            movwf   TBLPTRU,A           ; "
            movlw   high disp_dat       ; "
            movwf   TBLPTRH,A           ; "
            movlw   low disp_dat        ; "
            movwf   TBLPTRL,A           ; "
            movlw   lp_cnt
            movwf   PRODL,A             ; use PRODL as the loop count
send_lp     bcf     PORTB,RB4,A         ; enable data to be shifted into MAX7221
            tblrd*+                     ; read the upper 16 bits
            movf    SSPBUF,W,A          ; clear the BF flag
            movff   TABLAT,SSPBUF       ; send to MAX7221
wait1       btfss   SSPSTAT,BF,A        ; wait until 8 bits are shifted into MAX7221
            goto    wait1
            movf    SSPBUF,W,A          ; clear the BF flag
            tblrd*+                     ; read the lower 8 bits
            movff   TABLAT,SSPBUF       ; send to MAX7221
wait2       btfss   SSPSTAT,BF,A        ; wait until 8 bits are shifted into MAX7221
            goto    wait2
            bsf     PORTB,RB4,A         ; latch the lower 8 bits of data
            decfsz  PRODL,F,A           ; decrement the loop count
            goto    send_lp             ; not done yet and continue
            return
```

The C language version of the above subroutines will be left for you as an exercise. Cascading multiple MAX7221 is easy. The procedure is as follows:

1. Wire the $\overline{\text{CS}}$ signal of all MAX7221s together and connect them to an output pin of the PIC18 microcontroller.

2. Wire the SCK pins of all MAX7221s together and connect them to the SCK pin of the PIC18 microcontroller.

3. Wire the DIN pin of the first MAX7221 to the SDO pin of the PIC18 microcontroller, connect the DOUT pin of the first MAX7221 to the DIN pin of its adjacent MAX7221, and so on.

The circuit for cascading two MAX7221s is shown in Figure 10.13. In this example, the RB4 pin is used as a general-purpose output pin to drive the $\overline{\text{CS}}$ pin. Again, the R_{SET} is set to 62 KΩ.

Example 10.4

▼

Write a program to configure the MSSP module and display the value

80.5 °F 06 04 01 20 03

on the 16 seven-segment displays shown in Figure 10.13. Assume that the PIC18F8720 is driven by a 16-MHz crystal oscillator.

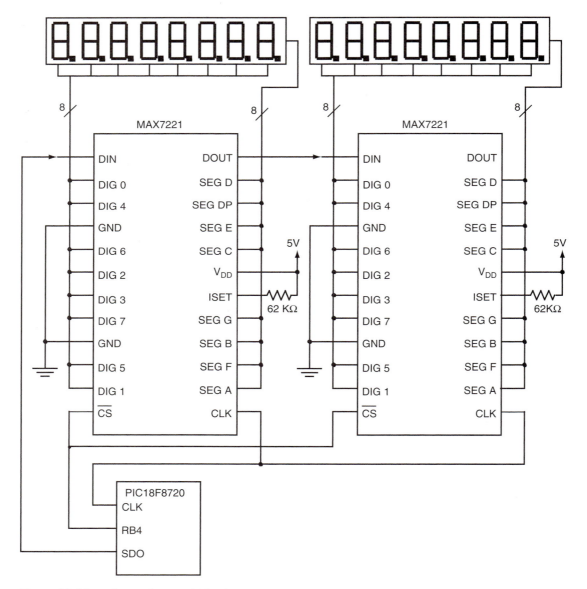

Figure 10.13 ■ Cascading two MAX7221 to display 16 BCD digits

Solution: The same configuration for the MSSP module in the previous example can be used here. The given digit string will be divided into two halves so that each is displayed by one MAX7221:

80.5 °F 06
04 01 20 03

The first MAX7221 will display seven characters, whereas the second MAX7221 will display eight digits. These two halves will be paired as shown in Figure 10.14.

digit	7	6	5	4	3	2	1	0	
		8	0	5	o	F	0	6	◄——— first MAX7221
	0	4	0	1	2	0	0	4	◄——— second MAX7221

Figure 10.14 ■ Data arrangement for transfer

In this circuit configuration, the user needs to send 13 32-bit values to set up display data. The values to be transferred are as follows:

Scan-limit register—The first and the second MAX7221 will display seven and eight digits, respectively. For this setup, the user needs to send the values 0x0B06 and 0x0B07 to the first and the second MAX7221, respectively.

Decode mode register—For the first MAX7221, no-decode mode will be used for digits 2 and 3 and code-B decode mode for other digits. For the second MAX7221, all digits will be displayed in decode mode. For this setup, the user should send the value 0x09F3 to the first MAX7221 and send the value 0x09FF to the second MAX7221.

Intensity register—For this example, the maximum intensity will be used to display the values. For this setup, the value 0x0A0F should be sent to both MAX7221s.

Shutdown register—Both chips should perform normally. Therefore, the value to be stored in this register is 0x01, and the value 0x0C01 should be sent to both MAX7221s.

Display test register—Both chips should perform normal operation. Therefore, the value to be stored in this register is 0x00 (bit 0 must be 0), and the value 0x0F00 should be sent to both MAX7221s.

Digit 7 register—The first MAX7221 will not display anything on digit 7, whereas the second MAX7221 will display the value 0 on digit 7. For this value, the user should send 0x0000 to the first MAX7221 and send the value 0x0800 to the second MAX7221.

Digit 6 register—The first and the second MAX7221 will display 8 and 4 on this digit, respectively. Therefore, the values 0x0708 and 0x0704 should be sent to the first and the second MAX7221, respectively.

Digit 5 register—Both MAX7221s will display 0 on digit 5. Therefore, the value 0x0600 should be sent to both MAX7221s.

Digit 4 register—The first and the second MAX7221 will display 5 and 1 on digit 4, respectively. Therefore, the values to be sent to the first and the second MAX7221 are 0x0505 and 0x0501, respectively.

Digit 3 register—The first MAX7221 will display °, whereas the second MAX7221 will display 2 on this digit. No-decode mode will be used in this digit for the first MAX7221. Therefore, the values to be sent to the first and the second MAX7221s are 0x0463 and 0x0402, respectively.

Digit 2 register—The first and the second MAX7221 will display F and 0 on digit 2, respectively. To display letter F, the first MAX7221 selects no-decode mode for this digit. Therefore, the values to be sent to the first and the second MAX7221s are 0x0347 and 0x0300, respectively.

Digit 1 register—Both MAX7221s will display 0 on this digit. Therefore, the value 0x0200 will be sent to both MAX7221s.

Digit 0 register—The first and the second MAX7221 will display 6 and 4 on this digit, respectively. Therefore, the values to be sent to the first and the second MAX7221s are 0x0106 and 0x0104, respectively.

The configuration used in Example 10.3 for the SPI subsystem can also be used in this example. The data to be transferred to the MAX7221s is defined as follows:

```
; *************************************************************
; The data to be sent to two MAX7221s are paired together. The data for the second
; MAX7221 are listed in the front because they should be sent out first.
; *************************************************************
max_data    db      $0B,$07,$0B,$06      ; data for scan limit registers
            db      $09,$FF,$09,$F3      ; data for decode mode registers
            db      $0A,$0F,$0A,$0F      ; data for intensity registers
            db      $0C,$01,$0C,$01      ; data for shutdown registers
            db      $0F,$00,$0F,$00      ; data for display test registers
            db      $08,$00,$00,$00      ; data for digit 7 registers
            db      $07,$04,$07,$08      ; data for digit 6 registers
            db      $06,$00,$06,$00      ; data for digit 5 registers
            db      $05,$01,$05,$05      ; data for digit 4 registers
            db      $04,$02,$04,$63      ; data for digit 3 registers
            db      $03,$00,$03,$47      ; data for digit 2 registers
            db      $02,$00,$02,$00      ; data for digit 1 registers
            db      $01,$04,$01,$06      ; data for digit 0 registers
```

The program for transferring this data to MAX7221s is straightforward and hence is left for you as an exercise.

▲

10.8 Digital Temperature Sensor TC72

The TC72 from Microchip is a digital temperature sensor with the SPI interface. The TC72 has 10-bit resolution; that is, it uses 10 bits to represent the ambient temperature. The pin assignment and functional block diagram are shown in Figure 10.15.

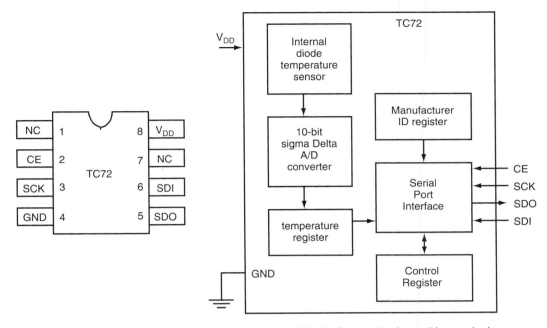

Figure 10.15 ■ TC72 pin assignment and functional block diagram (redraw with permission of Microchip)

10.8.1 Functioning of TC72

The TC72 is capable of reading temperatures from –55°C to +125°C. The T72 can be used either in the *continuous temperature conversion mode* or the *one-shot conversion mode*. The continuous conversion mode measures temperature approximately every 150 ms and stores the data in temperature registers. The TC72 has an internal clock generator that controls the automatic temperature conversion sequence. In contrast, the one-shot mode performs a single temperature measurement and returns to the power-saving shutdown mode.

10.8.2 Temperature Data Format

Temperature data is represented by a 10-bit two's complement word with a resolution of 0.25°C per least significant bit. The analog-to-digital converter is scaled from –128°C to +127°C with 0°C as 0x0000, but the operating range of the TC72 is specified from –55°C to +125°C. The 10-bit temperature value is stored in two 8-bit registers. Because the TC72 A/D converter is scaled from –128°C to +127°C and the conversion is represented in 10 bits, the temperature value is equal to the A/D conversion result divided by 4. The lowest two bits of the temperature are stored in the highest two bits of the lower byte of the temperature register. The temperature value is represented in the two's complement format. Whenever the most significant bit of the upper byte of the temperature register is 1, the temperature is negative. The magnitude of a negative temperature can be found by taking the two's complement of the temperature reading. After this is done, the upper eight bits become the integer part of the temperature, whereas the least significant two bits become the fractional part of the temperature. A sample of the temperature readings and their corresponding temperature values are shown in Table 10.8.

Binary high byte/low byte	Hex	Temperature
0010 0001/0100 0000	2140	33.25° C
0100 1010/1000 0000	4A80	74.5° C
0001 1010/1100 0000	1AC0	26.75° C
0000 0001/1000 0000	0180	1.5° C
0000 0000/0000 0000	0000	0° C
1111 1111/1000 0000	FF80	-0.5° C
1111 0010/1100 0000	F2C0	-13.25° C
1110 0111/0000 0000	E700	-24° C
1100 1001/0100 0000	C900	-54.75° C

Table 10.8 ■ TC72 Temperature output data

10.8.3 Serial Bus Interface

The serial interface consists of chip enable (CE), serial clock (SCK), serial data input (SDI), and serial data output (SDO). The CE input is used to select TC72 when there are multiple devices connected to the microcontroller. TC72 can operate as an SPI slave only.

The SDI input writes data into the TC72's control register, while the SDO output pin reads the temperature data from the temperature register and the status of the *shutdown bit* of the control register.

The T72 can shift data in/out using either the rising or the falling edge of the SCK input. The CE signal is active high. The SCK idle state is detected when the CE signal goes high. As shown in Figure 10.16, the clock polarity (CP) of SCK determines whether data is shifted on the rising or the falling edge. The highest SCK frequency (f_{SCK}) is 7.5 MHz.

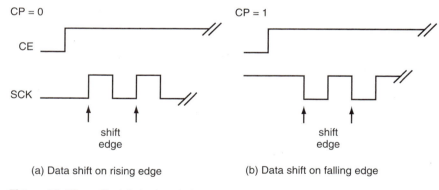

(a) Data shift on rising edge   (b) Data shift on falling edge

Figure 10.16 ■ Serial clock polarity

Data transfer to and from the TC72 consists of one address byte followed by one or multiple data (two to four) bytes. The most significant bit (A7) of the address byte determines whether a read (A7 = 0) or a write (A7 = 1) operation will occur. A multiple-byte read operation will start from a high address toward lower addresses. The user can send in the temperature result high byte address and read the temperature result high byte, low byte, and the control register. The timing diagrams for single-data-byte write, single-byte read, and three-data-byte read are shown in Figures 10.17a, 10.17b, and 10.17c.

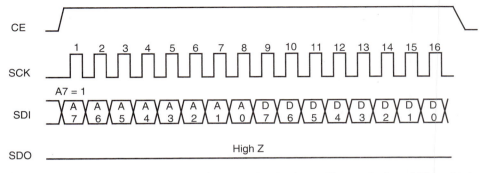

Figure 10.17a ■ Single data byte write operation (redraw with permission of Microchip)

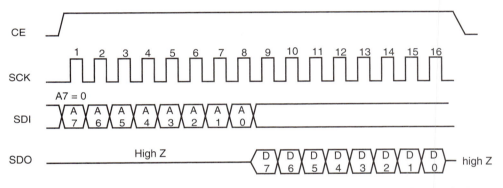

Figure 10.17b ■ Single data byte read operation (redraw with permission of Microchip)

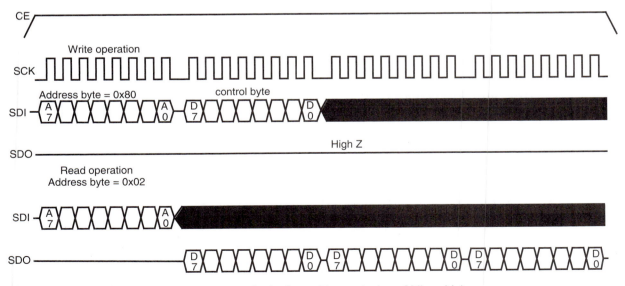

Figure 10.17c ■ SPI multiple data byte transfer (redraw with permission of Microchip)

The procedure for reading temperature result is as follows:

Step 1
Pull CE pin to high to enable SPI transfer.

Step 2
Send the temperature result high byte read address (0x02) to the TC72. Wait until the SPI transfer is complete.

Step 3
Read the temperature result high byte. The user needs to write a dummy byte into the SSPBUF register to trigger eight pulses to be sent out from the SCK pin so that the temperature result high byte can be shifted in.

Step 4
Read the temperature result low byte. Likewise, the user also needs to write a dummy byte into the SSPBUF register to shift in the temperature low byte.

Step 5
Pull CE pin to low so that a new transfer can be started.

10.8.4 Internal Register Structure

The TC72 has four internal registers: control register, LSB temperature, MSB temperature, and manufacturer ID. The contents of these are shown in Table 10.9.

Register	Read address	Write address	Bit 7	Bit 6	Bit 5	Bit 4	Bit 3	Bit 2	Bit 1	Bit 0	Value on POR/BOR
Control	0x00	0x80	0	0	0	OS	0	0	0	SHDN	0x05
LSB temperature	0x01	N/A	T1	T0	0	0	0	0	0	0	0x00
MSB temperature	0x02	N/A	T9	T8	T7	T6	T5	T4	T3	T2	0x00
Manufacturer ID	0x03	N/A	0	1	0	1	0	1	0	0	0x54

Note.
1. OS is One-Shot
2. SHDN is Shutdown

Table 10.9 ■ Register for TC72 (redraw with permission of Microchip)

Control Register

The control register is read/writable and is used to select the shutdown, continuous, or one-shot conversion operating mode. The temperature conversion mode selection logic is shown in Table 10.10.

Operation mode	One-Shot bit	Shutdown bit
Continuous temperature conversion	0	0
Shutdown	0	1
Continuous temperature conversion	1	0
One-shot	1	1

Table 10.10 ■ Control register temperature conversion mode selection

At power-up, the SHDN bit is 1. Thus, the T72 is in the shutdown mode at startup. The shutdown mode disables the temperature conversion circuitry; however, the serial I/O communication port remains active.

If the SHDN bit is 0, the TC72 will perform a temperature conversion approximately every 150 ms. In normal operation, a temperature conversion will be initialized by a write operation to the control register to select either the continuous temperature conversion or the one-shot operating mode. The temperature data will be available in the upper byte and lower byte of the temperature register approximately 150 ms after the control register is written into.

The one-shot mode performs a single temperature measurement and returns to the power-saving mode. After completion of the temperature conversion, the one-shot bit is reset to 0. The user must set the one-shot bit to 1 to initiate another temperature conversion.

TEMPERATURE REGISTER

The temperature register is a read-only register and contains a 10-bit two's complement representation of the temperature measurement. Bit 0 through bit 5 are always read as 0. After reset, the temperature register is reset to 0.

MANUFACTURING ID REGISTER

This register is read-only and is used to identify the temperature sensor as a Microchip component.

Example 10.5

Describe the circuit connection between the PIC18 microcontroller and TC72 for digital temperature reading and write a C program to read the temperature every 200 ms. Convert the temperature value into a string so that it can be displayed in an appropriate output device. A pointer to the buffer to hold the string will be passed to this function. The crystal oscillator of the demo board is assumed to be 16 MHz.

Solution: The circuit connection is shown in Figure 10.18. A 0.1-μF to 1.0-μF capacitor should be added between the VDD and the GND pins to filter out power noise.

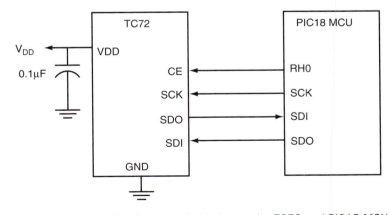

Figure 10.18 ■ Circuit connection between the TC72 and PIC18 MCU in the SSE8720 demo board

The C function (**read_temp( )**) that starts a temperature measurement and converts the temperature reading into a string is as follows:

```c
#include <p18F8720.h>
#include <spi.h>
#include <stdlib.h>
#include <timers.h>
void wait_200ms(void);
void read_temp (char *ptr);
char buf[10];
void main (void)
{
     TRISHbits.TRISH0 = 0;              /* configure RH0 pin for output */
     OpenSPI(SPI_FOSC_16, MODE_01, SMPEND);
     read_temp(&buf[0]);
}
void read_temp (char *ptr)
{
     char hi_byte, lo_byte, temp, *bptr;
     unsigned int result;
     bptr = ptr;
     PORTHbits.RH0 = 1;                 /* enable TC72 data transfer */
     SSPBUF = 0x80;                     /* send out TC72 control register write address */
     while(!SSPSTATbits.BF);            /* wait until data is shifted out */
     temp = SSPBUF;                     /* clear the BF flag */
     SSPBUF = 0x11;                     /* perform one shot conversion */
     while(!SSPSTATbits.BF);            /* wait until data is shifted out */
     PORTHbits.RH0 = 0;                 /* disable TC72 data transfer */
     temp = SSPBUF;                     /* clear the BF flag */
     wait_200ms();                      /* wait until temperature conversion is complete */
     PORTHbits.RH0 = 1;                 /* enable TC72 data transfer */
     SSPBUF = 0x02;                     /* send MSB temperature read address */
     while(!SSPSTATbits.BF);            /* wait until SPI transfer is complete */
     temp = SSPBUF;                     /* clear BF flag */
     SSPBUF = 0x00;                     /* read the temperature high byte */
     while(!SSPSTATbits.BF);            /* wait until SPI transfer is complete */
     hi_byte = SSPBUF;                  /* save temperature high byte and clear BF */
     SSPBUF = 0x00;                     /* read the temperature low byte */
     while(!SSPSTATbits.BF);            /* wait until SPI transfer is complete */
     lo_byte = SSPBUF;                  /* save temperature low byte and clear BF */
     PORTHbits.RH0 = 0;                 /* disable TC72 data transfer */
     lo_byte &= 0xC0;                   /* make sure the lower 6 bits are 0s */
     result = hi_byte * 256 + lo_byte;
     if (hi_byte & 0x80) {
             result = ~ result + 1;     /* take the two' complement of result */
             result >> = 6;
             temp = result & 0x0003;    /* place the lowest two bits in temp */
             result >> = 2;             /* get rid of fractional part */
             *ptr++ = 0x2D;             /* store the minus sign */
             itoa(result, ptr);
     }
     else {
```

```
            result >> = 6;
            temp = result & 0x0003;          /* save fractional part */
            result >> = 2;                   /* get rid of fractional part */
            itoa(result, ptr);               /* convert to ASCII string */
            }
            while(*bptr){                    /* search the end of the string */
            bptr++;
            };
            switch (temp){                   /* add fractional digits to the temperature */
            case 0:
                        break;
            case 1:     /* fractional part is .25 */
                        *bptr++ = 0x2E;      /* add decimal point */
                        *bptr++ = 0x32;
                        *bptr++ = 0x35;
                        *bptr = '\0';
                        break;
            case 2:     /* fractional part is .5 */
                        *bptr++ = 0x2E;      /* add decimal point */
                        *bptr++ = 0x35;
                        *bptr = '\0';
                        break;
            case 3:     /* fractional part is .75 */
                        *bptr++ = 0x2E;      /* add decimal point */
                        *bptr++ = 0x37;
                        *bptr++ = 0x35;
                        *bptr = '\0';
                        break;
                        default:
                        break;
            }
            }
void wait_200ms(void)
{
            OpenTimer0(TIMER_INT_OFF & T0_16BIT & T0_SOURCE_INT & T0_PS_1_16);
            TMR0 = 15535; /* let Timer0 overflow in 50000 clock cycles */
            INTCONbits.TMR0IF = 0;
            while(!INTCONbits.TMR0IF);
}
```

▲

10.9 The DS1306 Alarm Real-Time Clock

Keeping track of *time-of-day* is important for many applications. The DS1306 from DALLAS-MAXIM is a chip that can keep track of seconds, minutes, hours, date of the month, month, day of the week, and year with leap-year compensation valid up to 2100. The DS1306 supports the SPI and three-wire interface methods and provides two programmable alarms.

The DS1306 provides a full BCD clock calendar. All time information, including seconds, minutes, hours, day, date, month, and year, are in BCD format for easy reading. At the end of the month, date is automatically adjusted for months with fewer than 31 days. The clock operates in either the 24-hour or the 12-hour format with AM/PM indicator. In addition, the DS1306 provides 96 bytes of nonvolatile RAM for data storage.

10.9.1 Signal Descriptions

The pin assignment and block diagram of the DS1306 are shown in Figure 10.19. The function of each signal is as follows:

- V_{CC1}. Primary power supply.
- V_{CC2}—*secondary power pin.* In systems using the trickle charger, the rechargeable energy source is connected to this pin.
- V_{BAT}. Battery input for any standard 3-V lithium cell or other energy source.
- V_{CCIF}—*interface logic power supply input.* The V_{CCIF} pin allows the DS1306 to drive the SDO and 32-KHz output pins to a level that is compatible with the interface logic, thus allowing an easy interface to 3-V logic in mixed supply systems.
- *SERMODE*—*serial interface mode input.* This pin selects the three-wire mode when set to ground and selects the SPI mode when it is connected to VCC.
- *SCLK*—*serial clock input.* SCLK is used to synchronize data movement on the serial interface for either the SPI or the three-wire interface. The highest allowable frequency for the SCLK signal is 2 MHz.
- *SDI*—*serial data input.* When the SPI communication is selected, the SDI pin is the serial data input for the SPI bus. When the three-wire communication is selected, this pin must be tied to the SDO pin.
- *SDO*—*serial data output.* When the SPI communication is selected, the SDO pin is the serial data output for the SPI bus. When the three-wire communication is selected, this pin must be tied to the SDI pin.

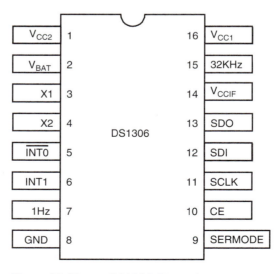

Figure 10.19a ■ DS1306 Pin assignment

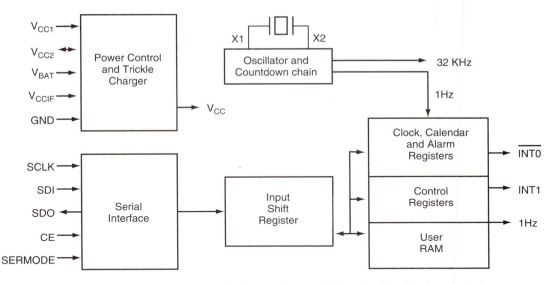

Figure 10.19b ■ DS1306 Block diagram (redraw with permission of Dallas Semiconductor)

- *CE—chip enable.* The chip enable signal must be asserted high during a read or a write for both the three-wire and the SPI communication. This pin has an internal 55-K pull-down resistor.
- *$\overline{INT0}$—interrupt 0 output.* The $\overline{INT0}$pin is active low and can be used as an interrupt to a processor. The $\overline{INT0}$pin can be programmed to be asserted by Alarm 0. This pin remains low as long as the status bit causing the interrupt is present and the corresponding interrupt enable bit is set. This pin is an open-drain output and requires an external pull-up resistor.
- *1Hz—1-Hz clock output.* The 1-Hz pin provides a 1-Hz square wave output. This output is active when the 1-Hz bit in the control register is 1.
- *INT1—interrupt 1 output.* The INT1 is active high and can be used as an interrupt input to a processor. The INT1 signal can be programmed to be asserted by Alarm1. When an alarm condition is present, the INT1 pin generates a 62.5-ms active-high pulse. The INT1 pin operates only when the DS1306 is powered by V_{CC2} or V_{BAT}.
- *32KHz—32.768-KHz clock output.* This signal is always present and provides a 32768-Hz output.
- *X1, X2—connection for a standard 32768-Hz quartz crystal.* The internal oscillator is designed for operation with a crystal having a specified load capacitance of 6 pF. The DS1306 can also be driven by an external oscillator signal. In this configuration, the X1 pin is connected to the external oscillator signal, and the X2 pin is floated.

10.9.2 RTC Registers

The time counting and alarm generating are controlled by a large set of registers. Some registers are read-only, whereas others are write-only. They are mapped to different addresses. The address map for RTC registers is shown in Figure 10.20.

The time, calendar, and alarm are set by writing to appropriate register bytes. Note that some bits are set to 0. Those bits always read 0 regardless of how they are written. Registers 0x12 to 0x1F and 0x92 to 0x9F are reserved. These registers always read 0.

Hex address		Bit 7	Bit 6	Bit 5	Bit 4	Bit 3	Bit 2	Bit 1	Bit 0	Range
Read	Write									
0x00	0x80	0	10 Sec			Sec				00-59
0x01	0x81	0	10 Min		Min					00-59
0x02	0x82	0	12 / 24	P / A / 10	10-HR	Hours				01-12 + P/A / 00-23
0x03	0x83	0	0	0	0	0	Day			01-07
0x04	0x84	0	0	10-Date		Date				1-31
0x05	0x85	0	0	10-Month		Month				01-12
0x06	0x86	10-Year				Year				00-99
0x07	0x87	M	10-Sec Alarm 0			Sec Alarm 0				00-59
0x08	0x88	M	10-Min Alarm 0		Min Alarm 0					00-59
0x09	0x89	M	12 / 24	P / A / 10	10-HR	Hour Alarm 0				01-12 + P/A / 00-23
0x0A	0x8A	M	0	0	0	0	Day Alarm 0			01-07
0x0B	0x8B	M	10-Sec Alarm 1			Sec Alarm 1				00-59
0x0C	0x8C	M	10-Min Alarm 1		Min Alarm 1					00-59
0x0D	0x8D	M	12 / 24	P / A / 10	10-HR	Hour Alarm 1				01-12 + P/A / 00-23
0x0E	0x8E	M	0	0	0	0	Day Alarm 1			01-07
		—								
0x0F	0x8F	Control Register								—
0x10	0x90	Status Register								—
0x11	0x91	Trickle Charger Register								—
0x12-1F	0x92-9F	Reserve								—
0x20-7F	0xA0-FF	96-Bytes User RAM								—

Note. Range for alarm registers does not include mask 'm' bits

Figure 10.20 ■ RTC registers and address map

The internal time and date registers continue to increment during write operations. However, the countdown chain is reset when the seconds register is written. This ensures consistent data. Terminating a write before the last bit is sent aborts the write for that byte.

Buffers are used to copy the time and the date register at the beginning of a read. When reading in burst mode, the user copy is static, while the internal registers continue to increment. During a *burst mode read,* the processor sends in the starting address of a block of registers and that whole block is read. *Burst mode writes* are also supported.

12-Hour or 24-Hour Mode

The DS1306 can be run in either the 12-hour or the 24-hour mode. When bit 6 of the hours register is 1, the 12-hour mode is selected. Otherwise, the 24-hour mode is selected. In the 12-hour mode, bit 5 of the hours register indicates AM (0) or PM (1). In the 24-hour mode, bit 5 and bit 4 are used as the 10-hour bits (20–23 hours).

Alarms

The DS1306 contains two time-of-day alarms. The time-of-day Alarm 0 is set by writing into registers 0x87 through 0x8A. The time-of-day Alarm 1 is set by writing into registers 0x8B through 0x8E. Bit 7 of the alarm register is a mask bit. Alarm time is stored in four registers. When all four mask bits of an alarm are 0s, a time-of-day alarm occurs only once per week when the values stored in timekeeping registers 0x00 through 0x03 match the values stored in the time-of-day alarm registers. An alarm is generated every day when bit 7 of the day alarm register is set to 1. An alarm is generated every hour when bit 7 of the day and hour alarm registers is set to 1. Similarly, an alarm is generated every minute when bit 7 of the day, hour, and minute alarm registers is set to 1. When bit 7 of the day, hour, minute, and seconds alarm registers is set to 1, an alarm occurs every second.

During each clock update (every second), the RTC compares the Alarm 0 and Alarm 1 registers with the corresponding clock registers. When a match occurs, the corresponding alarm flag in the status register is set to 1. If the corresponding alarm interrupt enable bit is set, an interrupt will be requested.

Control Register

This register has four functions: enable/disable write protection, enable/disable 1-Hz clock output, enable/disable Alarm 0 interrupt, and enable/disable Alarm 1 interrupt. The contents of this register are shown in Figure 10.21.

7	6	5	4	3	2	1	0
0	WP	0	0	0	1 Hz	AIE1	AIE0

WP: Write protect

   0 = no write protect

   1 = inhibits write operation

1 Hz: 1 Hz Output Enable

   0 = disables 1 Hz clock output

   1 = enables 1 Hz clock output

AIE1: Alarm Interrupt Enable 1

   0 = the setting of IRQF1 bit of the status register does not assert INT1 pin

   1 = the setting of IRQF1 bit of the status register asserts the INT1 pin

AIE0: Alarm Interrupt Enable 0

   0 = the setting of IRQF0 bit of the status register does not assert INT0 pin

   1 = the setting of IRQF0 bit of the status register asserts the INT0 pin

Figure 10.21 ■ RTC Control register

Status Register

The RTC status register records the status of alarm time comparisons. As shown in Figure 10.22, only two bits are used. When the IRQF1 flag is set to 1 and the AIE1 bit is also set to 1, the INT1 pin will output a 62.5-ms pulse.

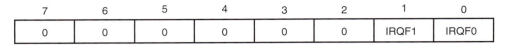

7	6	5	4	3	2	1	0
0	0	0	0	0	0	IRQF1	IRQF0

IRQF1: Interrupt 1 Request Flag

    0 = current time does not match the alarm time 1

    1 = current time has matched the alarm time 1

IRQF0: Interrupt 0 Request Flag

    0 = current time does not match the alarm time 0

    1 = current time has matched the alarm time 0

Figure 10.22 ■ RTC Status register

The IRQF0 flag is cleared when the address pointer goes to any of the Alarm 0 registers during a read or a write. When the IRQF0 flag is set to 1 and the AIE0 bit is also set to 1, the INT0 pin will go low. IRQF0 is activated when the device is powered by V_{CC1}, V_{CC2}, or V_{BAT}. IRQF1 is cleared when the address pointer goes to any of the Alarm 1 registers during a read or a write. IRQF1 is activated only when the device is powered by V_{CC2} or V_{BAT}.

TRICKLE CHARGE REGISTER

The DS1306 supports rechargeable power source by providing a trickle charger. The trickle charge register controls the trickle charger characteristics of the DS1306. A simplified diagram of the trickle charger is shown in Figure 10.23. In order to utilize the trickle charging capability, a rechargeable power source should be connected to V_{CC2} pin.

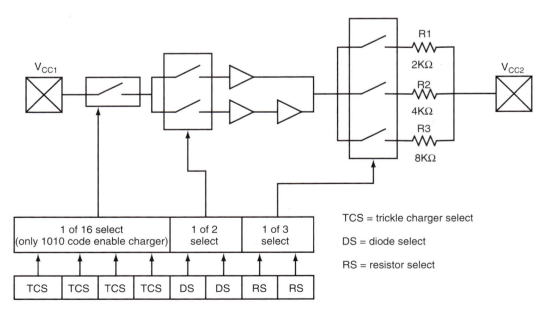

Figure 10.23 ■ Programmable trickle charger

The contents of the trickle charge register are shown in Figure 10.24. The user can select to have one or two diodes in the recharging path and select among three different resistor values to control the magnitude of the recharging current. The charging current can be calculated accordingly. For example, when one diode and R1 is chosen, the maximum charging current is

$$I_{MAX} = (5.0 \text{ V} - \text{diode drop})/R1 = 4.3 \text{ V}/2.0 \text{ K} = 2.2 \text{ mA}$$

7	6	5	4	3	2	1	0
TCS3	TCS2	TCS1	TCS0	DS1	DS0	RS1	RS0

TCS3..TCS0: Trickle charge select bits

DS1..DS0: diode select bits

RS..RS0: register select bits

T C S 3	T C S 2	T C S 1	T C S 0	D S 1	D S 0	R S 1	R S 0	
x	x	x	x	x	x	0	0	Disabled
x	x	x	x	0	0	x	x	Disabled
x	x	x	x	1	1	x	x	Disabled
1	0	1	0	0	1	0	1	1 Diode, 2KΩ
1	0	1	0	0	1	1	0	1 Diode, 4KΩ
1	0	1	0	0	1	1	1	1 Diode, 8KΩ
1	0	1	0	1	0	0	1	2 Diode, 2KΩ
1	0	1	0	1	0	1	0	2 Diode, 4KΩ
1	0	1	0	1	0	1	1	2 Diode, 8KΩ

Figure 10.24 ■ RTC Trickle charge register

10.9.3 Serial Interface of DS1306

The DS1306 can communicate with the SPI interface or with a standard three-wire interface. The SERMODE input selects the interface method. When the SERMODE input is high, the SPI method is selected. Otherwise, the three-wire communication method is selected. The SPI and three-wire communications differ in two aspects:

1. In SPI, SDO and SDI are separate pins. In three-wire, there is only one I/O pin for data coming in and going out.

2. In SPI, each byte is shifted most significant bit first. In three-wire, each byte is shifted least significant bit first.

In either method, an address byte is written to the device followed by a single data byte or multiple data bytes. The most significant bit of the address byte determines if a read or a write takes places. For a multiple byte transfer, each read or write cycle causes the RTC register or RAM address to automatically increment. Incrementing continues until the device is disabled (CE signals goes low). Since three-wire method is not supported in the PIC18 microcontroller, only the SPI method is explored in this section.

Like the TC72, the DS1306 can shift data on either clock edge, depending on the idle state of the SCLK clock signal. The DS1306 determines the clock polarity by sampling the SCLK input when the CE signal becomes active.

10.9.4 Power Supply Configurations

Since the DS1306 has three power input pins, power supply can be configured for the needs of applications. Figure 10.25 shows three power supply configurations that are allowed.

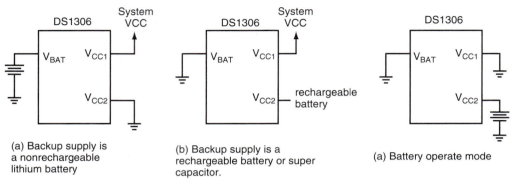

(a) Backup supply is a nonrechargeable lithium battery

(b) Backup supply is a rechargeable battery or super capacitor.

(a) Battery operate mode

Figure 10.25 ■ DS1306 power supply configurations

10.9.5 Applications

The main applications of the DS1306 are to keep track of the time of day and to remind people what needs to be done by sending the preset alarms. A circuit connection that configures the DS1306 in SPI mode is shown in Figure 10.26.

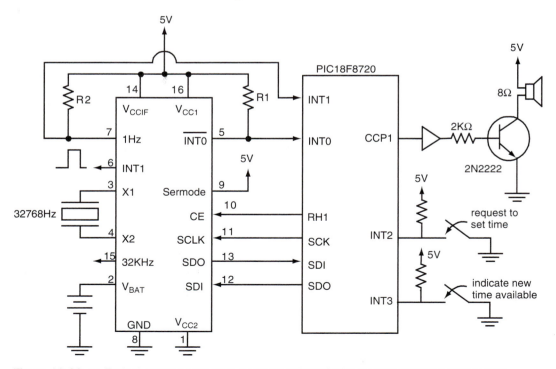

Figure 10.26 ■ Typical circuit connection between PIC18F8720 and the DS1306 (SPI mode)

In Figure 10.26, the user can program the DS1306 to remind the PIC18 microcontroller (via INT0 interrupt) to turn on the alarm when the alarm time is reached. The user can use DIP switches or a keypad to enter the time of day and alarm times to the PIC18 to be sent to the DS1306.

The 1-Hz output (when enabled) will interrupt the microcontroller once per second so that the time of day can be updated in a timely manner. The user can choose to use LCD or seven-segment displays to display the time-of-day information.

An embedded system that employs a DS1306 should contain at least the following components:

1. *Startup routine.* This part is executed whenever the microcontroller is reset. It will configure the SPI function and the DS1306 properly and display the current time of day so that the user can see if there is a need to update the time.

2. *Time update and display.* The 1-Hz square waveform interrupts the microcontroller once per second. If the user wants to change the current time of day, he or she presses the debounced key switch connected to the RB2/INT2 pin for at least 1 second, and it will be detected. If the user did not press the key, then the interrupt service routine simply reads the current time of day and displays it on the LCD. If the user has pressed the key, the microcontroller reads the current time from the DIP switches and then sends it to the DS1306.

3. *Alarm generation.* Whenever the alarm time is reached, the DS1306 interrupts the microcontroller (using the INT0 signal). In response, the PIC18 microcontroller uses the CCP1 to make a loud noise (at 2 KHz) with the speaker. The alarm will last for 1 minute. At the end of 1 minute, the PIC18 microcontroller turns off the speaker.

Example 10.6

▼

For the SSE8720 demo board, write a startup routine to configure the DS1306 to operate with the following features:

- Enable Alarm 0 interrupt and 1-Hz output
- Select 2 KΩ and one diode for trickle charger

Solution: The value to be written into the DS1306 control register to enable Alarm 0 interrupt and enable 1-Hz output is 0x05. The value to select the 2-KΩ resistor and one diode for the trickle charger is 0xA5.

The following C function will configure DS1306 accordingly:

```
void Open_DS1306 (void)
{
        unsigned char temp;
/* configure control register of the DS1306 */
        PORTH |= 0x02;              /* enable SPI transfer to DS1306 */
        SSPBUF = 0x8F;             /* send out the address of the control register */
        while (!SSPSTATbits.BF);
        temp = SSPBUF;             /* clear the BF flag */
        SSPBUF = 0x05;             /* enable alarm 0 interrupt & 1Hz output */
        while (!SSPSTATbits.BF);
        temp = SSPBUF;             /* clear the BF flag */
        PORTH &= 0xFD;             /* disable SPI transfer to DS1306 */
/* configure trickle charge register of the DS1306 */
        PORTH |= 0x02;              /* enable SPI transfer to DS1306 */
        SSPBUF = 0x91;             /* send out the address of trickle register */
        while (!SSPSTATbits.BF);
```

```
temp = SSPBUF;                    /* clear the BF flag */
SSPBUF = 0xA5;                    /* select 2K & 1 diode for trickle charger */
while (!SSPSTATbits.BF);
temp = SSPBUF;                    /* clear the BF flag */
PORTH &= 0xFD;                    /* disable SPI transfer to DS1306 */
}
```

The real-time clock system needs to read the current time of day from the DS1306. According to Figure 10.20, time of day has seven components (year, month, and so on), and each *alarm time* has four components. Since both alarm times are identical, only Alarm 0 is discussed in this section. For convenience, an array (**cur_time[0 . . . 10]**) is allocated to store the current time of day and Alarm 0. The order of time components follows that in Figure 10.20.

In a real-time clock, time of day is updated once per second. The following examples read and send time of day from/to DS1306.

Example 10.7

Write a function to read time of day from the DS1306 and save it in the array **cur_time [0 . . . 6]**.

Solution: The following C function will read the time of day from the DS1306:

```
char cur_time[11];                /* array to hold time-of-day and alarm time */
void read_time_of_day(void)
{
    char i, temp2;
    PORTHbits.RH1 = 1;            /* enable SPI transfer to/from DS1306 */
    SSPBUF = 0x00;               /* send read address of second to DS1306 */
    while (!SSPSTATbits.BF);      /* wait until address is shifted out */
    temp2 = SSPBUF;              /* clear the BF flag */
    for (i = 0; i < 7; i++) {
        SSPBUF = 0x00;           /* trigger 8 clock pulses */
        while(!SSPSTATbits.BF);
        cur_time[i] = SSPBUF;    /* read time component i and clear BF */
    }
    PORTHbits.RH1 = 0;           /* disable SPI transfer to/from DS1306 */
}
```

Example 10.8

Write a C function to send time of day stored in the array **cur_time[0..6]** to the DS1306.

Solution: The following C function will perform accordingly:

```
void send_cur_time(void)
{
    char i, temp1;
    PORTHbits.RH1 = 1;            /* enable SPI transfer to DS1306 */
    SSPBUF = 0x80;               /* send out the write address of seconds */
```

```
    while(!SSPSTATbits.BF);
    temp1 = SSPBUF;              /* clear the BF flag */
    for (i = 0; i < 7; i++) {
        SSPBUF = cur_time[i];
        while(!SSPSTATbits.BF);
        temp1 = SSPBUF;          /* clear the BF flag */
    }
    PORTHbits.RH1 = 0;           /* disable SPI transfer to DS1306 */
}
```

The user can write C functions to read and send alarm time from/to the DS1306. These two C functions are similar to those in Examples 10.7 and 10.8.

Before the current time of day can be displayed, it must be formatted. The following example formats the current time of day (stored in the **array cur_time[0 . . . 6]**) into two strings to be displayed on the first and second rows of the LCD display.

▲

Example 10.9

▼

Write a C function to convert the current time of day stored in the array **cur_time[0 . . . 6]** to two strings to be saved in two character arrays: **hms[0 . . . 11]** and **dmy[0 . . . 11]**. The array **hms[0 . . . 11]** holds the hours, minutes, and seconds, whereas the array **dmy[0 . . . 11]** holds month, date, and year. The format of display is as follows:

In 24-hour mode,

hh:mm:ss:xx

mm:dd:yy

where xx stands for day of week (can be "SU," "MO," "TU," "WE," "TH," "FR," and "SA").

In 12-hour mode,

hh:mm:ss:ZM

xx:mm:dd:yy

where Z can be "A" or "P" to indicate AM or PM and xx stands for day of week.

The following C function will format the current time of day into two strings:

```
void format_time(void)
{
    char                    temp3;
    temp3 = cur_time[3]& 0x07;
    if (cur_time[2] & 0x40) {    /* if 12-hour mode is used */
        hms[0] = 0x30 + ((cur_time[2] & 0x10) >> 4);   /* ten's hour digit */
        hms[1] = 0x30 + (cur_time[2] & 0x0F);          /* one's hour digit */
        hms[2] = ':';
        hms[3] = 0x30 + ((cur_time[1] & 0x70) >> 4);   /* ten's minute digit */
        hms[4] = 0x30 + (cur_time[1] & 0x0F);          /* minute's digit */
        hms[5] = ':';
        hms[6] = 0x30 + ((cur_time[0] & 0x70) >> 4);   /* ten's second digit */
        hms[7] = 0x30 + (cur_time[0] & 0x0F);          /* second's digit */
        hms[8] = ':';
        if (cur_time[2]& 0x20)
        hms[9] = 'P';
```

```
                          else
                          hms[9] = 'A';
                          hms[10] = 'M';
                          hms[11] = '\0';     /* terminate the string with a NULL */
                          switch (temp3) {    /* convert to day of week */
                              case 1:      dmy[0] = 'S';
                                           dmy[1] = 'U';
                                           break;
                              case 2:      dmy[0] = 'M';
                                           dmy[1] = 'O';
                                           break;
                              case 3:      dmy[0] = 'T';
                                           dmy[1] = 'U';
                                           break;
                              case 4:      dmy[0] = 'W';
                                           dmy[1] = 'E';
                                           break;
                              case 5:      dmy[0] = 'T';
                                           dmy[1] = 'H';
                                           break;
                              case 6:      dmy[0] = 'F';
                                           dmy[1] = 'R';
                                           break;
                              case 7:      dmy[0] = 'S';
                                           dmy[1] = 'A';
                                           break;
                              default:     dmy[0] = 0x20; /* space */
                                           dmy[1] = 0x20;
       }

                          dmy[2] = ':';
                          dmy[3] = 0x30 + ((cur_time[5] & 0x30) >> 4);        /* month */
                          dmy[4] = 0x30 + (cur_time[5] & 0x0F);
                          dmy[5] = ':';
                          dmy[6] = 0x30 + ((cur_time[4] & 0x30) >> 4);        /* date */
                          dmy[7] = 0x30 + (cur_time[4] & 0x0F);
                          dmy[8] = ':';
                          dmy[9] = 0x30 +(cur_time[6] >> 4);      /* year */
                          dmy[10] = 0x30 + cur_time[6] & 0x0F;
                          dmy[11] = '\0';                        /* NULL character */
                          } else {                              /* 24-hour mode */
                          hms[0] = 0x30 + ((cur_time[2]& 0x30) >> 4);        /* hours */
                          hms[1] = 0x30 + (cur_time[2]& 0x0F);
                          hms[2] = ':';
                          hms[3] = 0x30 + ((cur_time[1]& 0x70) >> 4);        /* minutes */
                          hms[4] = 0x30 + (cur_time[1]& 0x0F);
                          hms[5] = ':';
                          hms[6] = 0x30 + ((cur_time[0]& 0x70) >> 4);        /* seconds */
                          hms[7] = 0x30 + (cur_time[0]& 0x0F);
```

```
                    hms[8] = ':';
                    switch (temp3) {
                        case 1:        hms[9]              = 'S';
                                       hms[10]             = 'U';
                                       break;
                        case 2:        hms[9]              = 'M';
                                       dmy[10]             = 'O';
                                       break;
                        case 3:        hms[9]              = 'T';
                                       hms[10]             = 'U';
                                       break;
                        case 4:        hms[9]              = 'W';
                                       hms[10]             = 'E';
                                       break;
                        case 5:        hms[9]              = 'T';
                                       hms[10]             = 'H';
                                       break;
                        case 6:        hms[9]              = 'F';
                                       hms[10]             = 'R';
                                       break;
                        case 7:        hms[9]              = 'S';
                                       hms[10]             = 'A';
                                       break;
                        default:       hms[9]              = 0x20; /* space */
                                       hms[10]             = 0x20;
                    }
                    hms[11] = '\0';
                    dmy[0] = 0x30 + ((cur_time[5] & 0x30)>> 4);         /* month */
                    dmy[1] = 0x30 + (cur_time[5] & 0x0F);               /* month */
                    dmy[2] = ':';
                    dmy[3] = 0x30 + ((cur_time[4] & 0x30) >> 4);        /* date */
                    dmy[4] = 0x30 + (cur_time[4] & 0x0F);               /* date */
                    dmy[5] = ':';
                    dmy[6] = 0x30 +(cur_time[6] >> 4);        /* year */
                    dmy[7] = 0x30 + cur_time[6] & 0x0F;       /* year */
                    dmy[8] = '\0';                            /* NULL character */
            }
    }
```

The following C function will display the previous two strings on the LCD:

```
void disp_time(void)
{
            LCD_cmd(0x83);                    /* set cursor to the 3rd column of first row */
            LCD_putsram(&hms[0]);             /* display hours, minutes, and seconds */
            LCD_cmd(0xC3);                    /* set cursor to the 3rd column of 2nd row */
            LCD_putsram(&dmy[0]);             /* display month, date, and year */

}
```

Example 10.10

Write the main program that controls the overall operation. The main program will initialize the SPI, configure LCD, and set up the DS1306 operation and then stay in a while loop to wait for the flag **set_time** to be set true. When the **set_time** flag is true, the main program calls the **get_cur_time( )** function and the **send_cur_time( )** function to allow the user to enter the new time of day and the alarm time using the DIP switches and send it to the DS1306.

Solution: The main program is as follows:

```
void main (void)
{
    OpenSPI(SPI_FOSC_16, MODE_00, SMPMID);   /* configure SPI */
    LCD_init();                              /* configure LCD settings */
    TRISH = 0x00;                            /* configure port H for output */
    TRISF = 0xFF;                            /* configure port F for input */
    ADCON1 = 0x0A;                           /* configure RF7 through RF0 for digital input */
    TRISB = 0x0F;                            /* configure INT3..INT0 for input */
    Open_DS1306( );                          /* configure DS1306 */
                                             /* initialize set_time to 0 */
    set_time = 0;
                                             /* enable priority interrupts */
    INTCON2 = 0xBE;                          /* disable PORTB pullup, select active edges for INT0..INT3*/
    INTCON3 = 0x48;                          /* set INT1 in high priority, enable INT1 interrupt */
    RCON |= 0x80;                            /* enable priority interrupt */
    INTCON = 0xD0;                           /* enable global and INT0 interrupts */
                                             /* infinite loop to wait for set time request */
    while (1) {
        if (set_time){
            INTCON3bits.INT3IE = 1;          /* enable INT3 interrupt */
            get_cur_time( );                 /* get new time-of-day from keypad */
            send_cur_time( );                /* send new time-of-day to DS1306 */
            get_alarm_time( );               /* get new alarm time from keypad */
            send_alarm_time( );              /* send new alarm time to DS1306 */
            set_time = 0;                    /* done with set time */
            INTCON3bits.INT3IE = 0;          /* disable INT3 interrupt */
            INTCONbits.INT0IE = 1;           /* re-enable alarm interrupt */
            INTCON3bits.INT1IE = 1;          /* re-enable 1Hz interrupt */
        }
    }
}
```

Toward the end of the main program, the interrupts INT0 and INT1 are reenabled because these two interrupts are disabled when the user presses the RB3/INT3 pin to request a new time of day.

The main program calls the function **get_cur_time( )** to obtain the new time of day from the DIP switches. This function will use LCD to inform the user what time component to enter. Two separate functions are used to get user enter new time of day and alarm times. Because setting up DIP switches is a relatively slow process, it is better send the new time of day to the DS1306 immediately after the new seconds are entered so that DS1306 can start to count the new time. The resultant time will be more accurate. The following example illustrates the function.

Example 10.11

▼

Write the function get_cur_time.

Solution: This function will ask the user to enter time components starting from year toward seconds. Once seconds are entered, the function returns to the main program, and main program sends the new time of day to the DS1306 immediately. For each time component to be entered, this function clears the **get_it** flag to 0 and waits for the user to enter the specified time component. Once the user sets up the new time component using the DIP switches, he or she press the key switch connected to the RB3/INT3 pin and the microcontroller is interrupted. The INT3 service routine sets **get_it** to 1, reads the time component, and saves it in a temporary buffer, **time_buf.** When interrupts returns, the **get_it** flag is 1, and the function **get_cur_time( )** places the entered time component in the appropriate slot of the **cur_time[ ]** array. The **get_cur_time( )** function is as follows:

```c
void get_cur_time (void)
{
    LCD_cmd(0x80);              /* set cursor to row 1 column 1 */
    LCD_putstr(msg1);          /* remind user to enter year */
    get_it = 0;
    while(!get_it);            /* wait for the user to set up new year in DIP switches */
    cur_time[6] = time_buf;    /* save year */
    LCD_cmd(0x80);             /* set cursor to row 1 column 1 */
    LCD_putstr(msg2);          /* remind user to enter month */
    get_it = 0;
    while(!get_it);
    cur_time[5] = time_buf;    /* save month */
    LCD_cmd(0x80);             /* set cursor to row 1 column 1 */
    LCD_putstr(msg3);          /* remind user to enter date of month */
    get_it = 0;
    while(!get_it);
    cur_time[4] = time_buf;    /* save date */
    LCD_cmd(0x80);             /* set cursor to row 1 column 1 */
    LCD_putstr(msg4);          /* remind user to enter day of week */
    get_it = 0;
    while(!get_it);
    cur_time[3] = time_buf;    /* save day-of-week */
    LCD_cmd(0x80);             /* set cursor to row 1 column 1 */
    LCD_putstr(msg5);          /* remind user to enter hours */
    get_it = 0;
    while(!get_it);
    cur_time[2] = time_buf;    /* save hours */
    LCD_cmd(0x80);             /* set cursor to row 1 column 1 */
    LCD_putstr(msg6);          /* remind user to enter minutes */
    get_it = 0;
    while(!get_it);
    cur_time[1] = time_buf;    /* save minutes */
    LCD_cmd(0x80);             /* set cursor to row 1 column 1 */
    LCD_putstr(msg7);          /* remind user to enter seconds */
    get_it = 0;
    while(!get_it);
    cur_time[0] = time_buf;    /* save seconds */
}
```

The new alarm time can be set up using the same method. The function to obtain the new alarm time is called **get_alarm_time( )**. The messages that are used to inform the user to enter appropriate time components are as follows:

```
rom     char    *msg1     = "enter year:";
rom     char    *msg2     = "enter month:";
rom     char    *msg3     = "enter date of month:";
rom     char    *msg4     = "enter day of week:";
rom     char    *msg5     = "enter hours:";
rom     char    *msg6     = "enter minutes:";
rom     char    *msg7     = "enter seconds:";
rom     char    *msg8     = "enter alarm day:";
rom     char    *msg9     = "enter alarm hours:";
rom     char    *msg10    = "enter alarm minutes:";
rom     char    *msg11    = "enter alarm seconds:";
```

The real-time clock application also needs to trigger an alarm whenever the alarm time is reached. The following example illustrates how to trigger an alarm.

Example 10.12

Write a C function to trigger an alarm (2-KHz square waveform) using the PWM1 channel, assuming that the crystal oscillator frequency for the PIC18 microcontroller is 16 MHz.

Solution: This function will use Timer2 as the base timer for PWM1. Both Timer2 and Timer3 must be configured properly. By setting the prescale factor to Timer2 to 16, the value to be written into the period register is found to be 124, and the duty cycle value to be written into the CCPR1 register is 250. Since the alarm is to be lasted for 1 minute, a counter (**alarm_remain**) is used to keep track of the duration for the alarm. The counter **alarm_remain** is initialized to 60 and will be decremented by 1 each time the 1-Hz clock interrupts the microcontroller. The 1-Hz interrupt service routine will decrement this count value if the **in_alarm** flag is set to 1 (set by the INT0 service routine). When the count value is decremented to 0, the alarm will be stopped:

```c
void trigger_alarm (void)
{
    char temp4;
    OpenTimer2 (TIMER_INT_OFF & T2_PS_1_16 & T2_POST_1_1);
    OpenTimer3 (TIMER_INT_OFF & T3_16BIT_RW & T3_SOURCE_INT & T3_PS_1_1 & T12_SOURCE_CCP);
    OpenPWM1 (124);                 /* generate a PWM output at 2KHz */
    SetDCPWM1 (250);                /* set duty cycle to 50% */
    alarm_remain = 60;              /* let alarm to last for 1 minute */
                                    /* read alarm day to clear the IRQF0 flag */
    PORTHbits.RH1 = 1;              /* enable SPI transfer */
    SSPBUF = 0x0A;                  /* send alarm day read address */
    while(!SSPSTATbits.BF);
    temp4 = SSPBUF;                 /* clear the BF flag */
    SSPBUF = 0x00;                  /* read alarm day */
    while(!SSPSTATbits.BF);
    temp4 = SSPBUF;
    PORTHbits.RH1 = 0;             /* disable SPI transfer */
}
```

In this real-time clock application, three interrupts are used:

- *INT0 interrupt.* The Alarm 0 interrupt of the DS1306 is connected to this pin. Whenever this pin is asserted (low), the microcontroller will trigger an alarm that lasts for 1 minute. This service routine also sets the **in_alarm** flag to 1. The user can use any one of the available PWM channels to generate the alarm. Example 10.12 illustrates how to generate an alarm.

- *INT1 interrupt.* This interrupt pin is connected to the 1-Hz output of the DS1306. The service routine of this interrupt will check to see if the user press RB2/INT2 pin. If the RB2 pin is pressed, then the service routine will set a flag (called **set_time**) and return. If the user does not press the RB2/INT2 pin, then the INT1 service routine reads the current time of day from the DS1306, formats the data, and displays the result on the LCD screen.

- *INT3 interrupt.* This interrupt informs the microcontroller that the user has set up a time component using the DIP switches. The microcontroller will then read the data and place it in the appropriate slot in the array **cur_time[ ]**.

Example 10.13

▼

Write the interrupt service routines for INT0, INT1, and INT3.

Solution: In reality, the service routines for these three interrupt sources must be placed in the same function because the PIC18 microcontroller has only two interrupt priority levels. This example will enable priority interrupt:

```
#pragma code
#pragma interrupt high_ISR
void high_ISR(void)                    /* This ISR services only INT0..INT3 interrupts */
{
        if (INTCON3bits.INT3IF) {          /* caused by user pressed on RB3/INT3 pin */
            time_buf = PORTF;              /* read a time component */
            get_it = 1;                    /* informs get_cur_time that a time component is entered */
            INTCON3bits.INT3IF = 0;
            _asm
            retfie 0
            _endasm
        }
        if (INTCON3bits.INT1IF) {          /* caused by 1Hz interrupt */
            INTCON3bits.INT1IF = 0;
            if (!PORTBbits.RB2) {          /* check to see if user requests to enter new times */
                set_time = 1;
                INTCONbits.INT0IE = 0;     /* temporarily disable alarm interrupt */
                INTCON3bits.INT1IE = 0;    /* temporarily disable 1Hz interrupt */
                _asm
                retfie 0
                _endasm
            }
        else {                             /* no user request to enter new time-of-day */
            read_time_of_day( );
            format_time( );
```

```
                              disp_time( );
                       }
                       if (in_alarm) {
                              alarm_remain--;
                              if (alarm_remain == 0){
                                     ClosePWM1( );              /* stop alarm */
                                     in_alarm = 0;              /* clear alarm flag */
                              }
                       }
                }
                if(INTCONbits.INTOIF){                          /* interrupt caused by alarm at INT0 pin */
                       trigger_alarm();                        /* start PWM1 to drive the speaker */
                       in_alarm = 1;                           /* indicate alarm is on */
                       INTCONbits.INTOIF = 0;                   /* clear the INTOIF flag */
                }
                _asm                                           /* ignore all other interrupt */
                retfie 0
                _endasm
        }
```

Examples 10.6 to 10.13 illustrate that a real-time clock application can be very complicated. These examples also illustrate how to coordinate normal programs and interrupt service routines by using flags.

10.10 Summary

When high-speed data transfer is not necessary, using serial data transfer enables us to make the most use of the limited number of I/O pins available on the microcontroller device. Serial data transfer can be performed asynchronously or synchronously. The SPI is a synchronous protocol created by Motorola for serial data exchange between peripheral chips and microcontrollers.

In the SPI format, a device (must be a microcontroller) must be responsible for initiating the data transfer and generating the clock pulses for synchronizing data transfer. This device is referred to as the SPI master. All other devices in the same system are referred to as SPI slaves. The master device needs three signals to carry out the data transfer:

- SCK: A clock signal for synchronizing data transfer
- SDO: Serial data output from the master (also called MOSI in Motorola literature)
- SDI: Serial data input to the master (also called MISO in Motorola literature)

To transfer data to one or more SPI slaves, the microcontroller writes data into the SPI data register, and eight clock pulses will be generated to shift out the data in the SPI data register from the SDO pin. If the SDI pin of the microcontroller is also connected (to the slave), then eight data bits will also be shifted into the SPI data register. To read data from the slave, the microcontroller also needs to write data into the SPI data register to trigger clock pulses to be sent out from the SCK pin. However, the value written into the SPI data register is unimportant in this case.

When configuring as a slave device, the PIC18 microcontroller also needs the fourth signal, called slave select ($\overline{SS}$). The $\overline{SS}$ signal enables the PIC18 slave to respond to an SPI data transfer. Most slave peripheral devices have signals called CE (chip enable) or CS (chip select) to enable/disable the SPI data transfer.

Multiple peripheral devices with the SPI interface can be interfaced with a single PIC18 microcontroller simultaneously. There are many different methods for interfacing multiple peripheral devices (with SPI interface) to the microcontroller. Two popular connection methods can be identified as follows:

1. *Parallel connection.* In this method, the SDI, SDO, and SCK signals of the peripheral devices are connected to the same signals of the PIC18 device. The PIC18 microcontroller also needs to use certain unused I/O pins to control the CS (or CE) inputs of each individual peripheral devices. Using this method, the PIC18 microcontroller can exchange data with any selected peripheral device without affecting other peripheral devices.

2. *Serial connection.* In this method, the SDI input of a peripheral device is connected to the SDO pin of its predecessor, whereas the SDO output of a peripheral device is connected to the SDI input of its successor. The SDI input of the peripheral device that is closest (in terms of connection) to the microcontroller is connected to the SDO output of the microcontroller. The SDO output of the last peripheral device (in the loop) is connected to the SDI input of the microcontroller. The SCK inputs of all peripheral devices are tied to the SCK pin of the microcontroller. Using this method, the data to send to the last device in the loop will need to go through every other peripheral device. The CE (or CS) signal of all peripheral devices is either tied to ground or tied to high.

Among all peripheral devices with the SPI interface, shift registers such as 74HC595 and 74HC589 can be used to add parallel output and input ports to the microcontroller. A digital temperature sensor is useful for displaying the ambient temperature. The TC72 and TC77 from Microchip and the LM74 from National Semiconductor are examples of digital temperature sensors with the SPI interface. The SPI interface is also often added to LED and LCD display drivers. The MAX7221 from Dallas-Maxim can drive up to eight seven-segment displays. Motorola MC14489 can drive up to five seven-segment displays. Both driver chips can be cascaded to drive more displays. The DS1306 can keep track of time of day and allow the user to set up alarm times.

10.11 Exercises

E10.1 What signal pins are involved in the SPI transfer?

E10.2 Write an instruction sequence to configure SPI interface so that it can carry out data transfer with the following features, assuming that f_{OSC} = 32 MHz:

- SCK idle high and data transmitted on falling edge
- Input data sampled at middle of data output time
- Master mode
- Shift data at 8 MHz
- SPI enabled

E10.3 Suppose that there is an SPI-compatible peripheral output device that has the following characteristics:

- Has a CLK input pin that is used as the data shifting clock signal
- Has an SI pin to shift in data on the falling edge of the CLK input
- Has a $\overline{CE}$ pin that enables the chip to shift data when it is low
- The highest data shift rate is 500 KHz
- The most significant bit is shifted in first

Describe how to connect the SPI pins for the PIC18 microcontroller and this peripheral device and write an instruction sequence to configure the SPI subsystem properly for data transfer. Assume that f_{OSC} = 32 MHz.

E10.4 The 74HC589 is another SPI-compatible shift register. This chip has both serial and parallel inputs and is often used to expand the number of parallel input ports. The block diagram of the 74HC589 is shown in Figure 10E.4. The functions of these pins are as follows:

- *A, . . ., H—parallel data inputs.* Data on these pins is loaded into the data latch on the rising edge of the LC signal.

- *SA—serial data input.* Data on this pin is shifted into the shift register on the rising edge of the SC signal if the SS/$\overline{PL}$ pin is high. Data on this pin is ignored when the SS/$\overline{PL}$ signal is low.

- *SC—serial shift clock.* A low-to-high transition on this pin shifts data on the serial data input into the shift register; data on stage H is shifted out from Q_H, where it is replaced by the data previously stored in stage G. The highest shift clock rate is 5 MHz.

- *LC—latch clock.* A rising edge on this pin loads the parallel data on inputs A . . . H into the data latch.

- *$\overline{OE}$—output enable.* A high level on this pin forces the Q_H output into a high-impedance state. A low level enables the output. This control does not affect the state of the input latch or the shift register.

- *Q_H—serial data output.* This is a three-state output from the last stage of the shift register.

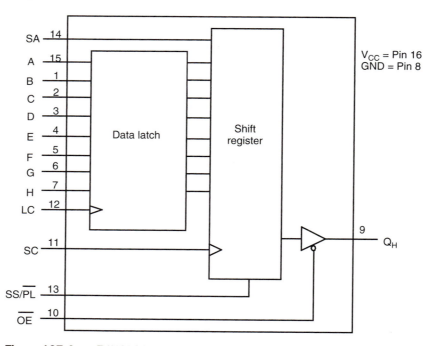

Figure 10E.4 ■ 74HC589 block diagram and pin assignment

Multiple 74HC589s can be cascaded. To cascade, the Q_H output is connected to the SA input of its adjacent 74HC589. All 74HC589s should share the same clock signal SC. Suppose that an application requires you to use two 74HC589s to interface with two DIP switches so that the circuit can read four hex digits into the PIC18 microcontroller in two SPI transfers. Describe the circuit connection and write an instruction sequence to read data from these two DIP switches. Leave these two bytes in PRODL and PRODH.

E10.5 The Motorola MC14489 is a five-digit seven-segment display driver chip. The datasheet is in the accompanying CD (also can be downloaded from Motorola Web site). Describe how to interface one MC14489 with the PIC18 microcontroller using the SPI interface. Write a program to display the value 13579 on the seven-segment displays driven by the MC14489.

E10.6 Write an instruction sequence to display the character string 10.50.30.PH on the seven-segment displays driven by the MAX7221 in Figure 10.12.

E10.7 Write an instruction sequence to display the character string abcdeFGH on the seven-segment displays driven by the MAX7221 in Figure 10.12.

E10.8 Use an MAX7221 to drive five seven-segment displays using a circuit connection similar to that in Figure 10.12. Write a program to display the following patterns:

```
        0
       10
      210
     3210
    43210
    54321
    65432
    76543
    87654
    98765
    09876
    10987
    21098
    32109
    43210

      . . .
```

Each pattern will be displayed for half a second. Repeat this operation forever.

E10.9 The LM74 is an SPI-compatible 12-bit plus sign digital temperature sensor made by National Semiconductor. The datasheet of this chip is included in the companion CD of this text. Show the circuit connection between the LM74 and the PIC18 microcontroller. Write a program to configure the LM74 in continuous conversion mode that reads the temperature once every second and convert the temperature reading into a string so that it can be displayed on the LCD. The LM74 has only a bidirectional data pin SI/O.

E10.10 Write the **get_alarm_time( )** function by referring to Example 10.11.

E10.11 Write the **format_alarm_time( )** function (similar to the function **format_time( )**), which will format the alarm time into one string in the format of

hh:mm:ss:PM (or AM):DD

where DD can be "SU" through "SA."

E10.12 Suppose you want to set the current time of day to

    10:30:50:AM   (hh:mm:ss)
    TU:06:17:03   (mm:dd:yy)

Write down the seven bytes to be sent to DS1306 starting from the write address 0x80.

E10.13 Suppose you want to set the new alarm time 0 to

    05:45:30:AM

which will alarm every morning. Write down the four bytes to be sent to the DS1306.

E10.14 Suppose you want to set the current time of day in 24-hour mode to

    14:30:10:TU   (hh:mm:ss)
    06:17:03   (mm:dd:yy)

Write down the seven bytes to be sent to DS1306 starting from the write address 0x80.

10.12 Lab Exercises and Assignments

L10.1 *Data shifting practice.* Use an MAX7221 to drive six seven-segment displays using a circuit connection similar to that in Figure 10.12. Write a program to display the following patterns:

    0
    10
    210
    3210
    43210
    543210
    654321
    765432
    876543
    987654
    098765
    109876
    210987
    321098
    432109
    543210

      . . .

Each pattern will be displayed for about half a second. Repeat this pattern forever.

L10.2 *Real-time clock and alarm practice.* Use the DS1306 chip and the required chips on the SSE8720 (or SSE8680) demo board to enter a new time of day and alarm time 0 and go through the displaying and alarm triggering process.

L10.3 *Digital thermometer.* Use the TC72 and the LCD kit on the SSE8720 (or SSE8680) to display the room temperature. Write a program that reads the temperature once per second and display it on the LCD. Press your finger on the TC72 chip to measure your hand temperature and see the result on the LCD.

11

Interintegrated Circuit Interface

11.1 Objectives

After completing this chapter, you should be able to

- Understand the characteristics of the I^{2}C protocol
- Describe the I^{2}C protocol in general
- Explain the I^{2}C signal components
- Explain the I^{2}C bus arbitration method
- Explain the I^{2}C data transfer format
- Use I^{2}C C library functions to perform data communications with I^{2}C slaves
- Use the DS1631A to measure ambient temperature
- Store and retrieve data in/from the 24LC08B

11.2 The I²C Protocol

The inter-integrated circuit (I²C) serial interface protocol was developed by Philips in the late 1980s. Version 1.0 was published in 1992. This version supports the following:

- Both the 100-Kbps (standard mode) and the 400-Kbps (fast mode) data rate
- 7-bit and 10-bit addressing
- Slope control to improve electromagnetic compatibility (EMC) behavior

After the publication of version 1.0, I²C became well received by the embedded application developers.

By 1998, the I²C protocol had become an industrial standard—it has been licensed to more than 50 companies and implemented in over 1000 different integrated circuits. However, many applications require a higher data rate than that provided by the I²C protocol. Version 2.0 was published in which the high-speed mode (with a data rate of 3.4 Mbps) was added to address this requirement. Since the PIC18 devices do not support the high-speed mode, it is not discussed in this chapter.

11.2.1 Characteristics of I²C Protocol

The I²C protocol has the following characteristics:

- *Synchronous in nature.* A data transfer is always initiated by a master device. A clock signal (SCL) synchronizes the data transfer. The clock rate can vary without disrupting the data. The data rate will simply change along with the changes in the clock rate.
- *Master slave model.* The master device controls the clock line (SCL). This line dictates the timing of all data transfers on the I²C bus. Other devices can manipulate this line, but they can only force the line low. By forcing the line low, it is possible to clock more data into any device. This is known as "clock stretching."
- *Bidirectional data transfer.* Data can flow in any direction on the I²C bus.
- *Serial interface method.* I²C uses only signals SCL and SDA. The SCL signal is the serial clock signal, whereas the SDA signal is known as serial data. In reality, the SDA signal can carry both the address and data.

11.2.2 I²C Signal Levels

I²C can have only two possible electrical states: *float high* and *driven low*. A master or a slave device drives the I²C bus using an open-drain (or open-collector) driver. As shown in Figure 11.1, both the SDA and the SCL lines are pulled up to V_{DD} via pull-up resistors. Because the driver circuit is open drain, it can pull the I²C bus only to low. When the clock or data output is low, the NMOS transistor is turned off. In this situation, no current flows from/to the bus to/from the NMOS transistor, and hence the bus line will be pulled to high by the pull-up resistor. Otherwise, the NMOS transistor is turned on and pulls the bus line to low.

The designer is free to use any resistor value for various speeds. But the calculation of what value to use will depend on the capacitance of the driven line and the speed of the I²C communication. In general, the recommended values for the pull-up resistors are 2.2 KΩ and 1 KΩ for standard mode and fast mode, respectively. These values are found to work frequently. When the data rate is below 100 Kbps, the pull-up resistor should be set to 4.7 KΩ.

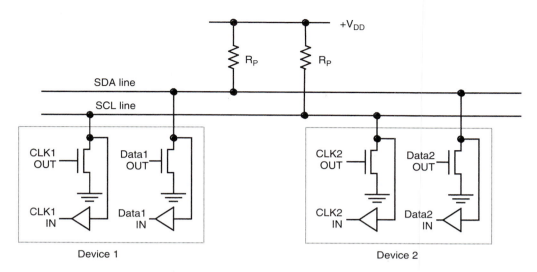

Figure 11.1 ■ Connecting standard- and fast-mode devices to the I²C bus

11.2.3 I²C Data Transfer Signal Components

An I²C data transfer consists of the following fundamental signal components:

- Start (**S**)
- Stop (**P**)
- Repeated start (**R**)
- Data
- Acknowledge (**A**)

START (S) CONDITION

A start condition indicates that a device would like to transfer data on the I²C bus. As shown in Figure 11.2, a start condition is represented by the SDA line going low when the clock (SCL) signal is high.

The start condition will initialize the I²C bus. The timing details for the start condition will be taken care of by the microcontroller that implements the I²C bus. Whenever the user wants to initiate a data transfer using the I²C bus, he or she must tell the microcontroller that he or she wants a start condition.

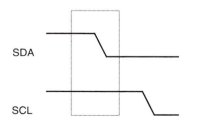

Figure 11.2 ■ I²C Start condition

Stop (P) Condition

A stop condition indicates that a device wants to release the I²C bus. Once released (the driver is turned off), other devices may use the bus to transmit data. As shown in Figure 11.3, a stop condition is represented by the SDA signal going high when the clock (SCL) signal is high.

Once the stop condition completes, both the SCL and the SDA signals will be high. This is considered to be an *idle bus.* After the bus is idle, a start condition can be used to send more data.

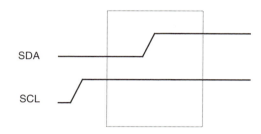

Figure 11.3 ■ Stop (P) condition

Repeated Start (R) Condition

A repeated START signal is a START signal generated without first generating a STOP signal to terminate the communication. This is used by the master to communicate with another slave or with the same slave in different mode (transmit/receive mode) without releasing the bus. A repeated start condition indicates that a device would like to send more data instead of releasing the line. This is done when a start must be sent but a stop has not occurred. It prevents other devices from grabbing the bus between transfers. The timing diagram of a repeated start condition is shown in Figure 11.4. In the figure, there is no stop condition occurring between the start condition and the restart condition.

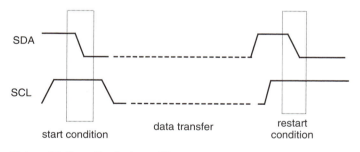

Figure 11.4 ■ Restart condition

Data

The data block represents the transfer of eight bits of information. The data is sent on the SDA line, whereas clock pulses are carried on the SCL line. The clock can be aligned with the data to indicate whether each bit is a "1" or a "0".

Data on the SDA line is considered valid only when the SCL signal is high. When SCL is not high, the data is permitted to change. This is how the timing of each bit works. Data bytes are used to transfer all kinds of information. When communicating with another I²C device, the eight bits of data may be a control code, an address, or data. An example of 8-bit data is shown in Figure 11.5.

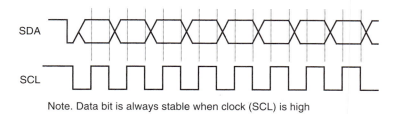

Note. Data bit is always stable when clock (SCL) is high

Figure 11.5 ■ I²C bus data elements

ACKNOWLEDGE (ACK) CONDITION

Data transfer in the I²C protocol needs to be acknowledged either positively (A) or negatively (NACK). As shown in Figure 11.6, a device can acknowledge (A) the transfer of each byte by bringing the SDA line low during the ninth clock pulse of SCL.

If the device does not pull the SDA line to low and instead allows the SDA line to float high, it is transmitting a negative acknowledge (NACK). This situation is shown in Figure 11.7.

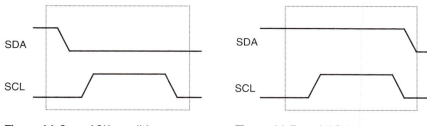

Figure 11.6 ■ ACK condition   **Figure 11.7** ■ NACK condition

11.2.4 Synchronization

All masters generate their own clocks on the SCL line to transfer messages on the I²C bus. Data is valid only during the high period of the clock. A defined clock is therefore needed for the bit-by-bit arbitration procedure to take place. For most microcontrollers (including the PIC18 devices), the SCL clock is generated by counting down a programmable reload value using the instruction clock signal.

Clock synchronization is performed using the wired-AND connection of I²C interfaces to the SCL line. This means that a high-to-low transition on the SCL line will cause the devices concerned to start counting off their low period, and once a device clock has gone low, it will hold the SCL line in that state until high state is reached (CLK1 in Figure 11.8). However, the transition from low to high of this clock may not change the state of the SCL line if another clock (CLK2) is still within its low period. The SCL line will therefore be held low by the device with the longest low period. Devices with shorter low periods enter a high *wait state* during this time.

When all devices concerned have counted off their low period, the clock line will be released and go high. There will then be no difference between the device clocks and the state of the SCL line, and all the devices will start counting their high periods.

In this way, a synchronized SCL clock is generated with its low period determined by the device with the longest clock low period and its high period determined by the one with the shortest clock high period.

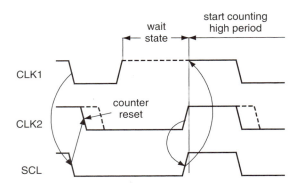

Figure 11.8 ■ Clock synchronization during the arbitration procedure

Clock synchronization occurs when multiple masters attempt to drive the I²C bus and before the arbitration scheme can decide which master is the winner. The I²C bus arbitration process is discussed in the next section.

11.2.5 Arbitration

I²C allows multiple master devices to coexist in the system. In the event that two or more master devices attempt to begin a transfer at the same time, an arbitration scheme is employed to force one master to give up the bus. The master devices continue transmitting until one attempts a high while the other transmits a low. Since the bus driver is open drain, the bus will be pulled low. The master attempting to transfer a high signal will detect a low on the SDA line and give up the bus by switching off its data output stage. The winning master continues its transmission without interruption; the losing master becomes a slave and receives the rest of the transfer. This arbitration scheme is nondestructive: one device always wins, and no data is lost.

An example of arbitration procedure is shown in Figure 11.9, where data 1 and data 2 are data driven by device 1 and device 2 and SDA is the resultant data on the SDA line. The moment there is a difference between the internal data level of the master generating data 1 and the actual level on the SDA line, its data output is switched off, which means that a high output level is then connected to the bus. This will not affect the data transfer initiated by the winning master.

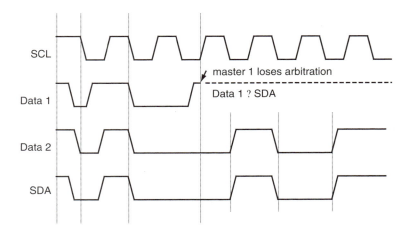

Figure 11.9 ■ Arbitration procedure of two masters

11.2.6 Data Transfer Format

I²C allows a device to use either the 7-bit or the 10-bit address to specify a slave device for data transfer. The following are the possible I²C data transfer formats:

- *Master transmitter to slave receiver.* The transfer direction is not changed. An example of this format using the 7-bit addressing is shown in Figure 11.10.
- *Master reads slave immediately after the first byte (address byte).* At the moment of the first acknowledgment, the master transmitter becomes a master receiver, and the slave receiver becomes a slave transmitter. The first acknowledgement is still generated by the slave. The stop condition is generated by the master, which has previously sent a negative acknowledgment ($\overline{A}$). An example of this format using the 7-bit addressing is shown in Figure 11.11.
- *Combined format.* During a change of direction within a transfer, both the start condition and the slave address are repeated, but with the R/$\overline{W}$ bit reversed. If a master receiver sends a repeated start condition, it has previously sent a negative acknowledgment. An example of this format in the 7-bit addressing is shown in Figure 11.12.

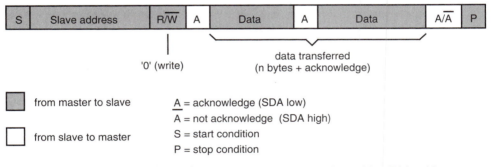

Figure 11.10 ■ A master-transmitter addressing a slave receiver with a 7-bit address. The transfer direction is not changed.

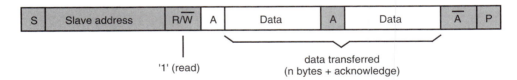

Figure 11.11 ■ A master read a slave immediately after the first byte

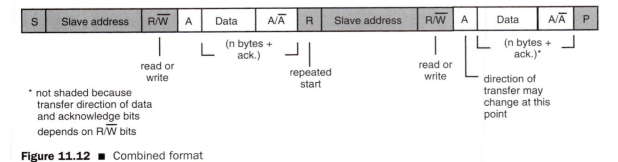

Figure 11.12 ■ Combined format

11.2.7 7-Bit Addressing

The addressing procedure for the I²C-bus is such that the first byte after the start condition usually determines which slave will be selected by the master. The exception is the *general call address* which can address all devices. When this address is used, all devices should, in theory, respond with an acknowledgment. However, devices can be made to ignore this address. The second byte of the general call address then defines the action to be taken.

The first byte after the start condition carries the 7-bit address and the direction of the message. The meaning of this byte is shown in Figure 11.13. When the least significant bit is "1," the master device will read information from the selected slave. Otherwise, the master will write information to a selected slave.

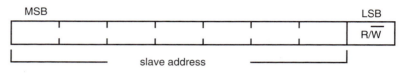

Figure 11.13 ■ The first byte after the start condition

When an address is sent, each device in a system compares the first seven bits after the start condition with its address. If they match, the device considers itself addressed by the master as a slave receiver or slave transmitter, depending on the R/$\overline{W}$ bit.

The I²C bus committee coordinates the allocation of I²C addresses. Two groups of eight addresses (0000xxx and 1111xxx) are reserved for the purpose shown in Table 11.1. The bit combination 11110xx of the slave address is reserved for 10-bit addressing.

Slave address	R/$\overline{W}$ bit	Description
0000 000	0	general call address
0000 000	1	start byte[1]
0000 001	x	CBUS address[2]
0000 010	x	Reserved for different bus format[3]
0000 011	x	Reserved for future purpose
0000 1xx	x	Hs-mode master code
1111 1xx	x	Reserved for future purposes
1111 0xx	x	10-bit slave addressing

Notes.
1. No device is allowed to acknowledge at the reception of the start byte.
2. The CBUS address has been reserved to enable the inter-mixing of CBUS compatible and I²C-bus compatible devices in the same system. I²C-bus compatible devices are not allowed to respond to on reception of this address.
3. The address reserved for a different bus format is included to enable I²C and other protocols to be mixed. Only I²C-bus compatible devices that can work with such formats and protocols are allowed to respond to this address.

Table 11.1 ■ Definition of bits in the first byte

GENERAL CALL ADDRESS

The general call address is for addressing every device connected to the I²C bus. However, if a device does not need any of the data supplied within the general call structure, it can ignore this address by not issuing an acknowledgment. If a device does require data from a general call address, it will acknowledge this address and behave as a slave receiver. The second and following bytes will be acknowledged by every slave receiver capable of handling this data. A slave that cannot process one of these bytes must ignore it by not acknowledging. The meaning of the general call address is always specified in the second byte as shown in Figure 11.14.

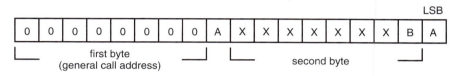

Figure 11.14 ■ General call address format

When bit B in Figure 11.14 is 0, the second byte has the following definitions:

- *00000110 (0x06)—reset and write programmable part of slave address by hardware.* On receiving this 2-byte sequence, all devices designed to respond to the general call address will reset and take in the programmable part of their address. Precautions have to be taken to ensure that a device is not pulling down the SDA or SCL line after applying the supply voltage since these low levels would block the bus.

- *00000100 (0x04)—write programmable part of slave address by hardware.* All devices that define the programmable part of their addresses by hardware will latch this programmable part at the reception of this 2-byte sequence. The device will not reset.

- *00000000 (0x00).* This code is not allowed to be used as the second byte.

Sequences of programming procedure are published in the appropriate device datasheets. The remaining codes have not been fixed, and devices must ignore them.

When bit B is 1, the 2-byte sequence is a *hardware general call.* This means that the sequence is transmitted by a hardware master device, such as a keyboard scanner, which cannot be programmed to transmit a desired slave address. Since a hardware master does not know in advance to which device the message has to be transferred, it can only generate this hardware general call and present its own address to identify itself to the system.

The seven bits remaining in the second byte contain the address of the hardware master. This address is recognized by an intelligent device (e.g., a microcontroller) connected to the bus, which will then direct the information from the hardware master. If the hardware master can also act as a slave, the slave address is identical to the master address.

In some systems, an alternative could be that the hardware master transmitter is set in the slave receiver mode after the system reset. In this way, a system-configuring master can tell the hardware master transmitter (which is now in slave receiver mode) to which address data must be sent (see Figure 11.15). After this programming procedure, the hardware master remains in the master transmitter mode.

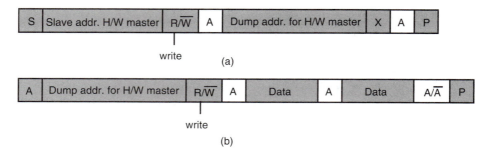

Figure 11.15 ■ Data transfer by a hardware-transmitter capable of dumping data directly to slave devices. (a) Configuring master sends dump address to hardware master (b) hardware master dumps data to selected slave

11.2.8 10-Bit Addressing

Ten-bit addressing is compatible with and can be combined with 7-bit addressing. Using 10-bit addressing exploits the reserved combination 1111xxx for the first seven bits of the first byte following a start (S) or repeated start (R) condition. The 10-bit addressing does not affect the existing 7-bit addressing. Devices with 7-bit and 10-bit addresses can be connected to the same I²C bus, and both 7-bit and 10-bit addressing can be used in standard- and fast-mode systems.

Although there are eight possible combinations of the reserved address bits 1111xxx, only the four combinations 11110xx are used for 10-bit addressing. The remaining four combinations 11111xx are reserved for future I²C bus enhancements.

Definitions of Bits in the First Two Bytes

The 10-bit slave address is formed from the first two bytes following a start (S) condition or repeated start (R) condition. The first seven bits of the first byte are the combination 11110xx, of which the last two bits (XX) are the two most significant bits of the 10-bit address. The eighth bit of the first byte is the R/$\overline{\text{W}}$ bit that determines the direction of the message. A 0 in the least significant bit of the first byte means that the master will write information to the selected slave. A 1 in this position indicates that the master will read information from the selected slave.

Formats with 10-Bit Address

The following data formats are possible in 10-bit addressing:

- *Master transmitter transmits to slave receiver with a 10-bit slave address.* The transfer direction is not changed (see Figure 11.16). When a 10-bit address follows a start condition, each slave compares the first seven bits of the first byte of the slave address (11110xx) with its own address and tests if the eighth bit is 0. It is possible that more than one device will find a match and generate an acknowledgment (A1).

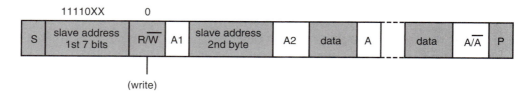

Figure 11.16 ■ A master-transmitter addresses a slave-receiver with a 10-bit address

All slaves that found a match will compare the eight bits of the second byte of the slave address with their own addresses, but only one slave will find a match and generate an acknowledgment (A2). The matching slave will remain addressed by the master until it receives a stop condition or a repeated start condition followed by a different slave address.

- *Master receiver reads slave transmitter with a 10-bit address.* The transfer direction is changed after the second R/$\overline{\text{W}}$ bit (see Figure 11.17). Up to and including the acknowledgment bit A2, the procedure is the same as that described for a master transmitter addressing a slave receiver. After the repeated start condition, a matching slave remembers that it was addressed before. This slave then checks if the first seven bits of the first byte of the slave address following repeated start (R) are the same as they were after the start condition (S) and tests if the eighth (R/$\overline{\text{W}}$) bit is 1. If there is a match, the slave considers that it has been addressed as a transmitter and generates acknowledgment bit A3. The slave transmitter remains addressed until it receives a stop condition (P) or until it receives another repeated start (R) condition followed by a different slave address. After a repeated start (R) condition, all other slave devices will also compare the first seven bits of the first byte of the slave address (11110xx) with their own addresses and test the eighth bit. However, none of them will be addressed because R/$\overline{\text{W}}$ = 1 (for 10-bit address) or the 11110xx slave address (for 7-bit address) does not match.

- *Combined format.* A master transmits data to a slave and then reads data from the same slave (see Figure 11.18). The same master occupies the bus all the time. The transfer direction is changed after the second R/$\overline{\text{W}}$ bit.

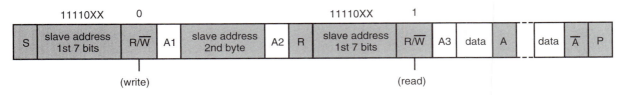

Figure 11.17 ■ A master-receiver addresses a slave-transmitter with a 10-bit address

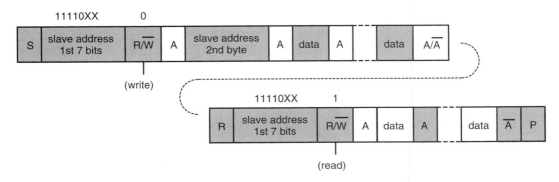

Figure 11.18 ■ **Combined format.** A master addresses a slave with a 10-bit address, then transmit data to this slave and reads data from this slave.

- *Combined format.* A master transmits data to one slave and then transmits data to another slave (see Figure 11.19). The same master occupies the bus all the time.
- *Combined format.* Ten-bit and 7-bit addressing combined in one serial transfer (see Figure 11.20). After each start (S) condition or each repeated start condition, a 10-bit or 7-bit slave address can be transmitted. Figure 11.20 shows how a master transmits data to a slave with a 7-bit address and then transmits data to a second slave with a 10-bit address. The same master occupies the bus all the time.

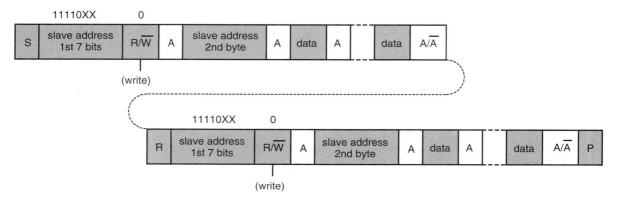

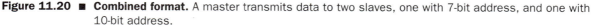

Figure 11.19 ■ **Combined format.** A master transmits data to two slaves, both with 10-bit addresses.

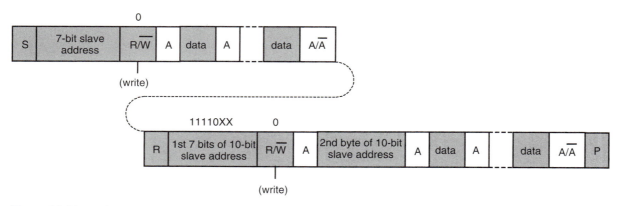

Figure 11.20 ■ **Combined format.** A master transmits data to two slaves, one with 7-bit address, and one with 10-bit address.

11.3 PIC18 MSSP Module in I²C Mode

The MSSP module in the I²C mode fully implements all master and slave functions (including general call support) and provides interrupts on start and stop bits in hardware to determine if a bus is free (multimaster function). Both 7-bit and 10-bit addressing are supported.

Two pins are used for data transfer:

- Serial clock (SCL)—RC3/SCK/SCL
- Serial data (SDA)—RC4/SDI/SDA

The user must configure these two bits for input or output through the TRISC<4:3> bits.

The PIC18 MSSP module also supports the system management bus (SMBus). The SMBus is an extension of the I²C bus that was developed to extend the benefits of two-wire querying and control to a breadth of applications not envisioned by the Philips I²C bus. A detailed discussion of SMBus is outside the scope of this text.

11.4 Registers for I²C Operation

The MSSP has six registers to support the I²C operations:

- MSSP control register 1 (SSPCON1)
- MSSP control register 2 (SSPCON2)
- MSSP status register (SSPSTAT)
- Serial receive/transmit buffer (SSPBUF)
- MSSP shift register (SSPSR)—not directly accessible
- MSSP address register (SSPADD)

11.4.1 The SSPCON1 Register

The SSPCON1 is a readable and writable control register. Its contents are shown in Figure 11.21. The WCOL bit will be set if the microcontroller attempts to write into the SSP-BUF register while one of the following operations is proceeding:

1. The SSPSR register is shifting out data.
2. The SSPSR register is shifting in data.
3. The START condition is not complete.
4. The STOP condition is not complete.
5. The ACK condition is not complete.

	7	6	5	4	3	2	1	0
value after reset	WCOL	SSPOV	SSPEN	CKP	SSPM3	SSPM2	SSPM1	SSPM0
	0	0	0	0	0	0	0	0

WCOL: Write collision detect bit

In master transmit mode:

   0 = no collision

   1 = A write to the SSPBUF register was attempted while I2C
      conditions were not valid for a transmission to be started.

In slave transmit mode:

   0 = no collision

   1 = The SSPBUF register is written while it is still transmitting
      the previous word.

In receive mode: don't care

SSPOV: Receive overflow bit

In receive mode:

   0 = no overflow

   1 = A byte is received while the SSPBUF register is still holding
      the previous byte.

In transmit mode: don't care

SSPEN: Synchronous serial port enable

   0 = Disable serial port and configure SDA and SCL as I/O pins

   1 = Enables the serial port and configures these two pins as serial port pins

CKP: SCK release control bit

In slave mode:

   0 = Holds clock low (clock stretch), used to ensure data setup time

   1 = Release clock

In master mode: unused in this mode

SSPM3..SSPM0: Synchronous serial port mode select bits

   0110 = I2C slave mode, 7-bit address

   0111 = I2C slave mode, 10-bit address

   1000 = I2C master mode, clock = $F_{OSC}/(4 * (SSPADD + 1))$

   1011 = I2C firmware controlled master mode (slave IDLE)

   1110 = I2C slave mode, 7-bit address with start and stop bit interrupt enabled

   1111 = I2C slave mode, 10-bit address with start and stop bit interrupt enabled

Figure 11.21 ■ The SSPCON1 register (redraw with permission of Microchip)

11.4.2 The SSPCON2 Register

This register is readable and writable. The contents of this register are shown in Figure 11.22. By programming the SSPCON2 register, the user can do the following:

■ Enable the interrupt when a general call address 0x00 is received by setting bit 7

■ Initiate an acknowledgment sequence by setting bit 4

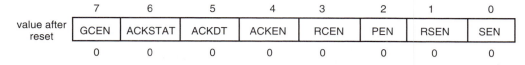

	7	6	5	4	3	2	1	0
value after reset	GCEN	ACKSTAT	ACKDT	ACKEN	RCEN	PEN	RSEN	SEN
	0	0	0	0	0	0	0	0

GCEN: General call enable bit
    0 = General call address disabled
    1 = Enable interrupt when a general call address (0x00) is received in the SSPSR
ACKSTAT: Acknowledge status bit (master transmit mode only)
    0 = Acknowledge was received from slave
    1 = Acknowledge was not received from slave
ACKDT: Acknowledge data bit (value to be transmitted when ACK is initiated)
    0 = Acknowledge
    1 = Not acknowledge
ACKEN: Acknowledge sequence enable bit (master receive mode only)
    0 = Acknowledge sequence idle
    1 = Initiate acknowledge sequence on SDA and SCL pins, and sends ACKDT bit
RCEN: Receive enable bit
    0 = Receive idle
    1 = Enables receive mode for I²C
PEN: Stop condition enable bit (master mode only)
    0 = Stop condition idle
    1 = Initiate stop condition on SDA and SCL pins (cleared by hardware
       automatically)
RSEN: Repeated start condition enabled bit (master mode only)
    0 = Repeated start condition idle
    1 = Initiate repeated start condition (automatically cleared by hardware)
SEN: Start condition enabled/stretch enabled bit
  In master mode:
    0 = Start condition idle
    1 = Initiate start condition on SDA and SCL pins (cleared by hardware
       automatically)
  In slave mode:
    0 = Clock stretching is disabled
    1 = Clock stretching is enabled for both slave transmit and slave receive

Figure 11.22 ■ The SSPCON2 register (redraw with permission of Microchip)

- Enable receive mode by setting bit 3
- Generate a stop condition by setting bit 2
- Generate a repeated start condition by setting bit 1
- Generate a start condition by setting bit 0

The start, the stop, and the repeated start conditions can be asserted only by a master device, whereas an acknowledgment can be generated by either a master or a slave.

11.4.3 The SSPSTAT Register (I²C mode)

This register records the status of all I²C data transmission/reception activities. The contents of this register are shown in Figure 11.23.

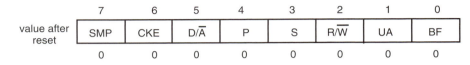

	7	6	5	4	3	2	1	0
value after reset	SMP	CKE	D/$\overline{\text{A}}$	P	S	R/$\overline{\text{W}}$	UA	BF
	0	0	0	0	0	0	0	0

SMP: Slew rate control bit

  0 = Slew rate control enabled for high speed mode (400 KHz)

  1 = Slew rate control disabled for standard speed mode (100 KHz and 1 MHz)

CKE: SMbus select bit

  0 = Disable SMbus specific inputs

  1 = Enable SMbus specific inputs

D/$\overline{\text{A}}$: Data/address bit

 In master mode: reserved

 In slave mode:

  0 = Indicates that the last byte received or transferred was address

  1 = Indicates that the last byte received or transmitted was data

P: Stop bit

  0 = A stop bit was not detected last

  1 = Indicates that a stop bit has been detected last

S: Start bit

  0 = Start bit was detected last

  1 = Indicates that a start bit has been detected last

R/$\overline{\text{W}}$: Read/Write bit information (I²C mode only)

 In slave mode:

  0 = Write

  1 = Read

 In master mode

  0 = Transmit is not in progress

  1 = Transmit is in progress

UA: Update address bit (10-bit slave mode only)

  0 = Address does not need to be updated

  1 = Indicates that the user needs to update the address in the SSPADD register

BF: Buffer full status bit

 In receive mode:

  0 = Receive not complete, SSPBUF empty

  1 = Receive complete, SSPBUF is full

 In transmit mode:

  0 = Data transmit complete, SSPBUF is empty

  1 = Data transmit in progress, SSPBUF is full

Figure 11.23 ■ The SSPSTAT register (redraw with permission of Microchip)

11.4.4 MSSP Address Register

The SSPADD register holds the slave device address when the SSP is configured in I²C slave mode. When configured in 10-bit slave mode, the user must write the upper byte of the 10-bit address into this register before a data transfer is started and then write the lower byte of the address after the upper byte address is matched. When the MSSP module is configured in master mode, the lower seven bits of the SSPADD register act as the baud rate generator reload value.

11.5 The PIC18 I²C Master Mode

The master mode is enabled by setting and clearing the appropriate bits in the SSPCON1 register. In master mode, the SCL and SDA signals are manipulated by the MSSP hardware. The user can operate the master mode in either the interrupt-enabled mode (SSPM3 . . . SSP0 = 1000) or the firmware-controlled mode (SSPM3 . . . SSPM0 = 1011). The interrupt-enabled master mode is used more often than the firmware-controlled master mode.

In the interrupt-enabled master mode, data transfer operation is supported by interrupt generation on the detection of the start and the stop conditions. The START (S) and the STOP (P) bits are cleared from a reset or when the MSSP module is disabled. The following events will cause the SSP interrupt flag bit, SSPIF, to be set:

- START condition
- STOP condition
- Data transfer byte transmitted/received
- Acknowledge transmit
- Repeated START

In firmware-controlled mode, user code conducts all I²C bus operations based on the START and the STOP bit conditions.

11.5.1 Master Mode Operation

Once the master mode is enabled, the user can perform the following six operations:

- Assert a START condition
- Assert a REPEATED START condition
- Write to the SSPBUF register to initiate transmission of data/address
- Configure the I²C port to receive data
- Assert an ACK condition at the end of a received byte of data
- Assert a STOP condition

ASSERT A **START** CONDITION

In assembly language, the START condition can be asserted as follows:

```
bsf    SSPCON2,SEN
```

In C language, this is done by the following statement:

```
SSPCON2bits.SEN = 1;
```

Usually, the user needs to use the following instruction sequence to make sure that the START condition is completed:

```
loop1   btfsc   SSPCON2,SEN    ; wait until start condition is complete
        bra     loop1          ; "
```

In C language, the same operation is achieved by the following statement:

```
while (SSPCON2bits.SEN);
```

ASSERT A REPEATED START CONDITION

The user can assert a repeated start condition by using the following instruction:

```
bsf      SSPCON2,RSEN
```

In C language, this is implemented by the following statement:

```
SSPCON2bits.RSEN = 1;
```

The user also needs to make sure that the repeated start condition is completed:

```
loop2    btfsc    SSPCON2,RSEN
         bra      loop2
```

In C language, the same operation is achieved by the following statement:

```
while (SSPCON2bits.RSEN);
```

ASSERT A STOP CONDITION

In assembly language, the STOP condition can be asserted as follows:

```
bsf      SSPCON2,PEN
```

In C language, this is done by the following statement:

```
SSPCON2bits.PEN = 1;
```

The user also needs to use the following instruction sequence to make sure that the STOP condition is completed:

```
loop1    btfsc    SSPCON2,PEN
         bra      loop1
```

In C language, the same operation is implemented by the following statement:

```
while(SSPCON2bits.PEN);
```

FINDING OUT IF I²C PORT IS IDLE

Before writing data into the I²C port, one must make sure that the I²C port is idle to avoid *write collision*. The I²C bus is idle if and only if none of the following conditions is active:

- Start condition
- Stop condition
- Repeated start
- Acknowledge
- Data reception
- Data transmission

The following routine will make sure that the I²C port is idle before returning to the caller:

```
i2c_idle   movlw    0x1F
           andwf    SSPCON2,W      ; make sure the first 5 busy conditions don't exist
           bz       chk_wr
           bra      i2c_idle       ; wait until bus is not busy
chk_wr     btfsc    SSPSTAT,R_W    ; check to see if SSP is transmitting data
           bra      chk_wr
           return   0
```

The C language version of the routine is as follows:

```
void i2c_idle (void)
{
            while ((SSPCON2 & 0x1F) | (SSPSTATbits.R_W));
}
```

WRITING DATA TO I²C BUS

The following subroutine will write the value in WREG to the I²C bus:

```
writeI2C   pushr   WREG        ; save WREG in the stack
           call    i2c_idle,0  ; make sure I²C bus is idle
           popr    WREG        ; retrieve data to be written
           movwf   SSPBUF      ; send out the byte
again      btfsc   SSPSTAT,BF  ; BF will be cleared when 8 bits are shifted out
           bra     again
           return  0
```

The C language version of the same function is as follows:

```
void    write_i2c (unsigned char dat)
{
            i2c_idle ( );
            SSPBUF = dat;
            while(SSPSTATbits.BF);      /* wait until data is shifted out */
}
```

ENABLE RECEIVING A BYTE

The user can use the following function to receive a byte from the I²C port and return it in WREG:

```
read_i2c   bsf     SSPCON2,RCEN   ; enable receive mode
loop3      btfss   SSPSTAT,BF     ; wait until one byte has been shifted in
           bra     loop3
           movf    SSPBUF,W       ; place data in WREG
           return  0
```

The C language version of the function is as follows:

```
unsigned char ReadI2C (void)
{
            SSPCON2bits.RCEN = 1;
            while(!SSPSTATbits.BF);     /* wait until a byte is received */
            return (SSPBUF);
}
```

ACKNOWLEDGING RECEPTION OF DATA

The successful reception of a byte should be acknowledged. The following instruction sequence will send an acknowledgment to the sender:

```
bcf   SSPCON2,ACKDT   ; set acknowledge bit state for ACK
bsf   SSPCON2,ACKEN   ; initiate bus acknowledge sequence
```

In C language, a function can be written to acknowledge the reception of data:

```
void AckI2C (void)
{
          SSPCON2bits.ACKDT = 0;
          SSPCON2bits.ACKEN = 1;
}
```

The following instruction sequence will make a negative acknowledgment:

```
bsf     SSPCON2,ACKDT,A      ; set acknowledge bit state for ACK
bsf     SSPCON2,ACKEN,A      ; initiate bus acknowledge sequence
```

In C language, the same function is written as follows:

```
void Not_AckI2C (void)
{
          SSPCON2bits.ACKDT = 1;
          SSPCON2bits.ACKEN = 1;
}
```

11.5.2 Master Mode Transmit Sequence

A typical transmit sequence would go as follows:

1. The user generates a START condition by setting the START enable bit SEN.
2. The SSPIF flag bit is set. The MSSP module will wait for the required start time before any other operation takes place.
3. The user loads the SSPBUF with the slave address to transmit.
4. The MSSP module shifts out the address from the SDA pin.
5. The MSSP module shifts in the ACK bit from the slave device and writes its value into the SSPCON2 register (bit 6).
6. The MSSP module generates an interrupt at the end of the ninth clock cycle by setting the SSPIF flag.
7. The user loads the SSPBUF with eight bits of data.
8. The MSSP module shifts out data from the SDA pin.
9. The MSSP module shifts in the ACK bit from the slave device and writes its value into the SSPCON2 register (bit 6).
10. The MSSP module generates an interrupt at the end of the ninth clock cycle by setting the SSPIF flag.
11. The user generates a stop condition by setting the STOP enable bit PEN (bit 2).
12. The MSSP module generates an interrupt once the stop condition is complete.

11.5.3 Baud Rate Generator

In the I^{2}C master mode, the baud rate generator (BRG) reload value is placed in the lower seven bits of the SSPADD register (shown in Figure 11.24). When a write occurs to SSPBUF, the baud rate generator will automatically begin counting. The BRG counts down to 0 and stops until another reload has taken place. The BRG count is decremented twice per instruction cycle (T_{CY}).

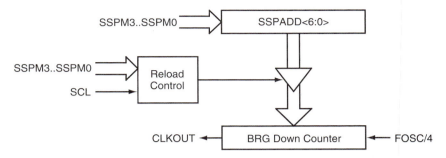

Figure 11.24 ■ Baud rate generator block diagram (redraw with permission of Microchip)

Once the given operation is complete (i.e., transmission of the last data bit is followed by ACK), the internal clock will automatically stop counting, and the SCL pin will remain in its last state.

The baud rate of I²C is given by the following expression:

baud rate = $F_{OSC} \div [4 \times (BRG + 1)]$

The value (BRG) to be written into the SSPADD register in order to achieve certain baud rate is as follows:

BRG = $F_{OSC} \div (4 \times baud\_rate) - 1$

A table of I²C clock rates versus reload value is given in Table 11.2. In Table 11.2, f_{CY} (= $f_{OSC}/4$) is the frequency of the instruction cycle clock.

f_{CY}	2 * f_{CY}	reload value (BRG)	buad rate (2 rollovers of BRG)
10 MHz	20 MHz	0x18	400 KHz[1]
10 MHz	20 MHz	0x31	200 KHz
10 MHz	20 MHz	0x63	100 KHz
8 MHz	16 Mhz	0x13	400 KHz[1]
8 MHz	16 MHz	0x27	200 KHz
8 MHz	16 MHz	0x4F	100 KHz
4 MHz	8 MHz	0x09	400 KHz[1]
4 MHz	8 MHz	0x13	200 KHz
4 MHz	8 MHz	0x27	100 KHz
2 MHz	4 MHz	0x04	400 KHz[1]
2 MHz	4 MHz	0x09	200 KHz
2 MHz	4 MHz	0x13	100 KHz
1 MHz	2 MHz	0x02	333 KHz[1]
1 MHz	2 MHz	0x04	200 KHz
1 MHz	2 MHz	0x09	100 KHz

Note
1. The I²C interface does not conform to the 400 KHz I²C specification (which applies to rates greater than 100 KHz) in all details, but may be used with care where higher rates are required by the application.

Table 11.2 ■ I²C clock rates versus reload values

11.5.4 Start Condition Timing

The user sets the SEN bit of the SSPCON2 register to initiate a START condition. If the SDA and SCL pins are sampled high, the baud rate generator is reloaded with the contents of SSPADD<6:0> and starts its counting. If SCL and SDA are both sampled high when the baud rate generator times out (T_{BRG}), the SDA pin is driven low. The action also sets the S bit of the SSPSTAT register. Following this, the baud rate generator is reloaded with the contents of SSPADD<6:0> and resumes its counting. When the baud rate generator times out (T_{BRG}), the SEN bit of the SSPCON2 register will be cleared. After this, the baud rate generator is suspended, leaving the SDA line held low, and the START condition is complete. The timing diagram of the START condition is illustrated in Figure 11.25. The parameter T_{BRG} is equal to $2 \times T_{OSC} \times (BRG + 1)$, where T_{OSC} is the period of the crystal oscillator output.

If at the beginning of the START condition the SDA and SCL pins are already sampled low or if during the START condition the SCL line is driven low, a bus collision occurs. The bus collision interrupt flag BCLIF is set, and the I²C module is reset to IDLE state.

If the user writes the SSPBUF register when a START sequence is in progress, the WCOL flag will be set, and the contents of the buffer are unchanged.

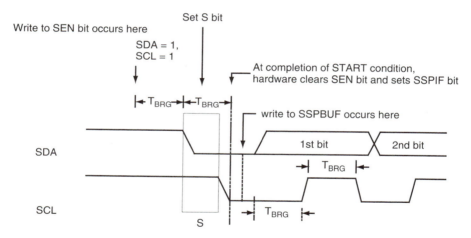

Figure 11.25 ■ Start condition timing (redraw with permission of Microchip)

11.5.5 Repeated Start Condition Timing

A repeated start condition occurs when the RSEN bit of the SSPCON2 register is set to 1 and the I²C module is in the idle state (the SCL pin is low). When the SCL pin is sampled low, the baud rate generator is loaded with the contents of SSPADD<6:0> and begins counting. The SDA pin is released (brought high) for one baud rate generator count (T_{BRG}). When the baud rate generator times out, if the SDA pin is sampled high, the SCL pin will be deasserted (brought high). When the SCL pin is sampled high, the baud rate generator is reloaded with the contents of SSPADD<6:0> and begins counting. The SDA and SCL pins must be sampled high for one T_{BRG}. This action is then followed by assertion of the SDA pin (SDA = 0) for one T_{BRG} while SCL is high. After this, the RSEN bit will be automatically cleared, and the baud rate generator will not be reloaded, leaving the SDA pin held low. As soon as a START condition is detected, the S bit will be set. The SSPIF bit will not be set until the baud rate generator has timed out. The detailed timing of a repeated start condition is illustrated in Figure 11.26.

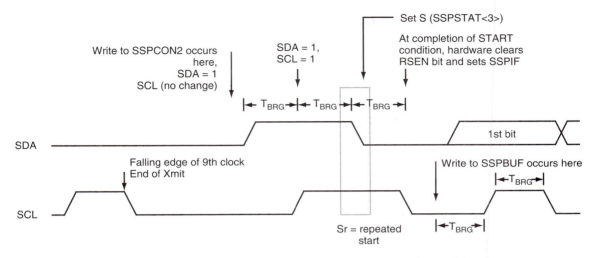

Figure 11.26 ■ Repeat start condition waveform (redraw with permission of Microchip)

If the user writes the SSPBUF when a repeated start sequence is in progress, the WCOL bit is set, and the contents of the buffer are not changed.

11.5.6 Master Mode Transmission

Transmission of a data byte, a 7-bit address, or the other half of a 10-bit address is accomplished by simply writing a value to the SSPBUF register. This action will set the buffer full flag BF and allow the baud rate generator to begin counting and start the next transmission. Each bit of address/data will be shifted out of the SDA pin on the falling edge of the SCL signal. The SCL signal is held low for one baud rate generator rollover count (T_{BRG}). Data should be valid before the SCL signal is released high. The SCL signal will stay at high for T_{BRG}. Data should remain stable for that duration and some hold time after the falling edge of SCL. After the eighth bit is shifted out, the BF flag is cleared, and the master releases the SDA pin. This allows the slave device being addressed to respond with an ACK bit during the ninth bit time if an address match occurred or if data was received correctly. The status of the ACK bit is written into the ACKSTAT bit on the falling edge of the ninth clock. If the master receives an acknowledgment, the acknowledge status bit, ACKSTAT, is cleared. If not, the bit is set. After the ninth clock, the SSPIF bit is set, and the master clock is suspended until the next data byte is loaded into the SSPBUF. SCL is held low, and SDA is allowed to float.

BF Status Flag

In transmit mode, the BF flag is set when the CPU writes to SSPBUF and is cleared when all eight bits are shifted out.

WCOL Status Flag

If the user writes into the SSPBUF when a transmission is already in progress, the WCOL is set, and the contents of the buffer are unchanged. WCOL must be cleared in software.

ACKSTAT Status Flag

In transmit mode, the ACKSTAT bit is cleared when the slave has sent acknowledgment (ACK = 0) and is set when the slave does not acknowledge (ACK = 1). A slave sends an acknowledgment when it recognizes its address or when the slave has properly received its data.

11.5.7 Master Mode Reception

Master mode reception is enabled when the RCEN bit of the SSPCON2 register is set to 1. The baud rate generator begins counting, and on each rollover the state of the SCL pin changes and data is shifted into the SSPSR register. After the falling edge of eighth clock, the receive enable flag is automatically cleared, the contents of the SSPSR are loaded into the SSPBUF, the BF flag is set, the SSPIF flag is set, and the baud rate generator is suspended from counting, holding SCL low. The MSSP module is now in idle state, waiting for the next command. When the buffer is read by the CPU, the BF flag is cleared. The user can then send an acknowledgment at the end of reception by setting the ACKEN bit in the SSPCON2 register.

The following bits are affected by the receive operation:

- *BF status flag.* In receive operation, the BF bit is set when an address or data byte is loaded into SSPBUF from SSPSR. It is cleared when the SSPBUF register is read.
- *SSPOV status flag.* In receive operation, the SSPOV flag is set when eight bits are shifted into the SSPSR and the BF is already set from a previous reception.
- *WCOL status flag.* If the user writes the SSPBUF register when a reception is already in progress, the WCOL bit is set, and the contents of the buffer are not changed.

11.5.8 Acknowledge Sequence Timing

An acknowledge sequence is enabled by setting the ACKEN bit of the SSPCON2 register. When this bit is set, SCL is pulled low, and the content of the ACKDT bit of the SSPCON2 register is applied to the SDA pin. If the user wishes to generate an acknowledgment, then the ACKDT bit should be cleared. If not, the user should set the ACKDT bit before starting an acknowledgment sequence. The baud rate generator then counts for one rollover period (T_{BRG}), and the SCL pin is then pulled high. When the SCL pin is sampled high, the baud rate generator counts for another T_{BRG} period. The SCL pin is then pulled low. Following this, the ACKEN bit is automatically cleared, the baud rate generator is turned off, and the MSSP module then goes into idle state. The detailed timing is illustrated in Figure 11.27.

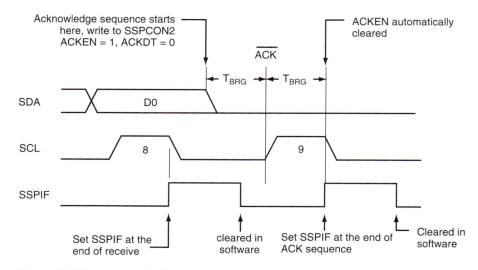

Figure 11.27 ■ Acknowledge sequence timing (redraw with permission of Microchip)

If the user writes into the SSPBUF register when an acknowledgment sequence is in progress, then the WCOL bit is set, and the contents of the buffer are not changed.

11.5.9 Stop Condition Timing

A STOP sequence is started by setting the PEN bit of the SSPCON2 register. At the end of a receive/transmit, the SCL line is held low after the falling edge of the ninth clock. When the PEN bit is set, the master will assert the SDA line low. When the SDA line is sampled low, the baud rate generator is reloaded and counts down to 0. When the baud rate generator times out, the SCL pin will be brought high, and one T_{BRG} interval later, the SDA pin will be deasserted. When the SDA pin is sampled high while SCL is high, the P bit is set. A T_{BRG} later, the PEN bit is cleared, and the SSPIF bit is set as shown in Figure 11.28.

If the user writes the SSPBUF when a STOP sequence is in progress, then the WCOL bit is set, and the contents of the buffer are unchanged.

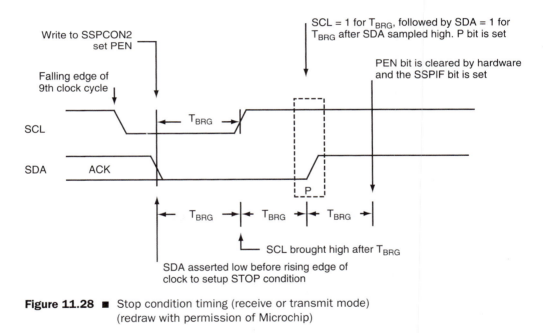

Figure 11.28 ■ Stop condition timing (receive or transmit mode) (redraw with permission of Microchip)

11.6 PIC18 I²C Slave Mode

The slave mode allows the PIC18 microcontroller to operate in a multimaster I²C environment. An address is assigned (by the user) to the PIC18 device in slave mode. The SCL and SDA pins must be configured as inputs. The MSSP module will override the input state with the output data when required (slave transmitter).

The I²C slave mode hardware will always generate an interrupt on an address match. Through the mode select bits, the user can also choose to interrupt on START and STOP bits.

When an address is matched or the data transfer after an address match is received, the hardware automatically generate the ACK pulse and load the SSPBUF register with the received value in the SSPSR register.

The MSSP module may not generate the ACK pulse if one of the following two conditions occurs:

- The buffer full bit BF was set before the transfer was received.
- The overflow bit SSPOV was set before the transfer was received.

In these cases, the SSPSR register is not loaded into the SSPBUF, but bit SSPIF is set. The BF bit is cleared by reading the SSPBUF register, whereas the SSPOV bit must be cleared by software.

11.6.1 Slave Addressing

Once the MSSP module is enabled in one of the slave modes, it waits for a START condition to occur. Following the START condition, eight bits are shifted into the SSPSR register. All incoming bits are sampled on the rising edge of the SCL signal. The value of the register SSPSR<7:1> is compared with the SSPADD register on the falling edge of the eighth SCL pulse. If the address match and the BF and SSPOV bits are clear, the following events will occur:

- The SSPSR register value is loaded into the SSPBUF register.
- The buffer full bit BF is set.
- An ACK pulse is generated.
- MSSP interrupt flag bit SSPIF is set on the falling edge of the ninth SCL pulse.

An address value must be written into the SSPADD register by the user before the MSSP is configured to operate in slave mode.

In 10-bit address mode, two address bytes need to be received by the slave. The five most significant bits of the first address byte specify if this is a 10-bit address. Bit R/$\overline{W}$ (SSPSTAT<2>) must specify a write so that the slave device will receive the second address byte. The first address byte must be 11110 A9 A8 0, where A9 and A8 are the two most significant address bits.

The UA bit of the SSPSTAT register is used in 10-bit address mode. When the high byte of the 10-bit address is matched, the UA bit will be set to 1. When the UA bit is 1, the user must load the lower byte of the 10-bit address into the SSPADD register so that it can be compared with the incoming lower byte of the 10-bit address.

The following occur when configuring in 10-bit slave mode:

1. The user first loads the value 1110 A9 A8 0 into the SSPADD register to be compared with the incoming higher address byte.
2. When the high address bytes match, UA bit is set to 1. The MSSP will acknowledge this situation to the master device.
3. The user makes sure that the UA bit is 1 and loads the lower address byte into the SSPADD register.
4. If the lower address byte is also matched, the MSSP will acknowledge again and wait for the following data byte or transmit data byte to the master device.

11.6.2 Reception in Slave Mode

The MSSP module expects to receive data when the R/$\overline{W}$ bit of the address is clear and an address match occurs. This condition will also cause the R/$\overline{W}$ bit of the SSPSTAT register to be cleared. The received address is loaded into the SSPBUF, and the SDA line is held low to acknowledge.

If the address byte overflow condition exists, then no ACK pulse will be generated. An overflow condition is defined as either the BF bit or the SSPOV bit being set. An MSSP interrupt is generated for each data transfer byte.

An MSSP interrupt is generated (by setting the SSPIF flag bit) for each data received. The SSPIF (PIR1<3>) flag bit must be cleared in software. The SSPSTAT register is used to determine the status of the byte.

If the SEN bit is set (clock stretching enabled), the RC3/SCL pin will be held low (clock stretch) following each data transfer. The clock is released by setting the CKP bit (SSPCON1<4>).

11.6.3 Transmission in Slave Mode

When the R/$\overline{\text{W}}$ bit of the incoming address byte is set and an address match occurs, the R/$\overline{\text{W}}$ bit of the SSPSTAT register is set. To prepare data for transmission, the user may need to stretch the SCL clock by clearing the CKP bit of the SSPSTAT register to low. The received address is loaded into the SSPBUF register. The ACK pulse will be sent on the ninth bit, and pin SCL is held low regardless of the setting of the SEN bit. The low state of SCL prevents the master from asserting another SCL pulse. When the slave has written the data byte into the SSPBUF register, it should set the CKP bit to release the SCL signal so that the master device can pulse the SCL pin again to shift out the data byte in the SSPSR register.

The ACK pulse from the master receiver is latched on the rising edge of the ninth SCL input pulse. If the SDA line is high (NACK), then data transfer is complete. In this case, when the NACK is latched by the slave, the slave logic is reset, and the slave monitors for another occurrence of the START condition. If the master device acknowledges, the slave device should load the next data byte into the SSPBUF register for transmission. Again the slave must set the CKP bit to release the SCL pin.

The MSSP interrupt is generated (setting the SSPIF flag) for each data transfer byte. The SSPIF bit is set on the falling edge of the ninth clock pulse.

11.6.4 Clock Synchronization and the CKP Bit

When the CKP bit is cleared, the SCL pin is forced to low. However, setting the CKP bit will not assert the SCL pin low until the SCL output is already sampled low. Therefore, the CKP bit will not assert the SCL pin until an external I²C master device has already asserted the SCL line. The SCL signal will remain low until the CKP bit is set and all other devices on the IC bus have deasserted SCL. This ensures that a write to the CKP bit will not violate the minimum high time requirement for the SCL signal.

11.6.5 General Call Address Support

The addressing procedure for the I²C bus is such that the first byte after the START condition usually determines which slave device that the master will address. The exception is the general call address, which can address all devices. When this address is used, all devices should, in theory, respond with an ACK.

The general call address consists of all 0s with R/$\overline{\text{W}}$ = 0 and is recognized when the general call enable bit (GCEN) of the SSPCON2 register is set. If the general call address after a START condition appears, the SSPSR register is transferred to the SSPBUF register, and the BF flag bit will be set on the eighth clock cycle. The SSPIF flag bit is set on the falling edge of the ninth clock bit.

When the interrupt is serviced, the source of the interrupt can be identified by reading the contents of the SSPBUF register. The value can be used to determine if the address was device specific or a general call address.

In the 10-bit mode, the SSPADD register is required to be updated for the second half of the address to match, and the UA bit is set. When in 10-bit addressing mode and a general call address occurs, the user need not load the lower byte of the address into the SSPADD register for comparison. In this situation, the UA bit will not be set, and the slave will begin receiving data after the acknowledgment.

11.6.6 Clock Stretching

Both 7- and 10-bit slave modes implement automatic clock stretching during a transmit sequence.

The SEN bit (SSPCON2<0>) allows clock stretching to be enabled during reception operations. Setting the SEN bit will cause the SCL pin to be held low at the end of each data receive sequence.

Clock Stretching for 7-bit Slave Receive Mode

In 7-bit slave mode, on the falling edge of the ninth clock at the end of the ACK sequence, if the BF bit is set, the CKP bit in the SSPCON1 register is automatically cleared, forcing the SCL output to be held low. The CKP being cleared to 0 will assert the SCL line low. The CKP bit must be set in the user's interrupt service routine (ISR) before reception is allowed to continue. By holding the SCL signal low, the user has time to service the interrupt and read the contents of the SSPBUF register before the master device can initiate another receive sequence (for the slave). This will prevent buffer overruns from occurring.

Clock Stretching for 10-bit Slave Receive Mode

In the 10-bit slave receive mode, during the address sequence, clock stretching automatically takes place but CKP is not cleared. During this time, if the UA bit is set after the ninth clock, clock stretching is initiated. The UA bit is set after receiving the upper byte of the 10-bit address. The release of the clock line occurs upon updating the SSPADD register. Clock stretching will occur on each data receive sequence as in the 7-bit address mode.

Clock Stretching for 7-bit Slave Transmit Mode

The 7-bit slave transmit mode implements clock stretching by clearing the CKP bit after the falling edge of the ninth clock if the BF bit is clear. This occurs regardless of the state of the SEN bit.

The user's ISR must set the CKP bit before transmission is allowed to continue. By holding the SCL signal low, the user has time to execute the ISR and load the contents of the SSPBUF register before the master device can initiate another transmit sequence.

Clock Stretching for 10-bit Slave Transmit Mode

In the 10-bit slave transmit mode, clock stretching is controlled during the first two address sequences by the state of the UA bit, just as it is in 10-bit slave receive mode. The first two address bytes are followed by a third address sequence, which contains the high order, which contains the high order bits of the 10-bit address and the R/$\overline{\text{W}}$ bit set to 1. After the third address sequence is performed, the UA bit is not set, the module is now configured in transmit mode, and clock stretching is controlled by the BF flag, as in 7-bit slave transmit mode.

Clock Synchronization and the CKP bit

If a user clears the CKP bit, the SCL output is forced to 0. Setting the CKP bit will not assert the SCL output low until the SCL output is already sampled low. If the user attempts to drive the SCL signal to low, the CKP bit will not assert the SCL signal until an external I²C master device has already asserted the SCL line. The SCL signal will remain low until the CKP bit is set and all other devices on the I²C bus have deasserted the SCL signal. This ensures that a write to the CKP bit will not violate the minimum high time requirement for the SCL signal.

11.7 Multimaster Mode

In multimaster mode, the interrupt generation on the detection of the START and STOP conditions allows the determination of whether the bus is free. Bus arbitration is often needed in multimaster mode. The SDA line is monitored for arbitration to see if the signal level is the expected output level.

A master can lose in arbitration under the following states:

- Address transfer
- Data transfer
- A START condition
- A repeated START condition
- An ACK condition

The bus collision flag (BCLIF) will be set to 1 if a master loses in I²C bus arbitration. In a bus collision, a master sends out 1 but detects 0 on the SDA line. In this situation, the bus master sets the BCLIF flag and resets the I²C port to its reset state. The bus collision timing diagram for transmit and acknowledge is shown in Figure 11.29.

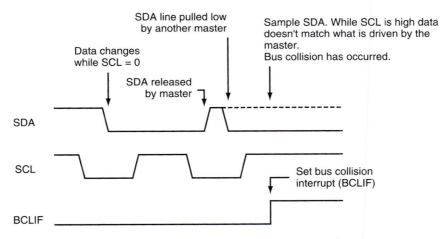

Figure 11.29 ■ Bus collision timing diagram for transmit and acknowledge

If a transmission was in progress when the bus collision occurred, the transmission is halted, the BF flag is cleared, the SDA and SCL lines are deasserted, and the SSPBUF can be written to. When the microcontroller executes the bus collision interrupt service routine and if the I²C bus is free, the user can resume communication by asserting a START condition.

If a START, repeated START, STOP, or ACK condition was in progress when the bus collision occurred, the condition is aborted, the SDA and SCL lines are deasserted, and the respective control bits in the SSPCON2 register are cleared.

In multimaster mode, the interrupt generation on the detection of START and STOP conditions allows the detection of when the bus is free. Control of the I²C bus can be taken when the P bit of the SSPSTAT register is set or the bus is idle and the S and P bits are cleared.

Example 11.1

▼

Write a function to initialize the MSSP module to operate in the I²C mode with the following parameters, assuming that f_{OSC} = 32 MHz:

- I²C master mode
- Enable slew rate control for high-speed mode
- Set baud rate to 400 KHz
- Configure SDA and SCL pins as serial port pins

Solution: This function is straightforward.

```
I2C_init   movlw    0x3F
           andwf    SSTSTAT,F      ;clear the upper two bits of SSPSTAT
           movlw    0x13           ;set baud rate to 400 KHz
           movwf    SSPADD         ; "
           clrf     SSPCON1
           clrf     SSPCON2
           bcf      SSPSTAT,SMP    ;enable slew rate control for high speed mode
           bsf      SSPSTAT,CKE    ;enable SMBus inputs
           bsf      TRISC,SCL      ;configure SCL pin for input
           bsf      TRISC,SDA      ;configure SDA pin for input
           movlw    0x28
           iorwf    SSPCON1,F      ;select master mode, enable SDA and SCL pins
           return   0
```

The C language version of the function is as follows:

```
void   I2C_init (void)
{
       SSPSTAT &= 0x3F;              /* enable slew rate control in high speed */
       SSPADD = 0x13;               /* set baud rate to 400 KHz */
       SSPCON1 = 0x00;              /* clear to reset state */
       SSPCON2 = 0x00;              /* clear to reset state */
       SSPCON1 |= 0x28;             /* enable serial port and select I2C master mode */
       TRISCbits.RC3 = 1;           /* configure SCL for input */
       TRISCbits.RC4 = 1;           /* configure SDA for input */
}
```

When the I²C module is not used, it should be closed so that the user can configure the MSSP module for other purposes. To close the I²C module, simply clear bit 5 of the SSPCON1 register:

```
bcf    SSPCON1,SSPEN
```

▲

11.8 Microchip C18 Library Functions for I²C

The I²C peripheral is supported with the functions listed in Table 11.3. The user must include the **i2c.h** header file in order to use these library functions. The prototype definitions and arguments of each function are described in the following:

void **AckI2C** (void);

This function generates an I²C bus ACK condition.

Function	Description
AckI2C	Generate I²C bus Acknowledge condition.
CloseI2C	Disable the SSP module.
DataRdyI2C	Is the data available in the I²C buffer?
getcI2C	Read a single byte from the I²C bus.
getsI2C	Read a string from the I²C bus operating in master I²C mode.
IdleI2C	Loop until I²C bus is idle.
NotAckI2C	Generate I²C bus Not Acknowledge condition.
OpenI2C	Configure the SSP module.
putcI2C	Write a single byte to the I²C bus.
putsI2C	Write a string to the I²C bus operating in either Master or Slave mode.
ReadI2C	Read a single byte from the I²C bus.
RestartI2C	Generate an I²C bus repeated start condition.
StartI2C	Generate an I²C bus start condition.
StopI2C	Generate an I²C bus stop condition.
WriteI2C	Write a single byte to the I²C bus.

Table 11.3 ■ The I²C peripheral library functions

void **CloseI2C** (void);

This function disables the SSP module.

unsigned char **DataRdyI2C** (void);

This function determines if there is a byte to be read in the SSP buffer. This function returns a 1 if a byte is in the buffer. Otherwise, it returns a 0.

unsigned char **getcI2C** (void);

unsigned char **ReadI2C** (void);

These two functions are identical and read a single byte from the I²C bus.

unsigned char **getsI2C** (unsigned char ***rdptr,** unsigned char **length**);

This function reads a fixed length string from the I²C bus operating in master mode. The argument **rdptr** points to the buffer to hold the string, and the argument **length** specifies the number of characters to be read from I²C device. A 1 is returned if a bus collision occurred. Otherwise, this function returns a 0.

void **IdleI2C** (void);

This function loops until I²C bus is idle.

void **NotAckI2C** (void);

This function generates an I²C bus NACK condition.

void **OpenI2C** (unsigned char **sync_mode,** unsigned char **slew**);

This function configures the SSP module. It has two arguments:

sync_mode may have three values:

SLAVE_7	I²C slave mode, 7-bit address
SLAVE_10	I²C slave mode, 10-bit address
MASTER	I²C master mode

slew may have two values:

SLEW_OFF	Slew rate disabled for 100-KHz mode
SLEW_ON	Slew rate enabled for 400-KHz mode

unsigned char **putI2C** (unsigned char **data_out**);

unsigned char **WriteI2C** (unsigned char **data_out**);

> These two functions are identical and write the byte **data_out** to the I²C bus. A 1 is returned if a bus collision occurred. Otherwise, a 0 is returned.

unsigned char **putsI2C** (unsigned char ***wrptr**);

> This function writes a data string pointed to by **wrptr** to the I²C bus until a null character is reached. This function can operate in both master and slave modes. This function returns the following:
>
> 0: if NULL character was reached in the data string
>
> –2: if the slave I²C device responded with a NACK
>
> –3: if a write collision occurred

void **RestartI2C** (void);

> This function asserts an I²C bus repeated START condition.

void **StartI2C** (void);

> This function asserts an I²C bus START condition.

void **StopI2C** (void);

> This function asserts an I²C bus STOP condition.

11.9 Interfacing the Digital Thermometer and Thermostat DS1631A with I²C

Many embedded products, such as network routers and switches, are used in larger systems, and their failures due to overheating can severely damage the functioning or even cause the total failure of the larger system. Using a thermostat to warn of potential system overheating is indispensable for the proper functioning of many embedded systems.

The digital thermostat device DS1631A from Dallas Semiconductor is one such product. The DS1361A will assert a signal (T_{OUT}) whenever the ambient temperature exceeds the *trip point* preestablished by the user.

11.9.1 Pin Assignment

As shown in Figure 11.30, the DS1631A is an 8-pin package. The SDA and SCL pins are used as data and clock lines so that the DS1631A can be connected to an I²C bus. Pins A2 . . .

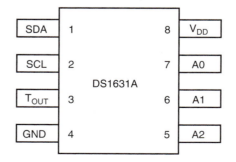

Figure 11.30 ■ Pin assignment of DS1631A

A0 are address inputs to the DS1631A. The T_{OUT} pin is the thermostat output, which is asserted whenever the ambient temperature is above the trip point set by the user.

11.9.2 Functional Description

The block diagram of the DS1631A is shown in Figure 11.31. The DS1631A converts the ambient temperature into 9-, 10-, 11-, or 12-bit readings over a range of –55° C to +125° C. The thermometer accuracy is ±0.5 °C from 0° C to +70° C with 3.0 V ≤ V_{DD} ≤ 5.5 V.

The thermostat output T_{OUT} is asserted whenever the converted ambient temperature is equal to or higher than the value stored in the T_H register. After being asserted, T_{OUT} will be deasserted only when the converted temperature reading drops below the value stored in the T_L register. The DS1631A automatically begins taking temperature measurements at power-up, which allows it to function as a stand-alone thermostat. The DS1631A conforms to the I²C bus specification.

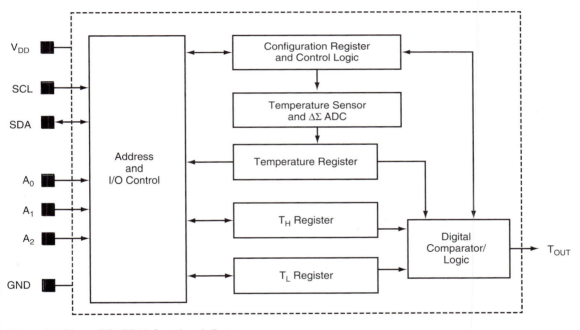

Figure 11.31 ■ DS1631A functional diagram

11.9.3 DS1631A Registers

Four registers support the operations of the DS1631A: Config, TH, TL, and Temperature. The SRAM-based Temperature register is a 2-byte register that holds the converted temperature value in two's complement format. The converted temperature is stored in the upper bits of the Temperature register with lower bits filled with 0s. When the most significant bit of this register is 1, the temperature is negative.

Both the T_H and the T_L are EEPROM-based 2-byte registers. T_H holds the upper alarm temperature value in two's complement format. Whenever the converted temperature value is equal to or higher than the value in T_H, the T_{OUT} signal is asserted. Once being asserted, T_{OUT} can be deasserted only when the converted temperature is lower than the value in the T_L register. The value in the T_L register is also represented in two's complement format.

The contents of the Config register are shown in Figure 11.32. The lower two bits of the Config register are EEPROM based, whereas the upper six bits are SRAM based. The meaning of each bit is well defined.

The Config register allows the user to program various DS1631A options, such as conversion resolution, T_{OUT} polarity, and operation mode. It also provides information to the user about conversion status, EEPROM activity, and thermostat activity. This register can be read from and written into using the **Access Config** command. When writing to the Config register, conversions should first be stopped using the **Stop Convert T** command if the device is in continuous conversion mode. Since the POL and 1SHOT bits are stored in EEPROM, they can be programmed prior to installation if desired. All other bits are in SRAM and are powered up in the state shown in Figure 11.32.

7	6	5	4	3	2	1	0
DONE	THF	TLF	NVB	R1	R0	POL*	1SHOT*

*NV (EEPROM)

power-up value

1	0	0	0	1	1	X	X

Done: Temperature conversion done (read-only)

    0 = Temperature conversion is in progress.

    1 = Temperature conversion is complete. Will be cleared when the Temperature
        register is read.

THF: Temperature high flag (read/write)

    0 = The measured temperature has not exceeded the value in T_H register.

    1 = The measured temperature has exceeded the value in T_H register. THF
        remains at 1 until it is overwritten with a 0 by the user, the power is
        recycled, or a software POR command is issued.

TLF: Temperature low flag (read/write)

    0 = The measured temperature has not been lower than the value in T_L register.

    1 = At some point after power up, the measured temperature is lower than the
        value stored in the T_L register. TLF remains at 1 until it is overwritten with a
        0 by the user, the power is recycled, or a software POR command is issued.

NVB: Nonvolatile memory busy (read only)

    0 = NV memory is not busy.

    1 = A write to EEPROM memory is in progress.

R1:R0 : Resolution bits (read/write)

    00 = 9-bit resolution (conversion time is 93.75 ms)

    01 = 10-bit resolution (conversion time is 187.5 ms)

    10 = 11-bit resolution (conversion time is 375 ms)

    11 = 12-bit resolution (conversion time is 750 ms)

POL: T_{OUT} polarity (read/write)

    0 = T_{OUT} active low

    1 = T_{OUT} active high

1SHOT: Conversion mode (read/write)

    0 = Continuous conversion mode. The Start Convert T command initiates
        continuous temperature conversions.

    1 = One-shot mode. The Start Convert T command initiates a single temperature
        conversion and then the device enters a low-power standby mode.

Figure 11.32 ■ DS1631A Configuration register

11.9.4 The DS1631A Operation

The DS1631A begins conversions automatically at power-up. It can be configured to perform continuous consecutive conversions (continuous conversion mode) or single conversions on command (one-shot mode).

The default resolution of the DS1631A is 12-bit. However, it can also be configured to 9-, 10-, or 11-bit. The resolution is changed via the R1:R0 bits of the Config register. A few samples of temperatures and their converted values are shown in Table 11.4. The lowest four bits are always 0s in the table because the resolution is 12-bit.

Temperture (°C)	Digital Output (binary)	Digital Output (hex)
+125	0111 1101 0000 0000	0x7D00
+25.0625	0001 1001 0001 0000	0x1910
+10.125	0000 1010 0010 0000	0x0A20
+0.5	0000 0000 1000 0000	0x0080
0	0000 0000 0000 0000	0x0000
-0.5	1111 1111 1000 0000	0xFF80
-10.125	1111 0101 1110 0000	0xF5E0
-25.0625	1110 0110 1111 0000	0xE6F0
-55	1100 1001 0000 0000	0xC900

Table 11.4 ■ 12-bit Resolution temperature/data relationship

Both the T_H and the T_L registers are in EEPROM. Their resolutions match the output temperature resolution and are determined by the R1:R0 bits. Writing to and reading from these two registers are achieved by using the **Access TH** and **Access TL** commands. When making changes to the T_H and T_L registers, conversions should first be stopped using the **Stop Convert T** command if the device is in continuous conversion mode.

Since the DS1631A automatically begins taking temperature measurements at power-up, it can function as a stand-alone thermostat. For stand-alone operation, the nonvolatile T_H and T_L registers and the POL and 1SHOT bits in the Config register should be programmed to the desired values prior to installation.

Table 11.4 shows that every 16 units (after ignoring the lowest four bits) of conversion result correspond to 1°C. The conversion result cannot be higher than 0x7D00 and lower than 0xC900 because the range of temperature that can be handled by DS1631A cannot be higher than 125 °C and lower than –55 °C. An easy way for finding the corresponding temperature of a certain conversion result is as follows

Positive Conversion Result

Step 1
Truncate the lowest four bits.

Step 2
Divide the upper 12 bits by 16.

For example, the conversion result 0x7000 corresponds to 0x700/16 = 112 °C. The conversion result 0x6040 corresponds to 0x604/16 = 96.25 °C.

Negative Conversion Result

Step 1
Compute the two's complement of the conversion result.

Step 2

Truncate the lowest four bits.

Step 3

Divide the upper 12 bits of the two's complement of the conversion result.

For example, the conversion result 0xE280 corresponds to –0x1D8/16 = –29.5 °C.

11.9.5 DS1631A Command Set

The DS1631A supports the following commands:

- ***Start Convert T [0x51]***—This command initiates temperature conversions. If the part is in one-shot mode, only one conversion is performed. In continuous mode, continuous temperature conversions are performed until a **Stop Convert T** command is issued.

- ***Stop Convert T [0x22]***—This command stops temperature conversions when the device is in continuous conversion mode.

- ***Read Temperature [0xAA]***—This command reads the last converted temperature value from the 2-byte Temperature register.

- ***Access TH [0xA1]***—This command reads or writes the 2-byte T_H register.

- ***Access TL [0xA2]***—This command reads or writes the 2-byte T_L register.

- ***Access Config [0xAC]***—This command reads or writes the 1-byte Config register.

- ***Software POR [0x54]***—This command initiates a software power-on-reset operation, which stops temperature conversions and resets all registers and logic to their power-up states. The software POR allows the user to simulate cycling the power without actually powering down the device.

11.9.6 I²C Communication with DS1631A

A typical circuit connection between the PIC18 microcontroller and a DS1631A is shown in Figure 11.33. The address input of the DS1631A is arbitrarily set to 001 in Figure 11.33.

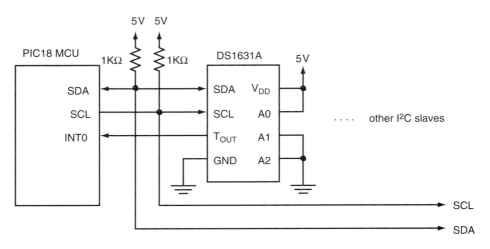

Figure 11.33 ■ Typical circuit connection between the PIC18 MCU and DA1631A

To initiate I²C communication, the PIC18 microcontroller asserts a START condition followed by a control byte containing the DS1631A slave address. The R/$\overline{\text{W}}$ bit of the control byte must be a 0 since the PIC18 microcontroller next will write a command byte to the DS1631A. The format for the control byte is shown in Figure 11.34. The DS1631A responds with an ACK after receiving the control byte. This must be followed by a command byte from the master, which indicates what type of operation is to be performed. The DS1631A again responds with an ACK after receiving the command byte. If the command byte is a **Start Convert T** or a **Stop Convert T** command, the transaction is finished, and the master must issue a STOP condition to signal the end of communication sequence. If the command byte indicates a write or read operation, additional actions must occur.

Figure 11.34 ■ Control byte for DS1631A

Write Data to DS1631A

The master can write data to the DS1631A by issuing an **Access Config, Access TH,** or **Access TL** command following the control byte. Since the read/write bit in the control byte was a 0, the DS1631A is already prepared to receive data. Therefore, after receiving an ACK in response to the command byte, the master device can immediately begin transmitting data. When writing to the Config register, the master must send one byte of data, and when writing to the T_H or T_L register, the master must send two bytes of data. After receiving each data byte, the DS1631A responds with an ACK, and the transaction is finished with a STOP from the master.

Read Data from DS1631A

The master can read data from the DS1631A by issuing an **Access Config, Access TH, Access TL,** or **Read Temperature** command following the control byte. After receiving an ACK in response to the command, the master must generate a repeated START condition followed by a control byte with the same slave address as the first control byte. However, this time the R/$\overline{\text{W}}$ bit must be set to 1 to inform the DS1631A that a "read" is being performed. After the DS1631A sends an ACK in response to this control byte, it begins transmitting the requested data on the next clock cycle. One byte of data will be transmitted when reading from the Config register, after which the master must respond with a NACK followed by a STOP condition. For 2-byte reads (i.e., from the Temperature, T_H, or T_L register), the master must respond to the first data byte with an ACK and to the second byte with a NACK followed by a STOP condition. If only the most significant byte of data is needed, the master can issue a NACK followed by a STOP condition after reading the first data byte.

Example 11.2

▼

Write an instruction sequence to configure the DS1631A in Figure 11.33 to operate in continuous conversion mode and set the T_{OUT} polarity to active high. Assume that the I²C has only one master and that there is no possibility of bus collision.

Solution: The procedure for configuring the DS1631A is as follows:

Step 1
Assert a START condition. Make sure that the I²C bus is idle before doing this.

Step 2
Send a control byte (0x92) to the DS1631A with R/$\overline{\text{W}}$ = 0.

Step 3
Wait until I²C bus is idle and check to see if the DS1631A acknowledges. If not, exit.

Step 4
Send an **Access Config** command to the DS1631A.

Step 5
Wait until I²C bus is idle and check to see if the DS1631A acknowledges. If not, exit.

Step 6
Write a command byte to configure the DS1631A to operate in continuous conversion mode and set the T_{OUT} polarity to active high.

Step 7
Wait until I²C bus is idle and check to see if the DS1631A acknowledges. If not, exit.

Step 8
Assert a STOP condition to complete the whole process.

The assembly function that performs the desired configuration is as follows:

```
open_DS1631A
            call        i2c_idle            ; ensure I2C module is idle
            bsf         SSPCON2,SEN         ; generate a START condition
loop1       btfsc       SSPCON2,SEN         ; wait until start condition is completed
            bra         loop1               ; "
            movlw       0x92                ; send out control byte to DS1631A
            movwf       SSPBUF              ; "
loop2       btfsc       SSPSTAT,BF          ; wait until 8 bits have been shifted out
            bra         loop2               ; "
            call        i2c_idle            ; ensure I2C module is idle
            btfsc       SSPCON2,ACKSTAT     ; check ACK bit
            goto        bad1                ; receive NACK
            movlw       0xAC                ; sends Access Config command
            movwf       SSPBUF              ; "
loop3       btfsc       SSPSTAT,BF          ; wait until 8 bits have been shifted out
            bra         loop3               ; "
            call        i2c_idle
            btfsc       SSPCON2,ACKSTAT     ; check ACK bit
            goto        bad1
            movlw       0x02                ; select continuous conversion mode and set
            movwf       SSPBUF              ; Tout polarity to active high
loop4       btfsc       SSPSTAT,BF
            bra         loop4
            call        i2c_idle
            btfsc       SSPCON2,ACKSTAT
            goto        bad1
            bsf         SSPCON2,PEN         ; generate a stop condition
```

```
loop5    btfsc    SSPCON2,PEN        ; wait until stop condition is complete
         bra      loop5
         movlw    0x00               ; set return code 0
         return   0                  ; "
bad1     movlw    0xFF               ; set return code to –1
         return   0                  ; "
```

The C language version of the function is as follows:

```
char open_ds1631A (void)
{
         IdleI2C( );
         StartI2C();               /* generate a START condition */
         while(SSPCON2bits.SEN);   /* wait until START condition is complete */
         if (WriteI2C(0x92))       /* send control code with R/W = 0 */
           return –1;
         IdleI2C( );
         if (SSPCON2bits.ACKSTAT)  /* received ACK ? */
           return –1;
         if (WriteI2C(0xAC))       /* send Access Config command */
           return –1;
         IdleI2C();
         if (SSPCON2bits.ACKSTAT)
           return –1;
         if(WriteI2C(0x02))        /* configure DS1631A in continuous conversion */
           return –1;             /* mode and select TOUT assertive high */
         IdleI2C();
         if (SSPCON2bits.ACKSTAT)
           return –1;
         StopI2C();                /* generate a STOP condition */
}
```

▲

Example 11.3

▼

Write a function to start temperature conversion.

Solution: The procedure for starting temperature conversion using the DS1631A is as follows:

Step 1
Assert a START condition. Make sure that the I²C module is idle before doing this.

Step 2
Send a control byte (0x92) to the DS1631A with R/$\overline{W}$ = 0.

Step 3
Wait until I²C bus is idle and check to see if the DS1631A acknowledges. If not, exit.

Step 4
Send a **Start Convert T** command to the DS1631A.

Step 5
Wait until I²C bus is idle and check to see if the DS1631A acknowledges. If not, exit.

Step 6
Assert a STOP condition to complete the whole process.

The assembly program that implements this procedure is as follows:

```
start_conv    call     i2c_idle                  ; ensure that I²C module is idle
              bsf      SSPCON2,SEN               ; generate a START condition
loop1         btfsc    SSPCON2,SEN               ; wait until start condition is completed
              bra      loop1                     ; "
              movlw    0x92                      ; send out control byte to DS1631A
              movwf    SSPBUF                    ; "
loop2         btfsc    SSPSTAT,BF                ; wait until 8 bits have been shifted out
              bra      loop2                     ; "
              call     i2c_idle                  ; ensure I²C bus is idle
              btfsc    SSPCON2,ACKSTAT           ; check ACK bit
              goto     bad2                      ; receive NACK
              movlw    0x51                      ; sends Start Convert T command
              movwf    SSPBUF                    ; "
loop3         btfsc    SSPSTAT,BF                ; wait until 8 bits have been shifted out
              bra      loop3                     ; "
              call     i2c_idle
              btfsc    SSPCON2,ACKSTAT           ; check ACK bit
              goto     bad2
              bsf      SSPCON2,PEN               ; generate a stop condition
loop5         btfsc    SSPCON2,PEN               ; wait until stop condition is complete
              bra      loop5
              movlw    0x00
              return   0
bad2          movlw    0xFF                      ; set return code to –1
              return   0                         ; "
```

The C language version of the function is as follows:

```c
char start_conv_T (void)
{
        IdleI2C( );                     /* ensure the I²C module is idle */
        StartI2C();                     /* generate a START condition */
        while(SSPCON2bits.SEN);         /* wait until START condition is complete */
        if (WriteI2C(0x92))             /* send control code with R/W = 0 */
          return –1;
        IdleI2C( );
        if (SSPCON2bits.ACKSTAT)        /* received ACK ? */
          return –1;
        if (WriteI2C(0x51))             /* send Start Convert T command */
          return –1;
        IdleI2C();
        if (SSPCON2bits.ACKSTAT)
          return –1;
        StopI2C( );                     /* generate a STOP condition */
          return 0;

}
```

Example 11.4

Write a function to set the high thermostat temperature. The upper and lower bytes of the high thermostat temperatures are passed in PRODH and PRODL.

Solution: The procedure for setting up the high thermostat temperature is as follows:

Step 1
Assert the START condition.

Step 2
Send a control byte 0x92 with $R/\overline{W} = 0$.

Step 3
Wait until I²C bus is idle and check to see if the DS1631A acknowledges. If not, exit.

Step 4
Send an **Access TH** command to DS1631A.

Step 5
Wait until I²C bus is idle and check to see if the DS1631A acknowledges. If not, exit.

Step 6
Send the upper byte of the thermostat temperature to the DS1631A.

Step 7
Wait until I²C bus is idle and check to see if the DS1631A acknowledges. If not, exit.

Step 8
Send the lower byte of the thermostat temperature to the DS1631A.

Step 9
Wait until I²C bus is idle and check to see if the DS1631A acknowledges. If not, exit.

Step 10
Assert the STOP condition.

The low thermostat temperature can be set using the same procedure.
The assembly program that implements this procedure is as follows:

```
set_TH    call     i2c_idle
          bsf      SSPCON2,SEN        ; generate a START condition
loop1     btfsc    SSPCON2,SEN        ; wait until start condition is completed
          bra      loop1              ;"
          movlw    0x92               ; send out control byte to DS1631A
          movwf    SSPBUF             ;"
loop2     btfsc    SSPSTAT,BF         ; wait until 8 bits have been shifted out
          bra      loop2              ;"
          call     i2c_idle           ; ensure I2C bus is idle
          btfsc    SSPCON2,ACKSTAT    ; check ACK bit
          goto     bad1               ; receive NACK
          movlw    0xA1               ; sends "Access TH" command
          movwf    SSPBUF             ;"
loop3     btfsc    SSPSTAT,BF         ; wait until 8 bits have been shifted out
          bra      loop3              ;"
          call     i2c_idle
          btfsc    SSPCON2,ACKSTAT    ; check ACK bit
          goto     bad1
```

```
              movff      PRODH,SSPBUF        ; sends upper byte of TH
loop4         btfsc      SSPSTAT,BF
              bra        loop4
              call       i2c_idle
              btfsc      SSPCON2,ACKSTAT     ; check ACK bit
              goto       bad1
              movff      PRODL,SSPBUF        ; send lower byte of TH
loop5         btfsc      SSPSTAT,BF
              bra        loop4
              call       i2c_idle
              btfsc      SSPCON2,ACKSTAT     ; check ACK bit
              goto       bad1
              bsf        SSPCON2,PEN         ; generate a stop condition
loop6         btfsc      SSPCON2,PEN         ; wait until stop condition is complete
              bra        loop5
              movlw      0x00
              return
bad1          movlw      0xFF                ; set return code to –1
              return
```

The C language version of the function is straightforward and hence is left for you as an exercise problem.

▲

Example 11.5
▼

Write a function to read the Config register from the DS1631A and return its value in the PRODL register. Assume that there is only one I²C bus master and that there is no possibility of bus collision.

Solution: The procedure for reading the Config register is as follows:

Step 1
Assert the START condition.

Step 2
Send a control byte 0x92 with R/$\overline{\text{W}}$ = 0.

Step 3
Wait until I²C bus is idle and check to see if the DS1631A acknowledges. If not, exit.

Step 4
Send an **Access Config** command (0xAC) to the DS1631A.

Step 5
Wait until I²C bus is idle and check to see if the DS1631A acknowledges. If not, exit.

Step 6
Assert a repeat START condition.

Step 7
Send the control byte 0x93 with R/$\overline{\text{W}}$ = 1, which tells the DS1631A that a read is being performed.

Step 8
Wait until I²C bus is idle and check to see if the DS1631A acknowledges. If not, exit.

Step 9
Read in the Config register.

Step 10

Make sure the bus is idle and send back a NACK to the DS1631A.

Step 11

Assert the STOP condition.

The assembly program that implements this procedure is as follows:

```
; **************************************************************
; The following function reads the Config register and returns the result in PRODL.
; **************************************************************
read_conf   bsf      SSPCON2,SEN          ; assert a START condition
lop1        btfsc    SSPCON2,SEN          ; wait until start condition is completed
            bra      lop1                 ; "
            movlw    0x92                 ; send control byte to DS1631A with R/W = 0
            movwf    SSPBUF               ; "
lop2        btfsc    SSPSTAT,BF           ; wait until 8 bits have been shifted out
            bra      lop2                 ; "
            call     i2c_idle             ; ensure I²C bus is idle
            btfsc    SSPCON2,ACKSTAT      ; check ACK bit
            goto     bad2                 ; receive NACK
            movlw    0xAC                 ; sends Access Config command
            movwf    SSPBUF               ; "
lop3        btfsc    SSPSTAT,BF           ; wait until 8 bits have been shifted out
            bra      lop3                 ; "
            call     i2c_idle
            btfsc    SSPCON2,ACKSTAT      ; check ACK bit
            goto     bad2
            bsf      SSPCON2,RSEN         ; assert a Repeat START condition
lop4        btfsc    SSPCON2,RSEN         ; wait until Repeat START condition is complete
            bra      lop4                 ; "
            movlw    0x93                 ; send control byte to the DS1631A with
            movwf    SSPBUF               ; R/W = 1
lop5        btfsc    SSPSTAT,BF           ; wait until 8 bits are shifted out
            bra      lop5
            call     i2c_idle             ; make sure I²C is idle
            btfsc    SSPCON2,ACKSTAT      ; check ACK bit
            goto     bad2
            bsf      SSPCON2,RCEN         ; enable reception
lop6        btfss    SSPSTAT,BF           ; wait until 8 bits have been shifted in
            bra      lop6                 ; "
            movff    SSPBUF,PRODL         ; transfer Config register contents to PRODL
            call     i2c_idle             ; make sure I2C is idle
            bsf      SSPCON2,ACKDT        ; send NACK
            bsf      SSPCON2,ACKEN        ; "
lop7        btfsc    SSPCON2,ACKEN        ; wait until NACK is complete
            bra      lop7
            bsf      SSPCON2,PEN          ; assert STOP condition
lop8        btfsc    SSPCON2,PEN
            bra      lop8
            movlw    0x00                 ; set return code to 0
            return
bad2        movlw    0xFF                 ; set return code to –1
            return
```

The C language version of the function is as follows:

```c
unsigned char read_conf (void)
{
        unsigned char return_val;
        StartI2C( );
        while(SSPCON2bits.SEN);      /* wait until START condition is complete */
        if (WriteI2C(0x92))          /* send control code with R/W = 0 */
          return -1;
        IdleI2C( );
        if (SSPCON2bits.ACKSTAT)     /* received ACK ? */
          return -1;
        if(WriteI2C(0xAC))           /* send Access Config command */
          return -1;
        IdleI2C();
        if (SSPCON2bits.ACKSTAT)
          return -1;
        RestartI2C( );               /* assert a repeated start condition */
        while(SSPCON2bits.RSEN);     /* wait until restart is complete */
        if(WriteI2C(0x93))           /* send control byte with R/W = 1 */
          return -1;
        IdleI2C( );
        if (SSPCON2bits.ACKSTAT)
          return -1;
        return_val = ReadI2C( );     /* read Config Register */
        IdleI2C();
        NotAckI2C();
        StopI2C();
        return return_val;
}
```

Example 11.6

Write a subroutine to read the converted temperature and return the upper and lower bytes in PRODH and PRODL, respectively. Assume that the temperature conversion has been started but that this function needs to make sure that the converted temperature value is resulted from the most recent **Start Convert T** command.

Solution: This subroutine will call the previous subroutine to make sure that the DONE bit is set to 1 and reads the converted temperature from the DS1631A. The assembly subroutine is as follows:

```
read_temp    call      read_conf
             btfss     PRODL,7        ; wait until a new conversion
             bra       read_temp      ; result is available
             call      i2c_idle       ; wait until I²C bus is idle
             bsf       SSPCON2,SEN    ; assert a START condition
lp1          btfsc     SSPCON2,SEN    ; wait until start condition is completed
             bra       lp1            ; "
```

```
              movlw      0x92                     ; send control byte to DS1631A with R/W = 0
              movwf      SSPBUF                   ; "
lp2           btfsc      SSPSTAT,BF               ; wait until 8 bits have been shifted out
              bra        lp2                      ; "
              call       i2c_idle                 ; ensure I²C bus is idle
              btfsc      SSPCON2,ACKSTAT          ; check ACK bit
              goto       bad3                     ; receive NACK
              movlw      0xAA                     ; sends "Read Temperature" command
              movwf      SSPBUF                   ; "
lp3           btfsc      SSPSTAT,BF               ; wait until 8 bits have been shifted out
              bra        lp3                      ; "
              call       i2c_idle
              btfsc      SSPCON2,ACKSTAT          ; check ACK bit
              goto       bad3
              bsf        SSPCON2,RSEN             ; assert a Repeated START condition
lp4           btfsc      SSPCON2,RSEN             ; wait until Repeated START is complete
              bra        lp4                      ; "
              movlw      0x93                     ; send control byte to the DS1631A with
              movwf      SSPBUF                   ; R/W = 1
lp5           btfsc      SSPSTAT,BF               ; wait until 8 bits are shifted out
              bra        lp5                      ; "
              call       i2c_idle                 ; make sure I²C is idle
              btfsc      SSPCON2,ACKSTAT          ; check ACK bit
              bra        bad3
              bsf        SSPCON2,RCEN             ; enable reception
lp6           btfss      SSPSTAT,BF               ; wait until 8 bits have been shifted in
              bra        lp6                      ; "
              movff      SSPBUF,PRODH             ; transfer converted temperature hi-byte to PRODH
              call       i2c_idle                 ; make sure I²C is idle
              bcf        SSPCON2,ACKDT            ; sends ACK
              bsf        SSPCON2,ACKEN            ; "
lp7           btfsc      SSPCON2,ACKEN            ; wait until ACK is done
              bra        lp7
              bsf        SSPCON2,RCEN             ; enable another reception
lp8           btfss      SSPSTAT,BF               ; wait until 8 bits are received
              bra        lp8                      ; "
              movff      SSPBUF,PRODL             ; copy converted temperature low byte to PRODL
              call       i2c_idle
              bsf        SSPCON2,ACKDT            ; send NACK
              bsf        SSPCON2,ACKEN            ; "
lp9           btfsc      SSPCON2,ACKEN            ; wait until NACK is done
              bra        lp9
              bsf        SSPCON2,PEN              ; assert STOP condition
lp10          btfsc      SSPCON2,PEN
              bra        lp10
              movlw      0x00                     ; set return code to 0
              return
bad3          movlw      0xFF
              return
```

The C language version of the function is straightforward and hence is left for you as an exercise.

In a system with multiple I²C masters, bus collision can occur. When a bus collision occurs, the BCLIF flag of the PIR2 register will be set to 1. The user should clear this flag and retry the same operation if a bus collision occurs. The functions described in Examples 11.2 to 11.6 can be modified slightly to return the error code for bus collision to the caller. The callers of these functions can clear the BCLIF and retry the same operation.

▲

11.10 Interfacing the Serial EEPROM 24LC08B with I²C

Some applications require the use of large amount of nonvolatile memory because these applications are powered by batteries and may be used in the field for an extended period of time. Many semiconductor manufacturers produce serial EEPROMs with a serial interface. Both serial EEPROMs with SPI and I²C interfaces are available.

The 24LC08B is a serial EEPROM with the I²C interface from Microchip. This device is an 8-Kbit EEPROM organized as four blocks of 256- × 8-bit memory. A low-voltage design permits operation down to 2.5 V with standby and active currents of only 1 µA and 1 mA, respectively. The 24LC08B also has a page write capability for up to 16 bytes of data.

11.10.1 Pin Assignment and Block Diagram

The pin assignment and the block diagram of the 24LC08B are shown in Figures 11.35 and 11.36, respectively.

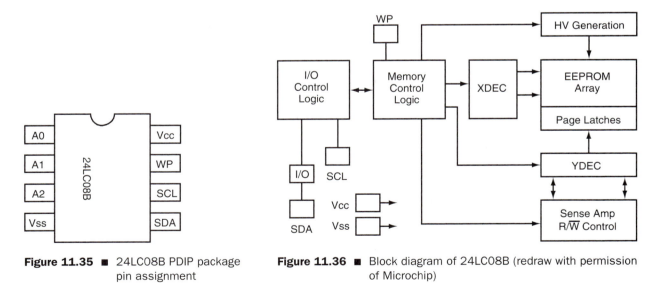

Figure 11.35 ■ 24LC08B PDIP package pin assignment

Figure 11.36 ■ Block diagram of 24LC08B (redraw with permission of Microchip)

The pins SCL and SDA are for I²C bus communications. The SCL line is a clock input that can be as high as 400 KHz. The WP pin is used as the write protection input. When this pin is high, the 24LC08B cannot be written into. Pins A2 . . . A0 are not used and can be left floating, grounded, or pulled to high.

11.10.2 Device Addressing

Like any other I²C slave, the first byte sent to the 24LC08B after the START condition is the control byte. The contents of the control byte for the 24LC08B are shown in Figure 11.37.

The upper four bits are the device address of the 24LC08B, whereas B1 . . . B0 are the block addresses of the memory location to be accessed. For any access to the 24LC08B, the master must also send an 8-bit byte address after the control byte. There is an address pointer inside the 24LC08B. After the access of each byte, the address pointer is incremented by 1.

7	6	5	4	3	2	1	0
1	0	1	0	X	B1	B0	R/$\overline{W}$

Figure 11.37 ■ 24LC08B Control byte contents

11.10.3 Write Operation

The 24LC08B supports byte write and page write operations. In a *byte write* operation, the following events will occur:

1. The master asserts the START condition.
2. The master sends the control byte to the 24LC08B.
3. The 24LC08B acknowledges the data transmission.
4. The master sends the byte address to the 24LC08B.
5. The 24LC08B acknowledges the data transmission.
6. The master sends the data byte to the 24LC08B.
7. The 24LC08B acknowledges the data transmission.
8. The master asserts the STOP condition.

In a *page write* operation, the master can send up to 16 bytes of data to the 24LC08B. The write control byte, byte address, and the first data byte are transmitted to the 24LC08B in the same way as in a byte write. But instead of asserting a STOP condition, the master transmits up to 16 data bytes to the 24LC08B that are temporarily stored in the on-chip page buffer and will be written into the memory after the master has asserted a STOP condition. After the receipt of each byte, the four lower address pointer bits are internally incremented by 1. The higher-order six bits of the byte address remain constant. If the master should transmit more than 16 bytes prior to generating the STOP condition, the address counter will roll over, and the previously received data will be overwritten.

11.10.4 Acknowledge Polling

When the 24LC08B is writing the data held in the write buffer into the EEPROM array, it will not acknowledge any further write operation. This fact can be used to determine when the cycle is complete.

Once the STOP condition for a write command has been issued from the master, the device initiates the internal write cycle. The ACK polling can be initiated immediately. This involves the master sending a START condition followed by the control byte for a write command (R/$\overline{W}$ = 0). If the 24LC08B is still busy, then no ACK will be returned. If the cycle is complete, then the device will return the ACK, and the master can then proceed with the next read or with a write command. The polling process is illustrated in Figure 11.38.

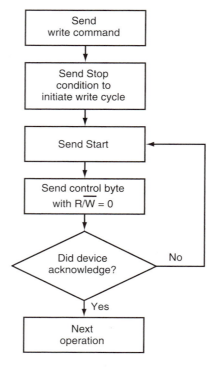

Figure 11.38 ■ Acknowledge polling flow

11.10.5 Read Operation

The 24LC08B supports three types of read operations: *current address read, random read,* and *sequential read.*

CURRENT ADDRESS READ

As was explained earlier, the internal address counter is incremented by 1 after each access (read or write). The current address read allows the master to read the byte immediately following the location accessed by the previous read or write operation. On receipt of the slave address with R/$\overline{\text{W}}$ bit set to 1, the 24LC08B issues an acknowledgment and transmits an 8-bit data byte. The master will not acknowledge the transfer but asserts a STOP condition, and the 24LC08B discontinues transmission.

RANDOM READ

Random read operations allow the master to access any memory location in a random manner. To perform this type of read operation, the user must send the address of the memory location to be read. The procedure for performing a random read is as follows:

Step 1
The master asserts a START condition.

Step 2
The master sends the control byte to the 24LC08B.

Step 3
The 24LC08B acknowledges the control byte.

Step 4
The master sends the address of the byte to be read to the 24LC08B.

Step 5
The 24LC08B acknowledges the byte address.

Step 6
The master asserts a repeated START condition.

Step 7
The master sends the control byte with $R/\overline{W} = 1$.

Step 8
The 24LC08B acknowledges the control byte and sends the data to the master.

Step 9
The master asserts NACK to the 24LC08B.

Step 10
The master asserts the STOP condition.

SEQUENTIAL READ

If the master acknowledges the data byte returned by the random read operation, the 24LC08B will transmit the next sequentially addressed byte as long as the master provides the clock signal on the SCL line. The master can read the whole chip using sequential read.

11.10.6 Circuit Connection between the I²C Master and the 24LC08B

The circuit connection of a 24LC08B and the PIC18 microcontroller is similar to that in Figure 11.33. The diagram of a system with one DS1631A and one 24LC08B and the PIC18 microcontroller is shown in Figure 11.39.

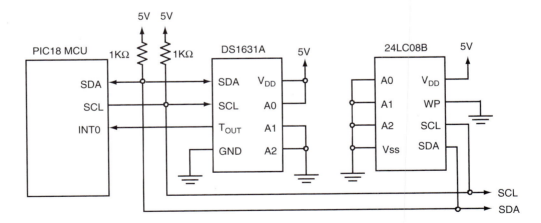

Figure 11.39 ■ Circuit connection of the PIC18 MCU, 24LC08B, and DA1631A

Example 11.7

▼

Write a function to read a byte from the 24LC08B. Pass the control byte and address in PRODH and PRODL to this subroutine. Return the data byte and error code in PRODL and PRODH, respectively. This function should check two errors:

- Bus collision (to make this function usable in a multimaster environment)
- Not ACK (implies that 24LC08B may be in error)

Solution: The following assembly subroutine implements the procedure for random read described earlier:

```
rand_read   call    i2c_idle            ; make sure the I²C bus is idle
            bsf     SSPCON2,SEN         ; assert START condition
eloop1      btfsc   SSPCON2,SEN         ; wait until START condition is done
            bra     eloop1             ; "
            btfss   PIR2,BCLIF         ; test bus collision error
            goto    next1
            movlw   0xFF                ; set error code to -1
            movwf   PRODH              ; "
            return                      ; "
next1       call    i2c_idle            ; avoid write collision
            movff   PRODH,SSPBUF        ; send out control byte
eloop2      btfsc   SSPSTAT,BF         ; wait until data is shifted out
            bra     eloop2             ; "
            call    i2c_idle            ; wait until MSSP module is not busy
            btfsc   SSPCON2,ACKSTAT    ;
            goto    not_ack            ; NACK error
            call    i2c_idle            ; make sure MSSP module is idle
            movff   PRODL,SSPBUF        ; send address to 24LC08B
eloop3      btfsc   SSPSTAT,BF         ; wait until data is shifted out
            bra     eloop3
            call    i2c_idle
            btfsc   SSPCON2,ACKSTAT    ; check NACK error
            goto    not_ack
            call    i2c_idle
            bsf     SSPCON2,RSEN       ; assert repeated start condition
eloop4      btfsc   SSPCON2,RSEN       ; wait until repeated START condition
            goto    eloop4             ; is complete
            btfss   PIR2,BCLIF         ; test bus collision error
            goto    next2
            movlw   0xFF                ; set error code to -1
            movwf   PRODH              ; "
            return                      ; "
next2       movlw   0x01
            addwf   PRODH,W            ; set the R/W bit of control byte to 1
            movwf   SSPBUF             ; send out the control byte to read
eloop5      btfsc   SSPSTAT,BF         ; wait until 8 bits have been shifted out
            bra     eloop5             ; "
            call    i2c_idle
            btfsc   SSPCON2,ACKSTAT    ; check NACK error
```

```
                goto        not_ack
                bsf         SSPCON2,RCEN        ; enable receive
eloop6          btfss       SSPSTAT,BF         ; wait until 8 bits have been shifted in
                bra         eloop6
                bsf         SSPCON2,ACKDT      ; assert NACK condition
                bsf         SSPCON2,ACKEN      ; "
eloop7          btfsc       SSPCON2,ACKEN      ; "
                bra         eloop7
                bsf         SSPCON2,PEN        ; assert STOP condition
eloop8          btfsc       SSPCON2,PEN        ; wait until STOP condition is complete
                bra         eloop8             ; "
                btfss       PIR2,BCLIF         ; test bus collision error
                goto        next3              ; "
                movlw       0xFF
                movwf       PRODH              ; "
                return                         ; "
next3           movff       SSPBUF,PRODL       ; place data in PRODL
                clrf        PRODH              ; set error code to 0 (no error)
                return
not_ack         movlw       0xFE               ; return –2 as the error code
                movwf       PRODH              ; "
                return
```

The C language version of the function is as follows:

unsigned int **EERandomRead**(unsigned char **control,** unsigned char **address**)
{

```
        IdleI2C( );                        // make sure I²C module is idle
        StartI2C( );                       // assert a START condition
        while (SSPCON2bits.SEN);           // wait until start condition is over
        if (PIR2bits.BCLIF)                // test for bus collision
                return (-1);               // return with Bus Collision error
        IdleI2C( );                        // make sure I²C module is idle
        WriteI2C(control);
        IdleI2C();                         // make sure I²C module is idle
        if (SSPCON2bits.ACKSTAT)           // test for NACK condition
                return (-2);
        IdleI2C( );
        WriteI2C(address);                 //send out read address
        IdleI2C( );                        // ensure module is idle
        if (SSPCON2bits.ACKSTAT)           // test for ACK condition, if received
                return (-2);               // return NACK error
        RestartI2C( );                     // assert I²C bus restart condition
        while ( SSPCON2bits.RSEN);         // wait until re-start condition is over
        if (PIR2bits.BCLIF)                // test for bus collision
                return (-1);               // return with Bus Collision error
        IdleI2C( );
        WriteI2C(control+1);               // write 1 byte – R/W bit should be 1
        IdleI2C( );                        // ensure module is idle
        if (SSPCON2bits.ACKSTAT)           // test for NACK condition
                return (-2);
        SSPCON2bits.RCEN = 1;              // enable master for 1 byte reception
```

```
        while (SSPCON2bits.RCEN);        // check that receive sequence is over
        NotAckI2C( );                    // send NACK condition
        while (SSPCON2bits.ACKEN);       // wait until ACK sequence is over
        StopI2C( );                      // send STOP condition
        while (SSPCON2bits.PEN);         // wait until stop condition is over
        if (PIR2bits.BCLIF)              // test for bus collision
            return (–1);                                 // return with Bus Collision error
        return ((unsigned int) SSPBUF);  // return with data
    }
```

Sequential read and current address read are quite similar and hence are left for you as exercise problems.

Example 11.8

Write a subroutine that writes a byte into the 24LC08B. The control byte, address, and data to be written are passed to this routine in the software stack.

Solution: The assembly subroutine that writes a byte into the 24LC08B to the specified address is as follows:

```
ctl_ byte       equ         –5              ; offset from top of software stack
addr            equ         –4              ; "
dat             equ         –3              ; "
; *************************************************************
; The following subroutine writes a byte into the EEPROM. The control byte, address,
; and data byte are pushed onto the stack in that order.
; *************************************************************
;
wr_byte_ee      pushr       FSR2L           ; save callers frame pointer
                pushr       FSR2H           ; "
                movff       FSR1L,FSR2L     ; set up new stack frame
                movff       FSR1H,FSR2H     ; pointer
                call        i2c_idle
                bsf         SSPCON2,SEN     ; assert the START condition
weloop1         btfsc       SSPCON2,SEN     ; wait until START condition is complete
                goto        weloop1
                btfss       PIR2,BCLIF      ; check bus collision error
                goto        next1
                popr        FSR2H           ; restore FSR2 before return
                popr        FSR2L           ; "
                movlw       0xFF            ; return bus collision error code (–1)
                return                      ; "
next1           movlw       ctl_byte        ; send out control byte to 24LC08B
                movff       PLUSW2,SSPBUF   ; "
weloop2         btfsc       SSPSTAT,BF      ; wait until control byte is shifted out
                bra         weloop2         ; "
                call        i2c_idle
                btfsc       SSPCON2,ACKSTAT ; check NACK error
                goto        nack_err
                movlw       addr
                movff       PLUSW2,SSPBUF   ; send out memory address to be accessed
```

```
weloop3       btfsc      SSPSTAT,BF            ; wait until address is shifted out
              goto       weloop3              ; "
              call       i2c_idle
              btfsc      SSPCON2,ACKSTAT      ; check NACK error
              goto       nack_err
              movlw      dat                  ; send out the write data to 24LC08B
              movff      PLUSW2,SSPBUF        ; "
weloop4       btfsc      SSPSTAT,BF,A
              bra        weloop4
              call       i2c_idle
              btfsc      SSPCON2,ACKSTAT      ; check NACK error
              bra        nack_err
              bsf        SSPCON2,PEN          ; assert the STOP condition
weloop5       btfsc      SSPCON2,PEN          ; wait until STOP condition is complete
              bra        weloop5
              popr       FSR2H
              popr       FSR2L
              movlw      0x00
              return
nack_err      popr       FSR2H
              popr       FSR2L
              movlw      0xFE                 ; NACK error (–2)
              return
```

This subroutine invokes several macros defined in Chapter 4. The C language version of the program is as follows:

```c
unsigned char EEByteWrite(unsigned char control, unsigned char address, unsigned char data)
{
        IdleI2C( );                        // ensure I²C bus is idle
        StartI2C( );                       // initiate START condition
        while (SSPCON2bits.SEN );          // wait until start condition is over
        if (PIR2bits.BCLIF)                // test for bus collision
                return (-1);               // return with Bus Collision error
        WriteI2C(control);                 // write byte - R/W bit should be 0
        IdleI2C( );                        // ensure module is idle
        if (SSPCON2bits.ACKSTAT)           // test for NACK condition received
                return (-2);
        IdleI2C( );
        WriteI2C(address);                 // write word address for EEPROM
        IdleI2C( );                        // ensure module is idle
        if (SSPCON2bits.ACKSTAT)           // test for NACK condition received
                return (-2);
        WriteI2C(data);                    // data byte for EEPROM
        IdleI2C( );                        // ensure module is idle
        StopI2C( );                        // send STOP condition
        while (SSPCON2bits.PEN);           // wait until stop condition is over
        if (PIR2bits.BCLIF)                // test for bus collision
                return (-1);               // return with Bus Collision error
        return (0);                        // return with no error
}
```

▲

Example 11.9

Write a C function that performs a page write operation. The control byte, the starting address of the destination, and the pointer to the data in RAM to be written are passed to this function.

Solution: This C function will invoke the library function **putsI2C( )** to output the string to the 24LC08B. The EEPageWrite function is as follows:

```
unsigned char EEPageWrite(unsigned char control, unsigned char address, unsigned char *wrptr)
{
            IdleI2C( );                           // ensure module is idle
            StartI2C( );                          // initiate START condition
            while (SSPCON2bits.SEN);              // wait until start condition is over
            if (PIR2bits.BCLIF )                  // test for bus collision
              return (-1);                        // return with Bus Collision error
            IdleI2C( );
            WriteI2C(control);                    // write control byte - R/W bit should be 0
            IdleI2C( );                           // ensure module is idle
            if(SSPCON2bits.ACKSTAT)               // test for NACK condition, if received
              return (-2);                        // NACK error
            WriteI2C(address);                    // write address byte to EEPROM
            IdleI2C( );                           // ensure module is idle
            if(SSPCON2bits.ACKSTAT)               // test for ACK condition, if received
              return(-2);
            if (putsI2C(wrptr))
              return(-4);                         // bus device responded possible error
            IdleI2C( );                           // ensure module is idle
            StopI2C( );                           // send STOP condition
            while (SSPCON2bits.PEN);              // wait until stop condition is over
            if (PIR2bits.BCLIF)                   // test for Bus collision
              return (-1);                        // return with Bus Collision error
            return (0);                           // return with no error
}
```

In some applications, the user program may need to write several blocks of data to different locations. In this situation, the user program will need to poll the EEPROM in order to find out if the internal write operation has been completed. The following example implements the algorithm illustrated in Figure 11.38.

Example 11.10

Write a C function to implement the algorithm described in Figure 11.38.

Solution: The function **EEAckPolling( )** will poll until the EEPROM is not busy or until certain errors occur:

```
unsigned char EEAckPolling(unsigned char control)
{
            IdleI2C( );  // ensure module is idle
            StartI2C( );// initiate START condition
```

```
    while (SSPCON2bits.SEN);            // wait until start condition is over
    if (PIR2bits.BCLIF)                 // test for bus collision
        return (-1);                    // return with Bus Collision error
    IdleI2C( );
    WriteI2C(control);                  // write byte - R/W bit should be 0
    IdleI2C( );                         // ensure module is idle
    if (PIR2bits.BCLIF)                 // test for bus collision
        return (-1);                    // return with Bus Collision error
    while (SSPCON2bits.ACKSTAT)         // test for ACK condition received
    {
        RestartI2C();                   // initiate Restart condition
        while (SSPCON2bits.RSEN);       // wait until re-start condition is over
        if (PIR2bits.BCLIF)             // test for bus collision
            return (-1);                // return with Bus Collision error
        IdleI2C( );
        WriteI2C(control);              // write byte - R/W bit should be 0
        IdleI2C( );
    }
    StopI2C();                          // send STOP condition
    while(SSPCON2bits.PEN);             // wait until stop condition is over
    if (PIR2bits.BCLIF)                 // test for bus collision
        return (-1);                    // return with Bus Collision error
    return ( 0 );                       // return with no error
}
```

▲

11.10.7 C Library Functions for I²C EEPROM

Microchip PIC18 compiler provides a handful library functions to support the use of the EEPROMs with the I²C interface. These library functions can be applied to 24LC01B/02B/04B/08B devices from Microchip. The user needs to include the header file **i2c.h** in order to use them. The prototype declarations and functions of these functions are as follows:

unsigned **EEAckPolling** (unsigned char **control**);

> This function generates the acknowledge polling sequence for Microchip EEPROM I²C devices. If there is no error, the return value of this function will be 0. If bus collision occurred during the polling, this function returns –1 to the caller. If there was a NACK error, the return value is –2. If there was a write collision error, this function returns –3.

unsigned char **EEByteWrite** (unsigned char **control,** unsigned char **address,** unsigned char **data**);

> This function writes the **data** byte to the memory location specified by **address** in the EEPROM device. This function returns the following:

> 0:   if there was no error
>
> –1:   if there was a bus collision error
>
> –2:   if there was a NACK error
>
> –3:   if there was a write collision error

unsigned int **EECurrentAddRead** (unsigned char **control**);

The only argument to this function is the device select address byte. This function reads a byte located at the current pointer in the EEPROM device. The return value is the following:

−1:   if a bus collision error occurred

−2:   if a NACK error occurred

−3:   if there was a write collision error

unsigned 16-bit integer that contains the data in the lower byte if there was no error.

unsigned char **EEPageWrite** (unsigned char **control**, unsigned char **address**, unsigned char ***wptr**);

This function writes a string of data to the EEPROM device from the I²C bus. The first argument is the device select address. The second parameter specifies an EEPROM internal location, whereas the third parameter points to the string in RAM to be output.

unsigned int **EERandomRead** (unsigned char **control**, unsigned char **address**);

This function reads a byte from the I²C bus at the location specified by **address.** The return value contains the value read in the least significant byte and error condition in the most significant byte. The error condition is the following:

0:      if there was no error

−1:   if there was a bus collision error

−2:   if there was a NACK error

−3:   if there was a write collision error

unsigned char **EEsequentialRead** (unsigned char **control,** unsigned char **address,** unsigned char ***rdptr,** unsigned char **length**);

This function reads in a predefined string length of data from the I²C bus. The meaning of returned error code is identical to those in **EERandonRead.**

Example 11.11

▼

Write a C program to illustrate the use of the I2C EEPROM library functions.

Solution:

```
#include <p18F452.h>
#include <i2c.h>
unsigned char array_2wr [ ] = {10, 11, 12, 13, 14, 15, 16, 17, 18, 19, 0};
unsigned char array_rd [10];
void main (void)
{
        char value;
        OpenI2C(MASTER, SLEW_ON);
        SSPADD = 9;                              /* set baud rate to 400 KHz at 16 MHz */
        EEByteWrite(0xA0, 0x20, 0x33);           /* write 0x33 to location at 0x20 */
        EEAckPolling (0xA0);                     /* poll until EEPROM is ready */
        value = EERandomRead (0xA0, 0x20);       /* read back the written value */
        EEPageWrite (0xA0, 0x50, array_2wr);     /* write the 9-byte array into EEPROM */
        EEAckPolling (0xA0);
        EESequentialRead (0xA0, 0x50, array_rd, 9); /* read back 9 bytes written before */
}
```

11.11 Summary

The I²C serial protocol is an alternative to the SPI serial interface protocol. Compared with the SPI protocol, the I²C bus offers the following advantages:

- No chip enable or chip select signal for selecting slave devices
- Allows multiple master devices to coexist in a system because it provides easy bus arbitration
- Allows many more devices in the same I²C bus
- Allows resources to be shared by multiple master (microcontrollers) devices

However, the SPI has the following advantages over the I²C interface:

- Higher data rates (no longer true for high-speed mode)
- Much lower software overhead to carry out data transmission

The data transfer over the I²C bus requires the user to generate the following signal components:

1. START condition
2. STOP condition
3. ACK
4. Repeated START condition
5. Data

Whenever there are multiple master devices attempting to send data over the I²C bus, bus arbitration is carried out automatically. The loser is decided whenever it attempts to drive the data line to high, whereas another master device drives the same data line to low.

To select the slave device without using the chip select (or chip enable) signal, address signals are used. Both 7- and 10-bit addresses are supported in the same I²C bus. Ten-bit addressing will be used in a system that consists of many slave devices. I²C bus supports three speed rates:

- 100 KHz
- 400 KHz
- 3.4 MHz

Current PIC18 devices do not support the 3.4-MHz clock rate.

Each data transfer starts with a START condition and ended with the STOP condition. One or two control bytes will follow the START condition, which specifies the slave device to receive or send data. For each data byte, the receiver must assert either the ACK or the NACK condition to acknowledge or unacknowledge the data transfer, respectively.

The DS1631A is a digital thermometer that allows the user to read the ambient temperature using the I²C bus. It has an output signal (T_{OUT}) that will become active whenever the ambient temperature reaches the value preset by the user. This feature is useful to provide early warning about the potential overheating of a system. The DS1631A can be configured to operate in one-shot mode or continuous conversion mode. The default temperature conversion resolution is 12-bit. However, it can also be reduced to 9-, or 10-, or 11-bit if high resolution is not so important compared to the required conversion time.

The 24LC08B is an EEPROM with an I²C interface. The capacity of the 24LC08B is 8 K bits. This chip has an internal address pointer that will increment automatically after each access. This feature can increase the access efficiency when the access patterns are sequential. Library functions are available for general read and write for an I²C environment and for EEPROM chips produced by Microchip.

Functions for performing basic read and write access to the I²C slave devices are provided in this chapter. Users can add their own processing functions to provide further processing or data formatting to make the data more user friendly.

11.12 Exercises

E11.1 Does the I²C clock frequency need to be exactly equal to 100 KHz or 400 KHz? Why?

E11.2 How many slave devices are supported in a 7-bit address scheme? Try to review the datasheets of peripheral devices with I²C interface and provide a realistic number of devices supported in an I²C bus system.

E11.3 Suppose that the 7-bit address of an I²C slave is B'10101 A1 A0' with A1 tied to low and A0 pulled to high. What is the 8-bit hex write address for this device? What is the 8-bit hex read address?

E11.4 Write an instruction sequence to configure the MSSP module to operate in the I²C slave mode using the 7-bit addressing and operating at 100 KHz.

E11.5 What value should be written into the SSPADD register so that the I²C module can operate at 300 KHz for an 8-MHz instruction cycle clock?

E11.6 What is the value of T_{BRG}, assuming that the instruction cycle clock frequency is 8 MHz and the value written into the SSPADD register is 19?

E11.7 Write a subroutine to configure the DS1631A to operate in one-shot mode, 10-bit resolution, and low active polarity for T_{OUT} output.

E11.8 Assuming that the DS1631A has been configured to operate in one-shot mode with 12-bit resolution, write a program to display the converted temperature in the format of three integer digits and two fractional digits in the LCD.

E11.9 For the circuit shown in Figure 11.33, add an alarm speaker to the CCP1 output and set the high temperature trip point to 50 °C. Whenever temperature reaches 50 °C or higher, turn on the alarm until temperature drops down to 23 °C.

E11.10 Write an assembly routine to implement the algorithm described in Figure 11.38.

E11.11 MAX5812 is a DAC with 12-bit resolution and I²C interface. The MAX5821 datasheet can be downloaded from the Web site www.maxim-ic.com. What is the 7-bit address of this device? What is the highest operating frequency of this chip? How many commands are available to this chip? How can you connect this device to the I²C bus?

E11.12 For the MAX5812 mentioned in Problem E11.11, write a program to generate a sine wave from the OUT pin.

E11.13 The MCP23016 is an I/O expander from Microchip. This chip has an I²C interface and can add 16 I/O pins to the microcontroller. Download the datasheet of this device from Microchip's Web site. Show the circuit connection of this chip to the microcontroller and write a sequence of instructions to configure the upper eight pins for input and configure the lower eight pins for output. Configure input polarity to active low.

E11.14 What are the corresponding temperatures for the conversion results 0x6800, 0x7200, 0x4800, 0xEE60, and 0xF280 output by the DS1631A?

E11.15 What will be the conversion results sent out by the DS1631A for temperatures 40 °C, 50 °C, 80.5 °C, −10.25 °C, and −20.5 °C?

E11.16 A sequential read can be performed after a read operation has been performed to the EEPROM. Write a few instructions to perform the sequential read and return the data in PRODL. (Hint: The I²C master sets the RCEN bit of the SSPCON2 register to perform the sequential read.)

E11.17 Write a function that performs current address read to the 24LC08B. Pass the control byte in PRODH to this subroutine. Return the data byte and error code in PRODL and PRODH, respectively. This function should check three errors:

- Bus collision (to make this function usable in a multimaster environment)
- Write collision
- Not ACK (implies that 24LC08B may be in error)

11.13 Laboratory Exercises and Assignments

L11.1 Write a program to store 0 to 255 in the first block of the 24LC08B of your demo board and then, staying in an infinite loop, read out one value from 24LC08B sequentially every half second and display the value on eight LEDs of the demo board.

L11.2 Connect the DS1631A to the I²C bus of your PIC18 demo board. Write a program to set up the high and low temperature trip points to 40 °C and 20 °C, respectively. Whenever the temperature goes above 40 °C, turn on the alarm (speaker). Turn off the alarm when the temperature drops below 20 °C. Display the temperature on the LCD display and update the display once every second.

12

Analog-to-Digital Converter

12.1 Objectives

After completing this chapter, you should be able to

- Explain the A/D conversion process
- Describe the resolution, the various channels, and the operation modes of the PIC18 A/D converter
- Interpret the A/D conversion results
- Describe the procedure for using the PIC18 A/D converter
- Configure the A/D converter for the application
- Use the temperature sensor TC1047A
- Use the humidity sensor IH-3605
- Use the barometric pressure sensor BPT

12.2 Basics of A/D Conversion

Many embedded applications deal with nonelectric quantities, such as weight, humidity, pressure, mass or airflow, temperature, light intensity, and speed. These quantities are *analog* in nature because they have a continuous set of values over a given range, in contrast to the discrete values of digital signals. To enable the microcontroller to process these quantities, they need to be represented in digital form; thus, an *analog-to-digital* (A/D) converter is required.

12.2.1 A Data Acquisition System

An A/D converter can deal only with electrical voltage. A nonelectric quantity must be converted into a voltage before A/D conversion can be performed. The conversion of a nonelectric quantity to a voltage requires the use of a *transducer*. In general, a transducer is a device that converts the quantity from one form to another. For example, a temperature sensor is a transducer that can convert the temperature into a voltage. A *load cell* is the transducer that can convert a weight into a voltage.

A transducer may not generate an output voltage in the range suitable for A/D conversion. A voltage *scaler* (or *amplifier*) is often needed to amplify the transducer output voltage into a range that can be handled by the A/D converter. The circuit that performs the scaling and shifting of the transducer output is called a *signal-conditioning circuit*. The overall A/D process is illustrated in Figure 12.1.

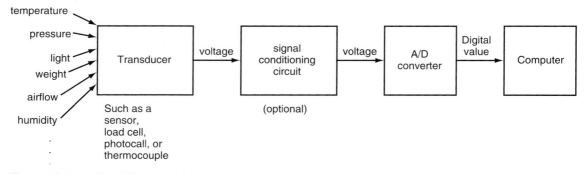

Figure 12.1 ■ The A/D conversion process

12.2.2 Analog Voltage and Digital Code Characteristic

An ideal A/D converter should demonstrate the linear input/output relationship as shown in Figure 12.2.

The output characteristic shown in Figure 12.2 is unrealistic because it requires the A/D converter to use infinite number of bits to represent the conversion result. The output characteristic of an ideal A/D converter using n bits to represent the conversion result is shown in Figure 12.3. An n-bit A/D converter has 2^n possible output code values. The area between the dotted line and the staircase is called the *quantization error*. The value of $V_{DD}/2^n$ is the resolution of this A/D converter. Using n bits to represent the conversion result, the average *conversion error* is $V_{DD}/2^{n+1}$ if the converter is perfectly linear. For a real A/D converter, the output characteristic may have *nonlinearity* (the staircase may have unequal steps in some values) and *nonmonotonicity* (higher voltage may have smaller code).

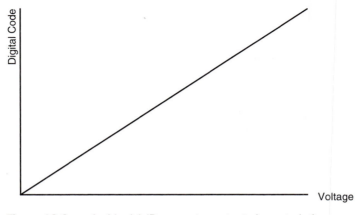

Figure 12.2 ■ An ideal A/D converter output characteristic

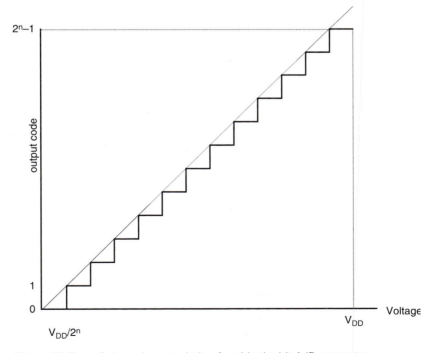

Figure 12.3 ■ Output characteristic of an ideal n-bit A/D converter

Obviously, the more bits used in representing the A/D conversion result, the smaller the conversion error will be. Most microcontrollers used 8 bits, 10 bits, or 12 bits to represent the conversion result. Some microcontrollers (mainly 8051 variants from Silicon Laboratory, TI, and Analog Devices) use 16 bits or even 24 bits to represent conversion results. Whenever the on-chip A/D converter cannot provide the required accuracy, an external A/D converter should be considered.

12.2.3 A/D Conversion Algorithms

Many A/D conversion algorithms have been invented in the past. These algorithms can be divided into four categories:

1. Parallel (flash) A/D converter
2. Slope and double-slope A/D converter
3. Sigma-delta A/D converter
4. Successive approximation A/D converter

PARALLEL (FLASH) A/D CONVERTERS

In this type of A/D converter, 2^n comparators are used. One of the inputs to each comparator is the input voltage to be converted, whereas the other input corresponds to the voltage that represents one of the 2^n combinations of n-bit values. The comparator output will be high whenever the analog input (to be converted) is higher than the voltage that represents one of the 2^n combinations of the n-bit value. The largest n-bit value that causes the comparator output to become true is selected as the A/D conversion value. It is obvious that this type of A/D converter will be very fast. However, they require a lot of hardware resources to implement and therefore are not suitable for implementing high-resolution A/D converters. Over the years, several variations to this approach have been proposed to produce high-speed A/D converters. The most commonly used technique is to pipeline a flash A/D converter, which will reduce the amount of hardware required while still achieving high conversion speed.

SLOPE AND DOUBLE-SLOPE A/D CONVERTERS

This type of A/D converter is used in PIC14000 microcontrollers in which the charging and discharging of a capacitor is used to perform the A/D conversion. This type of A/D converter requires relatively simple hardware and is popular in low-speed applications. In addition, high resolution (10-bit to 16-bit) can be achieved.

SIGMA-DELTA A/D CONVERTERS

This type of A/D converter uses the *oversampling* technique to perform A/D conversion. This type of converter has good noise immunity and can achieve high resolution. Sigma-delta A/D converters are becoming more popular in implementing high-resolution A/D converters. The only disadvantage is its conversion speed. However, this weakness is improving because of the improvement in CMOS technology.

SUCCESSIVE-APPROXIMATION A/D CONVERTERS

This method was first detailed in the December 1975 issue of the IEEE Journal of Solid State Circuit. This method utilizes charge redistribution over an array of ratioed capacitors to perform A/D conversion. The block diagram of this method is shown in Figure 12.4.

The successive-approximation method approximates the analog signal to n-bit code in n steps. It first initializes the successive-approximation register (SAR) to 0 and then performs a series of guessing, starting with the most significant bit and proceeding toward the least significant bit. The algorithm of the successive-approximation method is illustrated in Figure 12.5. For every bit of the SAR, the algorithm does the following:

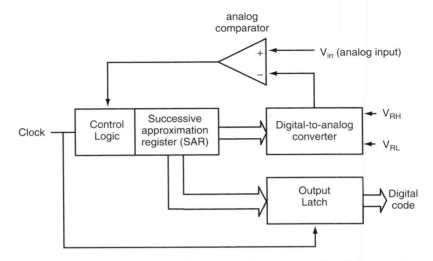

Figure 12.4 ■ Block diagram of a successive approximation A/D converter

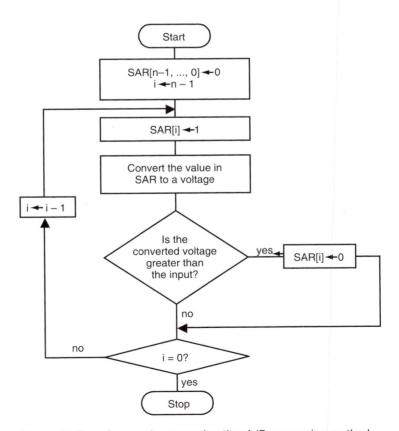

Figure 12.5 ■ Successive approximation A/D conversion method

- Guesses the bit to be a 1
- Converts the value of the SAR register to an analog voltage
- Compares the D/A output with the analog input and clears the bit to 0 if the D/A output is larger (which indicates that the guess is wrong)

Because of its balanced speed and precision, this method has become one of the most popular A/D conversion methods. Most microcontrollers use this method to implement the A/D converter. The PIC18 microcontrollers also use this technique to implement A/D converters.

12.2.4 Optimal Voltage Range for A/D Conversion

An A/D converter needs a *negative reference voltage* (V_{REF-}) and a *positive reference voltage* (V_{REF+}) to perform the conversion. The V_{REF-} voltage is often set to ground, whereas the V_{REF+} voltage is often set to V_{DD}. Some microcontrollers simply tie V_{REF-} to the ground voltage and leave only the high-reference voltage programmable. Most A/D converters are *ratiometric* because of the following:

- A 0-V analog input is converted to the digital code of n 0s.
- A V_{DD} (or V_{REF+}) analog input is converted to the digital code of $2^n - 1$.
- A k-V input will be converted to the digital code of $k \times (2^n - 1) \div V_{DD}$.

Here, n is the number of bits used to represent the A/D conversion result.

By common sense, the A/D conversion result would be most accurate if the value of analog signal covers the whole voltage range from V_{REF-} to V_{REF+}. The A/D conversion result k corresponds to an analog voltage V_k given by the following equation:

$$V_k = V_{REF-} + (range \times k) \div (2^n - 1) \tag{12.1}$$

where range = $V_{REF+} - V_{REF-}$.

Example 12.1

▼

Suppose that there is a 10-bit A/D converter with V_{REF-} = 1 V and V_{REF+} = 4 V. Find the corresponding voltage values for the A/D conversion results of 25, 80, 240, 500, 720, 800, and 900.

Solution: Range = $V_{REF+} - V_{REF-}$ = 4 V – 1 V = 3 V

The voltage corresponding to the A/D conversion results of 25, 80, 240, 500, 720, 800, and 900 are as follows:

$$1\,V + (3 \times 25) \div (2^{10} - 1) = 1.07\,V$$
$$1\,V + (3 \times 80) \div (2^{10} - 1) = 1.23\,V$$
$$1\,V + (3 \times 240) \div (2^{10} - 1) = 1.70\,V$$
$$1\,V + (3 \times 500) \div (2^{10} - 1) = 2.47\,V$$
$$1\,V + (3 \times 720) \div (2^{10} - 1) = 3.11\,V$$
$$1\,V + (3 \times 800) \div (2^{10} - 1) = 3.35\,V$$
$$1\,V + (3 \times 900) \div (2^{10} - 1) = 3.64\,V$$

▲

12.2.5 Scaling Circuit

Some of the transducer output voltages are in the range of $0 \sim V_Z$, where $V_Z < V_{DD}$. Because V_Z sometimes can be much smaller than V_{DD}, the A/D converter cannot take advantage of the available full dynamic range, and therefore conversion results can be very inaccurate. The voltage scaling (amplifying) circuit can be used to improve the accuracy because it allows the A/D converter to utilize its full dynamic range. The diagram of a voltage scaling circuit is shown in Figure 12.6. Because the OP AMP has an infinite input impedance, the current that flows through the resistor R_2 will be the same as the current that flows through R_1. In addition, the voltage at the inverting input terminal (same as the voltage drop across R_1) would be the same as that at the noninverting terminal (V_{IN}). Therefore, the voltage gain of this circuit is given by the following equation:

$$A_V = V_{OUT} \div V_{IN} = (R_1 + R_2) \div R_1 = 1 + R_2/R_1 \tag{12.2}$$

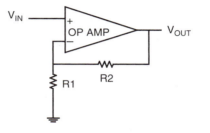

Figure 12.6 ■ A voltage scaler

Example 12.2

Suppose that the transducer output voltage ranges from 0 V to 200 mV. Choose the appropriate values for R_1 and R_2 to scale this range to $0 \text{ V} \sim 5 \text{ V}$.

Solution: $5 \text{ V} \div 200 \text{ mV} = 25$

$\therefore R_2/R_1 = 24$

By choosing 240 KΩ for R_2 and 10 KΩ for R_1, we obtain a R_2/R_1 ratio of 24 and achieve the desired scaling goal.

12.2.6 Voltage Translation Circuit

Some transducers have output voltage in the range of $V_1 \sim V_2$ (V_1 can be negative and V_2 can be unequal to V_{DD}) instead of $0 \text{ V} \sim V_{DD}$. The accuracy of A/D conversion can be improved by using a circuit that shifts and scales the transducer output so that it falls in the full range of $0 \text{ V} \sim V_{DD}$.

An OP AMP circuit that can shift and scale the transducer output is shown in Figure 12.7c. This circuit consists of a summing circuit (Figure 12.7a) and an inverting circuit (Figure 12.7b). The voltage V_{IN} comes from the transducer output, whereas V_1 is an adjusting voltage. By choosing appropriate values for V_1 and resistors R_1, R_2, and R_f, the desired voltage shifting and scaling can be achieved.

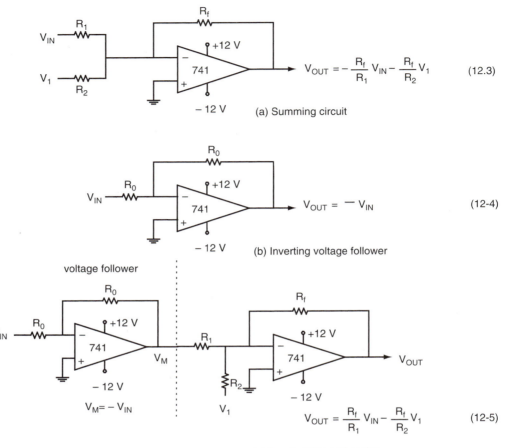

Figure 12.7 ■ Level shifting and scaling circuit

Example 12.3

Choose appropriate resistor values and the adjusting voltage so that the circuit shown in Figure 12.7c can shift the voltage from the range of –1.2 V ~ 3.0 V to the range of 0 V ~ 5 V.

Solution: Applying Equation 12.5,

$$0 = -1.2 \times (R_f/R_1) - (R_f/R_2) \times V_1$$
$$5 = 3.0 \times (R_f/R_1) - (R_f/R_2) \times V_1$$

By choosing $R_0 = R_1 = 10$ K, $R_2 = 50$ K, $R_f = 12$ K, and $V_1 = -5$ V, one can translate and scale the voltage to the desired range. This example tells us that the selection of resistors and the voltage V_1 is a trial-and-error process at best.

12.3 The PIC18 A/D Converter

A PIC18 microcontroller has a 10-bit A/D converter and may have from 5 to 16 analog inputs. The assignment of analog inputs and pins can be found in the datasheet of the appropriate PIC18 microcontroller.

12.3.1 Registers Associated with A/D Converter

The A/D converter has five registers that control the overall operation of the module:

- A/D control register 0 (ADCON0)
- A/D control register 1 (ADCON1)
- A/D control register 2 (ADCON2)
- A/D result high register (ADRESH)
- A/D result low register (ADRESL)

ADCON0 REGISTER

The contents of the ADCON0 register are shown in Figures 12.8a, 12.8b, and 12.8c. Earlier PIC18 members (e.g., PIC18FXX2 and PIC18FXX8) have only two control registers, whereas later PIC18 members (e.g., PIC18FXX20, PIC18F8X2X, PIC18F6X2X, PIC18F6X8X, and PIC18F8X8X) have three control registers.

The contents of ADCON0 for the PIC1320/1220 are shown in Figure 12.8c.

	7	6	5	4	3	2	1	0
	ADCS1	ADCS0	CHS2	CHS1	CHS0	GO/$\overline{\text{DONE}}$	--	ADON
value after reset	0	0	0	0	0	0	0	0

ADCS1:ADCS0: A/D conversion clock select bits
   (used along with ADCS2 of the ADCON1 register (shown in Table 12.1)
CHS2:CHS0: Analog channel select bits
   000 = channel 0, (AN0)
   001 = channel 1, (AN1)
   010 = channel 2, (AN2)
   011 = channel 3, (AN3)
   100 = channel 4, (AN4)
   101 = channel 5, (AN5)
   110 = channel 6, (AN6)
   111 = channel 7, (AN7)
GO/DONE: A/D conversion status bit
  when ADON = 1
   0 = A/D conversion not in progress
   1 = A/D conversion in progress (setting this bit starts the A/D conversion.
     This bit will be cleared by hardware when A/D conversion is done)
ADON: A/D on bit
   0 = A/D converter module is shut-off
   1 = A/D converter module is powered up

Figure 12.8a ■ ADCON0 register (PIC18FXX2 and PIC18FXX8) (redraw with permission of Microchip)

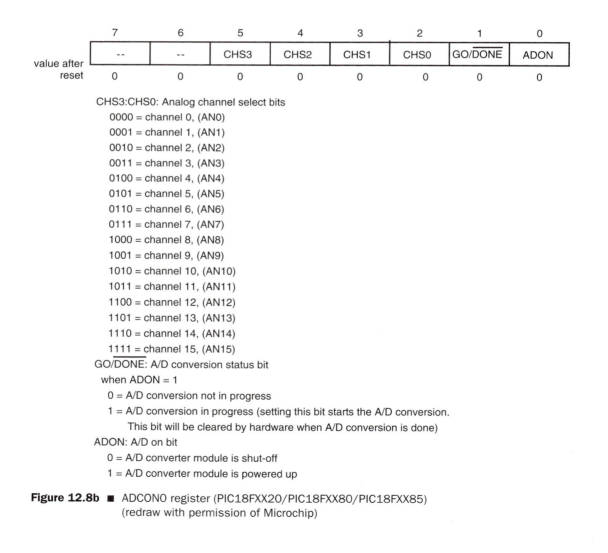

ADCS2: ADCS0	Clock conversion
000	FOSC/2
001	FOSC/8
010	FOSC/32
011	FRC (clock derived from RC oscillator)
100	FOSC/4
101	FOSC/16
110	FOSC/64
111	FRC (clock derived from RC oscillator)

Table 12.1 ■ A/D conversion clock source select bits

7	6	5	4	3	2	1	0
--	--	CHS3	CHS2	CHS1	CHS0	GO/$\overline{\text{DONE}}$	ADON

value after reset: 0 0 0 0 0 0 0 0

CHS3:CHS0: Analog channel select bits

0000 = channel 0, (AN0)
0001 = channel 1, (AN1)
0010 = channel 2, (AN2)
0011 = channel 3, (AN3)
0100 = channel 4, (AN4)
0101 = channel 5, (AN5)
0110 = channel 6, (AN6)
0111 = channel 7, (AN7)
1000 = channel 8, (AN8)
1001 = channel 9, (AN9)
1010 = channel 10, (AN10)
1011 = channel 11, (AN11)
1100 = channel 12, (AN12)
1101 = channel 13, (AN13)
1110 = channel 14, (AN14)
1111 = channel 15, (AN15)

GO/$\overline{\text{DONE}}$: A/D conversion status bit

when ADON = 1

0 = A/D conversion not in progress
1 = A/D conversion in progress (setting this bit starts the A/D conversion.
This bit will be cleared by hardware when A/D conversion is done)

ADON: A/D on bit

0 = A/D converter module is shut-off
1 = A/D converter module is powered up

Figure 12.8b ■ ADCON0 register (PIC18FXX20/PIC18FXX80/PIC18FXX85)
(redraw with permission of Microchip)

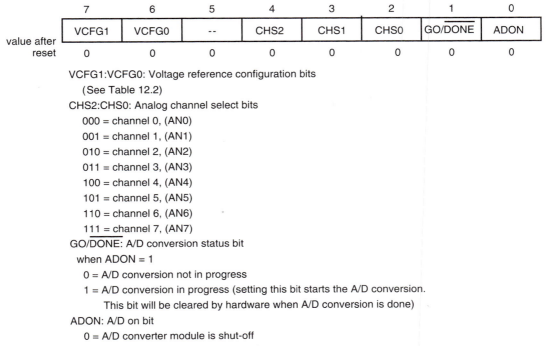

7	6	5	4	3	2	1	0
VCFG1	VCFG0	--	CHS2	CHS1	CHS0	GO/$\overline{\text{DONE}}$	ADON

value after reset: 0   0   0   0   0   0   0   0

VCFG1:VCFG0: Voltage reference configuration bits
(See Table 12.2)
CHS2:CHS0: Analog channel select bits
   000 = channel 0, (AN0)
   001 = channel 1, (AN1)
   010 = channel 2, (AN2)
   011 = channel 3, (AN3)
   100 = channel 4, (AN4)
   101 = channel 5, (AN5)
   110 = channel 6, (AN6)
   111 = channel 7, (AN7)
GO/$\overline{\text{DONE}}$: A/D conversion status bit
  when ADON = 1
   0 = A/D conversion not in progress
   1 = A/D conversion in progress (setting this bit starts the A/D conversion.
     This bit will be cleared by hardware when A/D conversion is done)
ADON: A/D on bit
   0 = A/D converter module is shut-off
   1 = A/D converter module is powered up

Figure 12.8c ■ ADCON0 register (PIC18F1220/1320) (redraw with permission of Microchip)

VCFG1:VCFG0	A/D V$_{\text{REF}}+$	A/D V$_{\text{REF}}-$
00	AV$_{\text{DD}}$	AV$_{\text{SS}}$
01	External V$_{\text{REF}}+$	AV$_{\text{SS}}$
10	AV$_{\text{DD}}$	External V$_{\text{REF}}-$
11	External V$_{\text{REF}}+$	External V$_{\text{REF}}-$

Table 12.2 ■ Voltage reference configuration bits

The user selects the analog input channel to be converted and the clock source to control the conversion process by programming the ADCON0 register. The ADON bit must be set to 1 in order to enable the A/D module, whereas the GO/$\overline{\text{DONE}}$ bit actually starts the conversion. The GO/$\overline{\text{DONE}}$ bit will be cleared automatically once the A/D conversion is done. The user has the option to use the power supply or the external supplied voltages as the high reference voltage and the low reference voltage.

ADCON1 Register

This register is used mainly to configure I/O pins as analog or digit port pins. The contents of this register are shown in Figures 12.9a, 12.9b, and 12.9c.

When a pin is connected to a signal to be converted, the pin must be configured as an analog input. Otherwise, the pin will be read as 0 and cannot serve the purpose of analog signal.

The ADFM bit in Figure 12.9a allows the user to select the result format to be left or right justified to better match the requirement of a specific application.

7	6	5	4	3	2	1	0
ADFM	ADCS2	--	--	PCFG3	PCFG2	PCFG1	PCFG0

value after reset: 0   0   0   0   0   0   0   0

ADFM: A/D result format select bit

    0 = left justified. Six least significant bits of ADRESL are 0s.

    1 = right justified. Most significant bits of ADRESH are 0s.

ADCS2: A/D conversion clock select.

    This bit along with the ADCS1:ADCS0 bits of ADCON0 are used to select

    clock source for A/D conversion.

PCFG3:PCFG0: A/D port configuration control bits.

    (see Table 12.3)

Figure 12.9a ■ ADCON1 register (PIC18FXX2 and PIC18FXX8) (redraw with permission of Microchip)

PCFG<3:0>	AN7	AN6	AN5	AN4	AN3	AN2	AN1	AN0	$V_{REF}+$	$V_{REF}-$	C/R
0000	A	A	A	A	A	A	A	A	VDD	VSS	8/0
0001	A	A	A	A	VREF+	A	A	A	AN3	VSS	7/1
0010	D	D	D	A	A	A	A	A	VDD	VSS	5/0
0011	D	D	D	A	VREF+	A	A	A	AN3	VSS	4/1
0100	D	D	D	D	A	D	A	A	VDD	VSS	3/0
0101	D	D	D	D	VREF+	D	A	A	AN3	VSS	2/1
011x	D	D	D	D	D	D	D	D	−	−	0/0
1000	A	A	A	A	VREF+	VREF−	A	A	AN3	AN2	6/2
1001	D	D	A	A	A	A	A	A	VDD	VSS	6/0
1010	D	D	A	A	VREF+	A	A	A	AN3	VSS	5/1
1011	D	D	A	A	VREF+	VREF−	A	A	AN3	AN2	4/2
1100	D	D	D	A	VREF+	VREF−	A	A	AN3	AN2	3/2
1101	D	D	D	D	VREF+	VREF−	A	A	AN3	AN2	2/2
1110	D	D	D	D	D	D	D	A	VDD	VSS	1/0
1111	D	D	D	D	VREF+	VREF−	D	A	AN3	AN2	1/2

A = analog input D = digital input
C/R = # of analog input channels/# of A/D voltage references

Table 12.3 ■ A/D port configuration control bits

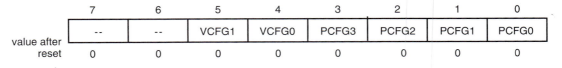

	7	6	5	4	3	2	1	0
	--	--	VCFG1	VCFG0	PCFG3	PCFG2	PCFG1	PCFG0
value after reset	0	0	0	0	0	0	0	0

VCFG1:VCFG0: Voltage reference configuration bits
  (see Table 12.2)
PCFG3:PCFG0: A/D port configuration control bits

	AN15	AN14	AN13	AN12	AN11	AN10	AN9	AN8	AN7	AN6	AN5	AN4	AN3	AN2	AN1	AN0
0000	A	A	A	A	A	A	A	A	A	A	A	A	A	A	A	A
0001	D	D	A	A	A	A	A	A	A	A	A	A	A	A	A	A
0010	D	D	D	A	A	A	A	A	A	A	A	A	A	A	A	A
0011	D	D	D	D	A	A	A	A	A	A	A	A	A	A	A	A
0100	D	D	D	D	D	A	A	A	A	A	A	A	A	A	A	A
0101	D	D	D	D	D	D	A	A	A	A	A	A	A	A	A	A
0110	D	D	D	D	D	D	D	A	A	A	A	A	A	A	A	A
0111	D	D	D	D	D	D	D	D	A	A	A	A	A	A	A	A
1000	D	D	D	D	D	D	D	D	D	A	A	A	A	A	A	A
1001	D	D	D	D	D	D	D	D	D	D	A	A	A	A	A	A
1010	D	D	D	D	D	D	D	D	D	D	D	A	A	A	A	A
1011	D	D	D	D	D	D	D	D	D	D	D	D	A	A	A	A
1100	D	D	D	D	D	D	D	D	D	D	D	D	D	A	A	A
1101	D	D	D	D	D	D	D	D	D	D	D	D	D	D	A	A
1110	D	D	D	D	D	D	D	D	D	D	D	D	D	D	D	A
1111	D	D	D	D	D	D	D	D	D	D	D	D	D	D	D	D

1. AN15:AN12 are available only in PIC18F8X8X devices
2. AN12 is also available in PIC18F2X20/PIC18F4X20 devices
3. AN5 through AN7 are not available in PIC18F2X20 devices

Figure 12.9b ■ ADCON1 register (PIC18FXX20/PIC18FXX80/PIC18FXX85) (excluding PIC18F1320/1220) (redraw with permission of Microchip)

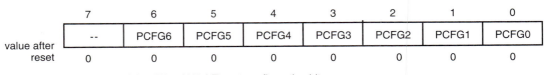

	7	6	5	4	3	2	1	0
	--	PCFG6	PCFG5	PCFG4	PCFG3	PCFG2	PCFG1	PCFG0
value after reset	0	0	0	0	0	0	0	0

PCFG6..PFCG0: AN6..AN0 A/D port configuration bit
  0 = Pin configured as an analog channel -- digital input disabled and read as 0
  1 = Pin configured as a digital input

Figure 12.9c ■ ADCON1 register (PIC18F1220/1320) (redraw with permission of Microchip)

ADCON2 REGISTER

This register is not available in earlier devices, such as PIC18FXX2 or PIC18FXX8. The main function of this register is to allow the user to program the A/D conversion clock source and A/D acquisition time. The newer PIC18 devices, such as PIC18F8X8X and PIC18F2X20/ 4X20, allow the user to program the data acquisition time to reduce the software overhead. For other devices that do not allow the user to program the data acquisition time, the user will need to use software delay to provide the required data acquisition time. To select an appropriate value for the data acquisition time, one needs to take clock source into consideration at the same time. The contents of this register are shown in Figures 12.10a and 12.10b.

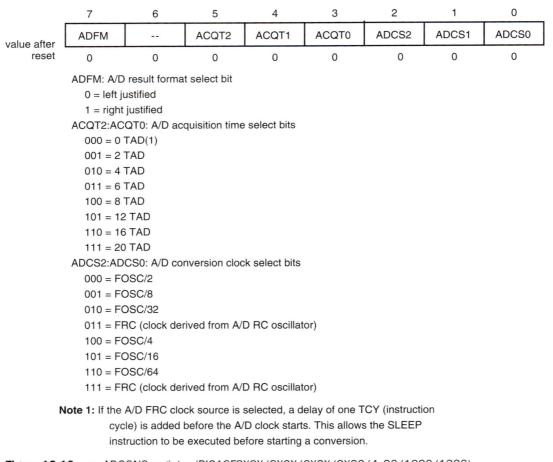

ADFM: A/D result format select bit
  0 = left justified
  1 = right justified
ACQT2:ACQT0: A/D acquisition time select bits
  000 = 0 TAD(1)
  001 = 2 TAD
  010 = 4 TAD
  011 = 6 TAD
  100 = 8 TAD
  101 = 12 TAD
  110 = 16 TAD
  111 = 20 TAD
ADCS2:ADCS0: A/D conversion clock select bits
  000 = FOSC/2
  001 = FOSC/8
  010 = FOSC/32
  011 = FRC (clock derived from A/D RC oscillator)
  100 = FOSC/4
  101 = FOSC/16
  110 = FOSC/64
  111 = FRC (clock derived from A/D RC oscillator)

Note 1: If the A/D FRC clock source is selected, a delay of one TCY (instruction
       cycle) is added before the A/D clock starts. This allows the SLEEP
       instruction to be executed before starting a conversion.

Figure 12.10a ■ ADCON2 register (PIC18F8X8X/8X2X/6X2X/2X20/4x20/1220/1320)
    (redraw with permission of Microchip)

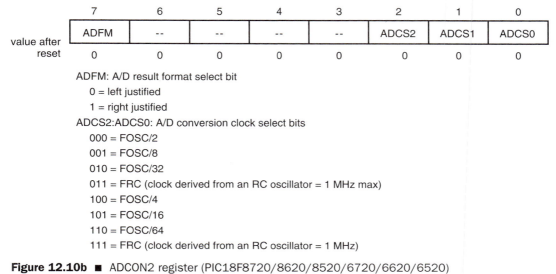

	7	6	5	4	3	2	1	0
	ADFM	--	--	--	--	ADCS2	ADCS1	ADCS0
value after reset	0	0	0	0	0	0	0	0

ADFM: A/D result format select bit
   0 = left justified
   1 = right justified
ADCS2:ADCS0: A/D conversion clock select bits
   000 = FOSC/2
   001 = FOSC/8
   010 = FOSC/32
   011 = FRC (clock derived from an RC oscillator = 1 MHz max)
   100 = FOSC/4
   101 = FOSC/16
   110 = FOSC/64
   111 = FRC (clock derived from an RC oscillator = 1 MHz)

Figure 12.10b ■ ADCON2 register (PIC18F8720/8620/8520/6720/6620/6520) (redraw with permission of Microchip)

12.3.2 A/D Acquisition Requirements

In order to ensure that the analog input voltage is stable during the A/D conversion process, the PIC18 microcontroller has a sample-and-hold circuit at the input. The analog input model is shown in Figure 12.11. The capacitor C_{HOLD} holds the voltage to be converted into digital code. For the A/D converter to meet its specified accuracy, the charge-holding capacitor (C_{HOLD}) must be allowed to fully charge to the input channel voltage level. The source impedance (R_S) and the internal sampling switch impedance (R_{SS}) directly affect the time required to charge the capacitor C_{HOLD}. The sampling switch impedance varies over the device voltage (V_{DD}). The source impedance affects the offset voltage at the analog input (because of pin leakage current). The maximum recommended impedance for analog sources is 2.5 KΩ. The voltage acquisition must be done before the conversion is started.

The minimum *acquisition time* (T_{ACQ}) is computed by the following equation:

T_{ACQ} = amplifier settling time + holding capacitor charging time + temperature coefficient
    = $T_{AMP} + T_C + T_{COFF}$

The *amplifier settling time* (T_{AMP}) is about 2 μs. The *temperature coefficient* is given by the following equation:

T_{COFF} = (temp – 25° C) × (0.05 μs/°C).

Temperature coefficient is equal to 0 at temperatures below 25°C.
The *holding capacitor charging time* is given by the following equation:

T_C = –120 pF × (1 K + R_{SS} + R_S) log(1/2047)
    = 120 pF × (1 KΩ + R_{SS} + R_S) log(2047)

Since R_S = 2.5 K and R_{SS} is 7 K in Figure 12.11, T_C is computed to be 9.61 μs. The acquisition time at 50°C is calculated to be 12.86 μs.

For earlier PIC18 members, when the GO/$\overline{DONE}$ bit is set, sampling is stopped, and a conversion begins. The user is responsible for ensuring that the required acquisition time has passed between selecting the desired input channel and setting the GO/$\overline{DONE}$ bit.

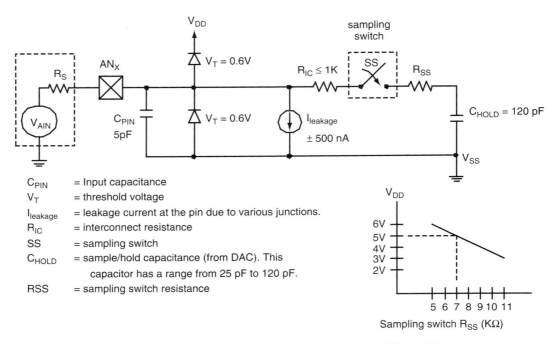

C_PIN = Input capacitance
V_T = threshold voltage
$I_{leakage}$ = leakage current at the pin due to various junctions.
R_{IC} = interconnect resistance
SS = sampling switch
C_HOLD = sample/hold capacitance (from DAC). This capacitor has a range from 25 pF to 120 pF.
RSS = sampling switch resistance

Figure 12.11 ■ Analog input circuit model (redraw with permission of Microchip)

For those devices that have automatic acquisition time, if ACQT2:ACQT0 bits (ADCON2<5:3>) are programmed to be a nonzero value, the A/D module will continue to sample after the GO/$\overline{\text{DONE}}$ bit is set for the selected acquisition time and then automatically begins a conversion. Once the conversion is started, the holding capacitor (C_HOLD in Figure 12.11) will be disconnected from the analog input. The holding capacitor will be reconnected to the analog input at the end of conversion. This feature will be very useful if the user wants to collect samples continuously. Since the acquisition time is programmed, there is no need to wait for an acquisition time between selecting a channel and setting the GO/$\overline{\text{DONE}}$ bit.

12.3.3 Selecting the A/D Conversion Clock

The A/D conversion time per bit is defined as T_{AD}. Each A/D conversion takes 12 T_{AD} for 10-bit resolution. There are seven possible options for T_{AD}:

- 2 T_{OSC}
- 4 T_{OSC}
- 8 T_{OSC}
- 16 T_{OSC}
- 32 T_{OSC}
- 64 T_{OSC}
- Internal RC oscillator period

For correct A/D conversion, the A/D conversion clock (T_{AD}) must be selected to be no smaller than 1.6 μs.

Table 12.4 shows the resultant T_{AD} times derived from the device operating frequencies and the A/D clock source selected. The RC clock source is used mainly in sleep mode. When the device is in sleep mode, all clock signals are stopped. Therefore, RC clock source is the only available clock source for the A/D converter.

AD clock source (T_{AD})		maximum device frequency	
Operation	ADCS2:ADCS0	Normal MCU	Low power MCU
2 T_{OSC}	000	1.25 MHz	666 KHz
4 T_{OSC}	100	2.50 MHz	1.33 MHz
8 T_{OSC}	001	5.00 MHz	2.66 MHz
16 T_{OSC}	101	10.0 MHz	5.33 MHz
32 T_{OSC}	010	20.0 MHz	10.65 MHz
64 T_{OSC}	110	40.0 MHz	21.33 MHz
RC	x11	1.00 MHz	1.00 MHz

Note: Lower power device has a letter L in its name. For example, PIC18LF8720

Table 12.4 ■ T_{AD} vs. device operating frequencies

12.3.4 A/D Conversion Process

Figure 12.12 shows the operation of the A/D converter after the GO bit has been set and the ACQT2:ACQT0 bits are cleared. A conversion is started after the following instruction to allow entry into sleep mode before the conversion begins.

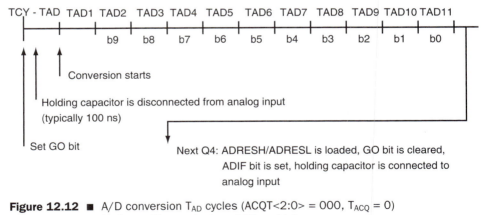

Figure 12.12 ■ A/D conversion T_{AD} cycles (ACQT<2:0> = 000, T_{ACQ} = 0)
(redraw with permission of Microchip)

Figure 12.13 shows the operation of the A/D converter after the GO bit has been set, the ACQT2:ACQT0 bits have been set to 010, and a 4-T_{AD} acquisition time has been set before the conversion starts.

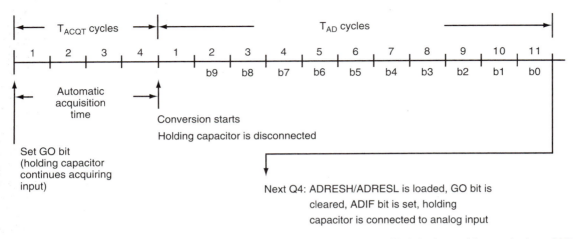

Figure 12.13 ■ A/D conversion T_{AD} cycles (ACQT<2:0> = 010, T_{ACQ} = 4T_{AD}) (redraw with permission of Microchip)

Clearing the GO/$\overline{\text{DONE}}$ bit during a conversion will abort the current conversion. The A/D result register pair will not be updated with the partially completed A/D conversion sample. This means that the ADRESH:ADRESL registers will continue to contain the value of the last completed conversion.

After the A/D conversion is completed or aboted, a minimum wait time of 2 T_{AD} is required before the next acquisition can be started. After this wait, acquisition on the selected channel is automatically started. The GO/$\overline{\text{DONE}}$ bit can then be set to start the conversion.

12.3.5 Use of the CCP2 Register

An A/D conversion can be started by the special event trigger of the CCP2 module. This requires that the CCP2M3:CCP2M0 bits (CCP2CON<3:0>) be programmed to "1011" and that the A/D module is enabled. When the trigger occurs, the GO/$\overline{\text{DONE}}$ bit will be set, starting the A/D conversion, and the Timer1 (or Timer3) counter will be reset to zero. Timer1 (or Timer3) is reset to automatically repeat the A/D acquisition period with minimal software overhead. The appropriate analog input channel must be selected and the minimum acquisition done before the special event trigger sets the GO/$\overline{\text{DONE}}$ bit.

If the A/D module is not enabled (ADON bit is cleared), the special event trigger will be ignored by the A/D module but will still reset the Timer1 (or Timer3) counter.

Example 12.4

Write an instruction sequence to configure the A/D converter of the PIC18F8680 to operate with the following parameters:

- Conversion result right justified
- f_{OSC} = 32 MHz
- Highest ambient temperature may reach 60°C
- Use V_{DD} and V_{SS} as the high and low reference voltages
- Convert channel AN0
- Enable A/D module

Solution: The temperature coefficient for the input circuit is $(60 - 25) \times 0.05 = 1.75$ μs. The required acquisition time is $(9.61 + 2 + 1.75)$ μs = 13.36 μs.

By setting the clock source to 64 f_{OSC}, T_{AD} is calculated to be 2 μs. The data acquisition time must be at least 8 T_{AD} to make it greater than 13.36 μs. Since channel AN0 is the channel to be converted, it must be configured for analog input.

The following instruction sequence will configure the A/D module of the PIC18F8680 as specified:

```
movlw 0x01        ; select channel AN0 and enable A/D
movwf ADCON0,A    ; "
movlw 0x0E        ; configure only channel AN0 as analog port,
movwf ADCON1,A    ; select VDD and VSS as reference voltage
movlw 0xA6        ; set A/D result right justified, set acquisition
movwf ADCON2,A    ; time to 8 TAD, clock source FOSC/64
```

In C language, the following statements will achieve the same configuration:

```
ADCON0 = 0x01;
ADCON1 = 0x0E;
ADCON2 = 0xA6;
```

▲

Example 12.5

▼

Write an instruction sequence to configure the A/D converter of the PIC18F452 to operate with the following parameters:

- Conversion result right justified
- f_{OSC} = 32 MHz
- Highest ambient temperature may reach 60° C
- Use V_{DD} and V_{SS} as the high and low reference voltages
- Convert channel AN0
- Enable A/D module

Solution: There are only two control registers to be programmed for the PIC18F452:

```
movlw 0x81        ; select FOSC/64 as conversion clock source,
movwf ADCON0,A    ; select channel AN0, enable A/D module
movlw 0xCE        ; A/D conversion result right justified, configure
movwf ADCON1,A    ; only channel AN0 as analog, VDD & VSS as
                  ; reference voltages for conversion
```

▲

12.4 Procedure for Performing A/D Conversion

Before the PIC18 can perform A/D conversion, the A/D module must be connected and configured properly. The procedure for performing an A/D conversion is as follows:

1. Configure the A/D module:
   - Configure the analog pins, reference voltages, and digital/analog I/O pins
   - Select the A/D input channel
   - Select the A/D acquisition time (if available)

- Select the A/D conversion clock source
- Enable the A/D module

2. Configure the A/D interrupt (if desired):
   - Clear the ADIF bit
   - Set the ADIE bit
   - Set the GIE bit

3. Wait for the desired acquisition time (if required).

4. Start the conversion by setting the GO/$\overline{\text{DONE}}$ bit.

5. Wait for the A/D conversion to complete by either waiting for the GO/$\overline{\text{DONE}}$ (DONE) bit to be cleared or waiting for the A/D interrupt.

6. Read the A/D result registers; clear the ADIF flag.

7. For the next conversion, go to Step 1 or Step 2 as required. The A/D conversion time per bit is defined as T_{AD}. A minimum waiting time of 2 T_{AD} is required before next acquisition starts.

Example 12.6

▼

Assume that the AN0 pin of a PIC18F8680 running with a 32-MHz crystal oscillator is connected to a potentiometer. The voltage range of the potentiometer is from 0 V to 5 V. Write a program to measure the voltage applied to the AN0 pin, convert it, retrieve the conversion result, and place it in PRODH:PRODL.

Solution: The program that follows the procedure described earlier is as follows:

```
          #include      <p18F8680.inc>
          org           0x00
          goto          start
          org           0x08
          retfie
          org           0x18
          retfie
start     movlw         0x01              ; select channel AN0 and enable A/D
          movwf         ADCON0,A          ; "
          movlw         0x0E              ; use VDD & VSS as reference voltages &
          movwf         ADCON1,A          ; configure channel AN0 as analog input
          movlw         0xA6              ; select FOSC/64 as conversion clock,
          movwf         ADCON2,A          ; 8 TAD for acquisition time, right-justified
          bsf           ADCON0,GO,A       ; start A/D conversion
wait_con  btfsc         ADCON0,DONE,A     ; wait until conversion is done
          bra           wait_con
          movff         ADRESH,PRODH      ; save conversion result
          movff         ADRESL,PRODL      ; "
          end
```

12.5 Microchip A/D Converter C Library Functions

The A/D converter is supported with six library functions. One needs to include the **adc.h** header file in order to use these library functions. The prototype declarations of these functions are as follows:

char **BusyADC** (void);

This function checks whether the A/D module is currently performing a conversion. The return value is 1 if the A/D module is busy performing a conversion. Otherwise, the return value is 0.

void **CloseADC** (void);

This function disables the A/D converter and A/D interrupt mechanism.

void **ConvertADC** (void);

This function starts an A/D conversion. The **BusyADC( )** function may be used to detect completion of the conversion.

void **OpenADC** (unsigned char **config**, unsigned char **config2**);

For the PIC18F1X20/2X20/4X20, the **OpenADC** function has a third argument **portconfig**. Because different PIC18 members may have different number of control registers, the setting of the parameters **config** and **config2** may be different:

PIC18FXX2, PIC18FXX2, and PIC18FXX8

The first argument, **config,** may have the following values:

A/D clock source

ADC_FOSC_2	$F_{OSC}/2$
ADC_FOSC_4	$F_{OSC}/4$
DC_FOSC_8	$F_{OSC}/8$
ADC_FOSC_16	$F_{OSC}/16$
ADC_FOSC_32	$F_{OSC}/32$
ADC_FOSC_64	$F_{OSC}/64$
ADC_FOSC_RC	Internal RC oscillator

A/D result justification

ADC_RIGHT_JUST	Result in least significant bits
ADC_LEFT_JUST	Result in most significant bits

A/D voltage reference source

ADC_8ANA_0REF	$V_{REF+} = V_{DD}, V_{REF-} = V_{SS}$, all analog channels
ADC_7ANA_1REF	$AN3 = V_{REF+}$, all analog channels except AN3
ADC_6ANA_2REF	$AN3 = V_{REF+}, AN2 = V_{REF-}$
ADC_6ANA_0REF	$V_{REF+} = V_{DD}, V_{REF-} = V_{SS}$
ADC_5ANA_1REF	$AN3 = V_{REF+}, V_{REF-} = V_{SS}$
ADC_5ANA_0REF	$V_{REF+} = V_{DD}, V_{REF-} = V_{SS}$
ADC_4ANA_2REF	$AN3 = V_{REF+}, AN2 = V_{REF-}$
ADC_4ANA_1REF	$AN3 = V_{REF+}$

ADC_3ANA_2REF	AN3 = V_{REF+}, AN2 = V_{REF-}
ADC_3ANA_0REF	V_{REF+} = V_{DD}, V_{REF-} = V_{SS}
ADC_2ANA_2REF	AN3 = V_{REF+}, AN2 = V_{REF-}
ADC_2ANA_1REF	AN3 = V_{REF+}
ADC_1ANA_2REF	AN3 = V_{REF+}, AN2 = V_{REF-}, AN0 = A
ADC_1ANA_0REF	AN0 is analog input
ADC_0ANA_0REF	All digital I/O

The second argument, **config2,** may have the following values:

Channel

ADC_CH0	Channel 0
ADC_CH1	Channel 1
ADC_CH2	Channel 2
ADC_CH3	Channel 3
ADC_CH4	Channel 4
ADC_CH5	Channel 5
ADC_CH6	Channel 6
ADC_CH7	Channel 7

A/D Interrupts

ADC_INT_ON	Interrupts enabled
ADC_INT_OFF	Interrupts disabled

PIC18C658/858, PIC18C601/801, PIC18F6X20, PIC18F8X20

The first argument, **config,** may have the following values:

A/D clock source

ADC_FOSC_2	$F_{OSC}/2$
ADC_FOSC_4	$F_{OSC}/4$
ADC_FOSC_8	$F_{OSC}/8$
ADC_FOSC_16	$F_{OSC}/16$
ADC_FOSC_32	$F_{OSC}/32$
ADC_FOSC_64	$F_{OSC}/64$
ADC_FOSC_RC	Internal RC oscillator

A/D result justification

ADC_RIGHT_JUST	Result in least significant bits
ADC_LEFT_JUST	Result in most significant bits

A/D port configuration

ADC_0ANA	All digital	
ADC_1ANA	Analog AN0	Digital AN1-AN15
ADC_2ANA	Analog AN0-AN1	Digital AN2-AN15
ADC_3ANA	Analog AN0-AN2	Digital AN3-AN15
ADC_4ANA	Analog AN0-AN3	Digital AN4-AN15

ADC_5ANA	Analog AN0-AN4	Digital AN5-AN15
ADC_6ANA	Analog AN0-AN5	Digital AN6-AN15
ADC_7ANA	Analog AN0-AN6	Digital AN7-AN15
ADC_8ANA	Analog AN0-AN7	Digital AN8-AN15
ADC_9ANA	Analog AN0-AN8	Digital AN9-AN15
ADC_10ANA	Analog AN0-AN9	Digital AN10-AN15
ADC_11ANA	Analog AN0-AN10	Digital AN11-AN15
ADC_12ANA	Analog AN0-AN11	Digital AN12-AN15
ADC_13ANA	Analog AN0-AN12	Digital AN13-AN15
ADC_14ANA	Analog AN0-AN13	Digital AN14-AN15
ADC_15ANA	All analog	

The second argument, **config2,** has the following values:

Channel

ADC_CH0	Channel 0
ADC_CH1	Channel 1
ADC_CH2	Channel 2
ADC_CH3	Channel 3
ADC_CH4	Channel 4
ADC_CH5	Channel 5
ADC_CH6	Channel 6
ADC_CH7	Channel 7
ADC_CH8	Channel 8
ADC_CH9	Channel 9
ADC_CH10	Channel 10
ADC_CH11	Channel 11
ADC_CH12	Channel 12
ADC_CH13	Channel 13
ADC_CH14	Channel 14
ADC_CH15	Channel 15

A/D Interrupts

| ADC_INT_ON | Interrupts enabled |
| ADC_INT_OFF | Interrupts disabled |

A/D voltage configuration

ADC_VREFPLUS_VDD	$V_{REF+} = AV_{DD}$
ADC_VREFPLUS_EXT	V_{REF+} = external
ADC_VREFMINUS_VSS	$V_{REF-} = AV_{SS}$
ADC_VREFMINUS_EXT	V_{REF-} = external

PIC18F6X8X, PIC18F8X8X, PIC18F1X20, PIC18F2X20, PIC18F4X20

The first argument, **config**, may have the following values:

A/D clock source

ADC_FOSC_2	$F_{OSC}/2$
ADC_FOSC_4	$F_{OSC}/4$
ADC_FOSC_8	$F_{OSC}/8$
ADC_FOSC_16	$F_{OSC}/16$
ADC_FOSC_32	$F_{OSC}/32$
ADC_FOSC_64	$F_{OSC}/64$
ADC_FOSC_RC	Internal RC oscillator

A/D result justification

ADC_RIGHT_JUST	Result in least significant bits
ADC_LEFT_JUST	Result in most significant bits

A/D acquisition time select

ADC_0_TAD	$0\ T_{AD}$
ADC_2_TAD	$2\ T_{AD}$
ADC_4_TAD	$4\ T_{AD}$
ADC_6_TAD	$6\ T_{AD}$
ADC_8_TAD	$8\ T_{AD}$
ADC_12_TAD	$12\ T_{AD}$
ADC_16_TAD	$16\ T_{AD}$
ADC_20_TAD	$20\ T_{AD}$

The second argument, **config2**, has the following values:

Channel

ADC_CH0	Channel 0
ADC_CH1	Channel 1
ADC_CH2	Channel 2
ADC_CH3	Channel 3
ADC_CH4	Channel 4
ADC_CH5	Channel 5
ADC_CH6	Channel 6
ADC_CH7	Channel 7
ADC_CH8	Channel 8
ADC_CH9	Channel 9
ADC_CH10	Channel 10
ADC_CH11	Channel 11
ADC_CH12	Channel 12
ADC_CH13	Channel 13
ADC_CH14	Channel 14
ADC_CH15	Channel 15

A/D Interrupts

ADC_INT_ON   Interrupts enabled

ADC_INT_OFF   Interrupts disabled

A/D voltage configuration

ADC_VREFPLUS_VDD $V_{REF+} = AV_{DD}$

ADC_VREFPLUS_EXT V_{REF+} = external

ADC_VREFMINUS_VSS $V_{REF-} = AV_{SS}$

ADC_VREFMINUS_EXT V_{REF-} = external

The third argument, **Portconfig,** is any value from 0 to 127 for the PIC18F1220/1320 and 0 to 15 for the PIC18F2220/2320/4220/4320/6X8X/8X8X, inclusive. This is the value of bits 0 through 6 or bits 0 through 3 of the ADCON1 register, which are the port configuration bits.

int **ReadADC** (void);

This function reads the result of an A/D conversion and returns it as a 16-bit signed value. Depending on the configuration, the conversion result may be right or left justified.

void **SetChanADC** (unsigned char **channel**);

This function selects the pin used as input to the A/D converter. The channel value can be ADC_CH0 through ADC_CH15.

Example 12.7

▼

Write a C program to configure the A/D module of the PIC18F452 with the following characteristics and take one sample, convert it, and store the result in a memory location:

- Clock source set to $F_{OSC}/64$
- Result right justified
- Set port A AN0 pin for analog input, others for digital
- Use V_{DD} and V_{SS} as high and low reference voltages
- Select AN0 to convert
- Disable interrupt

Solution: The C program that performs the configuration, takes one sample, and performs the conversion is as follows:

```
#include <p18F452.h>
#include <adc.h>
#include <stdlib.h>
#include <delays.h>
int result;
void main (void)
{
    OpenADC(ADC_FOSC_64 & ADC_RIGHT_JUST & ADC_1ANA_0REF, ADC_CH0 & ADC_INT_OFF);
    Delay10TCYx(20);        // provides 200 instruction cycles of acquisition time
    ConvertADC( );          // start A/D conversion
    while(BusyADC( ));      // wait for completion
    result = ReadADC( );    // read result
    CloseADC( );
}
```

The statement **Delay10TCYx(20)** is not needed for those devices with programmable acquisition time.

▲

12.6 Using the Temperature Sensor TC1047A

The TC1047A from Microchip is a three-pin temperature sensor whose voltage output is directly proportional to the measured temperature. The TC1047A can accurately measure temperatures from –40°C to 125°C with a power supply from 2.7 V to 5.5 V.

The output voltage range for these devices is typically 100 mV at –40°C, 500 mV at 0°C, 750 mV at +25°C, and +1.75 V at +125°C. As shown in Figure 12.14, the TC1047A has a 10 mV/°C voltage slope output response.

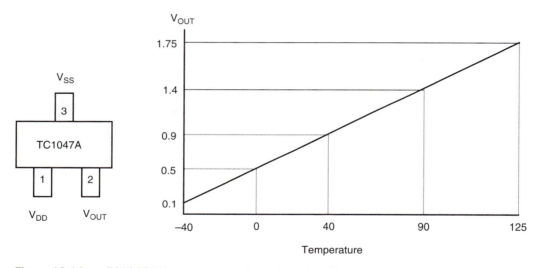

Figure 12.14 ■ TC1047A V_{OUT} vs. temperature characteristic

Example 12.8

Describe a circuit connection and the required program to build a digital thermometer. Display the temperature in three integral and one fractional digits using the LCD. Measure and display the temperature over the whole range of TC1047A, that is, –40°C to +125°C. Update the display data 10 times per second and assume that the PIC18F8680 operates with a 32-MHz crystal oscillator.

Solution: Since the voltage output of the TC1047A is from 0.1 V to 1.75 V for the temperature range of –40°C through 125°C, we will need to use the circuit shown in Figure 12.7c to perform the translation and scaling for V_{OUT}. The circuit connection is shown in Figure 12.15. This circuit will use V_{DD} and V_{SS} as the reference voltages required in the process of A/D conversion. The range of the voltage connected to the AN0 pin will be between 0 V and 5V for the given temperature range.

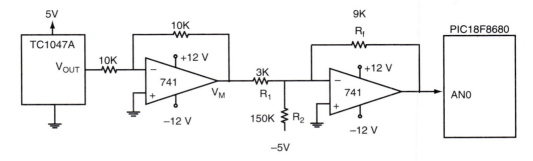

Figure 12.15 ■ Circuit connection between the TC1047A and the PIC18 MCU

To convert the A/D conversion result to the corresponding temperature, we need to do the following:

1. Divide the conversion result by 6.2
2. Subtract the quotient by 40

Since the PIC18 microcontroller cannot handle floating-point numbers directly, the operation of "dividing by 6.2" can be implemented by multiplying the conversion result by 10 and then dividing the product by 62.

The procedure for starting an A/D conversion, reading the conversion result, calculating the corresponding temperature, and converting the temperature value into an ASCII string every 200 ms is as follows:

Step 1
Configure the A/D converter and Timer0 properly. Timer0 is configured to overflow every 200 ms.

Step 2
Start an A/D conversion.

Step 3
Wait until the A/D conversion is complete.

Step 4
Multiply the conversion result by 10.

Step 5
Divide the product resulted in Step 4 by 62 to obtain the temperature reading. Use the variables **quo** and **rem** to hold the quotient and remainder.

Step 6
Subtract 40 from the variable **quo** to obtain the actual temperature.

Step 7
If **quo** > 0, go to Step 9; otherwise, replace quo with its two's complement.

Step 8
If **rem** ≠ 0, then

1. decrement **quo** by 1
2. rem ← 62 – rem

Step 9

Compute the fractional digit by multiplying **rem** by 10 and then dividing the resulting product by 62.

Step 10

Compute the integral digits by performing repeated division by 10 to **quo.**

Step 11

Wait until Timer0 overflows and then go to Step 2.

The assembly program that implements this algorithm is as follows:

```
#include <p18F8680.inc>
; ---
; include macro definitions for pushr, push_dat, popr, alloc_stk, dealloc_stk here.
; ---
; ****************************************************************
; The following definitions are used by the 16-bit multiplication routine
; ****************************************************************
loc_varm    equ     4           ; number of bytes used for local variable
pd0         equ     7           ; offset of PD3 from frame pointer
pd1         equ     6           ; offset of PD2 from frame pointer
pd2         equ     5           ; offset of PD1 from frame pointer
pd3         equ     4           ; offset of PD0 from frame pointer
MD_lo       equ     -8          ; offset of MD_lo from frame pointer
MD_hi       equ     -7          ; offset of MD_hi from frame pointer
ND_lo       equ     -6          ; offset of ND_lo from frame pointer
ND_hi       equ     -5          ; offset of ND_hi from frame pointer
buf_lo      equ     -4
buf_hi      equ     -3
ptr_hi      equ     0x01        ; address of buffer to hold product
ptr_lo      equ     0x00        ; "
; ****************************************************************
; The following definitions are used by the 16-bit unsigned divide routine
; ****************************************************************
loc_var     equ     2           ; local variable size
lp_cnt      equ     1           ; loop count
temp        equ     2           ; temporary storage
quo_hi      equ     -7          ; offset for quotient and dividend from frame
quo_lo      equ     -8          ; pointer
rem_hi      equ     -5          ; offset for remainder from frame pointer
rem_lo      equ     -6          ; "
dsr_hi      equ     -3          ; offset for divisor from frame pointer
dsr_lo      equ     -4          ; "
quo         set     0x02        ; memory location to hold the quotient
rem         set     0x04        ; memory location to hold the remainder
; ****************************************************************
; variables for holding temperature conversion result and string
; ****************************************************************
int_pt      set     0x06        ; space to hold integer part of the temperature
temp_buf    set     0x08        ; reserve 6 bytes to hold the temperature
                                ; digits, sign, period, and NULL
```

```
;****************************************************************
; Reset vector is here.
;****************************************************************
;
                org       0x00          ; reset vector
                goto      start
                org       0x08
                retfie
                org       0x18
                retfie
start           lfsr      FSR1,0xC00    ; set up stack pointer
                call      a2d_init      ; configure and turn on A/D module
; initialize the temperature string to ^^0.0
                movlw     0x20          ; store space character
                movwf     temp_buf      ; "
                movwf     temp_buf+1    ; "
                movlw     0x30          ; store a 0 digit
                movwf     temp_buf+2    ; "
                movwf     temp_buf+4    ; "
                movlw     0x2E          ; store a period
                movwf     temp_buf+3    ; "
                clrf      temp_buf+5    ; terminate the string with a NULL character
                call      OpenTmr0      ; initialize and enable Timer0
forever         bsf       ADCON0,GO,A   ; start A/D conversion
wait_a2d        btfsc     ADCON0,DONE,A ; wait until A/D conversion is complete
                bra       wait_a2d      ; "
                pushr     ADRESL        ; push the A/D conversion result in
                pushr     ADRESH        ; stack
                push_dat  0x0A          ; push 10 in the stack for multiplier
                push_dat  0x00          ; "
                push_dat  ptr_lo        ; pass buffer pointer to the subroutine
                push_dat  ptr_hi        ; "
                call      mul_16U,FAST  ; multiply A/D conversion result by 10
                dealloc_stk 6           ; deallocate space used in the stack
                movlw     ptr_lo        ; place the address of product in FSR0
                movwf     FSR0L,A       ; "
                movlw     ptr_hi        ; "
                movwf     FSR0H,A       ; "
                movf      POSTINC0,W,A  ; push (A/D result x 10)
                pushr     WREG          ; "
                movf      INDF0,W,A     ; "
                pushr     WREG          ; "
                alloc_stk 2
                push_dat  0x3E          ; push 62 into the stack as the divisor
                push_dat  0             ; "
                call      div16u,FAST   ;
                dealloc_stk 2
                popr      rem+1         ; retrieve remainder high byte (should be 0)
                popr      rem           ; retrieve remainder low byte
                popr      int_pt+1      ; retrieve integer part of the (should be 0)
                popr      int_pt        ; temperature
```

```
; subtract 40 from integer part to obtain the actual temperature
                movlw       0x28            ; calculate the actual temperature
                subwf       int_pt,F,A      ; "
                bnn         non_minus       ; if non-minus, no need for further check
                negf        int_pt,A        ; find the magnitude of temperature
                movlw       0x2D
                movlw       temp_buf        ; store the minus sign
                movf        rem,W,A         ; check the fractional part before divide
                bz          separate_dd     ; branch to separate integer digits if rem = 0
                decf        int_pt,F,A      ; fractional digit ≠ 0, decrement integer part
                movlw       0x3E            ; need to find the complement of the fractional
                subwf       rem,F,A         ; digit
                negf        rem,A           ; "
; calculate the fractional digit
non_minus
                movlw       0x0A            ; multiply the remainder by 10
                mulwf       rem             ; "
                pushr       PRODL           ; push (remainder × 10)
                pushr       PRODH           ; "
                alloc_stk   2
                push_dat    0x3E            ; push 62 into the stack as the divisor
                push_dat    0               ; "
                call        div16u,FAST     ;
                dealloc_stk 2
                popr        rem+1           ; retrieve remainder high byte (= 0)
                popr        rem             ; retrieve remainder low byte
                popr        quo+1           ; retrieve quotient high byte (= 0)
                popr        quo             ; retrieve quotient low byte
; round the fractional digit
                movlw       0x1F            ; is remainder >= 31?
                cpfslt      rem,A           ; smaller, then skip
                incf        quo,A
                movlw       0x0A            ; is quo equal to 10?
                cpfseq      quo,F,A
                goto        no_round
                clrf        quo,A
                incf        int_pt,A
no_round        movlw       0x30            ; convert to ASCII of BCD digit
                addwf       quo,F,A
                movwf       temp_buf+4      ; save the ASCII of the fractional digit
; separate the integral digits using repeated division by 10
separate_dd
                pushr       int_pt          ; push integer part
                push_dat    0               ; "
                alloc_stk   2
                push_dat    0x0A            ; push 10 as the divisor
                push_dat    0               ; "
                call        div16u,FAST
                dealloc_stk 2
```

```
                popr        rem+1
                popr        rem
                popr        quo+1
                popr        quo
                movlw       0x30
                addwf       rem,W,A
                movwf       temp_buf+2   ; save the one's digit
                movf        quo,W,A      ; check the quotient
                bz          next_time    ; wait to perform next conversion
; prepare to separate ten's digit
                movlw       0x0A         ; is the quotient >= 10?
                cpfslt      quo,A        ; "
                goto        yes_ge
                movlw       0x30
                addwf       quo,W,A
                movwf       temp_buf+1   ; save the ten's digit
                goto        next_time    ; "
yes_ge          movlw       0x31         ; save "1" as the hundred's digit
                movwf       temp_buf,A   ; "
                movlw       0x0A         ; separate the ten's digit and place it
                subwf       quo,W,A      ; in WREG
                addlw       0x30         ; convert ten's digit to ASCII and
                movwf       temp_buf+1,A ; save ten's digit
next_time       btfss       INTCON,TMR0IF,A    ; wait until 200 ms is over
                goto        next_time
; ---
; add instructions to update display here
; ---
                goto        forever      ; prepare to perform next A/D conversion
; ***********************************************************************
; This routine will place 15535 in TMR0 so that it overflows in 50000 count.
; When prescaler is set to 32 with f_OSC = 32MHz, it will overflow in 200 ms.
; ***********************************************************************
OpenTmr0        movlw       0x3C         ; place 15535 in TMR0
                movwf       TMR0H        ; so that it overflows in
                movlw       0xAF         ; 200 ms
                movwf       TMR0L        ; "
                movlw       0x84         ; enable TMR0, select internal clock,
                movwf       T0CON        ; set prescaler to 32
                bcf         INTCON,TMR0IF      ; clear TMR0IF flag
                return
; ***********************************************************************
; This routine initializes the A/D converter to select channel AN0 as
; analog input other pins for digital pin. Select VDD and VSS as A/D
; conversion reference voltages, result right justified, FOSC/64 as
; A/D clock source, 8 TAD for acquisition time.
; ***********************************************************************
a2d_init        movlw       0x01         ; select channel AN0
                movwf       ADCON0       ; and enable A/D module
                movlw       0x0E         ; use VDD & VSS as A/D reference voltage
```

```
                movwf       ADCON1      ; & configure AN0 as analog others for digital
                movlw       0xA6        ; result right justified, FOSC/64 as A/D clock
                movwf       ADCON2      ; source and set acquisition time to 8 TAD
                return
; ——-
; include subroutines div16u and mul_16U here.
; ——-
                end
```

The C language version of the program is as follows:

```c
#include         <p18F8680.h>
#include         <timers.h>
#include         <adc.h>
unsigned         char temp_buf[6];
void main (void)
{
    int a2d_val;
    unsigned int quo, rem;
    char fd1, fdr;
    ADCON0 = 0x01;          //select channel AN0, enable A/D module
    ADCON1 = 0x0E;          //use VDD, VSS as reference and configure AN0 for analog
    ADCON2 = 0xA6;          //result right justified, 8TAD acquisition time, FOSC/64
    OpenTimer0 (TIMER_INT_OFF & T0_16BIT & T0_SOURCE_INT & T0_PS_1_32); /*start Timer0 and make it
                overflow in 200 ms*/
    while (1) {
        temp_buf[5] = '\0';
        temp_buf[0] = 0x20; //set to space
        temp_buf[1] = 0x20; //set to space
        temp_buf[2] = 0x30; //set to digit 0
        temp_buf[3] = 0x2E; //store the decimal point
        temp_buf[4] = 0x30; //set to digit 0
        ConvertADC( );          //start an A/D conversion
        while(BusyADC());       //wait until A/D conversion is done
        a2d_val = 10 * ReadADC();
        quo = a2d_val / 62;  //convert to temperature
        rem = a2d_val % 62;
        if (quo < 40)           //is temperature minus?
        {
            quo = 40 - quo;
            temp_buf[0] = 0x2D;             //set sign to minus
            if (rem != 0)
            {
                quo—;
                rem = 62 - rem;
            }
        }
        fd1 = (rem * 10) / 62;              //fd1 will be between 0 and 9
        fdr = (rem * 10) % 62;
        if (fdr >= 31)
            fd1 ++;
```

```
    if (fd1 == 10) {          //fractional digit can only be between 0 and 9
        quo++;
        fd1 = 0;
    }
    temp_buf[4] = 0x30 + fd1;          //store the ASCII code of fractional digit
    temp_buf[2] = quo % 10 + 0x30;     //store ASCII code of one's digit
    quo = quo / 10;
    if (quo != 0)
    {
        temp_buf[1] = (quo - 10) + 0x30;       //ten's digit of temperature
        quo /= 10;
    }
    if (quo == 1)
        temp_buf[0] = 0x31; //hundred's digit of temperature
    while(!INTCONbits.TMR0IF); //wait until Timer0 overflows
    INTCONbits.TMR0IF = 0; //clear the TMR0IF flag
}
}
```

12.7 Using the IH-3606 Humidity Sensor

The IH-3605 is a humidity sensor made by Honeywell. The IH-3605 humidity sensor provides a linear output from 0.8 V to 3.9 V in the full range of 0% to 100% *relative humidity* (RH) when powered by a 5-V power supply. It can operate in a range of 0% to 100% RH, –40° to 185° F.

The pins of the IH-3605 are shown in Figure 12.16. The specification of the IH-3605 is listed in Table 12.5. The IH-3605 is light sensitive and should be shielded from bright light for best results.

Specification	Description
Total accuracy	± 2% RH, 0-100% TH @25° C
Interchangeability	± 5% RH up to 60% RH, ±8% RH at 90% RH
Operating temperature	–40 to 85° C (–40 to 185° F)
Storage temperature	–51 to 110° C (–60 to 223° F)
Linearity	±0.5% RH typical
Repeatability	±0.5% RH
Humidity Stability	±1% RH typical at 50% RH in 5 years
Temp. effect on 0% RH voltage	±0.007% RH° C (negligible)
Temp. effect on 100% RH voltage	–0.22% RH/° C
Output voltage	$V_{OUT} = (V_S)(0.16$ to $0.78)$ nominal relative to supply voltage for 0-100% RH; i.e., 1-4.9V for 6.3V supply; 0.8 – 3.9V for 5V supply; Sink capability 50 microamp; drive capability 5 microamps typical; low pass 1 KHz filter required. Turn on time < 0.1 sec to full output.
VS Supply requirement	4 to 9V, regulated or use output/supply ratio; calibrated at 5V
Current requirement	200 microamps typical @5V, increased to 2mA at 9V

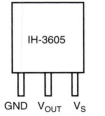

Figure 12.16 ■ Honeywell IH-3605 humidity sensor

Table 12.5 ■ Specifications of IH-3605

The IH-3605 can resist contaminant vapors, such as organic solvents, chlorine, and ammonia. It is unaffected by water condensate as well. Because of this capability, the IH-3605 has been used in refrigeration, drying, instrumentation, meteorology, and many other applications.

Example 12.9

▼

Construct a humidity measurement system that consists of the PIC18F8680, an IH-3605 humidity sensor, and an LCD. The PIC18F8680 is running with a 32-MHz crystal oscillator.

Solution: Since the output of the IH-3605 is between 0.8 V and 3.9 V with a 5-V power supply, it would be beneficial to use a circuit to translate and then scale it to between 0 V and 5 V so that the best accuracy can be achieved. The circuit connection is shown in Figure 12.17. A set of resistor values and V1 voltage are given in Figure 12.17. A low-pass filter that consists of a 1-KΩ resistor and a 0.16-μF capacitor is added to meet the requirement. The user can use an LCD or four seven-segment displays to display the relative humidity.

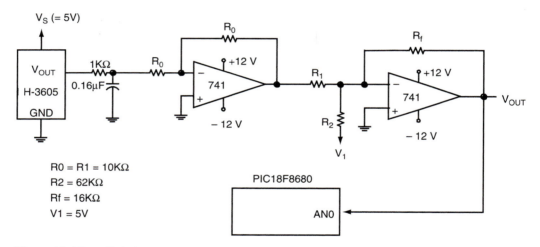

Figure 12.17 ■ Relative humidity measurement circuit

To translate from the conversion result to the relative humidity, divide the conversion result by 10.23. The following C program will configure the A/D module, start the A/D conversion, translate the conversion result to the relative humidity, and convert the humidity into an ASCII string to be output on the LCD.

The procedure for performing the measurement and display of relative humidity is similar to Example 12.8, and the C program that implements it is shown here:

```
#include    <p18F8680.h>
#include    <timers.h>
#include    <adc.h>
unsigned char hum_buf[6]; //buffer to hold relative humidity
void main (void)
{
    unsigned short long a2d_val;
```

```
unsigned short long quo, rem, temp1;
char fd1, i;
ADCON0 = 0x01;                    //select channel AN0, enable A/D module
ADCON1 = 0x0E;                    //use VDD, VSS as reference and configure AN0 for analog
ADCON2 = 0xA6;                    //result right justified, acquisition time = 8 TAD, F_OSC/64
OpenTimer0 (TIMER_INT_OFF & T0_16BIT & T0_SOURCE_INT & T0_PS_1_32);
          //start Timer0 and make it overflow in 200 ms
while (1) {
    hum_buf[0] = 0x20;            //set to space
    hum_buf[1] = 0x20;            //set to space
    hum_buf[2] = 0x30;            //set to digit 0
    hum_buf[3] = 0x2E;            //store the decimal point
    hum_buf[4] = 0x30;            //set to digit 0
    hum_buf[5] = '\0';           //terminate with a NULL character
    ConvertADC( );               //start an A/D conversion
    while (BusyADC( ));          //wait until A/D conversion is done
    a2d_val = 100 * ReadADC();
    quo = a2d_val / 1023;        //convert to relative humidity
    rem = a2d_val % 1023;        // "
    fd1 = (rem * 10) / 1023;     //compute the fractional digit
    temp1 = (rem * 10) % 1023;
    if (temp1 > 511)             //should round up the fractional digit
        fd1 ++;
    if (fd1 == 10) {             //if fractional digit becomes 10, zero it
        fd1 = 0;                 // and add 1 to integer part
        quo++;
    }
    hum_buf[4] = 0x30 + fd1;     //ASCII code of fractional digit
    hum_buf[2] = quo % 10 + 0x30; //ASCII code of one's digit
    quo = quo / 10;
    if (quo != 0)
    {
        hum_buf[1] = (quo % 10) + 0x30;
        quo /= 10;
    }
    if (quo == 1)
        hum_buf[0] = 0x31;
    while (!INTCONbits.TMR0IF);  //wait until Timer0 overflows
    for (i = 0; i < 4; i++) {    //wait for Timer0 to overflow four more times
        INTCONbits.TMR0IF = 0;   //clear the TMR0IF flag
        while(!INTCONbits.TMR0IF); //wait for Timer0 to overflow
    }
    INTCONbits.TMR0IF = 0
    }
}
```

This C program performs the A/D conversion and translates the result into an ASCII string so that it can be displayed in an LCD or a terminal.

▲

12.8 Measuring Barometric Pressure

Barometric pressure refers to the air pressure existing at any point within the earth's atmosphere. This pressure can be measured as an absolute pressure (with reference to absolute vacuum) or can be referenced to some other value or scale. The meteorology and avionics industries traditionally measure the absolute pressure and then reference it to a sea-level pressure value. This complicated process is used in generating maps of weather systems.

Mathematically, atmosphere pressure is exponentially related to altitude. Once the pressure at a particular location and altitude is measured, the pressure at any other altitude can be calculated.

Several units have been used to measure the barometric pressure: in-Hg, kPa, mbar, or psi. A comparison of barometric pressure using four different units at sea level up to 15000 feet is shown in Table 12.6.

Altitude (ft)	Pressure (in-Hg)	Pressure (mbar)	Pressure (kPa)	Pressure (psi)
0	29.92	1013.4	101.4	14.70
500	29.38	995.1	99.5	14.43
1000	28.85	977.2	97.7	14.17
6000	23.97	811.9	81.2	11.78
10000	20.57	696.7	69.7	10.11
15000	16.86	571.1	57.1	8.28

Table 12.6 ■ Altitude versus pressure data

There are three forms of pressure transducer: *gauge, differential,* and *absolute.* Both the gauge pressure (psig) and differential (psid) transducers measure pressure differentially. The acronym "psi" stands for "pounds per square inch," whereas the letters "g" and "d" stand for "gauge" and "differential," respectively. A gauge pressure transducer measures pressure against ambient air, whereas the differential transducer measures against a reference pressure. An absolute pressure transducer measures the pressure against a vacuum (0 psia), and hence it measures the barometric pressure.

The SenSym ASCX30AN is a 0- to 30-psia (psi absolute) pressure transducer. The range of barometric pressure is between 28 to 32 inches of mercury (in-Hg), or 13.75 to 15.72 psia or 948 to 1083.8 mbar. The transducer output is about 0.15 V/psi, which would translate to an output voltage from 2.06 V to 2.36 V. The complete specification of the SenSym ASCX30AN is shown in Table 12.7. Since the range of V_{OUT} is very narrow, the user will need to use a level shifting and scaling circuit in order to take advantage of the available dynamic range.

The pin assignment of the SenSym ASCX30AN is shown in Figure 12.18.

Characteristic	min	typ	max
Pressure	0 psia	–	30 psia
Zero pressure offset	0.205	0.250	0.295
Full-scale span[2]	4.455	4.500	4.545
Output at FS pressure	4.660	4.750	4.840
Combined pressure non-linearity and pressure hysteresis[3]	–	±0.1	±0.5
Temperature effect on span[4]	–	±0.2	±1.0
Temperature effect on offset[4]	–	±0.2	±1.0
Response time (10% – 90%)[5]	–	0.1	–
Repeatability	–	±0.05	–

Note

1. Reference conditions: TA = 25° C, supply voltage V_S = 5 V
2. Full scale span is the algebraic difference between the output voltage at full-scale pressure and the output at zero pressure. Full-scale span is ratiometric to the supply voltage.
3. Pressure non-linearity is based on the best-fit straight line. Pressure hysteresis is the maximum output difference at any point within the operating pressure range for increasing and decreasing pressure.
4. Maximum error band of the offset voltage or span over the compensated temperature range relative to the 25° C reading.
5. Response time for 0 psi to full-scale pressure step response.
6. If maximum pressure is exceeded, even momentarily, the package may leak or burst, or the pressure-sensing die may burst.

Table 12.7 ■ ASCX30AN performance characteristics[1]

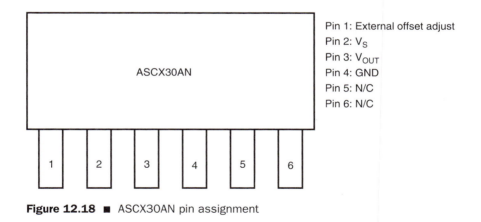

Pin 1: External offset adjust
Pin 2: V_S
Pin 3: V_{OUT}
Pin 4: GND
Pin 5: N/C
Pin 6: N/C

Figure 12.18 ■ ASCX30AN pin assignment

Example 12.10

▼

Describe the circuit connection of the ASCX30AN and the voltage level shifting and scaling circuit and write a program to measure and display the barometric pressure in units of mbar.

Solution: The circuit connection is shown in Figure 12.19. The circuit in Figure 12.19 will shift and scale V_{OUT} to the range of 0 V ~ 5 V. According to Table 12.7, the typical value for the offset adjustment is around 0.25 V and can be achieved by using a potentiometer.

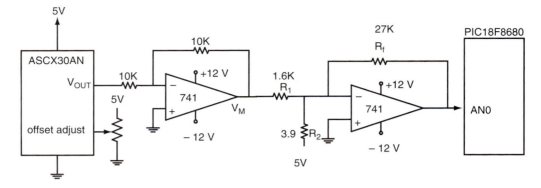

Figure 12.19 ■ Barometric pressure sensor output scaling and shifting circuit

The barometric pressure range is from 948 mbar to 1083.8 mbar. The A/D conversion result will be 0 and 1023 for these two values. Therefore, the user needs to divide the A/D conversion result by 7.53 in order to find the corresponding barometric pressure. The following equation shows the conversion:

Barometric pressure = 948 + A/D result/7.53
= 948 + (A/D result * 100)/753

The C program that performs A/D conversion, translates the result into barometric pressure, and then converts the barometric pressure into an ASCII string is as follows:

```
#include <p18F8680.h>
#include <timers.h>
#include <adc.h>
unsigned char bp_buf[7];
void main (void)
{
    unsigned short long a2d_val;
    unsigned short long quo, rem, temp1;
    char fd1, i;
    ADCON0 = 0x01;                      //select channel AN0, enable A/D module
    ADCON1 = 0x0E;                      //use V_DD, V_SS as reference and configure AN0 for analog
    ADCON2 = 0xA6;                      //result right justified, acquisition time = 8 TAD, FOSC/64
    OpenTimer0 (TIMER_INT_OFF & T0_16BIT & T0_SOURCE_INT & T0_PS_1_32);
                //start Timer0 and make it overflow in 200 ms
    while (1) {
        bp_buf[0] = 0x20;               //set to space
        bp_buf[1] = 0x20;               //set to space
        bp_buf[2] = 0x20;               //set to space
        bp_buf[3] = 0x30;               //set to digit 0
        bp_buf[4] = 0x2E;               //store the decimal point
        bp_buf[5] = 0x30;               //set to digit 0
        bp_buf[6] = '\0';               //terminate the string with a NULL character
        ConvertADC( );                  //start an A/D conversion
        while(BusyADC());               //wait until A/D conversion is done
```

```
        a2d_val = 100 * ReadADC();
        quo = a2d_val/753;                    //convert to barometric pressure
        rem = a2d_val%753;                    // "
        quo += 948;                           //add the barometric pressure at A/D
                                              //conversion result 0 to obtain the
                                              //actual barometric pressure
        fd1 = (rem * 10)/753;                 //compute the fractional digit
        temp1 = (rem * 10)%753;
        if (temp1 > 376)                      //should we round up the fractional digit?
           fd1 ++;
        if (fd1 == 10) {                      //if fractional digit becomes 10, zero it
           fd1 = 0;                           //and add 1 to integer part
           quo++;
        }
        bp_buf[5] = 0x30 + fd1;               //ASCII code of fractional digit
        bp_buf[3] = quo % 10 + 0x30;          //ASCII code of one's digit
        quo /= 10;
        bp_buf[2] = quo % 10 + 0x30;          //ten's digit
        quo /= 10;
        bp_buf[1] = quo % 10 + 0x30;          //hundred's digit
        quo /= 10;
        bp_buf[0] = quo + 0x30;               //thousand's digit
        while(!INTCONbits.TMR0IF);            //wait until Timer0 overflows
        for (i = 0; i < 4; i++) {             //wait for Timer0 to overflow four more times
           INTCONbits.TMR0IF = 0;             //clear the TMR0IF flag
           while(!INTCONbits.TMR0IF);         //wait for Timer0 to overflow
        }
        INTCONbits.TMR0IF = 0;
    }
}
```

▲

12.9 Summary

A data acquisition system consists of four major components: a transducer, a signal conditioning circuit, an A/D converter, and a computer. The transducer converts a nonelectric quantity into a voltage. The transducer output may not be appropriate for processing by the A/D converter. The signal conditioning circuit shifts and scales the output from a transducer to a range that can take advantage of the full capability of the A/D converter. The A/D converter converts an electric voltage into a digital value that will be further processed by the computer.

Because of the discrete nature of a digital system, the A/D conversion result has a quantization error. The accuracy of an A/D converter is dictated by the number of bits used to represent the analog quantity. The more bits are used, the smaller the quantization error will be.

There are four major A/D conversion algorithms:

- Parallel (flash) A/D converter
- Slope and double-slope A/D converters
- Sigma-delta A/D converters
- Successive-approximation A/D converters

The PIC18 microcontroller uses the successive-approximation algorithm to perform the A/D conversion. All A/D conversion parameters are configured via three A/D control registers: ADCON0, ADCON1, and ADCON2. Some PIC18 members have only the first two controller registers (ADCON0 and ADCON1). The following A/D conversion parameters need to be configured:

- Analog channel to be converted
- High and low reference voltages
- Analog or digital nature of the A/D port pins
- A/D result format (right or left justified)
- A/D acquisition time
- A/D clock source

For some earlier PIC18 members, the A/D acquisition time may not be programmable and must be provided by using software delay.

The TC1047A temperature sensor, the IH-3606 humidity sensor, and the ASCX30AN pressure sensor are used as examples to illustrate the A/D conversion process. The TC1047A can measure a temperature in the range from –40° C to 125° C. The IH-3606 can measure relative humidity from 0% to 100%. The ASCX30AN can measure a pressure in the range from 0 to 30 psi absolute. These three examples demonstrate the need for a good voltage shifting and scaling circuit.

There are applications that require A/D accuracy higher than that provided by the PIC18 microcontrollers. In this situation, the user has the option of using an external A/D converter with higher precision or of selecting a different microcontroller with higher A/D resolution. Many 8051 variants from Silicon Laboratory, TI, and Analog Devices have much higher A/D resolutions.

12.10 Exercises

E12.1 Design a circuit that can scale the voltage from the range of 0 mV ~ 100 mV to the range of 0 V ~ 5 V.

E12.2 Design a circuit that can shift and scale the voltage from the range of –80 mV ~ 160 mV to the range of 0 V ~ 5 V.

E12.3 Design a circuit that can shift and scale the voltage from the range of –50 mV ~ 75 mV to the range of 0 V ~ 5 V.

E12.4 Design a circuit that can shift and scale the voltage from the range of 2 V ~ 2.5 V to the range of 0 V ~ 5 V.

E12.5 Suppose that there is a 10-bit A/D converter with V_{REF-} = 2 V and V_{REF+} = 4 V. Find the corresponding voltage values for the A/D conversion results of 40, 100, 240, 500, 720, 800, and 1000.

E12.6 Suppose that there is a 12-bit A/D converter with V_{REF-} = 1 V and V_{REF+} = 4 V. Find the corresponding voltage values for the A/D conversion results of 80, 180, 480, 640, 960, 1600, 2048, 3200, and 4000.

E12.7 Write a few instructions to configure the PIC18F452 A/D converter with the following parameters:

- f_{OSC} = 16 MHz (you need to choose appropriate clock source)
- Channel AN3
- A/D result left justified

- Configure A/D channel AN3, AN1, and AN0 for analog input, other A/D port pins for digital I/O
- Select V_{DD} and V_{SS} as reference voltages

E12.8 Write an instruction sequence to configure the PIC18F4320 A/D converter with the following characteristics:

- f_{OSC} = 8 MHz
- Channel AN3
- A/D result right justified
- Configure A/D channel AN3 through AN0 for analog input, other A/D port pins for digital I/O
- Select V_{DD} and V_{SS} as reference voltages

E12.9 Write a few instructions to configure the PIC18F8720 A/D converter with the following parameters:

- f_{OSC} = 25 MHz (you need to choose appropriate clock source)
- Channel AN15
- A/D result right justified
- Configure A/D channel AN15 for analog input
- Select V_{DD} and V_{SS} as reference voltages

E12.10 The LM35 from National Semiconductor is a Centigrade temperature sensor with three external connection pins. The pin assignment and circuit connection for converting temperature are shown in Figure 12E.1 Use this device to construct a circuit to display the room temperature in Celsius. Assume that the temperature range is –27° C through 100° C. Use an LCD to display the temperature.

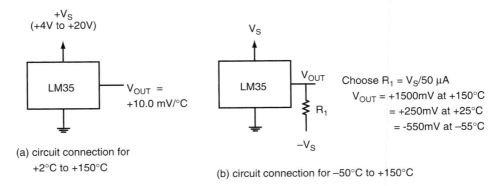

Figure 12E.1 ■ Circuit connection for the LM35

E12.11 The LM34 from National Semiconductor is a Fahrenheit temperature sensor with three external connection pins. The pin assignment and circuit connection for converting temperature are shown in Figure 12E.2. Use this device to construct a circuit to display the room temperature in Fahrenheit. Assume that the temperature range of interest is from –40° F to 215° F. Use an LCD to display the temperature.

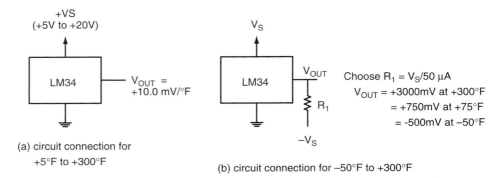

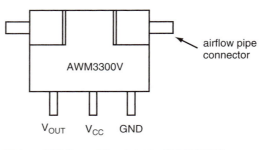

(a) circuit connection for
+5°F to +300°F

Choose $R_1 = V_S/50\ \mu A$
V_{OUT} = +3000mV at +300°F
        = +750mV at +75°F
        = -500mV at -50°F

(b) circuit connection for -50°F to +300°F

Figure 12E.2 ■ Circuit connection for the LM34

E12.12 The Microbridge AWM3300V is a mass airflow sensor manufactured by Honeywell. The block diagram of the AWM3300V is shown in Figure 12E.3. The Microbridge mass airflow sensor AWM3300V is designed to measure the airflow. Its applications include air-conditioning, medical ventilation/anesthesia control, gas analyzers, gas metering, fume cabinets, and process control. The AWM3300V operates on a single 10-V ± 10-mV power supply. The sensor output (from V_{OUT}) corresponding to the airflow rate of 0 ~ 1.0 liters per minute is 1.0 V to 5.0 V. The AWM3300V can operate in the temperature range of -25°C to 85°C. It takes 3 ms for the output voltage to settle after power-up. Design a circuit to measure and display the mass airflow using the AWM3300V. Write a program to configure the PIC18F8680 A/D module, start the A/D conversion, and display the mass airflow in an LCD display. Update the display 10 times per second.

AWM3300V

airflow pipe
connector

V_{OUT} V_{CC} GND

Figure 12E.3 ■ Microbridge AWM3300V

12.11 Lab Exercises and Assignments

L12.1 *A/D converter testing.* Perform the following steps:

Step 1
Set the function generator output (in the lab) to between 0 V and 5 V.

Step 2
Connect the AN1 pin to the functional generator output. Set the frequency to about 10 KHz.

Step 3

Write a program to perform the following operations:

- Configure the A/D module properly.
- Start the A/D conversion.
- Take 64 samples, convert them, and store the conversion results at appropriate SRAM locations.
- Compute the average value and store it in PRODL.

If your demo board has square waveform output, then select an output with frequency close to 16 KHz. Connect it to the AN1 pin.

L12.2 *Continuous voltage measurement.* Perform the following operations:

Step 1

Connect the AN0 pin to a potentiometer that can vary voltage from 0 V to 5 V.

Step 2

Configure the A/D module properly.

Step 3

Take one sample of the voltage from the potentiometer, convert the voltage, convert the A/D conversion result to the corresponding voltage, and display the voltage on the LCD.

Step 4

Repeat Step 3 every 100 ms.

L12.3 *Digital thermometer.* Use the LM34 temperature sensor, the 741 OP AMP, and the required resistors to construct a digital thermometer. The temperature should be displayed in three integral and one fractional digits. Use the LCD on your demo board to display the temperature. Update the temperature once every 100 ms.

L12.4 *Barometric pressure measurement.* The Motorola MPX4115A is a pressure sensor that can measure a pressure ranging from 15 kPa to 115 kPa (2.2–16.7 psi) and has a corresponding voltage output from 0.2 V to 4.8 V. The small outline package of this device has eight pins. Among these eight pins, only three pins carry useful signals:

Pin 2: V_S

Pin 3: Ground

Pin 4: V_{OUT}

This pressure sensor is often used in aviation altimeters, industrial controls, engine control, and weather stations and weather reporting devices.

Connect an MPX4115AC6U (or MPX4115A6U, case 482) device to your PIC18 demo board to measure the current ambient barometric pressure once every second and display the pressure on the LCD display. The datasheet of the MPX4115A can be found in the complimentary CD. Since the barometric pressure is in a narrow range (use the range mentioned in Example 12.10), one may need to construct a signal conditioning circuit to translate and scale the voltage output of the pressure sensor to the range of 0 V to 5 V.

13

Controller Area Network

13.1 Objectives

After completing this chapter, you should be able to

- Describe the layers of the CAN protocol
- Describe CAN's error detection capability
- Describe the formats of CAN messages
- Describe CAN message handling
- Explain CAN error handling
- Describe CAN fault confinement
- Describe CAN bit timing
- Explain CAN synchronization issue and methods
- Describe the CAN message structures
- Compute timing parameters to meet the requirements of your application
- Write subroutine to configure the PIC18 CAN module
- Write programs to transfer data over the CAN bus

13.2 Overview of Controller Area Network

The controller area network (CAN) was initially created by German automotive system supplier Robert Bosch in the mid-1980s for automotive applications as a method for enabling robust serial communication. The goal was to make automobiles more reliable, safe, and fuel efficient while at the same time decreasing wiring harness weight and complexity. Since its inception, the CAN protocol has gained widespread use in industrial automation and automotive/truck applications.

The description of CAN in this chapter is based on the CAN Specification 2.0 published in September 1991 by Bosch.

13.2.1 Layered Approach in CAN

The CAN protocol specified the lowest two layers of the ISO seven-layer model: *data link* and *physical* layers. The data link layer is further divided into two sublayers: logical link control (LLC) layer and medium access control (MAC) layer:

- The LLC sublayer deals with message acceptance filtering, overload notification, and error recovery management.
- The MAC sublayer presents incoming messages to the LLC sublayer and accepts messages to be transmitted forwarded by the LLC sublayer. The MAC sublayer is responsible for message framing, arbitration, acknowledgment, error detection, and signaling. The MAC sublayer is supervised by a self-checking mechanism, called *fault confinement*, which distinguishes short disturbances from permanent failures.

The physical layer defines how signal are actually transmitted, dealing with the description of bit timing, bit encoding, and synchronization. CAN bus driver/receiver characteristics and the wiring and connectors are not specified in the CAN protocol. These two aspects are not specified so that the implementers can choose the most appropriate transmission medium and hence optimize signal-level implementations for their applications. The system designer can choose from multiple available media technologies including twisted pair, single wire, optical fiber, radio frequency (RF), infrared (IR), and so on. The layered CAN protocol is shown in Figure 13.1.

13.2.2 General Characteristics of CAN

The CAN protocol was optimized for systems that need to transmit and receive relatively small amounts of information (as compared to Ethernet or USB, which are designed to move much larger blocks of data). The CAN protocol has the following features.

CARRIER SENSE MULTIPLE ACCESS WITH COLLISION DETECTION (CSMA/CD)

The CAN protocol is a CSMA/CD protocol. Every node on the network must monitor the bus (carrier sense) for a period of no activity before trying to send a message on the bus. Once this period of no activity occurs, every node on the bus has an equal opportunity to transmit a message (multiple access). If two nodes happen to transmit at the same time, the nodes will detect the *collision* and take the appropriate action. In the CAN protocol, a nondestructive bitwise arbitration method is utilized. Messages remain intact after arbitration is completed even if collisions are detected. Message arbitration will not delay higher-priority messages. To facilitate bus arbitration, the CAN protocol defines two bus states: **dominant** and **recessive.** The dominant state is represented by logic 0 (low voltage), whereas the recessive state is represented by logic 1 (high voltage). The dominant state will win over the recessive state.

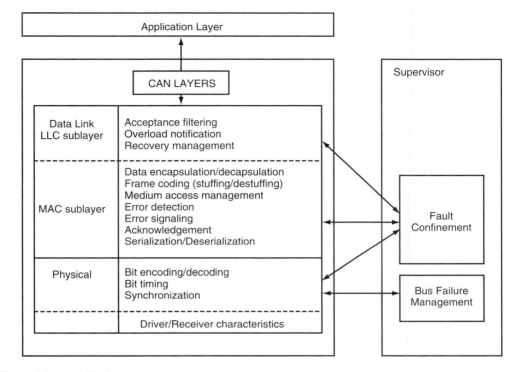

Figure 13.1 ■ CAN layers

MESSAGE-BASED COMMUNICATION

CAN is a message-based protocol, not an address-based protocol. Embedded in each message is an **identifier.** This identifier allows messages to arbitrate the use of the CAN bus and also allows each node to decide whether to work on the message. The value of the identifier is used as the priority of the message. The lower the value, the higher the priority. Each node in the CAN system uses one or more filters to compare the identifier of the incoming message. Once the identifier passes the filter, the message will be worked on by the node. The CAN protocol also provides the mechanism for a node to request data transmission from another node. Since address is not used in the CAN system, there is no need to reconfigure the system whenever a node is added to or deleted from a system. This capability allows the system to perform node-to-node or multicast communications.

ERROR DETECTION AND FAULT CONFINEMENT

The CAN protocol requires each node to monitor the CAN bus to find out if the bus value and the transmitted bit value are identical. For every message, cyclic redundancy check (CRC) is calculated, and the checksum is appended to the message. CAN is an asynchronous protocol, and hence clock information is embedded in the message rather than transmitted as a separate signal. Messages with long sequences of identical bits could cause synchronization problems. To resolve this, the CAN protocol requires the physical layer to use bit staffing to avoid a long sequence of identical bit values. With these measures implemented, the residual probability for undetected corrupted messages in a CAN system is as low as

message error rate $\times 4.7 \times 10^{-11}$

CAN nodes are able to distinguish short disturbances from permanent failures. Defective nodes are switched off from the CAN bus.

13.3 CAN Messages

The CAN protocol defines four different types of messages:

- *Data frame.* A data frame carries data from a transmitter to the receivers.
- *Remote frame.* A remote frame is transmitted by a node to request the transmission of the data frame with the same identifier.
- *Error frame.* An error frame is transmitted by a node on detecting a bus error.
- *Overload frame.* An overload frame is used to provide for an extra delay between the preceding and the succeeding data or remote frames.

Data frames and remote frames are separated from preceding frames by an interframe space.

13.3.1 Data Frame

As shown in Figure 13.2, a data frame consists of seven different bit fields: start-of-frame, arbitration, control, data, CRC, ACK, and end-of-frame.

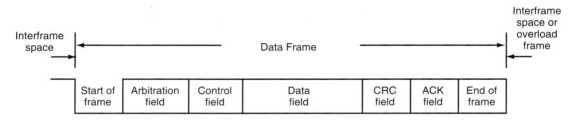

Figure 13.2 ■ CAN Data frame

START-OF-FRAME FIELD

This field is a single dominant bit that marks the beginning of a data frame. A node is allowed to start transmission only when the bus is idle. All nodes have to synchronize to the leading edge caused by this field of the node starting transmission first.

ARBITRATION FIELD

The format of the arbitration field is different for *standard format* and *extended format* frames, as illustrated in Figure 13.3. The identifier's length is 11 bits for the standard format and 29 bits for the extended format.

The identifier of the standard format corresponds to the base ID in the extended format. These bits are transmitted most significant bit first. The most significant seven bits cannot all be recessive.

The identifier of the extended format comprises two sections: an 11-bit base ID and an 18-bit extended ID. Both the base ID and the extended ID are transmitted most significant bit first. The base ID defines the base priority of the extended frame.

The *remote transmission request* (RTR) bit in data frames must be dominant. Within a remote frame, the RTR bit has to be recessive.

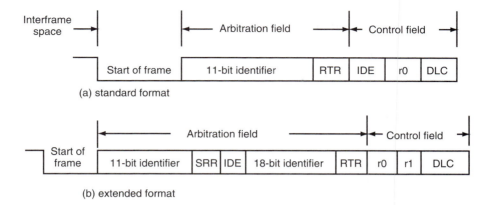

(a) standard format

(b) extended format

Figure 13.3 ■ Arbitration field

The *substitute remote request* (SRR) bit is a recessive bit. The SRR bit of an extended frame is transmitted at the position of the RTR bit in the standard frame and therefore substitutes for the RTR bit in the standard frame. As a consequence, collisions between a standard frame and an extended frame, where the base ID of both frames are identical, are resolved in such a way that the standard frame prevails over the extended frame.

The *identifier extension* (IDE) bit belongs to the arbitration field for the extended format and the control field for the standard format. The IDE bit in the standard format is transmitted dominant, whereas in the extended format the IDE bit is recessive.

CONTROL FIELD

The contents of this field are shown in Figure 13.4. The format of the control field is different for the standard format and the extended format. Frames in standard format include the data length code; the IDE bit, which is transmitted dominant; and the reserved bit **r0.** Frames in extended format include the data length code and two reserved bits, **r0** and **r1.** The reserved bits must be sent dominant, but the receivers accept dominant and recessive bits in all combinations. The data length code specifies the number of bytes contained in the data field. Data length can be zero to eight, as encoded in Table 13.1.

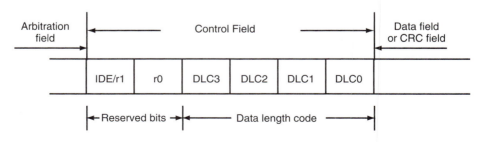

Figure 13.4 ■ Control field

Data length code				
DLC3	**DLC2**	**DLC1**	**DLC0**	**Data byte count**
d	d	d	d	0
d	d	d	r	1
d	d	r	d	2
d	d	r	r	3
d	r	d	d	4
d	r	d	r	5
d	r	r	d	6
d	r	r	r	7
r	d	d	d	8

d = dominant r = recessive

Table 13.1 ■ CAN Data length coding

DATA FIELD

The data field consists of the data to be transmitted within a data frame. It may contain from zero to eight bytes, each of which contains eight bits and are transferred most significant bit first.

CRC FIELD

The CRC field contains the CRC sequence followed by a CRC delimiter, as shown in Figure 13.5.

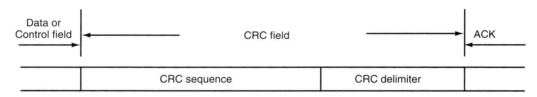

Figure 13.5 ■ CRC field

The frame-check sequence is derived from a cyclic redundancy code best suited to frames with bit counts less than 127. The CRC sequence is calculated by performing a polynomial division. The coefficients of the polynomial are given by the destuffed bit stream, consisting of the start of frame, arbitration field, control field, data field (if present), and 15 0s. This polynomial is divided (the coefficients are calculated using modulo-2 arithmetic) by the generator polynomial:

$$X^{15} + X^{14} + X^{10} + X^8 + X^7 + X^4 + X^3 + 1$$

The remainder of this polynomial division is the CRC sequence. In order to implement this function, a 15-bit shift register **CRC_RG(14:0)** is used. If **nxtbit** denotes the next bit of the bit stream, given by the destuffed bit sequence from the start of frame until the end of the data field, the CRC sequence is calculated as follows:

```
CRC_RG = 0;                        /* initialize shift register */
do {
    crcnxt = nxtbit ^ CRC_RG(14);  /* exclusive OR */
    CRC_RG(14:1) = CRC_RG(13:0);   /* shift left by one bit */
```

```
    CRC_RG(0) = 0;
    if crcnxt
        CRC_RG(14:0) = CRC_RG(14:0) ^ 0x4599;
} while (!(CRC SEQUENCE starts or there is an error condition));
```

After the transmission/reception of the last bit of the data field, **CRC_RG(14:0)** contains the CRC sequence. The *CRC delimiter* is a single recessive bit.

ACK Field

As shown in Figure 13.6, the ACK field is two bits long and contains the ACK slot and the ACK delimiter. A transmitting node sends two recessive bits in the ACK field. A receiver that has received a valid message reports this to the transmitter by sending a dominant bit in the ACK slot (i.e., it sends ACK).

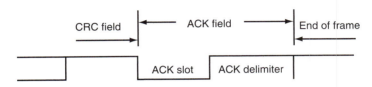

Figure 13.6 ■ ACK field

A node that has received the matching CRC sequence overwrites the recessive bit in the ACK slot with a dominant bit. This bit will be received by the data frame transmitter and learn that the previously transmitted data frame has been correctly received.

The *ACK delimiter* has to be a recessive bit. As a consequence, the ACK slot is surrounded by two recessive bits (the CRC delimiter and the ACK delimiter).

End-of-Frame Field

Each data frame and remote frame is delimited by a flag sequence consisting of seven recessive bits. This seven-bit sequence is the *end-of-frame* sequence.

13.3.2 Remote Frame

A node that needs certain data can request the relevant source node to transmit the data by sending a remote frame. The format of a remote frame is shown in Figure 13.7. A remote frame consists of six fields: start-of-frame, arbitration, control, CRC, ACK, and end-of-frame. The polarity of the RTR bit in the arbitration field indicates whether a transmitted frame is a *data frame* (RTR bit dominant) or a *remote frame* (RTR bit recessive).

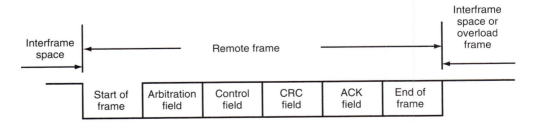

Figure 13.7 ■ Remote frame

13.3.3 Error Frame

The error frame consists of two distinct fields. The first field is given by the superposition of error flags contributed from different node. The second field is the error delimiter. The format of the error frame is shown in Figure 13.8. In order to terminate an error frame correctly, an *error-passive node* may need the bus to be idle for at least three bit times (if there is a local error at an error-passive receiver). Therefore, the bus should not be loaded to 100%. An error-passive node has an error count greater than 127 but no more than 255. An *error-active node* has an error count less than 127. There are two forms of error flags:

- *active-error flag.* This flag consists of six consecutive dominant bits.

- *passive-error flag.* This flag consists of six consecutive recessive bits unless it is overwritten by dominant bits from other nodes.

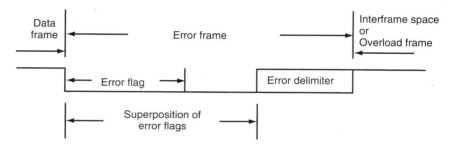

Figure 13.8 ■ Error frame

An error-active node signals an error condition by transmitting an active-error flag. The error flag's form violates the law of bit stuffing (to be discussed shortly) and applies to all fields from start-of-frame to CRC delimiter or destroys the fixed-form ACK field or end-of-frame field. As a consequence, all other nodes detect an error condition, and each starts to transmit an error flag. Therefore, the sequence of dominant bits, which actually can be monitored on the bus, results from a superposition of different error flags transmitted by individual nodes. The total length of this sequence varies between a minimum of 6 and a maximum of 12 bits.

An error-passive node signals an error condition by transmitting a passive-error flag. The error-passive node waits for six consecutive bits of equal polarity, beginning at the start of the passive-error flag. The passive-error flag is complete when these equal bits have been detected.

The *error delimiter* consists of eight recessive bits. After transmission of an error flag, each node sends recessive bits and monitors the bus until it detects a recessive bit. Then, it starts transmitting seven more recessive bits.

13.3.4 Overload Frame

The *overload frame* contains two bit fields: *overload flag* and *overload delimiter*. There are three different overload conditions that lead to the transmission of an overload frame:

1. The internal conditions of a receiver require a delay of the next data frame or remote frame.

2. At least one node detects a dominant bit during intermission.

3. A CAN node samples a dominant bit at the eighth bit (i.e., the last bit) of an error delimiter or overload delimiter. The error counters will not be incremented.

The format of an overload frame is shown in Figure 13.9. An overload frame resulting from Condition 1 is allowed to start only at the first bit time of an expected intermission, whereas an overload frame resulting from overload Conditions 2 and 3 starts one bit after detecting the dominant.

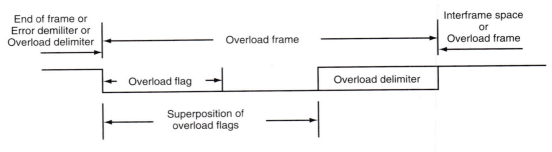
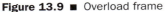

Figure 13.9 ■ Overload frame

No more than two overload frames may be generated to delay the next data frame or remote frame. The overload flag consists of six dominant bits. The format of an overload frame is similar to that of the active-error flag. The overload flag's form destroys the fixed form of the *intermission field*. As a consequence, all other nodes also detect an overload condition, and each starts to transmit an overload flag. In the event that there is a dominant bit detected during the third bit of *intermission* locally at some node, it will interpret this bit as the start-of-frame.

The overload delimiter consists of eight recessive bits. The overload delimiter has the same form as the error delimiter. After the transmission of an overload flag, the node monitors the bus until it detects a transition from a dominant to a recessive bit. At this point, every bus node has finished sending its overload flag, and all nodes start transmission of seven more recessive bits in coincidence.

13.3.5 Interframe Space

Data frames and remote frames are separated from preceding frames by a field called *interframe space*. In contrast, overload frames and error frames are not preceded by an interframe space, and multiple overload frames are not separated by an interframe space.

For nodes that are not error-passive or have been receivers of the previous message, the interframe space contains the bit fields of *intermission* and *bus idle*, as shown in Figure 13.10. The interframe space of an error-passive node consists of three subfields: *intermission, suspend transmission*, and *bus idle*, as shown in Figure 13.11.

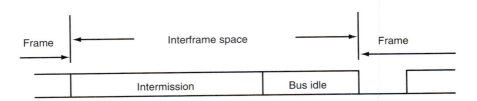

Figure 13.10 ■ Interframe space for non error-passive nodes or receiver of previous message

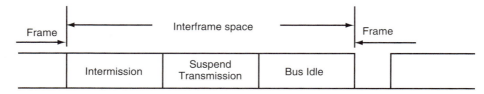

Figure 13.11 ■ Interframe space for error-passive nodes

The intermission subfield consists of three recessive bits. During intermission, no node is allowed to start transmission of the data frame or remote frame. The only action permitted is signaling of an overload condition.

The period of bus idle may be of arbitrary length. The bus is recognized to be free, and any node having something to transmit can access the bus. A message, pending during the transmission of another message, is started in the first bit following intermission. When the bus is idle, the detection of a dominant bit on the bus is interpreted as a start-of-frame.

After an error-passive node has transmitted a frame, it sends eight recessive bits following intermission before starting to transmit a new message or recognizing the bus as idle. If, meanwhile, a transmission (caused by another node) starts, the node will become the receiver of this message.

13.3.6 Message Filtering

A node uses filter(s) to decide whether to work on a specific message. Message filtering is applied to the whole identifier. A node can optionally implement mask registers that specify which bits in the identifier are examined with the filter.

If mask registers are implemented, every bit of the mask registers must be programmable; in other words, they can be enabled or disabled for message filtering. The length of the mask register can comprise the whole identifier or only part of it.

13.3.7 Message Validation

The point in time at which a message is taken to be valid is different for the transmitters and receivers of the message. The message is valid for the transmitter if there is no error until the end-of-frame. If a message is corrupted, retransmission will follow automatically and according to the rules of prioritization. In order to be able to compete for bus access with other messages, retransmission has to start as soon as the bus is idle. The message is valid for the receiver if there is no error until the last but one bit of the end-of-frame.

13.3.8 Bit Stream Encoding

The frame segments including start-of-frame, arbitration field, control field, data field, and CRC sequence are encoded by *bit stuffing*. Whenever a transmitter detects five consecutive bits of identical value in the bit stream to be transmitted, it automatically inserts a complementary bit in the actual transmitted bit stream.

The remaining bit fields of the data frame or remote frame (CRC delimiter, ACK field, and end-of-frame) are of fixed form and not stuffed. The error frame and overload frame are also of fixed form and are not encoded by the method of bit stuffing.

The bit stream in a message is encoded using the *non-return-to-zero* (NRZ) method. This means that during the total bit time, the generated bit level is either dominant or recessive.

13.4 Error Handling

There are five types of errors. These errors are not mutually exclusive.

13.4.1 Bit Error

A node that is sending a bit on the bus also monitors the bus. When the bit value monitored is different from the bit value being sent, the node interprets the situation as an error. There are two exceptions to this rule:

- A node that sends a recessive bit during the stuffed bit stream of the arbitration field or during the ACK slot detects a dominant bit.
- A transmitter that sends a passive-error flag detects a dominant bit.

13.4.2 Stuff Error

A stuff error is detected whenever six consecutive dominant or six consecutive recessive levels occur in a message field.

13.4.3 CRC Error

The CRC sequence consists of the result of the CRC calculation by the transmitter. The receiver calculates the CRC in the same way as the transmitter. A CRC error is detected if the calculated result is not the same as that received in the CRC sequence.

13.4.4 Form Error

A form error is detected when a fixed-form bit field contains one or more illegal bits. For a receiver, a dominant bit during the last bit of end-of-frame is not treated as a form error.

13.4.5 Acknowledgment Error

An acknowledgment error is detected whenever the transmitter does not monitor a dominant bit in the ACK slot.

13.4.6 Error Signaling

A node that detects an error condition signals the error by transmitting an error flag. An error-active node will transmit an active-error flag; an error-passive node will transmit a passive-error flag.

Whenever a node detects a bit error, a stuff error, a form error, or an acknowledgment error, it will start transmission of an error flag at the next bit time.

Whenever a CRC error is detected, transmission of an error flag will start at the bit following the ACK delimiter, unless an error flag for another error condition has already been started.

13.5 Fault Confinement

13.5.1 CAN Node Status

A node in error may be in one of three states: error-active, error-passive, or bus-off. An error-active node can normally take part in bus communication and sends an active-error flag when an error has been detected. An error-passive node must not send an active-error flag. It takes part in bus communication, but when an error has been detected, only a passive-error flag is sent. After a transmission, an error-passive node will wait before initiating further transmission. A bus-off node is not allowed to have any influence on the bus.

13.5.2 Error Counts

Two counts are implemented in every bus node to facilitate fault confinement: transmit error count and receive error count. These two counts are updated according to the following 12 rules:

1. When a receiver detects an error, the receive error count will be increased by 1, except when the detected error was a bit error during the sending of an active-error flag or an overload flag.

2. When a receiver detects a dominant bit as the first bit after sending an error flag, the receive error count will be increased by 8.

3. When a transmitter sends an error flag, the transmit error count is increased by 8. There are two exceptions to this rule. The first exception is when the transmitter is error-passive and detects an acknowledgment error because of not detecting a dominant ACK and does not detect a dominant bit while sending its passive error flag. The second exception is when the transmitter sends an error flag because a stuff error occurred during arbitration whereby the stuff bit is located before the RTR bit and should have been recessive and has been sent as recessive but monitored as dominant. The transmit error count is not changed under these two situations.

4. When an error-active transmitter detects a bit error while sending an active-error flag or an overload flag, the transmit error count is increased by 8.

5. When an error-active receiver detects a bit error while sending an active-error flag or an overload flag, the receive error count is increased by 8.

6. Any node tolerates up to seven consecutive dominant bits after sending an active error flag, a passive error flag, or an overload flag. After detecting the 14th consecutive dominant bit (in case of an active error flag or an overload flag) or after detecting the eighth consecutive dominant bit following a passive error flag, and after each sequence of additional eight consecutive dominant bits, every transmitter increases its transmit error count by 8, and every receiver increases its receive error count by 8.

7. After the successful transmission of a message (getting ACK and no error until end-of-frame is finished), the transmit error count is decremented by 1, unless it was already 0.

8. After the successful reception of a message (reception without error up to the ACK slot and the successful sending of the ACK bit), the receive error count is decremented by 1, if it was between 1 and 127. If the receive error count was 0, it stays 0, and if it was greater than 127, it will be set to a value between 119 and 127.

9. A node is error-passive when the transmit error count equals or exceeds 128 or when the receive error count equals or exceeds 128. An error condition letting a node become error-passive causes the node to send an active-error flag.

10. A node is bus-off when the transmit error count is greater than or equal to 256.

11. An error-passive node becomes error-active again when both the transmit error count and the receive error count are less than or equal to 127.

12. A node that is bus-off is permitted to become error-active (no longer bus-off) with its error counters both set to 0 after 128 occurrences of 11 consecutive recessive bits have been monitored on the bus.

An error count value greater than roughly 96 indicates a heavily disturbed bus. It may be advantageous to provide the means to test for this condition. If during system start-up only one node is online and if this node transmits some message, it will get no acknowledgment, detect an error, and repeat the message. It can become error-passive but not bus-off for this reason.

13.6 CAN Message Bit Timing

The setting of a bit time in a CAN system must allow a bit sent out by the transmitter to reach the far end of the CAN bus and allow the receiver to send back acknowledgment and reach the transmitter. In a CAN environment, the *nominal bit rate* is defined as the number of bits transmitted per second in the absence of resynchronization by an ideal transmitter.

13.6.1 Nominal Bit Time

The inverse of nominal bit rate is the *nominal bit time.* A nominal bit time can be divided into four nonoverlapping time segments as shown in Figure 13.12.

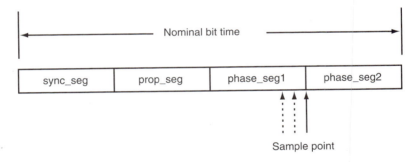

Figure 13.12 ■ Nominal bit time

The **sync_seg** segment is used to synchronize the various nodes on the bus. An edge is expected to lie within this segment. The **prop_seg** segment is used to compensate for the physical delay times within the network. It is twice the sum of the signal's propagation time on the bus line, the input comparator delay, and the output driver delay. The **phase_seg1** and **phase_seg2** segments are used to compensate for edge phase errors. These segments can be lengthened or shortened by synchronization. The *sample point* is the point in time in which the bus level is read and interpreted as the value of that respective bit. The sample point is located at the end of **phase_seg1.** A CAN controller may implement the three-samples-per-bit option in which the majority function is used to determine the bit value. Each sample is separated from the next sample by half a time quanta (CAN clock cycle). The *information processing time* is the time segment starting with the sample point reserved for calculation of the sample bit(s) level. The segments contained in a nominal bit time are represented in the unit of *time quantum.* The time quantum (TQ) is a fixed unit of time that can be derived from the oscillator period (T_{OSC}). TQ is expressed as a multiple of a *minimum time quantum.* This multiple is a programmable prescale factor. Thus, the time quantum can have the length of

time quantum = m × minimum time quantum

where m is the value of the prescaler.

13.6.2 Length of Time Segments

The segments of a nominal bit time can be expressed in the unit of time quantum as follows:

- **sync_seg** is 1 time quantum long.
- **prop_seg** is programmable to be 1, 2, . . ., 8 time quanta long.
- **phase_seg1** is programmable to be 1, 2, . . ., 8 time quanta long.
- **phase_seg2** is the maximum of **phase_seg1** and information processing time and hence will be programmable from 2 to 8 TQ.
- The information processing time is fixed at 2 TQ for the PIC18 CAN module.

The total number of time quanta in a bit time must be programmable over a range of at least 8 to 25.

13.7 Synchronization Issue

All CAN nodes must be synchronized while receiving a transmission; that is, the beginning of each received bit must occur during each node's **sync_seg** segment. This is achieved by synchronization. Synchronization is required because of phase errors between nodes, which may arise because of nodes having slightly different oscillator frequencies, or because of changes in propagation delay when a different node starts transmitting.

Two types of synchronization are defined: *hard synchronization* and *resynchronization.* Hard synchronization is performed only at the beginning of a message frame, when each CAN node aligns the **sync_seg** of its current bit time to the recessive-to-dominant edge of the transmitted start-of-frame. After a hard synchronization, the internal bit time is restarted with **sync_seg.** Resynchronization is subsequently performed during the remainder of the message frame whenever a change of bit value from recessive to dominant occurs outside the expected **sync_seg** segment. Resynchronization is achieved by implementing a *digital phase lock loop* (DPLL) function that compares the actual position of a recessive-to-dominant edge on the bus to the position of the expected edge.

13.7.1 Resynchronization Jump Width

There are three possibilities of the occurrence of the incoming recessive-to-dominant edge:

1. *After the* sync_seg *segment but before the sample point.* This situation is interpreted as a *late edge.* The node will attempt to resynchronize to the bit stream by increasing the duration of its **phase_seg1** segment of the current bit by the number of time quanta by which the edge was late, up to the resynchronization jump width limit.

2. *After the sample point but before the* **sync_seg** *segment of the next bit.* This situation is interpreted as an *early bit.* The node will now attempt to resynchronize to the bit stream by decreasing the duration of its **phase_seg2** segment of the current bit by the number of time quanta by which the edge was early, up to the resynchronization jump width limit. Effectively, the **sync_seg** segment of the next bit begins immediately.

3. *Within the **sync_seg** segment of the current bit time.* This is interpreted as no synchronization error.

As a result of resynchronization, **phase_seg1** may be lengthened, or **phase_seg2** may be shortened. The amount by which the phase buffer segments may be altered may not be greater than the *resynchronization jump width*, which is programmable to be between 1 and the smaller of 4 and **phase_seg1** time quanta.

Clocking information may be derived from transitions from one bit value to the other. The property that only a fixed maximum number of successive bits have the same value provides the possibility of resynchronizing a bus node to the bit stream during a frame.

The maximum length between two transitions that can be used for resynchronization is 29 bit times.

13.7.2 Phase Error of an Edge

The *phase error* of an edge is given by the position of the edge relative to **sync_seg**, measured in time quanta. The sign of phase error is defined as follows:

$e < 0$ if the edge lies after the sample point of the previous bit

$e = 0$ if the edge lies within **sync_seg**

$e > 0$ if the edge lies before the sample point

13.7.3 Synchronization Rules

Hard synchronization and resynchronization are the two forms of synchronization. They obey the following rules:

- Only one synchronization within one bit time is allowed.

- An edge will be used for synchronization only if the value detected at the previous sample point (previous read bus value) differs from the bus value immediately after the edge.

- Hard synchronization is performed whenever there is a recessive-to-dominant edge during bus idle.

- All other recessive-to-dominant edges (and optionally dominant-to-recessive edges in the case of low bit rates) fulfilling Rules 1 and 2 will be used for resynchronization with the exception that a node transmitting a dominant bit will not perform a resynchronization as a result of a recessive-to-dominant edge with a positive phase error, if only recessive-to-dominant edges are used for resynchronization.

13.8 PIC18 CAN Module

Among all PIC18 devices, the PIC18C658/858 devices, the PIC18F248/258/448/458 devices, and the PIC18F6585/8585/6680/8680 devices have an on-chip CAN module. Other PIC18 devices do not have on-chip CAN module. The CAN module. The CAN modules in the PIC18C658/858 devices and the PIC18F248/258/448/458 devices are identical. However, the CAN module in the PIC18F6585/8585/6680/8680 devices has been enhanced by adding mode 1 and mode 2. The mode 0 (also called *legacy* mode) of this group of devices is identical to that of the first two groups of devices.

The CAN module in the PIC18Cxx8 microcontroller and the PIC18Fxx8 microcontroller uses two pins (CANTX and CANRX) to transmit and receive data, whereas the PIC18F6585/8585/6680/8680 devices use three pins (CANTX1, CANTX2, and CANRX) to interface with the CAN bus. The CANTX1 and CANRX pins serve the same function as the CANTX and CANRX pins of earlier CAN modules. The CANTX2 pin makes available the CAN clock signal and hence can help debug the CAN baud rate problem. This pin can also be configured to carry the complement of the CANTX1 signal, which may be useful to drive those CAN transceivers that require this signal.

For other PIC18 devices without an on-chip CAN module, a CAN controller chip, such as the MCP2510, can be used to interface with the CAN bus. Most CAN controllers do not have enough driving capability and hence require a dedicated CAN bus transceiver chip, such as the MCP2551 (from Microchip) or the PCA82C250 (from Philips), to interface with the CAN bus. A typical CAN system that includes one or more PIC18 devices is shown in Figure 13.13.

Since the enhanced CAN module is a superset of the legacy CAN module, the discussion in this chapter focuses on the ECAN module contained in the PIC18F6585/8585/6680/8680 devices.

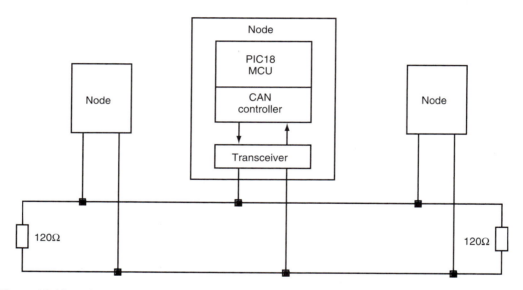

Figure 13.13 ■ A typical CAN bus system that comprises PIC18 devices

13.9 CAN Modes of Operation

The CAN module in the PIC18F6585/8585/6680/8680 device has six main modes of operation:

- Configuration mode
- Disable mode
- Normal operation mode
- Listen-only mode
- Loopback mode
- Error recognition mode

All modes except the error recognition mode are entered by programming the CANCON register. The contents of the CANCON register are shown in Figure 13.14. The error recognition mode is entered by programming the RXM bits of the receive buffer control registers RXB0CON and RXB1CON.

	7	6	5	4	3	2	1	0
Mode 0	REQOP2	REQOP1	REQOP0	ABAT	WIN2	WIN1	WIN0	--

	7	6	5	4	3	2	1	0
Mode 1	REQOP2	REQOP1	REQOP0	ABAT	--	--	--	--

	7	6	5	4	3	2	1	0
Mode 2	REQOP2	REQOP1	REQOP0	ABAT	FP3	FP2	FP1	FP0

REQOP2: REQOP0: Request CAN operation mode bits
    000 = Request normal mode
    001 = Request disable mode
    010 = Request loopback mode
    011 = Request listen only mode
    1xx = Request configuration mode
ABAT: Abort all pending transmission bits
    0 = Transmission proceeding as normal
    1 = Abort all pending transmission (in all transmit buffers)
WIN2:WIN0: Window address bits (Mode 0)
    This field selects which of the CAN buffers to switch into the access bank area
    000, 001, 110, 111 = Receive buffer 0
    010 = Transmit buffer 2
    011 = Transmit buffer 1
    100 = Transmit buffer 0
    101 = Receive buffer 1
FP3:FP0 FIFO read pointer bits (Mode 2)
    These bits point to the message buffer to be read
    0111:0000 = Message buffer to be read
    1111:1000 = Reserved

Figure 13.14 ■ PIC18F6585/8585/6680/8680 CANCON control register (redraw with permission of Microchip)

Only a subset of the CAN registers is in the access bank. The WIN2 . . . WIN0 bits select the CAN buffer to switch into the access bank. This allows access to the buffer registers from any data memory bank. After a frame has caused an interrupt, the ICODE3:ICODE0 bits can be copied to the WIN3:WIN0 bits to select the correct buffer.

When changing modes, the mode will not actually change until all pending transmissions are complete. Because of this, the user must verify that the device has actually changed into the requested mode before any further operations are executed. To verify whether a specific mode has been entered, the user must check the OPMODE bits of the CANSTAT register. The contents of this register are shown in Figure 13.15.

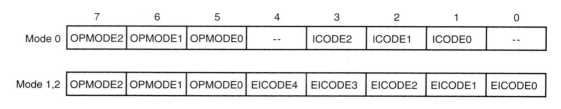

	7	6	5	4	3	2	1	0
Mode 0	OPMODE2	OPMODE1	OPMODE0	--	ICODE2	ICODE1	ICODE0	--

	7	6	5	4	3	2	1	0
Mode 1,2	OPMODE2	OPMODE1	OPMODE0	EICODE4	EICODE3	EICODE2	EICODE1	EICODE0

OPMODE2:OPMODE0: Operation mode status bits

000 = Normal mode

001 = Disable/SLEEP mode

010 = Loopback mode

011 = Listen only mode

100 = Configuration mode

101:111 = reserved

ICODE2:ICODE0 & EICODE4:EICODE0: Interrupt code bits

	Mode 0	Mode 1	Mode 2
No interrupt	00000	00000	00000
Error interrupt	00010	00010	00010
TXB2 interrupt	00100	00100	00100
TXB1 interrupt	00110	00110	00110
TXB0 interrupt	01000	01000	01000
RXB1 interrupt	01010	-----	-----
RXB0 interrupt	01100	-----	-----
Wake-up interrupt	01110	01110	01110
RXB0 interrupt	-----	10000	10000
RXB1 interrupt	-----	10001	10000
RX/TX B0 interrupt	-----	10010	10010
RX/TX B1 interrupt	-----	10011	10011[1]
RX/TX B2 interrupt	-----	10100	10100[1]
RX/TX B3 interrupt	-----	10101	10101[1]
RX/TX B4 interrupt	-----	10110	10110[1]
RX/TX B5 interrupt	-----	10111	10111[1]

Note 1. If buffer is configured as receiver, EICODE bits will contain
'10000' upon interrupt

Figure 13.15 ■ PIC18F6585/8585/6680/8680 CANSTAT status register (redraw with permission of Microchip)

When an interrupt occurs, a prioritized coded interrupt value will be present in these bits. This code indicates the source of the interrupt. By copying ICODE2:ICODE0 to WIN3:WIN0 (mode 0) or EICODE4:EICODE0 to EWIN4:EWIN0 (mode 1 and 2), it is possible to select the correct buffer to map into the *access bank* area.

Example 13.1

▼

Write an instruction sequence to set the CAN module to configuration mode.

Solution: The following instruction sequence will configure the CAN module in configuration mode:

```
           movlw   0x80
           movwf   CANCON      ; set to configuration mode
config_wt  movf    CANSTAT,W   ; read current mode state
           andlw   0x80
           bz      config_wt   ; if not configuration mode, branch
           . . .
```

▲

13.9.1 Configuration Mode

The CAN module has to be initialized before it can operate properly. The initialization of CAN module is only possible in the configuration mode. When the configuration mode is entered, the user can write into configuration registers, the acceptance registers, and the acceptance mask registers. After the CAN module is initialized properly, the user will want to clear the REQOP2:REQOP0 bits (of the CANCON register) to 0s in order to return to normal operation mode.

As long as a transmission or reception is taking place, the CAN module is not allowed to enter the configuration mode. This locking mechanism protects the following registers:

- Configuration registers
- Functional mode selection registers
- Bit timing registers
- Identifier acceptance filter registers
- Identifier acceptance mask registers
- Filter and mask control registers
- Mask selection registers

In configuration mode, the module will neither transmit nor receive. The error counters are cleared, and the interrupt flags remain unchanged.

13.9.2 Disable Mode

In disable mode, the CAN module will not transmit or receive. The module can still set the WAKIF flag (in PIR3 register) because of bus activity; however, any pending interrupts will remain, and the error counters will retain their value.

The disable mode causes the module internal clock to stop unless the module is active. If the module is active, the module will wait for 11 recessive bits on the CAN bus, detect that condition as an IDLE bus, and then accept the module disable command.

The WAKIF interrupt (wake-up interrupt) is the only CAN module interrupt that is still active in the module disable mode. If the WAKDIS bit (in BRGCON3 register) is cleared and the WAKIE bit (in PIE3 register) is set, the processor will receive an interrupt whenever the module detects a recessive-to-dominant transition. On wake-up, the CAN module will automatically be set to the previous mode of operation. For example, if the module was switched from normal to disable mode on bus activity wake-up, the module will automatically enter normal mode, and the first message that caused the module to wake up is lost. The module will not generate any error frame. The user software must detect this condition and make sure that retransmission is requested. If the processor receives a wake-up interrupt while it is sleeping, more than one message may get lost. The actual number of messages lost would depend on the processor oscillator start-up time and incoming message bit rate.

13.9.3 Normal Mode

This is the standard operation mode of the PIC18F6585/8585/6680/8680 devices. In this mode, the device actively monitors all bus messages and generates acknowledge bits, error frames, and so on. This is also the only mode in which the PIC18F6585/8585/6680/8680 devices will transmit messages over the CAN bus.

13.9.4 Listen-Only Mode

Listen-only mode allows the PIC18F6585/8585/6680/8680 devices to receive all messages, including messages with errors. This mode can be used for bus monitoring or for detecting the baud rate in *hot plugging* applications. For auto baud rate detection, there must be at least two other nodes communicating with each other. The baud rate can be detected empirically by testing different values until valid messages are received.

No messages, including error flags or acknowledgment, will be transmitted in the listen-only mode. However, the CAN module can use filters and masks to allow only particular messages to be loaded into the receive registers, or the filter masks can be set to all 0s to allow a message with any identifier to pass. The error counters are reset and deactivated in this state.

13.9.5 Loopback Mode

This mode allows internal transmission of messages from the transmit buffers to the receive buffers without actually transmitting messages on the CAN bus. This mode is used mainly in systems development and testing. In this mode, the ACK bit is ignored, and the device will allow incoming messages from itself, just as if they were coming from another node. In this mode, the CANTX pin will revert to port I/O. The filters and masks can be used to allow particular messages to be loaded into receive registers. The masks can be set to all 0s to provide a mode that accepts all messages.

13.9.6 Error Recognition Mode

In this mode, the CAN module can ignore all errors and receive any message. In functional mode 0, the error recognition mode is activated by setting the RXM<1:0> bit in the RXBnCON registers to 11. In this mode, the data that is in the message assembly buffer until the error time is copied into the receive buffer and can be read via the CPU interface.

13.10 CAN Module Registers

The CAN module has many control and data registers. These registers can be divided into seven groups:

- Control and status registers
- Dedicated transmit buffer registers
- Dedicated receive buffer registers
- Programmable TX/RX and automatic RTR buffers

- Baud rate control registers
- I/O control registers
- Interrupt status and control registers

These registers will be discussed wherever it is appropriate. In addition to the standard CAN control register discussed earlier, the ECAN module of the PIC18F6585/8585/6680/8680 devices also incorporates the ECANCON register to add additional features to the standard CAN module. The contents of this register are shown in Figure 13.16. This register allows the

7	6	5	4	3	2	1	0
MDSEL1	MDSEL0	FIFOWM	EWIN4	EWIN3	EWIN2	EWIN1	EWIN0

MDSEL1: MDSEL0: Mode select bits

    00 = Legacy mode

    01 = Enhanced legacy mode (mode 1)

    10 = Enhanced FIFO mode (mode 2)

    11 = Reserve

FIFOWM: FIFO high water mark bit (used in mode 2 only)

    0 = Will cause FIFO interrupt when four receive buffers remain

    1 = Will cause FIFO interrupt when one receive buffer remains (FIFO length

        of 4 or less will cause this bits to be set)

EWIN4:EWIN0: Enhanced window address bits (mode 1 and 2 only)

    00000 = Acceptance filters 0, 1, 2, and BRGCON3, 2

    00001 = Acceptance filters 3, 4, 5, and BRGCON1, CIOCON

    00010 = Acceptance filters mask, error and interrupt control

    00011 = Transmit buffer 0

    00100 = Transmit buffer 1

    00101 = Transmit buffer 2

    00110 = Acceptance filters 6, 7, 8

    00111 = Acceptance filters 9, 10, 11

    01000 = Acceptance filters 12, 13, 14

    01001 = Acceptance filters 15

    01010 - 01110 = reserved

    01111 = RXINT0, RXINT1

    10000 = Receive buffer 0

    10001 = Receive buffer 1

    10010 = TX/RX buffer 0

    10011 = TX/RX buffer 1

    10100 = TX/RX buffer 2

    10101 = TX/RX buffer 3

    10110 = TX/RX buffer 4

    10111 = TX/RX buffer 5

    11000 - 11111 = reserved

Figure 13.16 ■ Enhanced CAN control register ECANCON (redraw with permission of Microchip)

user to select the additional functional mode—FIFO watermark interrupt—and indicate the filters that caused the match.

The ECAN module provides additional communication status flags in the COMSTAT register shown in Figure 13.17.

	7	6	5	4	3	2	1	0
Mode 0	RXB0OVFL	RXB1OVFL	TXBO	TXBP	RXBP	TXWARN	RXWARN	EWARN

	7	6	5	4	3	2	1	0
Mode 1	--	RXBnVFL	TXBO	TXBP	RXBP	TXWARN	RXWARN	EWARN

	7	6	5	4	3	2	1	0
Mode 2	FIFOEMPTY	RXBnVFL	TXBO	TXBP	RXBP	TXWARN	RXWARN	EWARN

RXB0OVFL: Receive buffer 0 overflow bit (mode 0 only)
   0 = Receive buffer 0 has not overflowed
   1 = Receive buffer 0 has overflowed
$\overline{\text{FIFOEMPTY}}$: FIFO not empty bit (mode 2 only)
   0 = Receive FIFO is empty
   1 = Receive FIFO is not empty
RXB1OVFL: Receive buffer 1 overflow bit (mode 0 only)
   0 = Receive buffer 1 has not overflowed
   1 = Receive buffer 1 has overflowed
RXBnOVFL: (mode 1 and 2)
   0 = Receive buffer has not overflowed
   1 = Receive buffer has overflowed
TXBO: Transmitter bus off bit
   0 = Transmitter error counter ≤ 255
   1 = Transmitter error counter > 255
TXBP: Transmitter bus passive bit
   0 = Transmitter error counter ≤ 127
   1 = Transmitter error counter > 127
RXBP: Receiver bus passive bit
   0 = Receiver error counter ≤ 127
   1 = Receiver error counter > 127
TXWARN: Transmitter warning bit
   0 = Transmitter error ≤ 95
   1 = $95 <$ Transmitter error ≤ 127
RXWARN: Receiver warning bit
   0 = Receiver error ≤ 95
   1 = $95 <$ Receiver error ≤ 127
EWARN: Error warning bit
   0 = Neither the RXWARN nor the TXWARN bit is set
   1 = Either the RXWARN or the TXWARN bit is set

Figure 13.17 ■ PIC18F6585/8585/6680/8680 COMSTAT status register (redraw with permission of Microchip)

13.11 CAN Module Functional Modes

In additional to the CAN modes of operation, the ECAN module offers a total of three functional modes. These modes are referred to as mode 0, mode 1, and mode 2.

13.11.1 Mode 0—Legacy Mode

Mode 0 is designed to be fully compatible with CAN modules used in PIC18CXX8 and PIC18FXX8 devices. This is the default mode of operation on all reset conditions. As a result, programs written for the PIC18CXX8 and PIC18FXX8 can be run on the ECAN module without any changes.

The following resources are available in mode 0:

- Three transmit buffers: TXB0, TXB1, and TXB2
- Two receive buffers: RXB0 and RXB1
- Two acceptance masks, one for each receive buffer: RXM0 and RXM1
- Six acceptance filters, two for RXB0 and four for RXB1: RXF0, RXF1, RXF2, RXF3, RXF4, and RXF5

13.11.2 Mode 1—Enhanced Legacy Mode

Mode 1 is similar to mode 0 with the exception that more resources are available in mode 1. There are 16 acceptance filters and two acceptance mask registers. Acceptance filter 15 can be used as either an acceptance filter or an acceptance mask register. In addition to three transmit and two receive buffers, there are six more message buffers. One or more of these additional buffers can be programmed as transmit or receive buffers. These additional buffers can also be programmed to automatically handle RTR messages.

Fourteen out of 16 acceptance register filters can be dynamically associated to any receive buffer and acceptance mask register. One can use this capability to associate more than one filter to any one buffer.

When a receive buffer is programmed to use standard identifier messages, part of the full acceptance filter register can be used as data byte filter. The length of data byte filter is programmable from 0 to 18 bits. This functionality can simplify the implementation of high-level protocols, such as DeviceNet. The discussion of DeviceNet is outside the scope of this text.

The following resources are available in mode 1:

- Three transmit buffers: TXB0, TXB1, and TXB2
- Two receive buffers: RXB0 and RXB1
- Six buffers programmable as TX or RX: B0 to B5
- Automatic remote request (RTR) handling on B0 to B5
- Sixteen dynamically assigned acceptance filters: RXF0 to RXF15
- Two dedicated acceptance mask registers and one RXF15 programmable as a third: RXF0, RXF1, and RXF15
- Programmable data filter on standard identifier messages: SDFLC (useful for DeviceNet protocol)

13.11.3 Mode 2—Enhanced FIFO Mode

In this mode, two or more receive buffers are used to form the receive FIFO (first in, first out) buffer. There is no one-to-one relationship between the receive buffer and acceptance filter registers. Any filter that is enabled and linked to any FIFO receive buffer can generate acceptance and cause FIFO to be updated.

FIFO length is user programmable, from two to eight buffers deep. FIFO length is determined by the very first programmable buffer that is configured as a transmit buffer. For example, if buffer 2 (B2) is programmed as a transmit buffer, FIFO consists of RXB0, RXB1, B0, and B1, creating a FIFO length of four. If all programmable buffers are configured as receive buffers, FIFO will have the maximum length of eight.

The following resources are available in mode 2:

- Three transmit buffers: TXB0, TXB1, and TXB2
- Two receive buffers: RXB0 and RXB1
- Six buffers programmable as TX or RX; receive buffers form a FIFO: B0 to B5
- Automatic RTR handling on B0 to B5
- Sixteen acceptance filters: RXF0 to RXF15
- Two dedicated acceptance mask registers: RXM0 and RXM1; RXF15 programmable as a third mask
- Programmable data filter on standard identifier messages: SDFLC (useful for DeviceNet protocol)

13.12 CAN Message Buffers

The ECAN module has four sets of buffers:

- Dedicated transmit buffers
- Dedicated receive buffers
- Programmable transmit/receive buffers
- Programmable automatic RTR buffers

13.12.1 Dedicated Transmit Buffers

The PIC18 devices implement three dedicated transmit buffers: TXB0, TXB1, and TXB2. Each of these buffers occupies 14 bytes of SRAM and is mapped into the SFR memory map. CAN module in mode 0 can access only these three transmit buffers. Mode 1 and mode 2 can also access other additional buffers.

Each transmit buffer contains one control register (TXBnCON), four identifier registers (TXBnSIDL, TXBnSIDH, TXBnEIDL, and TXBnEIDH), one data length count register (TXBnDLC) and eight data byte registers (TXBnDm). The subscript n can be from 0 to 2, whereas the subscript m can be from 0 to 7.

The contents of the TXBnCON register are shown in Figure 13.18. The user can program the priority of a transmit buffer via this register. All transmission status flags are recorded in this register.

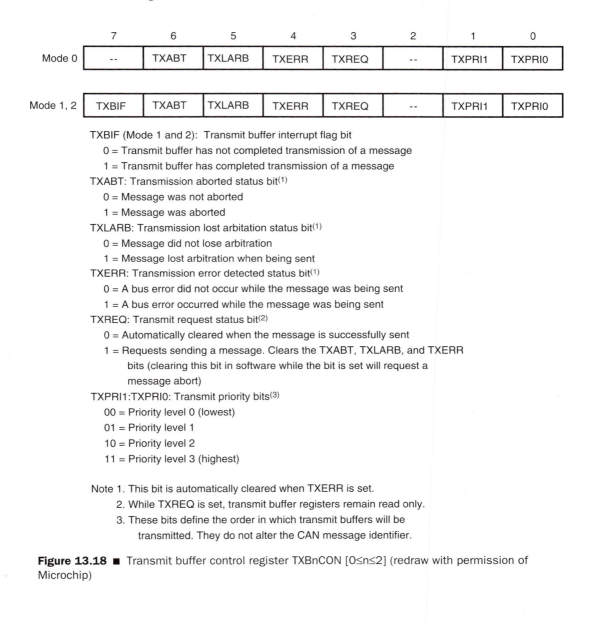

TXBIF (Mode 1 and 2): Transmit buffer interrupt flag bit

    0 = Transmit buffer has not completed transmission of a message

    1 = Transmit buffer has completed transmission of a message

TXABT: Transmission aborted status bit[1]

    0 = Message was not aborted

    1 = Message was aborted

TXLARB: Transmission lost arbitation status bit[1]

    0 = Message did not lose arbitration

    1 = Message lost arbitration when being sent

TXERR: Transmission error detected status bit[1]

    0 = A bus error did not occur while the message was being sent

    1 = A bus error occurred while the message was being sent

TXREQ: Transmit request status bit[2]

    0 = Automatically cleared when the message is successfully sent

    1 = Requests sending a message. Clears the TXABT, TXLARB, and TXERR
        bits (clearing this bit in software while the bit is set will request a
        message abort)

TXPRI1:TXPRI0: Transmit priority bits[3]

    00 = Priority level 0 (lowest)

    01 = Priority level 1

    10 = Priority level 2

    11 = Priority level 3 (highest)

Note 1. This bit is automatically cleared when TXERR is set.

      2. While TXREQ is set, transmit buffer registers remain read only.

      3. These bits define the order in which transmit buffers will be
         transmitted. They do not alter the CAN message identifier.

Figure 13.18 ■ Transmit buffer control register TXBnCON [0≤n≤2] (redraw with permission of Microchip)

A CAN node uses the identifier contained in the message to arbitrate for the use for the bus. To support both the standard and the extended formats, four registers are used to hold the identifier of a transmit buffer: TXBnSIDH, TXBnSIDL, TXBnEIDH, and TXBnEIDL. The contents of these four registers are shown in Figure 13.19.

	7	6	5	4	3	2	1	0
TXBnSIDH	SID10	SID9	SID8	SID7	SID6	SID5	SID4	SID3
TXBnSIDL	SID2	SID1	SID0	--	EXIDE	--	EID17	EID16
TXBnEIDH	EID15	EID14	EID13	EID12	EID11	EID10	EID9	EID8
TXBnEIDL	EID7	EID6	EID5	EID4	EID3	EID2	EID1	EID0

SID10..SID0: Standard identifier bits
EXIDE: Extended identifier enable bit
    0 = Message will transmit standard ID, EID17:EID0 are ignored
    1 = Message will transmit extended ID, SID10:SID0 becomes EID28:EID18
EID17:EID0: Extended identifier bits

Figure 13.19 ■ Transmit buffer identifier registers (redraw with permission of Microchip)

The *transmit buffer data length code register* specifies the number of bytes contained in a transmit buffer for a specific message. The contents of this register are shown in Figure 13.20.

7	6	5	4	3	2	1	0
--	TXRTR	--	--	DLC3	DLC2	DLC1	DLC0

TXRTR: Transmit remote frame transmission request bit

    0 = Transmit message will have RTR bit cleared

    1 = Transmit message will have RTR bit set

DLC3:DLC0: Data length code bits

    0000 = Data length is 0 bytes

    0001 = Data length is 1 byte

    0010 = Data length is 2 bytes

    0011 = Data length is 3 bytes

    0100 = Data length is 4 bytes

    0101 = Data length is 5 bytes

    0110 = Data length is 6 bytes

    0111 = Data length is 7 bytes

    1000 = Data length is 8 bytes

    others = unused

Figure 13.20 ■ TXBnDLC Transmit buffer n data length code register [$0 \leq n \leq 2$] (redraw with permission of Microchip)

13.12.2 Dedicated Receive Buffers

The PIC18 devices provide two dedicated receive buffers: RXB0 and RXB1. Each of these buffers consists of 14 bytes of SRAM and is mapped into SFR memory map. These are the only receive buffers available in mode 0. Mode 1 and mode 2 may access these and other additional buffers.

Each receive buffer contains one control register (RXBnCON), four identifier registers (RXBnSIDL, RXBnSIDH, RXBnEIDL, and RXBnEIDH), one data length count register (RXBnDLC) and eight data byte registers (RXBnDm).

There is also a separate message assembly buffer (MAB), which acts as an additional receive buffer. MAB is always committed to receiving the next message from the bus and is not directly accessible to user software. A message is transferred to appropriate receive buffers only if the corresponding acceptance filter criteria is met.

The contents of the receive buffer 0 and receive buffer 1 control registers are shown in Figures 13.21a and 13.21b. The RXFUL bit must be cleared in software after the receive buffer is read. No new message will be loaded into the receive buffer if this bit is 1. The lowest five bits (FILHIT4:FILHIT0) of the RXB0CON register tell the user which filter caused the message to be loaded into the dedicated received buffer (in modes 1 and 2).

Receive identifier registers are used to hold the identifier of the incoming message. The contents of these four receive identifier registers are shown in Figure 13.22.

The receive buffer data length code register (shown in Figure 13.23) specifies the number of bytes received in the receive buffer. The RXRTR bit indicates whether the received message is a remote transfer request.

The user application can use either interrupts or the polling method to find out if a message has been received. The following instruction sequence is a skeleton of a receive routine that uses the polling method:

```
rx_can      btfss    RXB0CON,RXFUL,A    ; receive buffer RXB0 is in access bank
            bra      done               ; no message, so return
            btfss    RXB0SIDL,EXID,A    ; Is this extended identifier?
            bra      std_msg            ; read all 29 bits of extended identifier
            . . .                       ; process extended identifier
std_msg     . . .                       ; process standard identifier
            lfsr     FSR0,can_rx_buf    ; use FSR0 as a pointer to the buffer
            lfsr     FSR1,0xF66         ; point to RXB0D0
            movf     RXB0DLC,W,A        ; save byte count in WREG
            bz       done               ; no data to copy
cpy_loop    movff    POSTINC1,POSTINC0  ; copy one byte
            decfsz   WREG,W,A           ; decrement the loop count
            bra      cpy_loop           ; continue
            bcf      RXB0CON,RXFUL      ; allow CAN to load new messages
done        return
```

13.12.3 Programmable Transmit/Receive Buffers

Although the CAN protocol was designed for applications that involved the transmission of a relatively small amount of information, the dedicated transmit buffers and receive buffers may still turn out to be inadequate under some situations. The programmable transmit/receive buffers are provided to remedy this problem.

The ECAN module implements six new buffers: B0 to B5. Each of these buffers can be programmed as either a transmit or a receive buffer. These buffers are available only in mode 1 and mode 2. Like dedicated transmit and receive buffers, each of these programmable buffers occupies 14 bytes and is mapped into SFR memory map.

Each buffer contains one control register (BnCON), four identifier registers (BnSIDL, BnSIDH, BnEIDL, and BnEIDH), one data length count register (BnDLC), and eight data byte registers (BnDm). Each of these registers contains two sets of control bits. Depending on whether the buffer is configured as transmit or receive, one would use the corresponding control bit set. By default, all buffers are configured as receive buffers. Each buffer can be individually configured as a transmit or a receive buffer by setting the corresponding TXENn bit in the BSEL0 register.

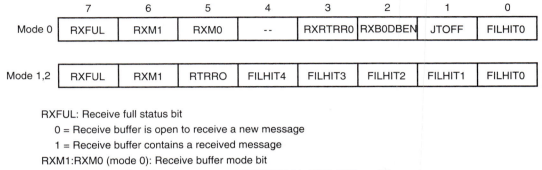

RXFUL: Receive full status bit

    0 = Receive buffer is open to receive a new message

    1 = Receive buffer contains a received message

RXM1:RXM0 (mode 0): Receive buffer mode bit

    00 = Receive all valid messages as per EXIDEN bit in RXFnSIDL register

    01 = Receive only valid messages with standard identifier (EXIDEN bit = 0)

    10 = Receive only valid messages with extended identifier (EXIDEN bit = 1)

    11 = Receive all messsages (including those with errors); filter critera ignored

RXM1 (mode 1, 2)

    0 = Receive all valid messages as per acceptance filters

    1 = Receive all messages (including those with errors); acceptance filters ignored

RTRRO: Remote transmission request bit for received message (read only)

    0 = A remote transmission request is not received

    1 = A remote transmission request is received

RXRTRRO: Remote transmission request bit for received message (read only)

    0 = A remote transmission request is not received

    1 = A remote transmission request is received

RXB0DBEN: Receive buffer 0 double-buffer enable bit

    0 = No receive buffer 0 overflow to receive buffer 1

    1 = Receive buffer 0 overflow will write to receive buffer 1

JTOFF: Jump table offset bit

    0 = Allows jump table offset between 1 and 0

    1 = Allows jump table offset between 6 and 7

FILHIT0 (mode 0): Filter hit bit 0

    This bit indicates which acceptance filter enabled the message reception into receive buffer 0.

    0 = Acceptance filter 0 (RXF0)

    1 = Acceptance filter 1 (RXF1)

FILHIT4:FILFIT0: Filter hit bits 4 to 0

    These five bits indicate which acceptance filter enabled the message reception into this receive buffer.

    00000 = Acceptance filter 0 (RXF0)

    00001 = Acceptance filter 1 (RXF1)

    ....

    01111 = Acceptance filter 15 (RXF15)

Figure 13.21a ■ PIC18 Receive buffer control register RXB0CON (redraw with permission of Microchip)

	7	6	5	4	3	2	1	0
Mode 0	RXFUL	RXM1	RXM0	--	RXRTRR0	FILHIT2	FILHIT1	FILHIT0

	7	6	5	4	3	2	1	0
Mode 1,2	RXFUL	RXM1	RTRRO	FILHIT4	FILHIT3	FILHIT2	FILHIT1	FILHIT0

RXFUL: Receive full status bit

    0 = Receive buffer is open to receive a new message

    1 = Receive buffer contains a received message

RXM1:RXM0 (mode 0): Receive buffer mode bit

    00 = Receive all valid messages as per EXIDEN bit in RXFnSIDL register

    01 = Receive only valid messages with standard identifier (EXIDEN bit = 0)

    10 = Receive only valid messages with extended identifier (EXIDEN bit = 1)

    11 = Receive all messages (including those with errors); filter criteria ignored

RXM1 (mode 1, 2): Receive buffer mode bit

    0 = Receive all valid messages as per acceptance filters

    1 = Receive all messages (including those with errors); acceptance filters ignored

RTRRO: Remote transmission request bit for received message (read only)

    0 = A remote transmission request is not received

    1 = A remote transmission request is received

RXRTRRO: Remote transmission request bit for received message (read only)

    0 = A remote transmission request is not received

    1 = A remote transmission request is received

FILHIT2:FILHIT0 (mode 0): Filter hit bit 0

    These bits indicate which acceptance filter enabled the message reception into receive buffer 1.

    000 = Acceptance filter 0 (RXF0), only possible when RXB0DBEN bit is set

    001 = Acceptance filter 1 (RXF1), only possible when RXB0DBEN bit is set

    010 = Acceptance filter 2 (RXF2)

    011 = Acceptance filter 3 (RXF3)

    100 = Acceptance filter 4 (RXF4)

    101 = Acceptance filter 5 (RXF5)

    others = reserved

FILHIT4:FILFIT0: Filter hit bits 4 to 0

    These five bits indicate which acceptance filter enabled the message reception into this receive buffer.

    00000 = Acceptance filter 0 (RXF0)

    00001 = Acceptance filter 1 (RXF1)

    ....

    01111 = Acceptance filter 15 (RXF15)

Figure 13.21b ■ PIC18 Receive buffer 1 control register RXB1CON (redraw with permission of Microchip)

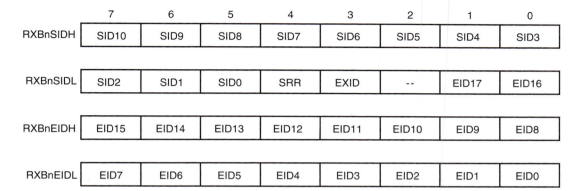

	7	6	5	4	3	2	1	0
RXBnSIDH	SID10	SID9	SID8	SID7	SID6	SID5	SID4	SID3
RXBnSIDL	SID2	SID1	SID0	SRR	EXID	--	EID17	EID16
RXBnEIDH	EID15	EID14	EID13	EID12	EID11	EID10	EID9	EID8
RXBnEIDL	EID7	EID6	EID5	EID4	EID3	EID2	EID1	EID0

SID10..SID0: Standard identifier bits

SRR: Substitute remote request bit

   This bit is always 0 when EXID = 1 or equal to the value of RXRTRRO when

   EXID = 0.

EXID: Extended identifier bit

   0 = Received message is a standard data frame

   1 = Received message is an extended data frame, SID10:SID0 are EID28:EID18

EID17:EID0: Extended identifier bits

Figure 13.22 ■ Receive buffer identifier registers (0 ≤ n ≤ 1) (redraw with permission of Microchip)

7	6	5	4	3	2	1	0
--	RXRTR	RB1	RB0	DLC3	DLC2	DLC1	DLC0

RXRTR: Transmit remote frame transmission request bit

   0 = No remote transfer request

   1 = Remote transfer request

RB1:RB0: Reserved bit 1 and 0

   Reserved by CAN spec and read as 0

DLC3:DLC0: Data length code bits

   0000 = Data length is 0 bytes

   0001 = Data length is 1 byte

   0010 = Data length is 2 bytes

   0011 = Data length is 3 bytes

   0100 = Data length is 4 bytes

   0101 = Data length is 5 bytes

   0110 = Data length is 6 bytes

   0111 = Data length is 7 bytes

   1000 = Data length is 8 bytes

   others = Invalid

Figure 13.23 ■ RXBnDLC Receive buffer n data length code register [0≤n≤1] (redraw with permission of Microchip)

When configured as transmit buffers, user software may access transmit buffers in any order similar to accessing dedicated transmit buffers. When configured in a receive buffer with mode 1 enabled, user software may also access receive buffers in any order required. But in mode 2, all receive buffers are combined to form a single FIFO. Actual FIFO length is programmable by user software. Access to FIFO must be done through the FIFO pointer bits in the CANCON register. There is no hardware protection against out-of-order FIFO reads.

The contents of the buffer control register (BnCON) are shown in Figures 13.24a and 13.24b. It is important that the user clears the RXFUL bit after reading the received buffer. No new messages will be loaded into the receive buffer if this bit is set. The function of the lowest five bits of the BnCON register is identical to that of the corresponding bits in the RXB0CON register. The user checks the RTRRO bit to decide if the received message is a *remote transmit request*. If this bit is 1, then the user software should send out a message with the same identifier associated with the remote transmit request.

7	6	5	4	3	2	1	0
RXFUL	RXM1	RTRRO	FILHIT4	FILHIT3	FILHIT2	FILHIT1	FILHIT0

RXFUL: Receive full status bit

    0 = Receive buffer is open to receive a new message

    1 = Receive buffer contains a received message

RXM1: Receive buffer mode bit

    0 = Receive all valid messages as per acceptance filters

    1 = Receive all messages including partial and invalid (acceptance filters are
        ignored.

RTRRO: Read only remote transmission request bit for received message

    0 = Received message is not a remote transmission request

    1 = Received message is a remote transmission request

FILHIT4:FILHIT0: Filter hit bits

    00000 = Acceptance filter 0 (RXF0)

    00001 = Acceptance filter 1 (RXF1)

    ....

    01111 = Acceptance filter 15 (RXF15)

Figure 13.24a ■ BnCON TX/RX Buffer n control register in receive mode [0≤n≤5] (These registers are available in mode 1 and 2 only) (redraw with permission of Microchip)

7	6	5	4	3	2	1	0
TXBIF	TXABT	TXLARB	TXERR	TXREQ	RTREN	TXPRI1	TXPRI0

TXBIF: Transmit buffer interrupt flag bit [1]

    0 = No message was transmitted

    1 = A message is successfully transmitted

TXABT: Transmission aborted status bit [1]

    0 = Message was not aborted

    1 = Message was aborted

TXLARB: Transmission lost arbitration status bit [2]

    0 = Message did not lose arbitration while being sent

    1 = Message lost arbitration while being sent

TXERR: Transmission error detected status bit [2]

    0 = A bus error did not occur while the message was being sent

    1 = A bus error occurred while the message was being sent

TXREQ: Transmit request status bit [3]

    0 = Automatically cleared when the message was being sent

    1 = Requests sending a message; clears the TXABT, TXLARB, and TXERR bits

RTREN: Automatic remote transmission request enable bit

    0 = When a remote transmission request is received, TXREQ will be unaffected.

    1 = When a remote transmission request is received, TXREQ will be set
       automatically.

TXPRI1:TXPRI0: Transmit priority bits

    00 = Priority 0 (lowest priority)

    01 = Priority 1

    10 = Priority 2

    11 = Priority 3 (highest priority)

Note 1. These registers are available in mode 1 and 2 only.

    2. This bit is automatically cleared when TXREQ is set.

    3. While TXREQ is set or transmission is in progress, transmit buffer
      registers remain read only.

Figure 13.24b ■ BnCON TX/RX Buffer n control register in transmit mode [0≤n≤5] (These registers are available in mode 1 and 2 only) (redraw with permission of Microchip)

The contents of the identifier registers are shown in Figures 13.25a and 13.25b. Since each buffer can be programmed as a transmit or a receive buffer, two sets of identifier bits are associated with each buffer.

	7	6	5	4	3	2	1	0
BnSIDH	SID10	SID9	SID8	SID7	SID6	SID5	SID4	SID3
BnSIDL	SID2	SID1	SID0	SRR	EXID	--	EID17	EID16
BnEIDH	EID15	EID14	EID13	EID12	EID11	EID10	EID9	EID8
BnEIDL	EID7	EID6	EID5	EID4	EID3	EID2	EID1	EID0

SID10..SID0: Standard identifier bits

SRR: Substitute remote request bit (only when EXID = 1)

    0 = No remote transmission request occurred

    1 = Remote transmission request occurred

EXID: Extended identifier bit

    0 = Received message is a standard data frame

    1 = Received message is an extended data frame, SID10:SID0 are EID28:EID18

EID17:EID0: Extended identifier bits

Figure 13.25a ■ TX/RX Buffer n Identifier registers in receive mode [0≤n≤5] (redraw with permission of Microchip)

	7	6	5	4	3	2	1	0
BnSIDH	SID10	SID9	SID8	SID7	SID6	SID5	SID4	SID3
BnSIDL	SID2	SID1	SID0	--	EXIDE	--	EID17	EID16
BnEIDH	EID15	EID14	EID13	EID12	EID11	EID10	EID9	EID8
BnEIDL	EID7	EID6	EID5	EID4	EID3	EID2	EID1	EID0

SID10..SID0: Standard identifier bits

EXIDE: Extended identifier enable bit

    0 = Message will transmit standard ID, EID10:EID0 are ignored

    1 = Message will transmit extended data frame, SID10:SID0 are EID28:EID18

EID17:EID0: Extended identifier bits

Figure 13.25b ■ TX/RX Buffer n Identifier registers in transmit mode [0≤n≤5] (redraw with permission of Microchip)

The contents of the data length code register (BnDLC) are shown in Figures 13.26a and 13.26b. In transmit mode, this register indicates the number of bytes that are being transmitted. In receive mode, this register indicates the number of bytes that are contained in the receive buffer.

7	6	5	4	3	2	1	0
--	RXRTR	RB1	RB0	DLC3	DLC2	DLC1	DLC0

RXRTR: Transmit remote frame transmission request bit

  0 = This is not a remote transfer request

  1 = This is a remote transfer request

RB1:RB0: Reserved bit 1 and 0

  Reserved by CAN spec and read as 0

DLC3:DLC0: Data length code bits

  0000 = Data length is 0 bytes

  0001 = Data length is 1 byte

  0010 = Data length is 2 bytes

  0011 = Data length is 3 bytes

  0100 = Data length is 4 bytes

  0101 = Data length is 5 bytes

  0110 = Data length is 6 bytes

  0111 = Data length is 7 bytes

  1000 = Data length is 8 bytes

  others = Reserved

Figure 13.26a ■ BnDLC TX/RX buffer n data length code register in receive mode [0≤n≤5] (redraw with permission of Microchip)

7	6	5	4	3	2	1	0
--	TXRTR	--	--	DLC3	DLC2	DLC1	DLC0

TXRTR: Transmit remote frame transmission request bit

  0 = Transmitted message will have RTR bit cleared

  1 = Transmitted message will have RTR bit set

DLC3:DLC0: Data length code bits

  0000 = Data length is 0 bytes

  0001 = Data length is 1 byte

  0010 = Data length is 2 bytes

  0011 = Data length is 3 bytes

  0100 = Data length is 4 bytes

  0101 = Data length is 5 bytes

  0110 = Data length is 6 bytes

  0111 = Data length is 7 bytes

  1000 = Data length is 8 bytes

  others = Reserved

Figure 13.26b ■ BnDLC TX/RX buffer n data length code register in transmit mode [0≤n≤5] (redraw with permission of Microchip)

The user can configure each of the TX/RX buffers as a transmit buffer or a receive buffer by programming the BSEL0 register. The contents of BSEL0 are shown in Figure 13.27.

7	6	5	4	3	2	1	0
B5TXEN	B4TXEN	B3TXEN	B2TXEN	B1TXEN	B0TXEN	--	--

B5TXEN:B0TXEN: Buffer 5 to buffer 0 transmit enable bit

    0 = Buffer is configured in receive mode

    1 = Buffer is configured in transmit mode

Figure 13.27 ■ Buffer select register 0 (BSEL0) (redraw with permission of Microchip)

13.12.4 Programmable Auto RTR Buffers

In mode 1 and mode 2, any one of the six programmable transmit/receive buffers may be programmed to automatically respond to predefined RTR messages without user software intervention. Automatic RTR handling is enabled by setting the BnTXEN bit in the BSEL0 register and the RTREN bit in the BnCON register. After this setup, when an RTR request is received, the TXREQ bit is automatically set, and current buffer content is automatically queued for transmission as an RTR response. As with all transmit buffers, once the TXREQ bit is set, the buffer becomes read only, and any writes to them will be ignored.

The procedure for handling automatic RTR messages is as follows:

Step 1
Set buffer to transmit mode by setting the BnTXEN bit of the BSEL0 register to 1.

Step 2
Preload at least one associated acceptance filter register with the expected RTR identifier.

Step 3
Set the RTREN bit in the corresponding BnCON register to 1.

Step 4
Preload the buffer with data to be sent as an RTR response.

The user must remember to keep the buffer data registers up to date. The user must also make sure to update the buffer when the automatic RTR response is not in the process of transmission.

13.13 CAN Message Transmission

For normal data transmission on the CAN bus, the user will follow this procedure:

Step 1
Load data into one of the transmit buffers (TXBnD0:TXBnD7 or BnD0:BnD7) when they are not busy (i.e., the associated TXREQ bit is clear).

Step 2
Set up the message identifier properly. This can be done by loading standard identifiers into the TXBnSIDH:TXBnSIDL (or BnSIDH: BnSIDL) or extended identifiers into TXBnEIDH:TXBnEIDL (or BnEIDH:BnEIDL).

Step 3

Set the TXREQ bit of the buffer to be transmitted. When the TXREQ bit is set, the TXABT, TXLARB, and TXERR bits will be cleared.

Setting the TXREQ bit does not initiate a message transmission; it merely flags a message buffer as ready for transmission. Transmission will start when the device detects that the bus is available. The device will then begin transmission of the highest-priority message that is ready.

When the transmission has completed successfully, the TXREQ bit (of the TXBnCON or BnCON register) will be cleared, the TXBnIF bit (in the PIR3 register) will be set, and an interrupt will be generated if the TXBnIE bit (in the PIE3 register) is set.

If the message transmission fails, the TXREQ bit will remain set, indicating that the message is still pending for transmission, and one of the condition flags (TXERR, IRXIF, and TXLARB) will be set. If the message started to transmit but encountered an error condition, the TXERR and the IRXIF bits (of the PIR3 register) will be set, and an interrupt will be generated. If the message lost arbitration, the TXLARB bit will be set.

Example 13.2

▼

Write an instruction sequence to send out the string T805H80 from transmit buffer 0. Use the standard identifier and set it to 0x100.

Solution: The standard identifier consists of 11 bits specified by the user and is 0010 0000 000. The first eight bits (0x20) will be loaded into the TXB0SIDH register. The remaining three bits in the identifier along with the don't care bit, the EXIDE bit, and the unused three extended identifier bits (set to 0s) will be loaded into the TXB0SIDL register. The following instruction sequence will set up the transmit buffer 0 and the associated identifier properly:

```
        movlw   'T'                 ; place the message in the transmit buffer
        movff   WREG,TXB0D0         ; "
        movlw   '8'                 ; "
        movff   WREG,TXB0D1         ; "
        movlw   '0'                 ; "
        movff   WREG,TXB0D2         ; "
        movlw   '5'                 ; "
        movff   WREG,TXB0D3         ; "
        movlw   'H'                 ; "
        movff   WREG,TXB0D4         ; "
        movlw   '8'                 ; "
        movff   WREG,RXB0D5         ; "
        movlw   '0'                 ; "
        movff   WREG,RXB0D6         ; "
; set up message identifier; no need to load TXB0EIDL:TXB0EIDH
        movlw   0x00
        movff   WREG,TXB0SIDL
        movlw   0x20
        movff   WREG,TXB0SIDH
; set up data length code
        movlw   0x07
        movff   WREG,TXB0DLC        ; prepare to transmit 7 bytes
```

```
; mark the buffer for transmission, set to priority 0
            movlw    0x08
            movff    WREG,TXB0CON
wait_tx     movff    TXB0CON,WREG
            btfsc    WREG,TXREQ
            bra      wait_tx
```

▲

13.13.1 Aborting Transmission

The PIC18 microcontroller can abort a message by clearing the TXREQ bit associated with the corresponding message buffer. The PIC18 microcontroller can abort messages in all transmit buffers by setting the ABAT bit (in the CANCON register). If the message has not yet started transmission or if the message transmission has been started but is interrupted by loss of arbitration or an error, the abort will be processed.

The abort of a message is indicated by the setting of the TXABT bit (TXBnCON<6> or BnCON<6>) for the corresponding buffer. If the message transmission has been started, the CAN module will attempt to transmit it fully. If the current message is transmitted fully and is not lost to arbitration or an error, the TXABT bit will not be set. Likewise, if a message is being transmitted during an abort request and the message is lost to arbitration or an error, the message will not be retransmitted, and the TXABT bit will be set, indicating that the message was successfully aborted.

Once an abort is requested by setting the ABAT or TXABT bits, it cannot be cleared to cancel the abort request. Only CAN module hardware or a power-on reset condition can clear it.

13.13.2 Transmit Priority

When there are multiple transmit buffers ready for transmission, the CAN module needs to decide which buffer to transmit first. The CAN module uses a priority scheme to make this decision. The transmit buffer with the highest priority will be sent first. There are four levels of transmit priority. If two buffers have the same priority setting, the buffer with the highest buffer number will be sent first. If the TXPR1 bits (of TXBnCON or BnCON) for a particular message buffer are set to 11, that buffer has the highest possible priority. If the TXPR1 bits are set to 00, that buffer has the lowest possible priority.

13.14 Message Reception

The message assembly buffer (MAB) of all receive buffers is always committed to receiving message from the bus. The entire contents of the MAB are moved into the receive buffer once a message is accepted. Once a message is loaded into a matching buffer, user software can determine which filter caused this reception by checking the filter hit bits in the RXBnCON or the BnCON register. In mode 0, the FILHIT<3:0> bits of the RXBnCON register serve as filter hit bits. In mode 1 and mode 2, the FILHIT<4:0> bits of the BnCON register serve as filter hit bits. The same registers also indicate whether the current message is an RTR frame. A received message is considered a message with a standard identifier if the EXID bit of the RXBnSIDL or the BnSIDL register is 0. Otherwise, the message is considered to have an extended identifier, and the user software should read SIDL, SIDH, EIDL, and EIDH registers. If the data length code is not 0, then the user program needs to copy data from either the RXBnDm or the BnDm register.

When a received message is a remote transmit request and if the current buffer is not configured for automatic RTR handling, user software must take appropriate action and respond manually.

13.14.1 Receive Modes

Each receive buffer contains RXM bits to set special receive modes. In mode 0, RXM<1:0> bits in RXBnCON define a total of four receive modes. In mode 1 and mode 2, the RXM1 bit in combination with the EXID mask and filter bit define the same four receive modes. These two bits are normally set to 00 to enable reception of all valid messages as determined by the appropriate acceptance filters. In this case, the determination of whether to receive messages with a standard or an extended identifier is determined by the EXIDE bit in the acceptance filter register (to be discussed later). In mode 0, if the RXM bits are set to 01 or 10, the receiver will accept only messages with a standard or an extended identifier, respectively. If an acceptance filter has the EXIDE bit set such that it does not correspond with the RXM mode, that acceptance filter is rendered useless. In mode 1 and mode 2, setting EXID in the SIDL mask register will ensure that only messages with a standard or an extended identifier are received. These two modes of RXM bits can be used in systems where it is known that only messages with a standard or an extended identifier will be on the bus. If the RXM bits are set to 11 (RXM1 = 1 in mode 1 and mode 2), the buffer will receive all messages regardless of the values of the acceptance filters. Also, if a message has an error before the end-of-frame, that portion of the message assembled in the MAB before the error frame will be loaded into the receive buffer. This mode may be a valuable debugging tool for a given CAN network. This setting should not be used in an actual system environment, as the actual system will always have some bus errors, and all nodes are expected to ignore them.

In mode 1 and mode 2, when a programmable buffer is configured as a transmit buffer and one or more acceptance filters are associated with it, all incoming messages matching this acceptance filter criteria will be discarded. To avoid this situation, the user should make sure that no acceptance filters are associated with a buffer configured as a transmit buffer.

13.14.2 Receive Priority

When in mode 0, RXB0 is the buffer with higher priority and has two message acceptance filters associated with it. RXB1 is the buffer with lower priority and has four acceptance filters associated with it.

The lower number of acceptance filters makes the match on RXB0 more restrictive and implies a higher priority for that buffer. Additionally, the RXB0CON register can be configured such that if RXB0 contains a valid message and another valid message is received, an overflow error will not occur, and the new message will be moved into RXB1 regardless of the acceptance criteria of RXB1. There are also two programmable acceptance filter masks available, one for each receive buffer.

In mode 1 and mode 2, there are a total of 16 acceptance filters available, and each can be dynamically assigned to any of the receive buffers. A buffer with a lower number has higher priority. Given this, if an incoming message matches with two or more receive buffer acceptance criteria, the buffer with the lower number will be loaded with that message.

13.14.3 Enhanced FIFO Mode

When configured for mode 2, two of the dedicated receive buffers in combination with one or more programmable transmit/receive buffers are used to create a maximum eight-buffer-deep FIFO. In this mode, there is no direct correlation between filters and receive buffer registers.

Any filter that has been enabled can generate an acceptance. When a message has been accepted, it is stored in the next available receive buffer register, and an internal write pointer is incremented. The entire FIFO must consist of contiguous receive buffers. The FIFO head begins at RXB0 buffer, and its tail spans toward B5. The maximum length of the FIFO is limited by the presence or absence of the first transmit buffer starting from B0. If a buffer is configured as a transmit buffer, the FIFO length is reduced accordingly. For instance, if B3 is configured as a transmit buffer, the actual FIFO will consist of RXB0, RXB1, B0, B1, and B2, a total of five buffers. If B0 is configured as a transmit buffer, the FIFO length will be two. If none of the programmable buffers is configured as a transmit buffer, the FIFO will be eight buffers deep. A system that requires more transmit buffers should try to locate transmit buffers at the very end of B0 to B5 buffers to maximize available FIFO length.

When a message is received in FIFO mode, the interrupt flag code bits (EICODE<4:0>) in the CANSTAT register will have a value of 10000, indicating that the FIFO has received a message. FIFO pointer bits FP<3:0> in the CANCON register point to the buffer that contain data not yet read. The user should use the FP bits to read the corresponding buffer data. When the received buffer data is no longer needed, the user should clear the RXFUL bit in the current buffer so that the CAN module can update the FP<3:0>.

To determine whether FIFO is empty, the user may use FP<3:0> bits to access the RXFUL bit in the current buffer. If the RXFUL bit is 0, then the buffer is empty. In mode 2, the module also provides a bit called the FIFO high-water mark (FIFOWM) in the ECANCON register. This bit can be used to cause an interrupt whenever the FIFO contains only one or four empty buffers. The FIFO high-water-mark interrupt can serve as an early warning to a full FIFO condition.

13.14.4 Time Stamping

In some applications, the arrival times of data are critical for decision making. The PIC18 CAN module has the capability of providing the arrival times of data. The CAN module can be programmed to generate a time stamp for every message that is received. When enabled, the module generates a capture signal for CCP1, which in turn captures the value of either timer 1 or timer 3. This value can be used as the message time stamp.

To use the time-stamp capability, the CANCAP bit (bit 4 of the CIOCAN register) must be set. This replaces the capture input for CCP1 with the signal generated from the CAN module. In addition, CCP1CON<3:0> must be set to 0011 to enable the CCP special event trigger for CAN events.

The following instruction sequence will perform the desired setting for CAN message time stamping:

```
bsf     CIOCAN,CANCAP    ; enable CAN message arrival time capture
movf    CCP1CON,W        ; set the lowest 4 bits of CCP1CON to
andlw   0xF3             ; 0011
iorlw   0x03             ; "
movwf   CCP1CON          ; "
```

13.15 Message Acceptance Filters and Masks

The CAN module uses message acceptance filters and masks to determine if a message in the MAB should be loaded into any receive buffers. Once a valid message has been received by the MAB, the identifier field of the message is compared to the filter values. If there is a match, that message will be loaded into the appropriate receive buffer. The filter masks are used to determine which bits in the identifier are examined with the filter. The truth table in Table 13.2 indicates how each bit in the identifier is compared to the masks and filters to determine if a message should be loaded into a receive buffer. If a mask bit is set to 0, then that bit in the identifier will be automatically accepted regardless of the filter bit.

Mask bit n	Filter bit n	Message identifier bit n001	Accept or reject bit n
0	x	x	Accept
1	0	0	Accept
1	0	1	Reject
1	1	0	Reject
1	1	1	Accept

Table 13.2 ■ Filter/mask truth table

In mode 0, acceptance filters RXF0 and RXF1 and filter mask RXM0 are associated with RXB0. Filters RXF2, RXF3, RXF4, and RXF5 and mask RXM1 are associated with RXB1.

In mode 1 and mode 2, there are 10 additional acceptance filters, RXF6 to RXF15, creating a total of 16 available filters. RXF15 can be used as either an acceptance filter or an acceptance mask register. For a message with standard identifier, the acceptance filter consists of two registers. If a message with extended identifier is expected, then two more registers will be added. These registers are shown in Figure 13.28.

Each of these acceptance filters can be individually enabled or disabled by setting or clearing the RXFnEN bit in the RXFCONn register. Any of these 16 acceptance filters can be dynamically associated with any of the receive buffers. Actual association is made by setting appropriate bits in the RXFBCONn register. Each RXFBCONn register contains a nibble for each filter. This nibble can be used to associate a specific filter to any of the available receive buffers. The user may associate more than one filter to any one specific receive buffer. The contents of the RXFCONn register and RXFBCONn are shown in Figures 13.29 and 13.30, respectively.

	7	6	5	4	3	2	1	0
RXFnSIDH	SID10	SID9	SID8	SID7	SID6	SID5	SID4	SID3

	7	6	5	4	3	2	1	0
RXFnSIDL	SID2	SID1	SID0	--	EXIDEN	--	EID17	EID16

	7	6	5	4	3	2	1	0
RXFnEIDH	EID15	EID14	EID13	EID12	EID11	EID10	EID9	EID8

	7	6	5	4	3	2	1	0
RXFnEIDL	EID7	EID6	EID5	EID4	EID3	EID2	EID1	EID0

SID10...SID0: Standard identifier filter bits (when EXIDEN = 0)
These 11 bits become EID28:EID18 bits if EXIDEN = 1.
EID17...EID0: Extended identifier filter bits
EXIDEN = 0
These bits are not used.
EXIDEN = 1
These bits are used as extended identifier filter bits
EXIDEN: Extended identifier filter enable bit
0 = Filter will only accept standard ID message
1 = Filter will only accept extended ID message
Note 1. Registers RXF6EIDH...RXF15EIDH and RXF6EIDL...RXF15EIDL are
available in mode 1 and 2 only.

Figure 13.28 ■ Receive acceptance filter **n** identifier filter registers [0 ≤n≤15] (redraw with permission of Microchip)

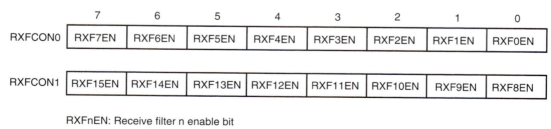

	7	6	5	4	3	2	1	0
RXFCON0	RXF7EN	RXF6EN	RXF5EN	RXF4EN	RXF3EN	RXF2EN	RXF1EN	RXF0EN

	7	6	5	4	3	2	1	0
RXFCON1	RXF15EN	RXF14EN	RXF13EN	RXF12EN	RXF11EN	RXF10EN	RXF9EN	RXF8EN

RXFnEN: Receive filter n enable bit
0 = Filter disabled
1 = Filter enabled

Figure 13.29 ■ Receive filter control register n [0≤n≤1] (redraw with permission of Microchip)

	7	6	5	4	3	2	1	0
RXFBCON0	F1BP_3	F1BP_2	F1BP_1	F1BP_0	F0BP_3	F0BP_2	F0BP_1	F0BP_0
RXFBCON1	F3BP_3	F3BP_2	F3BP_1	F3BP_0	F2BP_3	F2BP_2	F2BP_1	F2BP_0
RXFBCON2	F5BP_3	F5BP_2	F5BP_1	F5BP_0	F4BP_3	F4BP_2	F4BP_1	F4BP_0
RXFBCON3	F7BP_3	F7BP_2	F7BP_1	F7BP_0	F6BP_3	F6BP_2	F6BP_1	F6BP_0
RXFBCON4	F9BP_3	F9BP_2	F9BP_1	F9BP_0	F8BP_3	F8BP_2	F8BP_1	F8BP_0
RXFBCON5	F11BP_3	F11BP_2	F11BP_1	F11BP_0	F10BP_3	F10BP_2	F10BP_1	F10BP_0
RXFBCON6	F13BP_3	F13BP_2	F13BP_1	F13BP_0	F12BP_3	F12BP_2	F12BP_1	F12BP_0
RXFBCON7	F15BP_3	F14BP_2	F15BP_1	F15BP_0	F14BP_3	F14BP_2	F14BP_1	F14BP_0

FnBP_3...FnBP_0: Filter n buffer pointer nibble bits
    0000 = Filter n is associated with RXB0
    0001 = Filter n is associated with RXB1
    0010 = Filter n is assocaited with B0
    0011 = Filter n is associated with B1
    ....
    0111 = Filter n is assocaited with B5
    1111...1000 = Reserved

Figure 13.30 ■ Receive filter buffer control register **n** [0 ≤n≤7] (available in mode 1 and 2 only) (redraw with permission of Microchip)

The number of bits in the filter to be compared can be selected by programming the standard data bytes filter length count register (SDFLC). The contents of this register are shown in Figure 13.31.

In addition to the association of a dynamic filter to a buffer, in mode 1 and mode 2, each filter can also be dynamically associated to available acceptance mask registers. The FILn_m bits in the MSELn register can be used to link a specific acceptance filter to an acceptance mask

7	6	5	4	3	2	1	0
--	--	--	FLC4	FLC3	FLC2	FLC1	FLC0

FLC4...FLC0: Filter length count bits

Mode 0:

Not used; forced to 00000.

Mode 1 and 2:

00000...10010 = 0 ~ 18 bits are available for standard data byte filter. Actual number of bits used depends on DLC3...DLC0 bits (RXBnDLC<3...0> or BnDLC<3...0> if configured as RX buffer) of message being received

if DLC3...DLC0 = 0000 No bits will be compared with incoming data bits

if DLC3...DLC0 = 0001 Up to 8 data bits of <RXFnEID<7...0>, as determined by FLC2...FLC0, will be compared with the corresponding number of data bits of the incoming message.

if DLC3...DLC0 = 0010 Up to 16 data bits of RXFnEID<15...0>, as determined by FLC3...FLC0, will be compared with the corresponding number of data bits of the incoming message.

if DLC3...DLC0 = 0011 Up to 18 data bits of RXFnEID<17...0>, as determined by FLC4...FLC0, will be compared with the corresponding number of data bits of the incoming message.

Figure 13.31 ■ Standard data bytes filter length count registers (SDFLC) (redraw with permission of Microchip)

register. One can associate more than one mask to a specific acceptance filter. The contents of an acceptance mask are shown in Figure 13.32. There are four mask select registers, and their contents are shown in Figure 13.33.

When a filter matches and a message is loaded into the receive buffer, the filter number that enabled the message reception is loaded into the FILHIT bit (s). In mode 0 for the RXB1 buffer, the RXB1CON register contains the FILHIT<2:0> bits. They are coded as follows:

101 = acceptance filter 5 (RXF5)

100 = acceptance filter 4 (RXF4)

011 = acceptance filter 3 (RXF3)

010 = acceptance filter 2 (RXF2)

001 = acceptance filter 1 (RXF1)

000 = acceptance filter 0 (RXF0)

	7	6	5	4	3	2	1	0
RXMnSIDH	SID10	SID9	SID8	SID7	SID6	SID5	SID4	SID3
RXMnSIDL	SID2	SID1	SID0	--	EXIDEN	--	EID17	EID16
RXMnEIDH	EID15	EID14	EID13	EID12	EID11	EID10	EID9	EID8
RXMnEIDL	EID7	EID6	EID5	EID4	EID3	EID2	EID1	EID0

SID10...SID0: Standard identifier mask bits (when EXIDEN = 0)

These 11 bits become EID28:EID18 bits if EXIDEN = 1.

EID17...EID0: Extended identifier mask bits

EXIDEN = 0

These bits are not used.

EXIDEN = 1

These bits are used as extended identifier mask bits

EXIDEN: Extended identifier filter enable mask bit (available in mode 1 and 2

only)

0 = Filter will only accept standard ID message

1 = Filter will only accept extended ID message

Figure 13.32 ■ Receive acceptance mask **n** identifier mask registers [0 ≤n≤ 1] (redraw with permission of Microchip)

	7	6	5	4	3	2	1	0
MSEL0	FIL3_1	FIL3_0	FIL2_1	FIL2_0	FIL1_1	FIL1_0	FIL0_1	FIL0_0
MSEL1	FIL7_1	FIL7_0	FIL6_1	FIL6_0	FIL5_1	FIL5_0	FIL4_1	FIL4_0
MSEL2	FIL11_1	FIL11_0	FIL10_1	FIL10_0	FIL9_1	FIL9_0	FIL8_1	FIL8_0
MSEL3	FIL15_1	FIL15_0	FIL14_1	FIL14_0	FIL13_1	FIL13_0	FIL12_1	FIL12_0

FILn_1...FILn_0: Filter n select bits 1 and 0

00 = Acceptance mask 0

01 = Acceptance mask 1

10 = Filter 15

11 = No mask

Figure 13.33 ■ Mask select register n (MSELn) [0 ≤ n ≤3] (redraw with permission of Microchip)

The coding of the RXB0DBEN bit (of the RXB0CON register) enables the FILHIT<2:0> bits (of the RXB1CON register) to be used to distinguish a hit on filters RXF0 and RXF1 in either the RXB0 buffer or after a rollover into the RXB1 buffer:

> 111 = acceptance filter 1 (RXF1)
>
> 110 = acceptance filter 0 (RXF0)
>
> 001 = acceptance filter 1 (RXF1)
>
> 000 = acceptance filter 0

If the RXB0DBEN bit is clear, there are six codes corresponding to the six filters. If the RXB0DBEN bit is set, there are six codes corresponding to the six filters plus two additional codes corresponding to the RXF0 and the RXF1 filters that are associated with the rollover of the RXB0 buffer to the RXB1 buffer.

In mode 1 and mode 2, each buffer control register contains five filter hit bits FILHIT<4:0>. A value of 0 indicates a hit from RXF0, and a value of 15 indicates RXF15.

If more than one acceptance filter matches, the FILHIT bits will encode the binary value of the lowest-numbered filter that matched. In other words, if both the RXF2 filter and the RXF4 filter match, the FILHIT bits will be loaded with the value for the RXF2 filter. This essentially prioritizes the acceptance filters with a lower-number filter having higher priority. Messages are compared to filters in ascending order of the filter number.

The mask and filter registers can be modified only when the PIC18F6585/8585/6680/8680 devices are in configuration mode.

Associating a mask with the acceptance filter is easy. For example, the following instruction sequence associates mask 0 with acceptance filters 0 to 7 and associates mask 1 with filters 8 to 15:

```
movlw    0
movff    WREG,MSEL0    ; associates filters 0 to 7 with mask 0
movff    WREG,MSEL1    ; "
movlw    0x55          ; associates filters 8 to 15 with mask 1
movff    WREG,MSEL2    ; "
movff    WREG,MSEL3    ; "
```

The following C statements will perform the same associations:

```
MSEL0 = MSEL1 = 0;
MSEL2 = MSEL3 = 0x55;
```

13.16 Baud Rate Setting and Timing Parameters

All devices on the CAN bus must use the same bit rate. However, all devices are not required to have the same master oscillator clock frequency. For the different clock frequencies of the individual devices, the bit rate has to be adjusted by appropriately setting the baud rate prescaler and the number of time quanta on each segment.

The nominal bit rate is the number of bits transmitted per second, assuming an ideal transmitter with an ideal oscillator, in the absence of resynchronization. The nominal bit rate is defined to be a maximum of 1 Mb/s. The nominal bit time is defined as

T_{BIT} = 1/(nominal bit rate)

The nominal bit time is divided into four segments:

Synchronization segment (**sync_seg**)

Propagation time segment (**prop_seg**)

Phase buffer segment 1 (**phase_seg1**)

Phase buffer segment 2 (**phase_seg2**)

These time segments are in turn made up of integer units of time quanta (TQ). By definition, the nominal bit time is programmable from a minimum of 8 TQ to a maximum of 25 TQ. Therefore, the nominal bit time can be written as

$$T_{BIT} = TQ * (sync\_seg + prop\_seg + phase\_seg1 + phase\_seg2) \tag{13.1}$$

TQ is a fixed unit derived from the oscillator period. It is also defined by the programmable baud rate prescaler with integer values from 1 to 64 in addition to a fixed divide-by–2 prescaler for clock generation:

$$TQ = 2 * (BRP + 1) * T_{OSC} \tag{13.2}$$

where BRP is an integer from 0 to 63, represented by the binary value of the six bits BRGCON1<5:0>. The contents of the BRGCON1 register are shown in Figure 13.34. This register allows the user to select baud rate prescaler and synchronization jump width.

7	6	5	4	3	2	1	0
SJW1	SJW0	BRP5	BRP4	BRP3	BRP2	BRP1	BRP0

SJW1...SJW0: synchronized jump width bits

  00 = Synchronization jump width time = 1 x TQ

  01 = Synchronization jump width time = 2 x TQ

  10 = Synchronization jump width time = 3 x TQ

  11 = Synchronization jump width time = 4 x TQ

BRP5...BRP0: Baud rate prescaler bits

  000000 = TQ = (2 x 1) x T_{OSC}

  000001 = TQ = (2 x 2) x T_{OSC}

  000010 = TQ = (2 x 3) x T_{OSC}

  ....

  111110 = TQ = (2 x 63) x T_{OSC}

  111111 = TQ = (2 x 64) x T_{OSC}

Figure 13.34 ■ Baud rate control register 1 (BRGCON1) (redraw with permission of Microchip)

Suppose that F_{OSC} = 32 MHz, BRP<5:0> = 0x07, and nominal bit time = 8 TQ. Then

TQ = 2 * (BRP + 1)/F_{osc} = 2 * 8/32 MHz = 0.5 μs

T_{BIT} = 8 * TQ = 4 μs

Nominal bit rate = 1/T_{BIT} = 250 Kbits/s

In addition to the BRGCON1 register, two other registers (BRGCON2 and BRGCON3) are also involved in the CAN timing control. The contents of these two registers are shown in Figures 13.35 and 13.36.

7	6	5	4	3	2	1	0
SEG2PHTS	SAM	SEG1PH2	SEG1PH1	SEG1PH0	PRSEG2	PRSEG1	PRSEG0

SEG2PHTS: Phase segment 2 time select bit

   0 = Maximum of PHEG1 or information processing time (IPT), whichever is
      larger

   1 = Freely programmable

SAM: Sample of the CAN bus line

   0 = Bus line is sampled once at the sample point

   1 = Bus line is sampled three times prior to and at the sample point

SEG1PH2...SEG1PH0: Phase segment 1 bits

   000 = Phase segment 1 time = 1 x TQ

   001 = Phase segment 1 time = 2 x TQ

   010 = Phase segment 1 time = 3 x TQ

   011 = Phase segment 1 time = 4 x TQ

   100 = Phase segment 1 time = 5 x TQ

   101 = Phase segment 1 time = 6 x TQ

   110 = Phase segment 1 time = 7 x TQ

   111 = Phase segment 1 time = 8 x TQ

PRSEG2...PRSEG0: Propagation time select bits

   000 = Propagation time = 1 x TQ

   001 = Propagation time = 2 x TQ

   010 = Propagation time = 3 x TQ

   011 = Propagation time = 4 x TQ

   100 = Propagation time = 5 x TQ

   101 = Propagation time = 6 x TQ

   110 = Propagation time = 7 x TQ

   111 = Propagation time = 8 x TQ

Figure 13.35 ■ Baud rate control register 2 (BRGCON2) (redraw with permission of Microchip)

In the BRGCON2 register,

- The PRSEG bits set the length of the propagation segment in terms of TQ.
- The SEG1PH bits set the length of phase segment 1 in units of TQ.
- The SAM bit controls how many times the RXCAN pin is sampled. Setting this bit to 1 causes the bus to be sampled three times: twice (one at TQ and one at TQ/2 before the sample point) before the sample point and once at the normal sample point (at the end of phase segment 1). The value of the bus is determined by the majority function of these samples. Otherwise, only one sample is taken at the end of phase segment 1.
- The SEG2PHTS bit controls how the length of phase segment 2 is determined. If this bit is set to 1, then the length of the phase segment 2 is determined by the SEG2PH bits of the BRGCON3 register. Otherwise, the length of phase segment 2 is the greater of phase segment 1 and the information processing time.

7	6	5	4	3	2	1	0
WAKDIS	WAKFIL	--	--	--	SEG2PH2	SEG2PH1	SEG2PH0

WAKDIS: Wake-up disable bit

  0 = Enable CAN bus activity wakeup feature

  1 = Disable CAN bus activity wakeup feature

WAKFIL: Select CAN bus line filter for wakeup bit

  0 = CAN bus line filter is not used for wakeup

  1 = Use CAN bus line filter for wakeup

SEG2PH2...SEG2PH0: Phase segment 2 time select bits

  000 = Phase segment 2 time = 1 x TQ

  001 = Phase segment 2 time = 2 x TQ

  010 = Phase segment 2 time = 3 x TQ

  011 = Phase segment 2 time = 4 x TQ

  100 = Phase segment 2 time = 5 x TQ

  101 = Phase segment 2 time = 6 x TQ

  110 = Phase segment 2 time = 7 x TQ

  111 = Phase segment 2 time = 8 x TQ

Figure 13.36 ■ Baud rate control register 3 (BRGCON3) (redraw with permission of Microchip)

- The WAKDIS bit and the WAKFIL bit together allow the user to use the CAN bus activity to wake up the CAN module in the sleep mode.

The calculation of **prop_seg** requires the knowledge of the CAN transceiver delay times and hence is discussed in a later section.

13.17 Error Detection and Interrupts

The ECAN module detects all the errors required by the CAN protocol. The errors detected by the ECAN module include the following:

- *CRC error.* The ECAN transceiver computes and adds the CRC checksum to the message. The receiver node recalculates the CRC sequence using the same method and compares it with the received sequence. If a mismatch is detected, a CRC error has occurred, and an error frame will be generated.

- *Acknowledge error.* In the acknowledge field of a message, the transmitter checks if the acknowledge slot (sent out as a recessive bit) contains a dominant bit. If not, no other node has received the frame correctly. An acknowledge error has occurred; an error frame is generated, and the message has to be repeated.

- *Form error.* If a node detects a dominant bit in one of the four segments, including end-of-frame, interframe space, acknowledge delimiter, or CRC delimiter, then a form error has occurred, and an error frame is generated. The message is repeated.

- *Bit error.* A bit error occurs if a transmitter sends a dominant bit and detects a recessive bit or if it sends a recessive bit and detects a dominant bit when monitoring the actual bus level and comparing it to the just transmitted bit. In the case where the transmitter sends a recessive bit and a dominant bit is detected during the arbitration field and the acknowledge slot, no bit error is generated because normal arbitration is occurring.

- *Stuff error.* If six consecutive bits with the same polarity are detected between the start-of-frame and the CRC delimiter, the bit-stuffing rule has been violated. A stuff bit error occurs, and an error frame is generated. The message is repeated.

13.17.1 Error States

Detected errors are made public to all other nodes via error frames. The transmission of the erroneous message is aborted, and the frame is repeated as soon as possible. Furthermore, each CAN node is in one of the three error states—error-active, error-passive, or bus-off—according to the value of the internal error counters. In the error-active state, the bus node can transmit messages and active error frames. In the error-passive state, the bus node can transmit messages and passive error frames. The bus-off state prevents a node from participating in the bus communication temporarily. The error state transition is shown in Figure 13.37. The bus-off recovery sequence consists of 128 occurrences of 11 consecutive recessive bits. After that, the node will return to the error-active state.

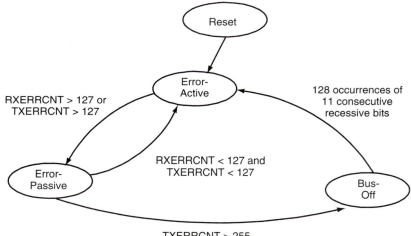

Figure 13.37 ■ Error modes state transition diagram

13.17.2 Error Modes and Error Counters

The ECAN module of the PIC18F6585/6680/8585/8680 devices contains two 8-bit error counters: the receive error counter (RXERRCNT) and the transmit error counter (TXERRCNT). The values of both counters can be read by the microcontroller. These counters are incremented and decremented in accordance with the CAN bus specification.

13.17.3 CAN Interrupts

The PIR3 register holds all the CAN interrupt flag bits, whereas the PIE3 holds all the CAN interrupt enable bits. The priority of the CAN interrupts can be set via the IPR3 register. The contents of these three registers are shown in Figures 13.38, 13.39, and 13.40, respectively. Each

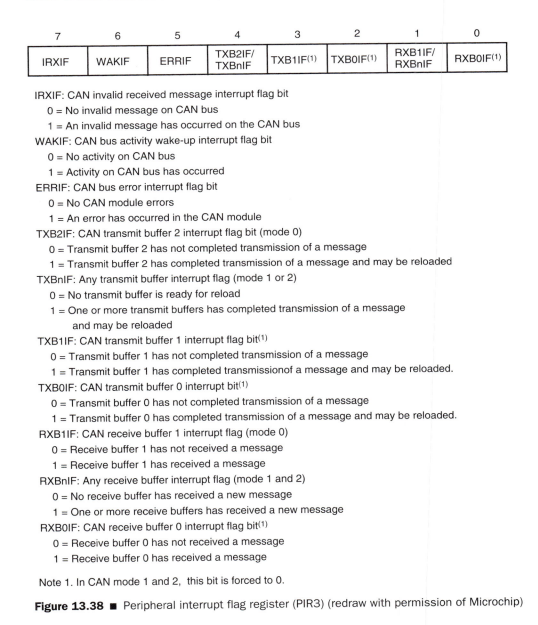

7	6	5	4	3	2	1	0
IRXIF	WAKIF	ERRIF	TXB2IF/ TXBnIF	TXB1IF[1]	TXB0IF[1]	RXB1IF/ RXBnIF	RXB0IF[1]

IRXIF: CAN invalid received message interrupt flag bit

    0 = No invalid message on CAN bus

    1 = An invalid message has occurred on the CAN bus

WAKIF: CAN bus activity wake-up interrupt flag bit

    0 = No activity on CAN bus

    1 = Activity on CAN bus has occurred

ERRIF: CAN bus error interrupt flag bit

    0 = No CAN module errors

    1 = An error has occurred in the CAN module

TXB2IF: CAN transmit buffer 2 interrupt flag bit (mode 0)

    0 = Transmit buffer 2 has not completed transmission of a message

    1 = Transmit buffer 2 has completed transmission of a message and may be reloaded

TXBnIF: Any transmit buffer interrupt flag (mode 1 or 2)

    0 = No transmit buffer is ready for reload

    1 = One or more transmit buffers has completed transmission of a message

      and may be reloaded

TXB1IF: CAN transmit buffer 1 interrupt flag bit[1]

    0 = Transmit buffer 1 has not completed transmission of a message

    1 = Transmit buffer 1 has completed transmissionof a message and may be reloaded.

TXB0IF: CAN transmit buffer 0 interrupt bit[1]

    0 = Transmit buffer 0 has not completed transmission of a message

    1 = Transmit buffer 0 has completed transmission of a message and may be reloaded.

RXB1IF: CAN receive buffer 1 interrupt flag (mode 0)

    0 = Receive buffer 1 has not received a message

    1 = Receive buffer 1 has received a message

RXBnIF: Any receive buffer interrupt flag (mode 1 and 2)

    0 = No receive buffer has received a new message

    1 = One or more receive buffers has received a new message

RXB0IF: CAN receive buffer 0 interrupt flag bit[1]

    0 = Receive buffer 0 has not received a message

    1 = Receive buffer 0 has received a message

Note 1. In CAN mode 1 and 2, this bit is forced to 0.

Figure 13.38 ■ Peripheral interrupt flag register (PIR3) (redraw with permission of Microchip)

IRXIE: CAN invalid received message interrupt enable bit

    0 = Disable invalid message received interrupt

    1 = Enable invalid message received interrupt

WAKIE: CAN bus activity wake-up interrupt enable bit

    0 = Disable bus activity wake-up interrupt

    1 = Enable bus activity wake-up interrupt

ERRIE: CAN bus error interrupt enable bit

    0 = Disable CAN bus error interrupt

    1 = Enable CAN bus error interrupt

TXB2IE: CAN transmit buffer 2 interrupt enable bit (mode 0)

    0 = Disable transmit buffer 2 interrupt

    1 = Enable transmit buffer 2 interrupt

TXBnIE: Transmit buffer interrupt enable bit (mode 1 or 2)

    0 = Disable transmit buffer interrupts

    1 = Enable transmit buffer interrupts; individual interrupt is enabled by the

        TXBIE and the BIE0 registers

TXB1IE: CAN transmit buffer 1 interrupt enable bit(1)

    0 = Disable transmit buffer 1 interrupt

    1 = Enable transmit buffer 1 interrupt.

TXB0IE: CAN transmit buffer 0 interrupt enable bit(1)

    0 = Disable transmit buffer 0 interrupt

    1 = Enable transmit buffer 0 interrupt.

RXB1IE: CAN receive buffer 1 interrupt enable bit (mode 0)

    0 = Disable receive buffer 1 interrupt

    1 = Enable receive buffer 1 interrupt

RXBnIE: CAN receive buffer interrupt enable bit (mode 1 and 2)

    0 = Disable all receive buffer interrupts

    1 = Enable receive buffer interrupts; individual interrupt is enabled by the BIE0

        register

RXB0IE: CAN receive buffer 0 interrupt enable bit(1)

    0 = Disable receive buffer 0 interrupt

    1 = Enable receive buffer 0 interrupt

Note 1. In CAN mode 1 and 2, this bit is forced to 0.

Figure 13.39 ■ Peripheral interrupt enable register (PIE3) (redraw with permission of Microchip)

7	6	5	4	3	2	1	0
IRXIP	WAKIP	ERRIP	TXB2IP/ TXBnIP	TXB1IP[1]	TXB0IP[1]	RXB1IP/ RXBnIP	RXB0IP[1]

IRXIP: CAN invalid received message interrupt priority bit

    0 = Low priority

    1 = High priority

WAKIP: CAN bus activity wake-up interrupt priority bit

    0 = Low priority

    1 = High priority

ERRIP: CAN bus error interrupt priority bit

    0 = Low priority

    1 = High priority

TXB2IP: CAN transmit buffer 2 interrupt priority bit (mode 0)

    0 = Low priority

    1 = High priority

TXBnIP: Transmit buffer interrupt priority bit (mode 1 or 2)

    0 = Low priority

    1 = High priority

TXB1IP: CAN transmit buffer 1 interrupt priority bit[1]

    0 = Low priority

    1 = High priority

TXB0IP: CAN transmit buffer 0 interrupt priority bit[1]

    0 = Low priority

    1 = High priority

RXB1IP: CAN receive buffer 1 interrupt priority bit (mode 0)

    0 = Low priority

    1 = High priority

RXBnIP: CAN receive buffer interrupt priority bit (mode 1 and 2)

    0 = Low priority

    1 = High priority

RXB0IP: CAN receive buffer 0 priority enable bit[1]

    0 = Low priority

    1 = High priority

Note 1. In CAN mode 1 and 2, this bit is forced to 0.

Figure 13.40 ■ Peripheral interrupt priority register (IPR3) (redraw with permission of Microchip)

dedicated transmit buffer and receive buffer and programmable TX/RX buffer has its own interrupt enable bit as shown in Figures 13.41 and 13.42.

7	6	5	4	3	2	1	0
--	--	--	TXB2IE	TXB1IE	TXB0IE	--	--

TXB2IE:TXB0IE: Transmit buffer 2-0 interrupt enable bit[2]

    0 = Transmit buffer interrupt is disabled

    1 = Transmit buffer interrupt is enabled

Note 1. This register is available in mode 1 and 2 only.

      2. TXBIE in PIE3 register must be set to get an interrupt

Figure 13.41 ■ Transmit buffers interrupt enable register (TXBIE)[1] (redraw with permission of Microchip)

7	6	5	4	3	2	1	0
B5IE	B4IE	B3IE	B2IE	B1IE	B0IE	RXB1IE	RXB0IE

B5IE: B0IE: Programmable transmit/receive buffer 5-0 interrupt enable bit[2]

RXB1IE:RXB0IE: Dedicated receive buffer 1-0 interrupt enable bit[2]

    0 = Interrupt is disabled

    1 = Interrupt is enabled

Note 1. This register is available in mode 1 and 2 only.

      2. Either TXBIE or RXBIE in PIE3 register must be set to get an interrupt

Figure 13.42 ■ Buffer interrupt enable register 0 (BIE0) (redraw with permission of Microchip)

13.17.4 Efficient Handling of CAN Interrupts

The two-level interrupt priority scheme sometimes makes the identification of interrupt source inefficient. Fortunately, the ECAN module encodes the CAN interrupt in a way that makes the identification easier. CAN interrupts are internally prioritized such that the higher-priority interrupts are assigned lower values. Once the highest-priority interrupt condition has been cleared, the code for the next-highest-priority interrupt that is pending will be reflected by the ICODE (mode 0) or EICODE (mode 1 and mode 2). Note that only those interrupt sources that have their associated interrupt enable bit set will be reflected in the ICODE (or EICODE). This scheme allows the user to use a jump table to handle CAN interrupt.

Transmit Interrupt

When the transmit interrupt is enabled, an interrupt will be generated when the associated transmit buffer becomes empty and is ready to be loaded with a new message. In mode 0, there are separate interrupt enable/disable and flag bits for each of the three dedicated transmit buffers. The TXBnIF bit will be set to indicate the source of the interrupt. In mode 1 and mode 2, all transmit buffers share one interrupt enable/disable and flag bits. The TXBIF flag indicates when a transmit buffer has completed transmission of its message. Individual transmit buffer interrupts can be enabled or disabled by setting or clearing TXBIE and BnIE register bits. When a shared interrupt occurs, user software must check the TXREQ bit of all transmit buffers to detect the source of interrupt.

RECEIVE INTERRUPT

When the receive interrupt is enabled, an interrupt will be generated when a message has been successfully received and loaded into the associated receive buffer. This interrupt is activated immediately after receiving the end-of-frame (EOF) field.

In mode 0, the RXBnIF bit is set to indicate the source of the interrupt. This flag must be cleared by the user software.

In mode 1 and mode 2, all receive buffers share RXBIE, RXBIF, and RXBIP in PIE3, PIR3, and IPR3, respectively. Individual receive buffer interrupts can be controlled by the TXBIE and BIE0 registers. In mode 1, when a shared receive interrupt occurs, user software must poll the RXFUL bit of each receive buffer to detect the source of interrupt. In mode 2, a receive interrupt indicates that the new message is loaded into the FIFO. The FIFO can be read by using the FIFO pointer bits, FP.

MESSAGE ERROR INTERRUPT

When an error occurs during the transmission or reception of a message, the message error flag, IRXIF, will be set. An interrupt will be generated if the IRXIE bit is set. This is intended to be used in facilitating baud rate determination when used in conjunction with the listen-only mode.

BUS ACTIVITY WAKE-UP INTERRUPT

When the PIC18 device is in sleep mode and the bus activity wake-up interrupt is enabled, an interrupt will be generated, and the WAKIF bit will be set when activity is detected on the CAN bus. This interrupt causes the PIC18 devices to exit sleep mode.

ERROR INTERRUPT

When the error interrupt is enabled, an interrupt is generated if an overflow condition occurs or if the error state of the transmitter or receiver has changed. The error flags in COMSTAT will indicate one of the following conditions:

- *Receiver overflow.* This condition occurs when the MAB buffer has assembled a valid message and the receive buffer associated with the filter is not available for loading of a new message. The associated RXnOVFL bit in the COMSTAT register will be set to indicate the overflow condition.
- *Receiver warning.* The warning interrupt will be generated when the receive error counter has reached the warning limit of 96.
- *Transmitter warning.* The transmitter error counter has reached the warning limit of 96.
- *Receiver bus passive.* The receive error counter has exceeded the error-passive limit of 127, and the device has entered the error-passive state.
- *Transmit bus passive.* The transmit error counter has exceeded the error-passive limit of 127, and the device has entered the error-passive state.
- *Bus off.* The transmit error counter has exceeded 255, and the device has entered the bus-off state.

13.17.5 CAN Module I/O Control Register

The voltage level for the CANTX2 pin and the CCP capture control can be programmed by configuring the CIOCON register. The contents of this register are shown in Figure 13.43. The CANCAP bit allows the user to capture the timer 1 value into CCP1 register when a message is received. This provides the time stamp to the message just received.

7	6	5	4	3	2	1	0
TX2SRC	TX2EN	ENDRHI	CANCAP	--	--	--	--

TX2SRC: CANTX2 pin data source bit

    0 = CANTX2 pin will output $\overline{CANTX1}$

    1 = CANTX2 pin will output the CAN clock

TX2EN: CANTX2 pin enable bit

    0 = CANTX2 pin will have digital I/O function

    1 = CANTX2 pin will output $\overline{CANTX1}$ or CAN clock as selected by TXSRC bit

ENDRHI: Enable drive high bit[1]

    0 = CANTX pin will be tri-state when recessive

    1 = CANTX pin will drive VDD when recessive

CANCAP: CAN message receive capture enable bit

    0 = Disable CAN capture, RC2/CCP1 input to CCP1 module

    1 = Enable CAN capture, CAN message receive signal replaces input on RC2/CCP1

Note 1. Always set this pin when using differential bus to avoid signal crosstalk in CANTX from other nearby pins.

Figure 13.43 ■ CAN I/O Control register (CIOCON) (redraw with permission of Microchip)

13.18 Physical CAN Bus Connection

The CAN protocol is designed for data communications over a short distance. A typical CAN bus system was shown in Figure 13.13, where both ends of the CAN bus must be terminated by a resistor with a resistance of about 120 Ω.

The CAN bus transceiver is connected to the CAN bus via two bus terminators, CAN_H and CAN_L, which provide differential receive and transmit capabilities. The nominal CAN bus levels are shown in Figure 13.44.

There are many semiconductor companies that produce CAN bus transceivers. The Microchip MCP2551, Philips PCA82C250/251, Texas Instruments SN65HVD251, and MAXIM MAX3050/3057 are among the most popular. All these chips are pin compatible. TI SN65HVD251, Philips PCA82C250, and Microchip MCP2551 are drop-in replaceable with each other.

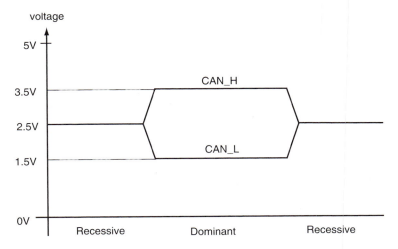

Figure 13.44 ■ Nominal CAN bus levels

13.18.1 The MCP2551 CAN Transceiver

The MCP2551 provides differential transmit and receive capability for the CAN protocol controller (or module). It operates at speeds up to 1 Mbps. The MCP2551 converts the digital signals generated by a CAN controller to signals suitable for transmission over the bus cabling (differential output). It also provides a buffer between the CAN controller and the high-voltage spikes that can be generated on the CAN bus by outside sources, such as EMI, ESD, electrical transients, and so on. The block diagram of the MCP2551 is shown in Figure 13.45.

The MCP2551 CAN transceiver outputs will drive a minimum load of 45 Ω, allowing a maximum of 112 nodes to be connected to the same CAN bus. The RxD pin reflects the differential bus voltage between CAN_H and CAN_L. The low and high states of the RxD output pin correspond to the dominant and recessive states of the CAN bus, respectively.

The R_S input allows the user to select one of the three operation modes:

- High speed
- Slope control
- Standby

The high-speed mode is selected by grounding the R_S pin. In this mode, the transmitter output drivers have fast output rise and fall times to support the high-speed CAN bus.

The slope-control mode further reduces electromagnetic interference (EMI) by limiting the rise and fall times of the CAN_H and CAN_L signals. The slope, or *slew rate* (SR), is controlled by connecting an external resistor (R_{EXT}) between the R_S pin and the ground pin. The slope is proportional to the current output at the R_S pin. Figure 13.46 illustrates typical slew rate values as a function of the slope-control resistance value.

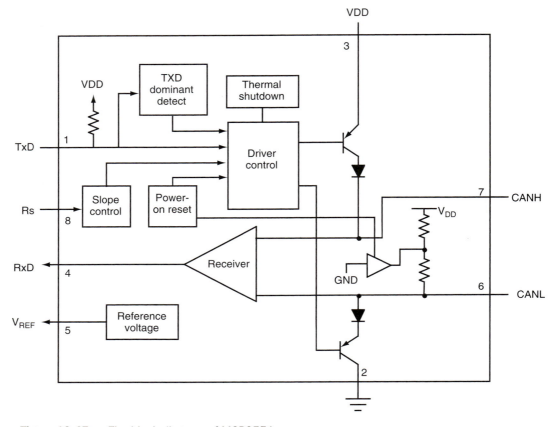

Figure 13.45 ■ The block diagram of MCP2551

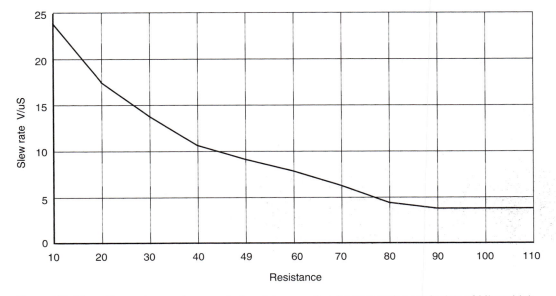

Figure 13.46 ■ Slew rate vs. slope control resistance value (redraw with permission of Microchip)

The MCP2551 can be placed in the standby or sleep mode by applying a high voltage at the R_S pin. In sleep mode, the transmitter is switched off, and the receiver operates at a lower current. The receive pin on the controller side (RxD) is still functioning but will operate at a slower rate. The attached microcontroller can monitor the RxD pin for CAN bus activity and place the transceiver into normal operation via the R_S pin.

13.18.2 Interfacing the MCP2551 to the PIC18F CAN Devices

A typical method of interfacing the MCP2551 transceiver to the PIC18 CAN module is shown in Figure 13.47. The CANTX2 pin is not used in this example. The CANTX2 pin can be used to find out the CAN baud rate in case problems caused by the baud rate mismatch arise or to provide the complementary version of the CANTX1 signal for CAN transceivers that require push and pull signals.

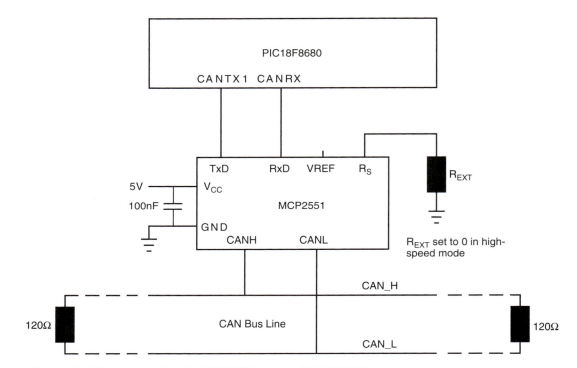

Figure 13.47 ■ Interfacing the MCP2551 with the PIC18F8680

The maximum achievable bus length in a CAN bus network is determined by the following physical factors:

1. The loop delay of the connected bus nodes (CAN controller and transceiver) and the delay of the bus line
2. The differences in bit time quantum length due to the relative oscillator tolerance between nodes
3. The signal amplitude drop due to the series resistance of the bus cable and the input resistance of bus node

The resultant equation after taking these three factors into account would be very complicated. The bus length that can be achieved as a function of the bit rate in the high-speed mode and with CAN bit timing parameters being optimized for maximum propagation delay is shown in Table 13.2.

Bit rate (kbit/s)	Bus length
1000	40
500	100
250	250
125	500
62.5	1000

Table 13.2 ■ CAN bus bit rate/bus length relation

The types and the cross sections suitable for the CAN bus trunk cable that has more than 32 nodes connected and span more than 100 meters are listed in Table 13.3.

Bus length/ Number of nodes	32	64	100
100 m	0.25 mm² or AWG 24	0.25 mm² or AWG 24	0.25 mm² or AWG 24
250 m	0.34 mm² or AWG 22	0.50 mm² or AWG 20	0.50 mm² or AWG 20
500 m	0.75 mm² or AWG 18	0.75 mm² or AWG 18	0.75 mm² or AWG 18

Table 13.3 ■ Minimum recommended bus wire cross-section for hte trunk cable

13.18.3 Setting the CAN Timing Parameters

The first component of the nominal bit time (NBT) in Equation 13.1 is **sync_seg.** The purpose of **sync_seg** is to inform all nodes on the CAN bus that a bit time is just started. The second component of NBT is the propagation delay segment, **prop_seg.** The existence of this segment is due to the fact that the CAN protocol allows nondestructive arbitration between nodes contending for access to the bus and the requirement for *in-frame acknowledgment.* In the case of nondestructive arbitration, more than one node may be transmitting during the arbitration field. Each transmitting node samples data from the bus in order to determine whether it has won or lost the arbitration and also to receive the arbitration field in case it loses arbitration. When each node samples a bit, the value sampled is the superposition (wired NAND is most common) of the bit values transmitted by each of the nodes arbitrating for bus access. In the case of the acknowledge field, the transmitting node transmits a recessive bit but expects to receive a dominant bit; in other words, a dominant value must be sampled at the sample time. The length of the **prop_seg** segment must be selected so that the earliest possible sample of the bit by a node until the transmitted bit values from all the transmitting nodes can reach all the nodes. To decide which node won in arbitration, the node that transmitted the earliest must wait until the last node transmits in order to find out if it won or lost. Suppose that node A and node B are the nodes that transmitted the earliest and the latest, respectively; then the worst-case value for t_{PROP_SEG} is given by

$$t_{PROP\_SEG} = t_{PROP(A,B)} + t_{PROP(B,A)} \tag{13.3}$$

where $t_{PROP(A,B)}$ and $t_{PROP(B,A)}$ are the propagation delays from node A to node B and from node B to node A, respectively. In the worst case, node A and node B are at the two ends of the CAN bus. Therefore, the propagation delay from node A to node B is given by

$$t_{PROP(A,B)} = t_{BUS} + t_{TX} + t_{RX} \tag{13.4}$$

where t_{BUS}, t_{TX}, and t_{RX} are the data traveling time on the bus, the transmitter propagation delay, and the receiver propagation delay, respectively.

Let node A and node B be two nodes at opposite ends of the CAN bus. Then the worst-case value for t_{PROP_SEG} is

$$t_{PROP\_SEG} = 2 \times (t_{BUS} + t_{TX} + t_{RX}) \tag{13.5}$$

The minimum number of time quanta (t_Q) that must be allocated to the **prop_seg** segment is therefore

$$prop\_seg = round\_up\ (t_{PROP\_SEG} \div t_Q) \tag{13.6}$$

where the **round_up( )** function returns a value that equals the argument rounded up to the next integer value.

In the absence of bus errors, bit stuffing guarantees a maximum 10-bit period between resynchronization edges (five dominant and five recessive bits, then followed by a dominant bit). This represents the worst-case condition for the accumulation of phase errors during normal communication. The accumulated phase error must be compensated for by resynchronization and therefore must be less than the programmed resynchronization jump width (t_{RJW}). The accumulated phase error is due to the tolerance in the CAN system clock, and this requirement can be expressed as

$$(2 \times \Delta f) \times 10 \times t_{NBT} < t_{RJW} \tag{13.7}$$

where Δf is the largest crystal oscillator frequency variation (in a percentage) of all CAN nodes in the network.

Real systems must operate in the presence of electrical noise, which may induce errors on the CAN bus. A node transmits an error flag after it has detected an error. In the case of a local error, only the node that detects the error will transmit the error flag. All other nodes receive the error flag and then transmit their own error flags as an echo. If the error is global, all nodes will detect it within the same bit time and will therefore transmit error flags simultaneously. A node can therefore differentiate between a local error and a global error by detecting whether there is an echo after its error flag. This requires that a node correctly samples the first bit after transmitting its error flag.

An error flag from an error-active node consists of six dominant bits, and there could be up to six dominant bits before the error flag if, for example, the error was a stuff error. A node must therefore correctly sample the 13th bit after the last resynchronization. This can be expressed as

$$(2 \times \Delta f) \times (13 \times t_{NBT} - t_{PHASE\_SEG2}) < min\ (t_{PHASE\_SEG1}, t_{PHASE\_SEG2}) \tag{13.8}$$

where, the function **min (arg1, arg2)** returns the smaller one of **arg1** and **arg2**.

The procedure for determining the optimum bit timing parameters that satisfy the requirements for proper bit sampling is as follows:

Step 1
Determine the minimum permissible time for the **prop_seg** segment using Equation 13.5.

Step 2
Choose the CAN system clock frequency, which will be half the CPU oscillator output frequency divided by a prescale factor. The CAN system clock is chosen so that the desired CAN bus nominal bit time is an integral multiple of the time quantum from 8 to 25.

Step 3
Calculate the **prop_seg** duration. The number of time quanta required for the **prop_seg** can be calculated by using Equation 13.6. If the result is greater than 8, go back to Step 2 and choose a lower CAN system clock frequency.

Step 4

Determine **phase_seg1** and **phase_seg2.** Subtract the **prop_seg** value and 1 from the time quanta contained in a bit time. If the difference is less than 3, then go back to Step 2 and select a higher CAN system clock frequency. If the difference is 3, then **phase_seg1** = 1 and **phase_seg2** = 2, and only one sample per bit may be chosen. Otherwise, divide the remaining number by 2 and assign the result to **phase_seg1** and **phase_seg2.**

Step 5

Determine the resynchronization jump width (RJW). RJW is the smaller one of four and **phase_seg1.**

Step 6

Calculate the required oscillator tolerance from Equations 13.7 and 13.8. If **phase_seg1** > 4, it is recommended that you repeat Steps 2 to 6 with a larger value for the prescaler. Conversely, if **phase_seg1** < 4, it is recommended that you repeat Steps 2 and 6 with a smaller value for the prescaler, as long as **prop_seg** < 8, as this may result in a reduced oscillator tolerance requirement. If the prescaler is already equal to 1 and a reduced oscillator tolerance is still required, the only option is to consider using a clock source with a higher frequency.

Example 13.3

▼

Calculate the CAN bit segments for the following constraints:

Bit rate = 800 Kbps

Bus length = 25 m

Bus propagation delay = 5×10^{-9} s/m

CAN transceiver plus receiver propagation delay = 150 ns at $85°\,C$

CPU oscillator frequency = 32 MHz

Solution: Follow the procedure described previously to do the calculation:

Step 1

Physical delay of the CAN bus = 25×5 = 125 ns:

$t_{PROP\_SEG} = 2 \times (125 + 150) = 550$ ns

Step 2

The minimum time quantum is obtained by setting the prescaler to 2, and the resultant time quantum is 2 ÷ (32 MHz/2) = 125 ns. One bit time is 1/800 Kbps = 1250 ns. Therefore, one bit time (NBT) corresponds to 10 (= 1250 ns ÷ 125) time quanta.

Step 3

Prop_seg = **round_up** (550 ns ÷ 125) = 5.

Step 4

Subtracting 5 (for **prop_seg**) and 1 (for **sync_seg**) from 10 time quanta per bit gives 4. Since the result is an even number, divide it by 2 and assign the quotient to **phase_seg1** and **phase_seg2.** Each receives two time quanta.

Step 5

Resynchronization jump width is the smaller one of 4 and **phase_seg1** and is 2.

Step 6

From Equation 13.7,

$\Delta f < RJW ÷ (20 \times NBT) = 2 ÷ (20 \times 10) = 1\%$

From Equation 13.8,

$\Delta f < \min(\text{phase\_seg1}, \text{phase\_seg2}) \div [2*(13 \times \text{NBT} - \text{phase\_seg2})]$
$= 2 \div 256 = 0.78\%$

The desired oscillator tolerance is the smaller one of these two values, that is, 0.78% over a period of 16 μs. Most crystal oscillators have tolerance smaller than 0.78%. Since **phase_seg1** is smaller than 4 and the prescaler is already set to 2 (prescaler 1 will make **prop_seg** to be 10 > 8 and is not allowed), there is no need to repeat Steps 2 to 6.

In summary:

Prescaler = 2
Nominal bit time = 10 (1.25 μs ÷ 125 ns)
prop_seg = 5
sync_seg = 1
phase_seg1 = 2
phase_seg2 = 2
RJW = 2
Oscillator tolerance = 0.78%

▲

Example 13.4

▼

Calculate the CAN bit segments for the following constraints:

Bit rate = 500 Kbps
Bus length = 70 m
Bus propagation delay = 5×10^{-9} s/m
CAN transceiver plus receiver propagation delay = 150 ns at 85° C
CPU oscillator frequency = 40 MHz

Solution: Solve the problem by following the procedure described previously:

Step 1
Physical delay of the bus = $70 \times 5 \times 10^{-9}$ sec/m = 350 ns:

$t_{\text{PROP\_SEG}} = 2 \times (350 + 150) = 1000$ ns

Step 2
Try the prescaler of 2 for the 40-MHz crystal oscillator. The resultant time quantum is 100 ns. This gives 2000/100 = 20 time quanta per bit.

Step 3
Prop_seg = **round_up** (1000 ÷ 100) = 10 > 8. Reset the prescaler to 4. The resultant new time quantum is 200 ns. One bit time corresponds to 10 time quanta. The new **prop_seg** = **round_up** (1000 ÷ 200) = 5.

Step 4

Subtracting 5 (for **prop_seg**) and 1 (for **sync_seg**) from 10 time quanta per bit gives 4. Since the result is even, divide it by 2 and assign the quotient to **phase_seg1** and **phase_seg2.**

Step 5

RJW is the smaller one of four and **phase_seg1** and is 2.

Step 6

From Equation 13.7,

$$\Delta f < RJW \div (20 \times NBT) = 2 \div (20 \times 10) = 1.0\%$$

From Equation 13.8,

$$\Delta f < \min(\text{phase\_seg1}, \text{phase\_seg2}) \div [2*(13 \times NBT - \text{phase\_seg2})]$$
$$= 2 \div 256 = 0.78\%$$

The desired oscillator tolerance is the smaller one of these two values, that is, 0.78% over a period of 25.6 μs. Most crystal oscillators have variations smaller than 0.78%. Since **phase_seg1** is smaller than 4 and the prescaler is already set to 4, there is no need to repeat Steps 2 to 6.

In summary:

Prescaler = 4

Nominal bit time = 10 (2 μs ÷ 200 ns)

Prop_seg = 5

Sync_seg = 1

Phase_seg1 = 2

Phase_seg2 = 2

RJW = 2

Oscillator tolerance = 0.78%

▲

13.19 CAN Programming

Data communications over the CAN bus requires the configuration and programming of the CAN module. The configuration of timing parameters should be done only once after power-on reset because all nodes over the CAN bus need to use the same baud rate in data communications. Other parameters, such as identifier filters and masks, may need to be changed when different data is needed for the application.

Only a small subset of CAN registers is resident in the access bank. When accessing those CAN registers residing outside the access bank, the user will need to switch the register bank. Using the **banksel** directive will simplify the register bank switching when writing assembly programs to access CAN-related registers. CAN registers that are located in the access bank are listed in Table 13.4. CAN registers located at banks 13, 14, and 15 are shown in Tables 13.5, 13.6, and 13.7, respectively. The user can also use the **movff** instruction to avoid the bank switching operation.

Address	Name	Address	Name	Address	Name
0xF60	RXB0CON	0xF6A	RXB0D4	0xF75	RXERRCNT
0xF61	RXB0SIDH	0xF6B	RXB0D5	0xF76	TXERRCNT
0xF62	RXB0SIDL	0xF6C	RXB0D6	0xF77	ECANCON
0xF63	RXB0EIDH	0xF6D	RXB0D7	0xF78	–
0xF64	RXB0EIDL	0xF6E	CANSTAT	0xF79	ECCP1DEL
0xF65	RXB0DLC	0xF6F	CANCON	0xF7A	–
0xF66	RXB0D0	0xF70	BRGCON1	0xF7B	–
0xF67	RXB0D1	0xF71	BRGCON2	0xF7C	–
0xF68	RXB0D2	0xF72	BRGCON3	0xF7D	–
0xF69	RXB0D3	0xF73	CIOCON	0xF7E	BAUDCON
		0xF74	COMSTAT	0xF7F	SPBRGH

Note. SPBRGH, BAUDCON, and ECCP1DEL are not in CAN module

Table 13.4 ■ CAN registers resident in the access bank

Address	Name	Address	Name	Address	Name
0xD60	RXF6SIDH	0xD78	RXF11SIDH	0xDD4	RXFCON0
0xD61	RXF6SIDL	0xD79	RXF11SIDL	0xDD5	RXFCON1
0xD62	RXF6EIDH	0xD7A	RXF11EIDH	0xDD8	SDFLC
0xD63	RXF6EIDL	0xD7B	RXF11EIDL	0xDE0	RXFBCON0
0xD64	RXF7SIDH	0xD80	RXF12SIDH	0xDE1	RXFBCON1
0xD65	RXF7SIDL	0xD81	RXF12SIDL	0xDE2	RXFBCON2
0xD66	RXF7EIDH	0xD82	RXF12EIDH	0xDE3	RXFBCON3
0xD67	RXF7EIDL	0xD83	RXF12EIDL	0xDE4	RXFBCON4
0xD68	RXF8SIDH	0xD84	RXF13SIDH	0xDE5	RXFBCON5
0xD69	RXF8SIDL	0xD85	RXF13SIDL	0xDE6	RXFBCON6
0xD6A	RXF8EIDH	0xD86	RXF13EIDH	0xDE7	RXFBCON7
0xD6B	RXF8EIDL	0xD87	RXF13EIDL	0xDF0	MSEL0
0xD70	RXF9SIDH	0xD88	RXF14SIDH	0xDF1	MSEL1
0xD71	RXF9SIDL	0xD89	RXF14SIDL	0xDF2	MSEL2
0xD72	RXF9EIDH	0xD8A	RXF14EIDH	0xDF3	MSEL3
0xD73	RXF9EIDL	0xD8B	RXF14EIDL	0xDF8	BSEL0
0xD74	RXF10SIDH	0xD90	RXF15SIDH	0xDFA	BIE0
0xD75	RXF10SIDL	0xD91	RXF15SIDL	0xDFC	TXBIE
0xD76	RXF10EIDH	0xD92	RXF15EIDH		
0xD77	RXF10EIDL	0xD93	RXF15EIDL		

Table 13.5 ■ CAN registers resident in the bank 13 (0xD80 ~ 0xDFC)

Address	Name	Address	Name	Address	Name
0xE20	B0CON	0xE40	B2CON	0xE60	B4CON
0xE21	B0SIDH	0xE41	B2SIDH	0xE61	B4SIDH
0xE22	B0SIDL	0xE42	B2SIDL	0xE62	B4SIDL
0xE23	B0EIDH	0xE43	B2EIDH	0xE63	B4EIDH
0xE24	B0EIDL	0xE44	B2EIDL	0xE64	B4EIDL
0xE25	B0DLC	0xE45	B2DLC	0xE65	B4DLC
0xE26	B0D0	0xE46	B2D0	0xE66	B4D0
0xE27	B0D1	0xE47	B2D1	0xE67	B4D1
0xE28	B0D2	0xE48	B2D2	0xE68	B4D2
0xE29	B0D3	0xE49	B2D3	0xE69	B4D3
0xE2A	B0D4	0xE4A	B2D4	0xE6A	B4D4
0xE2B	B0D5	0xE4B	B2D5	0xE6B	B4D5
0xE2C	B0D6	0xE4C	B2D6	0xE6C	B4D6
0xE2D	B0D7	0xE4D	B2D7	0xE6D	B4D7
0xE2E	CANSTAT_RO9	0x42E	CANSTAT_RO7	0xE6E	CANSTAT_RO5
0xE2F	CANCON_RO9	0x42F	CANCON_RO7	0xE6F	CANCON_RO5
0xE30	B1CON	0xE50	B3CON	0xE70	B5CON
0xE31	B1SIDH	0xE51	B3SIDH	0xE71	B5SIDH
0xE32	B1SIDL	0xE52	B3SIDL	0xE72	B5SIDL
0xE33	B1EIDH	0xE53	B3EIDH	0xE73	B5EIDH
0xE34	B1EIDL	0xE54	B3EIDL	0xE74	B5EIDL
0xE35	B1DLC	0xE55	B3DLC	0xE75	B5DLC
0xE36	B1D0	0xE56	B3D0	0xE76	B5D0
0xE37	B1D1	0xE57	B3D1	0xE77	B5D1
0xE38	B1D2	0xE58	B3D2	0xE78	B5D2
0xE39	B1D3	0xE59	B3D3	0xE79	B5D3
0xE3A	B1D4	0xE5A	B3D4	0xE7A	B5D4
0xE3B	B1D5	0xE5B	B3D5	0xE7B	B5D5
0xE3C	B1D6	0xE5C	B3D6	0xE7C	B5D6
0xE3D	B1D7	0xE5D	B3D7	0xE7D	B5D7
0xE3E	CANSTAT_RO8	0xE5E	CANSTAT_RO6	0xE7E	CANSTAT_RO4
0xE3F	CANCON_RO8	0xE5F	CANCON_RO6	0xE7F	CANCON_RO4

Table 13.6 ■ CAN registers resident in bank 14 (0xE20 ~ 0xE7F)

Address	Name	Address	Name	Address	Name
0xF00	RXF0SIDH	0xF20	TXB2CON	0xF40	TXB0CON
0xF01	RXF0SIDL	0xF21	TXB2SIDH	0xF41	TXB0SIDH
0xF02	RXF0EIDH	0xF22	TXB2SIDL	0xF42	TXB0SIDL
0xF03	RXF0EIDL	0xF23	TXB2EIDH	0xF43	TXB0EIDH
0xF04	RXF1SIDH	0xF24	TXB2EIDL	0xF44	TXB0EIDL
0xF05	RXF1SIDL	0xF25	TXB2DLC	0xF45	TXB0DLC
0xF06	RXF1EIDH	0xF26	TXB2D0	0xF46	TXB0D0
0xF07	RXF1EIDL	0xF27	TXB2D1	0xF47	TXB0D1
0xF08	RXF2SIDH	0xF28	TXB2D2	0xF48	TXB0D2
0xF09	RXF2SIDL	0xF29	TXB2D3	0xF49	TXB0D3
0xF0A	RXF2EIDH	0xF2A	TXB2D4	0xF4A	TXB0D4
0xF0B	RXF2EIDL	0xF2B	TXB2D5	0xF4B	TXB0D5
0xF0C	RXF3SIDH	0xF2C	TXB2D6	0xF4C	TXB0D6
0xF0D	RXF3SIDL	0xF2D	TXB2D7	0xF4D	TXB0D7
0xF0E	RXF3EIDH	0xF2E	CANSTAT_RO3	0xF4E	CANSTAT_RO1
0xF0F	RXF3EIDL	0xF2F	CANCON_RO3	0xF4F	CANCON_RO1
0xF10	RXF4SIDH	0xF30	TXB1CON	0xF50	RXB1CON
0xF11	RXF4SIDL	0xF31	TXB1SIDH	0xF51	RXB1SIDH
0xF12	RXF4EIDH	0xF32	TXB1SIDL	0xF52	RXB1SIDL
0xF13	RXF4EIDL	0xF33	TXB1EIDH	0xF53	RXB1EIDH
0xF14	RXF5SIDH	0xF34	TXB1EIDL	0xF54	RXB1EIDL
0xF15	RXF5SIDL	0xF35	TXB1DLC	0xF55	RXB1DLC
0xF16	RXF5EIDH	0xF36	TXB1D0	0xF56	RXB1D0
0xF17	RXF5EIDL	0xF37	TXB1D1	0xF57	RXB1D1
0xF18	RXM0SIDH	0xF38	TXB1D2	0xF58	RXB1D2
0xF19	RXM0SIDL	0xF39	TXB1D3	0xF59	RXB1D3
0xF1A	RXM0EIDH	0xF3A	TXB1D4	0xF5A	RXB1D4
0xF1B	RXM0EIDL	0xF3B	TXB1D5	0xF5B	RXB1D5
0xF1C	RXM1SIDH	0xF3C	TXB1D6	0xF5C	RXB1D6
0xF1D	RXM1SIDL	0xF3D	TXB1D7	0xF5D	RXB1D7
0xF1E	RXM1EIDH	0xF3E	CANSTAT_RO2	0xF5E	CANSTAT_RO0
0xF1F	RXM1EIDL	0xF3F	CANCON_RO2	0xF5F	CANCON_RO0

Table 13.7 ■ CAN registers resident in bank 15 (0xF00 ~ 0xF5F)

Example 13.5

▼

Write a program to configure the CAN module with the timing parameters in Example 13.3, disable all interrupts, and set the ECAN module to receive all messages with standard identifiers.

Solution: In order to configure the ECAN module, the user needs to set the ECAN module in configuration mode. To enable the ECAN to receive all messages with standard identifiers, the user can set all mask bits to 0 and set the EXIDEN bit of the RXFnSIDL registers to 0. The following assembly program does the desired configuration:

```
#include <p18F8680.inc>
org                 0x00
goto                start
org                 0x08
```

```
                retfie
                org             0x18
                retfie
start           movlw           0x80
                movwf           CANCON,A            ; set to configuration mode
conf_wait       movf            CANSTAT,W,A         ; read current mode state
                andlw           0x80
                bz              conf_wait           ; if not configuration mode, branch
                clrf            ECANCON,F,A         ; select mode 0
                movlw           0x41                ; set RJW to 2, prescaler set to 2
                movwf           BRGCON1,A           ; "
                movlw           0x89                ; take one sample, set phase_seg1 to 2 and
                movwf           BRGCON2,A           ; set prop_seg to 5
                movlw           0x41                ; enable wakeup feature, set phase_seg2 to 2
                movwf           BRGCON3,A           ; "
                movlw           0x20                ; RXB0 and RXB1 receive any standard messages
                movff           WREG,RXB0CON        ; "
                movff           WREG,RXB1CON        ; "
                movlw           0x00                ; clear receive mask registers
                movff           WREG,RXM0SIDH       ; to receive all standard format messages
                movff           WREG,RXM0SIDL       ; "
                movff           WREG,RXM1SIDH       ; "
                movff           WREG,RXM1SIDL       ; "
                clrf            CANCON,A            ; set to normal mode
conf_w1         movf            CANSTAT,W,A         ; read current mode state
                andlw           0xE0                ; check bits 7–5 for current mode
                bnz             conf_w1             ; if not normal mode, branch
                end
```

The C language version of the program is as follows:

```
#include <p18F8680.h>
void main (void)
{
    CANCON = 0x80;                  /* set to configuration mode */
    while (!(CANSTAT & 0x80));      /* wait until configuration mode is entered */
    ECANCON = 0x00;                 /* select mode 0 */
    BRGCON1 = 0x41;                 /* set RJW to 2, prescaler set to 2 */
    BRGCON2 = 0x8C;                 /* take one sample, set phase_seg1 to 2 and set prop_seg to 5 */
    BRGCON3 = 0x41;                 /* enable wakeup feature, set phase_seg2 to 2 */
    RXB0CON = 0x20;                 /* RXB0 receive all standard messages */
    RXB1CON = 0x20;                 /* RXB1 receive all standard messages */
    RXM0SIDH = 0;                   /* clear receive mask registers to receive all */
    RXM0SIDL = 0;                   /* standard format messages */
    RXM1SIDH = 0;
    RXM1SIDL = 0;
    CANCON = 0;                     /* set to normal mode */
    while (CANSTAT & 0xE0);         /* wait until normal mode is entered */
}
```

Example 13.6

▼

Write a program to configure the CAN module with the timing parameters computed in Example 13.4, disable all interrupts, and set the ECAN module to receive all extended messages started with extended identifier "SH". Use mode 0 and receive buffer 0 to receive the message.

Solution: Since we want to use receive buffer 0 to receive all the extended messages with identifiers started with "SH", we need to load the values of 0x53 and 0x49 (not 0x48) into the RXB0SIDH and RXB0SIDL registers, respectively. Since we don't care the rest of the identifier, the remainder bits can be cleared to 0. Therefore, we can load the value 0x00 into the RXB0EIDH and RXB0EIDL registers. The assembly program that performs the required configuration is as follows:

```
            #include <p18F8680.inc>
            org     0x00
            goto    start
            org     0x08
            retfie
            org     0x18
            retfie
start       movlw   0x80
            movwf   CANCON,A          ; set to configuration mode
conf_wait   movf    CANSTAT,W,A       ; read current mode state
            andlw   0x80
            bz      conf_wait         ; if not configuration mode, branch
            clrf    ECANCON,A         ; select mode 0
            movlw   0x43              ; set RJW to 2, prescaler set to 4
            movwf   BRGCON1,A         ; "
            movlw   0x8C              ; take one sample, set phase_seg1 to 2 and
            movwf   BRGCON2,A         ; set prop_seg to 5
            movlw   0x41              ; enable wakeup feature, set phase_seg2 to
            movwf   BRGCON3,A         ; 2
            movlw   0x53              ; load "SH" into receive acceptance filter
            movff   WREG,RXF0SIDH     ; registers
            movlw   0x49              ; "
            movff   WREG,RXF0SIDL     ; "
            movlw   0x00
            movff   WREG,RXF0EIDH     ; "
            movff   WREG,RXF0EIDL     ; "
            movlw   0xFF
            movff   WREG,RXM0SIDH     ; set up receive mask registers to receive all
            movlw   0xEB              ; extended messages with identifiers starting
            movff   WREG,RXM0SIDL     ; with "SH"
            movlw   0xE0              ; "
            movff   WREG,RXM1EIDH     ; "
            movlw   0x00
            movff   WREG,RXM1EIDL     ; "
            movff   WREG,CANCON       ; set to normal mode
conf_w1     movf    CANSTAT,W         ; read current mode state
            andlw   0xE0              ; check bits 7-5 for current mode
            bnz     conf_w1           ; if not normal mode, branch
            end
```

The C language version of the program is as follows:

```
#include <p18F8680.h>
void main (void)
{
    CANCON = 0x80;                  /* set to configuration mode */
    while(!(CANSTAT & 0x80));       /* wait until configuration mode is entered */
    ECANCON = 0x00;                 /* select mode 0 */
    BRGCON1 = 0x43;                 /* set RJW to 2, set baud rate prescaler to 4 */
    BRGCON2 = 0x8C;                 /* set phase_seg1 to 2, prop_seg to 5, one sample */
    BRGCON3 = 0x41;                 /* enable wakeup and set phase2_seg to 2 */
    RXFOSIDH = 0x53;                /* set receive acceptance filter to accept */
    RXFOSIDL = 0x49;                /* extended identifier started with "SH" */
    RXFOEIDH = 0x00;                /* " */
    RXFOEIDL = 0x00;                /* " */
    RXM0SIDH = 0xFF;                /* set up receive acceptance mask to compare */
    RXM0SIDL = 0xEB;                /* the first 19 bits */
    RXM0EIDH = 0xE0;                /* " */
    RXM0EIDL = 0x00;                /* " */
    CANCON = 0x00;                  /* enter normal mode */
    while (CANSTAT & 0xE0);         /* wait until normal mode is entered */
}
```

In some applications, a CAN node may need to input data from many other nodes to perform its functions. If the CAN node needs to process each data source separately, then using different receive acceptance filters for each data source would be a good idea. Mode 1 of the ECAN module is suitable for this application. The following example illustrates the configuration of this type of requirement.

▲

Example 13.7

▼

Assume that CAN node K needs to accept data from 15 other CAN nodes with one node sending one type of data that must be handled independently of other data sources. Write a subroutine to do the appropriate setup.

Solution: Assume that CAN node K and 15 nodes agree that they use the two-character identifier starting with K and followed with a hex digit from 0 to E as the identifier for the messages from each CAN node. Since there are 15 different identifiers, we will need 15 acceptance filters for making acceptance decision. Only mode 1 or mode 2 can support 15 acceptance filters. Mode 1 is chosen for this example. There are two fixed received buffers and six programmable buffers, B0 . . . B5. We will use five receive buffers to hold incoming messages: RXB0, RXB1, B0, B1, and B2. Three acceptance filters will be assigned to each receive buffer:

RXB0: RXF0, RXF1, RXF2

RXB1: RXF3, EXF4, RXF5

B0: RXF6, RXF7, RXF8

B1: RXF9, RXF10, RXF11

B2: RXF12, RXF13, RXF14

All these filters should be enabled. The following operations need to be performed by the configuration program:

1. Enter the configuration mode.
2. Select mode 1 for operation.
3. Enable and set up 15 receive acceptance filters.
4. Configure programmable buffers B0 . . . B2 for reception and configure buffers B3 . . . B5 for transmission.
5. Assign filters to receive buffers (three acceptance filter registers are assigned to one receive buffer).
6. Use receive acceptance mask register 0 and set up the mask register to compare the first 16 bits of the identifier for acceptance comparison.
7. Set up mask select registers (there are four mask select registers).

The following subroutine performs the desired configuration:

```
can_rcv_up      movlw     0x80
                movwf     CANCON,A            ; enter the configuration mode
conf_wait       movf      CANSTAT,W,A         ; wait until configuration mode is
                andlw     0x80                ; entered
                bz        conf_wait           ; "
                movlw     0x40                ; set CAN module to mode 1 operation
                movwf     ECANCON,A           ; "
; enable all receive acceptance filters
                movlw     0xFF                ; enable all receive acceptance filters 0 to 14
                movff     WREG,RXFCON0        ; "
                movlw     0x7F                ; "
                movff     WREG,RXFCON1        ; "
; associate receive acceptance filters with receive buffers
                movlw     0x00
                movff     WREG,RXFBCON0       ; associate filter 1 and 0 with RXB0
                movlw     0x10                ; associate filter 2 with RXB0 and associate
                movff     WREG,RXFBCON1       ; filter 3 with RXB1
                movlw     0x11                ; associate filter 4 and 5 with buffer RXB1
                movff     WREG,RXFBCON2       ; "
                movlw     0x22                ; associate filter 6 and 7 with buffer B0
                movff     WREG,RXFBCON3       ; "
                movlw     0x32                ; associate filter 8 with B0 and associate
                movff     WREG,RXFBCON4       ; filter 9 with B1
                movlw     0x33                ; associate filter 10 and 11 with buffer B1
                movff     WREG,RXFBCON5       ; "
                movlw     0x44                ; associate filters 12 and 13 with buffer B2
                movff     WREG,RXFBCON6       ; "
                movlw     0x04                ; associate filter 14 with buffer B2
                movff     WREG,RXFBCON7       ; "
; configure programmable buffers for receive and transmit
                movlw     0xE0                ; configure B5..B3 for transmit whereas
                movff     WREG,BSEL0          ; B2..B0 for receive
```

```
; associate all acceptance filters with mask 0
                movlw       0x00
                movff       WREG,MSEL0
                movff       WREG,MSEL1
                movff       WREG,MSEL2
                movff       WREG,MSEL3
; set up receive buffer control registers so that only valid messages with
; extended identifiers are received
                movff       WREG,RXB0CON         ; receive valid message per EXIDEN bit
                movff       WREG,RXB1CON         ; receive valid message per EXIDEN bit
                movff       WREG,B0CON           ; receive valid message per EXIDEN bit
                movff       WREG,B1CON           ; "
                movff       WREG,B2CON           ; "
; set up mask 0 identifier mask register
                movlw       0xFF
                movff       WREG,RXM0SIDH        ; set up acceptance mask 0 so that only the
                movlw       0xEB                 ; first 16 bits of the identifier are
                movff       WREG,RXM0SIDL        ; compared
                movlw       0xE0                 ; "
                movff       WREG,RXM0EIDH        ; "
                movlw       0x00
                movff       WREG,RXM0EIDL        ; "
; set up acceptance filter 0 to 5 (bank F)
                movlw       0x4B                 ; ASCII code of K
                movff       WREG,RXF0SIDH
                movff       WREG,RXF1SIDH
                movff       WREG,RXF2SIDH
                movff       WREG,RXF3SIDH
                movff       WREG,RXF4SIDH
                movff       WREG,RXF5SIDH
                movff       WREG,RXF6SIDH
                movff       WREG,RXF7SIDH
                movff       WREG,RXF8SIDH
                movff       WREG,RXF9SIDH
                movff       WREG,RXF10SIDH
                movff       WREG,RXF11SIDH
                movff       WREG,RXF12SIDH
                movff       WREG,RXF13SIDH
                movff       WREG,RXF14SIDH
                movlw       0x2A
                movff       WREG,RXF0SIDL
                movff       WREG,RXF1SIDL
                movff       WREG,RXF2SIDL
                movff       WREG,RXF3SIDL
                movff       WREG,RXF4SIDL
                movff       WREG,RXF5SIDL
                movff       WREG,RXF6SIDL
                movff       WREG,RXF7SIDL
```

```
          movlw      0x2B
          movff      WREG,RXF8SIDL
          movff      WREG,RXF9SIDL
          movlw      0x48
          movff      WREG,RXF10SIDL
          movff      WREG,RXF11SIDL
          movff      WREG,RXF12SIDL
          movff      WREG,RXF13SIDL
          movff      WREG,RXF14SIDL
          movlw      0x00
          movff      WREG,RXF0EIDH        ; hex digit 0
          movlw      0x20
          movff      WREG,RXF1EIDH        ; hex digit 1
          movlw      0x40
          movff      WREG,RXF2EIDH        ; hex digit 2
          movlw      0x60
          movff      WREG,RXF3EIDH        ; hex digit 3
          movlw      0x80
          movff      WREG,RXF4EIDH        ; hex digit 4
          movlw      0xA0
          movff      WREG,RXF5EIDH        ; hex digit 5
          movlw      0xC0
          movff      WREG,RXF6EIDH        ; ASCII code of hex digit 6
          movlw      0xE0
          movff      WREG,RXF7EIDH        ; ASCII code of hex digit 7
          movlw      0x00
          movff      WREG,RXF8EIDH        ; ASCII code of hex digit 8
          movlw      0x20
          movff      WREG,RXF9EIDH        ; ASCII code of hex digit 9
          movff      WREG,RXF10EIDH       ; ASCII code of hex digit A
          movlw      0x40
          movff      WREG,RXF11EIDH       ; ASCII code of hex digit B
          movlw      0x60
          movff      WREG,RXF12EIDH       ; ASCII code of hex digit C
          movlw      0x80
          movff      WREG,RXF13EIDH       ; ASCII code of hex digit D
          movlw      0xA0
          movff      WREG,RXF14EIDH       ; ASCII code of hex digit E
          movlw      0x00
          movff      WREG,RXF0EIDL        ; not needed byte
          movff      WREG,RXF1EIDL        ; not needed byte
          movff      WREG,RXF2EIDL        ; not needed byte
          movff      WREG,RXF3EIDL        ; not needed byte
          movff      WREG,RXF4EIDL
          movff      WREG,RXF5EIDL        ; not needed byte
          movff      WREG,RXF6EIDL        ; unneeded byte
          movff      WREG,RXF7EIDL        ; "
          movff      WREG,RXF8EIDL        ; "
          movff      WREG,RXF9EIDL        ; "
```

```
            movff      WREG,RXF10EIDL      ; "
            movff      WREG,RXF11EIDL      ; "
            movff      WREG,RXF12EIDL      ; "
            movff      WREG,RXF13EIDL      ; "
            movff      WREG,RXF14EIDL      ; "
            movff      WREG,CANCON         ; exit configuration mode and reenter normal mode
conf_w2     movf       CANSTAT,W           ; read current mode state
            andlw      0xE0
            bnz        conf_w2             ; if not normal mode, branch
            return     0
```

The C language version of the subroutine is straightforward and hence is left for you as an exercise.

Example 13.8

▼

Assume that one wants to set up an automatic response to any remote transmit request with an identifier starting with the string "TW" and send back the message "T is one" using buffer B5. Write an instruction to perform the required setup.

Solution: One should follow the procedure described in Section 13.12.4 to solve this problem. We will use the RXF14 as the acceptance filter and associate it with the RXM1. The acceptance filter will be set up to accept any extended identifiers starting with "TW". The instruction sequence to carry out the setup is as follows:

```
            movlw      0x80
            movwf      CANCON,A            ; enter configuration mode
conf_waitx  movwf      CANSTAT,A           ; wait until the configuration
            andlw      0x80                ; mode is entered
            bz         conf_waitx          ; "
            movff      BSEL0,WREG          ; enable B5 for transmission
            iorlw      0x80                ; "
            movff      WREG,BSEL0          ; "
            movff      B5CON,WREG          ; enable B5 to accept RTR and reply
            iorlw      0x04                ; "
            movff      WREG,B5CON          ; "
            movff      RXFCON1,WREG        ; enable the acceptance filter RXF14
            iorlw      0x40                ; "
            movff      WREG,RXFCON1        ; "
            movlw      0xF7                ; associate filter 14 with B5
            movff      WREG,RXFBCON7       ; "
            movlw      0xD0                ; select acceptance mask 1 for acceptance filter 14,
            movff      WREG,MSEL3          ; no mask for filter 15
            movlw      0x54                ; set "TW" as the acceptance
            movff      WREG,RXF14SIDH      ; filter
            movlw      0x4A                ; "
            movff      WREG,RXF14SIDL      ; "
            movlw      0xE0                ; "
            movff      WREG,RXF14EIDH      ; "
            movlw      0x00
```

```
        movff   WREG,RXF14EIDL    ; "
        movlw   0xFF
        movff   WREG,RXM1SIDH     ; set acceptance mask to compare
        movlw   0xEB              ; with the first 19 bits
        movff   WREG,RXM1SIDL     ; "
        movlw   0xE0              ; "
        movff   WREG,RXM1EIDH     ; "
        movlw   0x00
        movff   WREG,RXM1EIDL     ; "
        movlw   0x54              ; set up identifier ("TW") of response
        movff   WREG,B5SIDH       ; to be sent out
        movlw   0x4A              ; "
        movff   WREG,B5SIDL       ; "
        movlw   0xE0              ; "
        movff   WREG,B5EIDH       ; "
        movlw   0x00
        movff   WREG,B5EIDL       ; "
        movlw   'T'               ; store the response "T is one"
        movff   WREG,B5D0         ; in transmit buffer
        movlw   0x20              ; "
        movff   WREG,B5D1         ; "
        movff   WREG,B5D4         ; "
        movlw   'i'               ; "
        movff   WREG,B5D2         ; "
        movlw   's'               ; "
        movff   WREG,B5D3         ; "
        movlw   'o'               ; "
        movff   WREG,B5D5         ; "
        movlw   'n'               ; "
        movff   WREG,B5D6         ; "
        movlw   'e'               ; "
        movff   WREG,B5D7         ; "
        movlw   0x8               ; "
        movff   WREG,B5DLC        ; "
        clrf    CANCON,A          ; return to normal mode
conf_wx movf    CANSTAT,W,A       ; wait until normal mode is entered
        andlw 0xE0  ; "
        bnz conf_wx  ; "
```

The C function that performs the same setup is as follows:

```
#define    space        0x20
void rtr_auto_rply(void)
{
        CANCON = 0x80;                  /* enter configuration mode */
        while (0x80 & CANSTAT);         /* wait until configuration mode is entered */
        BSEL0bits.B5TXEN = 1;           /* enable B5 for transmission */
        B5CONbits.RTREN = 1;            /* enable B5 to accept RTR and reply */
        RXFCON1bits.RXF14EN = 1;        /* enable the acceptance filter RXF14 */
        RXFBCON7 = 0xF7;                /* associate filter 14 with B5 */
        MSEL3 = 0xD0;                   /* select acceptance mask 1 for acceptance filter 14 */
```

```
                RXF14SIDH = 0x54;              /* set "TW" as the acceptance filter string */
                RXF14SIDL = 0x4A;             /* " " */
                RXF14EIDH = 0xE0;             /* " " */
                RXF14EIDL = 0;               /* " " */
                RXM1SIDH = 0xFF;             /* set acceptance filter mask to compare the first 19
                RXM1SIDL = 0xEB;             /* bits */
                RXM1EIDH = 0xE0;            /* " " */
                RXM1EIDL = 0x00;           /* " " */
                B5SIDH = 0x54;            /* set up identifier ("TW") for response */
                B5SIDL = 0x4A;           /* " " */
                B5EIDH = 0xE0;          /* " " */
                B5EIDL = 0x00;         /* " " */
                B5D0 = 'T';           /* store the reply "T is one" in buffer B5 */
                B5D1 = space;        /* " " */
                B5D2 = 'i';         /* " " */
                B5D3 = 's';        /* " " */
                B5D4 = space;     /* " " */
                B5D5 = 'o';      /* " " */
                B5D6 = 'n';     /* " " */
                B5D7 = 'e';    /* " " */
                B5DLC = 8;        /* set data length count to 8 */
                CANCON = 0x00;            /* return to normal mode */
                while (CANSTAT & 0xE0);        /* wait until normal mode is entered */
}
```

How to handle the received message is dependent on the application. The minimum processing should include the copying of the message from the receive buffer to a data buffer set aside by the user and the clearing of the interrupt flag. The following example demonstrates how to write an interrupt service routine to handle CAN mode 0 interrupts.

Example 13.9

Write a skeleton of the C program that handles the CAN mode 0 receive interrupts, assuming that all CAN interrupts have been set to high priority.

Solution: The skeleton of this program is as follows:

```
#include <p18F8680.h>
void high_ISR (void);
void low_ISR (void);
unsigned char tempcanstat;
unsigned char rxb0_buf[14], rxb1_buf[14];
#pragma code high_vector = 0x08
void high_interrupt (void)
{
        _asm
        goto high_ISR
        _endasm
}
#pragma code low_vector = 0x18
```

```c
void low_interrupt (void)
{
        _asm
        goto low_ISR
        _endasm
}
#pragma code
#pragma interrupt high_ISR
void high_ISR (void)
{
        unsigned char i, *cpt;              /* pointer to unsigned char */
        tempcanstat = CANSTAT & 0x0E;       /* make a copy of the CANSTAT */
        switch (tempcanstat) {
        case 0: break;
        case 2: break;
        case 4: // handle TXB2 interrupt
                break;
        case 6: // handle TXB1 interrupt
                break;
        case 8: // handle TXB0 interrupt
                break;
        case 10: // handle RXB1 interrupt by copying the received message
                cpt = &RXB1D0;
                for (i = 0; i < RXB1DLC; i++)
                        rxb1_buf[i] = *cpt++;
                PIR3bits.RXB1IF = 0; // clear the interrupt flag RXB1IF
                break;
        case 12: // handle RXB0 interrupt by copying the received message
                cpt = &RXB0D0;
                for (i = 0; i < RXB0DLC; i++)
                        rxb0_buf[i] = *cpt++;
                PIR3bits.RXB0IF = 0; // clear the interrupt flag RXB0IF
                break;
        case 14: // handle wake-up interrupt
                break;
        default: break;
    }
}
#pragma code
#pragma interrupt low_ISR
void low_ISR (void)
{
     _asm
     retfie 0
     _endasm
}
void main (void)
{
}
```

13.20 Summary

The CAN bus specification was initially proposed as a data communication protocol for automotive applications. However, it can also fulfill the data communication needs of a wide range of applications, from high-speed networks to low-cost multiplex wiring.

The CAN protocol has gone through several revisions. The latest revision is 2.0A/B. The CAN 2.0A uses a standard identifier, whereas the CAN 2.0B specification accepts extended identifiers. Both the CAN and the ECAN modules in the PIC18 microcontrollers support the CAN 2A/B specifications.

The CAN protocol supports four types of messages: data frame, remote frame, error frame, and overload frame. Users need to transfer only data frames and remote frames. The other two types of frames are used by the CAN controller only to control data transmission and reception. Data frames are used to carry normal data, whereas remote frames are used to request other nodes to send messages with the same identifier as in the remote frame.

The CAN protocol allows all nodes on the bus to transmit simultaneously. When there are multiple transmitters on the bus, they arbitrate, and the CAN node transmitting a message with the highest priority wins. The simultaneous transmission of multiple nodes will not cause any damage to the CAN bus. CAN data frames are acknowledged in frame; that is, a receiving node sets a bit only in the acknowledge field of the incoming frame. There is no need to send a separate acknowledge frame.

The CAN bus has two states: **dominant** and **recessive.** The dominant state represents logic 0, and the recessive state represents logic 1 for most CAN implementations. When one node drives the dominant voltage to the bus while other nodes drive the recessive level to the bus, the resultant CAN bus state will be the dominant state.

Synchronization is a critical issue in the CAN bus. Each bit time is divided into four segments: **sync_seg, prop_seg, phase_seg1,** and **phase_seg2.** The sync_seg segment signals the start of a bit time. The sample point is between phase_seg1 and phase_seg2. At the beginning of each frame, every node performs a hard synchronization to align its sync_seg segment of its current bit time to the recessive-to-dominant edge of the transmitted start-of-frame. Resynchronization is performed during the remainder of the frame whenever a change of bit value from recessive to dominant occurs outside the expected sync_seg segment.

A CAN module (also called CAN controller) requires a CAN bus transceiver, such as the Philips PCA82C250 or the Microchip MCP2551, to interface with the CAN bus. Most CAN controllers allow more than 100 nodes to connect to the CAN bus. The CAN trunk cable could be a shielded cable, unshielded twisted pair, or simply a pair of insulated wires. It is recommended that you use shielded cable for high-speed transfer when there is a radio frequency interference problem. Up to a 1–Mbps data rate is achievable over a distance of 40 meters.

The CAN module of the PIC18 microcontroller has six major modes of operations:

    Configuration mode

    Disable mode

    Normal operation mode

    Listen-only mode

    Loopback mode

    Error recognition mode

All configuration operations can be performed only in the configuration mode. All data transfer and reception with other nodes are performed in normal operation mode.

In the PIC18 family, the PIC18C658/858, the PIC18F248/258/448/458, and the PIC18F6585/8585/6680/8680 have an on-chip CAN module. The CAN modules in the PIC18C658/858 and the PIC18F248/258/448/458 are identical. However, two additional modes (mode 1 and mode 2) were added to enhance the CAN modules in the PIC18F6585/8585/6680/8680 devices. Mode 0 of this group of devices is identical to that of the first two groups of devices.

Mode 0 of the CAN module has three transmit buffers, two receive modules, two acceptance masks, and six receive acceptance filters. Mode 1 and mode 2 ECAN have additional resources to provide more flexibility. The added resources include six programmable transmit/receive buffers and 10 additional receive acceptance filters.

Each transmit and receive buffer has eight data registers for holding data, four identifier registers for identification and arbitration purposes, one length count register, and one control register. In mode 1 and mode 2, any receive acceptance filter register can be associated with any receive buffer, and each filter register can select any one of the three mask registers.

13.21 Exercises

E13.1 Calculate the bit segments for the following system constraints, assuming that the Philips PCA82C250 transceiver is used:

Bit rate = 400 Kbps

Bus length = 50 m

Bus propagation delay = 5×10^{-9} s/m

PCA82C250 transceiver and receiver propagation delay = 150 ns at 85° C

Oscillator frequency = 32 MHz

E13.2 Calculate the bit segments for the following system constraints, assuming that the Philips PCA82C250 transceiver is used:

Bit rate = 200 Kbps

Bus length = 100 m

Bus propagation delay = 5×10^{-9} s/m

PCA82C250 transceiver and receiver propagation delay = 150 ns at 85° C

Oscillator frequency = 32 MHz

E13.3 Calculate the bit segments for the following system constraints, assuming that the Philips PCA82C250 transceiver is used:

Bit rate = 1 Mbps

Bus length = 20 m

Bus propagation delay = 5×10^{-9} s/m

PCA82C250 transmitter and receiver propagation delay = 150 ns at 85° C

Oscillator frequency = 32 MHz

E13.4 Calculate the bit segments for the following system constraints, assuming that the Philips PCA82C250 transceiver is used:

Bit rate = 400 Kbps

Bus length = 100 m

Bus propagation delay = 5×10^{-9} s/m

PCA82C250 transmitter and receiver propagation delay = 150 ns at 85° C

Oscillator frequency = 40 MHz

E13.5 Write a subroutine to configure the ECAN module of the PIC18F8680 with the bit timing segments computed in Example E13.2. Enable receive interrupt but disable transmit interrupt. Configure the CAN so that it accepts only messages with an 8-bit acceptance filter so that it accepts only messages with standard identifier H.

E13.6 Write a C function to perform the same setting as was done in Example 13.7.

E13.7 Write a few C statements to associate RXM0 with all the even-numbered acceptance filters and associate RXM1 with all the odd-numbered acceptance filters.

E13.8 Write a C function to set up the acceptance filters and masks to receive messages with standard identifiers starting with "T", "P", or "H" in mode 0.

E13.9 Write a C function to set up the acceptance filters and masks to receive messages with extended identifiers starting with "T0", "P0", or "H0" in mode 0.

E13.10 Write a C function to prepare data to be sent via the first available transmit buffer in mode 0. The identifier to be used is the string "H2", and the value to be send is 0x1001030121320003. One should prepare the identifier and data in a buffer, look for an available transmit buffer, copy data from the buffer to the transmit buffer, and enable the transmission.

13.22 Lab Exercises and Assignments

13.1 Practice data transfer over the CAN bus using the following procedures:

Step 1
Configure the CAN module to operate with the following setting:
1. Bit rate set to 500 Kbps with bus length less than 20 m.
2. Bus propagation delay = 5×10^{-9} sec/m
3. Oscillator frequency = 32 MHz
4. Mode 0 operation using transmit buffer 0 and receive buffer 0 with standard identifier.
5. The identifier of the message to be sent is the first letter of your last name only, i.e., eight bit only.
6. Enable recieve interrupt.
7. Recieve node use RXF0 as the filter and use RXM0 as the mask.

Step 2
Use a pair of insulated wires about 20 meters long. Connect the CAN_H pins of two demo boards (e.g., the SEE8680 boards) with one wire and connect the CAN_L pins of both demo boards with another wire. Connect both ends of these two wires together with 120-Ω resistores. The circuit connection is shown in Figure L13.1.

Figure L13.1 ■ CAN circuit connection for L13.1

Step 3

When started, your program will configure the CAN module and then wait for 30 seconds before start to send and receive data. Display the message "Wait 30 seconds. . . " on the LCD.

Step 4

The sending node performs A/D conversion, obtains the result in one integer and one fractional digit and send it out twice per second. The value sent out is displayed on the LCD. The message is "Send out x.x V".

Step 5

The receiving node will display the A/D result on the LCD. The format is "Received y.y V".

Step 6

Go to Step 4.

14

Internal and External Memory Programming and Expansion

14.1 Objectives

After completing this chapter, you should be able to

- Understand the overall memory system organization and mapping

- Erase and program the on-chip flash program memory

- Control the operation, programming, and protection of the on-chip flash program memory

- Erase and program the on-chip data EEPROM

- Understand the external memory expansion issue

- Make memory space assignment

- Design address decoders and memory control circuitry

- Perform timing analysis for the memory system

14.2 Overview of the PIC18 Memory System

The PIC18 microcontroller has three memory blocks:

- Program memory
- Data memory
- Data EEPROM

Internal program memory and data memory use separate buses, allowing for concurrent access to them.

Data memory is implemented in SRAM and is divided into special-function registers (SFRs) and general-purpose registers (GPRs). SFRs control the operations of peripheral functions. GPRs are used for data storage and scratch pad operations. SFRs start from 0xFFF (bank 15) and extend downward, whereas GPRs start from 0x000 (or bank 0) and extend upward. The access of data memory has been discussed in Chapters 1 and 2.

Data EEPROM is used mainly to store information that is changed frequently. The erasure, programming, and access are similar to those of the flash program memory.

The flash program memory is used mainly to store programs and tables of data. All the PIC18 devices with a letter F in its device name (e.g., PIC18F8720) have a certain amount of on-chip flash program memory. High-pin-count (mostly 80 pins and above) PIC18 members also support external memory. External memory can be SRAM, EPROM, EEPROM, or flash memory.

Adding external memory to the microcontroller will involve address decoder design, memory space assignment, and timing verification. These issues are discussed in appropriate sections.

14.3 On-Chip Flash Program Memory

The on-chip flash program memory is readable, writable, and erasable during the normal operation over the entire power supply (V_{DD}) range. A read from the program memory is proceeded one byte a time. A write to the program memory is executed on one block of eight bytes at a time. On-chip program flash memory is erased in a block of 64 bytes at a time. A bulk erase operation may not be issued from the user program. The on-chip program flash memory is 16 bits wide.

Since writing or erasing program flash memory takes much longer than one instruction cycle, it will stop instruction fetching until the operation is complete. An internal programming timer terminates program memory writes and erases.

The PIC18 microcontroller provides several versions of the following two instructions for accessing the program memory:

- Table read (TBLRD)
- Table write (TBLWT)

A table-read operation retrieves data from the program memory into the 8-bit TABLAT register. The user program can then copy the contents of this register to any other data register. A table-write operation stores the contents of the TABLAT register in a location in the program memory. One must place the data in the TABLAT register before writing into the program memory. In addition, one must place the address of the program memory in the TBLPTR register (consisting of the TBLPTRU, TBLPTRH, and TBLPTRL registers) before any read or write operation to the program memory is performed. Examples on reading from program memory have been discussed in Chapters 1, 2, and 4.

14.3.1 Control Registers of Program Flash Memory

One needs to deal with the following registers in order to read and write the program memory:

- EECON1
- EECON2
- TABLAT
- TBLPTR

EECON1 AND EECON2 REGISTERS

The EECON1 register controls the access of program flash memory. The EECON2 register is not a physical register. Reading the EECON2 register will return 0s. This register is used exclusively in the program memory erasing and writing sequences. The contents of the EECON1 register are shown in Figure 14.1.

7	6	5	4	3	2	1	0
EEPGD	CFGS	--	FREE	WRERR	WREN	WR	RD

EEPGD: Flash program or data EEPROM memory select bit
   0 = Access data EEPROM memory
   1 = Access flash program memory
CFGS: Flash program/data EEPROM or configuration select bit
   0 = Access flash program or data EEPROM memory
   1 = Access configuration registers
FREE: Flash row erase enable bit
   0 = Perform write only
   1 = Erase the program memory row addressed by TBLPTR on the next WR
      command (will be cleared by hardware when write is complete).
WRERR: Flash program/data EEPROM error flag
   0 = The write operation completed
   1 = A write operation is prematurely terminated.
WREN: Flash program/data EEPROM write enable bit
   0 = Inhibit write to the EEPROM
   1 = Allows write cycles
WR: Write control bit
   0 = write cycle to the EEPROM is complete
   1 = Initiate a data EEPROM erase/write cycle or a program memory erase
      cycle or write cycle.
RD: Read control bit
   0 = Does not initiate an EEPROM read
   1 = Initiates an EEPROM read

Figure 14.1 ■ EECON 1 register (redraw with permission of Microchip)

Both the flash program memory and the data EEPROM are controlled by the EECON1 register. The EEPGD bit determines whether the access is to the program memory (when equals 1) or the data EEPROM (when equals 0). The CFGS bit determines whether the access is to the configuration/calibration registers (when equals 1) or to the program memory/data memory (when equals 0).

The FREE bit must be set to 1 in order to erase the flash program memory. When this bit is 0, one can only write into but not erase the flash memory. The WREN bit enables/disables the write operation to the flash memory.

The WRERR bit will be set to 1 when a write operation is interrupted by a $\overline{MCLR}$ reset or a WDT time-out reset. The WR bit initiates the write operation to the flash memory. This bit cannot be cleared. It will be cleared when the write operation is completed.

TBLPTR Register

The table pointer (TBLPTR) addresses a byte within the program memory. As described in Chapter 1, this register consists of three registers. These three registers join to form a 22-bit-wide register. The lower 21 bits allow the user to access up to 2 MB of program memory space. The 22nd bit allows access to the device ID, the user ID, and the configuration bits.

TBLPTR is used in reads, writes, and erases of the flash program memory. When a TBLRD instruction is executed, all 22 bits of the table pointer determine the byte to be read from the program memory into the TABLAT register.

When the user writes data into the program memory, the data will be first written into the program memory holding register and then written into the flash memory. When a TBLWT instruction is executed, the three least significant bits of the TBLPTR register determine which of the eight holding registers of the program memory is written into. When the user sets the WR bit of the EECON1 register, the contents of these eight holding registers will be written into a block of the program memory. The block to be written into is determined by the upper 19 bits of the TBLPTR register.

14.3.2 Reading the Flash Program Memory

One uses the TBLRD instruction to retrieve data from the program memory and place it in the TABLAT register. One can then copy the contents of this register into any other general data register. The read operation takes only one instruction cycle (can be extended up to four instruction cycles) to complete. Examples on reading the flash program memory have been given in Chapters 1, 2, and 4.

14.3.3 Erasing Flash Program Memory

When executing the CPU instruction, one can only erase a block of 64 bytes in one operation. The most significant 16 bits of the TBLPTR register (TBLPTR<21:6>) select the block to be erased. The lowest six bits of the TBLPTR register are ignored. A larger block of program memory can be erased in one operation if one uses an external programmer or the in-circuit serial programming (ICSP) mechanism. One cannot erase a single word of the program memory at a time.

In order to erase a location in the flash memory, the following control bits (in EECON1) must be set up properly:

- EEPGD: set to 1
- WREN: set to 1
- FREE: set to 1

The flash memory starts a long write operation when an erase operation is started. The long write operation will be terminated by the internal programming timer. The sequence of events occurred during an erase operation is as follows:

1. Load the table pointer with the address of the row to be erased.
2. Set the EEPGD, WREN, and FREE bits of the EECON1 register to 1s for the erase operation. Clear the CFGS bit to select the flash memory.
3. Disable interrupts.
4. Write the value of 0x55 to the EECON2 register.
5. Write the value of 0xAA to the EECON2 register.
6. Set the WR bit of the EECON1 register to 1 to start the erase operation.
7. The CPU stalls for the duration of the erase cycle (about 2 ms using an internal timer).
8. Execute a NOP instruction.
9. Reenable interrupts.

The following instruction sequence will erase the row with the label **row_era:**

```
movlw    upper row_era      ; load TBLPTR with the value represented
movwf    TBLPTRU,A          ; by the label row_era
movlw    high row_era       ; "
movwf    TBLPTRH,A          ; "
movlw    low row_era        ; "
movwf    TBLPTRL,A          ; "
bcf      EECON1,CFGS,A      ; select flash program memory
bsf      EECON1,EEPGD,A     ; point to flash program memory
bsf      EECON1,WREN,A      ; enable write to flash memory
bsf      EECON1,FREE,A      ; enable erase operation
bcf      INTCON,GIE,A       ; disable interrupts
movlw    0x55
movwf    EECON2,A
movlw    0xAA
movwf    EECON2,A
bsf      EECON1,WR,A        ; start erase operation (CPU stall)
nop
bsf      INTCON,GIE,A       ; re-enable interrupts
```

The shaded instruction sequence is required for erasing the flash memory.

14.3.4 Writing to Flash Program Memory

The minimum programming unit for the flash memory is eight bytes. Byte or word programming is not supported.

To program the flash program memory, the user first loads the eight holding registers by executing the TBLWT instruction. After all eight holding registers have been loaded with data, the user executes a special required instruction sequence (to be shown shortly) and then sets the WR bit of the EECON1 register to start the write operation. After this, a long internal write operation is started, and the instruction execution is halted. The long write cycle will be terminated when the programming operation is completed.

In order to program the flash memory (or EEPROM) correctly, one must first erase the location to be programmed. Since the smallest size (64 bytes) for erasure is larger than the block size (eight bytes) for write operation, the user will need to do the following:

1. Copy the 64-byte block that contains the 8-byte block to be programmed into a buffer.
2. Write the block of data to be programmed into the flash memory into the appropriate 8-byte block in the buffer.
3. Erase the flash memory block.
4. Program the whole 64-byte buffer block into the corresponding flash memory block.

You have probably noticed that the lowest six bits of the table pointer (TBLPTR<5:0>) are not used to select the block to be programmed. The resultant block of this approach is called an *aligned block*. For a 2^k-byte aligned block, the least significant k address bits of the first byte in the block are all 0s.

It is possible that the eight locations (bytes) of the flash memory to be programmed straddle two 64-byte aligned blocks (the least significant six bits of the address of the first byte in an aligned 64-byte block are 0s). In this situation, the user will need to erase two 64-byte blocks and also program these two 64-byte blocks after the erasure. It is straightforward to test if these eight bytes straddle two aligned 64-byte blocks: Add 7 to the starting address of the flash memory to be programmed. If, after the addition, the lowest six bits of the sum is less than 7, then the 8-byte block straddle two 64-byte aligned blocks.

Suppose that among the eight bytes of data to be programmed, the first k bytes are to be programmed into the lower 64-byte block and then the remaining 8 - k bytes are to be programmed into the next-higher 64-byte block. The first k bytes of data would be written into the last k bytes of the first 64-byte aligned block, whereas the remaining 8 - k bytes of data are to be programmed into the first 8 - k bytes of the 64-byte aligned block. This is illustrated in Figure 14.2. A similar method can be used to find out if any k bytes (<64 bytes) of memory locations to be programmed are contained in a 64-byte aligned block.

The following example demonstrates a subroutine that programs k bytes ($1 \leq k \leq 8$) of data into the flash memory starting from a specified address. These k bytes are contained in an aligned 64-byte block.

Example 14.1
▼

Write a subroutine that programs k bytes ($1 \leq k \leq 8$) of data into the flash memory starting from the specified address. The k bytes of values to be programmed and the starting address of the flash memory are pushed into the stack. These k bytes are contained in a 2^6-byte aligned block.

Solution: The stack frame for this subroutine is shown in Figure 14.3. The buffer to save the 64-byte block and all local variables are allocated in the stack frame.

The subroutine and its calling sequence in assembly language are as follows:

```
          #include <p18F8680.inc>
byte_cnt    equ       -6        ; offset of byte_cnt from the frame pointer
addr_lo     equ       -5        ; offset of addr_lo from the frame pointer
addr_hi     equ       -4        ; offset of addr_hi from the frame pointer
addr_up     equ       -3        ; offset of addr_up from the frame pointer
count1      equ        7        ; offset of count1 from the frame pointer
count2      equ        8        ; offset of count2 from the frame pointer
buf0        equ        9        ; offset of buf0 from the frame pointer
```

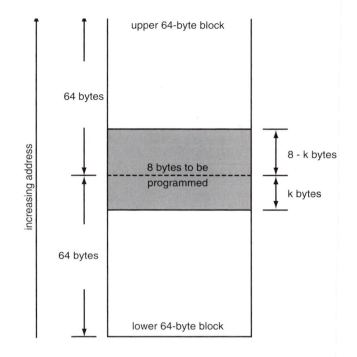

Figure 14.2 ■ Illustration of 8 bytes straddle two 64-byte aligned blocks

```
loc_varee    equ        66              ; size of local variables
pushr        macro      arg             ; macro to push the arg register into stack
             movff      arg,POSTINC1
             endm
popr         macro      arg             ; macro to pop the arg register from stack
             movff      POSTDEC1,arg    ; decrement the stack pointer
             movff      INDF1,arg       ; pop off a byte from the stack onto arg
             endm
alloc_stk    macro      n               ; this macro allocates n bytes in stack
             movlw      n
             addwf      FSR1L,F,A
             movlw      0x00
             addwfc     FSR1H,F,A
             endm
dealloc_stk  macro      n               ; this macro deallocate n bytes from stack
             movlw      n
             subwf      FSR1L,F,A
             movlw      0x00
             subwfb     FSR1H,F,A
             endm
             org        0x00
             goto       start
```

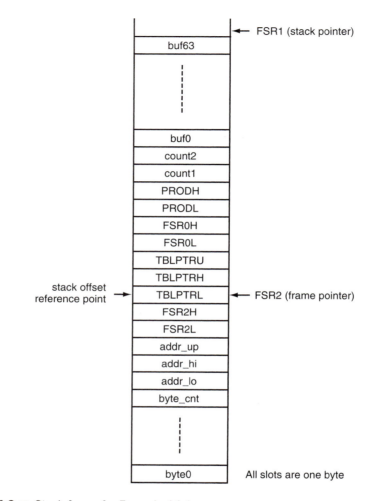

Figure 14.3 ■ Stack frame for Example 14.1

```
            org        0x08
            retfie
            org        0x18
            retfie
start       lfsr       FSR1,0xE00    ; set up software stack pointer
            movlw      0x31          ; push values (1,..,8) to be programmed
            pushr      WREG          ; "
            incf       WREG,W        ; "
            pushr      WREG          ; "
            incf       WREG,W        ; "
            pushr      WREG          ; "
            incf       WREG,W        ; "
```

```
                pushr       WREG            ; "
                incf        WREG,W          ; "
                pushr       WREG            ; "
                incf        WREG,W          ; "
                pushr       WREG            ; "
                incf        WREG,W          ; "
                pushr       WREG            ; "
                incf        WREG,W          ; "
                pushr       WREG            ; "
                movlw       0x08            ; push byte count
                pushr       WREG            ; "
                movlw       low addr        ; push address of flash memory to be
                pushr       WREG            ; programmed
                movlw       high addr       ; "
                pushr       WREG            ; "
                movlw       upper addr      ; "
                pushr       WREG            ; "
                call        prog_flash      ; call subroutine program flesh memory
                dealloc_stk 12              ; clean up the allocated stack space
forever         nop                         ; infinite loop
                bra         forever         ; "
; ************************************************************************
;
; The following subroutine programs 1 to 8 bytes of values into flash
; program memory. The memory locations to be programmed are contained in
; one aligned 64-byte block. The values to be programmed, the byte count,
; and the starting address to be programmed are pushed into the stack in that order.
; ************************************************************************
;
prog_flash      pushr       FSR2L           ; save caller's frame pointer
                pushr       FSR2H           ; "
                movff       FSR1L,FSR2L     ; set up new frame pointer
                movff       FSR1H,FSR2H     ; "
                pushr       TBLPTRL         ; save table pointer
                pushr       TBLPTRH         ; "
                pushr       TBLPTRU         ; "
                pushr       FSR0L           ; save FSR0
                pushr       FSR0H           ; "
                pushr       PRODL           ; save PRODL
                pushr       PRODH           ; save PRODH
                alloc_stk   loc_varee       ; allocate local variable space
                movf        FSR2L,W,A       ; set FSR0 to point to the start of the buffer
                addlw       buf0            ; in the stack frame
                movwf       FSR0L,A         ; "
                movlw       0               ; "
                addwfc      FSR2H,W,A       ; "
                movwf       FSR0H,A         ; "
                movlw       addr_lo         ; set TBLPTR to point to the start of the
                movf        PLUSW2,W        ; 64-byte aligned block to be saved
                andlw       0xC0            ; "
                movwf       TBLPTRL,A       ; "
```

```
                movlw      addr_hi              ; "
                movff      PLUSW2,TBLPTRH       ; "
                movlw      addr_up              ; "
                movff      PLUSW2,TBLPTRU       ; "
; back up the 64-byte block first
                movlw      64
                movwf      PRODL,A
cpy_lp1         tblrd*+                         ; read one byte from program memory
                movff      TABLAT, POSTINC0     ; into TABLAT and transfer to the buffer
                decfsz     PRODL,F,A            ; "
                bra        cpy_lp1              ; "
; set up offsets in buffer and data in stack for modifying buffer
                movlw      byte_cnt             ; place the number of bytes to be modified in
                movff      PLUSW2,PRODH         ; PRODH (also used as loop count)
                movlw      addr_lo              ; find out the offset in stack frame to
                movf       PLUSW2,W             ; be modified by programming data and
                andlw      0x3F                 ; leave it in count2
                addlw      buf0                 ; "
                movwf      PRODL,A              ; "
                movlw      count2               ; "
                movff      PRODL,PLUSW2         ; "
                movf       PRODH,W,A            ; place byte count in WREG
                sublw      byte_cnt             ; compute the stack offset of byte0 from FSR2
                movwf      PRODL,A              ; and leave it in count1
                movlw      count1               ; "
                movff      PRODL,PLUSW2         ; "
; write k bytes (k = 1..8) of data into the buffer before performing flash programming
mdy_lp          movlw      count1
                movf       PLUSW2,W             ; get the offset into WREG
                movff      PLUSW2,PRODL         ; place the programming data in PRODL
                movlw      count2               ; find the offset of the buffer slot to be
                movf       PLUSW2,W             ; modified
                movff      PRODL,PLUSW2         ; replace buffer value with programming data
                movlw      count1               ; increment the offset
                incf       PLUSW2,F             ; "
                movlw      count2               ; "
                incf       PLUSW2,F             ; "
                decfsz     PRODH,F,A            ; are k bytes all written in buffer?
                bra        mdy_lp               ; "
; erase the flash memory block
                movlw      addr_lo
                movff      PLUSW2,TBLPTRL
                movlw      addr_hi
                movff      PLUSW2,TBLPTRH
                movlw      addr_up
                movff      PLUSW2,TBLPTRU
                bsf        EECON1,EEPGD,A       ; point to flash memory
                bcf        EECON1,CFGS,A        ; access flash program memory
                bsf        EECON1,WREN,A        ; enable write to flash memory
```

```
                    bsf         EECON1,FREE,A           ; enable flash erase
                    bcf         INTCON,GIE,A            ; disable interrupts
                    movlw       0x55                    ; required sequence
                    movwf       EECON2                  ; "
                    movlw       0xAA                    ; "
                    movwf       EECON2                  ; "
                    bsf         EECON1,WR,A             ; start erase (CPU stall)
                    bsf         INTCON,GIE              ; re-enable interrupt
; perform flash programming and set up two layer of loops, each 8 times
                    movlw       8                       ; set outer loop count to 8
                    movwf       PRODL,A
                    movlw       count1
                    movff       PRODL,PLUSW2
                    movf        FSR2L,W,A
                    addlw       buf0
                    movwf       FSR0L,A
                    movlw       0
                    addwfc      FSR2H,W,A
                    movwf       FSR0H,A
prog_loop           movlw       8                       ; set inner loop count to 8
                    movwf       PRODL,A                 ; "
                    movlw       count2                  ; "
                    movff       PRODL,PLUSW2            ; "
write_wd            movf        POSTINC0, TABLAT        ; copy one byte to the holding register
                    TBLWT*+                             ; until buffers are full
                    movlw       count2                  ; "
                    decfsz      PLUSW2,F,A              ; "
                    bra         write_wd                ; "
                    bsf         EECON1,EEPGD,A          ; point to flash memory
                    bcf         EECON1,CFGS,A           ; access flash program memory
                    bsf         EECON1,WREN,A           ; enable write to memory
                    bcf         INTCON,GIE,A            ; disable interrupt
                    movlw       0x55                    ; required sequence
                    movwf       EECON2                  ; "
                    movlw       0xAA                    ; "
                    movwf       EECON2                  ; "
                    bsf         EECON1,WR,A             ; start programming (CPU stall)
                    nop
                    bsf         INTCON,GIE,A            ; enable interrupts
                    movlw       count1                  ; decrement outer loop count
                    decfsz      PLUSW2,F,A              ; "
                    bra         prog_loop               ; loop until done
                    bcf         EECON1,WREN,A           ; disable write to memory
done                dealloc_stk loc_varee
                    popr        FSR0H
                    popr        FSR0L
                    popr        TBLPTRU
                    popr        TBLPTRH
                    popr        TBLPTRL
```

```
            popr      FSR2H
            popr      FSR2L
            return    0
  addr      org       ($+0x3F) & 0x1fffc0      ;force to align with 64-byte boundary
            data      0x41,0x42,0x43,0x44
            data      0x45,0x46,0x47,0x48
            end
```

The shaded area in this program writes data into the programming holding registers (eight in total) and then starts the flash memory programming sequence. The CPU will be stalled during the programming process.

One of the applications for the flash memory programming is to write a *boot loader* for the PIC18 device. In this application, the boot loader is preprogrammed into the PIC18 microcontroller. After reset, the boot loader reads a hex file from the USART port and programs it into the flash memory and then jumps to the beginning of the downloaded program for execution. To perform this operation, one needs to run a terminal program on the PC. The boot loader will output a message to the terminal monitor to remind the user to download a hex file (simulated by the terminal program) and then wait for the hex file to come from the USART port.

Since the smallest erase block size is 64 bytes, one should always try to collect as many bytes to be programmed as possible in order to save the time spent on erasing flash memory. When there is a concern about the correctness of the flash memory programming, one should also perform a verification operation (read it out and compare it with what you just wrote into the same memory location) after programming is complete. This should be used in applications where excessive writes can stress bits near the application limit.

If a write operation is terminated by an unplanned event, such as loss of power or an unexpected RESET, the memory location just programmed should be verified and reprogrammed if needed. The WRERR bit (in the EECON1 register) is set when a write operation is interrupted by a $\overline{\text{MCLR}}$ reset or a WDT time-out reset during normal operation. In these situations, users can check the WRERR bit and rewrite the location.

▲

14.4 Data EEPROM Memory

The data EEPROM is readable and writable during the normal operation over the entire VDD range. The data EEPROM is not directly mapped in the register file space. Instead, it is indirectly addressed through the SFRs.

Five SFRs are involved in reading and writing of the data EEPROM memory:

- EECON1
- EECON2
- EEDATA
- EEADRH
- EEADR

Both the EECON1 and the EECON2 register have been discussed in Section 14.3.

The EEPROM data memory allows the byte-read and byte-write operations. The EEDATA register holds the 8-bit data for read and write. The EEADR and EEADRH registers hold the address of the EEPROM location being accessed. A PIC18 microcontroller may have 0, 256, or 1024 bytes of EEPROM data memory. A PIC18 microcontroller having only 256 bytes of data EEPROM does not have the EEADRH register. The EEADRH register has only two bits.

A data EEPROM write operation automatically erases the location and writes the new data. The write time is controlled by an on-chip timer. The write time may vary with voltage and temperature.

14.4.1 The EEADR and EEADRH Registers

This address register pair can address up to a maximum of 1024 bytes of data EEPROM. The two most significant address bits are stored in the EEADRH register, whereas the remaining eight least significant bits are stored in the EEADR register.

14.4.2 Reading the Data EEPROM Memory

To read a data memory location, one must do the following:

- Write the address into the EEADRH:EEADR (or simply EEADR for devices with only 256 bytes of data EEPROM) register pair
- Clear the EEPGD bit of the EECON1 register
- Clear the CFGS bit of the EECON1 register
- Set the RD bit of the EECON1 register

The data EEPROM read takes only one instruction cycle, and therefore the EEDATA register can be read by the next instruction. The EEDATA register will hold this value until another read operation or until it is written to by the user during a write operation. The following instruction sequence will read the contents of the data EEPROM location at **eedata_loc:**

```
movlw    high eedata_loc      ; set up upper address bits of EEPROM
movwf    EEADRH,A             ; data memory for reading
movlw    low eedata_loc       ; set up lower address byte of EEPROM
movwf    EEADR,A              ; data memory for reading
bcf      EECON1,EEPGD         ; point to EEPROM data memory
bcf      EECON1,CFGS          ; access EEPROM
bsf      EECON1,RD            ; read data EEPROM
movf     EEDATA,W             ; place data in WREG
```

14.4.3 Writing to the Data EEPROM Memory

To write data into an data EEPROM location, one must first place the address in the EEADRH:EEADR register pair (or EEADR if EEADRH does not exist) and place data in the EEDATA register.

The following instruction will write the value **ee_val** into the data EEPROM at **eedata_loc:**

```
movlw    high eedata_loc      ; set up address of data EEPROM
movwf    EEADRH,A             ; to be written
movlw    low eedata_loc       ; "
movwf    EEADR,A              ; "
movlw    ee_val               ; place data to be written in EEDATA
movwf    EEDATA,A             ; "
bcf      EECON1,EEPGD,A       ; point to data memory
bcf      EECON1,CFGS,A        ; access data EEPROM
bsf      EECON1,WREN,A        ; enable write to data EEPROM
bcf      INTCON,IE,A          ; disable interrupt
movlw    0x55                 ; required sequence
movwf    EECON2,A             ; "
movlw    0xAA                 ; "
```

```
        movwf     EECON2,A          ; "
        bsf       EECON1,WR,A       ; start write operation
        bsf       INTCON,GIE,A      ; re-enable interrupt
        . . .                       ; other user instructions
        . . .                       ; "
        bcf       EECON1,WREN       ; disable writes on write complete
```

At the end of the write cycle, the WR bit will be cleared by the hardware. The EEPROM write complete interrupt flag bit (EEIF) will be set. The EEIF flag can be cleared only by software.

In some applications where excessive writes may stress bits near the specification limit, one may want to verify the EEPROM location data against the original written value.

14.5 PIC18 External Memory Interface

At the time of this writing, a PIC18 device has no more than 128 KB of on-chip flash program memory. If an application requires more than 128 KB of program memory, one will need to consider adding external program memory supported by some PIC18 devices. The PIC18F8XXX devices and the PIC18C801 support up to 2 MB of external memory.

14.5.1 PIC18 Program Memory Modes

The PIC18C601/801 supports only external memory. All 80-pin PIC18F8XXX devices support external memory. These devices (PIC18F8X20, PIC18F8X8X, and PIC18F8X2X at the time of this writing) have four distinct operation modes:

- Microprocessor (MP)
- Microcontroller with boot block (MPBB)
- Extended microcontroller (EMC)
- Microcontroller

The program memory mode is determined by setting the two least significant bits of the CONFIG3L configuration byte, as shown in Figure 14.4.

7	6	5	4	3	2	1	0
WAIT	--	--	--	--	--	PM1	PM0

WAIT: External bus data wait enable bit

    0 = Wait programmable by WAIT1 and WAIT0 bits of MEMCOM register

    1 = Wait selections unavailable, device will not wait

PM1:PM0: Processor data memory mode select bits

    00 = Microcontroller mode

    01 = Microprocessor mode

    10 = Microcontroller with Boot block mode

    11 = Extended microcontroller mode

Figure 14.4 ■ CONFIG3L register (redraw with permission of Microchip)

The *microprocessor mode* allows access only to external program memory; the contents of the on-chip flash memory are ignored. The 21-bit program counter permits access to a 2-MB linear program memory space.

The *microprocessor with boot block mode* accesses on-chip flash memory within only the boot block. The size of the boot block is device dependent and is located at the beginning of the program memory. At the time of this writing, it is either 512 bytes (PIC18F8720/8620) or 2 KB (PIC18F8X8X and PIC18F852X). Beyond the boot block, external program memory is accessed all the way up to the 2-MB limit. Program execution automatically switches between the two memories as required.

The *microcontroller mode* accesses only the on-chip flash memory. External memory interface functions are disabled. Any attempt to read above the physical limit of the on-chip flash memory causes a read of all 0s.

The *extended microcontroller mode* allows access to both the internal and the external program memory as a single block. The device can access its entire on-chip flash memory; above this, the device accesses the external program memory up to the 2-MB program space limit. As with boot block mode, instruction execution automatically switches between the two memories as required.

In all modes, the microcontroller has complete access to the data RAM and the EEPROM. The microcontroller mode is the only mode that does not support external memory.

14.5.2 PIC18 External Memory Pins

The external memory interface (EMI) of the PIC18C601 uses 26 pins. The external memory interface of the PIC18C801 uses 38 pins. The PIC18F8XXX external memory interface uses 28 pins and is implemented across four I/O ports (D, E, H, and J). This chapter focuses mainly on the PIC18F8XXX external memory interface. The signal pins used in external memory interface are listed in Table 14.1.

The external memory interface can be enabled or disabled. The default read cycle to the external memory takes one instruction cycle time (T_{CY}). However, to allow slower memory devices and peripheral devices to be interfaced with the PIC18F8XXX, the read and write cycle times can be extended up to four instruction cycles (i.e., wait for up to 3 T_{CY}). The user program can read data from and write data into the external memory using the table-read and table-write instructions. The PIC18F8XXX EMI supports three different TBLWRT operations modes: *word-write*, *byte-select*, and *byte-write* modes. The operation of external memory is controlled by the MEMCON register. The contents of the MEMCON register are shown in Figure 14.5.

In both the microcontroller mode and the microprocessor mode, the EBDIS bit has no effect. In the other two modes, the EBDIS bit has no effect when the CPU is fetching instructions externally or executing table-read/table-write operations externally. While fetching instructions internally or executing table-read/table-write operations internally, the EBDIS control bit may change the pins from external memory to I/O port functions. When EBDIS = 0, the pins function as the external bus. When EBDIS = 1, the pins function as I/O ports.

If the device fetches or accesses external memory while EBDIS = 1, the pins will switch to the external bus. If the EBDIS bit is set by a program executing from external memory, the action of setting the bit will be delayed until the program branches into the internal memory. At that time, the pins will change from external bus to I/O ports.

When the device is executing out of the internal memory (with EBDIS = 0) in microprocessor with boot block mode or extended microcontroller mode, the control signals will be

Name	Function
RD0/AD0	EMI address bit 0 or data bit 0
RD1/AD1	EMI address bit 1 or data bit 1
RD2/AD2	EMI address bit 2 or data bit 2
RD3/AD3	EMI address bit 3 or data bit 3
RD4/AD4	EMI address bit 4 or data bit 4
RD5/AD5	EMI address bit 5 or data bit 5
RD6/AD6	EMI address bit 6 or data bit 6
RD7/AD7	EMI address bit 7 or data bit 7
RE0/AD8	EMI address bit 8 or data bit 8
RE1/AD9	EMI address bit 9 or data bit 9
RE2/AD10	EMI address bit 10 or data bit 10
RE3/AD11	EMI address bit 11 or data bit 11
RE4/AD12	EMI address bit 12 or data bit 12
RE5/AD13	EMI address bit 13 or data bit 13
RE6/AD14	EMI address bit 14 or data bit 14
RE7/AD15	EMI address bit 15 or data bit 15
RH0/A16	EMI address bit 16
RH1/A17	EMI address bit 17
RH2/A18	EMI address bit 18
RH3/A19	EMI address bit 19
RJ0/ALE	EMI address latch enable control pin
RJ1/$\overline{OE}$	EMI output enable ($\overline{OE}$) control pin
RJ2/$\overline{WRL}$	EMI write low ($\overline{WRL}$) control pin
RJ3/$\overline{WRH}$	EMI write high ($\overline{WRH}$) control pin
RJ4/BA0	EMI byte address bit 0
RJ5/$\overline{CE}$	EMI chip enable ($\overline{CE}$) control pin
RJ6/$\overline{LB}$	EMI lower byte enable ($\overline{LB}$) control pin
RJ7/$\overline{UB}$	EMI upper byte enable ($\overline{UB}$) control pin

Table 14.1 ■ PIC18F8XXX EMI Bus I/O port pins

inactive. They will go to a state where the AD<15:0> and A<19:16> signals are in high-impedance state; the $\overline{OE}$, $\overline{WRH}$, $\overline{WRL}$, $\overline{UB}$, and $\overline{LB}$ signals are high; and the ALE and BA0 signals are low.

14.5.3 Waveforms of Bus Signals

The waveform of a typical digital signal is shown in Figure 14.6. A bus signal cannot rise from low to high or drop from high to low instantaneously. The time needed for a signal to rise from 10% of the power supply voltage to 90% of the power supply voltage is referred to as the *rise time* (t_R). The time needed for a signal to drop from 90% of the power supply voltage to 10% of the power supply voltage is referred to as the *fall time* (t_F).

A single bus signal is often represented as a set of line segments (see Figure 14.7). The horizontal axis and vertical axis represent the time and the magnitude (in volts) of the signal, respectively. Multiple signals of the same nature, such as address and data, are often grouped together and represented as parallel lines with crossovers, as illustrated in Figure 14.8. A crossover in the waveform represents the point at which one or multiple signals change values.

Sometimes a signal value is unknown because the signal is changing. Hatched areas in the timing diagram, shown in Figure 14.9, represent single and multiple unknown signals.

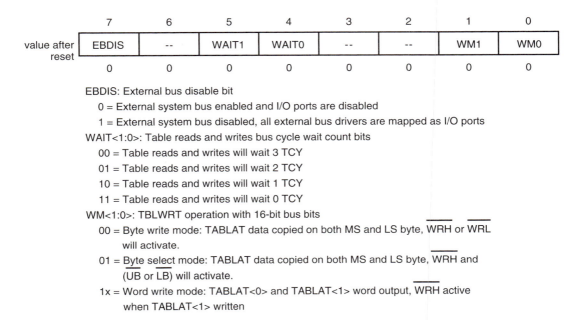

7	6	5	4	3	2	1	0
EBDIS	--	WAIT1	WAIT0	--	--	WM1	WM0
0	0	0	0	0	0	0	0

value after reset

EBDIS: External bus disable bit

    0 = External system bus enabled and I/O ports are disabled

    1 = External system bus disabled, all external bus drivers are mapped as I/O ports

WAIT<1:0>: Table reads and writes bus cycle wait count bits

    00 = Table reads and writes will wait 3 TCY

    01 = Table reads and writes will wait 2 TCY

    10 = Table reads and writes will wait 1 TCY

    11 = Table reads and writes will wait 0 TCY

WM<1:0>: TBLWRT operation with 16-bit bus bits

    00 = Byte write mode: TABLAT data copied on both MS and LS byte, $\overline{WRH}$ or $\overline{WRL}$ will activate.

    01 = Byte select mode: TABLAT data copied on both MS and LS byte, $\overline{WRH}$ and ($\overline{UB}$ or $\overline{LB}$) will activate.

    1x = Word write mode: TABLAT<0> and TABLAT<1> word output, $\overline{WRH}$ active when TABLAT<1> written

Figure 14.5 ■ MEMCON register (redraw with permission of Microchip)

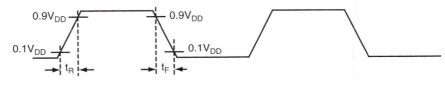

Figure 14.6 ■ A typical digital waveform

Figure 14.7 ■ Single signal waveform

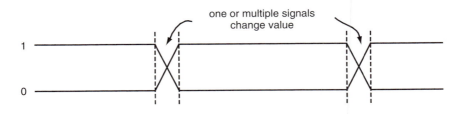

Figure 14.8 ■ Multiple-signal waveform

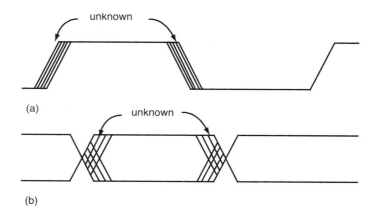

Figure 14.9 ■ Unknown signals. (a) single signal. (b) multiple signal

Sometimes one or multiple signals are not driven (because their drivers are not enabled) and hence cannot be received. An undriven signal is said to be *floating* or in the *high-impedance* state. Single and multiple floating signals are represented by a value between the high and low levels as shown in Figure 14.10.

In a microcontroller system, a bus signal falls into one of the three categories: *address, data,* or *control.*

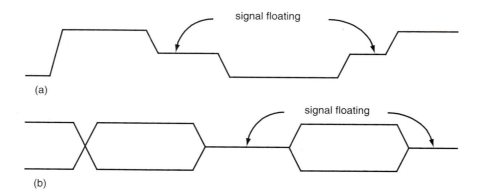

Figure 14.10 ■ Floating signals. (a) Single signal. (b) Multiple signals

14.5.4 Bus Transactions

A bus transaction includes sending the address and receiving or sending the data. A *read* transaction (also called a *read bus cycle*) transfers data from memory to either the CPU or the I/O device, and a *write* transaction (also called a *write bus cycle*) writes data to the memory. In a read transaction, the address is first sent down the bus to the memory, together with the appropriate control signals indicating a read. In Figure 14.11, this means pulling the read signal

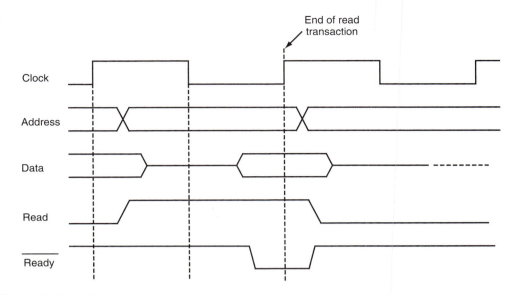

Figure 14.11 ■ A typical bus read transaction

to high. The memory responds by placing the data on the bus and driving the $\overline{\text{Ready}}$ signal to low. The $\overline{\text{Ready}}$ signal (asserted low) indicates that the data on the data bus is valid.

In Figure 14.11, a read bus cycle takes one clock cycle to complete. For some microcontrollers, the $\overline{\text{Ready}}$ signal is used to extend the bus cycle to more than one clock cycle to accommodate slower memory components. When data from external memory is not ready, the control circuitry does not pull the $\overline{\text{Ready}}$ signal to low. When the $\overline{\text{Ready}}$ signal is sampled to be low, the microcontroller copies the data into the CPU. The PIC18 microcontroller does not use the $\overline{\text{Ready}}$ signal to accommodate slower memory. In a write bus cycle, the CPU sends both the address and the data and requires no return of data.

In a bus transaction, there must be a device that can initiate a read or write transaction. The device that can initiate a bus transaction is called a *bus master*. A microcontroller is always a bus master. A device such as a memory chip that cannot initiate a bus transaction is called a *bus slave*.

In a bus transaction, there must be a signal to synchronize the data transfer. The signal that is used most often is a clock signal. The bus is *synchronous* when a clock signal is used to synchronize the data transfer. In a synchronous bus, the timing parameters of all signals use the clock signal as the reference. As long as all timing requirements are satisfied, the bus transaction will be successful.

An *asynchronous bus*, on the other hand, is not clocked. Instead, self-timed, handshaking protocols are used between the bus sender and receiver. Figure 14.12 shows the steps of a master performing a read on an asynchronous bus.

A synchronous bus is often used between the CPU and the memory system, whereas an asynchronous bus is often used to handle different types of I/O devices. A synchronous bus is usually faster because it avoids the overhead of synchronizing the bus for each transaction.

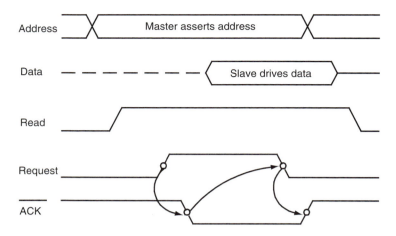

Figure 14.12 ■ Asynchronous read bus transaction

14.5.5 Bus Multiplexing

Designers of microcontrollers prefer to minimize the number of signal pins because it will make the chip less expensive to manufacture and use. By multiplexing the address bus and the data bus, many signal pins can be saved. The drawback of multiplexing bus signals is that the achievable bus transaction performance is compromised. Most 8-bit and many 16-bit microcontrollers multiplex their address and data buses.

For any bus transaction, the address signal input to the memory chips must be stable throughout the whole bus transaction. The memory system will need to use a circuit to latch the address signals so that they stay valid throughout the bus cycle. In a microcontroller that multiplexes the address and the data buses, the address signals are placed on the multiplexed bus first, along with certain control signals to indicate that the address signals are valid. After the address signals are on the bus long enough so that the external logic has time to latch them, the microcontroller stops driving the address signals and either waits for the memory devices to place data on the multiplexed bus (in a read bus cycle) or places data on the multiplexed bus (in a write bus cycle).

14.5.6 PIC18F8XXX External Memory Functions

The PIC18 external memory interface are mainly used for three functions:

- Program fetches
- Data reads (performed by the table-read instruction)
- Data writes (performed by the table-write instruction)

16-Bit Bus Overview

The PIC18F8XXX external memory interface is 16-bit. The data lines D15:D0 and the address lines A15:A0 are multiplexed on port D and port E pins. The highest three address pins are not multiplexed with data lines. The PIC18F8XXX provides 20 address signals to allow for access to 2 MB of memory space. Each address selects a 16-bit word. When access to the individual byte is needed, the BA0 pin provides the selection. Depending on the situation, the BA0 pin is driven by either the least significant bit of the program counter or the least significant bit of the table pointer.

There are seven control lines that are used in the external memory interface: $\overline{OE}$, $\overline{WRH}$, $\overline{WRL}$, $\overline{CE}$, $\overline{UB}$, $\overline{LB}$, and $\overline{ALE}$. The $\overline{OE}$ signal is used to enable external memory component to output data. The $\overline{WRH}$ and $\overline{WRL}$ signals may be used during fetches and writes. The $\overline{UB}$ and $\overline{LB}$ signals are used to select the upper byte and lower byte of a 16-bit word, respectively. The ALE signal is provided for external circuit to copy the address signals.

PROGRAM FETCHES

The PIC18 CPU pipelines the execution of two instructions. In each instruction cycle, a 2-byte instruction is executed while the external memory fetches the next 2-byte instruction. As long as the external memory system meets the timing requirement of the PIC18 instruction fetch cycle, there is no difference between the internal memory and the external memory accesses. Microchip uses the frequency of the crystal oscillator to express the external memory speed. At the time of this writing, the maximum speed of the external memory interface is 25 MHz for all except the PIC18F8X2X devices. The maximum speed of the external memory interface is 40 MHz for the PIC18F8X2X devices.

Each instruction cycle is divided into four Q cycles. The duration of each Q cycle is equal to the period of the external crystal oscillator. The program memory read timing is shown in Figure 14.13. The values of all timing parameters for the program memory read cycle are listed in Table 14.2. As shown in Figure 14.13, a program memory read cycle is completed in one

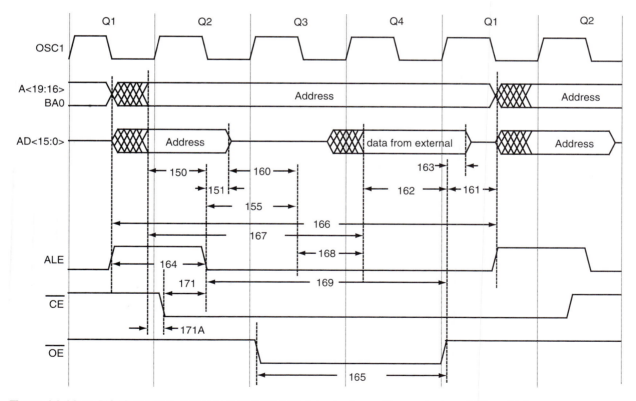

Figure 14.13 ■ PIC18 Program memory read timing diagram (redraw with permission of Microchip)

Param. No.	Symbol	Characteristics	Min	Typ	Max	Units
150	TadV2alL	Address out valid to ALE ↓ (address set up time)	0.25TCY - 10	–	–	ns
151	TalL2adL	ALE ↓ to address out invalid (address hold time)	5	–	–	ns
155	TalL2oeL	ALE ↓ to $\overline{OE}$ ↓	10	0.125TCY	–	ns
160	TadZ20eL	AD high-Z to $\overline{OE}$ ↓ (bus release to $\overline{OE}$)	0	–	–	ns
161	TOEH2adD	$\overline{OE}$ ↑ to AD driven	0.125TCY - 5	–	–	ns
162	TadV2oeH	LS data valid before $\overline{OE}$ ↑ (data setup time)	20	–	–	ns
163	ToeH2adl	$\overline{OE}$ ↑ to data in invalid (data hold time)	0	–	–	ns
164	Ta1H2a1L	ALE pulse width	–	0.25 TCY	–	ns
165	ToeL2oeH	$\overline{OE}$ pulse width	0.5TCY - 5	0.5 TCY	–	ns
166	Ta1H2a1H	ALE ↑ to ALE ↑ (cycle time)	–	TCY	–	ns
167	Tacc	Address valid to data valid	0.75TCY - 25	–	–	ns
168	Toe	$\overline{OE}$ ↓ to data valid	–	–	0.5 TCY - 25	ns
169	Ta1L2oeH	ALE ↓ to $\overline{OE}$ ↑	0.625TCY - 10	–	0.625TCY + 10	ns
171	Ta1H2csL	Chip enable active to ALE ↓	0.25TCY - 20	–	–	ns
171A	TubL2oeH	AD valid to chip enable active	–	–	10	ns

Note: TCY is instruction cycle time and is equal to four times the external oscillator period.

Table 14.2 ■ Program memory read cycle timing parameters (redraw with permission of Microchip)

instruction cycle or four crystal oscillator cycles. No wait states can be inserted. The least significant bit of the program counter is driven out from the BA0 pin, whereas the values of PC<20:17> and PC<16:1> are driven onto the pins A<19:16> and AD<15:0>, respectively. During the Q1 cycle, the ALE signal is enabled while the address signals A<15:0> become valid on the pins AD<15:0>. The falling edge of the ALE signal occurs slightly after the falling edge of the Q2 cycle. External circuitry utilizes this edge to copy the address signals A<15:0> into a latch and hold them valid throughout the program read cycle.

The $\overline{CE}$ signal goes low slightly after the rising edge of the Q2 cycle. This signal is used to enable the external memory to respond to the read request of the CPU. At about the middle of the Q3 cycle, the $\overline{OE}$ signal is asserted (goes low) to enable program memory to output data to the data bus. At the end of the Q4 cycle, the $\overline{OE}$ signal goes high, and the data is fetched from the memory at the low-to-high transition edge of the $\overline{OE}$ signal.

DATA READS (TABLE READS)

The user program fetches data from the external program memory by executing one of the table-read instructions. The timings are essentially the same, but unlike program fetching, reads are executed on a single-byte basis. During table-read operations, the least significant bit of the TBLPTR register is copied to the BA0 pin. The values of TBLPTR<20:1> appear on the address pins A<19:0>. Next, 16 bits of data are driven onto (by the memory device) the data bus. Circuitry in the TABLAT register will select either the high or the low byte of the data from the 16-bit bus based on the least significant bit of the address. When the BA0 signal is 0, the lower byte (D<7:0>) is selected; when the BA0 signal is 1, the upper byte (D<15:8>) is selected.

DATA WRITES (TABLE WRITES)

The user program writes data into the external program memory by executing one of the table-write instructions. When a table-write instruction is executed, the least significant bit of

TBLPTR is copied to the BA0 pin, and the values of TBLPTR<20:1> appear on the address pins A<19:0>. Depending on the external memory interface mode and the TBLPTR address, identical or different data may appear on the upper byte and the lower byte of the data bus. The EMI modes are discussed in the next section. The timing diagram and the values of timing parameters of the program memory write operation are shown in Figure 14.14 and Table 14.3, respectively.

In Figure 14.14, the write data is driven by the microcontroller at least TadV2wrH ns (timing parameter 156) before the rising edge of the $\overline{WRH}$ or the $\overline{WRL}$ signal. The interval in which either the $\overline{WRH}$ or the $\overline{WRL}$ signal is low is referred to as the *write pulse*. This interval will be at least half an instruction cycle period minus 5 ns long and is typically half an instruction cycle time. The data to be written will be valid on the data bus at least TadV2wrH ns before the rising edge of the $\overline{WRH}$ (or $\overline{WRL}$) signal. This parameter is often referred to as the *write data setup time*. The data to be written will remain valid for at least TwrH2adl ns after the rising edge of the $\overline{WRH}$ (or $\overline{WRL}$) signal. This parameter is referred to as the *write data hold time* and is also critical to the success of the write cycle.

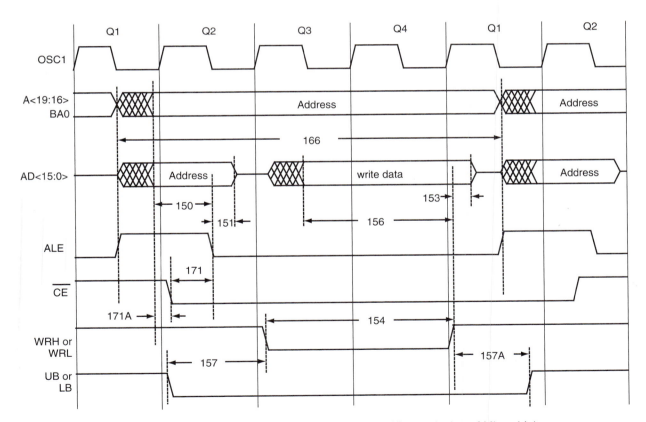

Figure 14.14 ■ PIC18 Program memory write timing diagram (redraw with permission of Microchip)

Param. No.	Symbol	Characteristics	Min	Typ	Max	Units
150	TadV2alL	Address out valid to ALE (address set up time)	0.25TCY - 10	–	–	ns
151	TalL2adL	ALE ↓ to address out invalid (address hold time)	5	–	–	ns
153	TwrH2ad1	$\overline{WRn}$ ↑ to data out invalid (data hold time)	5	–	–	ns
154	TwrL	$\overline{WRn}$ pulse width	0.5TCY - 5	0.5TCY	–	ns
156	TadV2wrH	Data valid before $\overline{WRn}$ ↑ (data set up time)	0.5TCY - 10	–	–	ns
157	TbsV2wrL	Byte select valid before $\overline{WRn}$) (byte select setup time)	0.25TCY	–	–	ns
157A	TwrH2bs1	$\overline{WRn}$ ↑ to byte select invalid (byte select hold time)	0.125TCY - 5	–	–	ns
166	Ta1H2a1H	ALE ↑ to ALE ↑ (cycle time)	–	TCY	–	ns
171	Ta1H2csL	Chip Ebable active to ALE	0.25TCY - 20	–	–	ns
171A	TubL2oeH	AD valid to chip enable active	–	–	10	ns

Note: $\overline{WRn}$ (n = H or L)

Table 14.3 ■ Program memory write cycle timing parameters

14.5.7 16-Bit EMI Operating Modes

The EMI operating modes are defined by the least significant two bits of the MEMCON register (shown in Figure 14.5). The setting of the EMI mode dictates the appropriate types of external memory available and the method for connection.

WORD-WRITE MODE

This mode is used for 16-bit-wide (i.e., in one location) memories that include some of the EPROM- and flash-type memories. This mode allows program fetches, table reads, and table writes from all forms of 16-bit memory.

When a table-write instruction writes to an even address (TBLPTR<0> = 0), the TABLAT data is transferred to a holding latch (referred to as TABLAT(0)), and the external address data bus changes to high-impedance state. When a table-write instruction writes to an odd address (TBLPTR<0> = 1), the TABLAT data is driven on the upper byte of the AD<15:0> bus. At the same time, the contents of the holding latch are presented on the lower byte of the AD<15:0> bus. The $\overline{WRH}$ signal is strobed for one write cycle; the $\overline{WRL}$ signal is not used. The signal on the BA0 pin indicates the least significant bit of TBLPTR but is not needed. Both the $\overline{UB}$ and the $\overline{UL}$ signal are active low to select these two bytes. The limitation to this method is that the table-write operation must be performed in a pair on a word boundary in order to correctly write to a memory word.

Suppose that we perform two consecutive table-write instructions to write the values of 0x36 and 0xA5 to two consecutive memory locations (one to 0x050020, the next to 0x050021). The simplified timing diagram for these two instructions is shown in Figure 14.15. In the first instruction, the value 0x028010 appears on the address pins A<19:0> during the Q1 and Q2 states, and then A<19:0> change to the high-impedance state during states Q3 and Q4. The least significant address signal BA0 will be 0 from Q1 through Q4. In the second instruction, the same value will appear on A<19:0> during Q1 and Q2. However, the value 0xA536 will be driven on the 16-bit data bus during Q3 and Q4. BA0 will be 1 in this instruction cycle.

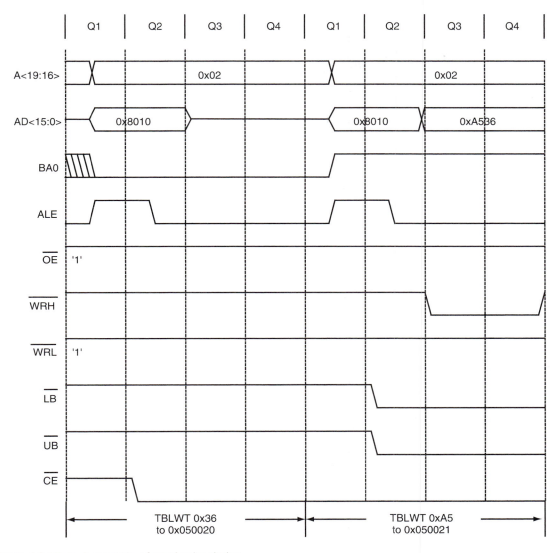

Figure 14.15 ■ A example of word write timing

BYTE-SELECT MODE

This mode allows one to use a table-write instruction to write into a byte of a 16-bit-wide external memory (one memory location has 16 bits) with byte selection capability. When a TBLWT instruction is executed, the contents of the TABLAT register are driven on both the upper and the lower bytes of the AD<15:0> bus. The $\overline{WRH}$ signal is pulsed in each write cycle, whereas the $\overline{WRL}$ signal is not used. Either the BA0 or the $\overline{UB}$ / $\overline{LB}$ signal is used to select the byte to be written based on the least significant bit of the TBLPTR register.

Figure 14.16 shows the timing diagrams of two TBLWT instructions. One writes the value 0x38 to the memory location at 0x092460. The other writes the value 0x92 to the memory location at 0x092461. These two instructions cannot be executed one after another because we need a **movlw 0x92** instruction to change the value in the TABLAT register to 0x92 before it can be written into the memory location at 0x92460. The least signal address bit is not included in A<19:0>. After dropping the least address bit, the value appears on A<19:0> becomes 0x049230. Their timing diagrams are placed side by side for ease of comparison.

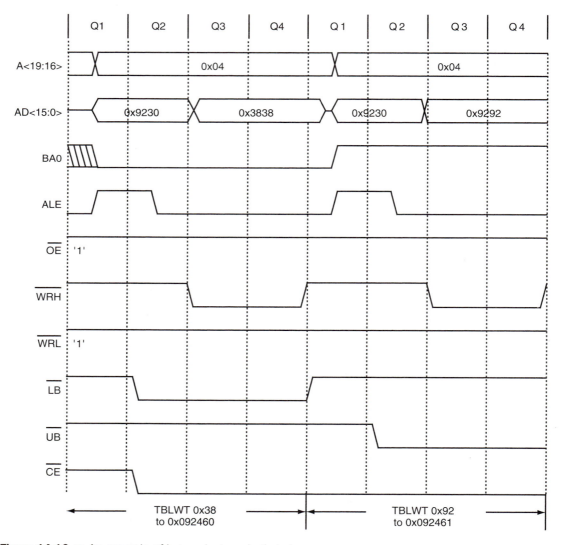

Figure 14.16 ■ An example of byte select mode timing

BYTE-WRITE MODE

This mode is used for two separate 8-bit-wide memories (each location has eight bits) connected for 16-bit operation. When a TBLWT instruction is executed, the TABLAT data is driven on both the upper and the lower bytes of the AD<15:0> bus. If the bit 0 of TBLPTR is 1, the $\overline{\text{WRH}}$ signal is asserted. Otherwise, the $\overline{\text{WRL}}$ control signal is pulsed.

Figure 14.17 illustrates the timing diagrams of two table-write instructions: one writes the value 0x28 to the memory location at 0x076460, whereas the other writes the value 0x45 to the memory location at 0x076461. In the timing diagram, both the $\overline{\text{LB}}$ and the $\overline{\text{UB}}$ signal are deasserted.

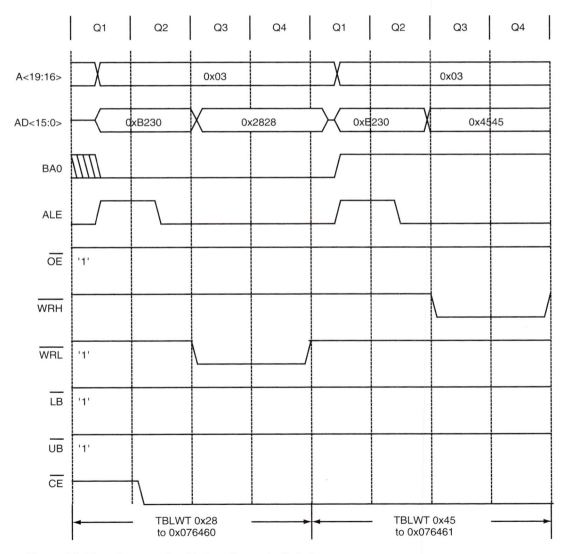

Figure 14.17 ■ An example of byte write mode timing

14.5.8 Wait States

The PIC18F8XXX devices provide the option of accessing external memory (or peripheral devices) using more than one instruction clock cycle. The user can choose to add up to three instruction cycles to the read or the write cycle. This can be beneficial for applications that need to use slower memory or peripheral devices. One such example is the EEPROM. The access time of the EEPROM device can range from 55 ns up to 200 ns. Wait states may need to be inserted in order for EEPROMs to be accessed correctly.

Figure 14.18 shows the timing diagram in which one wait state is added to a read cycle. In this figure, the $\overline{CE}$ and $\overline{OE}$ signals, once go low, will stay low until the end of the read bus cycle.

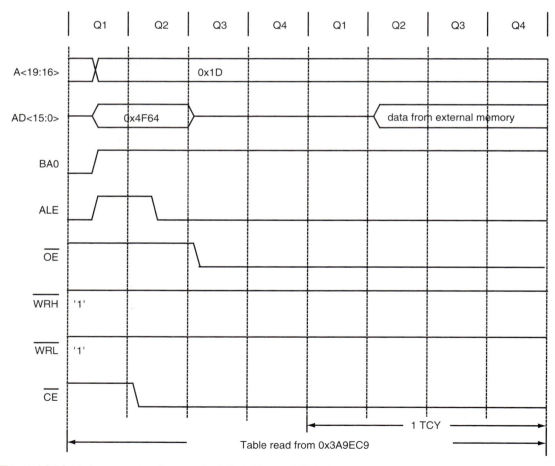

Figure 14.18 ■ An example of one-wait state table read timing

14.6　Issues Related to Adding External Memory

Adding external memory becomes necessary when the application gets larger than the on-chip program memory. When external memory is added, the designer should also consider treating external peripheral chips (when possible) as memory device because it will make the peripheral programming easier. When adding external memory, there are three issues that need to be considered:

- Memory space assignment
- Address decoder and control circuitry design
- Timing verification

14.6.1　Memory Space Assignment

Any space unoccupied by the on-chip flash memory can be assigned to external memory devices. When making memory space assignment, one has two options to choose from:

- *Equal size assignment.* In this method, the available memory space is divided into blocks of equal size, and then each block is assigned to a memory device without regard for the actual size of each memory-mapped device. A memory-mapped device could be a memory chip or a peripheral device. Memory space tends to be wasted using this approach because most memory-mapped peripheral chips need only a few bytes to be assigned to their internal registers.

- *Demand assignment.* In this approach, the user assigns the memory space according to the size of memory devices.

Example 14.2

▼

Suppose that a designer is assigned to design a PIC18F8680-based embedded product that requires 512 KB of external 16-bit SRAM, 256 KB of 16-bit EEPROM (used to hold a constant table), and a parallel peripheral interface (PPI) that requires only four bytes of address space. The only available SRAM to this designer is the 128-K × 8 SRAM chips (This chip has 128-K locations with each location containing eight bits). The only available EEPROM is the 128-K × 8 EEPROM chips. Suggest a workable memory space assignment.

Solution: The designer is assigned to design a 16-bit-wide memory system using the 8-bit-wide SRAM and EEPROM chips. Two 8-bit-wide memory chips are needed to construct a 16-bit memory module. Two 16-bit-wide SRAM modules are needed to provide the 512-KB capacity. One 16-bit-wide EEPROM module is needed to offer the 256-KB capacity.

The on-chip flash program memory occupies the space from 0x000000 to 0x00FFFF. The address space from 0x010000 to 0x1FFFFF is available for assignment. The following memory space assignment will be appropriate for this project:

```
SRAM0:   0x040000~0x07FFFF    ; 256 KB
SRAM2:   0x080000~0x0BFFFF    ; 256 KB
EEPROM:  0x0C0000~0x0FFFFF    ; 256 KB
PPI:     0x1FFFFC~0x1FFFFF    ; 4 bytes
```

▲

14.6.2 Address Decoder Design

The function of an address decoder is to make sure that there is no more than one memory device enabled to drive the data bus at a time. If there are two or more memory devices driving the same bus lines, *bus contention* will occur and could cause severe damage to the system. All memory devices or peripheral devices have control signals, such as *chip-enable* (CE), *chip-select* (CS), or *output-enable* (OE), to control their read and write operations. The address decoder outputs will be used as the chip-select or chip-enable signals of external memory devices.

Two address-decoding schemes have been used: *full decoding* and *partial decoding*. A memory device is said to be *fully decoded* when each of its addressable locations responds to only a single address on the system bus. A memory component is said to be *partially decoded* when each of its addressable locations responds to more than one address on the system bus. Memory components such as DRAM, SRAM, EPROM, EEPROM, and flash memory chips use the full address-decoding scheme more often, whereas peripheral chips or devices use the partial address-decoding scheme more often.

Address decoder design is closely related to memory space assignment. For the address space assignment made in Example 14.2, the higher address signals are used as inputs to the decoder, whereas the lower address signals are applied to the address inputs of memory devices.

Before *programmable logic devices* (PLDs) became popular and inexpensive, designers used transistor-transistor logic (TTL) chips, such as 74138, as address decoders. However, the off-the-shelf TTL decoders force designers to use equal-size memory space assignment. When PLDs became popular and inexpensive, designers started to use PLDs to implement address decoders. PLDs allow the designer to implement demand assignment.

One of the methods for implementing the address decoder is to use one of the hardware description languages (HDL), such as ABEL, CUPL, VHDL, or VERILOG. Low-density PLDs, such as GAL18V10, GAL20V8, GAL20V8, SPLD16V8, SPLD20V8, and SPLD20V8, are often used to implement address decoders for their ability to implement product terms of many variables. The *generic array logic* (GAL) devices are produced by Lattice Semiconductor. The *simple programmable logic devices* (SPLDs) are produced by Atmel. Both ABEL and CUPL are simple hardware description languages that are very suitable for describing circuit behaviors for address decoders. ABEL is supported by Lattice Semiconductor, whereas CUPL is supported by Atmel.

14.6.3 Timing Verification

When designing a memory system, the designer needs to make sure that timing requirements for both the microcontroller and the memory system are satisfied. In a read cycle, the most critical timing requirements are the *data setup time* (TadV2oeH, parameter #162) and *data hold time* (ToeH2adl, parameter #163) required by the PIC18 microcontroller. In addition, the designer must make sure that the address setup time and hold time requirements for the memory devices are met. The control signals needed by memory devices during a read cycle must be asserted at the appropriate times.

In a write cycle, the most critical timing requirements are the *write data setup time* and *write data hold time* required by the memory devices. Like a read cycle, the address setup time and address hold time must also be satisfied. Control signals required during a write cycle must also be generated at proper times.

14.7 Memory Devices

The control circuit designs for interfacing the SRAM, the EPROM, the EEPROM, and the flash memory to the PIC18 microcontroller are quite similar. In the following sections, we illustrate how to add SRAM chips with the 128-K × 8 organization to the PIC18F8680 microcontroller.

14.7.1 The CY7C1019

The CY7C1019 is a 128-K × 8 SRAM from Cypress that operates with a 5-V power supply. The CY7C1019 has a short access time in the range from 10 ns to 15 ns and three-state drivers. This device has an automatic power-down feature that can significantly reduce the power consumption when deselected. The pin assignment of the CY7C1019 is shown in Figure 14.19.

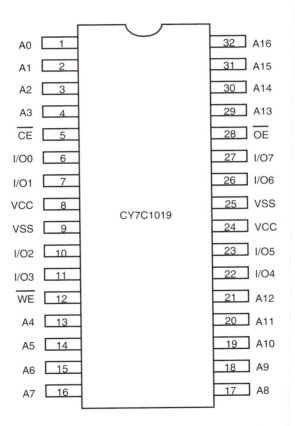

Figure 14.19 ■ The CY7C1019 Pin Assignment

The address signals A16 to A0 select one of the 128-K locations within the chip to be read or written. Pins I/O7 to I/O0 carry the data to be transferred between the chip and the microcontroller. The chip-enable ($\overline{CE}$) input allows/disallows the read/write access request to the CY7C1019. The $\overline{OE}$ signal is the output enable signal. When the $\overline{OE}$ signal is high, all eight I/O pins will be in the high-impedance state.

Depending on the assertion times of control signals, there are two timing diagrams for the read cycle and three timing diagrams for the write cycle (shown in Figures 14.20 and 14.21, respectively). The values of the related timing parameters for the read and write cycles are listed in Table 14.4.

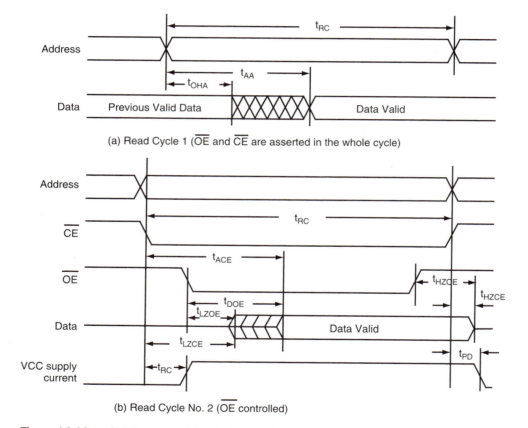

(a) Read Cycle 1 ($\overline{OE}$ and $\overline{CE}$ are asserted in the whole cycle)

(b) Read Cycle No. 2 ($\overline{OE}$ controlled)

Figure 14.20 ■ CY7C1019 Read cycle timing diagram

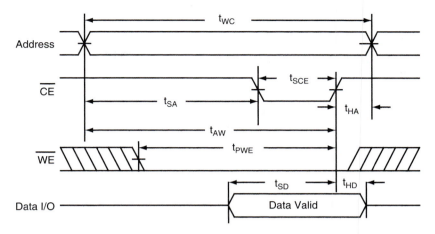

Figure 14.21a ■ TCY7C1019 Write cycle No. 1 ($\overline{CE}$ controlled)

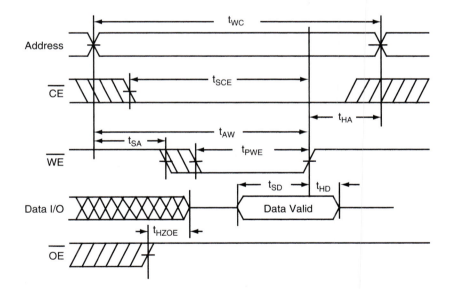

Figure 14.21b ■ CY7C1019 Write cycle No. 2 ($\overline{WE}$ controlled, $\overline{OE}$ high)

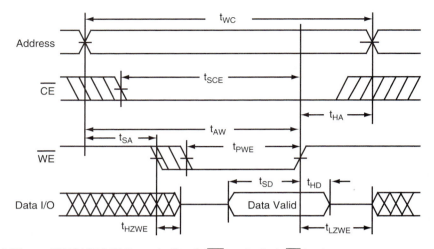

Figure 14.21c ■ CY7C1019 Write cycle No. 3 ($\overline{WE}$ controlled, $\overline{OE}$ low)

Parameter	Description	7C1019-10		7C1019-12		7C1019-15		Unit
		min	max	min	max	min	max	
Read Cycle								
t_{RC}	Read cycle time	10		12		15		ns
t_{AA}	Address to data valid		10		12		15	ns
t_{OHA}	Data hold from address change	3		3		3		ns
t_{ACE}	$\overline{CE}$ low to data valid		10		12		15	ns
t_{DOE}	$\overline{OE}$ low to data valid		5		6		7	ns
t_{LZOE}	$\overline{OE}$ low to low Z	0		0		0		ns
t_{HZOE}	$\overline{OE}$ high to high Z		5		6		7	ns
t_{LZCE}	$\overline{CE}$ low to low Z	3		3		3		ns
t_{HZCE}	$\overline{CE}$ high to high Z		5		6		7	ns
t_{PU}	$\overline{CE}$ low to power up	0		0		0		ns
t_{PD}	$\overline{CE}$ high to power down		10		12		15	ns
Write Cycle								
t_{WC}	Write cycle time	10		12		15		ns
t_{SCE}	$\overline{CE}$ low to write end	8		9		10		ns
t_{AW}	Address setup to write end	7		8		10		ns
t_{HA}	Address hold from write end	0		0		0		ns
t_{SA}	Address setup to write start	0		0		0		ns
t_{PWE}	$\overline{WE}$ pulse width	7		8		10		ns
t_{SD}	Data setup to write end	5		6		8		ns
t_{HD}	Data hold from write end	0		0		0		ns
t_{LZWE}	$\overline{WE}$ high to low Z	3		3		3		ns
t_{HZWE}	$\overline{WE}$ low to high Z		5		6		7	ns

Table 14.4 ■ CY7C1019 read and write timing parameters

In Figures 14.20a and 14.20b, the $\overline{CE}$ signal must be asserted for at least t_{RC} ns during a read cycle. The signal(s) that is asserted the latest determines the time that data will become available. For example, in Figure 14.20b, the $\overline{OE}$ signal is asserted the latest; therefore, data becomes valid t_{DOE} ns later. Data pins will go to high-impedance state t_{HZOE} ns after the $\overline{OE}$ signal goes to high or t_{HZCE} ns after the $\overline{CE}$ signal goes to high.

In a CE-controlled write cycle, the $\overline{CE}$ signal is asserted (going low) later than the $\overline{WE}$ signal. The write pulse width (t_{PWE} ns) is measured from the moment that the $\overline{WE}$ signal goes low until the $\overline{CE}$ signal goes high. The data to be written into the memory must be valid for t_{SD} ns before the $\overline{CE}$ signal goes high and must remain valid for t_{HD} ns after the $\overline{CE}$ signal goes high. In a WE-controlled write cycle, the $\overline{WE}$ signal is asserted later than the $\overline{CE}$ signal and becomes unasserted (goes high) earlier than the $\overline{CE}$ signal. The data to be written into the memory must be valid for $\mathbf{t_{SD}}$ ns before the $\overline{WE}$ signal goes high and must remain valid for $\mathbf{t_{HD}}$ ns after the $\overline{WE}$ signal goes high. The parameter t_{SD} is often called write data setup time, whereas the parameter t_{HD} is referred to as write data hold time. The timing diagrams in Figures 14.21b and 14.21c are quite similar except that if the $\overline{OE}$ signal is low (in Figure 14.21c), data I/O pins will get out of the high-impedance state when both the $\overline{CE}$ and $\overline{WE}$ signals are high.

14.7.2 The AT28C010 EEPROM

The AT28C010 is a 128-K × 8 electrically erasable, programmable, read-only memory. It needs only a 5-V power supply to operate and achieves access times ranging from 120 ns to 200 ns.

The AT28C010 supports a page-write operation that can write from 1 to 128 bytes. The device contains a 128-byte page register to allow writing up to 128 bytes simultaneously. During a write cycle, the address and 1 to 128 bytes of data are internally latched, freeing the address and data bus for other operations. Following the initiation of a write cycle, the device will automatically write the latched data using an internal control timer. The end of an internal write cycle can be detected by polling the I/O7 pin or checking whether the I/O6 pin has stopped toggling.

The device utilizes internal error correction for extended endurance and improved data retention. To prevent unintended write operation, an optional software data protection mechanism is available. The device also includes additional 128 bytes of EEPROM for device identification or tracking. The pin assignment for the AT28C010 is shown in Figure 14.22.

DEVICE OPERATION

The AT28C010 is accessed like a SRAM. When the $\overline{CE}$ and $\overline{OE}$ signals are low and the $\overline{WE}$ signal is high, the data stored at the location determined by the address pins is driven on the I/O pins. The I/O pins are put in the high-impedance state when either the $\overline{CE}$ or the $\overline{OE}$ signal is high.

A byte-write operation is started by a low pulse on the $\overline{WE}$ or the $\overline{CE}$ input with the $\overline{CE}$ or the $\overline{WE}$ input low, respectively, and the $\overline{OE}$ pin high. The address inputs are latched on the falling edge of the $\overline{CE}$ or the $\overline{WE}$ signal, whichever occurs the last. The data is latched by the first rising edge of the $\overline{CE}$ or the $\overline{WE}$ signal.

A page-write operation is initiated in the same manner as a byte-write operation; after the first byte is written, it can be followed by 1 to 127 additional bytes. Each successive byte must be loaded within 150 μs (t_{BLC}) of the previous byte. If the t_{BLC} limit is exceeded, the AT28C010 will cease accepting data and begin the internal programming operation. All bytes involved in a page-write operation must reside on the same page as defined by the state of the A16 to A7 inputs. For each high-to-low transition of the $\overline{WE}$ signal during the page-write operation, the address signals A16 to A7 must be the same. The inputs A6 to A0 are used to specify which

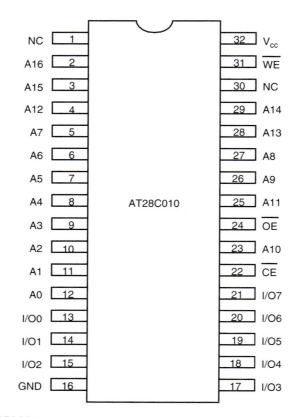

Figure 14.22 ■ The AT28C010 Pin Assignment

bytes within the page are to be written. The bytes may be loaded in any order and may be altered within the same load period. Only bytes that are specified for writing will be written.

The AT28C010 allows the user to poll the I/O7 pin to find out if an internal write operation has completed. Before an internal write operation has completed, a read of the last byte written will result in the complement of the written data to be presented on the I/O7 pin. Once the write cycle has been completed, true data is valid on all outputs, and the next write cycle can begin.

In addition to data polling, the AT28C010 provides another method for determining the end of a write cycle. During the internal write operation, successive attempts to read data from the device will result in I/O6 toggling between 1 and 0. Once the write operation has completed, the I/O6 pin will stop toggling, and valid data will be read.

DATA PROTECTION

Atmel has incorporated both hardware and software features to protect the memory against inadvertent write operations. The hardware protection method works as follows:

1. V_{CC} *sense.* If V_{CC} is below 3.8 V, the write function is inhibited.
2. V_{CC} *power-on delay.* Once V_{CC} has reached 3.8 V, the device automatically times out for 5 ms before allowing a write operation.

3. *Write inhibit.* Holding the $\overline{OE}$ signal low, the $\overline{CE}$ signal high, or the $\overline{WE}$ signal high inhibits write cycles.

4. *Noise filter.* Pulses of less than 15 ns on the $\overline{WE}$ or the $\overline{CE}$ input will not initiate a write cycle.

A software data protection (SDP) feature is included that can be enabled to prevent inadvertent write operations. The SDP is enabled by the host system issuing a series of three write commands; three specific bytes of data are written to three specific addresses. After writing the 3-byte command sequence and after the t_{WC} delay, the entire AT28C010 will be protected against inadvertent write operations. It should be noted that once protected, the host may still perform a byte write or page write to the AT28C010. This is done by preceding the data to be written by the same 3-byte command sequence used to enable the SDP.

Once set, the SDP will remain active unless the disable command sequence is issued. Power transitions do not disable the SDP, and the SDP will protect the AT28C010 during the power-up and power-down conditions. After setting the SDP, any attempt to write to the device without the 3-byte command sequence will start the internal write timer. No data will be written into the device.

The algorithm for enabling software data protection is as follows:

Step 1
Write the value of 0xAA to the memory location at 0x5555.

Step 2
Write the value of 0x55 to the memory location at 0x2AAA.

Step 3
Write the value of 0xA0 to the memory location at 0x5555. At the end of write, write-protect state will be activated. After this step, write operation is also enabled.

Step 4
Write any value to any location (1–128 bytes of data are written).

Step 5
Write the last byte to the last address.

Software data protection can be disabled anytime when it is undesirable. The algorithm for disabling software data protection is as follows:

Step 1
Write the value of 0xAA to the memory location at 0x5555.

Step 2
Write the value of 0x55 to the memory location at 0x2AAA.

Step 3
Write the value of 0x80 to the memory location at 0x5555.

Step 4
Write the value of 0xAA to the memory location at 0x5555.

Step 5
Write the value of 0x55 to the memory location at 0x2AAA.

Step 6
Write the value of 0x20 to the memory location at 0x5555. After this step, software data protection is exited.

Step 7
Write any value (s) to any location (s).

Step 8
Write the last byte to the last address.

Device Identification

An extra 128 bytes of EEPROM memory are available to the user for device identification. By raising the voltage at the A9 pin to 12 ± 0.5 V and using the address locations 0x1FF80 to 0x1FFFF, the bytes may be written to or read from in the same manner as the regular memory array.

Read and Write Timing

The read cycle timing diagram is shown in Figure 14.23.

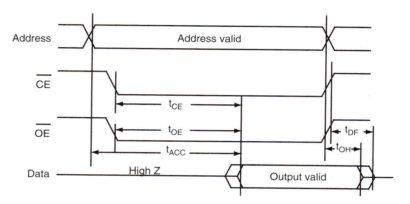

Figure 14.23 ■ AT28C010 read timing diagram

The AT28C010 has three read access times:

- Address access time t_{ACC}
- $\overline{CE}$ access time t_{CE}
- $\overline{OE}$ access time t_{OE}

Each of the read access times is measured by assuming that the other control signals and/or the address have been valid. For example, the address access time is the time from the moment that the address inputs to the AT28C010 become valid until data is driven out of data pins, assuming that the $\overline{CE}$ and $\overline{OE}$ signals have been valid (low) before the required moment (t_{CE} or t_{OE} ns before data become valid). After address change, data value will change immediately (data hold time is 0). The values of read timing parameters are listed in Table 14.5.

The write cycle timing diagram is shown in Figure 14.24. There are two write timing diagrams—WE-controlled and CE-controlled diagrams—depending on which signal is asserted the latest. In a WE-controlled write timing waveform, the $\overline{CE}$ signal is asserted (goes low) earlier than the $\overline{WE}$ signal and becomes inactive (goes high) after the $\overline{WE}$ signal goes high. In a CE-controlled write cycle, the $\overline{WE}$ signal goes low earlier and returns to high later than $\overline{CE}$ the signal.

In Figures 14.24a and 14.24b, the control signal that is asserted the latest must have a minimal pulse width of 100 ns. The address input must not be valid after the latest control signal

Symbol	Parameter	AT28C010-12		AT28C010-15		AT28C010-20		Unit
		Min.	**Max.**	**Min.**	**Max.**	**Min.**	**Max.**	
t_{ACC}	Address to output delay		120		150		200	ns
$t_{CE}^{(1)}$	CE to output delay		120		150		200	ns
$t_{OE}^{(2)}$	OE to output delay	0	50	0	55	0	55	ns
$t_{DF}^{(3)(4)}$	OE to output float	0	50	0	55	0	55	ns
t_{DH}	Output hold from OE, CE, or address, which occurred first	0		0		0		ns

Notes:
1. $\overline{CE}$ may be delayed up to t_{ACC}-t_{CE} after the address transition without impact on t_{ACC}.
2. $\overline{OE}$ may be delayed up to t_{CE}-t_{OE} after the falling edge of $\overline{CE}$ without impact on t_{CE} or by t_{ACC}-t_{OE} after an address change without impact on t_{ACC}.
3. t_{DF} is specified from $\overline{OE}$ or $\overline{CE}$ whichever occurs first (C_L = 5pF).
4. This parameter is characterized and is not 100% tested.

Table 14.5 ■ AT28C010 Read characteristics

Symbol	Parameter	Min.	Max.	Unit
t_{AS}, t_{OES}	Address, $\overline{OE}$ setup time	0		ns
t_{AH}	Address hold time	50		ns
t_{CS}	Chip select setup time	0		ns
t_{CH}	Chip select hold time	0		ns
t_{WP}	Write pulse width ($\overline{WE}$ or $\overline{CE}$)	100		ns
t_{DS}	Data setup time	50		ns
t_{DH}, t_{OEH}	Data, $\overline{OE}$ hold time	0		ns

Table 14.6 ■ AT28C010 Write characteristics

is asserted (goes low) and must remain valid for at least 50 ns after the same signal goes low. The values of write cycle timing parameters are shown in Table 14.6. The write data must be valid at least 50 ns before the latest control signal starts to rise. The write data need not be stable after the latest control signal (data hold time) rises.

The page mode write cycle timing diagram is shown in Figure 14.25, and the values of the timing parameters are shown in Table 14.7.

The write timing waveform illustrates only how the CPU writes data into the EEPROM. The EEPROM still needs to initiate an internal programming process to actually write data into the specified location. The CPU can find out whether the internal programming process has been completed by using the data polling or the toggle bit polling method.

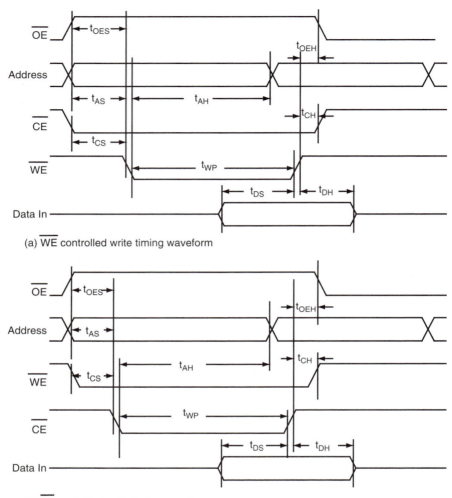

(a) $\overline{\text{WE}}$ controlled write timing waveform

(b) $\overline{\text{CE}}$ controlled write timing waveform

Figure 14.24 ■ AT28C010 Write cycle timing waveform

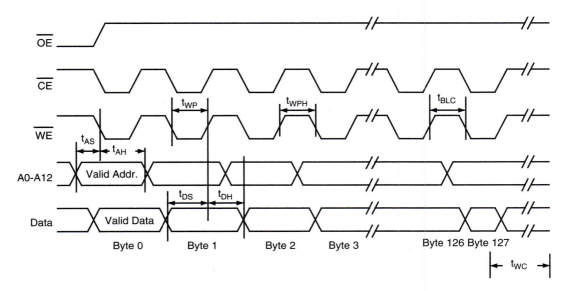

Figure 14.25 ■ AT28C010 Page mode write waveform

Symbol	Parameter	Min.	Max.	Unit
t_{WC}	Write cycle time		10	ms
t_{AS}	Address setup time	0		ns
t_{AH}	Address hold time	50		ns
t_{DS}	Data setup time	50		ns
t_{DH}	Data hold time	0		ns
t_{WP}	Write pulse width	100		ns
t_{BLC}	Byte load cycle time		150	μs
t_{WPH}	Write pulse width high	50		ns

Table 14.7 ■ AT28C010 Page-Write characteristics

14.8 Example of External Memory Expansion for the PIC18 Microcontroller

Suppose we want to design a demo board that has 256 KB of 16-bit external SRAM program memory and 256 KB of 16-bit EEPROM memory to store large amount of data. The design will allow the user to add an additional 256 KB of SRAM when it is needed. Two sockets will be provided but unpopulated to provide this possibility. In addition, this design will also provide chip-select signals for the remaining address space for users who want to add memory outside the demo board.

4.8.1 Memory Space Assignment

Because this design will be using two 128-KB SRAM chips (CY7C1019) to construct 256 KB of 16-bit wide memory, the lowest 17 address signals (A16 . . . A0) will be used as the address inputs for both SRAM chips. BA0 will not be needed. Because these two chips must be selected by the same chip-enable signal, the address space must be divided into 256-KB blocks when making memory space assignment to the SRAM module. The following assignment will be appropriate:

- *SRAM module 0.* 0x040000-0x07FFFF (the second 256 KB of the 2-MB memory space)
- *SRAM module 1.* 0x080000-0x0BFFFF (the third 256 KB of the 2-MB memory space)
- *EEPROM module.* 0x0C0000-0x0FFFFF (the fourth 256 KB of the 2-MB memory space)

14.8.2 Address Latch

Since the address signals A15 to A0 and data signals D15 to D0 are time-multiplexed, they need to be latched into a memory device such as the 74HC573 and held valid throughout the whole bus cycle. From Figures 14.13 and 14.14, it is easy to see that the falling edge of the ALE signal should be used to latch the address signals A15 to A0 into two 74HC573s. The pin assignment of the 74HC573 is shown in Figure 14.26.

In Figure 14.26, the values of pins D7 to D0 are latched into Q7 to Q0 on the falling edge of the $\overline{LE}$ signal (connected to the ALE signal directly). According to the datasheet provided by SGS-Thompson, D7 to D0 must be valid at least 15 ns (t_{SU}) before the falling edge of the $\overline{LE}$ signal and remain valid for 5 ns after the signal goes low in order to be correctly latched. Outputs Q7 to Q0 will be in the high-impedance state if the $\overline{OE}$ signal is high. The 74HC573 has a maximum propagation delay (from $\overline{LE}$'s falling edge to Q7 to Q0 valid) of 28 ns in the temperature range of –40°C to 85°C for a 5.0-V power supply. This delay is typically 15 ns at room temperature.

At 25 MHz, address signals A15 to A0 become valid $0.25T_{CY}$ – 10 ns (30 ns) before the falling edge of the ALE signal and remain valid for 5 ns after the ALE signal goes low and hence satisfy the D7 to D0 setup time and hold time requirements for the 74HC573. However, when the $\overline{LE}$ signal is high, the 74HC573 is transparent. That is, Q7 to Q0 will follow the value of D7 to D0 after the propagation t_{PLH} (or t_{PHL}). The value of t_{PHL} is load dependent. We will use the maximum value (28 ns) for a load of 50 pF in this analysis (the typical value of t_{PLH} or t_{PHL} is about 15 ns). Therefore, address signals A<15:0> will become available 60.5 ns in the worst case (47.5 ns typically) after the start of a bus cycle:

A<15:0> valid time (worst case)
= the delay of the rise edge of ALE from the start of Q1
+ pulse width of ALE - TadV2alL + t_{PHL}
= 22.5 ns + 40 ns – 30 ns + 28 ns (typical value is 15 ns)
= 60.5 ns

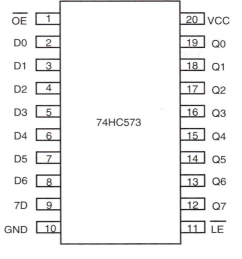

Figure 14.26 ■ Pin assignment of 74HC573

The address latching circuit is shown in Figure 14.27.

To facilitate timing verification, the timing parameters of read and write cycles for 25 MHz are shown in Table 14.8 and 14.9.

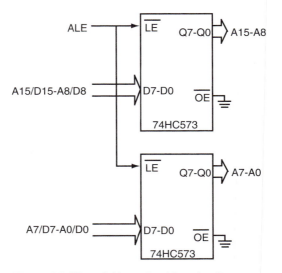

Figure 14.27 ■ Address latching circuit

Param. No.	Symbol	Characteristics	Min.	Typ	Max.	Units
150	TadV2a1L	Address out valid to ALE ↓ (address set up time)	30	–	–	ns
151	Ta1L2adL	ALE ↓ to address out invalid (address hold time)	5	–	–	ns
155	Ta1L2oeL	ALE ↓ to $\overline{OE}$ ↓	10	20	–	ns
160	TadZ2oeL	AD high-Z to $\overline{OE}$ ↓ (bus release to $\overline{OE}$)	0	–	–	ns
161	ToeH2adD	$\overline{OE}$ ↑ to AD driven	15	–	–	ns
162	TadV2oeH	LS data valid before $\overline{OE}$ ↑ (data setup time)	20	–	–	ns
163	ToeH2ad1	$\overline{OE}$ ↑ to data in invalid (data hold time)	0	–	–	ns
164	Ta1H2a1L	ALE pulse width	–	40	–	ns
165	ToeL2oeH	$\overline{OE}$ pulse width	75	80	–	ns
166	Ta1H2a1H	ALE ↑ to ALE ↑ (cycle time)	–	160	–	ns
167	Tacc	Address valid to data valid	95	–	–	ns
168	Toe	$\overline{OE}$ ↓ to data valid	–	–	55	ns
169	Ta1L2oeH	ALE ↓ to $\overline{OE}$ ↑	90	–	110	ns
171	Ta1H2csL	Chip enable active to ALE ↓	20	–	–	ns
171A	TubL2oeH	AD valid to chip enable active	–	–	10	ns

Note.
1. TCY is instruction cycle time and is 160 ns for 25 MHz crystal oscillator.
2. ALE signal and A15-A0 are valid 22.5 after the start of a bus cycle.

Table 14.8 ■ Program memory read cycle timing parameters at 25 MHz

Param. No.	Symbol	Characteristics	Min.	Typ	Max.	Units
150	TadV2a1L	Address out valid to ALE ↓ (address set up time)	30	–	–	ns
151	Ta1L2adL	ALE ↓ to address out invalid (address hold time)	5	–	–	ns
153	TwrH2ad1	$\overline{WRn}$ ↑ to data out invalid (data hold time)	5	–	–	ns
154	TwrL	$\overline{WRn}$ pulse width	75	80	–	ns
156	TadV2wrH	Data valid before $\overline{WRn}$ ↑ (data set up time)	70	–	–	ns
157	TbsV2wrL	Byte select valid before $\overline{WRn}$ ↓ (byte select setup time)	40	–	–	ns
157A	TwrH2bs1	$\overline{WRn}$ ↑ to byte select invalid (byte select hold time)	15	–	–	ns
166	Ta1H2a1H	ALE ↑ to ALE ↑ (cycle time)	–	160	–	ns
171	Ta1H2csL	Chip Enable active to ALE ↓	20	–	–	ns
171A	TubL2oeH	AD valid to chip enable active	–	–	10	ns

Note.
1. $\overline{WRn}$ (n = H or L)

Table 14.9 ■ Program memory write cycle timing parameters at 25 MHz crystal oscillator

14.8.3 Address Decoder Design

For this simple memory space assignment, the address decoding logic can be implemented using one 74F138 decoder. The address signals A19 to A17 will be decoded to select different memory modules. The chip select signal equations for these SRAM and EEPROM modules are (the exclamation character stands for "NOT"):

$$\overline{CS\_SRAM0} = !(!A19 * !A18 * A17 * \overline{CE})$$
$$\overline{CS\_SRAM1} = !(!A19 * A18 * !A17 * \overline{CE})$$
$$\overline{CS\_EEPROM} = !(!A19 * A18 * A17 * \overline{CE})$$

This design will also make the following chip-select signals available to the user:

$\overline{CS4}$ = !(A19 * !A18 * !A17 * $\overline{CE}$)	; space 0x100000-0x13FFFF
$\overline{CS5}$ = !(A19 * !A18 * A17 * $\overline{CE}$)	; space 0x140000-0x17FFFF
$\overline{CS6}$ = !(A19 * A18 * !A17 * $\overline{CE}$)	; space 0x180000-0x1BFFFF
$\overline{CS7}$ = !(A19 * A18 * A17 * $\overline{CE}$)	; space 0x1C0000-0x1FFFFF

As shown in Figures 14.13 and 14.14, the $\overline{CE}$ signal is asserted only after all address signals have become valid. Therefore, it should be used as a qualifying signal for the address decoder.

14.8.4 PIC18F8680 Timing Parameter Values at 25 MHz

To facilitate timing analysis and verification for the memory system to be designed, the values of all program memory read and write timing parameters at 25 MHz are calculated and listed in Tables 14.8 and 14.9.

14.8.5 Example PIC18F8680 Demo Board Memory System Diagram

As shown in Figure 14.28, the circuit required to interface with external SRAMs and EEPROMs is minimal. Two latches are needed to keep address signals A16 to A0 valid throughout the whole bus cycle, whereas one address decoder is used to generate the chip-select signals for selecting the SRAM and EEPROM modules. Since only the highest three address signals are used and the equal-size memory assignment is adopted, one can use the ubiquitous 74F138 decoder as the address decoder.

The $\overline{CE}$ signal from the PIC18F8680 is asserted (goes low) only when the address outputs are valid. It is used as the E0 input to the 74F138 to make sure that the decoder asserts one of its outputs to low when its address inputs A2 to A0 are valid.

The 74F138 has a maximum propagation delay (from A2 to A0 to On, t_{PHL} and t_{PLH}) about 9 ns in the temperature range from 0° C to 80° C. The typical propagation delay is about 5 ns at room temperature for a 50-pF load.

Example 14.3

▼

For the circuit shown in Figure 14.28, verify that the read and write timing requirements for the PIC18F8680 are satisfied when accessing SRAM modules, assuming that the PIC18F8680 is running with a 25-MHz crystal oscillator.

Solution: The timing verification consists of read cycle and write cycle and is based on the 25-MHz crystal oscillator.

▲

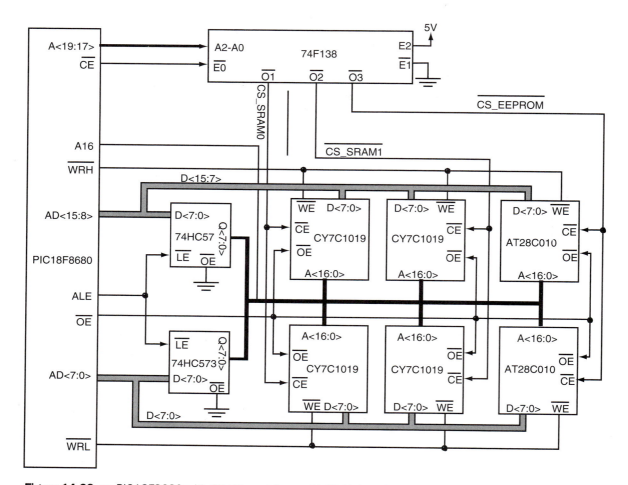

Figure 14.28 ■ PIC18F8680 with SRAM modules and LCD kit interfacing circuit

READ TIMING ANALYSIS

According to the earlier discussion, the address inputs to the SRAM will be valid 60.5 ns (worst case) after the read cycle starts (or 99.5 ns before the end of the read cycle if no-wait state is inserted).

The $\overline{OE}$ signal is valid (goes low) 75 ns before the end of the read cycle in no-wait state. The $\overline{CE}$ input to the CY7C1019 is valid 31.5 ns after the beginning of the read cycle (or 128.5 ns before the end of the read cycle):

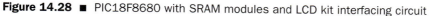

$\overline{CE}$ valid time = A<19:17> valid time from the start of the read cycle + 74F138 propagation delay
= 22.5 ns + 9 ns = 31.5 ns

Apparently, the $\overline{OE}$ signal becomes valid the latest, and hence it determines when the data output will be valid. Suppose that the CY7C1019-15 is chosen. The data will then become valid 60 ns (= 75 ns – 15 ns) before the end of the read cycle. The read data setup time requirement for the PIC18F8680 is 20 ns and hence is satisfied. In addition, the CY7C1019 holds data valid for 7 ns after the $\overline{OE}$ input goes high and hence also satisfies the PIC18F8680 read data hold time requirement (0 ns is required). The resultant read cycle timing diagram is shown in Figure 14.29.

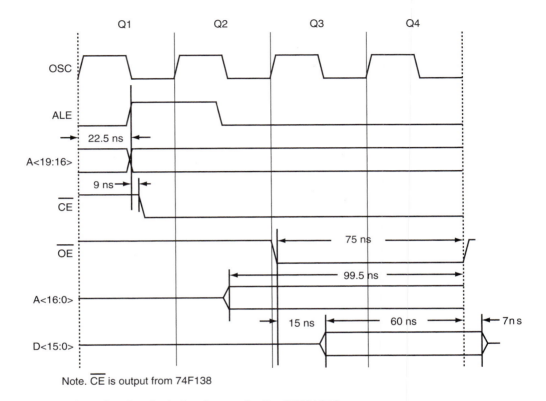

Note. $\overline{CE}$ is output from 74F138

Figure 14.29 ■ Read cycle timing diagram for the CY7C1019

WRITE TIMING ANALYSIS

The $\overline{WE}$ input is valid 75 ns before the end of the write cycle and hence is asserted later than the $\overline{CE}$ input (128.5 ns before the end of the write cycle). The write cycle for this demo board is therefore WE controlled. The CY7C1019 requires write data to be valid 8 ns before the $\overline{WE}$ signal rises and to remain valid for 0 ns after the rising edge of the $\overline{WE}$ signal.

The PIC18F8680 drives data onto the data bus 70 ns before the end of a write cycle and keeps it valid for 5 ns after the end of the write cycle and hence satisfies both the write data setup time and hold time requirements. In addition, the CY7C1019 requires the write pulse width ($\overline{WE}$) to be at least 10 ns. The $\overline{WE}$ input is driven by the $\overline{WRH}$ signal, which is 75 ns in width. Therefore, this requirement is also satisfied. The resultant write cycle timing diagram is shown in Figure 14.30.

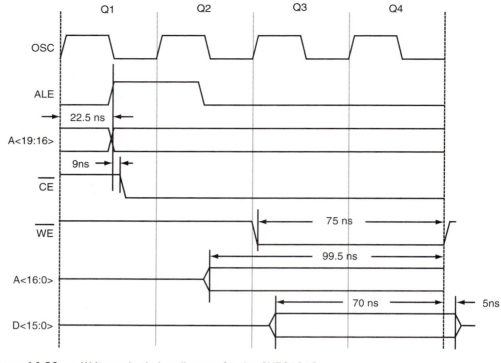

Figure 14.30 ■ Write cycle timing diagram for the CY7C1019

Example 14.4
▼

Since the CY7C1019 is much faster than what is needed, what is the slowest memory that can be successfully accessed without wait states at 25 MHz?

Solution: For read cycle, the CY7C1019 is 40 ns (= 60 ns – 20 ns) faster than what is actually needed. For write cycle, the PIC18F8680 drives data 55 ns (70 ns – 15 ns) earlier than what is actually needed by the CY7C1019. Therefore, we can choose a SRAM device with read access time equal to 55 ns (15 ns + 40 ns) or better and still can work correctly. However, most of today's SRAM chips are very fast. It is actually very difficult to find any company that manufactures slower SRAM chips today. However, sometimes we need to use external ROM memory (e.g., flash memory) to hold large application programs. This number tells us the slowest access time for the flash memory device that can be used at 25 MHz. For example, the Atmel 128-K × 8 flash memory AT49F001A-45 will work at this frequency without wait states.

▲

Example 14.5
▼

Perform an analysis to decide the minimum number of wait states that are needed for the different versions of the AT28C010 EEPROM. Discuss the pros and cons of using flash memory and EERPOM in a system.

Solution: *Read access timing.* Here we have three read access times to consider: t_{ACC}, t_{CE}, and t_{OE}. The address signals A<16:0>, $\overline{CE}$ signal, and $\overline{OE}$ signal are valid 99.5 ns, 128.5 ns (160 ns – 31.5 ns), and 75 ns before the end of the read cycle, respectively:

1. If the address access time is used (assuming that the other two control signals are valid earlier), then data from AT28C010 will be valid 20.5 ns (120 – 99.5), 50.5 ns (150 – 99.5), and 100.5 ns (200 – 99.5) after the end of an instruction cycle for the three versions of the AT28C010, respectively.

2. If the $\overline{CE}$ signal is used (assuming that A<16:0> and $\overline{OE}$ are valid already), then data will be valid 8.5 ns before the end of an instruction for the AT28C010-12 but will be valid 21.5 ns and 71.5 ns after the instruction cycle for the AT28C010-15 and AT28C010-20, respectively.

3. If the $\overline{OE}$ signal is used (assuming that A<16:0> and $\overline{CE}$ are valid earlier), then data will be valid 25 ns, 20 ns, and 20 ns before the end of an instruction cycle, respectively.

Apparently, address access time is the dominant access time. If one instruction cycle is added, then 160 ns will be added to the access time allowance. With two instruction cycles, the data from the AT28C010 will become valid 139.5 ns, 109.5 ns, and 59.5 ns before the end of the second instruction cycle and provide more than 20 ns of data setup time and hence would satisfy the read cycle data setup time for the PIC18F8680.

The AT28C010 provides 0 ns of read data hold time and satisfies the read data hold time requirement of the PIC18F8680. From this analysis, adding one wait cycle is adequate for three versions of the AT28C010.

Write cycle timing. Like the CY7C1019, the write access cycle of the AT28C010 is WE controlled. For the write access cycle, the AT28C010 requires write data to be valid 50 ns before the rising edge of the $\overline{WE}$ signal. With no wait state, the PIC18F8680 drives the write data 70 ns before the end of the instruction cycle. With one wait state, the write data is valid 230 ns before the rising edge of the $\overline{WE}$ signal. The PIC18F8680 keeps the write data valid for 5 ns after the rising edge of the $\overline{WE}$ signal, and hence both the write data setup time and the hold time requirements for the AT28C010 are satisfied.

The AT28C010 requires address inputs to be valid 0 ns before the falling edge of the $\overline{WE}$ signal and remains valid for 50 ns after the falling edge. A<16:0> is valid 99.5 ns before the end of the instruction cycle without wait state. The $\overline{WE}$ signal goes low 75 ns before the end of the instruction cycle. Therefore, A<16:0> is valid 24.5 ns before the falling edge of the $\overline{WE}$ signal. Since A<16:0> will remain valid for 60.5 ns after the end of the write cycle, it provides 295.5 ns of address hold time (75 ns + 160 ns + 60.5 ns). Both the address setup time and the hold time are satisfied for the AT28C010.

The AT28C010 requires the $\overline{WE}$ signal to be low (write pulse width) for at least 100 ns. The $\overline{WE}$ signal from the PIC18F8680 is low for 235 ns (= 75 ns + 160 ns) with one wait state, and hence this requirement is also satisfied.

Since all three versions of the AT28C010 can support one-wait state access, we would choose the AT28C010-20 because it should be the least expensive one.

When the contents of the AT28C010 need to be changed, one can write the new value(s) into the specified location(s) without erasing its (their) contents. This is not true for the flash memory. One needs to erase the flash memory either one block (a few kilobytes at a time) or the whole chip and hence is not efficient for changing a few bytes at a time. Therefore, EEPROM will be more suitable for holding tables that need to be changed frequently. However, the flash memory is cheaper. It is suitable for holding programs that the user normally erases the whole chip when there is a need to update them.

In order to insert wait states to the access cycle of the AT28C010, one needs to do the following:

- Configure the PIC18F8680 to extended microcontroller mode so that user can use both the on-chip flash program memory and external memory. This is done by clearing bit 7, bit 1, and bit 0 of the CONFIG3L register to 0, 1, and 1, respectively.
- Set the contents of the MEMCON register to 0x20 to enable the external bus, insert one T_{CY} to the table-read and the table-write operations, and select byte-write mode.

▲

14.9 Summary

A PIC18 microcontroller has a certain amount of on-chip flash program memory, data SRAM, and data EEPROM. Some of the data memory (referred to as special-function registers) is used to control peripheral functions, while the rest is used as scratch memory (referred to as general-purpose registers). On-chip data memory and flash program memory (or one-time programmable memory) are in separate memory spaces. A few PIC18 microcontrollers also support external memory. Users can add external SRAMs, EPROMs, or flash memory chips if the application requires that.

On-chip flash memory is used mainly to store program instructions. However, it also allows the user to use the table-read instruction to read from it and use the table-write instruction to write data into the program memory. In order to write data correctly into the flash memory, the user must erase the contents of the memory locations before programming to those locations. Certain procedures must be followed. One can erase flash program memory one block (64 bytes) at a time and can program into the flash memory also one block (eight bytes) at a time.

The data EEPROM is readable and writable during the normal operation over the entire V_{DD} range. This memory is not directly mapped in the register file space. Instead, it is indirectly addressed through the special-function registers EEADRH, EEADR, and EEDATA.

The PIC18 microcontroller has four program memory modes:

- Microprocessor
- Microcontroller with boot block
- Extended microcontroller
- Microcontroller

Only microcontroller mode is available for those PIC18 devices that do not have an external memory interface. The PIC18C601 and the PIC18C801 do not have on-chip memory and can operate only in microprocessor mode. For members that support external memory, all four modes are available.

For performance consideration, the PIC18 microcontroller was designed to fetch an instruction in one instruction clock cycle and cannot be extended. However, the PIC18 microcontroller allows the table-read and table-write operations to be extended up to four instruction cycles. This allows the user to use slower memory, such as EEPROM, as external memory to hold large tables.

There are three issues related to adding external memory to the PIC18 devices:

- Memory space assignment
- Address decoder and control circuitry design
- Timing verification

When making memory space assignment, it is common to assign memory space in blocks of equal size to memory components. However, this method would leave a lot of space unused when assigning space to peripheral devices that need only a few bytes.

All memory devices are attached to the common system bus. However, only one device should be allowed to drive the data bus at a time. Otherwise, bus contention will occur and could cause severe damage to the system. To avoid this problem, all memory devices and peripheral devices have control signals, such as chip-enable or chip-select, to allow/disallow a memory or peripheral device to access the system bus. The function of the address decoder is to generate chip-select (or chip-enable) signals.

When performing timing verification, the most important requirements are often the data setup and hold time requirements by the microcontroller and memory devices. Other requirements, such as address setup times and hold times, are also important. These requirements must be satisfied in order to make sure that the microcontroller can correctly access data from the memory system.

14.10 Exercises

E14.1   Write a program to erase the on-chip flash program memory from 0x8000 to 0xFFFF.

E14.2   Write a program to set the contents of the flash program memory from 0x8000 to 0xFFFF to 1, 2, 3, 4, 5, 6, 7, 8, 1, 2, 3, 4, 5, 6, 7, 8, and so on.

E14.3   Write a program to write the value from 1 to 100 to the first 100 bytes of the EEPROM data memory, read them back, and store them in general-purpose registers GPR100 to GPR199 for verification.

E14.4   Suppose that there is a PIC18F8680 demo board with 256 KB of external SRAM memory located at 0x40000 to 0x7FFFF. Write a function to sort an array of N 8-bit array located in the external SRAM. The array count N is 16-bit. Both the array count and the starting address of the array are passed to this function in the stack.

E14.5   Suppose that an application requires the designer to add 32 KB of EEPROM and 512 KB of SRAM (in two 256-KB modules) to the system. Make the memory space assignment and design the address decoder and the whole circuit similar to that in Figure 14.21. The 32-K 8 EEPROM AT28HC256 (from Atmel) and the 128-K 8 SRAM CY7C1019 are recommended for the design. The GAL16V8 device should be used to implement the address decoder.

E14.6   Perform a timing analysis to verify that all the timing requirements are satisfied. Are wait states needed for the EEPROM? If wait states are needed, can the EEPROM be used to hold the application program? Assume that a 25-MHz crystal oscillator is used in this design.

14.11 Lab Exercises and Assignments

L14.1   *Memory test for the demo board with external memory.* (Skip this exercise if a demo board with external SRAM is not available.) Write a program that tests the memory locations in the external SRAM. The program is written to test stuck-at-1 and stuck-at-0 faults and proper data connections:

1. To test stuck-at-0 faults, write 0xFF to each memory location and read it back to check.
2. To test stuck-at-1 faults, write 0x00 to each memory location and read it back to check.

3. To test for proper connection, write %10101010 and %01010101 to each memory location and read them back to check.

Output a message to the LCD kit to display the test result at the end of the program.

L14.2   *EEPROM programming.* Write a program to program the message "Long lives the microcontroller!" into the on-chip data EEPROM memory starting from address 0x000. Read back the message and echo it to the terminal port. A terminal program must be running in order for the echoed message to be seen.

L14.3   *Flash memory programming.* Write a program to compute the first 100 prime numbers and store these prime numbers in the flash program memory with each number stored in 16 bits. Use ICD2 and MPLAB to verify the program execution result.

15

System Configuraton
and Protection

15.1 Objectives

After completing this chapter, you should be able to

- Explain the function of each configuration register

- Set the value for each configuration register

- Explain the functioning of the watchdog timer

- Understand the reason for setting code protection

- Understand the reason for enabling power-down mode

- Understand the usefulness of in-circuit serial programming

15.2 Introduction

There are several PIC18 features that are important to the functioning of the microcontroller but cannot be included in an appropriate chapter for discussion: configuration registers, watchdog timer, power-down mode, and program memory protection. These features are discussed in this chapter.

15.3 Configuration Registers

Many of the settings and modes of the PIC18 devices are controlled by a group of configuration registers. The contents of these configuration registers are device dependent. A summary of the configuration bits of the PIC18FXX2, the PIC18FXX20, and the PIC18F8680/8585/6680/6585 is given in Appendix H. The contents of the configuration registers of these devices are discussed in this section.

The configuration registers are located from 0x300001 to 0x30000D, whereas the device IDs are stored at 0x3FFFFE to 0x3FFFFF. These addresses are beyond the user program memory space and can be accessed only by using the table-read and table-write instructions.

Programming the configuration registers is done in a manner similar to programming the flash memory. Setting the WR bit of the EECON1 register starts a self-timed write operation to the specified configuration register. The configuration registers are written one byte at a time.

15.3.1 The CONFIG1H Register

The contents of this register are shown in Figures 15.1a and 15.1b. The CONFIG1H register allows the user to select the type of oscillator crystal and enable the Timer1 external crystal oscillator. A 32-KHz crystal can be used with Timer1 when the bit 5 of the CONFIG1H register is cleared to 0.

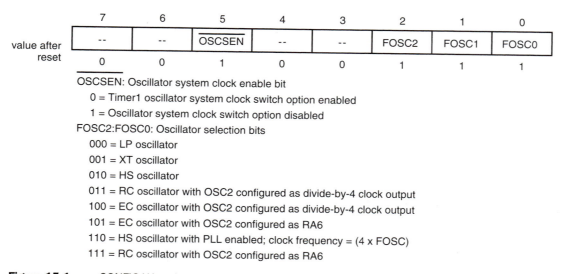

Figure 15.1a ■ CONFIG1H register of PIC18FXX2 and PIC18FXX20

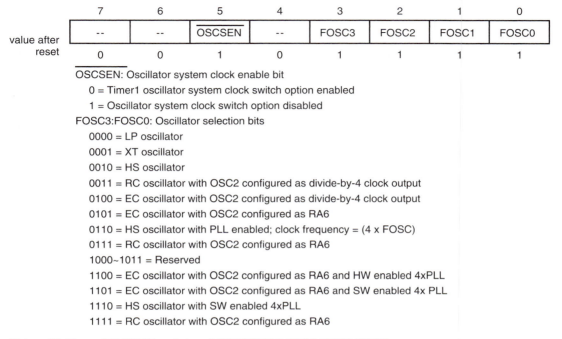

OSCSEN: Oscillator system clock enable bit

　0 = Timer1 oscillator system clock switch option enabled

　1 = Oscillator system clock switch option disabled

FOSC3:FOSC0: Oscillator selection bits

　0000 = LP oscillator

　0001 = XT oscillator

　0010 = HS oscillator

　0011 = RC oscillator with OSC2 configured as divide-by-4 clock output

　0100 = EC oscillator with OSC2 configured as divide-by-4 clock output

　0101 = EC oscillator with OSC2 configured as RA6

　0110 = HS oscillator with PLL enabled; clock frequency = (4 x FOSC)

　0111 = RC oscillator with OSC2 configured as RA6

　1000~1011 = Reserved

　1100 = EC oscillator with OSC2 configured as RA6 and HW enabled 4xPLL

　1101 = EC oscillator with OSC2 configured as RA6 and SW enabled 4x PLL

　1110 = HS oscillator with SW enabled 4xPLL

　1111 = RC oscillator with OSC2 configured as RA6

Figure 15.1b ■ CONFIG1H register of PIC18F8680/8585/6680/6585

15.3.2 The CONFIG2L Register

This register has three functions: select brown-out voltage level, enable/disable brown-out reset, and enable/disable power-up timer. *Brown-out* is a condition that the power supply drops below its normal operation level. The contents of this register are shown in Figure 15.2.

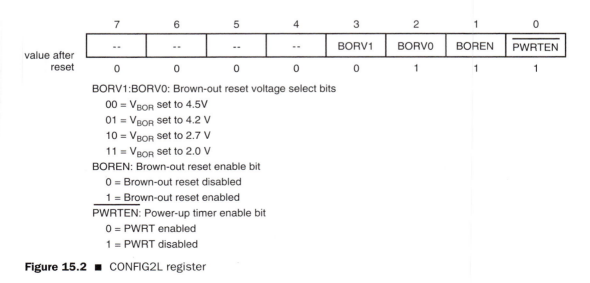

BORV1:BORV0: Brown-out reset voltage select bits

　00 = V_{BOR} set to 4.5V

　01 = V_{BOR} set to 4.2 V

　10 = V_{BOR} set to 2.7 V

　11 = V_{BOR} set to 2.0 V

BOREN: Brown-out reset enable bit

　0 = Brown-out reset disabled

　1 = Brown-out reset enabled

PWRTEN: Power-up timer enable bit

　0 = PWRT enabled

　1 = PWRT disabled

Figure 15.2 ■ CONFIG2L register

15.3.3 The CONFIG2H Register

This register allows the user to enable/disable the watchdog timer and also to select the appropriate postscale factor for the watchdog timer output. The postscale factor allows the user to enlarge the watchdog timer time-out interval. The contents of this register are shown in Figures 15.3a and 15.3b.

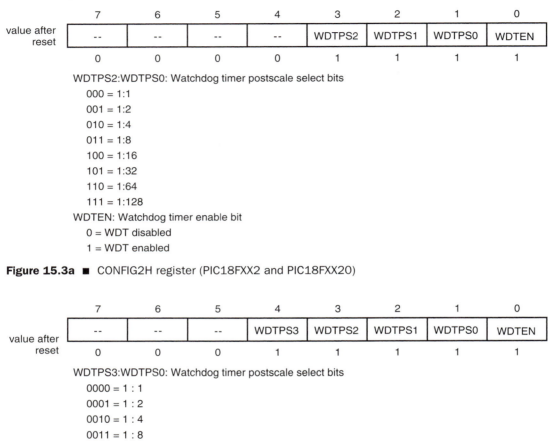

WDTPS2:WDTPS0: Watchdog timer postscale select bits
000 = 1:1
001 = 1:2
010 = 1:4
011 = 1:8
100 = 1:16
101 = 1:32
110 = 1:64
111 = 1:128

WDTEN: Watchdog timer enable bit
0 = WDT disabled
1 = WDT enabled

Figure 15.3a ■ CONFIG2H register (PIC18FXX2 and PIC18FXX20)

WDTPS3:WDTPS0: Watchdog timer postscale select bits
0000 = 1 : 1
0001 = 1 : 2
0010 = 1 : 4
0011 = 1 : 8
0100 = 1 : 16
0101 = 1 : 32
0110 = 1 : 64
0111 = 1 : 128
1000 = 1 : 256
1001 = 1 : 512
1010 = 1 : 1024
1011 = 1 : 2048
1100 = 1 : 4096
1101 = 1 : 8192
1110 = 1 : 16384
1111 = 1 : 32768

WDTEN: Watchdog timer enable bit
0 = WDT disabled
1 = WDT enabled

Figure 15.3b ■ CONFIG2H register (PIC18F8680/8585/6680/6585)

15.3.4 The CONFIG3L Register

This register allows the user to insert wait states to the external memory access cycles and also to select the microcontroller operation mode. The contents of this register are shown in Figure 15.4.

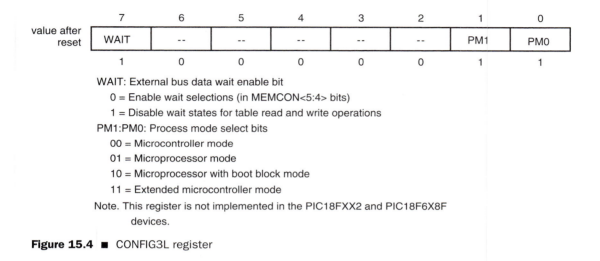

WAIT: External bus data wait enable bit

    0 = Enable wait selections (in MEMCON<5:4> bits)

    1 = Disable wait states for table read and write operations

PM1:PM0: Process mode select bits

    00 = Microcontroller mode

    01 = Microprocessor mode

    10 = Microprocessor with boot block mode

    11 = Extended microcontroller mode

Note. This register is not implemented in the PIC18FXX2 and PIC18F6X8F

       devices.

Figure 15.4 ■ CONFIG3L register

15.3.5 The CONFIG3H Register

This register allows the user to enable the $\overline{\text{MCLR}}$ pin and to configure the CCP2 pin multiplexing. The contents of this register are shown in Figures 15.5a, 15.5b, and 15.5c.

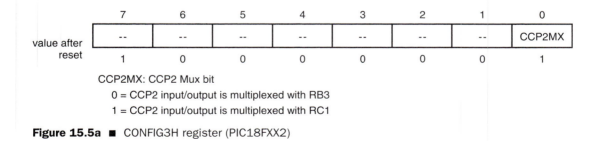

CCP2MX: CCP2 Mux bit

    0 = CCP2 input/output is multiplexed with RB3

    1 = CCP2 input/output is multiplexed with RC1

Figure 15.5a ■ CONFIG3H register (PIC18FXX2)

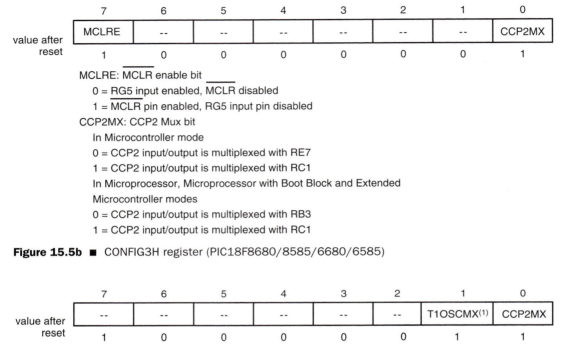

MCLRE: MCLR enable bit
   0 = RG5 input enabled, MCLR disabled
   1 = MCLR pin enabled, RG5 input pin disabled
CCP2MX: CCP2 Mux bit
   In Microcontroller mode
   0 = CCP2 input/output is multiplexed with RE7
   1 = CCP2 input/output is multiplexed with RC1
   In Microprocessor, Microprocessor with Boot Block and Extended
   Microcontroller modes
   0 = CCP2 input/output is multiplexed with RB3
   1 = CCP2 input/output is multiplexed with RC1

Figure 15.5b ■ CONFIG3H register (PIC18F8680/8585/6680/6585)

T1OSCMX: Timer1 Oscillator Mode bit
   0 = Low power Timer1 oscillator operation
   1 = Standard Timer1 oscillator operation
CCP2MX: CCP2 Mux bit
   In Microcontroller mode
   0 = CCP2 input/output is multiplexed with RE7
   1 = CCP2 input/output is multiplexed with RC1
   In Microprocessor, Microprocessor with Boot Block and Extended
   Microcontroller modes (PIC18F8X20 onoy)
   0 = CCP2 input/output is multiplexed with RB3
   1 = CCP2 input/output is multiplexed with RC1

Figure 15.5c ■ CONFIG3H register (PIC18FXX20)

15.3.6 The CONFIG4L Register

This register allows the user to enable/disable background debugger, low voltage in-circuit serial programming (ICSP), and return address stack overflow reset. The contents of this register are shown in Figure 15.6.

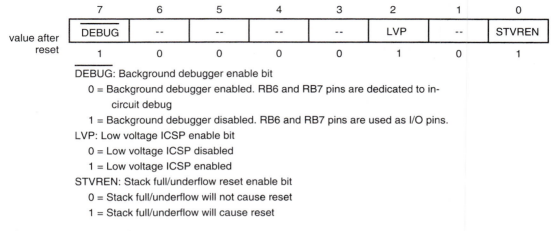

7	6	5	4	3	2	1	0
DEBUG	--	--	--	--	LVP	--	STVREN
1	0	0	0	0	1	0	1

value after reset

$\overline{\text{DEBUG}}$: Background debugger enable bit

  0 = Background debugger enabled. RB6 and RB7 pins are dedicated to in-circuit debug

  1 = Background debugger disabled. RB6 and RB7 pins are used as I/O pins.

LVP: Low voltage ICSP enable bit

  0 = Low voltage ICSP disabled

  1 = Low voltage ICSP enabled

STVREN: Stack full/underflow reset enable bit

  0 = Stack full/underflow will not cause reset

  1 = Stack full/underflow will cause reset

Figure 15.6 ■ CONFIG4L register

15.3.7 The CONFIG5L Register

The register allows the user to specify certain blocks of code to be protected against table-read and table-write from other blocks. The contents of this register are shown in Figures 15.7a, 15.7b, and 15.7c.

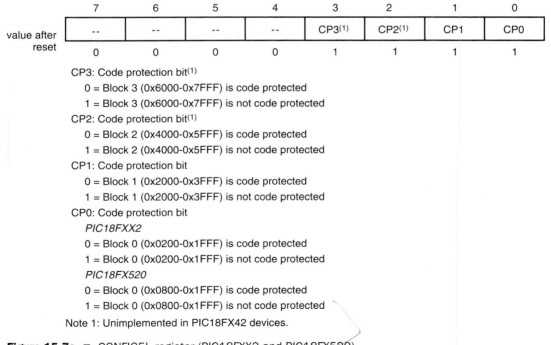

7	6	5	4	3	2	1	0
--	--	--	--	CP3[1]	CP2[1]	CP1	CP0
0	0	0	0	1	1	1	1

value after reset

CP3: Code protection bit[1]

  0 = Block 3 (0x6000-0x7FFF) is code protected

  1 = Block 3 (0x6000-0x7FFF) is not code protected

CP2: Code protection bit[1]

  0 = Block 2 (0x4000-0x5FFF) is code protected

  1 = Block 2 (0x4000-0x5FFF) is not code protected

CP1: Code protection bit

  0 = Block 1 (0x2000-0x3FFF) is code protected

  1 = Block 1 (0x2000-0x3FFF) is not code protected

CP0: Code protection bit

  PIC18FXX2

  0 = Block 0 (0x0200-0x1FFF) is code protected

  1 = Block 0 (0x0200-0x1FFF) is not code protected

  PIC18FX520

  0 = Block 0 (0x0800-0x1FFF) is code protected

  1 = Block 0 (0x0800-0x1FFF) is not code protected

Note 1: Unimplemented in PIC18FX42 devices.

Figure 15.7a ■ CONFIG5L register (PIC18FXX2 and PIC18FX520)

	7	6	5	4	3	2	1	0
value after reset	--	--	--	--	CP3[1]	CP2	CP1	CP0
	0	0	0	0	1	1	1	1

CP3: Code protection bit[1]
   0 = Block 3 (0xC000-0xFFFF) is code protected
   1 = Block 3 (0xC000-0xFFFF) is not code protected
CP2: Code protection bit
   0 = Block 2 (0x8000-0xBFFF) is code protected
   1 = Block 2 (0x8000-0xBFFF) is not code protected
CP1: Code protection bit
   0 = Block 1 (0x4000-0x7FFF) is code protected
   1 = Block 1 (0x4000-0x7FFF) is not code protected
CP0: Code protection bit
   0 = Block 0 (0x0800-0x3FFF) is code protected
   1 = Block 0 (0x0800-0x3FFF) is not code protected

Note 1: Unimplemented in PIC18FX585 devices.

Figure 15.7b ■ CONFIG5L register (PIC18F8680/8585/6680/6585)

	7	6	5	4	3	2	1	0
value after reset	CP7[1]	CP6[1]	CP5[1]	CP4[1]	CP3	CP2	CP1	CP0
	1	1	1	1	1	1	1	1

CP7: Code protection bit[1]
   0 = Block 7 (0x1C000-0x1FFFF) is code protected
   1 = Block 7 (0x1C000-0x1FFFF) is not code protected
CP6: Code protection bit[1]
   0 = Block 7 (0x18000-0x1BFFF) is code protected
   1 = Block 7 (0x18000-0x1BFFF) is not code protected
CP5: Code protection bit[1]
   0 = Block 7 (0x14000-0x17FFF) is code protected
   1 = Block 7 (0x14000-0x17FFF) is not code protected
CP4: Code protection bit[1]
   0 = Block 7 (0x10000-0x13FFF) is code protected
   ! = Block 7 (0x10000-0x13FFF) is not code protected
CP3: Code protection bit[1]
   0 = Block 3 (0xC000-0xFFFF) is code protected
   1 = Block 3 (0xC000-0xFFFF) is not code protected
CP2: Code protection bit
   0 = Block 2 (0x8000-0xBFFF) is code protected
   1 = Block 2 (0x8000-0xBFFF) is not code protected
CP1: Code protection bit
   0 = Block 1 (0x4000-0x7FFF) is code protected
   1 = Block 1 (0x4000-0x7FFF) is not code protected
CP0: Code protection bit
   0 = Block 0 (0x0800-0x3FFF) is code protected
   1 = Block 0 (0x0800-0x3FFF) is not code protected

Note 1: Unimplemented in PIC18FX620 devices.

Figure 15.7c ■ CONFIG5L register (PIC18FX620/PIC18FX720)

15.3.8 The CONFIG5H Register

This register allows the user to optionally protect the data EEPROM and boot block. The contents of this register are shown in Figure 15.8.

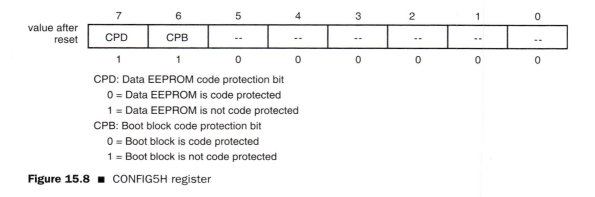

CPD: Data EEPROM code protection bit
 0 = Data EEPROM is code protected
 1 = Data EEPROM is not code protected
CPB: Boot block code protection bit
 0 = Boot block is code protected
 1 = Boot block is not code protected

Figure 15.8 ■ CONFIG5H register

15.3.9 The CONFIG6L Register

This register allows the user to provide write protection to the program memory. The contents of this register are shown in Figures 15.9a, 15.9b, and 15.9c.

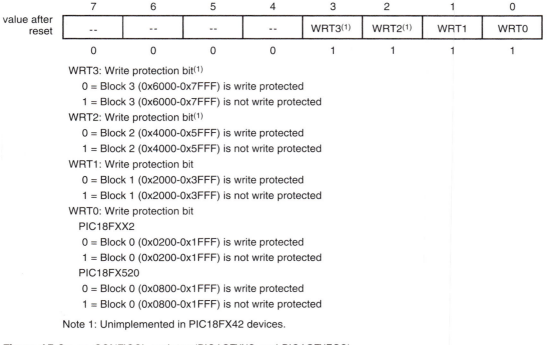

WRT3: Write protection bit[1]
 0 = Block 3 (0x6000-0x7FFF) is write protected
 1 = Block 3 (0x6000-0x7FFF) is not write protected
WRT2: Write protection bit[1]
 0 = Block 2 (0x4000-0x5FFF) is write protected
 1 = Block 2 (0x4000-0x5FFF) is not write protected
WRT1: Write protection bit
 0 = Block 1 (0x2000-0x3FFF) is write protected
 1 = Block 1 (0x2000-0x3FFF) is not write protected
WRT0: Write protection bit
PIC18FXX2
 0 = Block 0 (0x0200-0x1FFF) is write protected
 1 = Block 0 (0x0200-0x1FFF) is not write protected
PIC18FX520
 0 = Block 0 (0x0800-0x1FFF) is write protected
 1 = Block 0 (0x0800-0x1FFF) is not write protected

Note 1: Unimplemented in PIC18FX42 devices.

Figure 15.9a ■ CONFIG6L register (PIC18FXX2 and PIC18FX520)

value after reset	7	6	5	4	3	2	1	0
	--	--	--	--	WRT3[1]	WRT2	WRT1	WRT0
	0	0	0	0	1	1	1	1

WRT3: Write protection bit[1]
    0 = Block 3 (0xC000-0xFFFF) is write protected
    1 = Block 3 (0xC000-0xFFFF) is not write protected
WRT2: Write protection bit
    0 = Block 2 (0x8000-0xBFFF) is write protected
    1 = Block 2 (0x8000-0xBFFF) is not write protected
WRT1: Write protection bit
    0 = Block 1 (0x4000-0x7FFF) is write protected
    1 = Block 1 (0x4000-0x7FFF) is not write protected
WRT0: Write protection bit
    0 = Block 0 (0x0800-0x3FFF) is write protected
    1 = Block 0 (0x0800-0x3FFF) is not write protected

Note 1: Unimplemented in PIC18FX585 devices.

Figure 15.9b ■ CONFIG6L register (PIC18F8680/8585/6680/6585)

value after reset	7	6	5	4	3	2	1	0
	WRT7[1]	WRT6[1]	WRT5[1]	WRT4[1]	WRT3	WRT2	WRT1	WRT0
	1	1	1	1	1	1	1	1

WRT7: Write protection bit[1]
    0 = Block 7 (0x1C000-0x1FFFF) is write protected
    1 = Block 7 (0x1C000-0x1FFFF) is not write protected
WRT6: Write protection bit(1)
    0 = Block 6 (0x18000-0x1BFFF) is write protected
    1 = Block 6 (0x18000-0x1BFFF) is not write protected
WRT5: Write protection bit[1]
    0 = Block 5 (0x14000-0x17FFF) is write protected
    1 = Block 5 (0x14000-0x17FFF) is not write protected
WRT4: Write protection bit[1]
    0 = Block 4 (0x10000-0x13FFF) is write protected
    1 = Block 4 (0x10000-0x13FFF) is not write protected
WRT3: Write protection bit[1]
    0 = Block 3 (0xC000-0xFFFF) is write protected
    1 = Block 3 (0xC000-0xFFFF) is not write protected
WRT2: Write protection bit
    0 = Block 2 (0x8000-0xBFFF) is write protected
    1 = Block 2 (0x8000-0xBFFF) is not write protected
WRT1: Write protection bit
    0 = Block 1 (0x4000-0x7FFF) is write protected
    1 = Block 1 (0x4000-0x7FFF) is not write protected
WRT0: Write protection bit
    0 = Block 0 (0x0200-0x3FFF) is write protected
    1 = Block 0 (0x0200-0x3FFF) is not write protected

Note 1: Unimplemented in PIC18FX620 devices.

Figure 15.9c ■ CONFIG6L register (PIC18FX620/PIC18FX720)

15.3.10 The CONFIG6H Register

This register allows the user to provide write protection to the data EEPROM, the boot block, and the configuration register write protection bits. The contents of this register are shown in Figure 15.10.

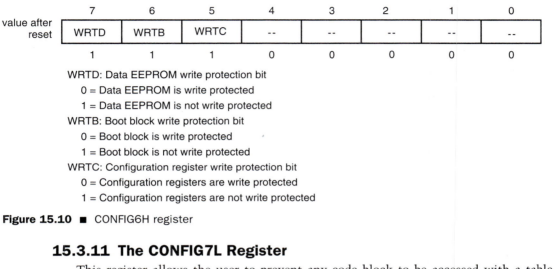

	7	6	5	4	3	2	1	0
value after reset	WRTD	WRTB	WRTC	--	--	--	--	--
	1	1	1	0	0	0	0	0

WRTD: Data EEPROM write protection bit
    0 = Data EEPROM is write protected
    1 = Data EEPROM is not write protected
WRTB: Boot block write protection bit
    0 = Boot block is write protected
    1 = Boot block is not write protected
WRTC: Configuration register write protection bit
    0 = Configuration registers are write protected
    1 = Configuration registers are not write protected

Figure 15.10 ■ CONFIG6H register

15.3.11 The CONFIG7L Register

This register allows the user to prevent any code block to be accessed with a table-read instruction from another block. The contents of this register are shown in Figures 15.11a, 15.11b, and 15.11c.

	7	6	5	4	3	2	1	0
value after reset	--	--	--	--	EBTR3[1]	EBTR2[1]	EBTR1	EBTR0
	0	0	0	0	1	1	1	1

EBTR3: Table read protection bit[1]
    0 = Block 3 (0x6000-0x7FFF) is protected from table-read from other blocks
    1 = Block 3 (0x6000-0x7FFF) is not protected from table-read from other blocks
EBTR2: Table read protection bit[1]
    0 = Block 2 (0x4000-0x5FFF) is protected from table-read from other blocks
    1 = Block 2 (0x4000-0x5FFF) is not protected from table-read from other blocks
EBTR1: Table read protection bit
    0 = Block 1 (0x2000-0x3FFF) is protected from table-read from other blocks
    1 = Block 1 (0x2000-0x3FFF) is not protected from table-read from other blocks
EBTR0: Table read protection bit
    PIC18FXX2
    0 = Block 0 (0x0200-0x1FFF) is protected from table-read from other blocks
    1 = Block 0 (0x0200-0x1FFF) is not protected from table-read from other blocks
    PIC18FX520
    0 = Block 0 (0x0800-0x1FFF) is protected from table-read from other blocks
    1 = Block 0 (0x0800-0x1FFF) is not protected from table-read from other blocks

Note 1: Unimplemented in PIC18FX42 devices.

Figure 15.11a ■ CONFIG7L register (PIC18FXX2 and PIC18FX520)

value after reset	7	6	5	4	3	2	1	0
	--	--	--	--	EBTR3[1]	EBTR2	EBTR1	EBTR0
	0	0	0	0	1	1	1	1

EBTR3: Table read protection bit[1]

   0 = Block 3 (0xC000-0xFFFF) is protected from table-read from other blocks

   1 = Block 3 (0xC000-0xFFFF) is not protected from table-read from other blocks

EBTR2: Table read protection bit

   0 = Block 2 (0x8000-0xBFFF) is protected from table-read from other blocks

   1 = Block 2 (0x8000-0xBFFF) is not protected from table-read from other blocks

EBTR1: Table read protection bit

   0 = Block 1 (0x4000-0x7FFF) is protected from table-read from other blocks

   1 = Block 1 (0x4000-0x7FFF) is not protected from table-read from other blocks

EBTR0: Table read protection bit

   0 = Block 0 (0x0800-0x3FFF) is protected from table-read from other blocks

   1 = Block 0 (0x0800-0x3FFF) is not protected from table-read from other blocks

Note 1: Unimplemented in PIC18FX585 devices.

Figure 15.11b ■ CONFIG7L register (PIC18F8680/8585/6680/6585)

value after reset	7	6	5	4	3	2	1	0
	EBTR7[1]	EBTR6[1]	EBTR5[1]	EBTR4[1]	EBTR3	EBTR2	EBTR1	EBTR0
	1	1	1	1	1	1	1	1

EBTR7: Table read protection bit[1]

   0 = Block 7 (0x1C000-0x1FFFF) is protected from table-read from other blocks

   1 = Block 7 (0x1C000-0x1FFFF) is not protected from table-read from other blocks

EBTR6: Table read protection bit[1]

   0 = Block 6 (0x18000-0x1BFFF) is protected from table-read from other blocks

   1 = Block 6 (0x18000-0x1BFFF) is not protected from table-read from other blocks

EBTR5: Table read protection bit[1]

   0 = Block 5 (0x14000-0x17FFF) is protected from table-read from other blocks

   1 = Block 5 (0x14000-0x17FFF) is not protected from table-read from other blocks

EBTR4: Table read protection bit[1]

   0 = Block 4 (0x10000-0x13FFF) is protected from table-read from other blocks

   1 = Block 4 (0x10000-0x13FFF) is not protected from table-read from other blocks

EBTR3: Table read protection bit[1]

   0 = Block 3 (0xC000-0xFFFF) is protected from table-read from other blocks

   1 = Block 3 (0xC000-0xFFFF) is not protected from table-read from other blocks

EBTR2: Table read protection bit

   0 = Block 2 (0x8000-0xBFFF) is protected from table-read from other blocks

   1 = Block 2 (0x8000-0xBFFF) is not protected from table-read from other blocks

EBTR1: Table read protection bit

   0 = Block 1 (0x4000-0x7FFF) is protected from table-read from other blocks

   1 = Block 1 (0x4000-0x7FFF) is not protected from table-read from other blocks

EBTR0: Table read protection bit

   0 = Block 0 (0x0200-0x3FFF) is protected from table-read from other blocks

   1 = Block 0 (0x0200-0x3FFF) is not protected from table-read from other blocks

Note 1: Unimplemented in PIC18FX620 devices.

Figure 15.11c ■ CONFIG7L register (PIC18FX620/PIC18FX720)

15.3.12 The CONFIG7H Register

This register allows the user to provide protection against table-read access to the boot block from other block. The content of this register is shown in Figure 15.12.

	7	6	5	4	3	2	1	0
value after reset	--	EBTRB	--	--	--	--	--	--
	0	1	0	0	0	0	0	0

EBTRB: Boot block table read protection bit
  0 = Boot block is protected against table-read from other blocks
  1 = Boot block is not protected against table-read from other blocks

Figure 15.12 ■ CONFIG7H register

15.3.13 The Device ID Registers

The user can use these two registers to identify the microcontroller and its revision number. There are two device ID registers: DEVID1 and DEVID2. The contents of these two registers are shown in Figures 15.13a and 15.13b.

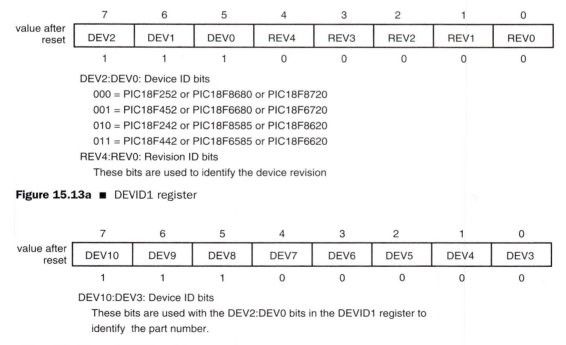

	7	6	5	4	3	2	1	0
value after reset	DEV2	DEV1	DEV0	REV4	REV3	REV2	REV1	REV0
	1	1	1	0	0	0	0	0

DEV2:DEV0: Device ID bits
  000 = PIC18F252 or PIC18F8680 or PIC18F8720
  001 = PIC18F452 or PIC18F6680 or PIC18F6720
  010 = PIC18F242 or PIC18F8585 or PIC18F8620
  011 = PIC18F442 or PIC18F6585 or PIC18F6620
REV4:REV0: Revision ID bits
  These bits are used to identify the device revision

Figure 15.13a ■ DEVID1 register

	7	6	5	4	3	2	1	0
value after reset	DEV10	DEV9	DEV8	DEV7	DEV6	DEV5	DEV4	DEV3
	1	1	1	0	0	0	0	0

DEV10:DEV3: Device ID bits
  These bits are used with the DEV2:DEV0 bits in the DEVID1 register to identify the part number.

Figure 15.13b ■ DEVID2 register

15.3.14 Programming the Configuration Bits

One can set the configuration bits by using the _ _**CONFIG** directive. The keyword CON-FIG is preceded by two underscore characters. The format for using the _ _**CONFIG** command is as follows:

_ _CONFIG <address>,<expression>

where **address** refers to the address of the configuration register of interest and **expression** is the value to be assigned to the configuration register. This command sets the processor configuration bits to the value represented by **expression.** Before this directive is used, the processor must be declared through the **processor** or **list** directive. The hex file output format must be set to INHX32 with the **list** directive when this directive is used with the PIC18 family.

To make the program more readable, it is recommended to use a symbolic name to refer to the configuration register. The include file supplied with each PIC18 microcontroller by the MPLAB® IDE has included the definitions for the address of each configuration register and the value of each configuration bit. One can simply use them in the program. For example,

_**CONFIG1L** refers to the address 0x300001 of the CONFIG1L register.

_**CONFIG1H** represents the address 0x300002 of the CONFIG1H register.

The MPLAB assembler defines the configuration bits in a way that the assembler will simply AND the expressions that represent configuration bits together and assign it to the specified configuration register.

For example, with the P18F8720.inc definitions

```
_EC_OSC_1H          equ        0xFC
_OSCS_OFF_1H        equ        0xFF
```

the assembler directive

_ _CONFIG _CONFIG1H, _EC_OSC_1H & _OSCS_OFF_1H

will assign the value 0xFC to the CONFIG1H register.

If one does not include a

_ _CONFIG _CONFIGxx, . . . ; select an option for each programmable bit

line for one of the configuration bytes in the source file, then the "1" option will be programmed for each programmable bit. Because of this, it is important to note which configuration bits may cause problems to the operation of the chip unless those bits are intended to be set to 1s.

In C language, the default linker script for each part contains a section named **CONFIG.** For example, the p18f8680.lkr script contains the following statements.

```
CODEPAGE   NAME=config    START=0x300000   END=0x30000D   PROTECTED
. . .
SECTION    NAME=CONFIG    ROM=config
```

The #pragma romdat CONFIG directive is used to set the current romdata section to the section named CONFIG. The configuration for the device can be specified using the _**CONFIG_DECL** macro and the **#defines** located in the processor-specific header file. The following example specifies the configuation of a PIC18F8660:

```
#include <18F8680.h>
#pragma romdata CONFIG
   _CONFIG_DECL (_OSC_EC_1H,     /*Select EC oscillator with OSC2 configured as divide-by-4 clock output */
   _BORV_45_2L,                  /*enable and set V_BOR set to 4.5 V */
   _WDT_OFF_2H,                  /* disable watchdog timer */
```

```
          _MODE_EM_3L,                    /* select microcontroller mode */
          _MCLRE_ON_3H,                   /* enable MCLR pin and multiplex CCP₂ pin with RC₁ */
          _LVP_OFF_4L,                    /* disable low volgate serial programming, disable background debug
                                             mode, enable stack overflow reset */
          _CONFIG5L_DEFAULT,              /* disable code protection */
          _CONFIG5H_DEFAULT,              /* disable data EEPROM protection, no boot block protection */
          _CONFIG6L_DEFAULT,              /* disable write protection to program memory */
          _CONFIG6H_DEFAULT,              /* no write protection to data EEPROM, boot block, &
                                             configuration registers */
          _CONFIG7L_DEFAULT,              /* No read protection to program memory */
          _CONFIG7H_DEFAULT               /* No table read protection to boot block */
          );
#pragma romdata
void main (void)
{
. . .
}
```

Every configuration register should be setup properly. Otherwise, default values will be used.

Example 15.1
▼

Write a few assembler directives to define the following configurations for the PIC18F8720:

- Disable oscillator switch for Timer1
- Select external oscillator with OSC2 pin configured as FOSC/4 output
- Enable power-up timer
- Disable brown-out reset
- Enable watchdog timer and set postscaler to 1:32
- Let CCP2 pin be multiplexed with the RE7 pin
- Select microcontroller mode
- Disable stack overflow reset, disable low-voltage reset, and disable background debugger
- Disable all protections to the program memory, the data EEPROM, and the boot block

Solution: The following assembler directives will achieve the desired requirements:

```
_ _CONFIG        _CONFIG1H, _OSCS_OFF_1H & _EC_OSC_1H
_ _CONFIG        _CONFIG2L, _BOR_OFF_2L & _PWRT_ON_2L
_ _CONFIG        _CONFIG2H, _WDT_ON_2H & _WDTPS_32_2H
_ _CONFIG        _CONFIG3L, _MC_MODE_3L
_ _CONFIG        _CONFIG3H, _CCP2MX_OFF_3H
_ _CONFIG        _CONFIG4L, _STVR_OFF_4L & _LVP_OFF_4L & _DEBUG_OFF_4L
```

The statements for CONFIG4H . . . CONFIG7H are not specified in this solution, which will disable all protections to the program memory, the data EEPROM, and the boot block.

▲

In C language, this can be done as follows:

```
#include <p18F8720.h>
#pragma romdata CONFIG
_CONFIG_DECL (_OSC_EC_1H,                    /* Select EC oscillator with OSC2 configured as
                                                divide-by-4 clock output */

        _PWRT_ON_2L &_BOR_OFF_2L,            /* enable power-on reset, no brown-out reset */
        _WDT_ON_2H &_WDTPS_32_2H,            /* enable watchdog timer with postscaler set to 32 */
        _MODE_EM_3L,                         /* select microcontroller mode */
        _CCP2MX_OFF_3H,                      /* disable MCLR pin and multiplex CCP2 pin with RE7 */
        _LVP_OFF_4L & STVR_OFF_4L & _DEBUT_OFF_4L,   /* disable low voltage serial programming, disable
                                                        background debug mode, disable stack overflow
                                                        reset */
        _CONFIG5L_DEFAULT,                   /* disable code protection */
        _CONFIG5H_DEFAULT,                   /* disable data EEPROM protection, no boot block protection */
        _CONFIG6L_DEFAULT,                   /* disable write protection to program memory */
        _CONFIG6H_DEFAULT,                   /* no write protection to data EEPROM, boot block, &
                                                configuration registers */
        _CONFIG7L_DEFAULT,                   /* No read protection to program memory */
        _CONFIG7H_DEFAULT                    /* No table read protection to boot block */
        );
#pragma romdata
```

15.4 Watchdog Timer

The watchdog timer (WDT) is a circuit that will reset the microcontroller when it times out (or overflows). The execution time of a well-written sequence of instructions, function, or program is predictable. Therefore, it is possible to add a few instructions to reset a timer periodically before it overflows (or referred to as "times out"). If a program has a certain bug, then it may prevent the program from resetting the timer before it times out.

The WDT is based on this concept. The main application of the WDT is to detect software bugs. By enabling the WDT and adding instructions to reset the WDT periodically, software bugs can be detected whenever the microcontroller is reset by the WDT. When using the WDT, the program should be written to repeat itself forever so that the instruction sequence that resets the WDT can be executed periodically to prevent the WDT from overflowing.

The clock signal to the PIC18 WDT comes from the output of a RC oscillator. This clock source enables the WDT to function even when the system clock has been stopped (e.g., stopped by the execution of a sleep instruction). Because the frequency of the RC oscillator varies with the fabrication process, the time-out period can range from 7 ms up to 33 ms (typical value is 18 ms). The WDT has a postscaler that can extend the WDT reset period. One can set the postscaler by programming the CONFIG2H register.

During a normal operation, a WDT time-out generates a reset to the microcontroller. If the microcontroller is in sleep mode, a WDT time-out causes the microcontroller to wake up and continue with normal operation. The $\overline{\text{TO}}$ bit in the RCON register will be cleared on a WDT time-out to indicate this condition.

The PIC18 microcontroller has a watchdog control register (WDTCON) that allows the user to override the WDT enable configuration bit if the configuration bit disabled the WDT. The WDTCON register has only the bit 0 (SWDTEN bit) implemented.

Example 15.2

▼

Assume that there is an embedded product designed with a PIC18F8680 and that it can complete all the processing in no more than 800 ms. Suppose that the designer wants to detect the software bug during the development stage and enables the WDT. What is the postscale factor to be set for this application?

Solution: In the worst case, the WDT will time out in 7 ms. When enabling the WDT, one must allow enough time for the application to complete the necessary processing. The smallest postscaler that will cause the WDT to time out in a period larger than 800 ms is 128. This postscaler will cause the WDT to time out in 896 ms. In order to more accurately detect the software bug, the user should add a 96-ms unnecessary delay to the application.

▲

15.5 Power-Down Mode (Sleep)

Many embedded products are powered by batteries. It is desirable to reduce the power consumption when it is not being used. The PIC18 devices have a power-down mode to dramatically reduce power consumption when the microcontroller is idle. The power-down mode is entered by executing the **sleep** instruction. The system clock signal will be stopped when the sleep mode is entered.

The PIC18 microcontroller can be waked up from sleep mode by the events listed in Table 15.1. External $\overline{\text{MCLR}}$ reset will cause a device reset. All other events are considered a continuation of program execution and will cause a wake-up. For the device to be waked up through an interrupt event, the corresponding interrupt enable bit must be set.

---

1. External reset input from $\overline{\text{MCLR}}$ pin
2. Watchdog timer wakeup
3. Interrupt from INT pin, RB port change
4. Parallel slave port read or write
5. TMR1 interrupt in asynchronous counter mode
6. TMR3 interrupt in asynchronous counter mode
7. CCP capture mode interrupt
8. Special event trigger (Timer1 in asynchronous mode using an external clock)
9. MSSP (start/stop) bit detect interrupt
10. MSSP transmit or receive in slave mode (SPI/I^{2}C)
11. USART RX or TX
12. A/D conversion complete (when the A/D clock source is RC clock)
13. EEPROM write operation complete
14. LVD interrupt

Table 15.1 ■ Events that can wake up PIC18 MCU from sleep mode

The application that includes entering the sleep mode should be written as an infinite loop. When the program starts, it accepts the user command and data and carries out all the necessary processing. After all the processing is done, the program initializes a time-out variable to certain value, enables the timer interrupt to decrement the variable, and continuously waits for the user command for more processing. If the user did not enter any new command and let the time-out variable to decrement to 0, the program will execute the **sleep** instruction to enter the low-power mode. The timer interrupt simply decrements the time-out variable by 1 and returns. If the user enters a new command before the time-out variable decrements to 0, the application performs the specified operation and resets the time-out variable to the starting value. This process repeats forever.

15.6 Program Code Protection

Some users may want to prevent their competitors from copying their software and even improve from it. If no protection is provided, then their competitors may be able to copy the software by using a universal programmer. Most of today's microcontrollers provide a protection to this type of software piracy. One must keep in mind that any such prevention aims to prevent reading or writing the user application from outside the chip. The protection mechanism does not aim to prevent the user program from reading from or writing into the program memory or data EEPROM.

To prevent piracy, the PIC18 microcontroller divides the user program memory into blocks on binary boundary. Each block has three separate code protection bits associated with it:

- Code protect bit (CP_n)
- Write protect bit (WRT_n)
- External block table read bit ($EBTR_n$)

The code protection bits are located in configuration registers 5L through 7H. Their locations within the registers are shown in Table 15.2. The sizes of the blocks to be protected are either 8 KB or 16 KB. The boot block is separated from block 0 and is protected separately.

Register Name	Bit 7	Bit 6	Bit 5	Bit 4	Bit 3	Bit 2	Bit 1	Bit 0
CONFIG5L	CP7*	CP6*	CP5*	CP4*	CP3	CP2	CP1	CP0
CONFIG5H	CPD	CPB	—	—	—	—	—	—
CONFIG6L	WRT7*	WRT6*	WRT5*	WRT4*	WRT3	WRT2	WRT1	WRT0
CONFIG6H	WRTD	WRTB	WRTC	—	—	—	—	—
CONFIG7L	EBTR7*	EBTR6*	EBTR5*	EBTR4*	EBTR3	EBTR2	EBTR1	EBTR0
CONFIG7H	—	EBTRB	—	—	—	—	—	—

Note* Unimplemented in PIC18FXX2, PIC18FX520, PIC18FX620, PIC18F6585/8585/8680/6680)

Table 15.2 ■ The PIC18 Code protection registers

15.6.1 Program Memory Code Protection

The user memory may be read from or written into using the table-read and table-write instructions. The device ID may be read with the table-read instructions. The configuration registers may be read and written with the table-read and table-write instructions.

In the user mode, the CP_n bits have no direct effect. The CP_n bits inhibit external reads and writes. A block of user memory may be protected from table writes if the WRT_n configuration bit is 0. The $EBTR_n$ bits control the table-read operation. For a block of user memory with the $EBTR_n$ bit set to 0, it can be read only from within the block but not from outside the block.

15.6.2 Data EEPROM Code Protection

The entire data EEPROM is protected from external reads and writes by two bits: CPD and WRTD. The CPD bit inhibits external reads and writes of data EEPROM. The WRTD bit prevents external writes to data EEPROM. The microcontroller can continue to read and write data EEPROM regardless of the settings of protection bits.

15.6.3 Configuration Register Protection

The configuration registers can be protected from writing. The WRTC bit controls the protection of the configuration registers. In user mode, the WRTC bit is readable only. This bit can be written only via an ICSP interface or an external programmer.

15.6.4 ID Locations

Users can store the checksum or other code identification numbers at eight designated memory locations from 0x200000 through 0x200007. These locations can be accessed during normal execution through the TBLRD and TBLWT instructions. The ID locations can be read when the device is code protected.

15.7 ICSP

All PIC18 microcontrollers with on-chip flash memory can be serially programmed while in the end-product circuit. This is done with two pins for clock and data and an additional three pins for power, ground, and the programming voltage. This allows users to manufacture printed circuit boards with unprogrammed devices and then program the microcontroller just before shipping the product. This also allows the most recent firmware (application software) to be programmed. To program the PIC18 microcontroller, one must follow the specific procedure published by Microchip. Documents on ICSP can be downloaded from Microchip Web site.

15.8 In-Circuit Debugger

In order to perform in-circuit debug, one must clear the DEBUG bit of the CONFIG4L register to 0. The MPLAB ICD 2 in-circuit debugger utilizes this feature to perform software debugging activities. A small amount of on-chip resources must be reserved for debugging activities and are not available for general use. The resources that are consumed by the debugger are listed in Table 15.3.

I/O pins	RB6, RB7
Stack	2 levels
Program memory	Last 576 bytes
Data memory	Last 10 bytes

Table 15.3 ■ PIC18 debugger resoruces

To use the in-circuit debugger function of the microcontroller, the design must implement ICSP connections to MCLR/V$_{PP}$, V$_{DD}$, GND, RB6, and RB7.

15.9 Low-Voltage ICSP Programming

When this feature is enabled, one can use the V$_{DD}$ voltage to program the PIC18 microcontroller. This allows the voltage level at the V$_{PP}$ pin to be left at the V$_{DD}$ level. In this mode, the RB5/PGM pin is dedicated to the programming function and cannot be used as a general I/O pin. During programming, V$_{DD}$ is applied to the $\overline{MCLR}$/V$_{PP}$ pin. To enter the programming mode, V$_{DD}$ must be applied to the RB5/PGM pin, provided that the LVP bit of the CONFIG4L register is set.

15.10 Summary

A group of configuration registers control the following important settings to the PIC18 normal operation:

- Type of crystal oscillator used
- WDT enable/disable and postscaler selection
- Microcontroller operation mode selection and external memory wait state enabling
- Background debugger enable/disable
- Low-voltage programming enable/disable
- Return address stack overflow/underflow reset enable/disable
- Program memory read/write protection
- Data memory protection
- Boot block protection

Setting the configuration bits can be done using the assembler directives provided by the MPASM.

The PIC18 microcontroller allows the user to reduce power consumption dramatically when the microcontroller is idle. The low-power mode is entered by executing the **sleep** instruction and exited by a reset or an external peripheral interrupt.

The PIC18 microcontroller also provides the ICSP capability to allow the user to program the PIC18 microcontroller on the end product if the interface is provided. This allows the user to upgrade the software whenever the new version of the software becomes available. The ICSP can be performed using the normal power supply level.

15.11 Exercises

E15.1 Write a sequence of assembler directives to configure the PIC18F452 with the following settings:

- Use RC oscillator with OSC2 pin configured as the RA6 pin
- Disable oscillator system clock switch option
- Enable power-up timer and set the brown-out voltage to 4.5 V
- Enable WDT and set the postscaler to 128

- Multiplex the CCP2 signal with the RC1 signal
- Disable background debugger, disable low-power programming, and disable stack overflow/underflow reset
- Enable code protection to block 0 and block 1
- Disable data EEPROM read and write protection
- Disable configuration register and boot block protection

E15.2 Write a sequence of assembler directives to configure the PIC18F8680 with the following settings:

- Use EC oscillator with OSC2 pin configured as $F_{OSC}/4$ output pin
- Disable oscillator system clock switch option
- Disable power-up timer and set the brown-out voltage to 4.5 V
- Select extended microcontroller mode and enable external memory wait state
- Disable WDT
- Multiplex the CCP2 signal with the RC1 signal
- Disable background debugger, disable low-power programming, and enable stack overflow/underflow reset
- Enable code protection to block 0 and block 1
- Enable data EEPROM read and write protection
- Disable configuration register and boot block protection

E15.3 Assume that there is a product that uses the PIC18F8680 as its controller. The application program of this product will take at most 440 ms to complete all the processing. Specify a postscaler to be used with the WDT to detect the software bug.

Glossary

Accumulator A register in a computer that contains an operand to be used in an arithmetic operation.

Acknowledgment error In CAN bus protocol, an acknowledgment error is detected whenever the transmitter does not monitor a dominant bit in the ACK slot.

Activation record Another term for stack frame.

Address access time The amount of time it takes for a memory component to send out valid data to the external data pins after address signals have been applied (assuming that all other control signals have been asserted).

Addressing The application of a unique combination of high and low logic levels to select a corresponding unique memory location.

Address multiplexing A technique that allows the same address pin to carry different signals at different times; used mainly by DRAM technology. Address multiplexing can dramatically reduce the number of address pins required by DRAM chips and reduce the size of the memory chip package.

Algorithm A set of procedure steps represented in pseudocode that is designed to solve certain computation issues.

ALU Arithmetic logic unit. The part of the processor in which all arithmetic and logical operations are performed.

Array An ordered set of elements of the same type. The elements of the array are arranged so that there is a zeroth, first, second, third, and so forth. An array may be one-, two-, or multidimensional.

ASCII American Standard Code for Information Interchange. A code that uses seven bits to encode all printable and control characters.

Assembler A program that converts a program in assembly language into machine instructions so that it can be executed by a computer.

Assembler directive A command to the assembler for defining data and symbols, setting assembler conditions, and specifying output format. Assembler directives do not produce machine code.

Assembly instruction A mnemonic representation of a machine instruction.

Assembly program A program written in assembly language.

Automatic variable A variable defined inside a function that comes into existence when the function is entered and disappears when the function returns.

Barometric pressure The air pressure existing at any point within the earth's atmosphere.

Binary coded decimal (BCD) A coding method that uses four binary digits to represent one decimal digit. The binary codes 0000_2 to 1001_2 correspond to the decimal digits 0 to 9.

Bit error In the CAN bus protocol, a node that is sending a bit on the bus also monitors the bus. When the bit value monitored is different from the bit value being sent, the node interprets the situation as a bit error.

Branch instruction An instruction that causes the program flow to change.

Break The transmission or reception of a low for at least one complete character time.

Breakpoint A memory location in a program where the user program execution will be stopped and the monitor program will take over the CPU control and display the contents of CPU registers.

Bubble sort   A simple sorting method in which an array or a file to be sorted is gone through sequentially several times. Each iteration consists of comparing each element in the array or file with its successor ($x[i]$ with $x[i + 1]$) and interchanging the two elements if they are not in proper order (either ascending or descending).

Bus   A set of signal lines through which the processor of a computer communicates with memory and I/O devices.

Bus cycle timing diagram   A diagram that describes the transitions of all the involved signals during a read or write operation.

Bus multiplexing   A technique that allows more than one set of signals to share the same group of bus lines.

Bus off   The situation in which a CAN node has a transmit error count above 256.

CAN transceiver   A chip used to interface a CAN controller to the CAN bus.

Central processing unit (CPU)   The combination of the register file, the ALU, and the control unit.

Charge pump   A circuit technique that can raise a low voltage to a level above the power supply. A charge pump is often used in A/D converter, in EEPROM and EPROM programming, and so on.

Comment   A statement that explains the function of a single instruction or directive or a group of instructions or directives. Comments make a program more readable.

Communication program   A program that allows a PC to communicate with another computer.

Computer   A computer consists of hardware and software. The hardware includes four major parts: the central processing unit, the memory unit, the input unit, and the output unit. Software is a sequence of instructions that control the operations of the hardware.

Contact bounce   A phenomenon in which a mechanical key switch will go up and down several times before it settles down when it is pressed.

Control unit   The part of the processor that decodes and monitors the execution of instructions. It arbitrates the use of computer resources and makes sure that all computer operations are performed in proper order.

Controller area network (CAN)   A serial communication protocol initially proposed to be used in automotive applications. In this protocol, data is transferred frame by frame. Each frame can carry up to eight bytes of data.

CRC error   Cyclic redundancy check error. In data communications, the CRC sequence consists of the result of the CRC calculation by the transmitter. The receiver calculates the CRC in the same way as the transmitter. A CRC error is detected if the calculated result is not the same as that received in the CRC sequence.

Cross assembler   An assembler that runs on one computer but generates machine instructions that will be executed by another computer that has a different instruction set.

Cross compiler   A compiler that runs on one computer but generates machine instructions that will be executed by another computer that has a different instruction set.

D/A converter   A circuit that can convert a digital value into an analog voltage.

Data hold time   The length of time over which the data must remain stable after the edge of the control signal that latches the data.

Datapath   The part of processor that consists of a register file and the ALU.

Data setup time   The amount of time over which the data must become valid before the edge of the control signal that latches the data.

DCE   Data communication equipment. DCE usually refers to equipment such as a modem, concentrator, router, and so on.

Demo board   A single board computer that contains the target microcontroller as the CPU and a monitor to help the user to perform embedded product development.

Direct mode   An addressing mode that uses an 8-bit value to represent the address of a memory location.

Dominant level   A voltage level in a CAN bus that will prevail when a voltage level at this state and a different level (recessive level) are applied to the CAN bus at the same time.

DTE   Data terminal equipment. DTE usually refers to a computer or terminal.

Dynamic memories   Memory devices that require periodic refreshing of the stored information, even when power is on.

EIA   Electronic Industry Association.

Electrically erasable programmable read-only memory (EEPROM)   A type of read-only memory that can be erased and reprogrammed using electrical signals. EEPROM allows each individual location inside the chip to be erased and reprogrammed.

Embedded system   A product that uses a microcontroller as the controller to provide the features. End users are interested in these features rather than the power of the microcontroller. A cell phone, a charge card, and a home security system are examples of the embedded system.

Erasable programmable read only memory (EPROM)   A type of read-only memory that can be erased by subjecting it to strong ultraviolet light. It can be reprogrammed using an EPROM programmer. A quartz window on top of the EPROM chip allows light to be shone directly on the silicon chip inside.

Error-active   A CAN node that has both transmit error count and receive error count lower than 127.

Error-passive   A CAN node that has either transmit error count or receive error count between 128 and 256.

Exception   A software interrupt, such as an illegal opcode, an overflow, division by zero, or an underflow.

Extended microcontroller mode   A PIC18 operation mode in which the PIC18 microcontroller can access both the internal and the external program memories as a single block.

Fall time   The amount of time a digital signal takes to go from logic high to logic low.

Fast register stack   A hardware circuit with three 8-bit registers intended for saving and restoring the BSR, STATUS, and WREG registers during the subroutine call and interrupt.

Floating signal   An undriven signal.

Form error   An error detected when a fixed-form bit field contains one or more illegal bits (in CAN bus protocol).

Frame pointer   A pointer used to facilitate access to parameters in a stack frame.

Framing error   A data communication error in which a received character is not properly framed by the start and stop bits.

Full decoding   An address-decoding scheme in which each memory location responds to only one address.

Full-duplex link   A four-wire communication link that allows both transmission and reception to proceed simultaneously.

General call address   The general call address is for addressing every device connected to the I²C bus.

Global memory   Memory that is available to all programs in a computer system.

Half-duplex link   A communication link that can be used for either transmission or reception but only in one direction at a time.

Hard synchronization   The synchronization performed by all CAN nodes at the beginning of a frame.

Hardware breakpoint   A hardware circuit that compares actual address and data values to predetermined data in setup registers. A successful comparison places the CPU in background debug mode or initiates a software interrupt.

Identifier acceptance filter   A group of registers that can be programmed to select those identifier bits of the incoming frames to be compared for acceptance.

Idle   A continuous logic high on the RxD line for one complete character time.

Illegal opcode   A binary bit pattern of the opcode byte for which an operation is not defined.

Immediate mode   An addressing mode that will be used as the operand of the instruction.

Indexable data structure   A data structure in which each element is associated with an integer that can be used to access it. Arrays and matrices are examples of indexable data structures.

Inline assembly instruction   Assembly instructions that are embedded in a high-level language program.

Input capture   The PIC18 function that captures the value of the 16-bit timer into a latch when the falling or rising edge of the signal connected to the input capture pin arrives.

Input handshake   A protocol that uses two handshake signals to make sure that the peripheral chip receives data correctly from the input device.

Input port   The part of the microcontroller that consists of input pins, input data register, and other control circuitry to perform input function.

Integrated development environment   A piece of software that combines a text editor, a terminal program, a cross compiler and/or cross assembler, and/or simulator that allows the user to perform program development activities without quitting any one of the programs.

Interframe space   A field in the CAN bus that is used to separate data frames or remote frames from the previous frames.

Interrupt   An unusual event that requires the CPU to stop normal program execution and perform some service to the event.

Interrupt overhead   The time spent on handling an interrupt. This time consists of the saving and restoring of registers and the execution of instructions contained in the service routine.

Interrupt priority   The order in which the CPU will service interrupts when all of them occur at the same time.

Interrupt service   The service provided to a pending interrupt by CPU execution of a program called a service routine.

Interrupt vector   The starting address of an interrupt service routine.

Interrupt vector table   A table that stores all interrupt vectors.

I/O synchronization   A mechanism that can make sure that CPU and I/O devices exchange data correctly.

ISO International Organization for Standardization.

Keyboard debouncing A process that can eliminate the key-switch bouncing problem so that the computer can detect correctly whether a key has indeed been pressed.

Keyboard scanning A process that is performed to detect whether any key has been pressed by the user.

Key wake-up A mechanism that can generate interrupt requests to wake up a sleeping CPU. The key wake-up ability is associated with I/O ports.

Label field The field in an assembly program statement that represents a memory location.

Linked list A data structure that consists of linked nodes. Each node consists of two fields: an information field and a next address field. The information field holds the actual element on the list, and the next address field contains the address of the next node in the list.

Load cell A transducer that can convert weight into a voltage.

Local variable Temporary variables that exist only when a subroutine is called. They are used as loop indices, working buffers, and so on. Local variables are often allocated in the system stack.

Low-power mode An operation mode in which less power is consumed. In CMOS technology, the low-power mode is implemented by either slowing down the clock frequency or turning off some circuit modules within a chip.

Machine instruction A set of binary digits that tells the computer what operation to perform.

Macro A method of grouping a sequence of instructions and assigning it a name in an assembly program and of commanding the assembler to regenerate the same sequence of instructions in the program wherever the macro name is encountered.

Mark A term used to indicate a binary 1.

Maskable interrupts Interrupts that can be ignored by the CPU. This type of interrupt can be disabled by setting a mask bit or by clearing an enable bit.

Masked ROM (MROM) A type of ROM that is programmed when it is fabricated.

Matrix A two-dimensional data structure that is organized into rows and columns. The elements of a matrix are of the same length and are accessed using their row and column numbers (i, j), where i is the row number and j is the column number.

Memory Storage for software and information.

Memory capacity The total amount of information that a memory device can store; also called memory density.

Memory organization A description of the number of bits that can be read from or written into a memory chip during a read or write operation.

Microcontroller A computer system implemented on a single, very large-scale integrated circuit. A microcontroller contains everything that is in a microprocessor and may contain memories, an I/O device interface, a timer circuit, an A/D converter, and so on.

Microcontroller mode The operation mode in which the PIC18 microcontroller functions without external address and data buses.

Microprocessor A CPU packaged in a single integrated circuit.

Microprocessor mode An operation mode for the PIC18 microcontroller in which the PIC18 microcontroller can access only external memory.

Microprocessor with boot block mode A PIC18 operation mode in which the PIC18 can access only the boot block of the on-chip flash memory (512 bytes or 2 KB) and the external memory.

Mode fault   An SPI error that indicates that there may have been a multimaster conflict for system control. Mode fault is detected when the master SPI device has its SS pin pulled low.

Modem   A device that can accept digital bits and change them into a form suitable for analog transmission (modulation) and can also receive a modulated signal and transform it back to its original digital representation (demodulation).

Multidrop   A data communication scheme in which more than two stations share the same data link. One station is designated as the master, and the other stations are designated as slaves. Each station has its own unique address, with the primary station controlling all data transfers over the link.

Multiprecision arithmetic   Arithmetic (add, subtract, multiply, or divide) performed by a computer that deals with operands longer than the computer's word length.

Multitasking   A computing technique in which CPU time is divided into slots that are usually 10 to 20 ms in length. When multiple programs are resident in the main memory waiting for execution, the operating system assigns a program to be executed to one time slot. At the end of a time slot or when a program is waiting for completion of I/O, the operating system takes over and assigns another program to be executed.

Nibble   A group of 4-bit information.

Nonmaskable interrupts   Interrupts that the CPU cannot ignore.

Nonvolatile memory   Memory that retains stored information even when power to the memory is removed.

Null modem   A circuit connection between two DTEs in which the leads are interconnected in such a way as to fool both DTEs into thinking that they are connected to modems. A null modem is used only for short-distance interconnections.

Object code   The sequence of machine instructions that results from the process of assembling and/or compiling a source program.

Output-compare   A PIC18 timer function that allows the user to make a copy of the value of the 16-bit main timer, add a delay to the copy, and then store the sum in a register. The output-compare function compares the sum with the main timer in each of the following clock cycles. When these two values are equal, the circuit can trigger a signal change on an output-compare pin and may also generate an interrupt request to the PIC18 microcontroller.

Output handshake   A protocol that uses two handshake signals to make sure that output device correctly receives the data driven by the peripheral chip (sent by the CPU).

Output port   The part of the circuit in a microcontroller that consists of output pins, data register, and control circuitry to send data to output devices.

Overflow   A condition that occurs when the result of an arithmetic operation cannot be accommodated by the preset number of bits (say, 8 or 16 bits); it occurs fairly often when numbers are represented by fixed numbers of bits.

Parallel slave port   A PIC18 parallel port (shared with PORTD) that allows the other microcontroller to read from and write into it after it is enabled.

Parameter passing   The process and mechanism of sending parameters from a caller to a subroutine where they are used in computations; parameters can be sent to a subroutine using CPU registers, the stack, or global memory.

Parity error   An error in which odd number of bits change value; it can be detected by a parity checking circuit.

Partial decoding   An address decoding method in which a memory location may respond to more than one address.

Phase_seg1 and Phase_seg2   Segments that are used to compensate for edge phase errors occurred in a CAN bus. These segments can be lengthened or shortened by synchronization.

Physical layer   The lowest layer in the layered network architecture. This layer deals with how signals are actually transmitted, the descriptions of bit timing, bit encoding, and synchronization.

Physical time   In the PIC18 timer system, the time is represented by the count in a 16-bit timer counter.

Point-to-point   A data communication scheme in which there are two stations communicate as peers.

Precedence of operators   The order in which operators are processed.

Program   A set of instructions that the computer hardware can execute.

Program counter   A register that keeps track of the address of the next instruction to be executed.

Program loops   A group of instructions or statements that are executed by the processor more than once.

Programmable logic device   A logic device to which the user can reconfigure its functionality by reprogramming the internal interconnection of the device.

PROM   Programmable read-only memory. A type of ROM that allows the end user to program it once and only once using a device called PROM programmer.

Prop_seg   The segment within a bit time used to compensate for the physical delay times within the CAN network.

Pseudocode   An expressive method that combines the use of plain English and statements similar to certain programming languages to represent an algorithm.

Pull   The operation that removes the top element from a stack data structure.

Pulse width modulation   A timer function that allows the user to specify the frequency and duty cycle of the digital waveform to be generated.

Push   The operation that adds a new element to the top of a stack data structure.

Queue   A data structure to which elements can be added at only one end and removed only from the other end. The end to which new elements can be added is called the **tail** of the queue, and the end from which elements can be removed is called the **head** of the queue.

RAM   Random-access memory. RAM allows read and write access to every location inside the memory chip. Furthermore, read access and write access takes the same amount of time for any location within the RAM chip.

Receiver overrun   A data communication error in which a character or a number of characters were received but not read from the buffer before subsequent characters were received.

Refresh   An operation performed on dynamic memories in order to retain the stored information during normal operation.

Refresh period   The time interval within which each location of a DRAM chip must be refreshed at least once in order to retain its stored information.

Register   A storage location in the CPU. It is used to hold data and/or a memory address during the execution of an instruction.

Remote request frame   A frame sent out by a CAN node to request another node to send data frames.

Repeated start condition   A START signal generated without first generating a STOP signal to terminate the communication in an I²C bus system. This is used by the master to communicate with another slave or with the same slave in different mode (transmit/receive mode) without releasing the bus. A repeated start condition indicates that a device would like to send more data instead of releasing the line.

Reset   A signal or operation that sets the flip-flops and registers of a chip or microprocessor to some predefined values or states so that the circuit or microprocessor can start from a known state.

Reset handling routine   The routine that will be executed when the microcontroller or microprocessor gets out of the reset state.

Resynchronization   All CAN nodes perform resynchronization within a frame whenever a change of bit value from recessive to dominant occurs outside of the expected **sync_seg** segment after the hard synchronization.

Resynchronization jump width   The amount of lengthening in **phase_seg1** or shortening in **phase_seg2** in order to achieve resynchronization in every bit within a frame. The resynchronization jump width is programmable to between 1 and 4.

Return address   The address of the instruction that immediately follows the subroutine call instruction (either JSR or BSR).

Return address stack   A set of registers organized as the stack structure to save and restore return address during subroutine calls in the PIC18 microcontroller.

Rise time   The amount of time a digital signal takes to go from logic low to logic high.

ROM   Read-only memory. A type of memory that is nonvolatile in the sense that when power is removed from ROM and then reapplied, the original data is still there. ROM data can only be read—not written—during normal computer operation.

Row address strobe   The signal used by DRAM chips to indicate that row address logic levels are applied to the address input pins.

RS232   An interface standard recommended for interfacing between a computer and a modem. This standard was established by the EIA in 1960 and has since then been revised several times.

Signal conditioning circuit   A circuit added to the output of a transducer to scale and shift the voltage output from the transducer to a range that can take advantage of the whole dynamic range of the A/D converter being used.

Simulator   A program that allows the user to execute microcontroller programs without having the actual hardware.

Simplex link   A line that is dedicated either for transmission or reception but not both.

Source code   A program written in either assembly language or a high-level language; also called a source program.

Source-level debugger   A program that allows the user to find problems in user code at the high-level language (such as C) or assembly language level.

Space   A term used to indicate a binary 0.

SPI   Serial peripheral interface. A protocol proposed by Motorola that uses three wires to perform data communication between a master device and a slave device.

Stack   A last-in, first-out data structure whose elements can be accessed only from one end. A stack structure has a top and a bottom. A new item can be added only to the top, and the stack elements can be removed only from the top.

Stack frame A region in the stack that holds incoming parameters, the subroutine return address, local variables, saved registers, and so on.

Start condition A signal combination in the I²C bus in which the data line goes to low when the clock line is at high. This condition is an indication that the bus master wants to start a data transfer.

Static memories Memory devices that do not require periodic refreshing in order to retain the stored information as long as power is applied.

Status register A register located in the CPU that keeps track of the status of instruction execution by noting the presence of carries, zeros, negatives, overflows, and so on.

Stepper motor A digital motor that rotates certain degrees clockwise or counterclockwise whenever a certain sequence of values is applied to the motor.

Stop condition A signal combination for the I²C bus master to indicate its intention to release the bus. The stop condition is represented by the data line going high when the clock line remains at high.

String A sequence of characters.

Structured programming A programming methodology in which the programmer begins with a simple main program whose steps clearly outline the logical flow of the algorithm and then assigns the execution details to subroutines. Subroutines may also call other subroutines. The structured programming methodology makes a complicated problem more manageable.

Subroutine A sequence of instructions that can be called from various places in the program and will return to the caller after its execution. When a subroutine is called, the return address will be saved on the stack.

Subroutine call The process of invoking the subroutine to perform the desired operations. The PIC18 has the CALL and RCALL instructions for making subroutine calls.

Successive approximation method A method for performing A/D conversion that works from the most significant bit toward the least significant bit. For every bit, the algorithm guesses the bit to be 1 and then converts the resultant value into the analog voltage and then compares it with the input voltage. If the converted voltage is smaller than the input voltage, the guess is right. Otherwise, the guess is wrong, and the bit is cleared to 0.

Switch statement A multiway decision based on the value of a control expression.

Sync_seg In the CAN format, the **sync_seg** segment is the segment within a bit time used to synchronize all CAN nodes.

Temperature sensor A transducer that can convert temperature into a voltage.

Text editor A program that allows the end user to enter and edit text and program files.

Thermocouple A transducer that converts a high temperature into a voltage.

Transducer A device that can convert a nonelectric quantity into a voltage.

Transpose An operation that converts the rows of a matrix into columns and vice versa.

Trap A software interrupt; an exception.

Union A variable that may hold (at different times) objects of different types and sizes, with the compiler keeping track of size and alignment requirements.

USART Universal synchronous asynchronous receiver and transmitter. An interface chip that allows the microprocessor to perform asynchronous serial data communication.

Vector A unidimensional data structure in which each element is associated with an index *i*. The elements of a vector are of the same length.

Volatile memory   Semiconductor memory that loses its stored information when power is removed.

Volatile variable   A variable that has a value that can be changed by something other than user code. A typical example is an input port or a timer register. These variables must be declared as volatile so that the compiler makes no assumptions on their values while performing optimizations.

Watchdog timer   A special timer circuit designed to detect software processing errors. If software is written correctly, then it should complete all operations within a certain amount of time. Software problems can be detected by enabling a watchdog timer so that the software resets the watchdog timer before it times out.

Write collision   The SPI error that occurs when an attempt is made to write to the SPDR register while data transfer is taking place.

Index